Cytotaxonomy and Vertebrate Evolution

Cytotaxonomy and Vertebrate Evolution

edited by A B CHIARELLI
Institute of Anthropology
University of Turin Italy

and E CAPANNA
Institute of
Comparative Anatomy
University of Rome Italy

1973

ACADEMIC PRESS London · New York
A Subsidiary of Harcourt Brace and Jovanovich, Publishers

ACADEMIC PRESS INC. (LONDON) LTD.
24/28 Oval Road,
London NW1

United States Editions published by
ACADEMIC PRESS INC.
111 Fifth Avenue
New York, New York 10003

Library of Congress Catalog Card Number: 72–7707
ISBN: 0 12 172535 9

PRINTED IN GREAT BRITAIN
BY W & J MACKAY LIMITED, CHATHAM

CONTRIBUTORS

M. BENAZZI	*Departimento di Zoologia, Università di Pisa, Via A. Volta 4, 56100 Pisa, Italia.*
AIMEE H. BAKKEN	*Biology Division, Oak Ridge National Laboratory, P.O. Box Y, Oak Ridge, Tennesse 37830, U.S.A.*
E. CAPANNA	*Istituto di Anatomia Comparata, "Battista Grassi", Via Alfonso Borelli 50, 00161 Roma, Italia.*
A. B. CHIARELLI	*Istituto di Antropologia, Università di Torino, Via Accademia Albertina 17, 10123 Torino, Italia.*
D. COLOMBERA	*Istituto Biologia Animale, Via Loredan 10, Padova, Italia.*
G. C. GORMAN	*College of Virgin Islands, U.S. Virgin Islands, St. Thomas VI 00801, U.S.A.*
L. KOULISCHER	*Institut de Morphologie Pathologigue, Alleé des Templiers, 29 bis, 6270 Loveral (Charleroi), Belgium.*
G. MANCINO	*Departimento di Istologia e Embriologia, Università di Pisa, Pisa, Italia.*
MARIA GABRIELLA MANFREDI ROMANINI	*Istituto di Antropologia, Università di Pavia, Palazzo Botta 10, 27100 Pavia, Italia.*
R. MATTHEY	*Institut de Biologie Animale, Université de Lausanne, Palais de Rumine, 1005 Lausanne, Switzerland.*
O. L. MILLER, JR.	*Biology Division, Oak Ridge National Laboratory, P.O. Box Y, Oak Ridge, Tennesse 37830, U.S.A.*
A. MORESCALCHI	*Istituto di Istologia e Embriologia, Università di Napoli, Via Mezzocannone 8, 80134 Napoli, Italia.*
S. OHNO	*Department of Biology and Pedology, City of Hope Medical Center, Duarte, California 91010, U.S.A.*
I. C. POTTER	*School of Biological Sciences, University of Bath, BA2 7AY, Somerset, England.*
R. RAY-CHAUDHURI	*Cytogenetics Laboratory, Department of Zoology Banaras Hindu University, Varanassi 5, India.*
E. S. ROBINSON	*School of Biological Sciences, Macquarie University North Ryde, New South Wales 2113, Australia.*
G. B. SHARMAN	*School of Biological Sciences, Macquarie University, North Ryde, New South Wales 2113, Australia.*
N. N. VORONTSOV	*Institute of Biology and Pedology, Far East Science Centre, Academy of Sciences of the U.S.S.R. Vladivostock 22, U.S.S.R.*
M. J. D. WHITE	*Department of Genetics, University of Melbourne, Australia.*

PREFACE

It was agreed during a meeting of the "Italian Study Group on Chordate Caryology and Cytotaxonomy" that there is a need for a comprehensive record of work in this field. Those of us most involved in vertebrate studies were asked to co-operate on a project to catalogue available information. However, we subsequently decided to ensure the success of such an endeavour by inviting some of the acknowledged authorities in this field to participate. The ones who responded enthusiastically, proposed to deal with some of the topics we had originally undertaken to document ourselves. Consequently our role was changed to that of editors, instead of authors. The difficulties posed by our undertaking should not be underrated: logistic problems and our determination to be concise, left us wary of shouldering so heavy a burden.

Where we were unable to find an author willing to cover a certain subject, we attempted to bridge the gap ourselves in an editors review.

Despite these problems and the trials of keeping pace with a subject of which our knowledge is rapidly expanding, we feel this book will give a useful insight into the phylogenetic relationships of vertebrates. Although studies of chromosome variation raise as many problems as they solve, the urge to uncover the truth has continually driven workers on to extend the boundaries of that knowledge.

We are obliged to Miss Chiara Bullo of the Institute of Anthropology, University of Turin, for her efficient assistance and the Academic Press for their kind co-operation.

A. B. CHIARELLI

The Institute of Anthropology
University of Turin.

E. CAPANNA

Institute of Comparative Anatomy
University of Rome.

CONTENTS

SECTION I—INTRODUCTION

SECTION II—THE PHYSICAL BASIS OF HEREDITY

SECTION III—CHROMOSOMES DYNAMIC

SECTION IV—PHYLOGENETIC AND TAXONOMIC INTERPRETATION OF THE KARYOLOGICAL DATA

8. The Chromosomes of Lower Chordates

9. The Chromosomes of the Cyclostomes

SECTION V—THE SEX CHROMOSOMES AND THE SEX DETERMINATION

16. The Evolution of the Sex Chromosomes

17. Evolution and the Mode of Action of the Genetic Regulatory Mechanisms which Determine Sex

18. Concluding Remarks

DEDICATION

In 1949 Professor Matthey's work "Les Chromosomes des Vertebrés" was published. It was the outcome of 20 years of passionate and highly qualified investigation in the field of vertebrate caryology.

For all of us who began to deal with the same subject around 1950 Matthey's book became the most reliable guide. We do not claim here to have attained further progress: no doubt information on vertebrate karyology has increased at a tremendous rate, but the problems so masterfully outlined by Matthey 20 years ago have remained unchanged. All of us are indebted to him for his untiring advice and encouragement. To this master, therefore, this book is dedicated.

Section I

INTRODUCTION

1

Cytotaxonomy and Evolution: General Remarks

MARIO BENAZZI
Department of Zoology, University of Pisa, Italy

It is no easy task indeed, in a work containing contributions of several authors, to outline the general aspects of the problems dealt with by each of them. However, I willingly agreed to accept the invitation by the Editors to write a few introductory remarks.

In the light of karyological investigations, taxonomy and evolution are not only rewarding fields of study, but they also represent a fascinating link between present day biology and the old trunk of natural history.

The bases of cytotaxonomy were laid down when the principle of chromosome individuality was accepted and the chromosome theory of heredity became established. It was noticed that the karyotype, characterized by chromosome number, size and morphology, is a definite and constant character of each species. Since the genetic material is carried by the chromosomes, it could be deduced that the genome has a specific pattern of its own. The karyotype exhibits more or less pronounced changes within different groups and may have, like other morphological features, phylogenetic implications.

Botanists and zoologists promptly realized that karyological information could be of use towards the elucidation of taxonomical problems; that is, it could contribute to our knowledge of the degree of affinity between species belonging to the same genus or family or even higher systematic categories.

Obviously, the karyotype elements available to the cytotaxonomist are largely variable for, in the organisms possessing complex structures such as giant chromosomes, an extremely refined analysis may be achieved. However, even where such "high resolution" structures are not available, as is the case in most eukaryotes, the mitotic karyotype, supplemented by data derived from meiotic processes (e.g. chiasma frequency and distribution) can provide fairly significant information.

Comparative analysis of chromosome sets is not, however, the universal answer to all the problems of systematics and evolution. Thus, priority with respect to other characters should not be ascribed to karyological data, however justifiable this may seem on the ground that the hereditary patrimony (hence the clue, so to speak, to the evolutionary history of the species) is carried by the chromosomes.

The karyotype, as far as its morphological features are concerned, is also part of the phenotype, as it is the result of an evolution whose course may have been very varied.

Evidence of this is definitely provided by the fact that in some groups the chromosome pattern has persisted substantially unaltered through long evolutionary stages, while in other groups it has undergone marked changes even in closely related species. In the Odonata, Diptera and Coleoptera, for instance, the chromosome number is fairly constant; whereas, in Lepidoptera and Trichoptera, in Scorpions as well as in Pisces, it displays marked variations. Genotypic and karyotypic variations may then arise independently and are not always the former reflected by the latter.

These observations aroused some doubts on the value of cytotaxonomy; The genetic material of related species obviously cannot present very marked differences. Therefore, the fact that relatively similar genic complexes were distributed in a very different way remained unclear. In the earliest decades of cytogenetics, the karyotype was in fact denied all evolutive significance by some eminent workers who maintained that it was not the chromosomes, but the genes they contained, that were of importance. This view was clearly expressed by E. B. Wilson in his famous treatise "The Cell in Development and Heredity" (1925, p. 886): "The number of chromosomes is *per se* a matter of secondary significance. Both cytological and genetic evidence prove that the chromosomes are compound bodies, containing many different components. So long as the sum-total of these remains the same, or nearly so, it seems immaterial whether they be grouped to form few or many aggregates."

But in the same year (1925), Sturtevant discovered the position effect and demonstrated that the effect of a locus is not dependent solely on the nature of the gene but also on its position with respect to the other genes. The evolutionary value of chromosomal variations (inversions, translocations etc.) was then brought to light by a large number of investigations. The physical bases of evolution are not, therefore, represented solely by genes, but also by the chromosomes as sharply defined morphological entities characterized by a determinate gene association and sequence, a determinate centromere position and by a determinate distribution of euchromatin and heterochromatin. This obviously lends value and significance to cytotaxonomy as envisaged in its traditional aspect; namely, that of the morphological analysis of the genome: "For the chromosome represents a higher level of organization and integration than that implied in a mere string of genes" (John and Lewis, 1968, p. 178).

All things considered, although the karyotype may be considered as a part of the phenotype, we may admit it has a lower adaptive value as compared to other morphological features (e.g. colouration, size, appendage morphology, cutaneous productions and so forth). For this reason, karyotype variations reflected general phylogenetic lines.

But we should now turn to a point of central concern, namely the question as to whether cytotaxonomy and evolution are a unitary problem. On the one hand, from a certain viewpoint, a positive answer may be given, since the interpretation of systematic relations on a chromosomal basis means indication of phyletic lines. But in this problem two aspects are implicit which should be considered separately. The first is inherent to speciation mechanisms, the second to the changes undergone by the karyotype during phylogeny. These two facets are in a certain sense related to the concepts of micro- and macroevolution of which they may perhaps express some essential contents, even though we admit that the differences between the two events are of a quantitative rather than a qualitative order.

It is true that in the term "evolution" the two concepts are indissolubly associated, for evolutionists since before Darwin's time have concentrated on the problem of how species originate and how they develop, step-by-step, their distinctive characteristics. Darwin himself called his fundamental work "The Origin of Species" though intending to refer to general phenomenon of evolution. As Mayr (1963, p. 424) rightly observes: "The true meaning of the term 'origin of species' was understood only rather recently. Not only were the pre-Darwinian evolutionists quite vague on this issue, but even Darwin himself seems to have considered 'origin of species' the same as evolution. Thus he confused two essentially different problems under the single heading 'origin of species'."

To put present day views on this question in clear and precise words, I quote the opening remarks written by White (1968) in a most interesting paper: "The modern or synthetic view of organic evolution regards speciation as a special and perhaps usually brief stage in evolutionary divergence, during which genetic isolating mechanisms develop to a level which makes the phyletic separation of incipient species irreversible (except for the special case of alloploidy, almost entirely restricted to plants)."

These concepts bring out, to my mind, what I have already stated about the advisability of considering speciation as an event *per se*, though correlated to the whole evolutionary phenomenon. This is valid particularly in its operative aspect, since speciation alone can be tackled by the experimental method. Genetics (either formal or of population and chromosomal), which is the matrix of this trend of research, enables the problem to be put in the following terms: in speciation processes, what is the respective importance of genic mutations and karyotype variations? The problem of the origin of species is thereby seen in a substantial perspective of which the great English naturalist could obviously have had no foresight.

Research carried out in the last decades has shown that the genetic processes lying at the base of speciation differ enormously, as can be readily understood, from case to case. The reproductive mechanisms which isolate one species from another are the most important attribute of a species, since they are by definition the species criteria. But the mechanisms of reproductive isolation are extremely varied and arise at most different times during the life cycle: from the preparatory events of gamete encounter, to fertilization, to the development of the new individual and to the maturation of its gametes. This makes up a long sequence of coordinated phases which are extremely varied in their essence and hence in their determinism. It may, therefore, be easily understood how they can be blocked or altered by most widely differing genic or chromosomal factors.

In the chapter "The genetics of speciation" of his well known treatise (1963, p. 535) Mayr, a student of speciation problems (although not personally specialized in karyology), stresses that many among the earliest cytogenetists believed that speciation is the result of a structural repatterning of geneticists believed that speciation is the result of a structural repatterning of tions: (1) that speciation is the origin of discontinuities, (2) that related species often differ in their chromosomal pattern and (3) that certain chromosomal changes, such as reciprocal translocations, are partially isolating on account of considerable sterility among the heterozygotes. These elements were combined into the theory that a structural chromosomal change might lead in a single step to the origin of an individual that is reproductively isolated from the members of the parental species. According to Mayr this viewpoint, however, is no longer acceptable: "Many species are polymorphic for the very chromosomal differences that in other cases differentiate closely related species. . . . Conversely, well-defined and reproductively isolated species may completely agree structurally in their chromosomes and differ only in their gene contents. It can no longer be doubted that speciation can be completed by genic differentiation without structural repatterning."

This conclusion cannot be rejected on any account. The significant example of some sibling *Drosophila* species showing no difference in the banding sequence of their polytenic chromosomes (*homosequential* species, according to the terminology proposed by Carson *et al.*, 1967) would be enough.

This, however, in no way excludes the possibility that chromosome variations themselves and *per se* are adequate to promote or merely initiate a speciation process. This problem was quite clearly synthetized by White (1969, p. 77) as follows: "The existence in the genus *Drosophila* of several strictly homosequential complexes of taxa which have undoubtedly reached the species level clearly proves that chromosomal rearrangement is not a *sine qua non* for speciation. But even in *Drosophila*, such homosequential complexes represent only a small fraction of the total number of species complexes that have been intensively studied. . . . The general situation in the animal kingdom seems to be, however, that the great majority of species are karyologically unique and differ from related species either in chromosome number or in other cytological features."

It may readily be understood that the cytogenetic approach of speciation presents considerable interpretative difficulties and that even painstaking investigations often lead to only partial solutions. In fact, the dynamics of speciation cannot be understood and elucidated through comparative karyological studies alone, but require different approaches which are not always feasible, particularly when studying meiosis in hybrids. John and Lewis (1965), in a study on genetic speciation in the grasshopper *Eyprepocnemis plorans*, put the question clearly as follows: "Mitotic complements, and individual chromosomes even more so, are distinguished by a few markers, whereas, the highly specific nature of quantitatively-limited meiotic pairing allows a recognition of differences which are otherwise concealed. Indeed, the mitotic complements of some species groups, like that to which the subjects of this paper belong, are so similar that distinctions are virtually impossible. But, as we shall see, one cannot on this evidence alone conclude that structural chromosome changes have been unimportant in their evolution."

Pertinent to our theme are also some recent considerations from Matthey (1970). He points out that chromosomal mutations have long been held to play a negligible role, since the genetic material remains unaltered except for some possible position effects; moreover, the species may have arisen without apparent karyotype changes, as for instance in *Felidae*, whereas enormous chromosome variations may be associated to minor morphological changes. But, he adds, revision of this viewpoint is dictated by recent works. According to this eminent scientist, Rodents, in full evolutionary explosion, illustrate in an exemplary way the correlations that may exist between karyotype differentiation and speciation. Chromosome evolution of a Robertsonian type stands out most clearly. As an example, in the Central African Republic, *Mus* (Leggada) populations belonging to the *minutoides-muscoloides* group with a number of arms (F.N.) = 38, Matthey has found all the diploid numbers from 18 to 34; namely, 16 out of the 17 possible Robertsonian combinations. But demarcation between species and subspecies and hence the exact significance of these karyological variations in speciation, are as yet unclear. This confirms the interpretative difficulties referred to above.

Obviously, the present general remarks do not permit illustration of single cases or detailed analyses otherwise available in large numbers in recent literature. However, an example will be given here, which in the writer's opinion is fairly significant to speciation processes, since it shows with particular evidence the different contributions afforded by chromosome variations in both number and structure. The main ancestor of the races of the domestic horse, as is well known, is *Equus przewalskii* Poliakoff, a small horse the size of a big pony which was largely widespread over Europe in the past (it was in fact pictured in the Lascaux caves by paleolithic artists), and which then became restricted to the Gobi desert at the boundary between China and Outer Mongolia, where the Russian colonel Przewalski discovered it in 1881. Despite its being extremely wild and hence difficult to capture and transport, a good number of specimens (up to 158 in 1969) have been bred in several Zoological Gardens, enabling cytogenetical studies to be carried out. Its diploid complement is made up of 66 chromosomes (Benirschke *et al.*, 1965); that is, two more than *E. caballus*, with $2n = 64$, as is well known. But crosses between the two species (or rather races) may occur, giving birth to a fertile hybrid with 65 chromosomes (Koulischer and Frechkop, 1966); meiosis, therefore, is not affected by the additional chromosome. Their karyotypes are otherwise alike in every respect, *przewalskii* having 13 pairs of metacentric autosomes and 19 pairs of acrocentric autosomes while *caballus* has 13 pairs of metacentrics and 18 pairs of acrocentrics; in both races chromosome X is a large metacentric and Y a small acrocentric. In the microevolutionary process leading to *caballus*, chromosomal structure was not altered or at least not to such an extent as to prevent the perfect pairing of homologous chromosomes with crossing-over.

On the other hand, the passage from *caballus* to *asinus* was quite different. These two species also differ by one chromosome pair, in that *E. asinus* displays $2n = 62$ and the mule $2n = 63$. However, as is well known, the latter is sterile (but there are some rare exceptions as regards females) on account of severe meiotic anomalies brought about by a series of inversions and translocations. *E. asinus* has in fact 19 pairs of metacentric autosomes and 11 pairs of acrocentric autosomes; furthermore, the sex chromosomes are morphologically different between the two species (Trujillo *et al.*, 1962; Benirschke *et al.*, 1962). There is little doubt that chromosomal rearrangements did play a major role, without excluding, however, the possibility that genic mutations may have occurred during the speciation process. Chromosome variations may also account for the separation, proposed by Bourdelle and Frechkop (1950), of the horse from the ass into two distinct genera: *Equus* and *Asinus*.

I wish now to touch briefly upon personal research by reporting some results obtained on fresh-water Triclads. Cytogenetic and cytotaxonomic investigations carried out by the writer and his colleagues have brought to light in various groups of species, some microevolutionary phenomena associated both to poliploidy related to gynogenesis (through mechanisms of meiotic and ameiotic oogenesis) and to structural chromosome variations. (General references in Benazzi, 1960, 1963 and Benazzi Lentati, 1970.)

The common European species *Dugesia lugubris s.l.* is a typical example of the above phenomena. Schmidt (1860) had described two species (*lugubris* and *polychroa*) that were afterwards accepted by some authors, but refuted by others who maintained it was a single species. Karyological studies on populations from several European localities have made possible the recognition of the existence of 7 biotypes, having a different significance as regards specific differentiation. The first 4 biotypes (A, B, C and D) make up a homogeneous series with polyploid evolution starting with biotype A which is diploid ($2n = 8$, $n = 4$) and amphigonic. Biotype B is triploid in the somatic line (12 chromosomes) and hexaploid in the female germ line (12 bivalents) through the premeiotic duplication of the complement. Biotype C is triploid in the somatic and female lines, since oocytes possess 12 univalents, i.e. they are asynaptic. Biotype D is like the latter though tetraploid instead of triploid. Biotypes B, C and D are pseudogamic (the egg develops through gynogenesis) and their male germ line is diploid and meiotic. Polyploid evolution has not promoted reproductive isolation (although crosses between diploid individuals acting as females and triplo-hexaploid individuals often fail to yield offspring) and the basic chromosome complement has persisted unchanged in all the 4 biotypes, consisting of one large submetacentric and 3 acrocentrics of different length.

Biotypes E, F and G are, on the contrary, chromosomally differentiated and reproductively isolated.

Biotype E is diploid ($2n = 8$, $n = 4$) but the oocyte and spermatocyte bivalents do not correspond in length and chiasma frequency to those of biotype A. This is confirmed in the corresponding karyotype, which shows that the haploid set is formed by three large acrocentric chromosomes of different length and by a very small one. Biotype F has $2n = 6$, $n = 3$ and in all probability originate from biotype E through a Robertsonian translocation, i.e. a centric fusion between two large acrocentrics. In fact, the haploid set is formed by a large metacentric, a medium length acrocentric and a very small chromosome.

Biotype G is also diploid ($2n = 8$, $n = 4$) but the haploid set is formed by two submetacentric chromosomes of different length, a medium length subtelocentric and a small chromosome. This chromosome set is more like that of biotype A, from which it may be derived through some chromosomal mutations, e.g. pericentric inversions and translocations.

These karyological and genetical results are of noteworthy taxonomical significance (Benazzi *et al.*, 1970). Biotypes A, B, C and D make up a single species, which on the basis of the morphological features of the copulatory system can be related to *D. polychroa* (O. Schmidt). On the contrary, biotype E corresponds to *D. lugubris* (O. Schmidt) in its morphological features and may also include biotype F, although, in the general case at least, the two biotypes are intersterile. Since they are chromosomally differentiated only by a "Robertsonian event", the latter may be assumed to have caused the initial stage of specific differentiation. Lastly, biotype G, has also reached the species rank, as testified by its reproductive isolation, which is clearly correlated with the structural rearrangement of chromosomes.

The superspecies *Dugesia lugubris*, therefore, displays a set of chromosomal variations with clearcut micro-evolutionary reverberations. Other groups of species, e.g. *D. gonocephala s.l.* and *Polycelis nigra-tenuis*, are analogous, though less conspicuous examples. On the other hand, the two well known American species *Dugesia dorotocephala* and *D. tigrina* exhibit identical karyotypes, thereby leading to the hypothesis that speciation may have occurred solely or prevalently through genic mutations (Benazzi *et al.*, unpublished).

Among speciation examples adequately studied from the karyological standpoint, I wish to draw the reader's attention also to *Jaera albifrons* (= *J. marina*). This intercotidal isopod, according to Staiger and Bocquet (1956) builds up a superspecies including five species, all characterized by a system of multiple sex chromosomes with female digamety (Z W_1W_2). However, each species possesses its own chromosome complement whose numbers reveal the existence of Robertsonian correlations. These also occur at the infraspecific level bringing about polymorphism in some species. Particularly clearcut is the case of *J.* (*a.*) *syei*, whose chromosome number shows a cline along the European Atlantic shores; its diploid number lying between 28 (Wilhelmshaven, North Germany) and 18 (Biarritz, Southern France). Lécher (1967) was able to confirm the occurrence of a Robertsonian mechanism by means of karyometric measurements and inspection of meiosis in hybrids from chromosomally different species.

I have dwelt at some length with the cytogenetics of speciation because of its central position in the problem of evolution and its being one of the more rewarding trends of study in present day Science.

However, I should like to add briefly some remarks on another aspect of cytotaxonomy; namely, on the transformations of the karyotype during phylogeny. In the writer's opinion, when karyotype transformations are considered, it should, first of all, be clear that they cannot reflect phylogenetic evolution in a congruous way. As has already been written (Benazzi, 1957) some time ago, the correlations between morphologic and karyologic evolution are hard to understand. Some phylogenetic trends are associated with chromosome variations, but we cannot establish, at least at present, to what extent and how these two events are linked to each other.

The karyotype is the fundamental physical basis of evolution, but it is also endowed with an evolution of its own dictated by intrinsic conditions and requirements, such as its functional mechanisms and the balance of the genic complex. The possibilities of karyotype transformation are obviously limited, nor can they adapt to the boundless series of morphological and functional variations realized by evolution within phyla, classes, orders and so forth. The karyotype might be compared to a set of pieces able to disassemble and reassemble although in a limited number of combinations.

Any attempt to rebuild the evolution of the karyotype in the major systematic taxa is met with overwhelming difficulties. Profitable analyses may be achieved solely at the level of genera or families. This assumption, already put forward by the writer (Benazzi, 1957), seems still valid, not disregarding the fact that the amplitude of the systematic categories in question varies from case to case.

When we consider the chromosome number, the variations of which represent the most conspicuous feature of the phylogenetic transformations of the karyotype, it may be seen that in some groups a "typical number" may be identified which, however, has mainly the value of a "modal number" without particular evolutionary significance (White, 1954). Furthermore, a typical number may be representative of more or less extended systematic categories (e.g. genus, family, order), that are always lower, at any rate, than the class or subclass.

The variations encountered in decapod crustaceans, scorpions or lepidopteres, for instance, are difficult to interpret. Identification of phylogenetic links between different karyotypes has been, on the contrary, possible within lower systematic taxa such as genera, subfamilies or even families. It was also possible to build up their evolutionary steps, at least partially.

Centric fusion, as is well known, represents a rather frequent mechanism of karyotype changes in several animal groups. At present, the converse, namely fission (or dissociation, in White's denomination) is also thought likely to occur. According to Capanna and Civitelli (1970) the direction of Robertsonian changes (fission or fusion) seems to move indifferently in both directions in the karyotypic evolution of Chriroptera.

Centric fusion satisfactorily explains the variations in chromosome number in the family Bovidae, for example, where the diploid number ranges from 60 (*Bos taurus*, *Bison*, *Capra*) with all acrocentric autosomes to 30 as in *Limnotragus speckii*, in which all the chromosomes, but for one pair of telocentric, are meta- or sub-metacentric (Koulisher *et al.*, 1967). Without any doubt, the Robertsonian type of fusion can occur among the Bovidae, both as interspecific and intraspecific variations, although this mechanism does not appear as the only one involved in speciation in this family.

Another reliable example is provided by Chen (1971), who carried out a comparative analysis of 20 species from the genus *Fundulus* and found that female karyotypes ranged in diploid number from 32 to 48 but in arm number from 48 to 52; an increase in complement by a pair of large bi-armed chromosomes was always accompanied by a decrease of two pairs of acrocentric chromosomes, indicating Robertsonian changes by centric fusion.

However, though widespread in some animal groups, fusion or fission mechanisms do not occur in others in which the arm number varies; this is well known in the genus *Peromiscus* (Cricetidae from North America) in which all the species and subspecies display a diploid number of 48, while the F.N. varies from 56 to 96, probably as a result of pericentric inversions (Hsu and Arrighi, 1966).

A general trend towards a decrease in the chromosome number is one of the most frequently observed evolutionary tendencies. This event, the significance of which is difficult to elucidate, is worthy of our attention since a decrease in chromosome number leads to a gene concentration; whereby the problem arises as to the possible advantages, from an evolutionary standpoint, of a different concentration of the gene complex. It may be assumed that the answer of selection to these changes varies from one group to another in relation to genic or chromosomal conditions that are not easy to analyse.

The decrease in the chromosome number may also be due to unequal reciprocal translocations. A rearrangement of this type was postulated by Morescalchi (1970) to explain the general tendency towards reduction of the chromosome number in many higher Anura, which seems to occur with the loss of small chromosomes, the material of which is possibly translocated onto larger elements.

It is certain, at all events, that even the most spectacular variations that have occurred in the karyotype of some groups cannot have elicited too drastic alterations of the gene complex. This concept was put forward by Ohno (1970, p. 41), according to whom the important parts of the gene have never changed in evolution and natural selection permits only trifling changes. The fact that in placental mammals the chromosome number varies from 84 (in the black rhinoceros) to 17 (in two rodent species) may indicate that these variations were not under strict control by natural selection; hence, casual chromosome reassortments have accompanied successive series of speciation without harm. These chromosome changes, though, were more apparent than real, since the original linkage groups have been retained to a large extent.

Another aspect of cytotaxonomy which deserves attention is that suggested by quantitative changes in the nuclear DNA, that is to say, the evolutionary significance of deletions and duplications. The latter were emphasized by Ohno (1970) in a quite suggestive manner, even though some perplexity may arise as regards some deductions.

In these introductory notes an attempt has been made to draw attention to only a few fundamental aspects of cytotaxonomy in relation to evolution.

Of course, more items of a great interest in this field will be adequately dealt with with regard to vertebrates in the proceeding chapters. The cytogenetics of sex deserve particular mention, since the evolutionary transformations of the sex chromosomes are convincing evidence of the methodologic value of the science we are interested in here. Moreover, we should not neglect the new acquisition about polyploidy which represents an evolutionary mechanism among Vertebrates also. Besides cases of polyploid parthenogenesis and gynogenesis, it is worth underlining that polyploidy is likely to have also been involved in the speciation of amphigonic forms. This was shown by the investigations conducted by Wolf *et al.* (1969) on *Barbus*, by Saez and Brum (1959), Beçak *et al.* (1966, 1970) and Bogart (1967) on some anuran amphibians.

It may, therefore, be maintained that the reckoning of our discipline is largely positive. Without a doubt, fresh fields are opening up in the ceaseless progress of Science; research work at a molecular level has already yielded some exciting suggestions. Scepticism about cytotaxonomy should be brushed aside, for this discipline has been, and will be for a long time, a most valuable tool and the leading thread in the historical and dynamic interpretation of the evolution of the living forms.

References

Beçak, M. L., Beçak, W. and Rabello, M. N. (1966). *Chromosoma* **19**, 188–193.

Beçak, M. L., Denaro, L. and Beçak, W. (1970). *Cytogenetics* **9**, 225–238.

Benazzi, M. (1957). *Boll. Zool.* **24**, 373–409.

Benazzi, M. (1960). Evoluzione e Genetica. Colloq. Intern. Quaderno 47 *Atti Accad. Naz. Lincei* **357**, 273–297.

Benazzi, M. (1963). *In* "The Lower Metazoa" pp. 405–422. Univ. of California Press, Berkeley, California.

Benazzi, M., Puccinelli, I. and Del Papa, R. (1970). *Atti Accad. Naz. Lincei, Rend.* **48**, 369–376.

Benazzi Lentati, G. (1970). *Int. Rev. Cytol.* **27**, 101–179.
Benirschke, K., Brownhill, L. E. and Beath, M. M. (1962). *J. Reprod. Fertil.* **4**, 319–326.
Benirschke, K., Malouf, N., Low, R. J. and Heck, H. (1965). *Science, N.Y.* **148**, 382–383.
Bogart, J. P. (1967). *Can. J. Genet. Cytol.* **9**, 531–542.
Bourdelle, E. and Frechkop, S. (1950). *Mammalia* **14**, 126–139.
Capanna, E. and Civitelli, M. V. (1970). *Caryologia* **23**, 79–111.
Carson, H. L., Claiton, F. E. and Stalker, H. D. (1967). *Proc. natn. Acad. Sci. U.S.A.* **57**, 1280–1285.
Chen, T. R. (1971). *Chromosoma* **32**, 436–453.
Hsu, T. C. and Arrighi, F. E. (1966). *Cytogenetics* **5**, 355–359.
John B. and Lewis, K. R. (1965). *Chromosoma* **16**, 308–344.
John, B. and Lewis, K. R. (1968). "The Chromosome Complement." Protoplasmatologia Bd. VI.
Koulischer, L. and Frechkop, S. (1966). *Science, N.Y.* **151**, 93–95.
Koulischer, L., Tijskens, J. and Mortelmans, J. (1967). *Acta Zool. Path. Antverpiensia.* **43**, 143–147.
Lécher, P. (1967). *Arch. Zool. Exp.* Gén. **108**, 633–698.
Matthey, R. (1970). *Revue Suisse Zool.* **77**, 625–629.
Mayr, E. (1963). "Animal Species and Evolution." Belknap Press. Harvard University Press. Cambridge, Mass.
Morescalchi, A. (1970). *Boll. Zool.* **37**, 1–28.
Ohno, S. (1970). "Evolution by Gene Duplication." Springer Verlag, Berlin.
Saez, F. A. and Brum, N. (1959). *An. Fac. Med. Montevideo* **44**, 414–423.
Schmidt, O. (1860). *Z. wiss. Zool.* **10**, 24–33.
Staiger, H. and Bocquet, C. (1956). *Bull. Biol. Fr. Belg.* **90**, 1–32.
Sturtevant, A. H. (1925). Genetics **10**, 117–147.
Trujillo, J. M., Stenius, C., Christian, L. C. and Ohno, S. (1962). *Chromosoma* **13**, 243–248.
White, M. J. D. (1954). "Animal Cytology and Evolution" 2nd Ed. Cambridge Univ. Press., London and New York.
White, M. J. D. (1968). *Science, N.Y.* **159**, 1065–1070.
White, M. J. D. (1969). *Ann. Rev. Genet.* **3**, 75–98.
Wilson, E. B. (1925). "The Cell in Development and Heredity." Macmillan, New York.
Wolf, U., Ritter, H., Atkin, N. B. and Ohno, S. (1969). *Humangenetik* **7**, 240–244.

Section II

THE PHYSICAL BASIS OF HEREDITY

2

The Ultrastructure of Chromosomes

EDITORIAL REVIEWS

In the last few years the control exerted by chromosomes over every cellular activity has given rise to a succession of important observations on their structural organization, whose variations are doubtlessly connected with the phenotypic variations, genetic differentiation and the biochemical activities of the cell.

Nevertheless, many questions are still unanswered. We do not know yet, for example, whether chromosomes are single-stranded or multistranded, whether longitudinal fibres run up and down the chromosome length. For the interpretation of the mechanisms involved in chromosome transformation, that are the basis of karyotype evolution, it would have been important beyond doubt to know with certainty what is the structure of the centromere, what is the structural aspect of the mechanism of centric fusion and whether the centromere is lost in this process or is retained. Mirsky and Ris were the first to suggest a model; they regarded the chromosome as a protein skeleton on which segments of DNA were attached. In the following years Kleinschmidt using the electron microscope (EM) was led to the assumption that the chromosome was made up of a bundle of irregular fibres. He did not recognize particular structures for the centromere or the telomere but only a variation in the density of the fibre packing.

The size of these fibres and their biochemical nature were then the object of many studies. DuPraw (1965, 1969) demonstrated in a series of important observations that the chromatin interphase and metaphase fibres had similar dimensions of about 150–230 Å and supposed that the fibre consisted of a single DNA double helix enclosed by a protein sheath. As the length of DNA was greater than the fibre length, he thought that the DNA was regularly supercoiled within the fibre. This fibre then, after being coiled again folded, would have formed the chromosome. In addition, he pointed out that these fibres, after gentle dissociation by trypsin, showed thin regions of 35–60 Å in which the DNA ratio to protein mass was 5:5

and that of the DNA length to fibre length was 5:1, and thicker regions of 200–300 Å in which the same ratios were respectively 3:7 and 50-1.

Ris (1967) demonstrated that the fibres of 250 Å separate into two fibres of 100 Å if they are treated with chelating agents and he asserted that the chromosome fibre was actually composed of two parallel fibres of 100 Å.

Comings and Okada (1970a) believe that the chromatid is made up of a single chromatin fibre which loops and folds upon itself. If two telocentric chromosomes, which have a single area of chromatid association, join to form a metacentric chromosome, the latter will have a quadripartite centromere consisting of two areas of chromatid association, an area of attenuation chromatin fibres and a central halo at the fusion site.

The biochemical hypotheses on the relationship between DNA and the various neclear proteins are also of interest. These proteins have two important functions: they control the genetic information and give the chromosome its structural conformation.

There is no explanation yet for the attachment process of different histones to DNA, we only know that lysine binds itself more easily to the A–T base pairs and arginine prefers the G–C base pairs. We also know that histones are responsible for the restrictions of the template activity, they are essential for the control of RNA synthesis, and they appear very early in evolution.

It would seem then (Hearst and Botchan 1970) that a functional specificity exists for the different histone fractions; the lysine-rich ones being involved in metaphase configuration, while the arginine-rich ones are important to the secondary structure of DNA. Furthermore, a larger proportion of lysine-rich histone is found at interphase, whereas the arginine-rich one is more abundant at metaphase.

On the other hand, the functional role of histones seems to come into play during interphase and doubling of both the non-histone proteins and RNA takes place at metaphase.

The aim of this brief and certainly incomplete synthesis of the presently available information on the structure of chromosomes is simply to introduce the problem to help the reader towards a better understanding of the following chapters.

References

Abuelo, J. G. and Moore, D. E., (1969). The human chromosome: electron microscopic observations on chromatin fiber organization. *J. cell Biol.* **41**, 78–83.

Comings, D. E. and Okada, T. A. (1970a). Whole-mount electron microscopy of the centromere region of metacentric and telocentric mammalian chromosomes. *Cytogenetics* **9**, 436–449.

Comings, D. E. and Okada, T. A. (1970b). Do half-chromatids exist?. *Cytogenetics* **9,** 450–459.

Comings, D. E. and Okada, T. A. (1971). Electron Microscopy of Chromosomes. *In* "Perspectives in Cytogenetics". (S. W. Wright, B. Cranoal and L. Boyer, eds), pp. 223–250, C. Thomas, Springfield, Illinois, U.S.A.

DuPraw, E. J., (1966). Evidence for a folded fibre organization in human chromosomes. *Nature Lond.* **209,** 577–581.

Hearst, J. E. and M. Botchan, (1970). The eukaryotic chromosome. *A. Rev. Biochem.* **39,** 151–182.

Hnilica, L. S., (1967). Proteins of the cell nucleus. *Prog. Nucleic Acid Res. Molecular Biology* 7, 25.

Hnilica, L. S. and Edwards, L. J., (1967). "The Evolution of Histones." Seventh International Congress of Biochemistry Colloquim IX, Histones.

Ris, H., (1967). Ultrastructure of the Animal Chromosome. "Regulation of Nucleic Acid and Protein Biosynthesis". (V. Koningsberer and L. Bosch, eds.) Elsevier Publishing Co. Amsterdam, pp. 11–22.

Stellwagen, R. H. and Cole, R. D. (1963). Chromosomal proteins. *A. Rev. Biochem.* **38,** 451–490.

Wolfe, S. L. and N. Grim, (1967). the relationship of isloated chromosome fibers to the fibers of the embedded nucleus. *J. Ultrastruc. Res.* **19,** 382–397.

3

Ultrastructure of Genetic Activity in Vertebrate Chromosomes

O. L. MILLER, Jr.
*Biology Division, Oak Ridge National Laboratory,**
Tennessee, U.S.A.
and
AIMÉE H. BAKKEN†
The University of Tennessee, Oak Ridge Graduate School of Biomedical Sciences, Oak Ridge, Tennessee, U.S.A.

I. INTRODUCTION

Several approaches currently are being utilized to elucidate the organization of genetic activity in chromosomes of higher organisms. A number of investigators have used DNA–DNA hybridization techniques to show that essentially all eukaryotic genomes contain nucleotide sequences ranging in incidence from highly repetitious (100 000 or more similar sequences present) to unique (one or only a few sequences present per haploid genome). The proportions of these various categories, however, may vary considerably from one species to the next (Britten and Davidson, 1971). In addition, RNA–DNA hybridization experiments have established that the main stable RNAs, i.e. the ribosomal RNAs and transfer RNAs, are transcribed from moderately repetitious nucleotide sequences (Bostock, 1972). Relatively labile cytoplasmic messenger RNAs and the highly labile heterogeneous nuclear RNAs, on the other hand, appear to be transcribed from both unique and repetitious sequences (Greenberg and Perry, 1971).

Although yielding a wealth of information on nuclear metabolism, the biochemical probes using nucleic acid hybridization have provided little

* Operated by Union Carbide Corporation for the U.S. Atomic Energy Commission.
† University of Tennessee Postdoctoral Trainee under contract No. 3322 with the Biology Division, Oak Ridge National Laboratory.

insight into the ultrastructural organization of genetically active chromosomes. We have taken a different approach toward understanding this organization—direct visualization of the structural aspects of genetic transcription. In this chapter we describe electron microscopic observations of nuclear RNA synthesis in two types of vertebrate cells, amphibian oocytes and cultured human (HeLa) cells. Previous reports of the oocyte studies can be found in Miller and Beatty (1969) and Miller *et al.* (1970) and current techniques are summarized in the legend for Fig. 2. The technical innovations that permit the visualization of active mammalian chromatin are described in the text and in the legend for Fig. 5.

II. RIBOSOMAL RNA GENES

The genes coding for ribosomal RNA (rRNA) are present in multiple copies in all organisms studied to date. In eukaryotes, these genes are arranged as tandem repeats within nucleolus organizer loci and the redundancy varies from around one hundred to several thousand per haploid genome, depending on the species (Attardi and Amaldi, 1970).

A. *Extrachromosomal Nucleoli of Amphibian Oocytes*

The amphibian oocyte has proven to be an excellent cell for studying structural aspects of nuclear RNA synthesis. In early oogenesis, the rRNA genes are amplified to several thousand times the number per haploid genome by an unknown mechanism (Brown and Dawid, 1968; Gall, 1969). As a result of this amplification, each oocyte nucleus contains many hundreds of active nucleoli, none of which show attachment to the chromosomes. In addition, the chromosomes are in the highly extended "lampbrush" state, actively synthesizing large amounts of RNA. The large size of amphibian oocyte nuclei (about 1 mm in diameter in mature oocytes of some species) allows rapid manual isolation of nuclear contents and sub-

Fig. 1. Thin section of extrachromosomal oocyte nucleolus showing a compact fibrous core surrounded by a granular cortex. *Triturus viridescens.*

Fig. 2. Portion of dispersed nucleolar core showing fibril gradients of matrix units and intercalated matrix-free axis segments. *T. viridescens.* Specimens are prepared for electron microscopy as follows. Nuclear contents of amphibian oocytes, containing both lampbrush chromosomes and extra-chromosomal nucleoli, are dispersed in water adjusted to pH 9, then centrifuged onto carbon-coated grids (3–4 min, 2350 X *g*) through 0.1 M sucrose plus 10% formalin (pH 9). The grids are rinsed in 0.4% Kodak photo-flo, dried, stained for 30 sec with 1% phosphotungstic acid in 70% ethanol (pH 2), rinsed in 95% ethanol and dried.

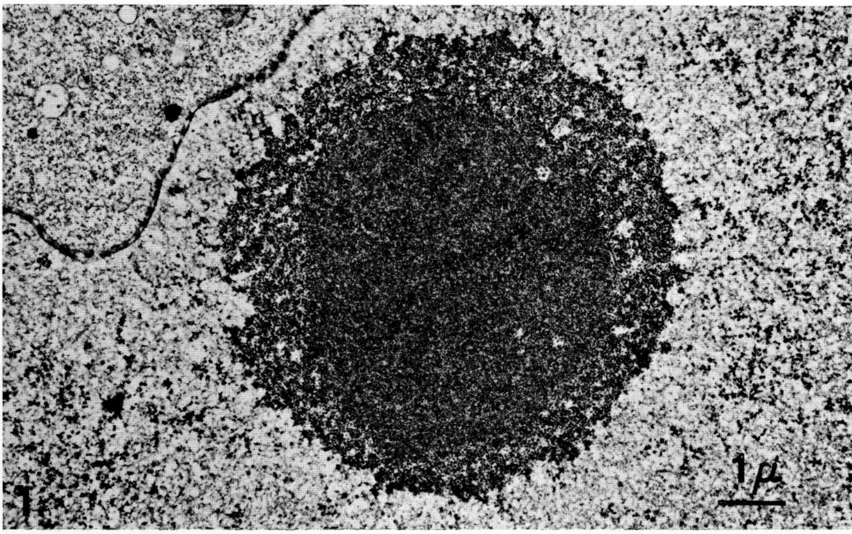
1μ
1

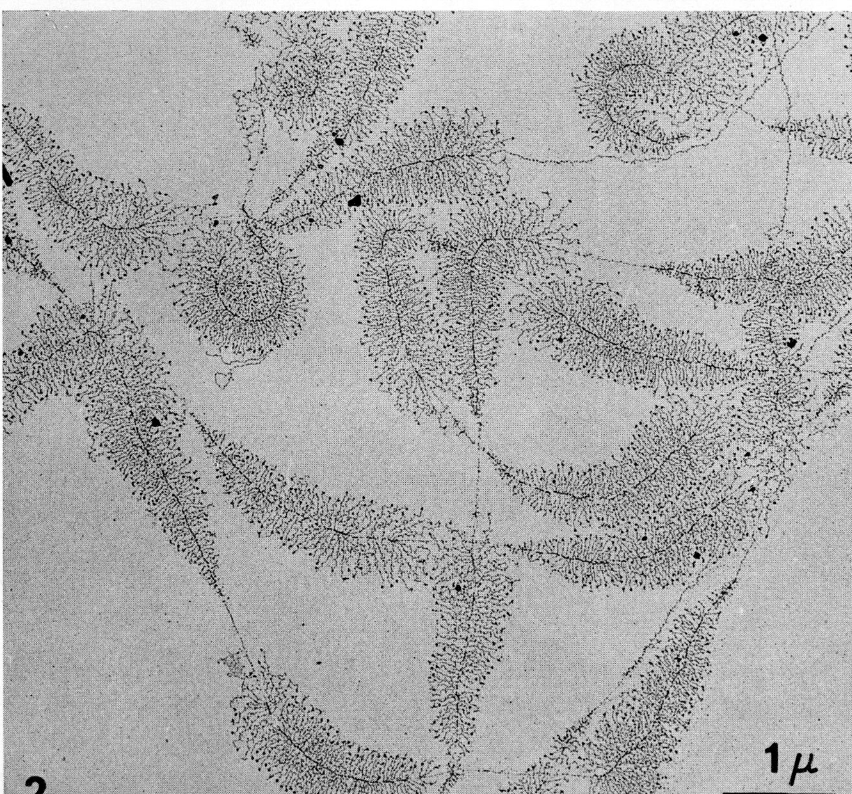
1μ
2

sequent preparation of both extrachromosomal nucleoli and lampbrush chromosomes for electron microscopy.

In thin section, each extrachromosomal nucleolus exhibits a dense fibrous core enclosed in a granular cortex (Fig. 1). When nucleoli are isolated in saline and immediately placed in distilled water adjusted to pH 7–9, the cortices disperse, allowing the compact cores to expand. Unwound cores prepared for electron microscopy as described in the legend for Fig. 2 are seen to consist of a thin axial fibre, periodically coated with matrix material. Each axis forms a circle, and the number of matrix units reflects the size of the nucleolus from which the core was isolated.

Each matrix unit is separated from its neighbours by matrix-free segments that typically are one-third or so the length of a matrix unit, but which may range up to 20 or more times the length of a matrix unit. Each matrix unit consists of a gradient of 80 to 100 closely spaced fibrils, each attached by one end to the core axis. Combined enzyme probes, specific protein staining and electron microscopic autoradiography show that the core axes are deoxyribonucleoprotein, the matrix fibrils are ribonucleoprotein (RNP) and that RNA synthesis is confined to the matrix units.

Several observations indicate that the RNA synthesized within the matrix segments consists of rRNA precursor molecules (rpRNA). First, most if not all of the amplified rRNA genes in amphibian oocytes are present in the extrachromosomal nucleoli (Evans and Birnstiel, 1968). Next, combined biochemical data (Gall, 1966) and autoradiographic results (Lane, 1967; Macgregor, 1967) indicate that synthesis of 40S rpRNA molecules is localized in the core regions of these nucleoli. Finally, the length of a matrix unit ($\sim$2.4 μ) is close to the B-conformation length of DNA ($\sim$ 2.6 μ) required to code for the 40S rpRNA of amphibians (molecular weight $\sim$ 2.6×10^6 daltons (Loening, 1970; Perry *et al.*, 1970)). Thus, it seems certain that the DNA segment within each matrix unit is a gene coding for rpRNA and that these genes repeat along a single DNA duplex, separated from neighbouring genes by nontranscribed "spacer" segments of DNA. The rRNA genes are easily visualized, because 80 to 100 RNA polymerase molecules simultaneously transcribe each gene, producing a gradient of fibrils containing rpRNA molecules in progressive stages of maturation (Fig. 3). The slight foreshortening of each gene from the expected B-conformation length may result from denaturation of the double helix at the transcription site within each of the polymerases.

The relatively shallow gradient of RNP fibrils on a gene suggests that rpRNA molecules are coiled within the protein that associates with the

Fig. 3. A single nucleolar gene showing RNA polymerase molecules on the DNA axis at the base of each matrix fibril. *T. viridescens*.
Fig. 4. Portion of dispersed HeLa cell genome.

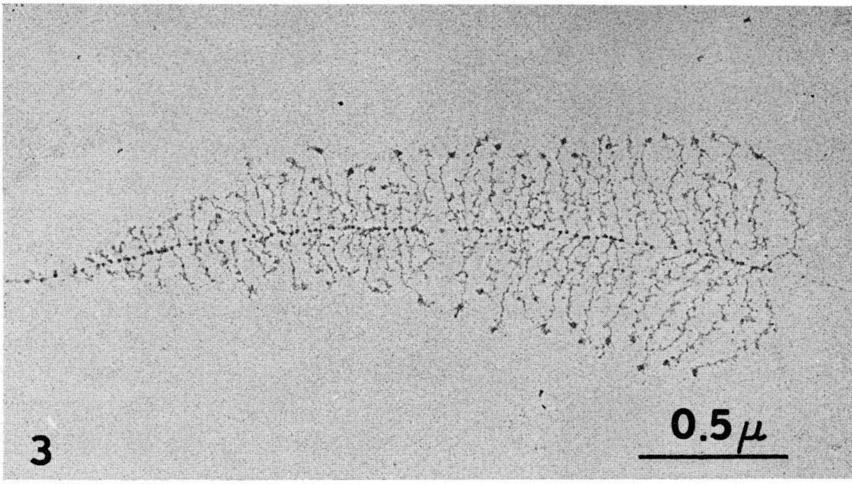
0.5 μ
3

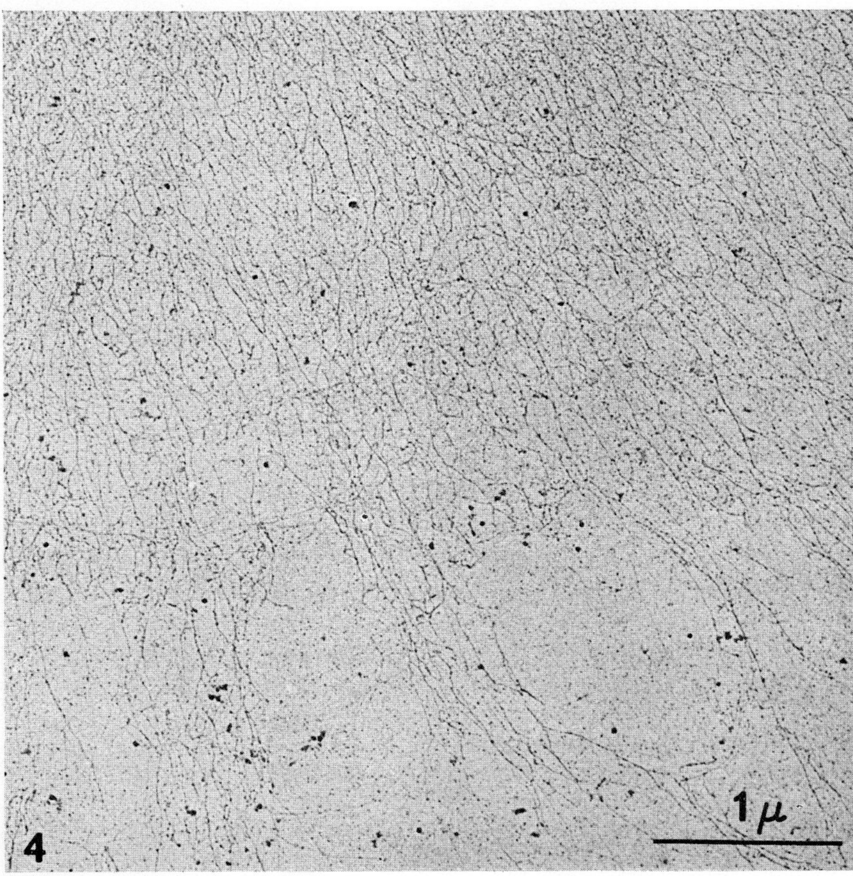
1 μ
4

molecules as they are synthesized. Fibrils at the termination end of a gene are about 0.5 μ long. A fully extended 40S rpRNA molecule would be about 5.2 μ long. Thus, as rpRNA molecules are synthesized, they are apparently foreshortened within RNP fibrils to only one-tenth or so the actual length of the nascent rpRNA molecules. Small granules appear on the ends of the nascent RNP fibrils about one-fifth of a gene length from the initiation end of each gene. Such granules do not appear to result from a rolling up of the ends of nascent RNP fibrils, since the fibrils actually appear longer, rather than shorter, when the granules first appear.

A rough estimate of the number of RNA polymerase molecules active in rpRNA synthesis per oocyte nucleus in the South African clawed toad, *Xenopus laevis*, can be obtained using the following information: (1) the amount of amplified DNA is 31 picograms (Macrgegor, 1968), which is about 9.3 m of duplex DNA (a picogram of duplex DNA is approximately 0.3 m long); (2) the repeating unit of an rRNA precursor gene plus spacer has a molecular weight of 9×10^6 daltons and this is equivalent to about 4.5 μ of duplex DNA (Dawid *et al.*, 1970) and (3) a minimum of 80 polymerases simultaneously transcribe each active rpRNA gene. If one assumes that all the amplified DNA is nucleolus organizer DNA and that every rpRNA gene is active, then at least 165×10^6 RNA polymerases are synthesizing rpRNA molecules per oocyte nucleus.

B. *HeLa Cell Cultures*

The large size of amphibian oocyte nuclei facilitates mechanical removal of the nuclear envelope and rapid dispersal of nuclear organelles prior to preparation for electron microscopy. The nuclei of typical vertebrate cells, however, are considerably smaller, usually ranging between 5 and 15 μ in diameter. Although such nuclei can be isolated by conventional homogenization procedures, the nuclear envelope is not ordinarily sensitive to osmotic shock and thus presents an obstacle to the dispersal and observation of nuclear organelles from single nuclei. Our initial efforts in the handling of small cells have focused on discovering chemical means to break down the mechanical barrier of the nuclear envelope without concomitant destruction of the structure of active genomes. These pilot studies have used cultured HeLa cells as a model cell type. Some success was obtained by exposing

Fig. 5. Putative HeLa rRNA genes showing gradients of nascent RNP fibrils containing RNA molecules in successive stages of maturation. Log-phase HeLa cells are diluted 1:9 with cold *Joy* detergent to give a final concentration of 0.3%. After 35 sec, the cells are fixed with 1 ml of cold 0.1 M sucrose plus 10% formalin (pH 9), then processed for electron microscopy as described in the legend for Fig. 2.

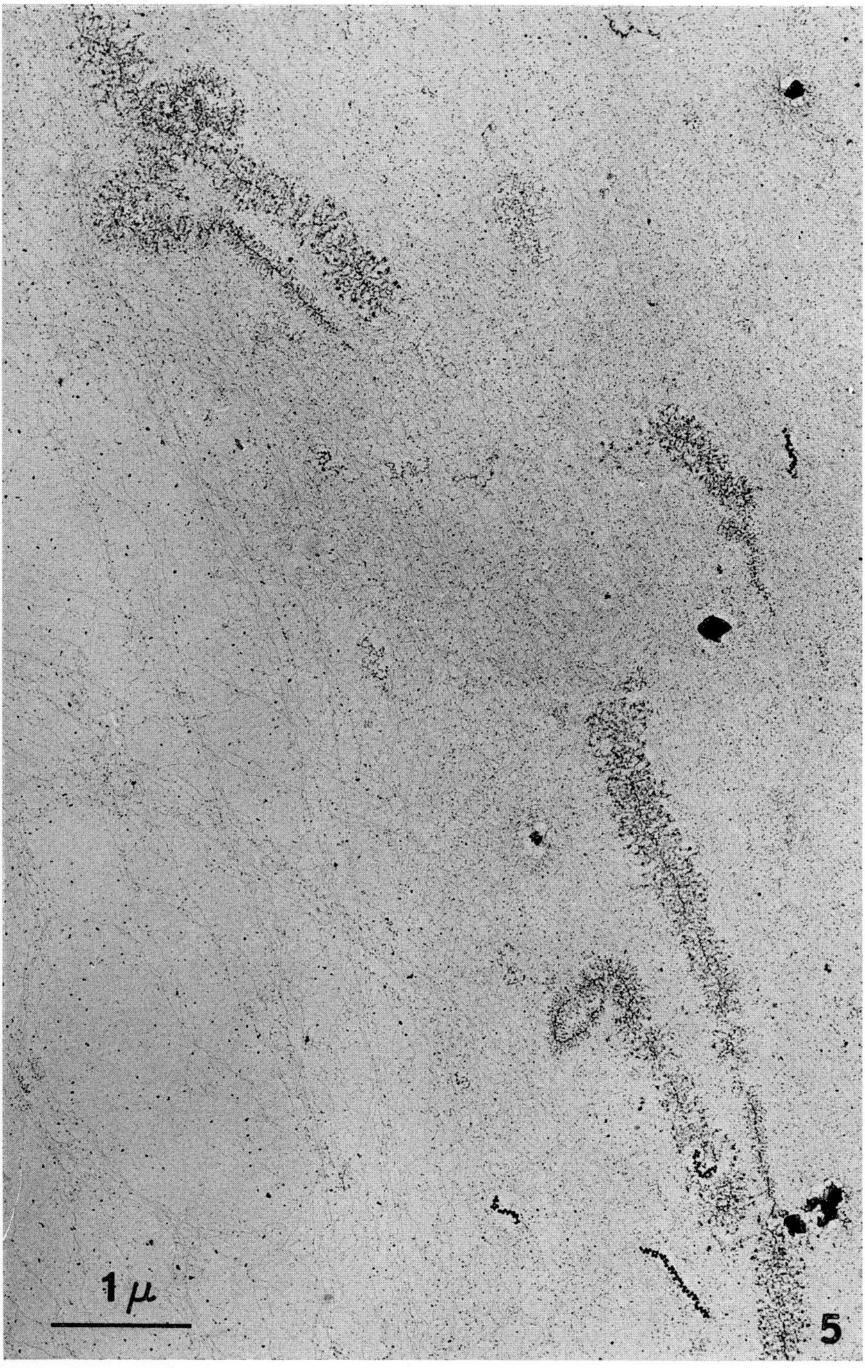
1 μ
5

nuclei to alkaline pH for a short time (30 sec, pH 11.5), by adding low concentrations of a mild commercial detergent to cultures (*Joy*, Proctor and Gamble) or by a combination of these two treatments.

Figure 4 shows an electron micrograph of a portion of the extruded contents of a HeLa cell nucleus, illustrating the complexity of the human genome. The task of unravelling such a quantity of DNA appears truly formidable. Figures 5 and 6 show segments of the HeLa genome tentatively identified as rpRNA genes because of the length of such segments and the striking similarity in the structure of the matrices to the well-characterized rRNA genes of amphibian oocytes.

The molecular weight of rpRNA molecules of HeLa cells is near 4.5×10^6 daltons (Loening, 1970). Thus, the B-conformation length of the rpRNA genes would be about 4.5 μ. The segments we observe are about 3.5 μ in length and we estimate that they have an average of 125 RNA polymerases active in RNA synthesis (Figs 6 and 7). As in the amphibian rRNA genes, we suggest that these loci are foreshortened from the expected B-conformation length by the large number of polymerases simultaneously transcribing each segment.

The gradient of nascent RNP fibrils on the putative HeLa rpRNA genes is shallower than that on the rRNA genes of amphibians (Fig. 6), suggesting that the RNA molecules are foreshortened to a greater degree within RNP fibrils. The RNP fibrils at the termination ends of these segments are about 0.25 μ long, or approximately one-thirtieth the length of a fully extended HeLa rpRNA molecule. At present, we do not know whether such an extreme foreshortening exists *in vivo* or results from the extraction of some protein from the nascent RNP fibrils by detergent or high pH treatments applied during isolation. As noted on amphibian rpRNP fibrils, granules appear on the nascent RNP fibrils of these presumptive HeLa rpRNA genes. In these cells, however, the granules appear approximately half a gene length from the initiation end.

It is likely that the putative rpRNA genes of HeLa cells are also separated by transcriptionally inactive spacers, even though we have not yet been able to visualize these directly because of the large numbers of deoxyribonucleoprotein fibrils forming a background on our grids. The presence of spacers can be surmised, however, from the fact that in well-dispersed clusters of these loci, distances up to 4 μ are observed separating single segments from other active segments of the cluster.

Fig. 6. Single HeLa rRNA gene.

Fig. 7. Portions of putative HeLa rRNA genes showing RNA polymerase molecules on the genes at the attachment points of nascent RNP fibrils.

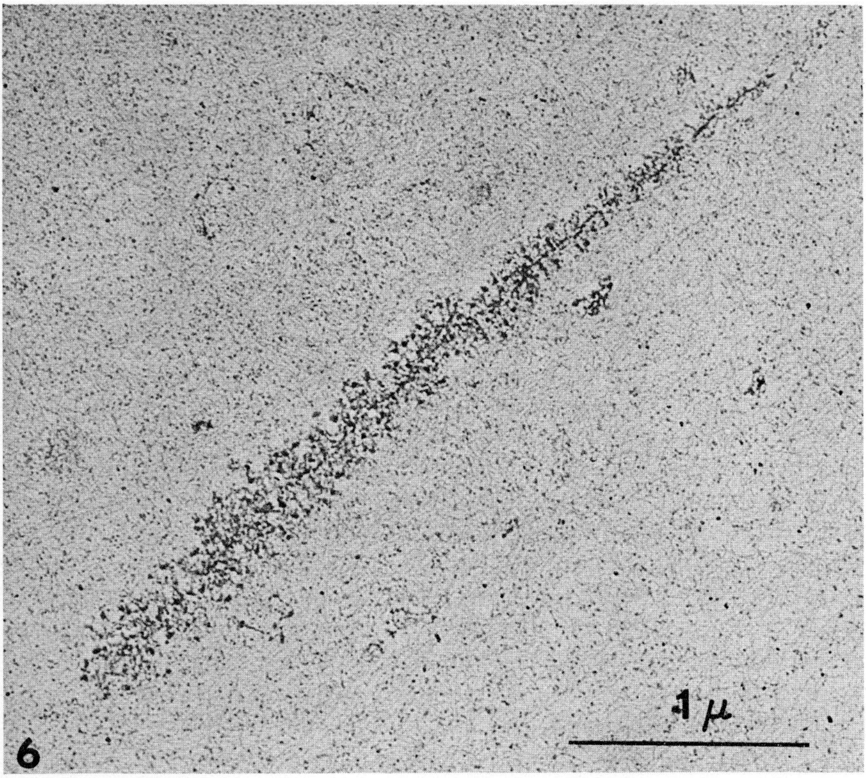
1 μ
6

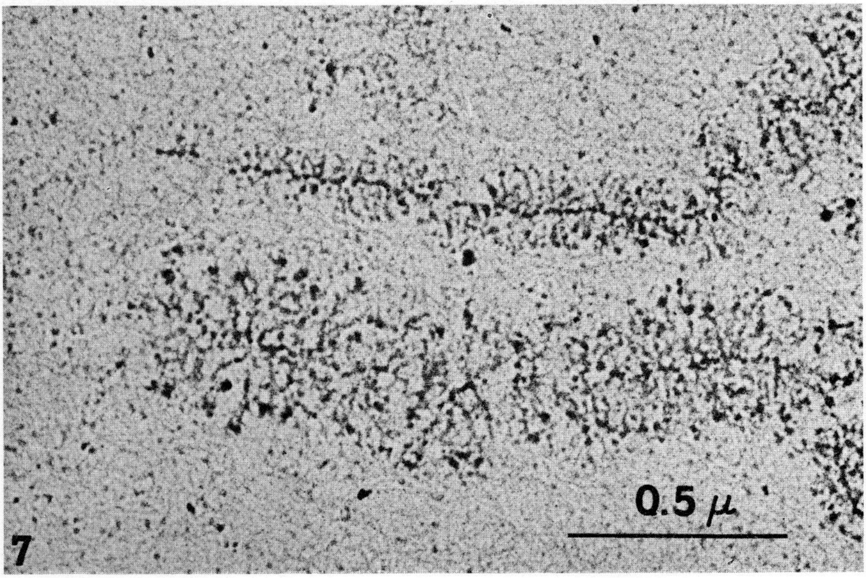
0.5 μ
7

C. *Discussion*

Although we have not provided definitive proof that segments of HeLa genome described in the preceding section are rpRNA genes, we believe that the weight of circumstantial evidence strongly supports this conclusion, and in this discussion we assume that such segments are active rpRNA genes.

In amphibians, the rpRNA molecule has a molecular weight of about 2.6×10^6 daltons, whereas the combined weight of the 18s and 28s rRNAs is 2.2×10^6 daltons. Thus, approximately 15% of each rpRNA molecule is discarded in the processing to form one 18s and one 28s rRNA. In HeLa cells, where the rRNAs have a combined weight of 2.45×10^6 daltons and the rpRNA has a molecular weight of 4.5×10^6 daltons, some 45% of the rpRNA is degraded during processing (Loening, 1970). In spite of these differences in rpRNA size and processing, active rRNA genes of amphibian and human cells exhibit an amazingly similar structural array. In both cell types, RNA polymerases cover virtually every active rRNA gene and the nascent rpRNA molecules are considerably foreshortened within their RNP fibrils. A third similarity between these genes in the two cell types is the appearance of small granules on the free ends of the fibrils. Finally, neighbouring rRNA genes of HeLa cells also appear to be separated by transcriptionally inactive spacers.

There is no known function for either the granules or spacer segments. However, data from amphibian cells (Reeder and Brown, 1970), mammalian cells (Siev *et al.*, 1969) and plant cells (Brown and Haselkorn, 1971) suggest that the 18s rRNA portion of the rpRNA molecule is synthesized before the 28S rRNA component. Therefore, the 18S rRNA probably is nearer the initiation end of the precursor molecule and also near, if not adjacent to, the granule-tipped ends of the rpRNP fibrils. It also has been shown that the 18s rRNA or its immediate precursor, migrates from the nucleolus to the cytoplasm much more rapidly than the pre-28s rRNA (Gall, 1966; Weinberg *et al.*, 1967). Thus, one is led to speculate that the terminal granules have a role in the processing and/or migration of the 18s rRNA.

Spacer segments between rRNA genes have been detected by electron microscopy in both amphibian and mammalian cells and may be present in all vertebrate cells. Brown *et al.* (1972) have shown that the rDNAs of two closely related *Xenopus* species have diverged considerably with respect to the base ratios of the spacers, whereas the rRNA cistrons are indistinguishable by DNA–RNA hybridization. However, spacer length in the two species is approximately the same. Consequently, any possible spacer function would not seem to require a stringent conservation of nucleotide sequences but may require maintenance of a minimal length. Spacer segments of dispersed nucleolus organizers typically lack matrix material. However, material has been found associated with spacers of isolated amphibian organizers under two abnormal conditions, after prior treatment of oocytes with actinomycin D and following storage of female animals at 12°C for several months. In both instances, fibro-granular material similar in structure to nucleolar cortex material was attached to spacer segments. One might speculate, therefore, that mature rpRNA fibrils have a labile association with spacers possibly in order to ensure proper cleavage during the first steps of processing the rRNAs.

III. CHROMOSOMAL RNA

In this section we consider structural aspects of the nuclear synthesis of large RNAs other than rRNAs. The inclusive term chromosomal RNA (chRNA) is used to denote these DNA-like RNAs.

A. *Amphibian Oocyte Chromosomes*

The structure and synthetic activity of lampbrush chromosomes of amphibian oocytes have been considered in a large number of studies (reviews by Callan, 1963; Callan and Lloyd, 1960; Gall, 1958; Hess, 1971). The general organization of such chromosomes as observed by phase-contrast light microscopy is shown in Fig. 8 and a schematic concept of the structure of lampbrush chromosomes is depicted in Fig. 9. These oocyte chromosomes are in an extended diplotene stage, with each pair of homologues joined by one to several chiasmata. The main axis of each chromosome consists of a series of closely spaced Feulgen-positive chromomeres joined by a submicroscopic fibre. Chromomeres typically exhibit pairs of Feulgen-negative lateral loops, each of which shows a thin and a thick insertion end at its chromomere. A variety of cytochemical techniques have been used to demonstrate that the loops consist of a DNase-sensitive axis coated with an RNP matrix and that each loop is synthesizing RNA.

Figure 10 shows an electron micrograph of a portion of a lateral loop near the thin insertion end. Similar to nucleolar genes, RNP fibrils are attached to RNA polymerase molecules that are closely spaced on the loop axis. Nascent RNA molecules again appear to be foreshortened within RNP fibrils, but in this case only to about one-fifth their full length. Although we have not been able to examine thick insertion ends of loops where the longest RNP fibrils would be expected, we have observed fibrils over 25 μ long at intermediate points along loops. If there are no nicks in the RNA, such fibrils must contain RNA molecules well over 100 μ long, or over 50×10^6 daltons in molecular weight.

An estimate of the number of RNA polymerases active in RNA synthesis on the lampbrush chromosomes of the spotted newt *Triturus viridescens* can be obtained with the following information. From our electron micrographs, we estimate the average centre-to-centre polymerase spacing to be 350 Å. Since the chromosome number of this species is eleven, there are 44 chromatids at the diplotene stage. An average chromatid bears approximately 1000 lateral loops, averaging about 50 μ in length (Gall, 1956). Thus, there are about 1400 polymerases per average loop or nearly 62×10^6 polymerase molecules simultaneously synthesizing chRNA in each oocyte of this species.

In spite of the large amount of information on the structure and synthetic activity of lampbrush chromosomes, little is known about the specific genetic activity of such chromosomes. At least some of the chRNA synthesized during oogenesis is stored in the oocyte and used for protein synthesis during early embryogenesis (Crippa *et al.*, 1967; Davidson *et al.*, 1966). Also it is known that a portion of the chRNA is synthesized on unique and a portion on repetitious nucleotide sequences of the genome (Davidson and Hough, 1969, 1971). However, it is not yet possible to identify cytologically the loci responsible for the stored mRNA or to locate unique and repetitious transcription sites.

B. *HeLa Cell Genomes*

The metabolism of chRNA in HeLa cells has been well characterized biochemically (Darnell, 1968). Excluding stable RNAs more than 95% of the RNA being synthesized at any instant in the HeLa cell nucleus is heterogeneous nuclear RNA (hnRNA) with sedimentation values of 20 to 100S. This RNA is contained in RNP complexes up to 2000S and is very labile, with an average half-life of 3 min. Since almost all non-nucleolar

Fig. 8. Phase-contrast photograph of a portion of a pair of lampbrush chromosomes isolated from an oocyte of *T. viridescens*. (Photograph courtesy Dr. J. G. Gall, Yale University).

Fig. 9. Upper left: diagrammatic sketch of a pair of lampbrush chromosomes joined by two chiasmata. Upper right: drawing of pairs of loops showing differences in morphology and length. Bottom: concept of continuity of chromatids in main chromosome axis with those forming axes of the lateral loops. (Reproduced with permission from Gall, 1956 after Swanson, 1957).

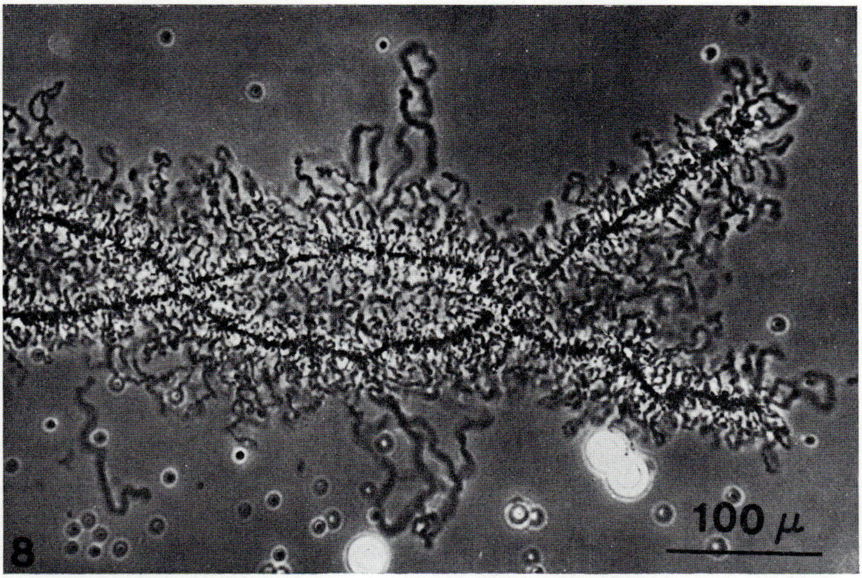
100 μ
8

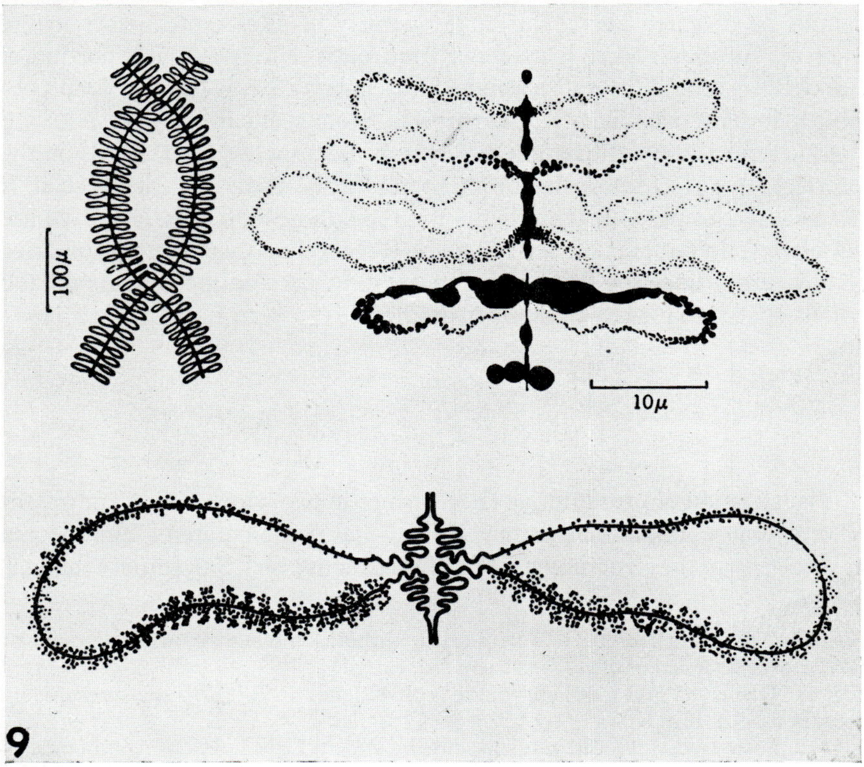
100μ
10μ
9

RNA synthesis per unit time is hnRNA synthesis, we would predict that essentially all synthetic activity, other than rRNA synthesis, visualized on isolated HeLa genomes would be hnRNA synthesis.

Figure 11 shows a portion of dispersed HeLa chromatin with attached fibrils. Although we have no direct proof, it is highly likely that these RNP fibrils contain nascent hnRNA molecules. Assuming little loss of material during isolation, RNA polymerases on active chromosomal loci in these cells are very widely spaced compared to those on lateral loops of amphibian lampbrush chromosomes. Unattached fibrils similar in structure to the fibrils attached to the genome are also present in extruded nuclear contents (Fig. 12).

C. *Discussion*

It is difficult to compare the mechanics of chRNA synthesis in the two studies described. No hnRNA synthesis has been reported for amphibian oocytes, whereas most of the RNA synthesized in HeLa cells is hnRNA. It is possible that hnRNA is synthesized on lampbrush chromosomes but cannot be detected above the huge quantity of rRNA precursors in the nucleus. Although there is evidence from both cell types that a portion of the chRNA enters the cytoplasm as messenger RNA, it has not been possible to localize messenger RNA synthesis cytologically in either case.

One can calculate the length of RNA within attached RNP fibrils on the lateral loops of lampbrush chromosomes fairly accurately, because it is possible to compare fibril gradients with length of loop axes. We have not yet observed gradients on active loci of HeLa cells other than the putative rRNA genes, since the RNA polymerases elsewhere on the genome are too widely spaced to form distinct fibril gradients.

IV. CONCLUSIONS

Visualization of presumptive rRNA genes in two widely divergent classes of vertebrates, amphibians, and mammals has demonstrated a high degree of similarity in the structural aspects of these active loci. Our initial examina-

Fig. 10. Thin insertion end of a lampbrush chromosome loop showing gradient of RNP fibrils attached to the loop axis. *T. viridescens*.

Fig. 11. Dispersed HeLa cell chromatin with attached RNP fibrils that presumably contain nascent hnRNA.

Fig. 12. Dispersed HeLa cell genome showing attached and free RNP complexes.

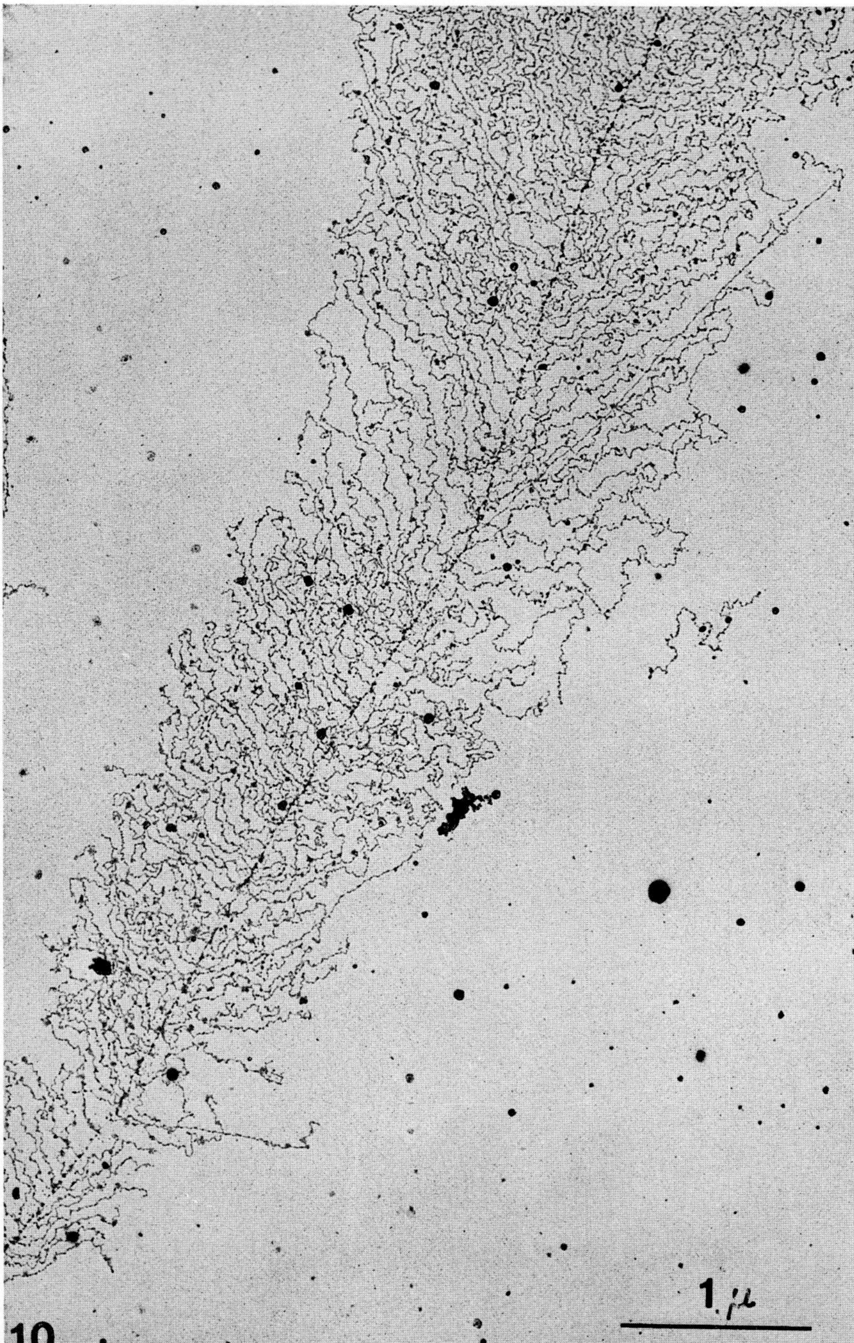
10
1 μ

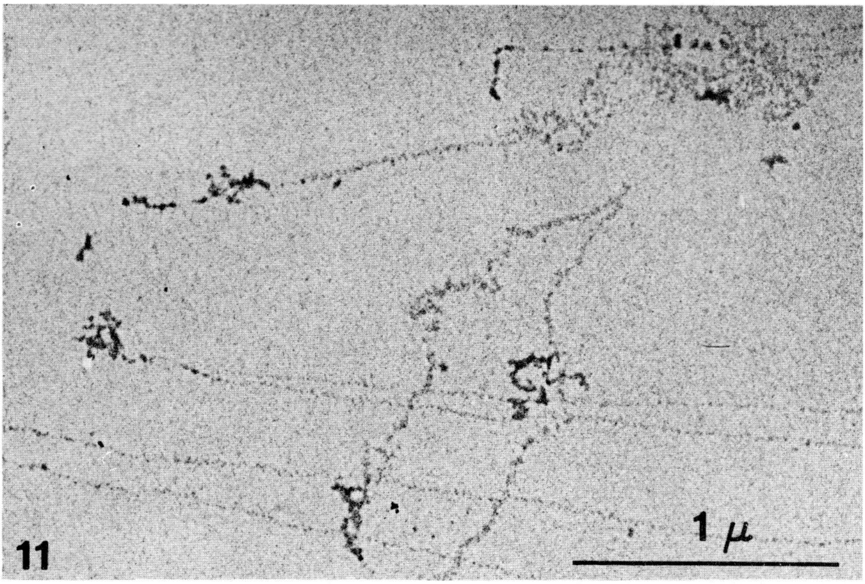

11

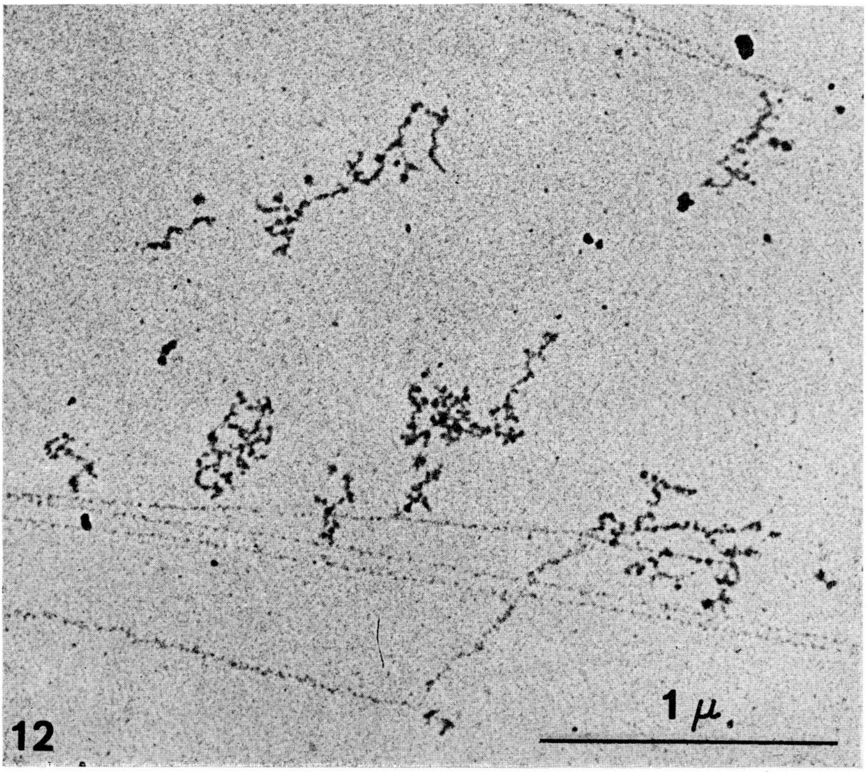

12

tions of the structural aspects of chRNA synthesis, however, present very different pictures for the two cell types studied.

It seems clear from these studies that to make valid comparisons between the structures of genetic activity in cells of different species it will be necessary to examine cells of the same type and, if they are dividing, at the same stage of the cell cycle.

Ultimately, innovations in techniques for dispersing nuclear contents of single vertebrate cells in known stages of the cell cycle should allow observation of specific genetic loci other than rRNA genes and open the way for comparative cytogenetics at the ultrastructural level.

Acknowledgement

We are grateful to Dr. R. B. Cumming of the Biology Division, Oak Ridge National Laboratory, for providing the HeLa cells and culture equipment used in these studies.

References

Attardi, G. and Amaldi, F. (1970). *A. Rev. Biochem.* **39**, 183–226.
Bostock, C. J. (1972). *Adv. Cell Biol.* in press.
Britten, R. J. and Davidson, E. H. (1971). *Rev. Biol.* **46**, 11–138.
Brown, D. D. and Dawid, I. B. (1968). *Science NY.* **160**, 272–280.
Brown, D. D., Wensink, P. C. and Jordon, E. (1972). *J. molec. Biol.* **63**, 57–73.
Brown, R. D. and Haselkorn, R. (1971). *J. molec. Biol.* **59**, 491–503.
Callan, H. G. (1963). *Int. Rev. Cytol.* **15**, 1–34.
Callan, H. G. and Lloyd. L. (1960). *Phil. Trans. R. Soc.* (*Lond*) Ser. B. **243**, 135–219.
Crippa, M., Davidson, E. H. and Mirsky, A. E. (1967). *Proc. natn. Acad. Sci. U.S.A.* **57**, 885–892.
Darnell, J. E. (1968). *Bact. Rev.* **32**, 262–290.
Davidson, E. H. and Hough, B. R. (1969). *Proc. natn. Acad. Sci. U.S.A.* **63**, 342–349.
Davidson, E. H. and Hough, B. R. (1971). *J. molec. Biol.* **56**, 491–506.
Davidson, E. H., Crippa, M., Kramer, F. R. and Mirsky, A. E. (1966). *Proc. natn. Acad. Sci. U.S.A.* **56**, 856–863.
Dawid, I. B., Brown, D. D. and Reeder, R. H. (1970). *J. molec. Biol.* **51**, 341–360.
Evans, D. and Birnstiel, M. L. (1968). *Biochim. biophys. Acta* **166**, 274–276.
Gall, J. G. (1956). *Brookhaven Symp. Biol.* **8**, 17–32.

Gall, J. G. (1958). *In* "The Chemical Basis of Development" (W. D. McElroy and B. Glass, eds), pp. 103–135. The Johns Hopkins Press, Baltimore.
Gall, J. G. (1966). *Natn. Cancer Inst. Monogr.* **23**, 475–488.
Gall, J. G. (1969). *Genetics* (*Suppl.*) **61**, 121–132.
Greenberg, J. R. and Perry, R. P. (1971). *J. Cell Biol.* **50**, 774–786.
Hess, O. (1971). "Handbuch der Allgemeinen Pathologie." Bd. II.
Lane, N. J. (1967). *J. Cell Biol.* **35**, 421–434.
Loening, U. E. (1970). *Symp. Soc. gen. Microbiol.* **20**, 77–106.
Macgregor, H. C. (1967). *J. Cell Sci.* **2**, 145–150.
Macgregor, H. C. (1968). *J. Cell Sci.* **3**, 437–444.
Miller, O. L., Jr. and Beatty, B. R. (1969). *Science, N.Y.* **164**, 955–957.
Miller, O. L., Jr., Beatty, B. R., Hamkalo, B. A. and Thomas, C. A., Jr. (1970). *Cold Spring Harb. Symp. quant. Biol.* **35**, 505–512.
Perry, R. P., Cheng, T–Y, Freed, J. J., Greenberg, J. R., Kelly, D. E. and Tartof, K. D. (1970). *Proc. natn. Acad. Sci. U.S.A.* **65**, 609–616.
Reeder, R. H. and Brown, D. D. (1970). *J. molec. Biol.* **51**, 361–377.
Siev, M., Weinberg, R. and Penman, S. (1969). *J. Cell Biol.* **41**, 510–520.
Swanson, C. P. (1957). "Cytology and Cytogenetics." Prentice Hall, Englewood Cliffs, New Jersey.
Weinberg, R. A., Loening, U., Willems, M. and Penman, S. (1967). *Proc. natn. Acad. Sci. U.S.A.* **58**, 1088–1095.

4

The DNA Nuclear Content and the Evolution of Vertebrates

MARIA GABRIELLA MANFREDI ROMANINI
Institute of Anthropology and Chair of Histology and Embryology, University of Pavia

I. INTRODUCTION

The whole of the DNA, that will make up the frame of the chromosomal structures characteristic of each individual and each species at mitosis, is present in the interphase nucleus in a particular chemical and physical state; that is, in a more or less marked uncoiled condition (eu- and heterochromatin). The cytochemical quantitative dosage of this DNA may therefore be taken as an indirect value of the "dimensions" which the karyotype will assume at metaphase and, above all, as the actual record of the amount of the substance which is directly or not involved in the coding of the gene in a definite cell; hence in a definite individual and in a definite species.

These concepts were already implicit in the so-called law of the constancy of the DNA content in the nuclei of all the somatic cells in individuals from the same species displaying the same degree of ploidy (Boivin *et al.*, 1948). In the following decades, through the shift from biochemical to cytochemical methods and the ever increasing data from both the animal and vegetable kingdoms, this assumption has attained more exact significance. In fact, once its limits are recognized to lie within those of a biological "constancy" correlated to the cell cycle (Fautrez, 1960), to particular metabolic requirements (Roels 1966, Pelc 1958, 1964, Pelc *et al.*, 1969), to hormonal stimuli (Alfert *et al.*, 1955; Desaive, 1967; Hymer *et al.*, 1970) and to major physiological crises (senescence etc.) (Fautrez, 1960, 1966a, b; Lesher *et al.*, 1961; Neurath *et al.*, 1970), the existence of a precise ratio between the DNA nuclear content (or "mass of genetic material" in Ohno's intuitively efficient words) and the karyotype characterizing a genus or even a species appears to be of the utmost importance.*

* All the more is this ratio important when the taxonomic allocation is dependent

Obviously, at the basis of such transpositions lies the concept of the validity of a correlation between the nuclear DNA content and the shape and size of the karyotype, or in more general terms, between the "genetic material" quantitatively determined and the corresponding formal genetic datum.

On these bases, the variability in DNA content between animal classes, genera and species, was readily viewed as a new tool for both an up-to-date approach to taxonomic problems and the study of the phylogenetic relationships among the various groups (Mirsky and Ris, 1951; Stebbins, 1966).

To accept the said ratio, however, we need accurately to clarify (a) possible biological and technical errors; (b) whether and to what extent the assumption of the existence of a ratio of the DNA nuclear content to the karyotype "size" is justifiable and (c) whether the ratio of the DNA nuclear content to nuclear and cellular dimensions may be established. Moreover, it is hoped that in the near future we shall be able to discuss this problem also with respect to (d) the possible significance in the frame of constant taxonomic values of the DNA content, of the correlation between "active" and "repetitive" DNA, since their ratio may differ at various taxonomic levels, and is also hypothesized to be endowed with evolutionary significance.

On these bases it will then be possible to outline the trend of distributions of the values of the DNA nuclear content among vertebrates in the light of the most recent phylogenetic interpretations. However, owing to the substantial inadequacy of such wide interpretations liable to criticism from many viewpoints, we are now compelled to restrict our observations to (a) the evaluation of the DNA nuclear content within classes, subclasses and superfamilies and (b) the attempts to utilize the studies on DNA nuclear content in taxonomically near orders and species, with the aim to contribute to the elucidation of the cytogenetic mechanisms that are deemed to be responsible for intrageneric, or even intraspecific, differentiation.

II. OUTLINE OF THE DISCUSSION

The value of the DNA content measurable in the interphase nucleus:
- (a) biological limits
- (b) technical limits and problems

DNA nuclear content and the karyotype

---

exclusively upon cytogenetic criteria (as for example in sympatric species neither reproductively isolated nor sufficiently differentiated phenotypically) and when as many discriminating items as possible are needed for a discussion on speciation mechanism.

DNA nuclear content and the meaning ascribable to repetitious DNA in different species.

DNA nuclear content and the major phylogenetic surveys proposed for the vertebrates.

DNA nuclear content and some phylogenetic fields referred to limited groups.

DNA nuclear content as a tool for elucidating cytogenetic problems.

A. *The DNA Content of the Interphase Nucleus*

(a) *Biological limits*

The DNA nuclear content (X) at the end of mitosis remains almost constant during the functional (specific secretive) phase of the cell cycle (G_1), doubles (2X) during synthesis (S) and so remains during the second period of stasis (G_2). As it is known, the length of these periods varies according to the cell type and its functional and nutritional state, as well as its volume (Killander and Zetterberg, 1965a, b) and also changes in different tissues and taxonomic levels. The synthesis may occur according to a different behaviour and in different moments during the phase S, for different chromosome groups of the same karyotype (Tagaki and Sandberg, 1968; Wright *et al.*, 1970). As no morphologic feature is associated with the transition from the first to the second and third phase, the cytochemical quantitative determinations of DNA, when obtained in the interphase nuclei are likely to fall in values X or from X to 2X, or 2X in proportion to the respective lengths of the various phases (Chévrémont and Bassleer, 1969). For this reason, cell cultures are utilized, which are synchronized by means of antimitotic drugs (Terasima *et al.*, 1963; Williams and Ockey, 1970) or, more frequently, those cells that may be considered to be consistently in phase G_1, e.g. cells from circulating blood. Lymphocytes have the advantage to possess a roundish nucleus, whereby the measurement of the area becomes easier. Although it is not possible to enter into a full discussion on this matter, I wish to recall that the measurements carried out on lymphocytes are 10% lower than in other cell types (Hale, 1963). Moreover, the importance of the steric configuration of the nucleoprotein complex in DNA cytophotometric determinations has been recently clarified by Mayall (1967, 1968, 1969, 1970 and Mayall *et al.*, 1967, 1970), not only as regards the protein component, but also when we assume that owing to the different degree of "condensation" of DNA in the chromatin, in the steric configuration of the molecule, some points become hydrolizable and therefore in a condition to yield dosable stained products.

In the same individual, differences in the DNA content in cell types displaying different nuclear volumes, may therefore be due to a different chromatin "condensation" rather than to absolute DNA content (Mayall, 1969; Noeske, 1968, 1969, 1971).

The validity of the constancy of the DNA nuclear content was also questioned when the possible existence of a so-called "metabolic" DNA fraction (Roels, 1966; Pelc, 1972), varying according to different cell requirements, was brought to light. That a given intensity level of physiological function may influence the synthetic activity of the genetic apparatus of the differentiated cells is at present largely accepted even in higher animals (Meerson, 1963, 1965; Pelc, 1969, and Pelc and co-workers, 1964 a, b). But recent interpretations (Noeske, 1971) are in contrast with this idea and attribute the variations of Feulgen values in cells variously stimulated to changes in the state of the chromatin. [For "metabolic" DNA in vegetables see Evans (1968a, b) and Miksche (1971) and for variations between wild and cultivated species Paroda and Rees (1971). For habitat influence of the DNA nuclear content in animals see Hauschteck-Jungen (1970)].

At any rate, the advisability of drawing a comparison only between morphologically and physiologically correspondent cell types from animals exhibiting as identical as possible physiological conditions, should be stressed.

(b) *Technical Limits and Problems*

The determination of the DNA nuclear quantity is a problem which has been studied from several points of view. Biochemically, i.e. with DNA dosages of known quantities of tissues followed by counts of the nuclei statistically present in the same tissue, Boivin *et al.*, 1948, and successively Vendrely and Vendrely, 1948, 1957 enunciated the first law of the DNA nuclear content constancy in cells pertaining to animals of the same species. Later on, in a parallel direction with this kind of "extra situm" research, dosages *in situ* of the nuclei, one at time, were utilized.

Owing to the delicacy of the instruments, we cannot, in the routine work, resort to the dosage of the absorption at 260 mμ of unstained material (Caspersson, 1936). Furthermore, this dosage is made on the total amount of purine and pyrimidine bases and it is not selective between DNA and RNA. It requires therefore concomitant enzymatic digestions (ribonuclease) which give rise to sundry problems (such as enzyme purity, optimum of the enzymatic activity and so on) not easy to be resolved in an unequivocal way. It is almost twenty years since coloured reactions (Methyl green (Brachet, 1942), naphtol yellow S. (Deitch, 1955), methylene blue (Deitch, 1964), fast green, Azur B. etc.) were utilized.

However none of these methods seem to avoid the risk of a possible final DNA association with competing proteins groups. (Devreux *et al.*, 1951; Godman and Deitch, 1957; Deitch, 1955, 1964; Flax and Himes, 1952; Bloch, 1958; Bloch and Godman, 1955a; Swift and Rasch, 1956; Alfert *et al.*, 1955; Bassleer, 1965).

The data obtained by these techniques are therefore to be judged with caution, all the more for being better acquainted today with the importance of DNA association with different proteins in different periods of the cell cycle and in different tissues (Auer *et al.*, 1970; Yateganas *et al.*, 1970).

In these latest years a notable favour was met with by the Gallocyanin chromalum staining method which, visualizing the phosphate groups of nucleic acids, is common to DNA and RNA but at a determined pH (1.65) seems to be exempt from the influence of the proteins and is a well reproducible and progressive stain (Sandritter *et al.*, 1966).

A DNA specific binding is obtained with acridine orange (Bertalanffy and Bickis, 1956, Ringertz *et al.*, 1969) but as yet such method has a problematic validity for quantitative researches (Ganter and Jolles, 1969) and it is used for the measurement of DNA template activity in various functional moments (Ringertz *et al.*, 1971; Noeske, 1971).

But most data on DNA nuclear content in animal and vegetal tissues were lastly obtained with the reactions derived from the classical one of Feulgen and Rossenbeck (1924).

In the meantime, changes were made in the type (Bloch and Godman, 1955a, b; Di Stefano, 1948) and modalities of the hydrolysis (Decosse and Ayello, 1968; Deitch *et al.*, 1968) and in the reactants revealing the presence of aldehydes (Ostergren, 1948; Van Duijn, 1956 etc.)

In this last field, some Schiff reagents with basic fluorochromes (Kasten, 1959; Prenna, 1964) were introduced in current use. The advantage of these fluorescent-Schiff reagents is due to the increased sensitivity (up to 10 times) and to the possibility of avoiding the distributional error (Prenna, 1968).

The Feulgen's type reactions are principally questioned on account of their poor reproducibility which compels comparisons to be made only among preparations which were simultaneously subjected to hydrolysis. Hence the need to use consistently when many comparisons are required, standard control preparations, which may involve complications in the understanding of the results (see for instance, Rasch, 1971). Furthermore, the influence of fixatives on the quantity of the stained product (Swift, 1955; Dutt, 1971) is well known. Critical surveys of the limits and amplitude of error inherent in different phases of the reaction are continuously reported in the literature and I refer to the same for better analysing this problem (Ghosh and Ghosh, 1969; Deitch, 1968; Sandritter *et al.*, 1966).

Moreover also the methods of quantitative photometric determination of the coloured product of the reaction and the technique of microfluorimetric dosage are ceaselessly revised (see Hale, 1966; Richards, 1966; Chayen and Denby, 1968; Ruch, 1966; Tonkelaar *et al.*, 1965; Mayall, 1969, 1970; Mayall *et al.*, 1970; Mendelsohn and Richards, 1958). Recently, the Feulgen reactions were studied on a critical basis also with reference to some contradictory results (Noeske, 1971).

According to Sandritter *et al.* (1965) and Bohm and Sandritter (1966) in the interphase nucleus two types of chromatin (condensed heterochromatin and diffuse euchromatin) give different responses to the Feulgen reaction, whereby constant results though not in the same ratio are obtained. This behaviour is explained by Ruch and Rosselet (1970) in terms of varying ratios between DNA and histonic proteins for heterochromatin and euchromatin. They believe that the histone influences the degree of condensation of the chromatin which, when the ratio DNA/histone exceeds a given value, condenses to a chromocentre and looses the function of a matrix for RNA synthesis (see also Noeske, 1968). According to Mayall's papers quoted above, the DNA in nuclei of greatly differing volumes, from the same animal, may show a different condensation which leads to different histophotometric results above all after the Feulgen reaction.

Garcia (1969 a) in his study of aploid and diploid cells "with different degrees of chromatin coiling" advocates the possibility that in the interphasic nucleus the more diffuse the chromatin the higher the DNA value recorded histophometrically.

We have been able to confirm Garcia's statements by some results obtained in lymphocytes subjected to hypotonic treatment. The treated cells have yielded a histophotometrically recorded DNA value 30% higher than normal cells to which the identical technique for metaphasic plates preparation had been applied. Only when the hypotonic procedure is more pronounced, loss of material sets in (Porcelli, 1971). In our opinion this last observation, which argues in favour of the invalidation of histophotometric results by some more or less marked "masking" of the active sites of the DNA molecule is of importance in the criticism specially with regard to the transfer of method such as the Feulgen reaction to isolated chromosomes (Gaillard *et al.*, 1968; Zirkin and Wolfe, 1970; Nitsch *et al.*, 1970).

Obviously, all the processing for DNA cytological dosage, in particular each step in Feulgen reaction, must always be analysed with the utmost care, in order to detect whether possible DNA loss is due either to prefixation (Dutt, 1971; Porcelli and Poggi, 1971), to hydrolysis (Porcelli and Farinetti, 1971) or to other causes.

Comparison between cells from different species may then be valid only provided the same cell type is studied, under the same conditions of chromatin condensation, interphase cycle time and functional state of the tissue.

B. *DNA Nuclear Content and the Karyotype*

Evidence has been largely provided, as previously reported, that the DNA nuclear content is statistically constant in all the cells exhibiting identical ploidy in the various tissues from the same animal or from animals of the same species, whereas it is significantly divergent in different species. According to this knowledge, in a given individual the correlation logically ensues between the nuclear DNA amount and the karyotype, namely the number and shape of the chromosomes characteristic of the species.

However, it must be emphasized that an explanation has failed to be given so far from either a chemical, physicochemical or structural standpoint, of the fact that a given amount of nuclear material DNA and proteins is stimulated at mitosis to become distributed into a particular number of chromosomes, above all of particular shape, attaining very precise dimensions when their coiling reaches a maximum.

Some attempts to discuss the above events are found in Fautrez, 1960, 1966 a, b (see also Ris, 1969; Luykx, 1970; Du Praw, 1966, 1970).

Failure to understand these phenomena is due to our imperfect knowledge of chromosome structure (Wolfe, 1969). As already pointed out, there is a correlation between the species, the DNA content, on the one hand, and, on the other, shape, dimension and structure of the chromosomes: whereby the existence of a precise relationship between DNA content and karyotype may be admitted.

In 1958, Kleinfeld and Chu had already shown that between the DNA content and the chromosome number no link exists nor, as it emerged in the vast majority of cases from our studies (Manfredi Romanini, 1971), does it exist between the DNA content and the autosome fundamental number (Matthey's F.N., 1949).

Therefore, the need arises to establish a value related to the actual dimensions of the karyotype instead of to discrete basic units such as individual chromosomes or a single chromosome arm that may change considerably in different karyotypes.

Ohno (see Atkin *et al.*, 1965) was the first to establish a correlation between DNA nuclear content and karyotype dimensions; he was able to measure the karyotype through enlarged photographs of metaphase plates that were cut out and weighed. Despite the notable causes of error inherent, in this type of measurement, the values obtained by him could be correlated in large scale comparisons, with those yielded by the cytophotometric determinations of the DNA nuclear content. Of course rather than being analytical, this comparison pertains to two patterns of behaviour, thereby indicating the possible existence of a ratio of the interphase nuclear DNA content to the volume of the chromatinic material at its maximum coiling in the metaphase plate, in which the DNA content has doubled and is about to divide.

Using Ohno's technique the size of the karyotype is considered as an "area", the constancy of the third dimension being taken for granted. A similar photographic method was utilized by Fox (1969) for measuring the chromosomes of *Dermestes*. He determined the length and the area of each chromosome or bivalent and from these values he calculated the volume, expressed in arbitrary units. Other techniques (Chiarelli *et al.*, 1959; Chiarelli, 1962, 1963a, b; Capanna and Civitelli, 1967; Martin and Hyman, 1965) go as far as to disregard two of the three spatial dimensions, by admitting that in the metaphase plate the chromosome calibre at its maximum coiling is constant, and just refer to the "genomic length" by measuring by means of more or less refined techniques the length values of individual chromosomes and summing them up.

The results thus obtained allow interesting comparisons among the various karyotypes to be made, especially as regards the assessment of varying extents in the loss of chromosome material. The validity of Capanna's method of chromosome measurement was indirectly confirmed by us (Capanna and Manfredi Romanini, 1971) in *Miniopterus*, where photometric evidence proved a relatively lower DNA content, as compared to the genus *Myotis*, in sharp agreement with Capanna and Civitelli's (1965) data on the genoma length.

Unfortunately, the genome length has not been measured on a sufficient number of karyotypes from species living at different taxonomic levels, so as to enable the compatibility, if any, between the DNA nuclear content and the fluctuations of the genome length among all of the vertebrates.

C. *The Nuclear DNA Content and the Nuclear and Cellular Volumes*

As early as 1957 the correlation between DNA nuclear content and nuclear volume in the erithrocytes from non-mammalian vertebrates was demonstrated by Vialli. He stressed, however, that this correlation is valid only within a relatively wide error limit, namely of species exhibiting DNA values almost twice that of other species (e.g. tench and carp) having similar nuclear volumes. Moreover, a number of "anomalies" in the nuclear DNA content were revealed that are to be ascribed not only to the particular features of a given tissue, but also to the cell type (capable or not of proliferation), to the cell cycle time (Bloch and Godman, 1955b; Prenna and Bianchi, 1964b) (in connection with DNA replication at the interphase) or even related to the conditions of the organisms to which the cell belongs (e.g. senescence).

Subsequent investigations have further clarified the above findings, and to-day it is known that the ratio of DNA content to nuclear volume may vary, in some cell types at least (Schnedl, 1969), since the doubling of the former is not exactly paralleled by that of the latter.

However, while a consistent direct correlation between DNA nuclear content and nuclear volume is advocated by some workers (Swift, 1950; Fautrez, 1960), others endorse Bloch's postulate, based on experimental assays by a large number of authors (Schrader and Leuchtenberger, 1950; Leuchtenberger and Schrader, 1951; Alfert *et al.*, 1955; Bassleer, 1965; Schreiber *et al.*, 1966, 1971; Bachmann and Cowden, 1965a, b) that in a great many cases a "heterosynthetic interphase" occurs in which the variations in the nuclear volume are more dependent on those of the protein content than on DNA duplication. The special role played by the nuclear proteins during mitotic rearrangements is being studied at present through different approaches and also through the refinement of DNA dosages (Auer *et al.*, 1970).

Moreover in senescence, evidence was provided that the ratio DNA/nuclear volume undergoes variations (Desaive, 1967; Neurath *et al.*, 1970).

In 1969 Alfert and Das proposed that, for the nuclear DNA, a real correlation ought to be established not with nuclear volume but with the nuclear surface area and formulated the hypothesis that the nuclear envelope may control DNA synthesis.

The idea is now generally accepted that comparison between DNA nuclear content and nuclear volume must be accomplished only between cells in the G_1 period, either destined or not to be followed by DNA synthesis (e.g. Bizzozzero's "labile" and "perennial" cell populations, respectively).

An interpretation, revolutionary under certain aspects, of the correlation between DNA nuclear content and nuclear volume, has been suggested quite recently by Szarsky (1970). He shifts the correlation DNA nuclear content/volume to DNA nuclear content/cell volume, adopting the well-known hypothesis of a correlation between nuclear volume and cell volume. Szarsky maintains that the cell volume should be regarded as an adaptative value, influenced by the environment and natural selection. According to him, in the larger cells, with slower metabolism, the degree of metabolic activity should be seen in relation to the ratio cell volume/cell surface. In the vertebrate cells where metabolism is speeded up, the cell volume would also have diminished in the course of evolution. This is in keeping with the decrease in DNA content paralleling the increase of the evolutionary level, as is generally agreed (Mirsky and Ris, 1951). Thus, the differences in DNA content would result from the different cell volume dictated by the evolutionary adaptation to the environment and life conditions.

The existence of a nuclear volume/cytoplasmic volume ratio has long been known; and the existence of some correlation between the three variables was also emphasized by Vialli, (1961). Recently it was postulated by Vialli, in his introductory remarks (Vialli and Sacchi Vialli, 1969) to an extensive biological and paleontological study on bone lacunae, the link between extant and fossil material as regards the DNA nuclear content in vertebrates on the assumption of the validity of a consistent correlation between DNA nuclear content and the cell volume, of which the bone lacunae represent the mould (see also Vialli, 1971).

Moreover, it has been demonstrated that the cell volume (Fox and Pardee, 1970) does not affect the degree of proliferative activity of a population of synchronized cells in culture, which seems to conflict somewhat with Szarsky's hypothesis however suggestive it may be.

In the light of present day knowledge the validity of the following correlations seems tenable:

(a) DNA nuclear content/karyotype size, or more precisely DNA nuclear content with the area of metaphase chromosomes (Ohno *et al.*, 1966) or the overall length of the chromosome complex;

(b) DNA nuclear content/nuclear volume provided the above mentioned exceptions are kept in mind and the possible different degree of chromatin coiling, liable to invalidate cytophotometric data in different cell types or differently treated cells, is not disregarded;

(c) nuclear DNA content/cell volume: for comparisons on a wide scale, namely, as concerns individuals highly diverging in their nuclear DNA content and cell volume.

D. *DNA Nuclear Content and the Meaning ascribable to Repetitive DNA in different species*

A completely different outlook of this problem may be in future obtained from the recent results reported by Britten and co-workers (Britten and Kohne, 1968; Britten and Davidson, 1969) of the existence in the DNA of a given species, of a correlation between total DNA and the bulk of repeated sequences of DNA, to which a general controlling role (the Housekeeper) (Walker *et al.*, 1969) and also the evolutive potentiality of the cell (namely, its capability of future adaptation, depending upon environmental changes) may be attributed. This correlation differs in different groups and perhaps different taxonomic levels. Obviously, the most exciting aspect of these new findings from the viewpoint of taxonomic rebuilding, lies in the possibility, already stated by the earliest authors who brought to light this condition, that "the extent of reassociation between DNA strands derived from different species was shown to be a measure of the evolutionary relations between species" (Hover *et al.*, 1965).

Although the ratio of DNA reassociation to the genome size (here considered as the haploid DNA cell content) has been demonstrated, we are still far from its utilization as a research tool. In all the species studied, DNA repeated sequences were seen to occur and their presence seems to be a general characteristic of higher organisms (Britten and Kohne). The pattern of amount of repeated sequences differs from species to species, but does not vary in cells belonging to different tissues of the same individual or to individuals from the same species, or even in cultured cell lines arising from the same cell strain. Thus, at least from this point of view, the amount of repetitious DNA and the length of the repeated sequences seem to be characteristics of the cell genetic content, as likewise are the parameters "total DNA nuclear content" and the other on which cytogenetics is based, namely the species-specific constancy of the characteristics of the karyotype.

Recently (Laird, 1971), the relationship between kinetics of strand reassociation of DNA and cytochemical nuclear DNA dosages have been discussed in order to define a model of chromatid structure (see addendum).

E. *The DNA Nuclear Content and Some Phylogenetic Surveys proposed for the Vertebrates: General Pattern of the Nuclear DNA Value in Vertebrates*

As the so-called law of the constancy of the DNA nuclear content in individuals from the same species was formulated by Vendrely and Vendrely, (1948), a large amount of biochemical results were also provided by them, bringing a light a striking variability of the DNA values in animals' cells, with particular reference to the vertebrates. Mirsky and Ris first envisaged in 1951 the possibility of lending an evolutionary value to the nuclear DNA and directed their attention to the whole animal kingdom by supplementing a number of data also concerning the invertebrates. Manton (1950), and more recently, Stebbins (1966) expanded their observations to the vegetable kingdom. They noticed some fluctuation in the nuclear DNA content in vegetables as well. For a general discussion on the variations of DNA

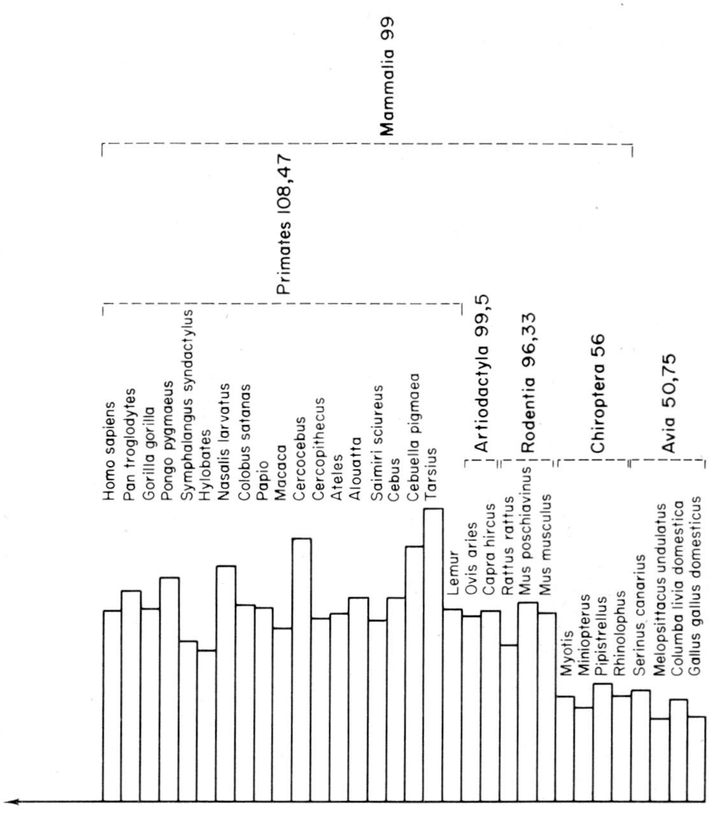
Mammalia 99
Primates 108,47
Artiodactyla 99,5
Rodentia 96,33
Chiroptera 56
Avia 50,75
Homo sapiens
Pan troglodytes
Gorilla gorilla
Pongo pygmaeus
Symphalangus syndactylus
Hylobates
Nasalis larvatus
Colobus satanas
Papio
Macaca
Cercocebus
Cercopithecus
Ateles
Alouatta
Saimiri sciureus
Cebus
Cebuella pigmaea
Tarsius
Lemur
Ovis aries
Capra hircus
Rattus rattus
Mus poschiavinus
Mus musculus
Myotis
Miniopterus
Pipistrellus
Rhinolophus
Serinus canarius
Melopsittacus undulatus
Columba livia domestica
Gallus gallus domesticus

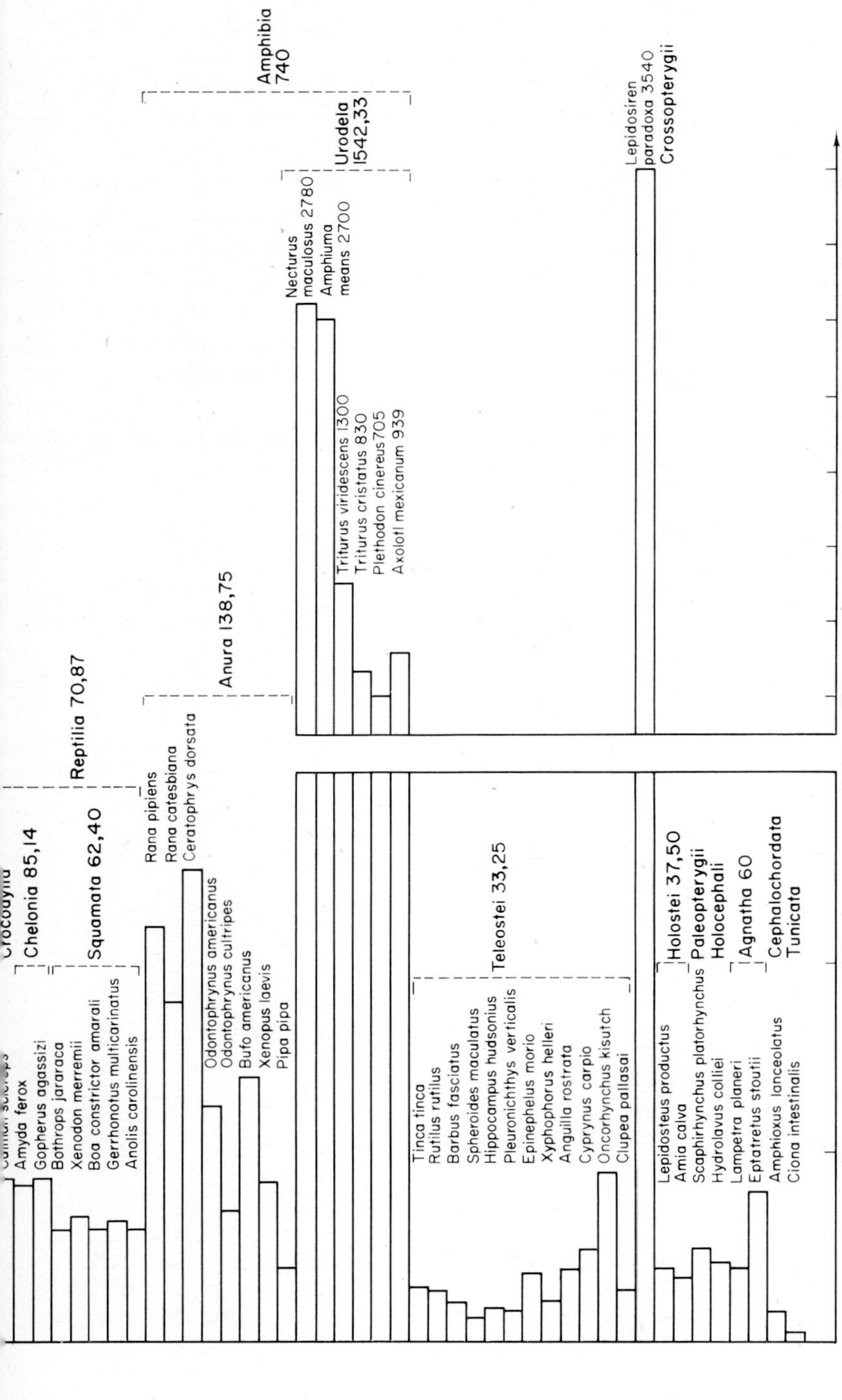

Fig. 1. DNA Nuclear Content (percentage of value in picograms in Man) in Vertebrata. (Ohno, 1970; Manfredi Romanini 1968, 1971; Capanna and Manfredi Romanini, 1971)

nuclear content in the animal and vegetable kingdom see also Schreiber (1971).

Among animals, the greatest variations are found in the vertebrates. In Fig. 1, the values (expressed in picograms) of the nuclear DNA content from specimens belonging to various vertebrate classes are reported. The data are expressed according to Ohno as a percentage of the value regarded as typical of mammals; that is to say, the volume of the man being reckoned as 100. It can be clearly seen that (1) a precise relationship exists between taxonomic position and DNA values, in the sense that the individuals within a given group (allowance being made for ample fluctuations) make up homogeneous complexes with respect to near groups (see Amphibia in particular); (2) no gradation can be established in moving from one group to another.

In Fig. 2, the data taken into consideration above are related to the

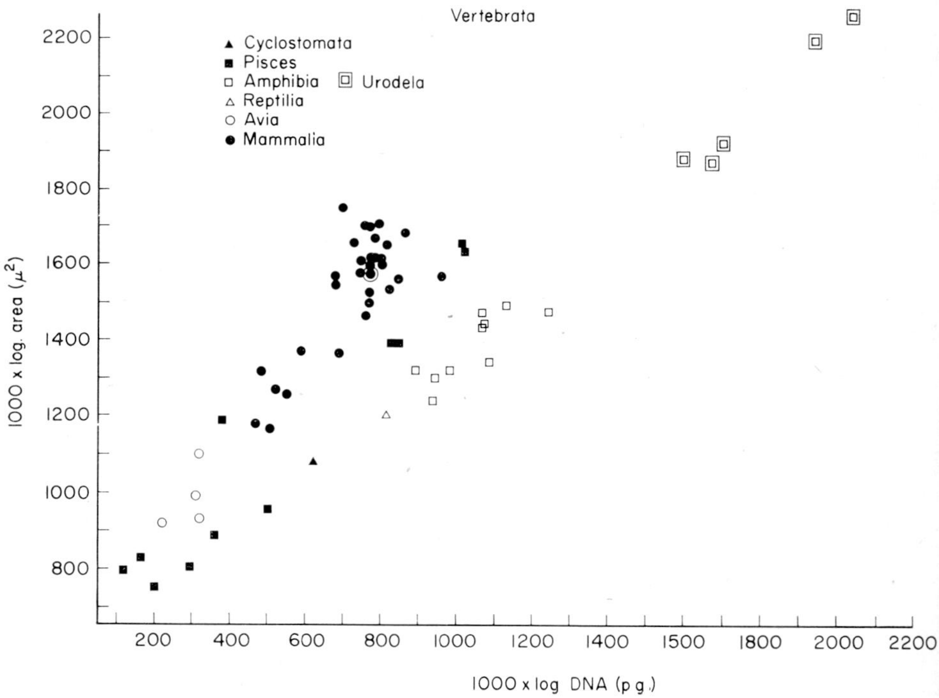

Fig. 2. Correlation between DNA Nuclear Content and Nuclear Area in Vertebrata. DNA Nuclear Content is Feulgen cytophotometrically recorded from Mammals Lymphocytes and Non Mammalian Vertebrata Erythrocytes; recalculated in picograms on the basis of the value of 6 picograms calculated for Man (Vendrely) and expressed in picogram logarithms × 1000. (From Vialli, 1957a; Manfredi Romanini, 1968, 1971; Capanna and Manfredi Romanini, 1971.)

nuclear area (which in the smears is known to be proportional to the volume), and the pattern of correlation proposed by Vialli appears expanded. It should be noted, however, that, unlike urodela, the other groups do not occupy sharply demarcated zones. Here not only the relationships between mammals, birds and the two major reptile groups (according to Ohno) seem clearer, but also the position of the two reptile groups, of the selachians, teleosts, dipnoa and in particular of the forms that can be regarded as primitive, such as the cyclostomes and ascidian as compared to gnatostomes and chordates (Atkin and Ohno, 1967) as well as the monotremes (Bick and Jackson, 1967), as compared to eutherian mammals.

Thus far, attempts to interpret in an evolutionary sense this pattern of nuclear DNA values are scanty. These attempts are logically and rightly tending to elucidate, on the basis of the variations in the DNA nuclear content, the mechanisms responsible for the transformation of the nuclear "genetic content" and through what means they have simultaneously affected the DNA nuclear content and reassessed the karyotype. In fact, on the assumption, discussed in the first section of this paper, that a precise even causal relationship between nuclear DNA content and the karyotype features exists, one may logically advocate the occurrence of strictly defined genetic mechanisms to be responsible for the vast transformations that have taken place passing from one group of vertebrates to another.

As is known, vertebrate evolution has occurred, according to Ohno, through gene duplication: "A new gene with a new function arose from a duplicate of the old gene" (Ohno, 1967). By comparing the most primitive extant chordates *Ciona intestinalis* with nuclear DNA $= 0.21$ mg $\times$ 10^{-9} equal to about 6% in eutherian mammals, karyotype $2n = 28$; *Amphioxus lanceolatus*, nuclear DNA $= 0.70$ mg $\times$ 10^{-9} equal to about 17% in eutherian mammals, $2n = 32$), he postulates a shift through pedomorphosis from *Ciona* to *Amphioxus*, by means of the duplication of the DNA nuclear content, may be through a tandem duplication and tetraploidy. In fishes, the nuclear DNA content from a value corresponding to that of amphioxus (17% of eutherian mammals) attains values even higher than those of extant mammals (*Lepidosiren paradoxa*, Dipnoa) and wide variations are found within the same genus (*Barbus*). Moreover, not one among the forms regarded as the earliest (holocephales, condrosteans, holosteans) or less evolved (cyclostomes) has very small genoma; *Lepidosiren*, thought of as a relic from the Dipnoa phylum that has survived unchanged from the middle Devonian, has a huge genome (3500% of eutherian mammals). Ohno believes that these differences may again be explained by a tandem duplication and poliploidy. Thus, he admits tandem duplication may explain the passage of the agnates to the various orders of magnitude of the genomes in the teleosts, while tetraploidy associated with genetic redundancy would account for the anomalies found, for instance, when ciprinid species with very divergent DNA contents and similar karyotypes are compared.

In *Lepidosiren* the huge genome might be viewed as an "evolutionary oddity" (Ohno, 1970) due to an abnormal increase through tandem duplication of the already conspicuous genome of the other primitive forms of the same group. The respective increase of so many tandemly duplicated copies of each cistron might have resulted in an enormous genetic redundancy paralleled by an equally high adaptive potential. This condition would have been followed by the fairly similar one found in most urodeles.

According to Ohno a "labyrinthodont lineage" would differentiate fishes, giving rise to anurans, from reptiles, birds and mammals and a "lepospondyls side branch" that after giving rise to urodela became extinct. Along the "labyrinthodont lineage" two more lines may somehow be recognized: (a) Diapsida and (b) Synapsida. On the basis of the nuclear DNA content (that is around 50% of that of eutherian mammals) and the characteristic occurrence of microsomes (Atkin *et al.*), the squamate reptiles and the whole class of birds (Diapsida line) may be fairly sharply recognized (Atkin *et al.*, 1965; Ohno, 1970). Another peculiar feature of this line is the mechanism of sex determination through female heterogamety. Ohno admits that the basic mechanism of these transformations is tetraploidization starting from those fish groups in which the nuclear DNA content averages 20% of that in eutherian mammals, but he explains the excess nuclear DNA content of squamate reptiles, compared to that of birds, as an excess of genetic redundancy due to tandem duplication following tetraploidization.

The other evolutionary line into which the labyrinthodont lineage divides is the synapsida line *sensu stricto*, which would lead to mammals, without the intermediate reptile forms. However, Ohno in this connection detaches the hydrosaurians from the "diapsida line" bringing them near the chelonian (anapsidae) on the basis of (1) the DNA nuclear content and (2) the lack of microchromosomes. For these reasons he links them to mammals and also here he admits the occurrence of a tetraploidization mechanism, though starting from those fish groups in which the DNA nuclear content is some 50% of that in eutherian mammals (e.g. Salmonida). Another feature which differentiated this evolutionary line from the preceding one is the mechanism of sex determination based on male heterogamety.

Lastly, it is noteworthy that according to Ohno the synapsida line is continued in all the three extant mammal groups (namely monotremes, marsupials and placentals) between which substantial differences in DNA content do not occur. Hence, a polyphyletic origin is advocated by the author both in the passage from fishes to the amphibians and, afterwards, separating the reptiles into two well demarcated groups essentially on the basis of the DNA nuclear content and the presence of microchromosomes (Ohno, 1970; Ohno, *et al.*, 1969).

Goin and Goin (1968) aim at correlating the greatest fluctuations in DNA nuclear content detected in the vertebrates with the great crises, which may be reflecting equally great ecological crises that were responsible for deep morphological changes (Simpson, 1953). In their opinion, two main periods of increase in the DNA nuclear content may be admitted: the first being associated to the passage of the earliest chordates to the vertebrates (see the comparison between the DNA nuclear content of *Ciona* and *Amphioxus*—already discussed by Ohno—and the average values for the vertebrates, among which the cyclostomes, nearer taxonomically, already have a value over twice that of *Amphioxus* and about 7 times that of *Ciona*); the second at the time of the passage from the great aquatic forms, such as Dypnoa to land life. From these high values one would then attain the actual distributions of values through many mechanisms of chromosome transformations involving the reduction of the total DNA amount in the genome.

Szarsky in a broader sense correlates (see Szarsky, 1970) the high cell volume and hence the high DNA content, with a slower metabolism, in relation to particular adjustments to the environment. Logically, this author also deals with the passage from aquatic to terrestrial life as a key event helping towards the understanding of phylogenetic relationships within the vertebrates. But, from his report, the idea emerges above all that the enormous DNA values of Dipnoa and Urodela correlated with the huge cell dimensions may be accounted for by their slow metabolism and the need of their cells to cope with the variations, often pronounced, in the composition of the body fluids. That is to say, rather than a phyletic correlation between the two groups he admits a convergence in their adjustment to similar environmental requirements. In relation to the adjustment of the individual to terrestrial life and by virtue of the body homothermy the cell would undergo a decrease in volume and DNA nuclear content as may be observed in mammals and, to a larger extent, in birds.

This interpretation in physiological terms, by Szarsky, is in generic agreement with Morescalchi's cytomorphological data (Morescalchi, 1970). The latter takes the lead from the fact that in all the tetrapod groups and in Actinopterigii a karyotype pattern is consistently found which exhibits the following characteristics: (a) high chromosomal number; (b) presence of many acrocentrics and; (c) presence of microchromosomes. Furthermore, in all the vertebrate classes a modal DNA nuclear value ranging from 4 to 5π per nucleus is found. Hence Morescalchi proposes that all the karyological changes of all the classes start from this ever present karyotype undergoing either DNA loss or gain, whereby the wide fluctuations of this value is accounted for.

He believes that some degree of differentiation may result in microchromosome loss (as is seen to occur in passing from actinopterigians to teleosts) and is generally coincident with a decrease in DNA nuclear content. Thus, he calls "conservative" the karyotype which displays many microchromosomes found by him in many reptiles, birds, monotremes and, possibly, in *Tarsius* (Primates). On the basis of some analogies, above all morphological ones, Morescalchi claims that the karyotype of the anura may be linked with that of marsupials and placentals; but he stresses that the great variability in nuclear DNA found in the amphibians is contrasted by its striking stability in mammals.

To sum up, a monophyletic evolution of the genome in the various tetrapod classes is claimed by Morescalchi following increase or decrease in the DNA nuclear content that would be due to environmental requirements (see Szarsky) or merely as a consequence of a greater or lesser degree of differentiation.

From any point of view, it is clear that the picture provided by cytological data and DNA nuclear content is of an extreme complexity and hard to interpret.

The dark points and the perplexity aroused by any of the above interpretations are not less than those encountered when aiming at an investigation into the relationship between the various groups on purely morphological grounds (Simpson, 1953).

Ohno's postulate of the existence of a correlation between some values of the DNA nuclear content, sufficiently stable in some major groups, may be accepted as evidence of the attainment of evolutionary stability (birds, mammals). We may recognize the validity of Morescalchi's postulate that it is necessary to go back to a DNA content (5 to 6π) frequently found in all the groups, as a reference point in order to understand the differentiation of the various genera or species in terms of loss or gain of genetic material. Moreover, the concept that this differentiation must have been markedly influenced by the need of adjusting to environmental changes is supported by Szarsky's explanation (adopted by Morescalchi) of the huge amount of nuclear DNA content characterizing the Dipnoa and urodela. Both authors view them as the result of adaptation to similar environmental requirements rather than admitting an artificial phyletic correlation between the two groups.

It is clear that questions are plentiful on these points as well. Present knowledge makes it premature and even risky to build up an unequivocal system to which the whole evolutionary pattern of vertebrates is referred.

At least two orders of knowledge are needed for such an attempt to be tried:

(1) a far higher number of data on both the DNA nuclear content (and on nuclear volume/cell volume) and the karyotype of species living in a known environment, and above all of the species lying at the key points of the major branchings of the evolutionary tree. Thus, the DNA nuclear content of *Latimeria*, the only extant relic of Chrossopterigians (Vialli, 1957b) that is quite near that of fishes and far from the extremely high values of Chrossopterigians has completely changed the interpretations of the phylogenetic relationships between Dypnoi and the Chrossopterigyans.

(2) Some knowledge, even indirect, is also needed about the possible genetic nuclear content of species now extinct, obtainable through both the analysis of the protein content (Prosser and Brown, 1961) and of the respiratory mechanism of fossil material (Raff and Raff, 1970) or by means of measurements of the areas of bone lacunae as indirect indication of the nuclear DNA content (Vialli and Sacchi Vialli, 1969).

We think that in this way, i.e. adding *the time* as a new parameter, we should obtain a tridimensional view of the problem which today is still studied only on the basis of two parameters: the DNA nuclear content (genetic material) and its patterning into the karyotype. In this way we should be able to understand the interrelations between the two mentioned parameters.

F. *DNA Nuclear Content and some Phylogenetic Pictures related to limited Groups*

With the actual research means at hand, it seems more logical and promising, leaving aside vast synthetic surveys, to go further into the analysis of small groups within which the characteristics of near genera or species are more readily grasped. Without descending to the finest microevolutionary levels (which we will discuss again later), many interesting features have emerged from an intensive study of the distribution of the DNA nuclear contents within a class (amphibians), a subclass (teleosts), a superfamily (primates) or a group within a family (microchiroptera palaeoarctica). Morescalchi *et al.*, in a quite recent discussion on the DNA nuclear content in the representatives of the three amphibians orders (apods, urodeles and anurans) have presented the following interesting comments:

(1) the high DNA nuclear content and its high variability sharply demarcate these from all the other vertebrate groups. This appears as

evidence of a common monophyletic origin of all the class. In a more analytical sense, a correlation between DNA nuclear content and rate of development or metabolism (Szarsky) seems admissible.

(2) As to the high values displayed by some urodeles the above authors view them as a sort of "hypertely" both on the ground that (*a*) most primitive forms of the same group (Hynobiidae) have a much lower DNA nuclear content and (*b*) that 90% of the DNA from the species with very high values (Dipnoa) instead of genetical significance would be merely functional may be in connection with cell division.

Hinegardner (1968) has studied over 200 teleost species, measuring the nuclear DNA content fluorometrically. In this group, DNA values lie between 0.4 and 4.4 picograms. The particular interest of his study lies in his having correlated the DNA nuclear values with some changes in body shape to be regarded as a greater or lesser degree of specialization. It clearly emerges from his results that fishes attaining a most specialized body shape possess a lower amount of DNA nuclear content, while in the most primitive ones this content is higher.

Thus, Mirsky and Ris's (1951) postulates of the fact that a high DNA content reflects a still operating adaptative ability are confirmed, while lower nuclear DNA values would point to higher specialization.

Some studies have been carried out by us on a mammalian family, namely the primates, whose evolutionary plasticity is still recognized today. The results obtained by us are in line with those of Hinegardner. Among primates the higher nuclear DNA content is found in *Tarsius* (Porcelli, 1971a) showing a marked degree of primitivity. Lower values are those of Hylobatinae (Fig. 3) (Porcelli, 1971b) which are envisaged, as is known, as the "cul de sac" of an evolutionary phylum extinguished today.

Within an even narrower context, the DNA nuclear content of a group of paleoarctic microchiroptera was studied in collaboration with Capanna. Three *Rhinolophidae* ($2n = 58$–56; N.F. $= 60$) and nine *Vespertilionidae* ($2n = 44$–36; N.F. $= 50$) were used. Throughout this group the mean DNA nuclear content is almost 40% lower than that of Primates (Fig. 4). This result is quite in line with the generally accepted view of the evolutionary stability reached by bats (Baker (1967), Hus *et al.* (1969) as compared to the considerable plasticity still characterizing the Primates.

Moreover, the most evolved species, e.g. *Miniopterus*, adapted to swift flight, have a very low DNA nuclear content (see Mirsky and Ris, 1951). While a correlation with the chromosome number and the autosomic fundamental number was not found in any case, the ratio of DNA nuclear content to genoma length has been confirmed precisely by comparing *Miniopterus* and *Myotis*.

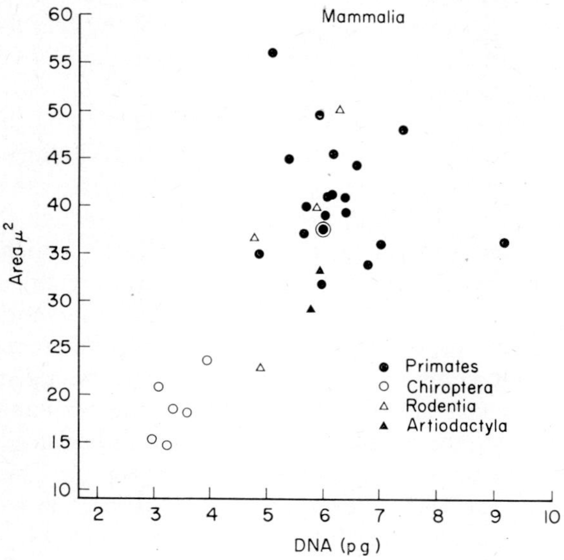

Fig. 3. Correlation between DNA Nuclear Content (picograms) and Nuclear Area in Mammals Lymphocytes.

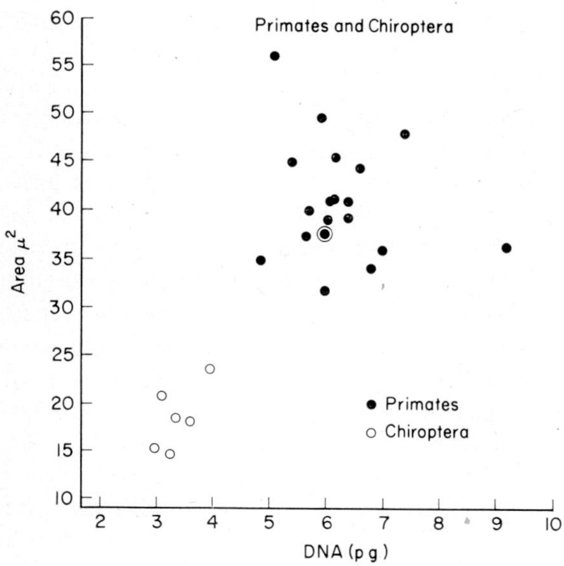

Fig. 4. Correlation between DNA Nuclear Content (picograms) and Nuclear Area in Primates and in Microchiroptera Palaeoarctica.

Analyses of this type, above all the association of the karyological datum with the quantitative cytochemical determination of the DNA nuclear content, are not frequent as yet.

If the exclusively karyological studies recently carried out by Beçak and Beçak on snakes and by Jovanic and Atkins (1969), and Ohno *et al.* (1964) on four fowl families have been supplemented by relative data on the DNA nuclear content, they would be of deeper significance, though already useful and interesting *per se*. And in fact, in contrast to Ohno's tendency to survey panoramically the behaviour of the DNA nuclear content according to classes and subclasses, I think that the DNA nuclear content datum is useful above all for inspection of finer karyotypic differences. For this reason it was planned by us to extend our study, among mammals, to groups displaying a well known extreme plasticity and adaptability, as is the case in rodents, and to the phylogenetically earliest eutherian group, e.g. the insectivores. At the same time comparison of European Chiroptera with the large American species has been planned in collaboration with Capanna. Confirmation was aimed at for all the Chiroptera, if possible, of the strikingly homogeneous behaviour observed in the first group already studied (*Microchiroptera palaeoarctica*), which consisted of ecologically fairly similar species.

G. *The DNA Nuclear Content as a Tool for understanding Cytogenetic Problems*

Though as yet sporadically distributed, much information is available on the DNA nuclear content in groups that are taxonomically so near one another and display such definite differences as to their karyotypes, that correlation of the latter with the variations in the nuclear DNA content appears feasible. Only a few instances closely related to the problem under study will be given here. Keyl's (1965) investigations on *Chironomus* and those by Rees and Jones (1967) on *Lalium* and *Allium* have made it possible to correlate differences in the DNA nuclear content, between near species, with a higher or lower amount of duplications of the chromosome segments excluding the possibility of ascribing them to differential polynemy. Grant (1966) has studied a series of taxa from *Betulla* with $2n = 28, 42, 56, 70$ and 84, showing a lower DNA nuclear content than that theoretically predictable in the species possessing a larger chromosome number. He has interpreted the above results as an adaptative phenomenon according to Darlington (1955), in relation to so high a degree of polyploidy.

Among the extremely numerous reports concerning the invertebrates, Schrader and Hughes-Schrader investigated various species of the genus

Thyanta (Hemiptera) under this aspect. They were able to ascertain, on the basis of the variations in the DNA nuclear content, that frequently the chromosome number of related species is the result of "fragmentation" rather than of polyploidy.

Fox (1969) carried out some studies in six species of the genus *Dermestes* (Coleoptera), reaching the conclusion that the high variations in DNA nuclear content from species to species are related to chromosome size variations (rather than number). He proposed this as a model in which longitudinal variations in structure of chromosomes are to be held as responsible for the differences between species.

Berendes and Keyl (1967) explained the differences in DNA nuclear content in a variety of cell models from *Drosophila* on the basis of the non-coincidence of DNA replication processes in euchromatin with respect to heterochromatin. Christensen (1966) succeeded in demonstrating, in various Enchytraeidae (Oligochaeta) species, the existence of DNA variations unrelated to variations in the chromosome number among species and in ascertaining the increase in DNA nuclear content both for differences among species due to polyploidy and for differential polyteny.

As regards vertebrates, recent karyological data (Chiarelli *et al.*, 1968) seem to have shed new light (Vialli, 1969) upon the complex problem of polyploidy relationships as was hypothesized and discussed for some cyprinid species (Meneghetti Gennaro, Conrieri, Manfredi Romanini *et al.*, Ohno and Atkin), while a keen interest was aroused by the studies performed by Schreiber and co-workers on the ratio of DNA to nuclear volume in polyploid amphibians (Schreiber *et al.*, 1966; Beçak *et al.*, 1967).

Ullerich (1966) has detected differences in the DNA nuclear content of three *Bufo* species and through the analysis of their karyotypes was able to exclude their dependence on differential polyteny. He explained them by admitting that during evolutionary processes a local increase in DNA occurred in all the chromosomes in *Bufo bufo* as compared to *Bufo calamita*, but only in the largest chromosomes in *Bufo viridis*. So far, many findings of intraspecific differences collected by Vialli and co-workers between 1950 and 1960 have not been paralleled by morphological studies on the karyotypes, hence their interpretation in cytogenetic terms is not feasible today.

Among the above studies, those by Vialli and Riccadonna (1957) on *Triturus cristatus* and *Triturus taeniatus* deserve mention, together with those of Vialli and Gerzeli (1955) on the ass and the horse. Furthermore, Vialli was the first to start a series of studies on hybrids (Vialli and Riccadonna and Vialli and Gerzeli) that today has been expanded along various lines and associated with karyologic observations, thereby already affording some relevant information (Ullerich, 1966).

Obviously, the fields already fairly well studied from a known cytogenetic point of view are likely to yield more rewarding results.

The wealth of cytogenetic data in mammals offers an ample choice of adequate material to provide evidence as to what extent the "genic content" changes concomitantly with determined speciation mechanisms.

The study of such a promising aspect of this problem has just been started in our laboratory. In some *Microchiroptera palaeoarctica*, besides suggesting some phylogenetic involvement in the behaviour of DNA values, the detailed analysis of single results has proved of remarkable interest in view of the association of quantitative DNA measurements with cytogenetics. Changes in the DNA nuclear content are not discernible between *Rhinolophus ferrumequinum* and *Rhinolophus hipposideros* (a centric fusion being claimed by Capanna and Civitelli, 1964 to have occurred in the transition from one species to the other) whereas changes in nuclear DNA content are observed in the transition from *Rhinolophus ferrumequinum* and *Rhinolophus euryale* viewed as dependent on two distinct non-Robertsonian mechanisms. DNA loss occurs when passing from *Myotis* (with a karyotype considered as primitive) and *Miniopterus* (more evolved) among which a variety of chromosomal mechanisms, with at least one centric fusion are believed to have occurred. DNA loss, however, cannot be clearly interpreted, owing to the complexity of the differences between the karyotypes of these two species.

Some differences in the karyotype length and a lack of correspondence between individual chromosomes, were pointed out by Capanna and Civitelli, 1970 between *Pipistrellus kuhli* and *Pipistrellus savii*, that display an identical chromosome and autosomal fundamental numbers. A more substantial difference has emerged from the discrepancy between the respective DNA contents, with DNA loss, if speciation is carried out in the direction *kuhlii* → *savii*. Lastly, the well known uniformity of the karyotype in the genus *Myotis* (Strelkov and Volobujev, 1969) seems merely morphological, since significant changes in DNA content were found to exist between *Myotis myotis* and *Myotis capaccinii*.

That no DNA loss occurs in particular in centric fusions seems further supported by the fact that no significative difference in the DNA nuclear content was found in the lymphocytes between two groups of *Rattus rattus* specimens displaying 38 and 42 chromosomes respectively (Manfredi Romanini and Capanna, 1971). Centric fusions between the two karyotypes are postulated (Capanna *et al.*, 1970).

Further evidence of the usefulness of some knowledge of DNA content at a cytogenetic level may be provided by a quite recent finding obtained by us on the material brought to notice by Gropp, namely on *Mus poschiavinus*. This species (tobacco mouse) is sympatric to *Mus musculus* from which it derived after adaptation to a particular habitat. It was found in fact exclusively in a restricted area of Poschiavo Valley, at about 1000 m above sea level, in the Grigioni Swiss Alps. *Mus musculus* and *Mus poschiavinus* display an identical N.F. = 40; the karyotype of *Mus musculus* consists of 40 acrocentrics; that of *Mus poschiavinus* is $2n = 26$ (14 metacentrics, 12 acrocentrics), seven centric fusions being involved in the mechanism of speciation. Comparison between the DNA contents from two males results in no significant difference. Comparison between two females, on the other hand, shows that in *Mus poschiavinus* the female has a higher DNA nuclear content; this discrepancy, difficult to explain, seems to invalidate the results yielded by males and raises a fresh problem. The mobilization of genetic material in speciation seems likely to have occurred. The comparison between sexes leads to the hypothesis of a different state of heterochromatinization of autosomic material in the two species, and probably in the two sexes. Obviously the above hypothesis needs further support, maybe by further research on the degree of inertia of the heterochromatinic material (Fujita, 1965) in *Mus poschiavinus*. Again comparison between *Ovis aries* and *Capra hircus*, two differing genera studied by us (Manfredi Romanini *et al.*) which exhibit almost identical karyotypes (Dain, 1970) with $2n = 56$ and $2n = 60$ respectively and two centric fusions (Taylor *et al.*, 1969) has yielded preliminary results similar to those from *Mus musculus* and *Mus poschiavinus*. Further investigation is therefore suggested, extended to other situations, by means of histochemical different methods, into the problem of centric fusions.

III. CONCLUDING REMARKS

(1) The DNA nuclear content may be considered as a constant value in all the somatic cells from all the individuals of a given species, provided it is measured in cells that are in the same interphase period, under comparable physiological conditions and with standardized technical processing.

(2) The DNA nuclear content, being a species-specific value, may accordingly be correlated with the morphological substrate of the definition of species, that is to say, to the karyotype regarded as its total dimension or as values correlated to the latter such as the area or total length of chromosomes.

(3) The ratio of the DNA nuclear content to the nuclear volume is valid only in fairly wide limits and must be discussed for each type of model under study. Moreover, the ratio of the DNA nuclear content to the nuclear volume may be correlated with the cell volume. The variations of these

three values may be taken as functional adaptations to various environmental requirements.

(4) An overall survey of the behaviour of the above values among vertebrates clearly indicates that they are stabilized at the various levels. At present, both the values of the DNA nuclear content and, within certain limits, those nuclear and cellular volumes which are more variable may be considered as characteristic of the various groups within the different classes (a maximum variability being found in the amphibians and a minimum in birds and placental mammals).

(5) The correlation existing between the DNA nuclear content and the karyotype has made it possible for the cytogenetists to construct some interpretations of the course of phylogeny in the vertebrates. We recall here the polyphyletic hypothesis (based on various mechanisms of gene duplication) advanced by Ohno, and the monophyletic interpretation recently forwarded by Morescalchi, in which DNA gains and losses at varying levels are hypothesized.

The lack of a sufficient body of information, above all at the level of the major branchings of the phylogenetic tree, and the fact that we cannot complete the present survey with the necessary information on the behaviour of the DNA nuclear content in the earliest epochs (information which it is hoped will be obtained in future from fossil materials), make any reconstruction of this kind questionable.

(6) More reliable, and hence of greater use for providing partial information, are investigations going deep into a taxonomically or ecologically homogeneous group, above all when the value of the DNA nuclear content may be correlated either to the karyotype value or to accurately carried out morphological variations.

(a) From the body of investigations performed in this sense at various levels the hypothesis of Mirsky and Ris (1951) remains valid in its main outline. Where a higher specialization degree is attained, with the ensuing reduction in evolutionary plasticity, the DNA nuclear content has decreased (Hinegardner (1968), in Teleosts; Manfredi Romanini (1971) in Primates; Capanna and Manfredi Romanini (1971) in Microchiroptera).

(b) In addition, the amphibians as a class are detached from all the other vertebrates for both the generally high level of their DNA nuclear content and the extremely great variability shown by its value (along with the related nuclear and cellular volumes). In the Amphibia therefore this appears as a model particularly suited to verify also the correlations between the values under examination and the environmental metabolic requirements according to Szarsky.

(7) According to present knowledge, the level at which the investigations

on the DNA content seem liable to contribute more definitely also in the field of genetic mechanisms of evolution, lies in the analytical study of the events of speciation. A number of examples are available in the literature concerning the invertebrates; less abundant as yet is information on the vertebrates. Recently our studies on Microchiroptera, on specimens of *Rattus rattus* with either 42 or 38 chromosomes, on the sympatric species *Mus musculus* and *Mus poschiavinus*, as well as on genera displaying a closely similar karyotype such as *Capra hircus* and *Ovis aries* have led to the following stimulating hypotheses:

(a) the absence of DNA loss in karyotypes arising from one another by centric fusion seems evident, while DNA loss occurs when speciation is due to other Robertsonian genetic mechanisms.

(b) Comparison of total nuclear DNA in males and females in species whose karyotypes are apparently correlated by centric fusions seems to suggest a different heterochromatinization of the autosomic material in the two species, with different patterns between sexes. Therefore, in this argument studies should be either extended to analogous situations or directed to the elucidation of the behaviour of the transferred material by means of histoautoradiography.

To date, in view of the ever increasing complexity of the events related to cytogenetics (see Mathey's chapter in this book) and the growing sensitivity of morphologic taxonomy (phenetic) to the critical requirements attending its phylogenetic reconstructions (Gilmour, 1961, Simpson, 1961, Sokal *et al.*, 1963), the expediency of quantitative cytochemical data on the DNA nuclear content is worth emphasizing, as a means of examining the mechanisms of speciation at all levels, on the basis of analytical comparisons.

These are in fact the key problems whose solution is mandatory for any phylogenetic interpretation.

References

Alfert, M. and Das, N. K. (1969) Evidence for control of the rate of nuclear DNA synthesis by the nuclear membranes in eukaryotic cells. *Proc. natn. Acad. Sci. U.S.A.* **63**, 123.

Alfert, M., Bern, H. A. and Kahn, R. H. (1955) Hormonal influence on nuclear synthesis. *Acta Anat.* **23**, 185.

Atkin, N. B. and Ohno, S. (1967) DNA values of four primitive cordates. *Chromosoma*, **23**, 10.

Atkin, N. B., Mattinson, G., Beçak, W. and Ohno, S. (1965) The comparative DNA content of 19 species of placental mammals, reptilis and birds. *Chromosoma*, **17**, 1.

Auer, G., Zetterberg, A. and Killander, D. (1970) Changes in binding between DNA and arginine residues in Histoche induced by cell crowding. *Expl. Cell Res.* **62**, 32.

Bachmann, K. and Cowden, R. R. (1965a) Quantitative cytophotometric studies on isolated liver cell nuclei of the bullfrog *Rana catesbiana*. *Chromosoma*, **17**, 22.

Bachmann, A. and Cowden, R. R. (1956b) Quantitative cytophotometric studies on polyploid liver cell nuclei of frog and rat. *Chromosoma*, **17**, 181.

Baker, R. J. (1967): Keryotypes of bats of the family Phyllostomidae and their texonomic implications. *Southwestern Nat.*, **12**, 407.

Bassleer, R. (1965) Determination de la teneur en protéines totales et en ADN dans les fibroblastes présentant diverses anomalies morphologiques des noyaux. *C. R. Hebd. Séanc. Acad. Sci. Paris*, **260**, 1777.

Beçak, W. and Beçak, M. L. (1969) Cytotaxonomy and chromosomal evolution. *Cytogenetics*, **8**, 247.

Beçak, W., Beçak, M. L., Levalle, D. and Schreiber, G. (1967) Further studies on Polyploid Amphibian (*Ceratophydidae*) II DNA content and nuclear volume. *Chromosoma*, **23**, 14.

Berendes, H. D. and Keyl, H. G. (1967) Distribution of DNA in heterochromatin and euchromatin of polytene nuclei of *Drosophila hidei*. *Genetics*, **57**, 1.

Bertalanffy, F. D. von and Bickis, I. (1956) Identification of cytoplasmic basophilia (Ribonucleic acid) by fluorescence microscopy. *J. Histochem. Cytochem*, **4**, 481.

Bick, Y. A. E. and Jackson, W. D. (1967) DNA content of Monotremes. *Nature Lond.* **215**, 192.

Bloch, D. P. (1958) Changes in desoxiribosonuclein complex during the all cycle. *In* "Frontiers of Cytology" (Palay, ed.) New Haven Univ. Press.

Bloch, D. P. and Godman, G. C. (1955a) A microspectrophotometric study of the synthesis of DNA and nuclear histone. *J. biophys. biochem. Cytol.*, **1**, 17.

Bloch, D. P. and Godman, G. C. (1955b) Evidence of difference in the deoxyribonucleoprotein complex of rapidly proliferating and non dividing cells. *J. biohpys. biochem. Cytol.*, **1**, 531.

Bohm, N. and Sandritter, W. (1956) Feulgen hydrolysis of normal cells and mouse ascites tumour cells. *J. Cell Biol.*, **28**, 1.

Boivin, A., Vendrely, E. and Vendrely, C. (1948) L'acide desoxiribonucléique du noyau cellulaire dépositaire des caractères héréditaires; arguments d'ordre analytique. *C.R. Hebd. Séanc. Acad. Sci. Paris*, **226**, 1061.

Brewen, J. C. and Peacock, W. J. (1969) Restricted rejoining of chromosomal subunits in aberration formation: a test for subunit dissimilarity. *Proc. natn. Acad. Sci. U.S.A.* **62**, 389.

Britten, R. J. and Davidson, E. H. (1969) Gene regulation for higher cells: a theory. *Science N.Y.*, **165**, 349.

Britten, R. J. and Kohne, D. E. (1968) Repeated sequences in DNA. *Science N.Y.*, **161**, 529.

Britten, R. J. and Kohne, D. E. (1969a) Repetition of nucleotide sequences in chromosomal DNA. *In* "Handbook of Molecular Cytology" (A. Lima de Faria, ed.) North Holland Publ. Co., Amsterdam and London.

Britten, R. J. and Kohne, D. E. (1969b) Implication of repeated nucleotide sequences. *In* "Handbook of Molecular Cytology" (A. Lima de Faria, ed.) North Holland Publ. Co., Amsterdam and London.

Bruce, V. G. (1965) Cell division Rhythms and the circadian clock. *In* "Circadian Clocks" (J. Aschoff, ed.) North Holland Publ. Co., Amsterdam.

Capanna, E. and Civitelli, M. V. (1964) Contributo alla conoscenza della cariologia dei Rinolofidi (Mammalia-Chiroptera). *Caryologia*, **17**, 361.

Capanna, E. and Civitelli, M. V. (1965) Cariologia e cariometria del Miniottero (Mammalia-Chiroptera). *Caryologia*, **18**, 541.

Capanna, E. and Civitelli, M. V. (1970) Chromosomal mechanism in the evolution of Chiropteran karyotype chromosomal tables of Cheroptera. *Caryologia*, **23**, 79.

Capanna, E. and Manfredi Romannini, M. G. (1971) Nuclear DNA content and morphology of the karyotype in certain paleoarctic microchiroptera. *Caryologia* **24**, 471.

Capanna, E., Civitelli, M. V. and Neser, R. (1970a) The karyotype of the black rat (*Rattus rattus* L.). Another population with a 38 chromosome complement. *Experientia*, **26**, 422.

Capanna, E., Savoia, G. and Manfredi Romanini, M. G. (1970b) Contenuto in DNA e area nucleare dei linfociti di alcune specie di microchirotteri. *In* Atti XXXIX Conv. V.2.I. Salice T.

Caspersson, T. (1936) Über den chemischen Aufbau der Strukturen des Zellkernes. *Scand. Arch. Physiol.*, **73**, suppl. 8, 1.

Caspersson, T., Lomakka, G. and Rigler, R. (1965) Regisrierender Fluoreszenzmikrospektrograph zur Bestimmung der Primär- und Sekundërfluoreszens verschiedener Zellsubstansen. *Acta Histochem.*, suppl. **6**, 123.

Chayen, J. and Denby, E. F. (1968) "Biophysical Technique as Applied to Cell Biology." Methuen and Co. Monographs, Worcester and London.

Chévrémont, M. and Bassleer, R. (1969) Recherches récentes sur la préparation à la mitose. *Acta Anat.*, suppl. **56**, 115.

Chiarelli, B. (1962) Comparative morphometric analysis of the Primate chromosome. I. *Caryologia*, **15**, 99.

Chiarelli, B. (1963a) Comparative morphometric analysis of the Primate chromosome. II. *Caryologia*, **15**, 401.

Chiarelli, B. (1963b) Comparative morphometric analysis of the Primate chromosome. III. *Caryologia*. **16**, 637.

Chiarelli, B. (1967) Statistical analysis of the relation between the length of the genome and the diploid chromosome number in *Cercopithecus*. *Experientia*, **23**, 672.

Chierelli B. (1968) Chromosome polymorphism in the species of the genus *Cereopithecus*. Cytologie **33**, 1–16.

Chiarelli, B., Nuzzo, F., de Carli, L. and Aulisa, B. (1959) Morphometric analysis of human chromosomes. *In* Atti A.G.I. **5**, 265–274.

Chiarelli, B., Ferrantelli, O. Cucchi, C. and Rossi, A. C. (1968) Dati preliminari sulla coryolopia depli Attinotterigi, *Boll. Zool.* **35**, 331.

Christensen, B. (1966) Cytophotometric studies on the DNA content in Diploid and Polipploid Enchytraeidae (Oliogochaeta). *Chromosoma*, **18**, 305.

Conrieri, C. (1957) Osservazioni citometriche ed istochimiche quantitative sul DNA negli eritrociti di *Tinca vulgaris* e *Barbus plebeius*. *Riv. Istochim. Norm. Pat.*, **3**, 253.

Dain, A. R. (1970) Separating the Sheep from the Goats on the basis of their chromosomes. *Nature Lond* **228**, 500.

Darlington, C. D. (1955), Nature (London), **175**, 1139.

Decosse, J. J. and Ayello, N. (1968) Feulgen hydrolysis: effect of acid and temperature. *J. Histochem. Cytochem.*, **16**, 371.

Deitch, A. D. (1955) Microspectrophotometric study of the binding of the anionic dye, naphthol yellow S by tissue sections and by purified proteins. *Lab. Invest.*, **4**, 324.

Deitch, A. D. (1964) A method for cytophotometric estimation of nucleic acids using methylene blue. *J. Histochem. Cytochem.*, **12**, 451.

Deitch, A. D., Wagner, D. and Richart, R. M. (1968) Conditions influencing the intensity of the Feulgen reaction. *J. Histochem. Cytochem.*, **16**, 371.

Desaive, C. (1967) Etude cytophotométrique des acides désoxyribonucléiques dans les noyaux de la glande de Loewenthal du rat albinos mâle à differents âges ou après modification de l'etat endocrinien. *Arch. Biol.*, **78**, 521.

Devreux, S., Johannson, M. and Errera, M. (1951) Affinité du vert de methy pour l'acide désoxyribonucléique. *Bull. Soc. Chim. biol.*, **33**, 800.

Di Stefano, H. S. (1948) A cytochemical study of the Feulgen nucleal reaction. *Chromosoma*, **3–4**, 282.

Du Praw, E. J. (1966) Evidence for a "folded fibre" organisation in human chromosomes. *Nature Lond* **209**, 577.

Du Praw, E. J. (1963) Cell and Molecular Biology. Academic Press, New York and London.

Dutt, M. K. (1971) Intensity of Feulgen staining of rat liver nuclei fixed with to different concentration of formalin. *Histochemie*, **25**, 60.

Einarson, L. (1951) On the theory of gallocyanin chromalum staining and its application for quantitative estimation of basophilia. A selective staining of exquisite progressivity. *Acta path. microbiol. scand.*, **28**, 82.

Evans, G. M. (1968a) Nuclear changes in flax. *Heredity*, **23**, 25.

Evans, G. M. (1968b) Induced chromosomal changes in flax. *Heredity*, **23**, 301.

Fautrez, J. (1960) Correlation between volume of the nucleus and its DNA content. *Z. mikrosk.-anat. Forsch.*, **66**, 210.

Fautrez, J. (1966a) "Le Dynamisme des Ecides Nucléiques." Gauthier-Villars, Paris.

Fautrez, J. (1966b) Variation de la teneur en acide désoxyribonucléique (ADN) des noyaux interphasiques en fonction de l'activité cellulaire et leurs relation avec les variations du volume nucléaire. *In* "Handbuch der Historchemie." Bd III/3 (W. Graumann and P. Neurath, eds.) Fischer, Stuttgart.

Feulgen, R. and Rossenbeck, H. (1924) Mikroskopisch-chemischer Nachweiss einer Nukleinsäure vom Typus Thymusnukleinsäure und die deranf beruhende elektive Färbung von Zellkernen in mikroskopischen Präparaten. Moppe-Seyler's *Z physiol. Chem.*, **135**, 203.

Flamm, W. G., Walker, P. M. B. and McCallum, M. (1969) Some properties of the single strands I selected from the DNA of the nuclear Satellite of the Mouse (*Mus musculus*). *J. Molec. Biol.*, **40**, 423.

Flax, M. H. and Himes, M. H. (1952) Microspectrophotometric analysis of metachromatic staining of nucleic acids. *Physiol. Zool.*, **25**, 297.

Fox, D. P. (1959) The relationship between DNA value and chromosome volume in the Coleopteran Genus *Dermestes*. *Chromosoma*, **27**, 130.

Fox, T. O. and Pardee, A. B. (1970) Animal cells: noncorrelation of length of G. Phase with size after mitosis. *Science N.Y.*, **167**, 80.

Fujita, S. (1965) Chromosomal organisation as a genetic basis of cytodifferentiation in multicellular organisms. *Nature Lond.* **206**, 742.

Gaillard, J. L. J., Van Duijn, P. and Schaberg, A. (1968) Photometric determination of DNA in human chromosomes. *Expl. Cell Res.*, **53**, 417.

Ganter, P. and Jolles, G. (1969) "Histochimie Normale et Pathologique." Gauthier-Villars, Paris.

Garcia, A. M. (1969a) Cytophotometric studies on haploid and diploid cells with different degrees of chromatin coiling. *Ann. N.Y. Acad. Sci.*, **157**, 1, 237.

Garcia, A. M. (1969b) Studies on deoxyribonucleic acid in leukocytes and related cells of mammals. VI. The Feulgen-deoxyribonucleic acid content of rabbit leukocytes after hypotonic treatment. *J. Histochem. Cytochem.*, **17**, 47.

Ghosh, S. and Ghosh, I. (1969) Some apects of the Feulgen reaction *in situ*. *Histochemie*, **19**, 165.

Gilmour, J. S. L. (1961) Taxonomy. *In* "Contemporary Botanical Thought" (A. M. McLead and L. S. Cobley, eds.) Oliver and Boyd, Edinburgh.

Gledhill, B. L., Gledhill, M. P., Rigler, R. and Ringertz, N. R. (1966) Changes in deoxyribonucleoprotein during spermiogenesis in the bull. *Expl. Cell Res.*, **41**, 652.

Godman, G. C. and Deitch, A. D. (1957) A cytochemical study of the L.E. bodies of systematic lupus erythematosus. I. Nucleic acids. *J. Expl. Med.* **106**, 575.

Goin, O. B. and Goin, C. J. (1968) DNA and the evolution of the vertebrates. *Am. Midl. Nat.*, **80**, 289.

Grant, W. F. (1966) Decreased DNA content of Birch (*Betula*) chromosomes at high ploidy as determined by cytophotometry. *Chromosoma*, **26**, 326.

Gropp, A. (1969) Cytologic mechanism of karyotype evolution in insectivores. *In* "Comparative Mammalian Cytogenetics" (K. Benirschke, ed.) Springer-Verlag, Berlin, Heidelberg and New York.

Gropp, A., Citoler, P. and Geisler, M. (1969) Karyotype variation and Heterochromatin-muster bei Igeln (*Erinaceus* und *Hemiechinus*). *Chromosoma*, **27**, 283.

Hale, A. J. (1963) The leucocyte as a possible exception to the theory of deoxyribonucleic acid constancy. *J. Path. Bact.*, **85**, 311.

Hale, A. J. (1966) Feulgen microspectrophotometry and its correlation with other cytochemical methods. *In* "Introduction to Quantitative Cytochemistry" (G. L. Wied, ed.) Academic Press, New York and London.

Hale, A. J. and Cooper, E. H. (1965) Studies on DNA replication in leukaemic and non leukaemic leukocytes. *In* "Current Research in Leukamia" (F. G. H. Hoyhoe, ed.) Cambridge Univ. Press. London.

Hall, A. V. (1968) Methods for aiding identification of critical groups in taxonomy. *Nature* (*Lond*), **218**, 203.

Hauschteck-Jungen, E. (1970) Quantitative DNS-Besimmungen von Ameisen-hirnzellen. II. Einfluss der Umwelt auf die Feulgen-Extinktions werte. *Chromosoma*, **32**, 79.

Hinegardner, R. (1968) Evolution of cellular DNA content in Teleost Fishes. *Am. Nat.*, **102**, 517.

Hover, B. H., Bolton, E. T., McCarthy, B. J. and Roberts, R. B. (1965) *In* "Evolving Genes and Proteins" (V. Bryson and H. Vogel, eds.) Academic Press, New York and London

Hsu, T. C. and Mead, R. A. (1969) Mechanism of chromosomal changes in mammalian speciation. *In* "Comparative Mammalian Cytogenetics" (K. Benirschke, ed.) Springer-Verlag, Berlin, Heidelberg and New York.

Hymer, W. C., Mastro, A. and Griwold, E. (1970) DNA synthesis in the anterior pituitary of the male rat: effect of castration and photoperiod. *Science N.Y.*, **167**, 1629.

Kasten, F. H. (1958) Additional Schiff-type reagents for use in cytochemistry. *Stain Technol.*, **33**, 39.

Kasten, F. H. (1959) Schiff-type reagents in cytochemistry. I. Theoretical and practical considerations. *Histochemie*, **1**, 466.

Kasten, F. H., Burton, V. and Glover, P. (1959) Fluorescent Schiff-type reagents for cytochemical detection of polyaldehyde moieties in sections and smears. *Nature Lond.*, **184**, 1797.

Keyl, H. G. (1965) Duplikationen von Untereinheiten der chromosomalen DNS Während der Evolution von Chironomus thummi. *Chromosoma*, **17**, 139.

Killander, D. and Zetterberg, A. (1965a) Quantitative cytochemical studies on interphase growth. I. Determination of DNA, RNA and mass content of age determined mouse fibroblasts in vitro and of intercellular variation in generation time. *Expl. Cell Res.*, **38**, 272.

Killander, D. and Zetterberg, A. (1956b) A quantitative cytochemical investigation of the relationship between cell mass and initiation of DNA synthesis in mouse fibroblasts *in vitro*. *Expl. Cell Res.*, **40**, 12.

Kleinfeld, R. and Chu, E. H. Y. (1958) DNA determination of kidney cell cultures of three species of Monkeys. *Cytologia*, **23**, 452.

Kubitschexk, H. E. (1971) The distribution of cell generation times. *Cell Tissue Kinet.*, **4**, 113.

Jovanovic, V. and Atkin, L. (1969) Karyotypes of four Passerine Birds Belonging to the families *Turdidae*, *Mimidae* e *Corvidae*. *Chromosoma*, **26**, 388.

Laird, C. D. (1971) Chromatid structure: Relationship between DNA content and nucleotide sequence diversity. *Chromosoma*, **32**, 378.

Lesher, S., Fry, R. J. and Kohn, H. I. (1961) Age and the generation time of the mouse duodenal epithelial cell. *Expl. Cell Res.*, **24**, 334.

Leuchtenberger, C. (1958) Quantitative determination of DNA in cells by Feulgen microspectrophotometry. *In* "General Cytochemical Methods" (J. F. Danielli, ed.)

Leuchtenberger, C. and Schrader, F. (1951) Relationship between nuclear volume, amount of intranuclear protein and DNA in various rat cells. *Biol. Bull.*, **101**, 95.

Lorick, G. (1970) Differential DNA synthesis in heterochromatic and euchromatic chromosome sets of *Planococcus citri. Chromosoma*, **32**, 11.

Luykx, P. (1970) Cellular mechanisms of chromosome distribution.. *Int Rev Cytol.*, Suppl. 2.

Manfredi Romanini, M. G. (1968) Quantitative relative determination of the nuclear DNA in the Old world Primates. *In* "Taxonomy and Phylogeny of Old World Primates with References to the Origin of Man". Rosenberg and Sellier, Torino.

Manfredi Romanini, M. G. (1970) The variations in nuclear genetic material content of Lymphocytes. *In* "Cytotaxonomic Picture of Primates". III Int. Congr. of Primatology, Zurich.

Manfredi Romanini, M. G. (1971) The nuclear DNA content and area of Primates Lymphocytes as a cytotoxonomical tool. *J. Hum. Evol.* (in press).

Manfredi Romanini, M. G. and Capanna, E. (1971) Contenuto in DNA nei nuclei postcinetici in due popolazioni cromosomicamente differenti di *Rattus rattus* (L.) Rend. Qcc. Naz. Lincei (Classe Scienze) Ser. VIII, **51**, 105.

Manfredi Romanini, M. G. and Metzar, A. (1957) Osservazioni sul contenuto in ADN dei nuclei degli eritrociti dei ciprinidi. *Boll. Zool.*, **24**, 565.

Manfredi Romanini, M. G., De Stefano, G. F. and Fontana, F. (1967) Ricerche sul contenuto nucleare in ADN e sulle aree dei linfociti in alcune specie di "Cercopithecoidea". *Riv. Antr.*, **54**, 3.

Manfredi Romanini, M. G., Minazza, M. and Capanna, E. (1971) DNA nuclear content in lymphocytes from *Mus musculus* and *Mus porchiavinus. Boll. Zool.*, **38**, 321.

Manfredi Romanini, M. G., Veneroni, S. and Porcelli, F. (1971) The DNA nuclear content in Lymphocytes of Sheep and Goats (in preparation).

Manton, I. (1950) "Problems of Cytology and Evolution in the Pteridophyta". Cambridge Univ. Press, London.

Martin, P. G. and Hyman, D. L. (1965) A quantitative method for comparing the karyotypes related species. *Evolution*, **19**, 157.

Matthey, R. (1949) "Les Chromosomes des Vertébrés". Libraire de l'Université F. Rouge, Lausanne.

Mayall, B. H. (1967) Variability in the stoichiometry of deoxiribonucleic stains. *J. Histochem. Cytochem.*, **15**, 762.

Mayall, B. H. (1968) Variability of DNA staining. III° Int. Congr. Histochem. and Cytochem., New York.

Mayall, B. H. (1969) Deoxyribonucleic acid cytophotometry of stained human leucocytes. I. Differences among cell types. *J. Histochem. Cytochem.*, **17**, 249.

Mayall, B. H. (1970) Deoxyribonuclei acid cytophotometry of stained human leucocytes. II. The mechanical scanner of CYDAC, the theory of scanning photometry and the magnitude of residual errors. *J. Histochem. Cytochem.*, **18**, 383.

Mayall, B. H. and Mendelsohn, M. L. (1967) Chromatin and chromosome compaction and the stoichiometry of DNA staining. *J. cell Biol.*, **35**, 182.

Mayall, B. H. and Mendelsohn, M. L. (1970) Errors in absorption cytophotometry: some theoretical and practical considerations. *In* "Introduction to Quantitative Cytochemistry" (G. L. Wied and G. F. Bahr, eds.) II, Academic Press, New York and London.

Meerson, F. Z. (1963) Interrelation of Physiological Function and Genetic Apparatus of the Cell. Moscow.

Meerson, F. Z. (1965) Intensity of function of structures of the differentiated cell as a determinant of activity of its genetic apparatus. *Nature Lond.*, **206**, 483.

Mendelsohn, M. L. and Richards, B. M. (1958) A comparison of scanning and two-wavelength microspectrophotometry. *J. biophys. biochem. Cytol.*, **5**, 707.

Meneghetti Gennaro, A. (1956) Osservazioni sul tenore relativo in ADN dei nuclei di eritrociti di Carpa e Tinca. Rend. Ist. Lombardo Sc. Lett., Cl. Sci., **90**, 172.

Miksche, J. P. (1971) Intraspecific variation of DNA per cell between *Picea sitchensis* (Bong.) Carr. Provenances. *Chromosoma*, **32**, 343.

Mirsky, A. E. and Ris (1951) The desossyribonucleic acid content of animal cells and its evolutionary significance. *J. gen. Physiol.*, **34**, 475.

Morescalchi, A. (1970) Karyology and vertebrate phylogeny. *Boll. Zool.*, **37**, 1.

Morescalchi, A., Olmo, E. and Gargiulo, G. (1970) Considerazioni sull 'evoluzione del DNA nucleare negli anfibi. *Boll. Zool.* (in press).

Muramoto, J. and Ohno, S. (1968) On the diploid state of the fish order Ostanophysi. *Chromosoma*, **24**, 59.

Nadler, C. F. (1969) Chromosomal evolution in rodents. *In* "Comparative Mammalian Cytogenetics" (K. Benirschke, ed.), Springer-Verlag, Berlin, Heidelberg and New York.

Neurath, P., De Romer, K., Bell, B. and Jarvik, L. (1970) Chromosome loss compared with chromosome size, age and sex of subject. *Nature Lond.*, **225**, 281.

Nitsch, B., Murken, J. D. and Brück, H. J. (1970) Determining Feulgen-DNA of individual chromosomes by fluorescence cytophotometry with incident light. *Histochemie*, **23**, 254.

Noeske, K. (1968) Die Zusammensetzung der Zellkernproteine und ihre Beziehung zum DNS-Gehalt im Verlauf der normalen Myelo- und Erythropoese beim Menschen Eine zytophotometrische Untersuchung. Thesis. Giessen.

Noeske, K. (1969) Diskrepanzen von Feulgen -wert und DNS-Gehalt. *Histochemie*, **20**, 322.

Noeske, K. (1971) Discrepancies between cytophotometric Feulgen values and deoxyribonucleic acid content. *J. Histochem. Cytochem.*, **19**, 169.

Ohno, S. (1967) Sex chromosomes and sex-linked genes. *In* "Monographs on Endocrinology" (A. Labhart, T. Mann and L. T. Samuels, eds.) Vol. I. Springer-Verlag, Berlin, Heidelberg and New York.

Ohno, S. (1969) The role of gene duplication in vertebrate evolution. *In* "The Biological Basis of Medicine" (E. E. Bittar and N. Bittar, eds.). Vol. IV. Academic Press, London and New York.

Ohno, S. (1970) "Evolution by Gene Duplication". Springer-Verlag, Berlin and New York.

Ohno, S. and Atkin, N. B. (1966) Comparative DNA values and chromosoma complements of eight species of fishes. *Chromosoma*, **18**, 455.

Ohno, S., Stenius, C., Christian, L. C., Beçak, W. and Beçak, M. L. (1964) Chromosomal uniformity in the avian subclass *Carinatae*. *Chromosoma*, **15**, 280.

Ohno, S., Muramoto, J., Christian, L. and Atkin, N. B. (1967) Diploid-tetraploid relationship among old-world members of the fish family Cyrpinidae. *Chromosoma*, **23**, 1.

Ohno, S., Wolf, U. and Atkin, N. B. (1968) Evolution from fish to mammals by gene duplication. *Hereditas* (*Lund*), **59**, 109.

Ohno, S., Muramoto, J., Stenius, C., Christian, L., Kittell, W. A. and Atkin, N. B. (1969) Microchromosomes in Holocephalian, Chondrostean and Holostean Fishes. *Chromosoma*, **26**, 35.

Ostergren, G. (1948) Chromatin stains of Feulgen type involving other dyes than fuchsin. *Hereditas*, **34**, 510.

Paroda, R. S. and Rees, H. (1971) Nuclear DNA variation in Eu-Sorghums. *Chromosoma*, **32**, 353.

Pelc, S. R. (1958) Nuclear uptake of labelled adenine in the seminal vesicle of the mouse. *Expl. Cell Res.*, **14**, 301.

Pelc, S. R. (1964) Labelling of DNA and cell division in so called non-dividing tissues. *J. cell. Biol.*, **22**, 21.

Pelc, S. R. (1972) Metabolic DNA in Ciliateol Protozoa Salivary Glands chromosomes, and mammolian cells, Int. Rev. Cyt., **32**, 327.

Pelc, S. R. and Viola-Magni, M. P. (1969) Decrease of labelled DNA in cells of adrenal medulla after intermittent exposure to cold. *J. cell Biol.*, **42**, 460.

Policard, A. (1964) "Cellules Viventes et Populations Cellulaires. Dynamique et Structure Moléculaire". Masson et Cie., Paris.

Porcelli, F. (1971a) Nuclear area and DNA content of lymphocytes in *Tarsius syrichta* L., Proc. 3rd. Congr. Primat., Zurich, 1970. Karker, Basel.

Porcelli, F. (1971b) Nuclear areas and DNA content of lymphocytes in Hylobatinae. Proc. 3rd .Congr. Primat., Zurich, 1970. Karger, Basel.

Porcelli, F. (1971c) Ricerche sui limiti di quantitatività della reazione di Feulgen. II. Influenza de trattamento ipotonico. *Rass. Med. Sper.*, **18**, 83.

Porcelli, F. and Farinetti, F. (1971) Ricerche sui limiti di quantitatività della reazione di Feulgen. I. L'idrolisi cloridrica: influenza dei tre parametri, concentrazione, temperatura, durata. *Riv. Istochem. norm. pat.* **17**, 77.

Porcelli, F. and Poggi, P. (1971) Ricerche sui limiti di quantitatività della reazione di Feulgen. III. Influenza della fissazione. *Riv. Istochem. norm. pat.* **17**, 61.

Prenna, G. (1964) Caratteristiche dell (2′-(p-aminofenil) 6-metil, 2–6′ bi-benzoatiazolo)-SO_2, un nuovo reagente tipo Schiff fluorescente ad elevata sensibilità. *Riv. Istochem. Norm. Pat.*, **10**, 469.

Prenna, G. (1968) Qualitative and quantitative application of fluorescent Schiff-type reagents. *Mikroskopie*, **23**, 150.

Prenna, G. and Bianchi, U. (1964a) Reazoni di Feulgen fluorescenti e loro possibilità citofluorometriche quantitative. IV. *Riv. Istochem. Norm. Pat.*, **10**, 645.

Prenna, G. and Bianchi, U. (1964b) Variazioni delle dimensioni nucleari dell'ADN in rapporto al ciclo riproduttivo cellulare. *Arch. Biol.*, **79**, 121.

Prenna, G. and De Paoli, A. M. (1964) La deidro-tio-p-toluidina-SO_2, un nuovo reagente tipo Schiff altamente fluorescente. *Riv. Istochem. Norm. Pat.*, **10**, 185.

Prescott, D. M. (1970) The structure and replication of eukaryotic chromosome. *Advs. Cell Biol.*, **1**, 57.

Prosser, C. L. and Brown, J. F. A. (1961) "Comparative Animal Physiology". 2nd Ed. Saunders, Philadelphia and London.

Rasch, E. M. (1971) The DNA content of sperm of *Drosophila melanogaster*. *Chromosoma*, **33**, 1.

Raff, R. A. and Raff, E. C. (1970) Respiratory mechanisms and the metazoan fossil redord. *Nature Lond.*, **228**, 1003.

Rees, H. and Jones, R. N. (1967) Structural basis of quantitative variation in nuclear DNA. *Nature Lond.*, **216**, 825.

Richards, B. M. (1966) Cytochemistry of nucleic acids. *In* "Chemistry and Cytochemistry of Nucleic Acids and Nuclexarproteins". Springer-Verlag, Vienna and New York.

Ringertz, N. R., Darzynkiewicz, Z. and Bolund, L. (1969) Actinomycin binding properties of stimulated human lymphocytes. *Expl. Cell Res.*, **56**, 411.

Ringertz, N. R., Bolund, L. and Darzynkiewicz, Z. (1971) AO-binding of intracellular nucleic acids in fixed cells in relation to cell growtt. *Expl. Cell Res.*, **63**, 233.

Ris, H. (1969) The molecular organisation of chromosomes. *In* "Handbook of Molecular Cytology". (A. Lima de Faria, ed.). North Holland Publ. Co., Amsterdam.

Roels, H. (1966) "Metabolic" DNA: A cytochemical study. *Int. Rev. Cytol.*, **19**, 1.

Ruch, F. (1966) Determination of DNA content by microfluorometry. *In* "Introduction to Quantitative Cytochemistry" (G. L. Wied, ed.) Academic Press, New York and London.

Ruch, F. and Rosselet, A. (1970) A cytochemical study of euchromatin and heterochromatin in roots of *Rhoeo discolor*. *Expl. Cell Res.*, **62**, 219.

Rudkin, G. T. (1966) Microspectrophotometry of chromosomes. *In* "Introduction to Quantitative Cytochemistry". (G. L. Weid, ed.) Academic Press, New York and London.

Sacerdote, M. (1960) La poliploidia somatica negli Urodeli. Ricerche istofotometriche in *Triturus cristatus* carnifex Laur., *Riv. Istochem. Norm. Pat.*, **6**, 193.

Sandritter, W. (1966) Methods and results in quantitative cytochemistry. *In* "Introduction to Quantitative Cytochemistry". (G. L. Wied, ed.) Academic Press, New York and London.

Sandritter, W., Jobst, K., Rakow, L. und Bosselman, K. (1965) Zur Kinetic der Feulgen reaktion bei verlangerter Hydrolysezeit. Cytophotometrische Missungen im sichtbaren und ultraviolet Licht. *Histochemie*, **4**, 420.

Sandritter, W., Kiefer, G. and Rick, W. (1966) Gallocianin chrome alum. *In* "Introduction to Quantitative Cytochemistry". (G. L. Wied, ed.) Academic Press, New York and London.

Schnedl, W. (1969) Sex chromatin und Kerngrosse im Zellzyklus. Kombinierte histophotometrische und autoradiographische Untersuchungen an Menschlichen Fibroblastenkulturen. *Chromosoma*, **27**, 421.

Schrader, F. and Leuchtenberger, C. (1950) A cytochemical analysis of the functional interrelations of various cell structures in *Arvelius albopunctatus* (De Geer.). *Expl. Cell Res.*, **1**, 421.

Schreiber, G. (1971) Evoluçao do conteúdo de DNA por núcleo Ciencia e Cultura, **24**, 428.

Schreiber, G. *et al.* (1966) Pesquisas de citologia quantitativa XIX DNA e volume nuclear nos recides somaticos dos vertebrados. *Mem. Inst. Butantan*, **1**. 135.

Schreiber, G. and Cavenaghi, T. M. C. M. (1971) Variations of DNA content during the Phylogenesis. 1 DNA content in Onycophora (in press).

Sibatani, A. (1953) Feulgen reaction and quantitative cytochemistry of desoxypentose nucleic acid. III. Effects of histone on the Feulgen reaction in vitro. *J. Biochem.* (*Tokyo*), **40**, 119.

Simpson, G. G. (1953) "The Major Features of Evolution". Columbia Univ. Press, New York.

Simpson, G. G. (1961) "Principles of Animal Taxonomy". Columbia Univ. Press, New York.

Sokal, R. R. (1961) Distance as a measure of taxonomic similarity. *Syst. Zool.*, **10**, 70.

Sokal, R. R. and Sneath, P. H. A. (1963) "Principles of Numerical Taxonomy". W. H. Freeman and Co., San Francisco and London.

Stebbins, G. L. (1966) Chromosomal variation and evolution. *Science N.Y.*, **152**, 1463.

Stowell, R. (1945) Feulgen reaction for thymonucleic acid. *Stain Technol.*, **20**, 45.

Strelkov, P. P. and Volobujev, V. T. (1969) Identity of Karyotypes in the genus Myotis. Proc. of the II Congr. of the Union Mammalogy, Moscow.

Stubblefield, E. and Waay, W. (1971) Architecture of the chinese hamster metaphase chromosome. *Chromosoma*, **32**, 262.

Szarsky, H. (1970) Changes in the amount of DNA in cell nuclei during vertebrate evolution. *Nature Lond.*, **226**, 651.

Swift, H. (1950) The desoxiribose nucleic acid content of animal nuclei. *Physiol. Zool.*, **23**, 169.

Swift, H. (1955) Cytochemical techniques for nucleic acid. *In* "The Nucleic Acids" (E. Chargaff and J. N. Davidson, eds.) Academic Press, New York and London.

Swift, H. and Rasch, E. M. (1956) Microphotometry with visible light. *In* "Physical Techniques in Biological Research" (G. Oster and A. W. Pollister, eds.) Vol. 3. Academic Press, New York and London.

Takagi, N. and Sandberg, A. A. (1968) Chology and pattern of human chromosome replication. VII. Cellular and chromosomal DNA behavior. *Cytogenetics*, 7, 118.

Taylor, K. M., Hungerford, D. A. and Snyder, R. L. (1969) Artiodactyl mammals: their chromosome cytology in relation to patterns of evolution. *In*

"Comparative Mammalian Cytogenetics" (K. Benirschke, ed.) Springer-Verlag, New York.

Terasima, T. and Tolmach, L. J. (1963) Growth and nucleic acid synthesis in synchronously dividing populations of He La cells. *Expl. Cell Res.*, **30**, 344.

Thaer, A. (1970) Anwendung der mikroskop photometrie in der cytochemie. *Anal. Chem.*, **252**, 82.

Tonkelaar, E. M. de and van Duijn, P. (1964) Photographic colorimetry as a quantitative cytochemical method. II. Determination of relative amounts of DNA in cell nuclei. *Histochemie*, **4**, 10.

Ullerich, F. H. (1966) Karyotyp und DNS-Gehalt von *Bufo bufo*, *B. viridis*, *B. bufo* × *B. viridis*, und *B. calamita* (Amphibia, Anura). *Chromosoma*, **18**, 316.

Van Duijn, P. (1956) A histochemical specific thionine-SO_2 reagent and its use in a bi-color method for deoxyribonucleic acid and periodic acid Schiff positive substances. *J. Histochem. Cytochem.*, **4**, 55.

Vendrely, R. and Vendrely, C. (1948) La teneur du noyau cellulaire en ADN à travers les organes, les individus et les espèces animales. *Experientia*, **4**, 434.

Vendrely, R. and Vendrely, C. (1957) "ADN, Substance Fondamentale de la Cellule Vivante" Legrand & C. ed. Paris.

Vialli, M. (1957a) Volume et contenu en ADN par noyau. *Expl. Cell Res.* suppl. **4**, 284.

Vialli, M. (1957b) La quantità di acido desossiribonucleico per nucleo negli eritrociti di Latimeria. *Ist. Lomb. Sci. Lett.*, **91**, 680.

Vialli, M. (1958) Anomalie quantitative dell'acido desossiribonucleico e situazioni anomale del nucleo. *Arch. It. Anat. Embriol.*, **63**, 309.

Vialli, M. (1961) Ricerche di istochimica quantitativa comparata sull'acido desossiribonucleico e sulle proteine. *Monit. Zool. It.*, **69**, 28.

Vialli, M. (1969) Contributo dei dati sul numero cromosomico alla soluzione negativa del discusso rapporto fra Carpa e Tinca. *Rend. Ist. Lomb. Sci. Lett.*, Cl. Sci., **103**, 47.

Vialli, M. and Gerzeli, G. (1955) Il tenore di acido desossiribonucleico nei linfociti dell'asino, del cavallo e del mulo. *Rend. Ist. Lomb. Sci. Lett.*, Cl. Sci., **88**, 273.

Vialli, M. and Riccadonna, M. (1957) Forma, grandezza e contenuto in ADN per nucleo negli eritrociti degli ibridi diretti e reciproci di *Triturus cristarus* e *Triturus taeniatus*. *Riv. Istochem. Norm. Pat.*, **3**, 127.

Vialli, M. and Sacchi Vialli, G. (1969) Morfometria delle lacune ossee di vertebrati attuali e fossili alla luce delle conoscenze di biologia cellulare. *Rend. Ist. Lomb. Sci. Lett.*, Cl. Sci., **103**, 234.

Vialli, M. (1971) Morfometria delle lacune ossee in Onfibi poliploidi. *Boll. Zool.* **38**, 67.

Whitehouse, H. L. K. (1967) A cycloid model for the chromosome. *J. Cell Sci.*, **2**, 9.

Williams, C. A. and Ockey, C. H. (1970) Distribution of DNA replicator sites in mammalian nuclei after different methods of cell synchromosation. *Expl. Cell Res.* **53**, 365.

Wolfe, S. L. (1969) Molecular organization of chromosomes. *In* "The Biological

Basis of Medicine" (E. Bittar and N. Bittar, eds.) Vol 4. Academic Press, London and New York.

Wolfe, S. L. and Martin P. G. (1968) The ultrastructure and strandedness of chromosomes from two species of *Vicia*. *Expl. Cell Res.*, **50**, 140.

Wolff, S. (1969) Strandedness of chromosomes. *Int. Rev. Cytol.*, **25**, 279.

Wright, W. C., Mukherjee, B. B., Mann, K. E., Ghasal, S. K. and Burkholder, G. D. (1970) Quantitative autoradiographic analysis of rates of DNA synthesis in X-chromosomes of bovine females. *Expl. Cell Res.*, **63**, 138.

Yataganas, X., Gahrton, G. and Thorell, B. (1970) DNA, RNA and hemoglobin during erythroblast maturation. *Expl. Cell Res.* **62**, 254.

Zirkin, B. R. and Wolfe, S. L. (1970) The chemical composition of nuclei and chromosomes isolated by the Langmuir trough technique. *Chromosoma*, **32**, 162.

ADDENDUM

While the present paper was in press a lot more information on repetitive DNA has become available. It has been clarified that mouse satellite DNA is made up of "highly repetitive sequences (Walker, 1969;). On the basis of some findings satellite DNA might be equated with heterochromatin. Early concepts about euchromatin and heterochromatin are changing at a quite fast rate (Rees and Jones, 1972, and also Berendes and Keyl, 1967; Lonck, 1970; Noeske, 1971; Ruch and Rosselet, 1970).

The relationships between repetitive DNA and chromosomes has recently been discussed by Flamm (Flamm; 1972), and new data on this subject have been critically reviewed by this worker, whereby many of the working hypotheses posited at the outset are now seen in a somewhat different light.

Perhaps the finding of major relevance to the present discussion is the demonstration that the percent ratio of the DNA nuclear content consisting of DNA with very highly repetitive sequences, to the total DNA nuclear content differs markedly even in closely related species (Majo, personal communication cited by Flamm, 1972; Hatch and Mazrimas, 1970; Arrighil F. E. *et al.* 1970; Hennig and Walker, 1970; Corneo *et al.*, 1970).

The highly repetitive sequences vary according to species not only as regards the "base composition" but also in terms of nucleotide sequence. It was confirmed that they are specific species. Comparison between the features of highly repetitive DNA in several *Mus* species leads to the assumption that this DNA represents the evolutionarily youngest genome fraction, and to the assignment of a "high evolutionary rate of turnover" to a very highly repetitive sequence.

On the other hand, confirmation has been reached of the fact that the highly repetitive component is consistently alike in all the cells of different organs, the sperm and cultures of different tissues from the same animal, both as concerns the percent of DNA nuclear content and its chemical and physical characteristics. In particular, this is true also for the comparison between different mouse strains.

The mechanism through which closely related species possessing very similar "DNA nuclear content" have acquired diversities in their highly repetitive component during admittedly recent evolution (a few million years) is beyond our understanding even today. Prerequisite for a discussion on this point is the clarification of the "function" of this DNA fraction. The following are the most recent hypotheses also discussed by Flamm.

1) DNA may be involved in the initiation of DNA synthesis or in its termination (Walker *et al.*, 1969, Smith 1970). However, the fact that in synchronized cells of mouse lymphoma the highly repetitive sequences cannot be evidentially demonstrated until about the end of the *S* phase, is at odds with this interpretation (Bostock and Prescott, 1971, Tobia *et al.*, 1971).

2) The highly repetitive sequence act as templates for RNA synthesis. This was shown *in vitro*; however, all the DNA-RNA hybridization experiments have consistently failed to reveal any complementarity between RNA and the mouse satellite DNA. Hence it may also be admitted that the repetitive sequence have the potential ability to act as templates for the synthesis of some type of RNA, though it may be hypothesized that this ability is somehow repressed *in vivo*.

3) The highly repetitive sequences would be involved in the mechanisms of self-strand association through the so-called "reversed sequences" which would serve as "recognition sites" by means of which "one duplex can recognize another". In this sense they would come into play in all those phenomena of pairing of homologous chromosomes in meiosis, of recombination, and so forth, postulated by Walker (1969) in his "housekeeper" concept.

That the highly repetitive sequences should govern all the organization of chromosomes was questioned by Walker himself (Walker 1971) on the basis of the fact that the satellite is localized in given small regions of the genome, such localization being at odds with so general a control function.

At any rate, this problem is raised once more (Flamm 1972) by the knowledge of their possible presence within chromocenters at interphase, in key sites for the refashioning of the chromosomal strands.

Moreover, the belief that the satellite sequences are present in the "distal arms of metaphase chromosomes" may suggest that they serve some function at the centromeric level, thereby lending support to the concept of a general regulatory function.

4) The idea that DNA acts as "spacer" between genes (Jones and Robert-

son, 1970) is considered valid by Rees and Jones (1972), being related in the end to the gene regulatory role proposed by Britten and Davidson (1969). Nevertheless, according to Flamm (1972) this is a very generic possibility.

5) Medvedev (Medvedev 1972) suggests a possible role in the determination of the cell life span as assignable to the highly repetitive sequences. In his opinion the repetitive DNA sequences would interfere with the accumulation of coded information derived from molecular accidents. Hence he believes that, under otherwise comparable conditions, the genomes possessing a more abundant fraction of repetitive DNA would have a longer life span. This hypothesis of some protection being exerted by repetitive DNA might also be taken into consideration to explain the different ageing rates in cells from different tissues.

6) Britten's view that repetitive DNA may represent a raw material potentially usable by future evolution has not, on the contrary, been confirmed so far, although hinted at by Flamm not without reservations about it as "not at all inconsistent with the properties of satellites as they are understood today." It appears evident, however, that thorough studies have been made on very few species, above all in mammals and rodents in particular. A more extensive comparative outlook may provide more adequate models helping towards clarification in this sense.

A new critical re-examination of information available today on the high variations in DNA nuclear content throughout living forms, animals and plants, is due to Rees and Jones (1972). They set out with the idea that a relationship exists between *changes in the base ratio* (demonstrated in closely related species, among bacteria, higher plants, insects, mammals) and *variations in DNA amount*. In this sense all the mechanisms more frequently coming into play in speciation (polyploidy, aneuploidy, deletion or repetition, differential polynemy viewed as differences in chromosome strandedness) may bring about variations in the DNA nuclear content by altering somehow the distribution of the base sequences.

From a general outlook on the distribution of DNA content in a huge number of animal and plant species, the authors turn to analytical inspection of instances in which differences between species from the same genus are found. In their discussion attention is given not only to the DNA nuclear content, but also to chromosome size, acknowledging the existence of a substantial relationship between these two values, in accordance with what reported by the present writer.

Some hypotheses are then formulated by them about the possible structural bases of quantitative changes in chromosomal DNA, namely, lateral multiplication or reduction, lengthwise repetition or deletion.

Their interpretation of DNA loss and gain observed in the overall picture, which as a rule are viewed as the result of major evolutionary changes, is grounded on the presence of repetitive DNA at all the higher taxonomic levels. Obviously, they base themselves on Britten's idea that repetitive DNA is to be held as raw material for long-term evolutionary transformation. Undoubted adaptative advantages would be secured by DNA increase, which may then be logically correlated by and large with the increase in genetic complexity from primitive to more highly evolved forms.

DNA loss, which predominates within the more highly evolved animals and plants is interpreted by these authors as a loss of potentially evolutionary material. Such a loss would occur in fact concomitantly with the acquisition of an advanced genetical configuration, hence no longer capable of further adaptation.

These conclusions by Rees and Jones find a place in the discussion on the role played by DNA in phylogenesis. (See the present paper.) It seems of importance, on the contrary, to point out that in Rees and Jones' work the first systematic attempt is made to discuss the findings related to DNA nuclear content in parallel with a) the karyotype morphology (genome total length, chromosome size; b) the schemes proposed by molecular morphology. Another integration should be programmed in a future time, namely, with fine cytochemical data, also at the ultrastructural level, on the chromosome proteinaceous components, in view of their importance in conditioning, if so, genetic mechanisms.

In Rees and Jones' work two unpublished data are presented on the comparison of the DNA nuclear content between inbreeding and outbreeding species. In *Lolium*, inbreeders possess larger DNA amounts than outbreeders; in *Lathyrus* the contrary is true. As far as I know, one single finding is reported in the literature about Eusorghum (Paroda and Rees, 1971). This type of research deserves attention, since the starting moment of inbreeding being known, it will be possible to calculate within which "time" DNA variations have occurred, thus introducing at least indirectly a third variable: "time" which is so important for any type of phylogenetic reconstruction.

Work is in progress (Manfredi Romanini and Capanna) to estimate the DNA content in some specimens from *Rattus norvegicus* (Winstar strain) whose inbreeding pattern is known, and in some from *Rattus norvegicus* in the wild state. On the basis of the first sets of measurements a significant DNA loss would seem to occur in inbreeders as compared to outbreeders. Should this decrease in DNA in inbreeders be confirmed, the first evidence would be gained that a DNA portion can be lost by the genome in correlation with a group advancement, hence concomitantly with a gradual loss of evolutionary capacity.

References

Arrighi, F. E. *et. al.* (1970). *Biochem. Genet.* **4**, 367.
Berendes and Keyl (1967). *Genetics* **57**, 1.
Bostock, C. G. and Prescott, D. M. (1971). *Expl. Cell Res.* **64**, 267, 481.
Britten, J. C. and Davidson, E. M. (1969). *Science N.Y.* **161**, 529.
Corneo, *et. al.* (1970). *J. molec. Biol.* **48**, 319.
Flamm. W. (1972). *Int. Rev. Cytol.* **32**, 1.
Hatch, F. T. and Mazrimas, J. A. (1970). *Biochem. biophys. Acta.* **224**, 291.
Hennig, W. and Walker, P. M. B. (1970). *Nature, Lond.* **225**, 915.
Jones, K. W. and Roberts, J. W. (1970). *Chromosoma* **31**, 331.
Lorick, (1970). *Chromosoma* **32**, 11.
Medvedev, Zh. A. (1972). *Nature, Lond.* **237**, 453.
Noeske, (1971). *J. Histochem. Cytochem.* **19**, 169.
Paroda, R. S. and Rees, H. (1971). Nuclear DNA variation in Eu-Sorghums. *Chromosoma* **32**, 353.
Rees, H. and Jones, R. N. (1972). *Int. Rev. Cytol.*, **32**, 53.
Ruch and Russelet, (1970). *Expl. Cell Res.* **56**, 411.
Smith, B. J. (1970). *J. molec. Biol.* **47**, 101.
Tobia, *et. al.* (1971). *J. molec. Biol.* (in press).
Walker, P. M. B. (1969). *Progr. Nucl. Acid. Res. molec. Biol.* **9**, 301.
Walker, P. M. B. (1971). *Nature, Lond.* **229**, 306.
Walker, P. M. B. *et al.* (1969). *In* "Handbook of Molecular Cytology".
(Lima de Faria, A. ed.) North Holland, Amsterdam.

5

Lampbrush Chromosomes in Cytotaxonomic and Cytogenetic Studies

GIORGIO MANCINO

Chair of Histology and Embryology, Faculty of Sciences, Pisa University
Pisa, Italy

The lampbrush chromosomes, present in the prophasic primary oöcytes several groups of animals, are regarded as "high resolution chromosomes", as they have cytogenetic markers (i.e. loops with various morphology, dimensions, nucleolus-organizing regions, sphere sites, centromeres as well as other axial and lateral differentiations, Figs 1 and 2) which enable us to draw maps, thus using a technique not unlike that successfully employed in the studies on giant polytene chromosomes in Diptera larvae. The lampbrush chromosomes are therefore useful too in karyological studies as they satisfy certain fundamental requisites in cytotaxonomic and cytogenetic research. In fact with the aid of the maps, each chromosome of the set can be clearly identified even if it belongs to a group of chromosomes which, as regards length and centromeric index are very close, and which are therefore interchangeable in the arrangement of the mitotic karyotype. Furthermore, it is possible to identify and interpret exactly the structural rearrangements which in the majority of cases would be easily overlooked analysing only the mitotic chromosomes. Examination of the lampbrush chromosomes also gives precise data on the meiotic behaviour, frequency and distribution of the chiasms, and contributes to a better understanding of the structure, organization and function of the chromosomes. Thus from the comparison of the chromosomic maps the degree of genetic relationship existing between various species can be established as well as the importance and role that possible mutations may have had in the speciation processes. Obviously the more limited the systematic category taken into consideration, and the closer the degree of affinity of the species under comparison, the greater the validity of the deductions of a cytotaxonomic and cytogenetic nature. It is to be hoped that this method will be used also in the study of hybrid sets, where it is possible to verify directly the degree of homology that exists

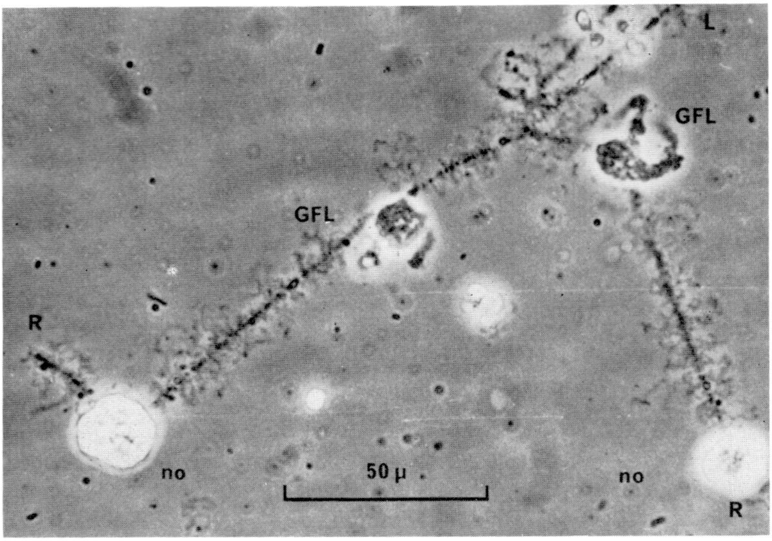

Fig. 1. Part of lampbrush chromosome X of *Triturus marmoratus* showing normal loops, giant fusing loops (GFL) and the nucleolus organizing regions (no). R = right, L = left.

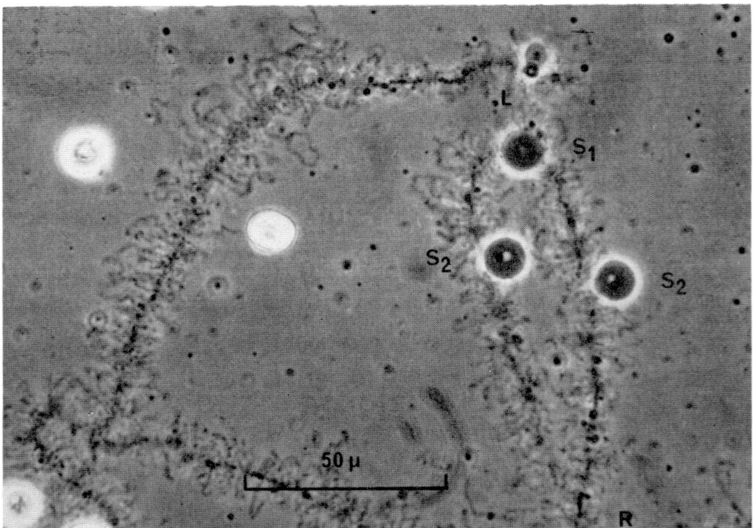

Fig. 2. Part of lampbrush chromosome V of *Triturus cristatus carnifex* showing three spheres (S). S_1 results from the fusion of the two homologous subterminal spheres. R = right, L = left.

between the chromosomes of related species which have given rise to hybrid individuals. This may help, on one hand, to resolve problems connected with chromosome differentiation, and on the other hand to supply chromosome evidence that may elucidate the systematic and phylogenetic position of the species.

As far as vertebrates are concerned, the bulk of the work done so far has been carried out on amphibians, in which it was possible to couple the analysis of the giant chromosomes of the oöcytes to the study of the mitotic chromosomes. Significant results have been forthcoming and it is worthwhile mentioning here the success obtained from analysis of the lampbrush karyotypes in the identification of sex chromosomes in species in which the traditional karyological analysis had always given negative results. Such is the case in the complements of *Tritunus cristatus* and *T. marmoratus* where it has been demonstrated that the partner chromosomes of bivalent I present a large area in which there are no chiasm, and where various pairs of heteromorphous loops and other structures in heterozygous conditions are inserted; in *T. marmoratus* the two partners of lampbrush bivalent I are constantly of different lengths (Fig. 3). An elucidation of the mechanism determining the sex in these species has emerged, where previously indirect tests and experiments on hybrids appeared to suggest a condition of male heterogamety.

Interesting data have emerged from investigations carried out on the polytypic species *Triturus cristatus* and on the systematically closely related species *T. cristatus* and *T. marmoratus*, and *Pleurodeles waltlii* and *P. poireti* which have given rise to the hypothesis that some morphological differences in the sequence of the loops and other chromosomic differentiations can be attributed to allelic changes intervened in the individual loci during speciation of these salamandrid urodeles. These conclusions appear to be valid, even when it is borne in mind that any modification in the morphology of single loops and other landmarks can often be attributed to the different physiological conditions of the oöcytes which, within certain limits, modify the morphology, the texture and the dimensions of these loops and of other chromosome structures. These conditions are perhaps the most important factors which make it difficult to obtain a morphological modification which can be related to a gene mutation, as was found, for example, in the study of the phenomonon of semi-albinism in newts. The discovery of gene mutations by analysing the "phenotype" of the lampbrush loops is hindered, above all, by the scarce knowledge of formal genetics and the difficulty in establishing the exact relationship between single structures which can be identified along the lampbrush chromosomes and single genetically defined units of inheritance.

More studies on chromosome mutations in the lampbrush karyotypes would appear to be worthwhile, as demonstrated by reports which deal with identification and interpretation of experimentally induced or spontaneous aberrations. The latter especially demonstrate the existence not only

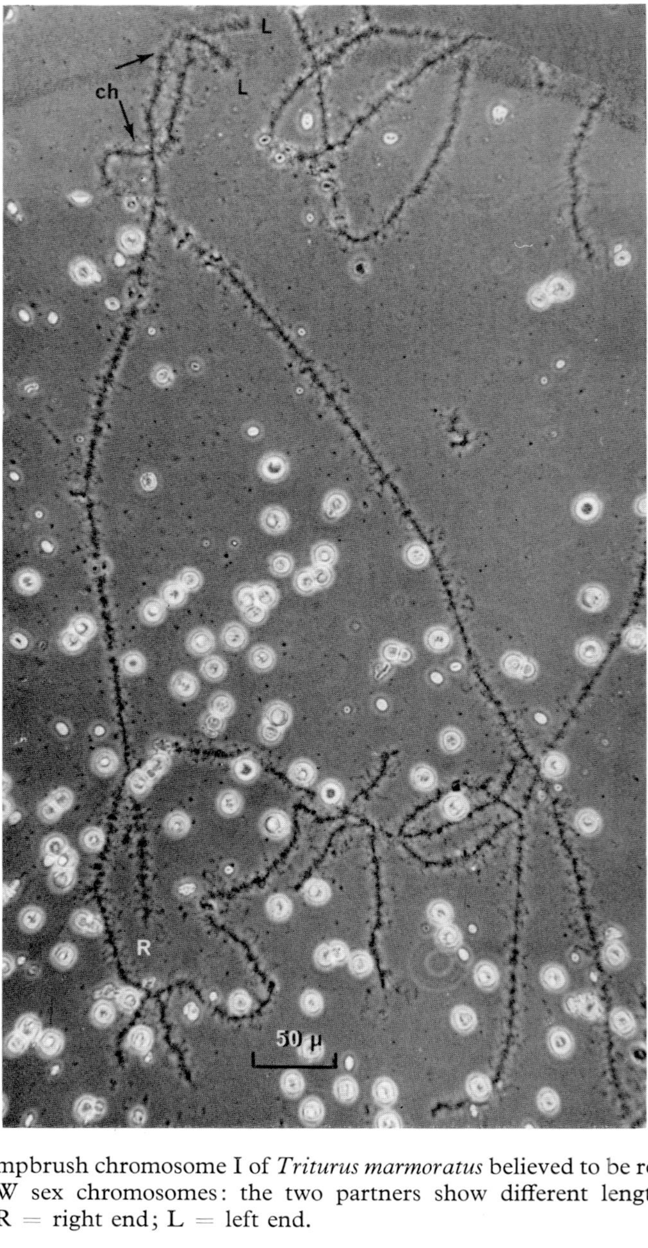

Fig. 3. Lampbrush chromosome I of *Triturus marmoratus* believed to be represented by the ZW sex chromosomes: the two partners show different lengths. ch = chiasma; R = right end; L = left end.

of genetic mutations, but also of more important structural rearrangements, such as large inversions, translocations, deficiencies and reduplications, which play a significant role in the karyological evolution of the Amphibians.

BIBLIOGRAPHY

A. *Occurrence of Lampbrush Chromosomes in Animal Groups*

Ahmad Sharif, M. (1970). Development, structure and composition of lampbrush chromosomes in domestic fowl. *Can. J. Genet. Cytol.* **12**, 728.

Barsacchi, G., Mancino, G. and Taddei, C. (unpublished). Lampbrush chromosomes of *Pseudoemis scripta elegans.*

Delobel, N. and Delavault, R. (1971). Chromosomes en écouvillon chez *Echinaster sepositus* (Echinoderme, Astéride). *Caryologia* **24**, 251.

Gaginskaya, E. R. (1972). The nuclear structure of oöcytes in adult birds. I. Chromosome behaviour during the oöcyte cytoplasm growth. *Tsitologiya* **14**, 426.

Gaginskaya, E. R. and Grouzova, M. N. (1969). Peculiarities of the oogenesis of *Fringilla. Tsitologiya* **11**, 1241.

Lécher, P. (1971). Évolution nucléaire méiotique chez *Jaera (albifrons) syei* (Isopodes Asellotes). *Archs. Zool. exp. gén.* **108**, 319.

Mancino, G. (1971). Citotassonomia di Gasteropodi Opistobranchi italiani e inglesi. *Accad. Naz. Lincei* **51**, 238.

Ribbert, D. and Kunz, W. (1969). Lampenbürstenchromosomen in den Oöcytenkernen von *Sepia officinalis. Chromosoma* **28**, 93.

Further Numerous References are Reviewed in:

Davidson, E. H. (1968). "Gene Activity in Early Development." Academic Press, New York and London.

Hess, O. (1971). Lampenbürstenchromosomen. *In* "Handbuch der Allgemeinen Pathologie" (Altmann, H.-W., Büchner, F., Cottier, H., Grundmann, E., Holle, G., Letterer, E., Masshoff, W., Meesen, H., Roulet, F., Seifert, G. and Seibert, G., eds), p. 215. Springer-Verlag Berlin Heidelberg, New York.

B. *Methods*

Gall, J. G. (1965). Techniques for the study of lampbrush chromosomes. *In* "Methods in Cell Physiology" (Prescott, D., ed.), v.2. p. 3. Academic Press, New York and London.

C. *Structure, Ultrastructure and Organization*

Bier, K., Kunz, W. and Ribbert, D. (1967). Structur und Funktion der

Oöcytenchromosomen und Nukleolen sowie der Extra- DNS während der Oögenese panoistischer und meroistischer Insekten. *Chromosoma* **23**, 214.

Callan, H. G. (1963). The nature of lampbrush chromosomes. *In* "International Review of Cytology" Vol. 15 (Bourne, G. H., and Danielli, J. F., eds) p. 1. Academic Press, New York and London.

Callan, H. G. (1963). The structure of lampbrush chromosomes. *Proc. XI Int. Congress Genetics* **2**, 323.

Callan, H. G. (1967). The organization of genetic units in chromosomes. *J. Cell Sci.* **2**, 1.

Duryee, W. R. (1950). Chromosomal physiology in relation to nuclear structure. *Ann. N.Y. Acad. Sci.* **50**, 920.

Fabergé, A. C. (1970). A doublet structure in the loops of lampbrush chromosomes of the newt, *Taricha granulosa*. *Caryologia* **23**, 241.

Gall, J. G. (1956). On the sub-microscopic structure of chromosomes. *Brookhaven Symp. Biol.* **8**, 17.

Gall, J. G. (1963). Kinetics of deoxyribonuclease action on chromosomes. *Nature Lond.* **198**, 36.

Gall, J. G. and Callan, H. G. (1963). Structure and function of lampbrush chromosomes. *Proc. XVI Int. Congress Zool.* **4**, 280.

Guyénot, E. and Danon, M. (1953). Chromosomes et ovocytes de Batraciens. *Rev. Suisse Zool.* **60**, 1.

Hess, O. (1966). Funktionelle und strukturelle Organisation der Lampenbürsten-Chromosomen. *In*: "Probleme der Biologischen Reduplikation" (P. Sitte, ed.), Springer-Verlag Berlin, Heidelberg and New York.

Hess, O. (1971). Lampenbürstenchromosomen. *In* "Handbuch der Allgemeinen Pathologie" (Altmann, H.-W., Büchner, F., Cottier, H., Grundmann, E., Holle, G., Letterer, E., Masshoff, W., Meesen, H., Roulet, F., Seifert, G. and Seibert, G., eds), p. 215, Springer-Verlag Berlin, Heidelberg and New York.

Kezer, J., Macgregor, H. C. and Schabtach, E. (1971). Observations on the membranous components of amphibian oocyte nucleoli. *J. Cell. Sci.* **8**, 1.

Lafontaine, J. G. and Ris, H. (1958). An electron microscope study of lampbrush chromosomes. *J. biophys. biochem. Cytol.* **4**, 99.

Mancino, G. (1963). La minuta struttura di un tratto del *lampbrush chromosome* XII di *Triturus cristatus cristatus*. *Accad. Naz. Lincei* **34**, 65.

Miller, O. L., Jr. (1965). Fine structure of lampbrush chromosomes. *Natn. Cancer Inst. Monogr.* **18**, 79.

Ris, H. and Kubai, D. F. (1970). Chromosome structure. *In* "Annual Reviews Genetics." Vol. 5. (Roman, H. L., Sandler, L. M. and Campbell, A., eds), p. 263, Annual Reviews Inc., Palo Alto California.

Snow, M. H. L. and Callan, H. G. (1969). Evidence for a polarized movement of the lateral loops of newt lampbrush chromosomes during oogenesis. *J. Cell Sci.* **5**, 1.

Ullerich, F.-H. (1970). DNS- Gehalt und Chromosomenstruktur bei Amphibien. *Chromosoma* **30**, 1.

Wischnitzer, S. (1957). A study of the lateral loop chromosomes of amphibian oocytes by phase contrast microscopy. *Am. J. Anat.* **101**, 135.

D. *DNA, RNA and Protein Synthesis*

Barsacchi, G. and Gall, J. G. (1972). Chromosomal localization of repetitive DNA in the newt, *Triturus. J. Cell. Biol* **54**, 580.

Brown, D. D. and Dawid, I. B. (1968). Specific gene amplification in oocytes. *Science N.Y.* **160**, 272.

Bucci, S., Nardi, I., Mancino, G. and Fiume, L. (1971). Incorporation of tritiated uridine in nuclei of *Triturus* oöcytes treated with α- amanitin. *Expl. Cell Res.* **69**, 462.

Callan, H. G. (1969). Biochemical activities of chromosomes during the prophase of meiosis. *In* "Handbook of molecular cytology" (A. Lima- de-Faria, ed.) p. 541, North-Holland Publ. Co., Amsterdam and London.

Davidson, E. H. (1968). Genomic function in amphibian oögenesis. Atti Convegno Fond. Antonio Baselli: Le basi molecolari del differenziamento, *Acc. Naz. Lincei, Quad.N.* **104**, 77.

Davidson, E. H., Crippa, M., Kramer, F. R. and Mirsky, A. E. (1966). Genomic function during the lampbrush chromosome stage of amphibian oögenesis. *Proc. natn. Acad. Sci.* **56**, 856.

Davidson, E. H. and Hough, B. R. (1969). High sequence diversity in the RNA synthesized at the lampbrush stage of oögenesis. *Proc. natn. Acad. Sci* **63**, 342.

Edström, J.-E. and Gall, J. G. (1963). The base composition of ribonucleic acid in lampbrush chromosomes, nucleoli, nuclear ap, and cytoplasm of *Triturus oocytes. J. cell Biol.* **19**, 279.

Fiume, L., Nardi, I., Bucci, S. and Mancino, G., (1972). Effects of cordycepin on morphology and RNA synthesis of amphibian lampbrush chromosomes. *Expl. Cell Res.* **75**, 11.

Gaginskaya, E. R. (1972). The nuclear structure of oöcytes in adult birds. II. Protein bodies and the karyosphere. *Tsitologiya* **14**, 568.

Gall, J. G. (1966). Nuclear RNA of the salamander oöcyte. *Natn. Cancer Inst. Monogr.* **23**, 475.

Gall, J. G. (1968). Differential synthesis of the genes for ribosomal RNA during amphibian oögenesis. *Proc. natn. Acad. Sci.* **60**, 553.

Gall, J. G. and Callan, H. G. (1962). H^3-uridine incorporation in lampbrush chromosomes. *Proc. natn. Acad. Sci.* **48**, 562.

Macgregor, H. C. (1965). The role of lampbrush chromosomes in the formation of nucleoli in amphibian oöcytes. *Q. J. Microsc. Sci.* **106**, 215.

Macgregor, H. C. (1968). Nucleolar DNA in oöcytes of *Xenopus laevis. J. Cell Sci.* **3**, 437.

Macgregor, H. C. and Callan, H. G. (1962). The actions of enzymes on lampbrush chromosomes. *Q. J. Microsc. Sci.* **103**, 173.

Makarov, V. B., Safronov, V. V., Kanel, C. I. (1972). Polarised non-replicative synthesis of DNA of gigantic granular loop of XIIth chromosome in oöcytes of *Triturus cristatus cristatus.* Doklady Akademii nauk SSSR. *Proc. Acad. Sci. USSR* **205**, 220.

Mancino, G., Barsacchi, G. and Nardi, I. (1967). Uno studio autoradiografico sulla sintesi di RNA nei *lampbrush chromosomes* di *Triturus* (Anfibi Urodeli). *Boll. Zool.* **34**, 134.

Mancino, G., Barsacchi, G. and Nardi, I. (1968). Incorporazione di fenilalanina tritiata nei *lampbrush chromosomes* e nei nucleoli di *Triturus vulgaris meridionalis* (Anfibi Urodeli). *Acc. Naz. Lincei* **44**, 840.

Mancino, G., Barsacchi, G. and Nardi, I. (1968). Ulteriori ricerche autoradiografiche sull'incorporazione di H^3-fenilalanina nei *lampbrush chromosomes* di *Triturus* (Anfibi Urodeli). *Acc. Naz. Lincei* **45**, 180.

Mancino, G., Barsacchi, G. and Nardi, I. (1968). Effetti della actinomicina D tritiata sui *lampbrush chromosomes* di *Triturus* (Anfibi Urodeli). *Acc. Naz. Lincei* **45**, 591.

Mancino, G., Nardi, I., Corvaja, N., Fiume, L. and Marinozzi, V. (1971). Effects of α- amanitin on *Triturus* lampbrush chromosomes. *Expl. Cell Res.* **64**, 237.

Miller, O. L., Jr., Beatty, B. R., Hamkalo, B. A. and Thomas, C. A., Jr. (1970). Electron microscopic visualization of transcription. *Cold Spring Harbor Symp. quant. Biol.* **35**, 505.

Miller, O. L., Jr. and Hamkalo, B. A. (1972). Visualization of R.N.A. synthesis on chromosomes. In "International Review of Cytology." Vol. (Bourne, G. H. and Danielli, J. F. eds.) p. 1. Academic Press, New York and London.

E *Maps of lampbrush chromosomes*

Barsacchi, G., Bussotti, L. and Mancino, G. (1970). The maps of the lampbrush chromosomes of *Triturus* (Amphibia Urodela). IV. *Triturus vulgaris meridionalis. Chromosoma* **31**, 255.

Callan, H. G. and Lloyd, L. (1960). Lampbrush chromosomes of crested newts *Triturus cristatus*. Phil. Trans. Roy. Soc. Lond., **B**, **243**, 135.

Callan, H. G. (1966). Chromosomes and nucleoli of the axolotl., *Ambystoma mexicanum. J. Cell Sci.* **1**, 85.

Delobel, N. (1971). Étude descriptive des chromosomes en écouvillon chez *Echinaster sepositus* (Échinoderme, Astéride). *Ann. Embr. Morph.* **4**, 383.

Gall, J. G. (1954). Lampbrush chromosomes from oocyte nuclei of the newt. *J. Morph.* **94**, 283.

Giogri, F. and Galleni, L. (1972). The lampbrush chromosomes of *Rana esculenta* L. (Amphibia, Anura). *Caryologia* **25**, 107.

Lacroix, J.-C. (1968). Étude descriptive des chromosomes en écouvillon dans le genre *Pleurodeles* (Amphibien, Urodèle). *Ann. Embr. Morph.* **1**, 179.

Mancino, G. and Barsacchi, G. (1965). Le mappe dei cromosomi *lampbrush* di *Triturus*. I. *Triturus alpestris apuanus. Caryologia* **18**, 637.

Mancino, G. and Barsacchi, G. (1966). The maps of the lampbrush chromosomes of *Triturus* (Amphibia Urodela). II. *Triturus helveticus helveticus. Riv. Biologia* **59**, 311.

Mancino, G. and Barsacchi, G. (1969). The maps of the lampbrush chromosomes of *Triturus*. III. *Triturus italicus. Ann. Embr. Morph.* **2**, 355.

Mancino, G., Barsacchi, G. and Nardi, I. (1969). The lampbrush chromosomes of *Salamandra salamandra* (L.) (Amphibia Urodela). *Chromosoma* **26**, 365.

Mancino, G., Nardi, I. and Ragghianti, M. (1972). Structural correspondence between nucleolus- and sphere-organizing regions of the lampbrush chromosomes and secondary constrictions of the mitotic chromosomes. *Experientia* **28**, 586.

Nardi, I., Ragghianti, M. and Mancino, G. (1972). Characterization of the lampbrush chromosomes of the marbled newt *Triturus marmoratus* (Latreille, 1800). *Chromosoma* **37**, 1.

Ragghianti, M., Baroni, R., Nardi, I. and Mancino, G. (in preparation). The lampbrush chromosomes of *Salamandrina terdigitata*.

F. *Other Papers having Cytotaxonomic Interest*

Mancino, G. (1965). Osservazioni cariologiche sull'Urodelo della Sardegna *Euproctus platycephalus*: morfologia dei bivalenti maschili e dei *lampbrush chromosomes*. *Acc. Naz. Lincei* **39**, 540.

Mancino, G. (1968). Sulla validità specifica di *Triturus italicus* (Peracca 1898) (Anfibi Urodeli). *Acc. Naz. Lincei* **44**, 697.

Mancino, G. and Barsacchi, G. (1966). Il corredo cromosomico di *Euproctus montanus*, Urodelo della Corsica. *Acc. Naz. Lincei* **41**, 396.

Ragghianti, M., Nardi, I. and Mancino, G. (1972). Completion of the morphology of the lampbrush chromosomes of the Italian alpine newt *Triturus alpestris apuanus* (Bonaparte). *Experientia* **28**, 588.

Srivastava, M. D. L. and Bhatnagar, A. N. (1962). Lampbrush chromosomes of *Rana cyanophlyctis*. *Cytologia (Tokyo)* **27**, 60.

G. *Sex Chromosomes*

Callan, H. G. and Lloyd, L. (1960). Lampbrush chromosomes. *In*: "New Approaches in Cell Biology" (P. M. B. Walker, ed.). Academic Press, New York and London.

Lacroix, J.-C. (1970). Mise en évidence sur les chromosomes en écouvillon de *Pleurodeles poireti* Gervais, Amphibien urodèle, d'une structure liée au sexe, identifiant le bivalent sexuel et marquant le chromosome W. *C. r. hebd. Séanc. Acad. Sc. Paris* **271**, 102.

Mancino, G. and Nardi, I. (1971). Chromosomal heteromorphism and female heterogamety in the marbled newt *Triturus marmoratus* (Latreille). *Experientia* **27**, 821.

Mancino, G., Ragghianti, M., Nardi, I. and Andreuccetti, P. (1972). Sex chromosomes in newts. *Boll. Zool.* **39**, (in press).

H. *Gene Mutations*

Callan, H. G. and Lloyd, L. (1956). Visual demostration of allelic differences within cell nuclei. *Nature Lond.* **178**, 355.

Mancino, G., Nardi, I. and Ragghianti, M. (1972). Lampbrush chromosomes from semi-albino crested newts, *Triturus cristatus carnifex* (Laurenti). *Experientia* **29**, 856.

I. *Spontaneous Chromosome Mutations*

Mancino, G., Nardi, I. and Barsacchi, G. (1970). Spontaneous aberrations in lampbrush chromosome XI from a specimen of *Triturus vulgaris meridionalis* (Amphibia, Urodela). *Cytogenetics* **9**, 260.

J. *Experimentally Induced Genome and Chromosome Mutations*

Gallien, L., Labrousse, M. and Lacroix, J.-C. (1966). Détection sur les chromosomes mitotiques et sur les chromosomes en écouvillon (*lampbrush*) d'anomalies provoquées par l'irradiation de l'oeuf, chez l'Amphibien Urodèle *Pleurodeles waltlii* Michah. *C. r. hebd. Seanc. Acad. Sc. Paris* **263**, 1984.
Lacroix, J.-C. (1967). Obtention de femelles trisomiques fertiles chez l'Amphibien Urodèle *Pleurodeles waltlii* Michah. *C. r. hebd. Séanc. Acad. Sci. Paris* **264**, 85.
Lacroix, J.-C. (1968). Variations expérimentales ou spontanées de la morphologie et de l'organisation des chromosomes en écouvillon dans le genre *Pleurodeles* (Amphibien, Urodèle). *Ann. Embr. Morph.* **1**, 205.
Lacroix, J.-C. and Loones, M. T. (1971). Fragmentation par les rayons X de l'organisateur d'une différenciation de chromosome en écouvillon (*Lampbrush*), chez *Pleurodeles waltlii*. *Chromosoma* **36**, 112.

Section III

CHROMOSOME DYNAMIC

6

Chromosomal Rearrangements in Mammalian Population Polymorphism and Speciation

M. J. D. WHITE
Department of Genetics, University of Melbourne
Australia

I. THE THEORY

It is only in the last few years that a study of the chromosomes has made any real contribution to the problems of evolutionary differentiation of populations, races and species in vertebrates. Thus when the field was reviewed by Matthey (1949) there seemed to be no clear instance of intrapopulation chromosomal polymorphism in vertebrates and no really critical studies had been carried out on the role (if any) of chromosomal rearrangements in race- and species-formation in the vertebrata.

Within the past two decades the picture has changed radically. We now have reliable karyotypes for about 1000 species of mammals, approximately one-quarter of the total, and for several hundred species of fishes, amphibia, reptiles and birds. More important than the mere bulk of the data, however, are the detailed studies that have been carried out on a number of critical genera and species-complexes, particularly in the mammalia and in the Urodeles, in the latter case making use of the giant lampbrush chromosomes in the oocyte nuclei.

We shall be concerned in this chapter with two forms of chromosomal variation: (1) intrapopulation chromosomal polymorphism, due to structural rearrangements such as inversions or translocations that have not reached fixation, but exist in a floating state, so that both heterozygotes and homozygotes occur within the natural populations, (2) cytotaxonomic differences between races of a species or between closely related or incipient species. In the latter case one or more chromosomal rearrangements, absent in one taxon, have reached fixation (i.e. homozygosity) in the other; where geographic overlaps occur some hybrids may be produced and if so will be heterozygous for the rearrangement(s), but this condition of chromosomal polymorphism is not geographically widespread in the second type of case, being confined to a zone of overlap a few kilometres or at most tens of kilometres in width.

The first type of situation is clearly one of balanced genetic polymorphism, whether due to simple heterosis (adaptive superiority of the heterozygote), frequency-dependent selection or niche-adaptation ("annidation"). It is thus in principle similar to the inversion polymorphisms that have been studied especially in *Drosophila* species and several groups of grasshoppers (references in Dobzhansky, 1970 and White, 1973). In those vertebrates that do not have lampbrush chromosomes suitable for detailed analysis paracentric inversions are virtually undetectable and many may hence escape notice. On the other hand pericentric inversions, duplications, deletions and all forms of translocations, including chromosomal fusions (evolutionary decreases in chromosome number) and dissociations (increases in chromosome number) will be detectable, unless they involve very short chromosome segments.

Heterozygosity for inversions, whether paracentric or pericentric, may involve reversed-loop formation at meiosis. If so, genetic crossovers may occur within the loop, resulting in chromatids bearing duplications and deletions. In the case of paracentric inversions these chromatids will also be dicentric or acentric. In *Drosophila*, because of the absence of crossing over in the male, this problem exists only in the females, which generally "excrete" the dicentric and acentric strands in the polar body nuclei. Heterozygotes for paracentric inversions thus do not have their fecundity reduced, and this category of rearrangements is a "permitted" one in *Drosophila*. Pericentric inversions, on the other hand, are a "forbidden" category in *Drosophila*, in the absence of any mechanism ensuring that the strands carrying a duplication and a deficiency regularly pass into the polar nuclei at meiosis.

Certain species of grasshoppers which are regularly polymorphic for pericentric inversions, seem to have got over this difficulty through replacement of reversed loop formation by "straight", non-homologous pairing at zygotene (certainly, in the males and presumably also in the females). Probably these species carry undetectable paracentric inversions as well as the detectable pericentric ones. Midges of the genus *Chironomus* (which have chiasmata in both sexes) seem to be intermediate in this regard—inversion heterozygotes form some reversed loops at meiosis, but not nearly as many as are theoretically possible (Martin, 1967); the populations of some Chironomid species are highly polymorphic for paracentric but not for pericentric inversions. Some dicentric and acentric chromatids are formed at meiosis in the males, but these impede anaphase separation so that only giant non-functional sperms are produced (Beermann, 1956).

The situation in vertebrates is not well known, but we may anticipate that in those species which are polymorphic for pericentric inversions, reversed-loop formation will be found to be rare or absent.

Individuals heterozygous for centric fusions or dissociations will, in general, show trivalents at meiosis, made up of two rod-shaped acrocentric chromosomes associated with a V-shaped metacentric. The fecundity of such individuals will not necessarily be reduced, provided that the orientation of the trivalents on the spindle is such that both acrocentrics regularly pass to one pole and the metacentric to the other. This orientation will depend in part on the relative dimensions of chromosomes, cell and spindle, and on the number and positions of the chiasmata. Thus certain species may be unable to tolerate heterozygosity for fusions or dissociations because the mechanics of their meiosis leads to frequent malorientation of trivalents. Such malorientation is, for example, frequent in laboratory produced hybrids between the mouse species *Mus musculus* and *M. poschiavinus*, which are heterozygous for seven centric fusions (Gropp *et al.*, 1970; Tettenborn and Gropp, 1970).

A controversy has arisen between the present author and some other cytogeneticists such as John and Lewis (1968) and John and Hewitt (1968) as to the mechanisms of centric fusion and dissociation. The latter workers believe that strictly telocentric chromosomes (i.e. ones with terminal centromeres, as opposed to acrocentrics with a long limb and a very short one on the other side of the centromere) exist in the normal karyotypes of some or many species and that simple fusions of telocentrics may give rise to metacentrics and, conversely, that fission through the centromeres of metacentrics may give rise to viable and stable telocentrics. They consequently deny (implicitly or explicitly) that special telomeres are necessary at the ends of every chromosome. This controversy would be unimportant in the context of the present article were it not for the fact that according to White's viewpoint a centric fusion always involves a loss of short chromosome segments while a dissociation involves a gain or duplication of centromeric and telomeric regions; whereas on the alternative hypothesis no net gain or loss of genetic material is necessarily involved.

Heterozygosity for ordinary mutual translocations (i.e. ones where the breakage points are not adjacent to the centromeres, and which leave the chromosome number unchanged) will generally lead to the formation of rings or chains of four chromosomes at meiosis; these must orientate with the adjacent centromeres passing to opposite poles at meiotic anaphase in order to give unimpaired fecundity. Fecundity is likely to be severely reduced if they do not form rings or chains of four with complete regularity (i.e. if they sometimes form two bivalents or a trivalent and a univalent). It is thus not surprising that there are relatively very few species of animals whose populations are polymorphic for this kind of chromosomal rearrangement.

We have dealt with the factors which tend to permit or forbid the establishment of certain types of karyotypic changes in natural populations because they are obviously of vital significance in determining the role of structural rearrangements of the karyotype both in population polymorphism and in the possible initiation of genetic isolating mechanisms in raciation and speciation. The machinery of meiosis is complex and varies from species to species in many quantitative ways, so that universal laws or regularities must not be expected in this field.

With regard to the cytotaxonomic differences between allopatric taxa or ones which overlap geographically only in narrow zones (the *parapatric* type of distribution), there seem to be two possibilities. Either these are the remains of former adaptive polymorphisms, or they are not. On the first alternative a widespread ancestral population was at some time in the past polymorphic for the two alternative chromosome types which have subsequently become monomorphic (i.e. homozygous) in the descendent taxa. This would imply that adaptive chromosomal polymorphisms may sometimes, in the course of evolution, cease being adaptive, either because the heterozygote is no longer heterotic or for some similar reason. There can be no doubt that this is what has happened in the case of many species of *Drosophila* which differ in respect of one or more paracentric inversions; in such cases it is highly likely that a former adaptive polymorphism has broken down into two monomorphisms in the course of speciation.

However, in *Drosophila*, in grasshoppers and in some vertebrates, we find a fairly large number of cytotaxonomic differences due to rearrangements of types which are not known (or only rarely known) to give rise to adaptive polymorphisms in the particular genera or species concerned. It has consequently been suggested (White, 1968, 1969, 1970, 1973) that these are instances where the rearrangements in question have established themselves in phylogeny without passing through a stage of balanced polymorphism. If this is possible it is obviously also possible that they may, on occasion, play a direct role, as divisive agents in speciation, generating genetic isolating mechanisms involving hybrid sterility. Such a view is, of course, reconcilable only with difficulty with the view of Mecham (1961) and Littlejohn (1969) according to which pre-mating isolating mechanisms (i.e. ethological ones) are the primary ones in speciation. We shall return to this question later, after we have considered some of the evidence, because the Mecham–Littlejohn dogma seems to have been widely accepted by vertebrate zoologists on account of its obvious plausibility.

A very important question in this connection is the simple one: how widespread are cytotaxonomic differences between closely related species? If they were universal there would obviously be a *prima-facie* case for concluding that chromosomal rearrangements did play a primary role in speciation; on the other hand if closely related species frequently had identical karyotypes, there would be little support for such a view.

It is now clear that even the most closely related species of animals generally have morphologically distinguishable karyotypes, which differ either in chromosome number or in the morphology of the individual chromosomes, i.e. in ways which must be ascribed to the fixation of different chromosomal rearrangements in the two lineages which have diverged from a common starting point, the ancestral population (in some cases, of course, fixation of rearrangements may have occurred in one lineage only and not in the other).

Speciation entirely without the occurrence of chromosomal rearrangements has been proved to occur in some instances in the genus *Drosophila*. Particularly in the Hawaiian archipelago, complexes of so-called homosequential species have been discovered, with karyotypes that seem to be identical or at least indistinguishable, even in the giant polytene nuclei in which the banding pattern of the individual chromosomes can be studied in great detail. But this is certainly not the usual state of affairs, even in *Drosophila*, where most pairs or groups of sibling species do seem to differ in karyotype, even if the differences consist only of a few inverted chromosome segments.

As far as vertebrates are concerned, we do not have polytene chromosomes which could be used to demonstrate the existence of homosequential species, if such exist in vertebrates. And the mapping of the giant lampbrush chromosomes of the Amphibia has not proceeded to a point where they can be used in this way. Thus all that we can conclude on this vitally important question is that the vast majority of closely related species in the vertebrata are *not* homosequential (because they show obvious differences in karyotype), but we cannot exclude the possibility that a few may be.

In view of the near-universality of cytotaxonomic differences between even the most closely-related species of vertebrates, it is remarkable how little attention has been paid by workers on vertebrate speciation to chromosomal rearrangements. In the past this was excusable. Thus Mayr, in his important work "Systematics and the Origin of Species" (1942), hardly discusses them at all and even in his later book (1963) karyotypic changes due to chromosomal rearrangements are treated as being, in general, not different to other types of genetic changes such as point mutations at single loci. In his latest work, however, Mayr (1970) fully recognizes and discusses the significance of chromosomal rearrangements in speciation, while still emphasizing that "numerous cases in various groups of organisms are now known of well-defined and reproductively isolated species that agree completely in their chromosome structure and differ only in their gene contents" (the word *numerous* seems incorrect here—such cases are extremely rare, even in *Drosophila*, and are hardly known at all in other genera of Diptera with polytene chromosomes; the alleged karyotypic uniformity in groups like Acridid grasshoppers is being increasingly shown to be spurious).

An under-estimation of the probable role of chromosomal rearrangements in speciation certainly pervades much ethological work on speciation, in which it is assumed or implied that pre-mating isolating mechanisms (songs, calls, display patterns, odours) are invariably primary in speciation, chromosomal rearrangements being regarded as absent, unimportant or secondary factors.

Chromosomal rearrangements seem to arise spontaneously in most species that have been studied with a frequency of about 1 in 500 individuals. Many of these will be of types that are inviable or so deleterious that they will be automatically eliminated by natural selection. Of the remainder, some will carry with them gene combinations which are adaptive; they may hence become established in the population as adaptive polymorphisms. A few, however, in spite of carrying adaptive gene combinations, will lead to mechanical difficulties at meiosis in the heterozygote, so that fecundity is reduced because of the formation of some aneuploid gametes. It is this last category, giving rise to well-adapted homozygotes, but negatively heterotic, which may be suspected of being important in generating genetic isolating mechanisms in speciation.

There seems to be an important difference in principle between chromosomal rearrangements and ordinary gene mutations. The latter can and do recur with a measurable frequency. If a mutation is adaptive under a particular set of environmental conditions, we may hence expect it to crop up at multiple points of origin over the whole of the area occupied by the population or species. Chromosomal rearrangements, on the other hand, depend on a minimum of two breaks (excluding for a moment the possibility of simple fission through the centromere, believed in by certain cytogeneticists such as B. John, K. R. Lewis and G. M. Hewitt). Now a break at a *particular* point along the length of a chromosome is a very rare event (assuming 10000 possible breakage points it would have a frequency of about 1 in 300000*); the coincident occurrence of two particular breaks in a cell will be the square of this or about 1 in 10^{11}. Thus, for all practical purposes, each chromosomal rearrangement will be unique and will have a single point of origin in space and time, in a particular individual. We are

* Assuming the frequency of single breaks to be 1 in $\sqrt{1000}$.

consequently faced, in the case of chromosomal rearrangements, with the problem of their spread, i.e. in the case of a geographically widespread rearrangement, it must have either made its way from its point of origin through a pre-existing population or its carriers must have invaded territory not previously occupied by the species.

Although the above propositions regarding two-break rearrangements seem theoretically irrefutable, there is the question how far two rearrangements due to very similar breakage points would be genetically equivalent. For example there can presumably be many different fusions between two particular acrocentric chromosomes A and B, to give a metacentric AB, each with slightly different breakage points very close to the centromeres of the original acrocentrics. These would be indistinguishable by the ordinary techniques of cytological analysis; we can only speculate as to how far they would exhibit the same genetic properties. What these considerations mean in practice is probably that wherever we find what appears to be the same chromosomal rearrangement occurring over a wide geographic area, occupied by a single race of a species, several races or even several closely related species, then we are probably justified in postulating that it had a single point of origin, in a particular ancestral individual. But where it appears that a rearrangement such as a fusion between two particular chromosomes has occurred in two different phyletic lineages, we should not insist on a common origin. In *Drosophila*, with polytene chromosomes available for cytogenetic analysis, there seem to be no known cases of the same inversion (i.e. with precisely the same breakage points) having arisen more than once in phylogeny.

Although most chromosomal rearrangements that have occurred in evolution have undoubtedly been 2-break ones, cases of spontaneous 3-break and even 4-break rearrangements have been recorded in grasshopper populations and in man. They undoubtedly occur with a frequency which is much higher than the theoretical expectation, perhaps because multiple breaks in a cell may be caused by abnormal conditions such as virus-infection.

One aspect that needs to be stressed (because it has been misunderstood by some writers) is that chromosomal rearrangements—like other types of genetic change—will be subject to selection in accordance with the totality of their genetic effects. To speak of them being selected solely as isolating mechanisms is hence unreal—their other (pleiotropic) effects, i.e. on viability under the particular conditions of the environment, will always be relevant. Failure to appreciate this point seems to be at the root of Littlejohn's (1969) insistence that post-mating isolating mechanisms can only arise incidentally *after* genetic isolation has already been achieved by ethological mechanisms. Another point that should probably be borne in mind is that a number of genetic systems are known in which inheritance deviates from mendelian expectation. These have been collectively designated *meiotic drive* mechanisms. There is as yet no direct evidence that they may play a part in speciation, but this is obviously a possibility in some cases. Speciation is a rare event in the life of a population and there is consequently no reason to believe that the primary events which give rise to it will necessarily obey deterministic mathematics. Pleiotropism, genetic drift and meiotic drive are all phenomena which should warn us against simplistic dogmas in the field of speciation theory.

The modern literature on vertebrate chromosomes deals largely with somatic karyotypes and is consequently rather deficient in information about the behaviour of the chromosomes at meiosis. It does, however, provide generally accurate data on chromosome numbers and the relative lengths of the chromosome arms. Many authors follow Matthey (1959) in referring to the number of major chromosome arms (i.e. excluding the minute "second arms" of acrocentrics) as the *nombre fondamental* or fundamental number (often abbreviated N.F.). The only criticism that one can make of this system is that it is sometimes difficult to decide just how short a chromosome arm has to be before it is excluded from the count. Where two individuals or related taxa have different N.F.'s but the same chromosome number, it has often been assumed that they differ in respect of one or more pericentric inversions. This may well be so in the majority of instances; but unless the chromosomes have been carefully measured or the DNA values determined, the possibility exists that in the form with the higher N.F. duplications of material, perhaps usually heterochromatic, have increased the length of the short arms so that they get included in the count. In general, it must be admitted that in spite of carefully established karyotypes, the high chromosome numbers of most vertebrates, and especially mammals, render difficult precise evolutionary interpretations in terms of individual structural rearrangements.

II. THE FACTS

In what follows we shall be concerned solely with the evidence from vertebrates and especially mammals, referring to other groups of organisms only where comparison seems essential. One advantage of vertebrate material is that we have, in general, far more information about population-structure than we do in the case of most invertebrates. The vagility of small rodents and insectivores is probably of the same order of magnitude as that of flightless grasshoppers and beetles, land isopods and many mollusca, but the factor of territoriality may introduce complications into their population structure which are absent in most invertebrates.

In the Marsupials, apart from two cases of supernumerary chromosomes (see p. 120) there is only one known case of intrapopulation chromosomal polymorphism. That is in the hare-wallaby *Lagorchestes conspicillatus*, where Martin and Hayman (1966) have described a polymorphism for a fusion between two autosomes. Although very careful chromosome measurements and determinations of DNA values have been carried out by Martin and Hayman (1965, 1967; see also Hayman and Martin, 1969), they do not provide much evidence on the role of chromosomal rearrangements in speciation, except insofar as they emphasize that in addition to the easily-identified evolutionary changes such as fusions, dissociations and pericentric inversions, there have been many small-scale rearrangements which, since they affect the total amount of DNA present, must have been duplications or deletions. This is a salutary warning, since in most work on mammalian karyotypes no DNA values have been determined and interpretations of karyotype evolution in terms of the more obvious structural changes is certainly only a first approximation to the complete truth.

The first example of chromosomal polymorphism in a vertebrate population to be seriously investigated was that of the chromosomal fusions in the small shrew *Sorex araneus* (Sharman, 1956; Ford *et al.*, 1957; Meylan, 1964, 1965; Ott, 1968; Ford and Graham, 1964; Ford, 1970; Kozlovsky, 1970). As recently as 1959 this was considered by Matthey to be the only proven case of balanced chromosomal polymorphism in the mammalia. The chromosome number of this species ranges from $2n♂ = 33$ to 21 due to the presence of six fusions or dissociations in the natural populations (all males have an XY_1Y_2 sex chromosome mechanism). Since some other species of *Sorex* have chromosome numbers ranging from $2n = 38$ to 66, it is likely that chromosomal evolution has been from higher to lower numbers in the *araneus* lineage and that we are dealing in this case with fusions rather than dissociations. *S. araneus* has a sibling species, *S. gemellus*, first clearly recognized by Ott, although earlier suggested on chromosomal grounds by Meylan. *Araneus* ranges from the British Isles to Siberia and is apparently chromosomally polymorphic throughout this vast area; while *S. gemellus* is chromosomally monomorphic with $2n = 23$ and the same XY_1Y_2 system; it occupies a much more restricted area from the Alps to Holland and the island of Jersey. There is a broad overlap zone between the two species in central Europe, but no natural hybrids seem to have been recorded and the karyotypic differences between the two species appear to be multiple and extensive (even between a 23-chromosome male of *araneus* and one of *gemellus*). Nevertheless, it is fairly clear that *gemellus* must have arisen as a structurally homozygous segregate from the chromosomally polymorphic *araneus*.

No information is, of course, available as to whether any particular chromosomal rearrangement played a role in initiating genetic isolation between the two species. The population structure of these small insectivores is surely of a type which would tend to loss of genetic polymorphism through inbreeding and genetic drift, so that the fact that an extensive system of chromosomal polymorphism has been preserved for a vast period of time in *S. araneus* argues for it having a very strong adaptive significance (powerful heterotic effects in the heterozygotes or some equivalent mechanism).

Some other insectivores also show considerable karyotypic differences between related species. In the genus *Tupaia* (sometimes included in the order Primates), six species from southeast Asia have chromosome numbers ranging from $2n = 52$ (N.F. 74) to 68 (N.F. 74) according to Arrighi *et al.* (1969); the karyotypes of *T. glis* and *T. longipes* are apparently indistinguishable ($2n = 60$, N.F. 60), but all the others are different. The closely-related *Urogale everetti* has $2n = 44$ (N.F. 80).

The tenrecs of Madagascar have been studied cytologically by Borgaonkar (1967) and Borgaonkar and Gould (1968), who have postulated a reciprocal translocation as the basis for speciation between *Microgale dobsoni* and *M. talazaci*; their interpretation is not entirely convincing, since it is not based on precise measurements of chromosome limbs and they have not considered alternative explanations.

In the European Hedgehogs, Gropp *et al.* (1969) have shown that *Erinaceus europeus* and *E. roumanicus* both have $2n = 48$, but show very different karyotypes. They have extensive heterochromatic and late-replicating segments in several autosomes, but these are quite differently arranged in the two species, suggesting a complex pattern of duplications and deletions of heterochromatin, and possibly (but not certainly) translocations as well, since the evolutionary separation of these two forms. *Hemiechinus auritus* and *H. megalotis* have indistinguishable karyotypes and similar DNA replication patterns; they are considered to be subspecies of a single species.

Of particular interest are processes of chromosomal evolution in small burrowing mammals, since these tend to form local colonies or demes between which migration may occur only rarely, due to low vagility of the individuals. The population structure of such species must resemble that of some invertebrates, e.g. wingless grasshoppers and snails, more than it does that of larger cursorial swimming or flying vertebrates. We may consider three well-investigated cases, the pocket gophers (*Thomomys*) of the United States, the mole-voles (*Spalax*) of eastern Europe and Palestine and the tuco-tucos (*Ctenomys*) of Argentina.

The morphological variability of *Thomomys bottae* is clear from the fact that more than 150 named races or subspecies have been recognized by taxonomists (whatever one may think of the utility or propriety of such nomenclatural abandon). The morphological–taxonomic situation is clearly a reflection of the population structure. Patton and Dingman (1970) studied 24 populations from at least 16 of the named races, in New Mexico, Arizona, southern California and northern Mexico. With the exception of one individual of *T.b. simulus* (the only one studied) which had $2n = 74$, all the others showed a uniform $2n = 76$. But the number of acrocentric chromosomes ranges from 0 (in five races) to 24 in *T.b. fulvus*, with a corresponding variation in the number of metacentric, submetacentric and subacrocentric chromosomes. Presumably at least 12 (probably many more) pericentric inversions have occurred in the phylogeny of this species. It seems usual for each population to have a constant karyotype; some named subspecies also exhibit karyotypic uniformity, but in other cases a named form may exhibit several karyotypes in different localities. Some allopatric subspecies appear to have identical karyotypes, but with such a high chromosome number it is difficult to be certain of this.

Intra-deme polymorphism for pericentric inversions is clearly not generally present in this species, although it does occur in some cases (see below).

The subspecies *Thomomys bottae grahamensis*, from the Graham Mountains of Arizona, exhibits an invariable chromosome number ($2n = 76$), but the number of acrocentric and metacentric chromosomes in the karyotype is variable, apparently due to intra-population polymorphism for several pericentric inversions (Patton, 1970). The number of acrocentric chromosomes increases steadily from an elevation of 2195 m to 3048 m; the populations at the lower elevations are all chromosomally polymorphic, but the one at 3048 m may be monomorphic.

The situation in the *Thomomys talpoides* complex differs in several respects from that in *T. bottae*. Thaeler (1968) studied 16 populations in Wyoming, Colorado, New Mexico and Arizona. Eight different karyotypes with the chromosome numbers $2n = 40, 44, 46, 48, 56, 58$ and 60 are known (the two populations with $2n = 56$ differed in the numbers of acrocentric and metacentric chromosomes). *T. t. pygmaeus* ($2n = 58$, autosomal *nombre fondamental* 76) co-exists at one locality with *T. t. ocius* ($2n = 56$, *nombre fondamental* 78) without interbreeding. *Pygmaeus* also coexists with *T. t. bridgeri* ($2n = 40$, *nombre fondamental* 70) without known interbreeding and points of contact without interbreeding are also reported for *T. t. ocius* and *T. t. bridgeri*. It is clear that these taxa are, in general, behaving as biological species, distinguished by cytotaxonomic differences based on chromosomal fusions (or dissociations) and pericentric inversions which have changed the *nombre fondamental*. The correlation between the chromosomal data and the currently accepted taxonomy of this complex is far from perfect. Thus "*T. t. fossor*" includes populations having $2n = 48$, 56 and 60.

T. t. ocius and *T. t. rostralis* south of Rawlins, Wyoming have indistinguishable karyotypes, but a population from the Medicine Bow Mountains currently regarded as *T. t. rostralis* has $2n = 48$. The absence of recorded interbreeding between taxa with different karyotypes at points of contact should not be interpreted as indicating that interbreeding *never* occurs, but only that if it takes place it is rare, so that introgression would be on a restricted scale. What barriers (ethological or other) prevent interbreeding is not known; it is conceivable that cross-mating does take place but that the hybrids are inviable. Extensive sympatry between different karyotypes does not seem to occur. It is interesting that at least ten chromosomal fusions (or dissociations) have reached fixation in the *talpoides* complex, whereas only one is known to have done so in *bottae*. No populations polymorphic for fusions or dissociations are known in either of these taxa. It seems clear that both are splitting up into a large number of biological species and that chromosomal rearrangements (pericentric inversions in *bottae*, pericentric inversions and fusions or dissociations in *talpoides*) are playing an active role in this multiple speciation process. It is, of course, quite possible that ethological isolating mechanisms have also been involved, but no information seems to exist on this point.

Karyotype evolution in burrowing rodents of the *Spalax ehrenbergi* complex in Palestine has been studied by Wahrman *et al.* (1969a, b). Four parapatric forms with $2n = 60$ (*nombre fondamental* 76), 58 (N.F. 76) 54 (N.F. 82) and 52 (N.F. 84) were encountered. As in the *Thomomys talpoides* complex the karyotypes differ in respect of both fusions (or dissociations) and pericentric inversions. A limited number of natural hybrids between the 52- and 54-chromosome taxa and also between the 58- and 60-chromosome forms were collected at points of contact. There is a somewhat irregular cline of chromosome numbers from the Negev ($2n = 60$) to Lebanon ($2n = 52$ and 54). Nevo (1969) has put forward evidence for some degree of ethological isolation between these taxa (more pronounced aggressive behaviour and lower frequency of copulation between heterogamic pairs, by comparison with homogamic pairs). He suggests that ethological isolating mechanisms "may have played an important role in speciation in *Spalax*, concomitant with cytologic mechanisms". This is one of the very few instances in the vertebrata where both types of phenomena have been investigated, but even so the evidence is still inadequate to define their respective roles or to decide which developed first.

Apparently *Spalax leucodon*, in the Balkans and eastern Europe, is similarly fragmented into taxa having $2n = 48$, 50, 54 and 56 (Matthey,

1959; Walknowska, 1963; Soldatović *et al.*, 1966, 1967; Raicu, *et al.*, 1968). *S. microphthalmus*, on the other hand, probably has a constant karyotype with $2n = 62$ (Raicu *et al.*, 1968).

Wahrman and Goitein (1972) have also studied karyotype evolution in the Spiny Mouse *Acomys cahirinus*. Material from Israel shows $2n = 38$ (16 pairs of metacentric autosomes, 2 pairs of acrocentrics), but a race with $2n = 36$ (all autosomes metacentric) exists in the Sinai Peninsula. A hybrid zone about 20 km wide, in which some 37-chromosome heterozygotes occur, has been found on the west coast of the Gulf of 'Aqaba. The *A. cahirinus* complex or superspecies has a vast distribution, from north-western India to North and East Africa, and is chromosomally unknown throughout most of its range. The 38- and 36-chromosome forms are believed to be the penultimate and ultimate members in a series with increasing numbers of fusions of acrocentric elements; but there is a possibility that the Israeli form may have been derived from the Sinaitic one through dissociation of one metacentric. Wahrman and Goitein believe that the hybrid zone is one of "secondary intergradation", but their reasons for this conclusion are not clear; absence of information regarding populations on the east coast of the Gulf of 'Aqaba prevents a satisfactory overall interpretation.

Forms of *A. cahirinus* occurring on the island of Crete were studied by Matthey (1963d); there were two races, one with $2n = 38$ (N.F. 66) and one with $2n = 40$ (N.F. 68). The karyotype was somewhat variable in spleen cells from the same individual, possibly due to non-disjunction. Matthey studied two female hybrids from a cross between a male Cretan *Acomys* ($2n = 40$) and a female *cahirinus* from Israel; he believed that two acrocentrics from the Cretan parent must have undergone centric fusion in the cytoplasm of *cahirinus* to produce a metacentric. The evidence on this point is unfortunately not really clear. Male hybrids between the same two forms showed some chains of chromosomes and univalents at diakinesis.

Two South African species of *Acomys* are distinguished from the rest of the genus by having "giant" metacentric X-chromosomes, about three times the usual length (Matthey, 1965b, 1968b); *A. selousi* has a unique XO condition in the female soma; the female germ-line has not been studied but it seems likely that it is XX and that the missing X becomes eliminated from the female somatic tissues or some of them, as in the case of the Marsupial Peramelidae studied by Hayman and Martin (1965). Altogether Matthey (1963d, 1965a, b, 1968b) has studied ten species of *Acomys* with the chromosome numbers 36 (N.F. = 68), 38 (N.F. = 68), 38 (N.F. = 68), 38 (N.F. = 70), 38–40 (N.F. = 68), 50 (N.F. = 66), 60 (N.F. = 70), 60 (N.F. = 76), 64 (N.F. = 66), 66 (N.F. = 66). Different centric fusions seem to have established themselves in the various lineages of this genus. Thus when

Wahrman (unpublished) crossed members of the superspecies *cahirinus* from Israel and Cyprus (both with $2n = 38$) the F_1 males showed at meiosis a chain of 28 chromosomes, a chain of 4, two autosomal bivalents and an XY bivalent. Intuitively, one might imagine that sequences of fusions would not arise in two phyletic lines from a common ancestor with all its chromosomes acrocentric in such a manner as to give a chain of 28 in the F_1 hybrid and that some special explanation of this regularity is required. However, a statistical analysis suggests that in this case the intuitive conclusion is probably wrong.

The burrowing rodents known as tuco-tucos in South America (genus *Ctenomys*) have been studied cytogenetically by Reig and Kiblisky (1969). The karyotypes of the eleven species so far investigated are very diverse and three species are chromosomally polymorphic. The diploid numbers range from 68 down to 22 and the N.F. from 122 down to 44. The very large number of chromosomal rearrangements which have established themselves in this genus must be a reflection of the population structure of these little rodents; Reig and Kiblisky estimate that at least 30 chromosomal rearrangements separate the karyotypes of two closely related species *C. talarum* and *C. porteousi*, which both, however, have $2n = 48$. On this basis, many hundreds, and perhaps thousands of structural changes in the karyotype must have occurred during the evolutionary radiation of this one genus.

The chromosomal races of the pocket-mouse *Perognathus goldmani*, which inhabit the states of Sonora and Sinaloa in northwestern Mexico, have a highly unusual type of distribution. Each occupies a territory of its own, but these are separated by the main rivers which traverse the area. The riparian vegetation fringing these rivers is not inhabited by either of the "parapatric" races but by a related species, *P. artus* (Patton, 1969a), which serves to isolate them. According to Patton's analysis, an ancestral population with $2n = 56$ (autosomal N.F. 54) existed in southern Sonora in pre-pluvial time. This is envisaged as having given rise allopatrically (i.e. by invasion of previously unoccupied territory) to race β (southward) by a pericentric inversion A_1 in chromosome 9 and to race ε (northward) by a centric fusion A_2 between acrocentrics 3 and 4. At this stage the mesically-adapted *P. artus*, which is derived from the same stock, is believed to have penetrated as a wedge between races β and ε. The latter is supposed to have given rise to races θ (northward) and γ (southward) by different fusions A_3 and A_4 (both these races also carry fusion A_2). Race γ then gave rise, by a pericentric inversion in the X (converting it into an acrocentric) to race α, which re-invaded most of the *artus*-occupied territory after the end of the pluvial period, and in turn gave rise by a further fusion A_5 (eastward) to race δ, which has all three fusions A_2, A_4 and A_5.

Patton's interpretation of this case, which rests largely on a hypothetical but plausible interpretation of Pleistocene phytogeography, is essentially an allopatric one, i.e. he explicitly rejects the "stasipatric" model of White (1968). There can be no doubt of the essential correctness of his cytogenetic analysis as such, i.e. the four fusions and the two pericentric inversions did actually occur in the differentiation of races β, ε, θ, γ, α and δ. Moreover, there is no evidence that any of these rearrangements is polymorphic in any area, and they may hence have never existed in a balanced polymorphic state. This is what we should expect if they had played a direct role as genetic isolating mechanisms. Introgression between the races seems to be minimal at the present time, since only three karyotypically hybrid individuals (hybrids between races α and δ) were found in the whole investigation (involving a study of 319 specimens).

Although Patton's chromosome phylogeny of *P. goldmani* seems certain, his geographic interpretation is obviously much more hypothetical. He points out that, on the supposition that fusions have occurred, the phylogenetically primitive races (or karyotypes) β and ε occupy the southern and northern margins respectively, of the total species-range, with the derived races γ, α and δ occupying the central part of the overall distribution (race θ occurs far to the north, in an area which is ecologically anomalous for the species, so that it may have recently invaded the territory it occupies). If dissociations rather than fusions had occurred the situation would, of course, be reversed, i.e. the most primitive races would occupy the centre of the species-range, with the derived ones on the periphery (but this interpretation is considered unlikely). Although Patton rejects the stasipatric model, partly because of considerations involving race θ (which may be somewhat anomalous), it will be apparent that there is an essential similarity between the geographic pattern in *P. goldmani* and the *viatica* group of morabine grasshoppers, from which the stasipatric model was originally derived (derived karyotypes central in both cases, ancestral ones peripheral). Neither picture is really clear and the interpretations in both cases are uncertain; perhaps both involve a combination of allopatric and stasipatric processes. One general objection that can be raised against strictly allopatric interpretations of cases such as these (and which seems particularly cogent in the case of the *Thomomys* species and species-groups referred to earlier, which have larger total ranges) is that they seem to require the ancestral forms to have occupied what seem unreasonably small areas of territory from which they invaded previously unoccupied regions, presumably at least in part owing to the adaptive properties of the new chromosome arrangements carried in homozygous form by the invaders.

Patton (1969b) has also studied karyotypic evolution in *Perognathus penicillatus*. All races of this species have $2n = 46$, but the N.F. ranges from 48 to 56 as a result of a number of pericentric inversions. The races *P. p.*

penicillatus and *P. p. angustirostris* from Arizona, southern California and Baja California, have what is believed to be the ancestral karyotype with NF = 54 (5 pairs of metacentrics). *P. p. eremicus* and *P. p. atrodorsalis* (New Mexico to the southern tip of the Mexican state of Tamaulipas) have a karyotype which differs by only one pericentric inversion (N.F. = 56, 5 pairs of metacentrics and one pair of submetacentrics). *P. p. pricei* (southern Arizona and Sonora) and *P. p. seri* (Tiburón Island) have a karyotype which differs from the supposed ancestral one by three pericentric inversions (N.F. = 48, only 2 pairs of metacentrics).

The karyotypes of 20 species of the North American cricetine rodent genus *Peromyscus* were studied by Hsu and Arrighi (1966, 1968). With the exception of one species, sometimes put in a separate genus by systematists, all of them had $2n = 48$ (the exceptional species had $2n = 52$). Chromosomal fusions and dissociations consequently do not seem to have occurred in the phylogeny of *Peromyscus sensu stricto*. The number of metacentric, submetacentric and acrocentric chromosomes in the karyotype is, however, highly variable, so that the N.F. (including the sex chromosomes) ranges from 56 (in *P. crinitus* and *P. boylei*) to 96 (in *P. eremicus* and *P. collatus*). A number of the species recognized by taxonomists (e.g. *P. truei*, *P. difficilis*) seem to include several different karyotypes and may be complexes of sibling species. A major difficulty in analysing karyotypes of members of this genus is the fact that the Y-chromosome seems to be variable in size and shape, in some cases within a species. The attempt of Kreizinger and Shaw (1970) to establish reliable karyotypes for seven subspecies of *P. maniculatus* was more or less unsuccessful for this reason. Some workers (e.g. Ohno *et al.*, 1966; Sparkes and Arakaki, 1966) have reported structural heterozygosity in *P. maniculatus*, which was interpreted as an indication of chromosomal polymorphism, presumably for pericentric inversions. In view of the difficulty of establishing karyotypes in this genus, their conclusions cannot be regarded as firmly established, although they may well be correct in some instances. Arakaki *et al.* (1970) demonstrated that karyotypic variability existed among siblings—but this was in a laboratory colony of *P. maniculatus* derived by intercrossing three different subspecies several years before. Clearly, pericentric inversions have played a major role in the karyotype evolution of *Peromyscus* and, on the whole, it seems as if most of the individuals caught in the wild that have been karyotyped are structurally homozygous, so that intra-population polymorphism, if it exists, seems to be on a relatively reduced scale, at any rate in most of the species. A few species pairs, such as *P. nudipes* and *P. megalops*, are stated to have indistinguishable karyotypes. It is obvious that no more than a beginning has been made in analysing the complex karyotype evolution of this genus, and its relation to race- and species-formation.

Studies on meiotic chromosomes should certainly be carried out to determine whether the alleged population polymorphism for pericentric inversions in some of the species is genuine or not.

The karyotypes of another North American cricetid genus, *Neotoma*, were studied by Baker and Mascarello (1969). Six out of seven species of the subgenus *Neotoma* had $2n = 52$, but their autosomal N.F. ranged from 50 to 66, possibly as a result of pericentric inversions, and in *N. micropus* and *M. lepida* the N.F. was variable, presumably due to "floating" inversions in the natural populations. *N. fuscipes* was anomalous, having $2n = 56$ and N.F. = 82. In the subgenus *Teonoma*, *N. cinerea* had $2n = 54$ (N.F. = 60) and in the subgenus *Teanopus*, *N. phenax* had $2n = 38$ (N.F. = 50). The chromosomal polymorphism of *N. micropus* has been studied in greater detail by Baker *et al.* (1970). One hundred individuals, from a total of four localities, were studied; the chromosome number was invariably $2n = 52$, but four different female karyotypes and three different male ones were encountered. Females had from 1 to 4 large metacentric chromosomes and males from 1 to 3; it appears probable that the polymorphism involves two chromosome pairs and that one of these is the X pair in the female (the Y appears to be an invariable subacrocentric). It may be suggested that metacentric and acrocentric X's both occur with a reasonably high frequency but that in the case of the autosomal pair the acrocentric type is relatively rare. If this is the case, individuals with no metacentric chromosomes probably occur in both sexes, but in very low frequency, so that their absence from a sample of 100 would not be remarkable. A peculiar feature of this case seems to be that the difference between the metacentric and acrocentric homologues appears not to be due either to fusions or dissociation (because the chromosome number is the same in all individuals)—or to pericentric inversions, because the metacentrics are much longer than any of the acrocentrics. Most probably the metacentrics have one arm composed of supplementary heterochromatin (although this is not the conclusion of Baker *et al.*, who speak rather vaguely of "both inversions and translocations", without considering what the genetic consequences would be).

The karyotypes of the African mice of the subgenus (or genus) *Leggada*, which are closely related to the house mouse, *Mus musculus*, have been studied by Matthey, in a long series of investigations. The taxonomy of the *Leggadas* is in a somewhat confused state and does not agree altogether with the karyological findings. In general, there seem to have been a large number of different chromosomal fusions in this group, and many populations are polymorphic for fusions; but pericentric inversions have hardly become established at all, except in *L. oubanguii*, which has a constant chromosome number of $2n = 28$, but in which the N.F. ranges from 30 to 34 (Jotterand, 1970).

These little *Leggada* mice fall into two groups, on the basis of their sex chromosomes. In what Matthey calls the PR group, the X and Y have the primitive acrocentric shape, as in *Mus musculus*. In the TR group the original X and Y have become fused or translocated to a pair of what were originally autosomes, so that the X has become metacentric and the Y submetacentric.

The PR group includes *L. tenellus*, *L. bufo*, *L. indutus* and *L. setulosus* and subspecies 1 of *L. minutoides* (from the Congo). The TR group includes *L. triton*, subspecies 2 (South Africa) and 4 (Ivory Coast) of *L. minutoides*, *L. minutoides musculoides* and *L. goundae*, from the Central African Republic (Matthey, 1964b; Jotterand, 1970). *L. minutoides* subspecies 3 (from Rhodesia) is in a sense intermediate between the PR and TR groups since it has the Y-autosome fusion but not the X-autosome and hence has X_1X_2Y males (Matthey, 1965c).

The primitive karyotype in the genus *Leggada* undoubtedly consisted of 36 acrocentrics (i.e. 17 pairs of acrocentric autosomes, an acrocentric X and a shorter acrocentric Y); this karyotype still persists in *L. bufo* (Congo) *L. setulosus* (central and west Africa), *L. tenellus* (Ghana), and *L. indutus* (South Africa).

L. bufo and *L. setulosus* seem to be chromosomally primitive species which have retained the ancestral karyotype and do not show any evident chromosomal polymorphism (Matthey, 1966a). But the *tenellus* group is a little more complex.

Material of this group from Ghana (probably *L. mattheyi*) showed the primitive karyotype of $2n = 36$ acrocentrics, but a closely related population from the Niger Republic (probably also *L. mattheyi*), was homozygous for one autosomal fusion and polymorphic for a second, so that individuals with 32, 33 and 33 chromosomes occurred; in addition one of the small autosomes has become metacentric in consequence of some kind of structural change (Matthey, 1968c, 1970a).

The majority of the TR section of the genus *Leggada* fall into the great *minutoides–musculoides* complex, which contains a bewildering number of forms, every chromosome number between $2n = 34$ and $2n = 18$ being now known (Matthey, 1963b, c, 1966c, 1970b). Most populations are polymorphic for fusions; in general only a single taxon of this complex occurs at each locality, but at Ippy, Central African Republic, two seem to co-exist, one with $2n = 26–32$ and one with $2n = 18–20$. It must be admitted that the interpretation of the data in relation to processes of speciation is not possible without far more data on the geographic distribution of the various forms, the existence of intergradation, sympatry or parapatry and the occurrence of ethological isolation or mechanisms of hybrid sterility. As in the case of the South American genus *Ctenomys*, Matthey's work on the African muridae has revealed a great wealth of evolutionary complexity, from which general conclusions are hard to draw at present.

In the 18-chromosome *L. m. minutoides* from South Africa the Y-chromosome is acrocentric, but its short arm still forms a chiasma with the metacentric X. In material from Angola, however, ("subspecies 6") the Y is a metacentric. It seems that in both forms the sex chromosomes are of the TR type, but that in the South African form one limb of the Y has been lost by deletion (Matthey, 1966e).

L. goundae, from the Central African Republic, differs from the other *Leggadas* in that it has apparently reduced its chromosome number below $2n = 18$, the lowest number that can be attained from $2n = 36$ acrocentrics solely by fusions. Actually, however, the chromosome number of *L. goundae* varies from 16 to 20, due to polymorphism for two fusions or dissociations. Presumably, at some stage in the ancestry of this species two metacentric chromosomes became re-converted into acrocentrics by pericentric inversion and then fused, reducing the chromosome number to 16 (Jotterand, 1970).

Leggada triton apparently includes two forms. Subspecies 1, from the Congo, has $2n = 32$, all autosomes being acrocentric, while the X is a large metacentric and the Y a small acrocentric (N.F. = 34). Subspecies 2, from Malawi and Tanzania, has approximately 20–22 chromosomes and N.F. = 32, there being a number of metacentric autosomes (Matthey, 1963a, 1964c, 1966f). However, in one population from the Congo, four females were found in which the shorter arm of one of the X's had undergone a deletion, so that the chromosome was an acrocentric. It is fairly clear that this is a balanced polymorphism, but it is not known whether males with deleted X's and females with two such chromosomes are viable or not. Altogether, 10 of 21 females from near Bukavu, Congo Republic, had one deleted X, but none were homozygous for the deleted X, which was not found in a total of 18 males (Matthey, 1966f, 1967c). A similar deletion, but in this case involving only part of one arm of the X, was found in the heterozygous state in four out of fourteen females of *L. minutoides musculoides* from the Ivory coast (Matthey, 1967a, b); four males all carried an undeleted X. Obviously, in both these cases there is reason to suspect that males with deleted X's are either not produced or are non-viable; this would, of course, automatically ensure that homozygous females would not occur. It is fairly obvious that this heterozygosity for deletions in the X, in both species, has something to do with the inactivation of one X-chromosome in the somatic cells of female mammals (the Lyon phenomenon); but we do not know whether the deletions are in the limb derived from the original X or in the one which was originally autosomal.

The karyotypes of the Australian Muridae have been inadequately studied, but Matthey (1968a) has reported that *Leggadina hermannsburgensis* differs from both *Mus* and *Leggada* by its higher chromosome number and N.F. ($2n = 48$, N.F. $= 56$, the X being a large metacentric). Kennedy (1969) has shown that *Hydromys chrysogaster*, *Pseudomys minnie*, two species of *Leggadina* and two species of *Notomys* all have rather similar (but not indistinguishable) karyotypes with $2n = 48$; this karyotypic uniformity of the native Australian Muridae contrasts strongly with the situation in the murid faunas of other continents. But in *Notomys* a number of the chromosomes which are acrocentric in the other genera have become subacrocentric.

The case of the "tobacco-mouse", *Mus poschiavinus*, is in many respects unique. In the year 1869 the zoologist V. Fatio described a new species of mouse from tobacco warehouses in Val Poschiavo, eastern Switzerland. Darker than the commensal *Mus musculus* and with other minor morphological differences, *M. poschiavinus* has never been found at any other locality. It stills exists in Poschiavo and was shown by Gropp *et al.* (1970) to have $2n = 26$, including 7 pairs of metacentric chromosomes, in place of the $2n = 40$ acrocentrics karyotype of *M. musculus*, i.e. it is homozygous for seven different chromosomal fusions. Laboratory hybrids between *poschiavinus* and *musculus* show seven trivalents at meiosis, each made up of one *poschiavinus* metacentric and two *musculus* acrocentrics. These trivalents are frequently mal-orientated on the meiotic spindle, so that much non-disjunction occurs and more than half of the gametes formed are aneuploid (Tettenborn and Gropp, 1970); fecundity in backcrosses is accordingly severely reduced.

We do not know what the situation in nature really is. Does the tobacco mouse exclude *M. musculus* from its limited range? Alternatively, if the two occur together, does natural hybridization take place? Is the distribution a parapatric one, with limited interaction in a narrow zone of overlap? Neither do we know the evolutionary history of the tobacco mouse, except that it must surely have arisen from *M. musculus* by some process that involved seven different chromosomal fusions. The really unique feature of this case is that *M. poschiavinus* must certainly have arisen from the widespread *M. musculus within* the range of the latter. This would seem to exclude a strictly allopatric origin, in the classical sense, and clearly suggests the stasipatric speciation model of White (1968). It is, however, possible that the tobacco mouse evolved from an isolated population of *M. musculus* cut off on a nunatak during the Pleistocene, when the alpine glaciers were far more extensive than at present. The frequent malorientation of the trivalents at meiosis in the hybrids between *musculus* and *poschiavinus* is somewhat unexpected, since trivalents apparently orientate regularly in several species of *Leggada*; and in a stock of *M. musculus* heterozygous for a spontaneously-arisen fusion (Evans *et al.* 1967) 98% of the second metaphases showed either a metacentric or the two acrocentrics.

That familiar pest and commensal of man, the black rat (*Rattus rattus*) has been shown to be a karyotypically complex species in recent years. One group of populations, in Japan, Europe and Malaysia, have $2n = 42$. However, Bianchi *et al.* (1969) have described two South American populations with $2n = 38$, and an apparently identical karyotype has been found by Capanna *et al.* (1970) in rats from Tuscany, the Argentario peninsula and two islands in the Tyrrhenian sea; the same karyotype appears to exist also in New Guinea, Australia and New Zealand (Yosida and Tsuchiya, 1969). Finally, Badr and Badr (1970) have claimed that near Cairo, Egypt, as many as three forms of *R. rattus* occur, the 42- and 38-chromosome races previously referred to and a form with about 54 chromosomes. It seems likely that the 38-chromosome race has arisen from the 42 one by fixation of two chromosomal fusions (or, alternatively, that two dissociations have undergone fixation in the ancestry of the 42-chromosome race); but the origin of the 54-chromosome race, if it really occurs, is not clear. In Asia the longest autosome of *R. rattus* shows a balanced polymorphism for a short pericentric inversion, and the No. 9 and No. 13 chromosomes are similarly polymorphic in certain populations (Yosida *et al.*, 1965, 1967; Yosida and Tsuchiya, 1969). A similar polymorphism exists in *Rattus norvegicus* (Yosida and Amano, 1965).

The karyotypes of numerous wild Malayan species of *Rattus* have been investigated by Yong (1968, 1969, 1972); they show a range of chromosome numbers from $2n = 36$ to 52 (N.F. range from 36 to 83). Chromosomal evolution has undoubtedly been quite complex in this genus. There are a number of native species of *Rattus* in Australia; a form with a 42-chromosome karyotype like that of *R. lutreolus* probably gave rise by two fusions to the karyotypes of *R. assimilis*, *R. fuscipes* and *R. greyi*, all of which have $2n = 38$—and by three further fusions to *R. conatus*, with $2n = 32$ (Kennedy, 1969; Martin, 1969; Dartnall, 1970). A change in the opposite direction, i.e. one involving a series of dissociations, must have taken place in the evolution of the Australian *R. villosissimus* ($2n = 50$, there being two pairs of large metacentrics).

An African subgenus of *Rattus*, *Mastomys* has been investigated cytotaxonomically by Matthey (1965a, 1966b, d). In South Africa a form with $2n = 36$ (N.F. either 56 or 58) is presumably *M. natalensis*. In West Africa there are two sympatric species, one with $2n = 32$, the other with $2n = 38$ (the latter is probably *M. erythroleucus*). The form with 32 chromosomes exists also at Fort-Lamy (Chad), where it is polymorphic for a pericentric inversion.

Karyotype evolution in ground squirrels of the super-species *Spermophilus townsendi* in the northwestern U.S.A. was studied by Nadler (1968) No chromosomally polymorphic populations were encountered, but five taxa (referred to as "subspecies" by Nadler, but perhaps biological species in some cases) showed karyotypes ranging from $2n = 36$ to 46. The N.F., on the other hand, only varied from 66 to 68. Chromosomal fusions (or dissociations) consequently seem to have been more frequent in the karyotype evolution of this complex than pericentric inversions. Nadler suggests that there was formerly an ancestral population which was polymorphic for chromosome number and included 46, 38 and 36 chromosome karyotypes (a view for which there is certainly no evidence). Alternatively, he supposes that an ancestral form with the karyotype of *canus* ($2n = 46$) gave rise southward and eastward to *mollis* and *idahoensis* (both with $2n = 38$) by a series of four fusions. Northward it is supposed to have given rise by four or five fusions to *townsendi* ($2n = 36$) and *nancyae* ($2n = 38$). The northernmost and southernmost populations *nancyae* and *mollis* and another more or less peripheral one, on the east, *idahoensis*, are stated to have indistinguishable karyotypes. On Nadler's interpretation *nancyae* and *mollis* would have acquired their karyotypes independently from a $2n = 46$ condition. Experimental hybridization between these taxa would reveal whether the four fusions are between the same chromosomes in each case; if they proved to be, Nadler's interpretation involving an independent origin of the 38-chromosome karyotypes would become very suspect. In that case we might have to consider the possibility of chromosomal dissociations rather than fusions having been involved; this would mean that the present distribution would have the primitive karyotypes on the periphery, as in the cases of *Perognathus goldmani* and the grasshoppers of the *viatica* group previously mentioned. We raise these possibilities here merely to show that Nadler's diagram of the course of karyotype evolution in this complex is certainly not the only possible one.

Experimental hybridization has been used to compare the karyotypes of various species of gerbils and jirds. Lay and Nadler (1969) crossed two partially sympatric jird species, *Meriones shawi*♀ and *M. lybicus*♂. Although both have $2n = 44$, the karyotypes differ in respect of several autosomal rearrangements and in the shape of the X-chromosome, which is acrocentric in *lybicus* and metacentric in *shawi*. Female hybrids were fertile in backcrosses, but male ones have much reduced spermatogenesis. Chromosome numbers range from $2n = 42$ to 72 in the genus *Meriones* (Matthey, 1957).

Gerbillus pyramidum is a superspecies which includes a number of forms with very different chromosome numbers. Populations in the Negev have $2n = 66$ or 62, one from Israel has $2n = 52$ and one from Algeria has $2n = 40$ (Wahrman and Zahavi, 1955, 1958; Zahavi and Wahrman, 1957;

Matthey, 1952, 1953). A male laboratory hybrid between a Negev female and an Algerian male showed 3 bivalents, 9 to 11 trivalents and 3 to 4 chains of 5 to 7 chromosomes at meiosis. This demonstrates, as in the case of the *Acomys* populations of Israel and Cyprus (p. 108) that some of the fusions that have occurred in the lineages leading to the Negev taxon are different from those that have reached fixation in the Algerian form.

The most extreme cytotaxonomic difference between closely-related species of mammals is possibly that between the deer *Muntiacus reevesi* ($2n = 46$) and *M. muntjac* ($2n♂ = 7$, $♀ = 6$, there being an XY_1Y_2 sex chromosome mechanism) described by Wurster and Benirschke (1970). Equally extreme is the case of the burrowing rodents of the genus *Ellobius*—*E. talpinus* from Tadzhikistan and Kirgizstan has races with $2n = 52$ and 54 (N.F. = 56 in both; sex chromosomes not identified) according to Vorontsov and Rajabli (1967) and Ivanov (1967); while the Persian *E. lutescens* has $2n♂♀ = 17$, there being an unpaired sex chromosome in both sexes (Matthey, 1954, 1958, 1962, 1963a, 1964a; Castro-Sierra and Wolf, 1967). Another case, this time involving sibling species not previously distinguished on external morphology is that of the cotton rats *Sigmodon hispidus* and *S. arizonae* (Zimmerman, 1970). The former, which extends from Tennessee to Texas and apparently as far south as Venezuela (Kiblisky, 1969a) has $2n = 52$, but *S. arizonae* from Arizona has $2n = 22$.

Sigmodon minimus, from New Mexico, was shown by Hsu (1969) to have a normal chromosome number of $2n = 30$; but an extraordinary type of chromosomal polymorphism was found in a population near Silver City, N.M., sampled in two successive years—namely six out of seven individuals had 29 chromosomes and were apparently carrying a fusion between two homologous acrocentrics (i.e. an isochromosome), since they regularly showed 15 chiasmate bivalents at diakinesis. Theoretically, such a rearrangement should lead to 50% sterility in crosses between two 29-chromosome individuals and total sterility in ones between 29-chromosome and 30-chromosome individuals. This is an example of a "deleterious" rearrangement which will increase in frequency in a local population if it has once reached a frequency of 50% as a result of drift, or a combination of drift and selection for viability.

Another species of *Sigmodon*, *S. fulviventer*, is polymorphic for a fusion between two autosomes and shows $2n = 28$ to 30 (Lee and Zimmerman, 1969).

If fossorial mammals show one extreme of vagility, flying ones should illustrate the other. It is hence of interest to examine the situation with regard to chromosomal polymorphism in bats (order Chiroptera). The literature is now extensive and has been reviewed by Capanna and Civitelli (1970). The large genus *Myotis* certainly shows great karyotypic uniformity

(Capanna *et al.*, 1968). Eighteen species appear to have almost identical karyotypes ($2n = 44$, autosomal N.F. $= 50$); in some instances, however, their sex chromosomes may differ (i.e. the X may be metacentric or submetacentric and the Y may be either acrocentric, submetacentric or very small). Two other species of *Myotis* have slightly different karyotypes. The closely-related genus *Pipistrellus* shows more karyotypic variation, since the following chromosome numbers ($2n$) are known: 26 (autosomal N.F. $= 44$), 28 (N.F. $= 46$), 30 (N.F. $= 56$), 34 (N.F. $= 46$), 36 (N.F. $= 48$), 42 (N.F. $= 48$) and 44 (three species with autosomal N.F. $= 48$, 50 and 50). *Rhinolophus* shows an intermediate condition; three species have $2n = 56$ (autosomal N.F. $= 60$), but three others have divergent karyotypes. The other genera, *Glossophaga* and *Tadarida* show considerable karyotypic uniformity. On the whole, it seems that there is some evidence for a greater degree of karyotypic uniformity within the genera of Chiroptera than is usual in other orders of mammals, where mobility is less highly developed.

The most complex instances of karyotype evolution in bats occur in the family Phyllostomatidae. Complex sex chromosome mechanisms of the XY_1Y_2 type seem to have evolved on at least three occasions in this group, once in the genus *Choeroniscus*, once in *Carollia* and once in the subfamily Stenoderminae (genera *Artibeus*, *Enchisthenes* and *Ametrida*). The evidence has been presented by Hsu *et al.* (1968), Yonenaga *et al.* (1969), Kiblisky (1969b) and Baker and Hsu (1970). Chromosome numbers in the Phyllostomatidae range from 16 to 46, and the N.F. is also highly variable (22 to 68), so we may assume that in addition to numerous centric fusions other types of structural rearrangements have occurred. *Mesophylla macconnelli* has 22 acrocentric or subacrocentric chromosomes in the female and 21 in the male; it may have XO males, as Baker and Hsu suggest, but is more probably X_1X_2Y as a result of a fusion between an originally minute Y and one member of a pair of autosomes.

The cats, another group of mammals in which the individuals are wide-ranging, so that small localized colonies of the type met with in rodents do not occur, seem to have a considerable degree of karyotypic uniformity. Almost all species of Felidae have $2n = 38$, except for the ocelot (*Felis pardalis*) and the marguay (*F. wiedii*), which have $2n = 36$, as a result of a fusion (Hsu *et al.*, 1963; Hsu and Rearden, 1965). Nevertheless, the karyotypes of the various species are certainly not completely indistinguishable except perhaps in a few instances, e.g. the jaguar (*Panthera onca*) and the black panther (*Panthera pardus*).

One group of mammals in which the karyotypes of all the species appear to be identical, by current techniques, is the Camelidae, with 2 palaearctic and either 2 or 4 neotropical species (according to whether the llama, alpaca and guanaco are considered separate species or not). According to

Taylor, *et al.* (1968) and Benirschke (1967) all these forms have $2n = 74$ and indistinguishable karyotypes. However the high chromosome number and consequent difficulties in observation and interpretation should restrain us from concluding that *no* chromosomal rearrangements have undergone fixation in their phylogeny. But if any mammalian species are homosequential in the way that certain species complexes in *Drosophila* are (see p. 99), the Camelidae are probably to be numbered among them.

The living species of horses, donkeys and zebras, seem to be taxonomically closer to one another than the species of Camels (they are all members of the genus *Equus*) and their population structure is presumably roughly similar; but they show an amazing degree of karyotypic diversity, with chromosome numbers ranging from $2n = 66$ in *E. przewalskii* down to 34 in *E. zebra* (Bernischke *et al.*, 1964, 1965; Benirschke, 1964). Palaeontologists who have compared the phylogenies of the horses and elephants may be interested to know that the karyotypes of the Indian and African elephants are, on the other hand, almost indistinguishable (Hungerford *et al.*, 1966).

The chromosomal evolution of the order Primates is of particular interest in view of the light which it might be expected to throw on problems of human evolution. We have already (p. 104) referred to the situation in the tree shrews, which occupy a place at the very base of primate phylogeny (it has been remarked that the Primates are a peculiar order of mammals because the most primitive forms have persisted alongside the more advanced types).

A number of species of Madagascan lemurs have been studied cytologically. Several appear to be complexes which include several biological species with quite different karyotypes. Thus *Hapalemur griseus* has "subspecies" showing $2n = 54$ and $2n = 58$; while *Lemur fulvus fulvus* includes forms with $2n = 60$ and $2n = 48$ (Chu and Swomley, 1961; Rumpler and Albignac, 1969a, b). Whether these taxa are reproductively isolated in nature does not seem to be known (counts of $2n = 58$ in *L. f. fulvus* were probably due to a minute pair of chromosomes having been missed). In *L. fulvus collaris* Rumpler and Albignac (1969b) have reported three different chromosome numbers ($2n = 48, 52, 60$), presumably due to fusions or dissociations. Since no heterozygous animals were encountered, it is probable that there are a number of chromosomally different populations, perhaps allopatric or parapatric. *Lemur catta*, on the other hand, is probably a karyotypically uniform species (Rumpler, 1970).

The Tarsiodea seem to have very high chromosome numbers; *Tarsius bancanus* has $2n = 80$ (Klinger, 1963).

In the family Cercopithecidae, the genera *Macaca* and *Papio* (macaques and baboons) are relatively conservative, karyotypically, with a uniform chromosome number of $2n = 42$. But in the genus *Cercopithecus* chromosome numbers range from 54 to 72 (Bender and Chu, 1963; Chiarelli, 1966, 1968). Several species show more than one chromosome number, probably in different races or subspecies (*C. l'hoesti* with $2n = 58$ and 60; *C. neglectus* with $2n = 58$ and 62; *C. mona* with $2n = 66$ and 68 and *C. nictitans* with $2n = 66$ and 70).

Comparisons between karyotypes of the anthropoid apes and that of man have been carried out by Hamerton *et al.* (1961) and Chiarelli (1961, 1962). The gorilla, chimpanzee and orang-utan all show $2n = 48$, whereas man has $2n = 46$. It seems fairly certain that a centric fusion occurred at some stage in the hominid phyletic line. Chiarelli (1961) has suggested that the chromosome produced by this fusion was probably number 2 or number 5 of the Denver classification. None of the anthropoid species seem to be chromosomally polymorphic; but a lack of precise correspondence between the karyotypes of the several species of anthropoids and between these and that of man prove that numerous rearrangements such as pericentric inversions and duplications or deletions must have occurred. *Homo sapiens* is an essentially monomorphic species, karyotypically. Apart from individuals with abnormal karyotypes due to spontaneous rearrangements or trisomy, individuals of all human individuals and races seem to be chromosomally alike, except for some minor variation in the sizes of the satellites on some of the acrocentrics and a rather more significant variation in the length of the Y-chromosomes, perhaps confined to the segment in the long arm which fluoresces with quinacrine dyes. Man has probably always been a wandering species, not liable to form small isolated colonies like most of the species of rodents we have considered earlier.

The gibbons seem to include two lineages, from the chromosomal standpoint. The various species of *Hylobates* all seem to have $2n = 44$ (Chiarelli, 1963) but the siamang (*Symphalangus syndactylus*) has $2n = 50$ (Bender and Chu, 1963; Klinger, 1963). The family Pongidae seem to be relatively conservative as far as their karyotypic evolution is concerned.

Supernumerary chromosomes (i.e. ones present in variable number in the karyotype and absent altogether in some members of the population) have only been recorded in a few species of mammals and seem to be rare in vertebrates as a whole. Why this form of chromosomal polymorphism should be very frequent in certain groups, such as grasshoppers, and unknown in others, such as the genus *Drosophila*, is not clear. In Marsupials, Hayman and Martin (1965) have described a case in the Australian *Schoenobates volans* and Hayman *et al.* (1969) report the presence of mitotically unstable supernumerary chromosomes in *Echymipera kalabu* in New Guinea. In rodents Shellhammer (1969) has recorded 0 to 7 supernumeraries in *Reithrodontomys megalotis* in North America.

There seems no reason to suppose that supernumeraries existing in a state of balanced polymorphism have been of any significance in speciation. Extra chromosome fragments consisting of a centromere and short adjacent segments "capped" by telomeres may, however, have been important in the process of chromosomal dissociation in many groups.

Where material from zoos, parks or domestic herds has been found to be chromosomally polymorphic, it may be difficult or impossible to determine whether this condition was present in the original wild population of the species. This is the case with the asiatic deer *Cervus nippon*, in which Gustavsson and Sundt (1968, 1969) have described polymorphism for several centric fusions in material from European deer parks. In cattle, the centric fusion studied by Gustavsson (1966) in two breeds in central and southern Sweden must have arisen as a spontaneous rearrangement; it is clearly not deleterious under the artificial conditions of a domestic herd.

III. THE CONCLUSIONS

Chromosome studies may be directed at an understanding of the differentiation of populations, races and species. This is what we have tried to discuss in this article. Some authors have been concerned with more general questions of karyotypes in relation to the phylogeny of orders, classes and phyla. The latter is necessarily a much more speculative field, in which certain writers have allowed their imagination full rein. The views of Ohno and Atkin (1966), Atkin and Ohno (1967), Taylor (1967) and Ohno *et al.* (1968), according to which a series of polyploidizations occurred in the phylogeny of the lower vertebrates, are in this category and are rightly regarded with suspicion by Dobzhansky (1970). The present writer shares his scepticism on this point and fully agrees with the conclusion of Morescalchi (1970) according to which "There seems no indication that it (i.e. polyploidy) has played any extensive evolutionary role in the phylogeny of the subphylum." Certainly, it does not enter into the picture, as far as the Mammalia are concerned. Another fanciful theory, which does nevertheless contain a germ of truth within it, is the "karyotypic fissioning" hypothesis of Todd (1970) according to which the evolution of karyotypes in the Carnivora and some other mammalian orders is believed to have come about through a progressive "fissioning" of metacentric chromosomes (apparently conceived of as single breaks through the centromere), with few or no rearrangements of other types. The germ of truth is, of course, that over a long period of evolutionary time there must be some kind of equilibrium between increases and decreases of chromosome number. The simplest type of equilibrium would seem to be an equal number of fusions and dissociations. But the mechanism of fusion seems an easier one than that of dissociation (at least as conceived of by the present author) although it might be otherwise if simple breaks through the centromere could frequently give rise to functional teleocentric chromosomes, according to the views of John and Lewis (1968), John and Hewitt (1968) and Todd (1970). Thus most students of vertebrate and, especially, mammalian karyotypes have interpreted differences in chromosome numbers in terms of evolutionary fusions rather than dissociations, probably rightly. Possibly this dilemma may be resolved if species with lower chromosome numbers (i.e. in which a number of fusions have undergone fixation) have a lower evolutionary potential, either because they are dead ends, destined to eventual extinction, or because they are less likely to undergo speciation, giving rise to two or more descendent species.

Much more significant and essentially factual ideas on long-term trends of karyotype evolution in vertebrate phylogeny than those of Todd have been put forward by Morescalchi (1970). This author's most distinctive contribution in this field has been to suggest that primitive amphibia, reptiles, birds and monotremes all share karyotypes characterized by high chromosome numbers and a definite dichotomy between large macrochromosomes and numerous minute microchromosomes. The karyotypes of the modern groups of amphibia, marsupials and eutherian mammals are thought to have been derived by a general reduction of chromosome number with loss of the microchromosomes (presumably by fusion rather than by actual loss).

With the karyotypes of one-quarter to one-third of all species of mammals now worked out, it is obvious that karyotyping by itself will soon enter on a period when it will yield diminishing returns, if indeed this stage has not already been reached. What are urgently needed now are synthetic studies of mammalian speciation in which detailed investigations of geographical, ethological and cytogenetic isolation will be carried out simultaneously. Only thus can we get further insights into speciation as it actually occurs in nature, rather than as it is in theoretical models, however helpful these may be in visualizing the possibilities. At the same time it would be most helpful to have actual information as to how mechanisms of balanced cytogenetic polymorphism actually operate in vertebrates, since almost all of our present knowledge on the role of these polymorphisms comes from studies on *Drosophila*.

Clearly, the restrictions on the establishment of chromosomal rearrangements (whether as balanced polymorphisms or as isolating mechanisms) do not operate in the same way in different groups of organisms. Thus in one genus, species-group or evolutionary lineage we find numerous pericentric inversions establishing themselves, in another it is fusions that have been favoured by selection and in yet another dissociations or additions of extra heterochromatic chromosome limbs. These regularities have been referred to as the principle of karyotypic orthoselection and are more fully discussed elsewhere (White, 1973); they are certainly evident in many groups of mammals.

Nevo (1969) has written of concomitant evolution of ethological and cytologic isolating mechanisms in *Spalax*. It seems probable that this is a very usual situation in vertebrates. But what precise models of this concomitance can we construct? What is the genetic basis of ethological isolating mechanisms, anyhow, in particular cases? Littlejohn (1969) has suggested that it must be polygenic, but he had amphibian calls in mind. Mammalian ethological isolation may be based largely on odours (pheromones) which may have a simpler biochemical and hence genetic basis. A first approach to a "concomitance model" would involve the assumption of a chromosomal rearrangement occurring at a single point (i.e. in a single individual) and spreading from there; and an ethological isolating mechanism arising independently in many individuals at many points of origin. But a complete model would involve locating these events in time and space, i.e. in relation to the evolution of the species' geographic range. The essence of the stasipatric model is the assumption that a chromosomal rearrangement which produces negative heterosis (i.e. which acts as at least a partial isolating mechanism) can, under certain circumstances, spread through the pre-existing population of the species. The only alternative, since it is clear that spreading of such rearrangements *has* occurred on numerous occasions, is the strictly allopatric dogma that they can only have spread into territory previously unoccupied by the species. The greatest need at the present time would seem to be precise, intensive and extensive studies of "concomitance" in critical cases such as the species of *Spalax*, *Thomomys*, *Perognathus* and others, with precise measurements of the strength of ethological and cytogenetic isolation between different populations taken from areas remote from, adjacent to, and actually located in the zones of narrow overlap. It is easy enough to envisage the various possible models of "concomitance", but we need far more facts to provide some evidence as to which of them are at all realistic.

There is, however, also a need for much more detailed karyotyping than has usually been attempted up till now, with careful measurements of chromosome arms, use of tritiated thymidine autoradiography to locate late-labelling segments and of DNA value determinations and other techniques to identify, as far as possible, evolutionary increases and decreases in the total amount of genetic material present. It might be illuminating to apply such techniques in the relatively few instances where a group of related species seem to have indistinguishable karyotypes when studied by the usual methods. The processes of karyotype evolution are sufficiently complicated that we can be sure that this field will not be worked out in the near future and that many fundamental problems will still be unsolved for a considerable time.

References

Arakaki, D. T., Veomett, I. and Sparkes, R. S. (1970). *Experientia* **26**, 425–426.

Arrighi, F., Sorenson, M. W. and Shirley, L. R. (1969). *Cytogenetics* **8**, 199–208.

Atkin, N. B. and Ohno, S. (1967). *Chromosoma* **23**, 10–13.

Badr, F. M. and Badr, R. S. (1970). *Chromosoma* **30**, 465–475.

Baker, R. J. and Hsu, T. C. (1970). *Cytogenetics* **9**, 131–138.

Baker, R. J. and Mascarello, J. T. (1969). *Cytogenetics* **8**, 187–198.

Baker, R. J., Mascarello, J. T. and Jordan, R. G. (1970). *Experientia* **26**, 426–428.

Beermann, W. (1956). *Chromosoma* **8**, 1–11.

Bender, M. A. and Chu, E. H. Y. (1963). *In* "Evolutionary and Genetic Biology of Primates" Vol 1 (J. Buettner-Janusch, ed.) pp. 261–310. Academic Press, New York and London.

Benirschke, K. (1964). *Chromosoma* **15**, 300.

Benirschke, K. (1967). *In* "Comparative Aspects of Reproductive Failure" (K. Benirschke, ed.). Springer Verlag, New York.

Benirschke, K., Low, R. J., Brownhill, L. E., Caday, L. B. and De Venecia-Fernandez, J. (1964). *Chromosoma* **15**, 1–13.

Benirschke, K., Maloef, N., Low, R. J. and Heck, H. (1965). *Science* **148**, 382–383

Bianchi, N. O., Paulete-Vandrell, J. and De Vidal Rioja, L. A. (1969). *Experientia* **25**, 1111–1112.

Borgaonkar, D. S. (1967). *J. Hered.* **58**, 211–213.

Borgaonkar, D. S. and Gould, E. (1968). *Experientia* **24**, 506–509.

Capanna, E. and Civitelli, M. V. (1970). *Caryologia* **23**, 79–111.

Capanna, E., Civitelli, M. V. and Nezer, R. (1970). *Experientia* **26**, 422–425.

Capanna, E., Civitelli, M. V. and Spagnuolo, C. (1968). *Caryologia* **21**, 225–240.

Castro-Sierra, E., and Wolf, U. (1967). *Cytogenetics* **6**, 268–275.

Chiarelli, B. (1961), *Nature* (*Lond*) **192**, 285.

Chiarelli, B. (1962). *Caryologia* **15**, 99–121.

Chiarelli, B. (1963). *Caryologia* **16**, 637–648.

Chiarelli, B. (1966). *Am. J. Phys. Anthrop.* **24**, 155–170.

Chiarelli, B. (1968). *Cytologia* **33**, 1–16.

Chu, E. H. Y. and Swomley, B. A. (1961). *Science* **133**, 1925–1926.

Dartnall, J. A. (1970). *Pap. Proc. R. Soc. Tasmania* **104**, 79–80.

Dobzhansky, T. (1970). "Genetics of the Evolutionary Process." Columbia Univ. Press.

Evans, E. P., Lyon, M. F. and Daglish, M. (1967). *Cytogenetics* **6**, 105–119.

Ford, C. E. (1970), "Human Population Cytogenetics" Pfizer Medical Monographs 5, pp. 221–239. Edinburgh Univ. Press.
Ford, C. E., Hamerton, J. L. and Sharman, G. B. (1957). *Nature (Lond)* **180**, 392.
Ford, P. J. and Graham, C. F. (1964). *Bull. Mammal Soc. Br. Isl.* **22**, 10.
Gropp, A., Citoler, P. and Geisler, M. (1969). *Chromosoma* **27**, 288–307.
Gropp, A., Tettenborn, U. and Von Lehmann, E. (1970). *Cytogenetics* **9**, 9–23.
Gustavsson, I. (1966). *Nature (Lond)* **211**, 865–866.
Gustavsson, I. and Sundt, C. O. (1968). *Hereditas* **60**, 233–248.
Gustavsson, I. and Sundt, C. O. (1969). *Chromosoma* **28**, 245–254.
Hamerton, J. L., Fraccaro, M., de Carli, L., Nuzzo, F., Klinger, H. P., Hullinger, L., Taylor, A. and Lang, E. M. (1961). *Nature (Lond)* **192**, 225–228.
Hayman, D. L. and Martin, P. G. (1965). *Aust. J. Biol. Sci.* **18**, 1081–1082.
Hayman, D. L. and Martin, P. G. (1969). *In* "Comparative Mammalian Cytogenetics." (K. Benirschke, ed.). Springer Verlag, Berlin and New York.
Hayman, D. L., Martin, P. G. and Waller, P. F. (1969). *Chromosoma* **27**, 371–380.
Hsu, T. C. (1969). *Experientia* **25**, 205–206.
Hsu, T. C. and Arrighi, F. E. (1966). *Cytogenetics* **5**, 355–359.
Hsu, T. C. and Arrighi, F. E. (1968). *Cytogenetics* **7**, 417–446.
Hsu, T. C. and Reardon, H. H. (1965). *Chromosoma* **16**, 365–371.
Hsu, T. C., Reardon, H. H. and Luquette, G. F. (1963). *Am. Nat.* **97**, 225–234.
Hsu, T. C., Baker, R. J. and Utakoji, T. (1968). *Cytogenetics* **7**, 27–38.
Hungerford, D. A., Chandra, H. A., Snyder, R. L. and Ulmer, F. A. Jr. (1966). *Cytogenetics* **5**, 243–246.
Ivanov, V. G., (1967). *Tsitol. Moscow.* **9**, 879–883.
John, B. and Hewitt, G. M. (1968). *Chromosoma* **25**, 40–74.
John, B. and Lewis, K. R. (1968), "The Chromosome Complement. Protoplasmatologia", VIa. Springer Verlag, Vienna.
Jotterand, M. (1970). *Experientia* **26**, 1360–1361.
Kennedy, J. A. (1969). *Aust. J. Zool.* **17** 465–471.
Kiblisky, P. (1969a). *J. Mammal.* **50**, 810–811.
Kiblisky, P. (1969b). *Experientia* **25**, 1203–1204.
Klinger, H. P. (1963). *Cytogenetics* **2**, 140–151.
Kozlovsky, A. N. (1970). *Tsitol. Moscow* **12**, 1459–1464.
Kreizinger, J. D. and Shaw, M. W. (1970). *Cytogenetics* **9**, 52–70.
Lay, D. M. and Nadler, C. F. (1969). *Cytogenetics* **8**, 35–38.
Lee, M. R. and Zimmerman, E. G. (1969). *J. Mammal.* **50**, 333–339.
Littlejohn, M. J. (1969). Systematic Biology Internat. Conf., U.S. Nat. Acad. Sci. pp. 459–482.
Martin, J. (1967). *Can. J. Genet. Cytol.* **9**, 255–268.
Martin, J. H. D. (1969). *Qd. J. Agric. Sci.* **26** 125–141.
Martin, P. G. and Hayman, D. L. (1965). *Evolution* **19**, 157–161.
Martin, P. G. and Hayman, D. L. (1966). *Chromosoma* **19**, 159–175.
Martin, P. G. and Hayman, D. L. (1967). *Chromosoma* **20**, 290–310. See also *Errata* in *Chromosoma* **22**, 390.
Matthey, R. (1949). "Les Chromosomes des Vertébrés." Rouge, Lausanne.

Matthey, R. (1952). *Arch. Julius Klaus-Stift. VererbForsch* **27**. 163–166.
Matthey, R. (1953). *Revue Suisse Zool.* **60**, 225–283.
Matthey, R. (1954). *Caryologia* **6**, 1–44.
Matthey, R. (1957). *Säugetierk. Mitt.* **5**, 145–150.
Matthey, R. (1958). *Experientia* **14**, 240–241.
Matthey, R. (1959). *Revue Suisse Zool.* **66**, 175–209.
Matthey, R. (1962). *Cytogenetics* **1**, 180–195.
Matthey, R. (1963a). *Cytogenetics* **2**, 290–322.
Matthey, R. (1963b). *Revue Suisse Zool.* **70**, 173–190.
Matthey, R. (1963c). *Experientia* **19**, 145–146.
Matthey, R. (1963d). *Chromosoma* **14**, 468–497.
Matthey, R. (1964a). *Revue Suisse Zool.* **71**, 401–410.
Matthey, R. (1964b). *Experientia* **20**, 657–665.
Matthey, R. (1964c). *Arch. Biol.* **75**, 169–206.
Matthey, R. (1965a). *Mammalia* **29**, 228–249.
Matthey, R. (1965b). *Revue Suisse Zool.* **72**, 119–144.
Matthey, R. (1965c). *Chromosoma* **16**, 351–364.
Matthey, R. (1966a). *Revue Suisse Zool.* **73**, 585–607.
Matthey, R. (1966b). *Mammalia* **30**, 105–119.
Matthey, R. (1966c). *Experientia* **22**, 400.
Matthey, R. (1966d). *Chromosoma* **18**, 188–200.
Matthey, R. (1966e). *Genetica* **37**, 171–180.
Matthey, R. (1966f). *Z. Vererb Lehre* **97**, 361–369.
Matthey, R. (1967a). *Cytogenetics* **6**, 168–177.
Matthey, R. (1967b). *Arch. Julius Klaus-Stift. VererbForsch* **42**, 21–30.
Matthey, R. (1967c). *Experientia* **23**, 133.
Matthey, R. (1968a). *Experientia* **24**, 1160–1161.
Matthey, R. (1968b). *Mammalia* **32**, 621–627.
Matthey, R. (1968c). *Genetics* **38**, 211–226.
Matthey, R. (1970a). *Experientia* **26**, 102–103.
Matthey, R. (1970b). *Revue Suisse Zool.* **77**, 625–629.
Mayr, E. (1942). "Systematics and the Origin of Species." Columbia Univ. Press.
Mayr, E. (1963). "Animal Species and Evolution." Belknap Press, Harvard Univ. Press.
Mayr, E. (1970). "Populations, Species and Evolution." Belknap Press, Harvard Univ. Press.
Mecham, J. S. (1961). *In* "Vertebrate Speciation" (W. F. Blair, ed.). Univ. Texas Press.
Meylan, A. (1964). *Revue Suisse Zool.* **71**, 903–983.
Meylan, A. (1965). *Revue Suisse Zool.* **72**, 636–646.
Morescalchi, A. (1970). *Boll. Zool.* **37**, 1–28.
Nadler, C. F. (1968). *Cytogenetics* **7**, 144–157.
Nevo, E. (1969). *Science* **163**, 484–486.
Ohno, S. and Atkin, N. B. (1966). *Chromosoma* **18**, 455–466.
Ohno, S., Weiler, C., Poole, J., Christian, L. and Stenius, C. (1966). *Chromosoma* **18**, 177–187.

Ohno, S., Wolf, U. and Atkin, N. B. (1968). *Hereditas* **59**, 169–187.
Ott, J. (1968). *Revue Suisse Zool.* **75**, 53–75.
Patton, J. L. (1969a). *Evolution* **23**, 645–662.
Patton, J. L. (1969b). *Caryologia* **22**, 351–358.
Patton, J. L. (1970). *Chromosoma* **31**, 41–50.
Patton, J. L. and Dingman, R. E. (1970). *Cytogenetics* **9**, 139–151.
Raicu, P., Bratosin, S. and Hamar, M. (1968). *Caryologia* **21**, 127–135.
Reig, O. A. and Kiblisky, P. (1969). *Chromosoma* **28**, 211–244.
Rumpler, Y. (1970). *Cytogenetics* **9**, 239–244.
Rumpler, Y. and Albignac, R. (1969a). *C. R. Séanc. Soc. Biol.* **163**, 1247–1250.
Rumpler, Y. and Albignac, R. (1969b). *C. R. Séanc. Soc. Biol.* **163**, 1989–1992.
Sharman, G. B. (1956). *Nature (Lond)* **177**, 941–942.
Shellhammer, H. S. (1969). *Chromosoma* **27**, 102–108.
Soldatović, B., Garzičić, B. and Živković, S. (1966). *Caryologia* **19**, 261–265.
Soldatović, B., Živković, S., Savić, J. and Milosević, M. (1967) *Z. Säugetierk.* **32**, 238–245.
Sparkes, R. S. and Arakaki, D. T. (1966). *Cytogenetics.* **5**, 411–418.
Taylor, K. M. (1967). *Chromosoma* **21**, 181–188.
Taylor, K. M. Hungerford, D. A., Snyder, R. L. and Ulmer, F. A. Jr. (1968). *Cytogenetics* **7**, 8–15.
Tettenborn, U. and Gropp, A. (1970). *Cytogenetics* **9**, 272–283.
Thaeler, C. S. Jr. (1968). *Chromosoma* **15**, 172–183.
Todd, N. B. (1970). *J. Theoret. Biol.* **26**, 445–480.
Vorontsov, N. N. and Rajabli, S. I. (1967). *Tsitol., Moscow* **9**, 848–852.
Wahrman, J. and Goitein, R. (1972). *Chromosomes Today* **3**, 228–237.
Wahrman, J. and Zahavi, A. (1955). *Nature (Lond)* **175**, 600–602.
Wahrman, J. and Zahavi, A. (1958). *Proc. 10th Internat. Congr. Genet.* Vol 2, 304–305.
Wahrman, J., Goitein, R. and Nevo, E. (1969a). In "Comparative Mammalian Cytogenetics" (K. Benirschke, ed.), pp. 30–48. Springer-Verlag, New York.
Wahrman, J., Goitein, R. and Nevo, E. (1969b). *Science* **164**: 82–84.
Walknowska, A. (1963). *Folia Biol.* **11**, 293–309.
White, M. J. D. (1968). *Science* **159**, 1065–1070.
White, M. J. D. (1969). *Ann. Rev. Genetics* **3**, 75–98.
White, M. J. D. (1970). *J. Aust. Ent. Soc.* **9**, 1–6.
White, M. J. D. (1973). "Animal Cytology and Evolution." 3rd Edn. Cambridge University Press.
Wurster, D. H. and Benirschke, K. (1970). *Science* **168**, 1364–1366.
Yonenaga, Y., Frota-Pessoa, O. and Lewis, K. R. (1969). *Caryologia* **22**, 63–79.
Yong, H. S. (1968). *Cytologia* **33**, 174–180.
Yong, H. S. (1969). *Chromosoma* **27**, 245–267.
Yong, H. S. (1972). *Chromosomes Today.* 3, 223–227.
Yosida, T. H. and Amano, K. (1965). *Chromosoma* **16**, 658–667.
Yosida, T. H. and Tsuchiya, K. (1969). *A. Rep. Natn. Inst. Genet.* **19**, 11–12.
Yosida, T. H., Nakamura, A. and Fukaya, T. (1965). *Chromosoma* **16**, 70–78.

Yosida, T. H., Moriguchi, Y., Kang, Y. S. and Shimakura, K. (1967). *A. Rep. Natn. Inst. Genet.* **17**, 61–63.

Zahavi, A. and Wahrman, J. (1967). *Mammalia* **21**, 341–380.

Zimmerman, E. G. (1970). "Karyology, Systematics and Chromosomal Evolution in the Rodent Genus, *Sigmodon*." Vol. 4, No. 9, 385–454. Publ. Mus. Michigan State Univ.

7

Common Patterns of Chromosome Evolution in Mammalian Cell Cultures or Malignant Tumors and Mammalian Speciation

L. KOULISCHER
Institut de Morphologie Pathologique
Loverval, Belgium

I. INTRODUCTION

The karyotype of somatic cells of most vertebrates, and especially of mammals, is indeed very stable. In normal conditions, from the first division after conception to death, somatic cells will show the same karyotype. This is very important and very useful to know in the field of cytogenetics: any somatic tissue collected after conception, for instance embryonic tissue collected *in situ* or recovered from abortusses, can be used for individual or specific karyotype studies. Thus, the chromosomes of goat/sheep hybrids, which are not viable but develop some time *in utero*, have been studied on embryonic tissue (Buttle and Hancock, 1966): the results correspond exactly to what one would expect to find if the hybrids had lived until birth. And it is now a common experience, in human cytogenetics, to study the chromosome of a foetus by culture of amniotic fluid and to observe the confirmation of diagnosis at birth.

This remarkable chromosome stability has been reported in human diploid cell strains cultivated *in vitro* (Hayflick and Moorhead, 1961; Hayflick, 1965). The life span of these cultures is however finite, and cell death occurs after more or less 50 divisions. Thus the conservation of "normality", including a normal karyotype, is limited in time.

In some special circumstances nevertheless, the karyotype stability of somatic mammalian cells is not observed any more. This is the case for tumoral cells (spontaneous as well as induced tumors) or cell lines grown *in vitro* by different tissue culture techniques and originated from normal or pathologic tissues. It is generally admitted that these changes correspond to the adaptation of the genome of the cell to a new environment, with new life requirements. Should the cell and its genome not be adapted, it would disappear. On the other hand, if the cell survives and adapts itself, it is capable of giving rise to a cell line which is "immortal" (in contrast with cell strains with finite life span): hundreds of generations can easily be obtained by means of adequate tissue cultures.

The study of karyotypes in malignant tumors is based on the well known and widely used concept of stemline, introduced by Makino (1957). The stemline is responsible for the development and the propagation of the tumor.

From a cytogenetic point of view, the stemline is characterized by a particular karyotype, observed in the majority of the dividing cells. In a given tumor, two or more stemlines can coexist. Tumoral stemlines however are not always stable: sidelines, with new karyotypes, may appear and supplant the original stemline. In favourable cases, marker chromosomes (i.e. chromosomes with a particular morphology making them easily recognizable) attest the parentship between the different karyotypes observed. The chromosome changes in the same cell system may occur spontaneously or be linked with different modifications of the environment, *in vivo* as well as *in vitro*. Repeated cytogenetic analysis allows this "evolution" of karyotypes to be followed and the evolutive patterns at the cell level to be reconstructed.

The purpose of this text is to compare cellular karyotype evolution with karyotype changes linked with speciation in mammals. Needless to say, mammalian somatic cells from a tumor or a tissue culture are very different in many respects from a whole living organism, with its many functions including the meiotic process before reproduction. However some basic common cytogenetic patterns of evolution are observed in both systems and will be discussed below.

II. MODELS OF KARYOTYPE EVOLUTION IN CELL CLONES

A. *Introduction*

The data that will be presented here are those available in literature. They concern the chromosomes of cells transformed into permanent cell lines by means of tissue culture or cells from malignant tumors. Indeed, these cells have acquired characteristic properties making them very different from the normal original cells. At first sight, the many well documented reports on karyotype clonal evolution convey the impression that each tumor or each cell culture cytogenetically studied has its own story. However, common patterns do exist; but before describing them, some precision about the selection of the data seem necessary.

(a) *Tumoral cells and cell lines maintained in vitro: is a distinction necessary?*

As it has been stated by Levan (1970): "The formation of a permanent tissue culture cell line may be looked upon as a genetic process of adaptation comparable to the change of a normal cell into a cancer cell." This has been experimentally demonstrated. Thus, normal diploid C57BL mouse cells are maintained *in vitro* and permanent cell lines established (Barski and Cassingena, 1963); one of the cell lines, after the sixteenth passage *in vitro* produced sarcomas when inoculated intramuscularly into isologous mice. Other similar experiments have shown the transformation of normal cells into malignant cells by tissue culture *in vitro* (see for instance Barski *et al.*, 1966; Evans *et al.*, 1967). In view of this evidence, the cytogenetic results found in tissue cultures and in malignant tumors will be considered in the present text as having the same signification.

(b) *Methodology of karyotype studies*

Two principal methods are used for the study of karyotype clonal evolution. The first one is the repeated analysis of chromosomes at regular intervals. This indeed is the best method for following and for understanding chromosome evolution. For practical reasons, this technique is used in tissue cultures *in vitro*, or when tumors are propagated *in vivo*. However, when cells are easily available, serial analyses can also be carried out in spontaneous tumors of animals and man, for instance in leukemias or cancerous effusions which can be repeatedly punctured.

The second method, mainly used in solid tumors, consists of examining only once the karyotype of cells, collected from a biopsy or by surgical means. Chromosome evolution is then more difficult to reconstruct and is based on hypotheses. The actual observed karyotype is the result of a long *in vivo* evolution which began when the initial cell (or cells) become cancerous. However, such single observations are very often of great interest and will be commented on here when necessary.

All reports are based on optical microscope observation and microphotographs. This technique has well known deficiencies. Small chromosome rearrangements may escape detection and the corresponding karyotype will be incorrectly described. On the other hand, the origin of marker chromosomes may be obscure. Nevertheless with all its imperfections, the technique has brought, and still brings, much valuable information.

(*c*) *Stemlines and sidelines*

Beside the stemline, characterized by a given diploid number of chromosomes and a particular karyotype, sidelines may appear. These sidelines can be of great importance in clonal evolution, or have no significance at all. Stemlines and sidelines could have their equivalents in paleontology: the successful and the fossil branches of evolution (Lejeune, 1965). For the sake of clarity, and without denying the interest of sidelines, only stemlines will be considered in the present text.

Seven different models of karyotype evolution will be commented upon; these are the "main" models: combination of them can explain every evolution observed at the cell level.

B. *No apparent change of the original karyotype*

When an attempt to establish a new cell line *in vitro* is made, the original karyotype is observed during a variable length of time (from some days to more than 9 months in a bone marrow C3H mouse cell line described by Hsu *et al.*, 1961). Sooner or later, a shift toward aneuploidy happens, and karyotype evolution begins: the genome adapts itself to its new environment. Some cell lines or some tumors, however, show a "normal" karyotype, or a karyotype identical to the one of the original cells, for years, so that this can be considered as characteristic of the stemline. The transformation of normal cells into a permanent cell line or a malignant tumor apparently takes place without any detectable change of the original karyotype.

This has been observed in cell lines derived from human leukocytes or human hematopoietic cells. In 12 out of 14 established cell lines from hematopoietic origin, Miles *et al.* (1968) reported a normal karyotype. In 5 out of 7 cell lines derived from peripheral blood of patients with leukemia, Zur Hausen (1967) observed also a normal karyotype. The same eudiploidy has been shown 3 to 23 months after the initiation of cell lines from hematopoietic human cells (Huang and Moore, 1969), and in a cell line from hematopoietic cells of a patient with melanoma, maintained *in vitro* for 15 months (Huang *et al.*, 1969) before being studied. In cell lines derived from individuals with sex chromosome anomalies (XXY and XYY), the original karyotype was maintained (Moore *et al.*, 1969). In more than 50% of cases of acute leukemia in man, no chromosome changes could be detected (Sandberg, 1966). In direct analysis of Burkitt's tumors, the karyotype was normal in 4 cases out of 16 (Jacobs *et al.*, 1963). At this stage, it is necessary to stress one point which will be discussed later; it would be tempting to postulate that human hematopoietic or lymphocytic cells show a tendency to keep their original karyotype whatever the modifications of their environment. However, one must be cautious with such generalizations, for the same cell systems may show, in similar conditions, a shift towards aneuploidy. *The examples here must be considered as specific for a given cell line, but not for all cell lines derived from the same tissue.*

The conservation of an apparently normal karyotype has been observed in fibroblast-like cultures from chinese hamster FAF, more than 10 months after initiating the culture, with a generation time of 14 hours (Yerganian and Leonard, 1961). Bergman and Reitalu (personal communication to Hsu *et al.*, 1961) could culture human skin 18 months *in vitro* without any detectable karyotype change. In the PTKI cell line started from kidney cells of a female rat kangaroo, Levan (1970) observed after more than 100 passages a karyotype almost identical to the normal, including the labelling pattern after tritiated thymidine incorporation.

Tumoral tissues other than human acute leukemias can also show a normal karyotype. In some experimental cancers, Bayreuther (1960) did not observe any chromosome change. Hellström *et al.* (1963) reported normal karyotypes in 1/5 primary and 6/18 transplanted mouse tumors induced by the polyoma virus. For these authors: "that 7 of the 23 polyoma tumors were diploid lends support to the view that such chromosomal changes detectable in telocentric mouse chromosomes by cytologic techniques are not required for malignant transformation as such. However, changes may have occurred even in the diploid tumors since only a small part of possible chromosomal rearrangements permit detection."

C. *Pseudodiploid cell lines*

In many of the cell systems mentioned above, besides lines with the original eudiploid karyotype, lines with pseudodiploid karyotypes were observed. The diploid original number remains normal; but the karyotypes are different from the original ones: chromosomes of some groups increase while chromosomes from other groups decrease in number, or marker chromosomes appear.

The example of the characteristic deleted G marker of chronic myeloid leukemia in man is well known (Nowell and Hungerford, 1960): in more than 90% of all cases of this form of leukemia, tumoral bone marrow cells are pseudodiploid and exhibit the marker, called "Philadelphia chromosome". This is however the only example of a specific marker in a human tumoral process.

In Burkitt's lymphoma direct analysis of the tumoral cells has shown the presence of pseudodiploid cell lines in some cases (Jacobs *et al.*, 1963), with 46 chromosomes and markers. The same Burkitt's tumor in long term tissue culture has shown a characteristic marker: a group C chromosome with a distal constriction of the long arm (Miles and O'Neill, 1967). In cell lines derived from leukocytes of 3 patients with infectious mononucleosis, Kohn *et al.* (1967) observed the same marker reported in Burkitt's tumors. In lines derived from human hematopoietic cells, Huang and Moore (1969) reported the presence of pseudodiploid stemlines, with or without markers, after 15 or 20 months of culture *in vitro*. A telocentric marker has been noted in pseudodiploid hematopoietic cell lines derived from leukocytes of a patient with myeloma (Huang *et al.*, 1969).

In tissue cultures of root vole (Ferno and Mandahl, 1970), pseudodiploidy was retained after 40–50 passages and still observed at the 143rd passage after 21 months of culture. In some Rous sarcoma in mice, Mark (1970) observed a pseudodiploid stemline 154 and 240 days after the beginning of the study, with a small telocentric marker.

In fact, the original diploid number seems very often to have a selective advantage, even if karyotypes differ from cell to cell in the same culture. In a detailed cytogenetic analysis of baby hamster kidney (BHK) cells, Hughes (1968) observed after 451 and 480 cell generations, different karyotypes from cell to cell, all showing nevertheless the normal diploid number. Hughes made the same observation with Burkitt's tumor cells *in vitro*: karyotype changes were linked with environmental changes. As no "classical" stemline could be observed, Hughes proposed the concept of "pseudostemline": the normal diploid number is retained, but this is the only common point between karyotypes of different cells. In these systems, cooperation between cells would be almost as important as competition. Owing to observation techniques, pseudodiploidy is perhaps more generalized than actually thought, and with improved methods some lines considered as eudiploid will in fact appear as pseudodiploid. For instance, Moore and Sandberg (1964) considered that a human tumor cell line of epithelial origin (RPMI 2650) had a normal karyotype after 35 generations *in vitro* for a period of 1 year. Moorhead (1965) studying the same line, showed that it was pseudodiploid. Should these observations be more frequent, the concept of a selective advantage of the original diploid number would be strengthened.

D. *Centromeric or Robertsonian fusion of chromosomes*

The importance of centric fusion of telocentric chromosomes in speciation was found by Robertson in 1916, and is often referred to as "Robertsonian" fusion. Indeed, the mechanism of the "fusion" is different depending on whether telocentric or acrocentric chromosomes are involved. In the case of telocentrics, with a terminal centromere, a real "fusion" may occur with a theoretically double centromere, behaving however at cell division as a single unit. If acrocentric chromosomes are involved, the mechanism of the "fusion" is that of a reciprocal symmetrical translocation (Rieger *et al.*, 1968), with production of two monocentric chromosomes. When the long arms of two translocated chromosomes form a "new" large chromosome and the short arms a very small one, very often (at least in Man, and perhaps in many other species) the small centric chromosome is lost. There is thus a decrease of the total number of chromosomes, a decrease in the number of acro- or telocentrics and an increase of the meta- or submetacentrics. Indeed, the total number of chromosome arms remains constant. This form of karyotype evolution is observed in many families and the total number of arms is the "nombre fondamental" or NF of Matthey (1949), which does not include the very small arms of acrocentric chromosomes.

The centromeric fusion of chromosomes is observed in many tumors and cell lines. In Man, this type of karyotype evolution has been postulated as occurring in tumors (Levan, 1967) or cancerous effusions (Muldal *et al.*, 1971), because the number of acrocentric chromosomes of the groups D and G shows a tendency to decrease, with a corresponding increase of meta- or submetacentric C like chromosomes. However, the Robertsonian mode is more clear in cell lines or cell tumors derived from species with many acro- or telocentrics: the presence of unusual metacentrics can be attributed to centromeric fusions. Nevertheless, these metacentrics can be formed by isochromosomes (Hsu *et al.*, 1961). But when the NF of a cell line is close to or identical with the NF of the species from which it derives, the most likely explanation is that centromeric fusion has occurred.

Makino (1963) observed, in spontaneous venereal sarcomas of the dog, cell lines with a diploid number of 59; 42 chromosomes were telocentrics and 17 metacentrics (NF of the tumor: 76). The normal karyotypes of the dog show $2n = 78$ chromosomes, with 76 acro- or telocentrics, the X chromosome being metacentric, the Y telocentric and the NF thus 79 or 80 (Wurster and Benirschke, 1968). The NF of the tumor and the NF of the normal cells being close to each other, Robertsonian fusion was postulated. Almost identical observations were reported by Weber *et al.* (1965) in primary and transplanted canine venereal sarcomas: the diploid number was $2n = 59$, with 44 telocentrics and 15 metacentrics. A year later, Barski and Cornefert-Jensen (1966) confirmed the karyotype evolution of this type of tumor; in personal communications to these authors, Miles and Moore reported that they had been able to show the same karyotype evolution by fusion of telocentrics in a cell line derived from the spleen of a dog with lymphosarcoma.

In tissue cultures derived from normal tissue of bovine origin (*Bos taurus*,

Capra hircus, *Ovis aries*), Nelson-Rees *et al.* (1967) reported an accumulation of centric fusions with time, the NF remaining constant. A dog cell line however did not show any chromosome fusion in the same technical conditions. This could indicate, in the dog, an increased tendency to fusion of chromosomes in venereal sarcoma and in lymphosarcoma, not found in cultures derived from normal tissues. This aspect of possibly "directed" chromosome evolution, shall be discussed later. Lithner and Ponten (1966) observed the same fusion events in bovine fibroblasts in long term tissue cultures. However, this happened in cell cultures with a finite life span; hypoploidy, severe aneuploidy and cell death began after 70 passages. The same chromosome evolution was observed with bovine fibroblasts exposed to the Rous sarcoma virus (Ponten and Lithner, 1966), while chromosome evolution after exposure to the SV 40 was quite different.

E. *Hyperdiploidy*

Hyperdiploidy is frequently observed in malignant tumors or cell cultures *in vitro*. The increase of the original diploid number varies from one supernumerary chromosome to nearhypotriploidy. In many cases, the supernumerary chromosomes of a hyperdiploid stemline do present a "normal" morphology: non-disjunction or selective duplication very likely happened to form the new karyotype. However, sometimes marker chromosomes appear; hyperdiploidy then is due to more complex rearrangements. Indeed, both mechanisms may occur in the same cell, showing together chromosomes of normal appearance and markers.

A remarkable example of hyperdiploidy has been given by Lejeune *et al.* (1963) in a case of acute leukemia in a trisomic 21 child. The constitutional diploid number was, following the terminology of the authors, $2n = 47$, tri 21. The leukemic stemline presented $2n = 54$ chromosomes. All intermediate numbers between 47 and 54 were found. Karyotype studies showed a very progressive transformation from the lower to the higher diploid number (Table I).

The increase of the chromosome number was essentially due to acquisition and duplication of supernumerary chromosomes. Intermediate karyotypes between the constitutional $2n = 47$ and the tumoral $2n = 54$ stemlines represented maladapted patterns almost completely eliminated. They were thought likely because the cytogenetic analysis was carried out at the beginning of the acute leukemic process. Lejeune (1965) reviews 9 other reports concerning acquisition and duplication of supernumerary chromosomes in tumors or tissue cultures from neoplastic origin. This duplication could represent a general mode of chromosome evolution in tumors, for

TABLE I

Progressive increase of the chromosome number in a case of acute leukemia in a trisomic 21 child (Lejeune et al., 1963)

Chromosome number	Karyotype	Number of cells
47	tri 21	22
48	tri 21 + 1G	3
49	tri 21 + 2G	1
50	tri 21 + 2G + 1D	1
51	tri 21 + 2G + 2D	1
52	tri 21 + 2G + 2D + 1C	3
53	tri 21 + 2G + 2D + 2C	7
54	tri 21 + 2G + 2D + 2C + 1F	36
55	tri 21 + 2G + 2D + 2C + 2F	3

Letters G, D, C and F represent chromosome groups following the Denver and the Chicago systems of classification of the human chromosomes.

even markers appear frequently in a duplicate state. Other general trends or "laws" should be found in clonal evolution, as pointed out by Lejeune. Indeed, modifications of the genetic balance must respect some fundamental requirements of the cell. For instance to ensure survival, a new karyotype cannot be linked with biochemical needs impossible to satisfy; on the other hand, in the case of tumors *in vivo*, the mutant cell must not be recognized by the immunological processes of the host organism. In summary, clonal evolution does not happen at random, but is strongly dependent upon environment.

Many cancerous processes (human or animal) show hyperdiploidy. In some favourable cases, intermediate karyotypes can be observed, as in Lejeune's patient described above or in blastic crisis of human chronic myeloid leukemia (de Grouchy *et al.*, 1968). In other cases, the origin of the hyperdiploid karyotypes cannot be traced. Hyperdiploidy can be due to the presence of only one or a few supernumerary chromosomes, representing a total increase of 2–5% of the original diploid number. Thus, in Burkitt's lymphoma, Jacobs *et al.* (1963) observed in some tumors a supernumerary C chromosome ($2n = 47$ instead of 46). In cell cultures derived from biopsies of Burkitt's lymphoma, the same observation was made (Miles and O'Neill, 1967; Huang *et al.*, 1969). In cell lines derived from blood of patients with leukemia, melanoma or myeloma, diploid numbers of 47 were reported (Huang *et al.*, 1969). In Rous sarcomas of the mouse, Mark (1970) noted stemlines with 41 to 43 chromosomes the normal diploid number of the mouse being $2n = 40$.

In other cases, hyperdiploidy represents an increase of 10–30% of the original chromosome number. Many human tumors show diploid numbers ranging from 51 to 60 chromosomes. In the Landschutz carcinoma of the mouse, after 200 intraperitoneal passages during 4 years, Leonard (1964) observed $2n = 46$ chromosomes, with one metacentric marker. The same diploid number had been reported for the same tumor by Tjio and Levan in 1955: the stability of this tumor is indeed remarkable. In mouse polyoma tumors, Hellström *et al.* (1963), besides lines with diploid karyotypes, observed hyperdiploid stemlines with chromosome numbers ranging from 41 to 50 (higher numbers were also observed but were in the range of triploidy and tetraploidy). In tissue cultures of KB cells of human origin, DeCarli and Chiarelli (1960) reported stemlines with a mean number of 62 chromosomes. In other cultures of human origin (EUE) Nuzzo *et al.* (1961) counted hyperdiploid numbers of 60 chromosomes. In one line derived from Ehrlich tumor of the mouse, Piccotti and Chiarelli (1967) observed $2n = 44$ chromosomes with the presence of a marker.

It would be interesting to determine, for each type of tumor, whether hyperdiploidy begins by the presence of the same supernumerary chromosome. No doubt that new fluorescence techniques will help to clarify this problem.

F. *Hypodiploidy*

Robertsonian fusion of telocentrics, with a reduction of the diploid number, is a particular case of hypodiploidy already described. This section concerns the presence, in malignant tumors or in tissue cultures, of hypodiploid cell lines without any detectable structural rearrangements. Hypodiploidy here seems due to an apparent "loss" of chromosomes. Indeed, hypodiploidy is less frequent than hyperdiploidy. This is theoretically unexpected: if hyperdiploidy is due to non-disjunction, corresponding hypodiploid cells should be observed. As it is not the case, it could be thought either that hyperdiploidy is more likely due to duplication of single chromosomes or that hypodiploid cells are disadvantaged in some way. This second opinion is more generally admitted. Thus, in a cell line originated from mouse tissue, Hsu *et al.* (1961) found an hypodiploid stemline number around 38; no important structural rearrangement could be found and this was puzzling the authors, for: "it seemed unlikely that a loss of 2 entire chromosomes from the normal diploid karyotype should result in increased growth capacity", so "cryptostructural reshuffling of genetic material" was assumed; one month later the cell line was hypotetraploid. On the other hand, although it may not be relevant here, it can be recalled that constitutional monosomies and even deletions have a more deleterious effect than trisomies in human pathology.

Stable hypodiploidy has been reported in human leukemias (for a review, see de Nava, 1970) and in hematopoietic cell lines of human origin, with one chromosome missing (Huang and Moore, 1969; zur Hausen, 1967). Transient hypodiploidy, soon followed by hypertriploidy, has been observed in an hematopoietic human cell line of a patient with myeloma (Huang *et al.*, 1969): the diploid number was 38–39 and then 71–74. Hypodiploidy followed by hypotetraploidy has been reported in 4 Rous sarcomas of the mouse (Mark, 1970) and in other mouse tissues (Hsu *et al.*, 1961).

G. *Triploidy and tetraploidy*

Triploidy and, more often, tetraploidy are very frequently observed in tissue cultures or tumors. Triploid or tetraploid stemlines seem to have a real selective advantage (Levan, 1967). However as far as tetraploidy is concerned, it must be pointed out that cells with $4n$ chromosomes may arouse spontaneously in tissue cultures such as blood lymphocytes. Tetraploid mitoses are found in cultures of many species and have been observed either in this laboratory or elsewhere (Sinha, 1967). Moreover tetraploidy or polyploidy seems physiological in some mammalian tissues.

Strict triploidy or tetraploidy, that is to say exactly $3n$ or $4n$ chromosomes, is not as common as the hypo- or hyper-state. In cultures of mouse bone marrow and peritoneum, Hsu *et al.* (1961) observed tetraploidy soon followed by hypotetraploidy or hypodiploidy soon followed by hypotetraploidy. The "two step" mechanism, aneuploidy around the diploid number followed by neartetraploidy, has been observed in other tissue cultures of mouse origin (Barski and Cassingena, 1963; Barski *et al.*, 1966; Pollack *et al.*, 1970) or in mouse induced tumors such as the Rous sarcoma (Mark, 1970) or the Ehrlich tumor (Piccotti and Chiarelli, 1967). However tetraploidy does not always ensure perennity of a cell strain; thus a human diploid cell strain infected by SV 40 showed hypotetraploid cells, but cell death happened 4 to 6 months after the viral infection (Moorhead and Saksela, 1965).

Triploidy or chromosome numbers near to triploidy have been reported in the human cell line LU 106 derived from a male embryo lung (Hampel and Levan, 1964), in another human cell line derived from a normal adult gingiva (Sonnenschein and Smilow, 1966), in human leukemias (for a review, see de Nava, 1970), in a culture from a human osteosarcoma (MacAllister *et al.*, 1971) and in a cell line derived from a mixed tumor of the salivary gland maintained *in vitro* for 3 years (Kondo *et al.*, 1971).

These are only a few examples. Indeed, it can be expected that sooner or later polypoid stemlines will be observed, especially in tissue cultures.

H. *Minute chromosomes*

Morphologically, minute chromosomes appear as very small double chromatin structures. It is not always possible to ascertain the presence of a centromere, and "minutes" could be small acentric fragments (Cox *et al.*, 1965) as well as tiny centric chromosomes.

Minute chromosomes have been repeatedly reported in human tumors of the nervous system such as neuroblastoma (Cox *et al.*, 1965; Levan *et al.*, 1968), glial tumor (Lubs *et al.*, 1966), medulloblastoma and rhabdomyosarcoma (Cox *et al.*, 1965) and subependymal glioma (Kuccheria, 1968). They have also been observed in some carcinomas (Spriggs *et al.*, 1962; Cox *et al.*, 1965), including a "brain carcinoma" (Yamada *et al.*, 1966), and in a meningioma (Zang and Singer, 1967). Pierre *et al.* (1971) noted the presence of "minutes" in several cases of human leukemia; especially of the myelomonocytic type. Sporadically, minutes have been observed in other human leukemias. Some induced Rous sarcomas in mice exhibited minute chromosomes (Mark, 1967). The origin of minute chromosomes or "double minutes" is still open to discussion. They may prove to be, like any marker, a stable characteristic of a stemline and persist for months in tissue cultures (White and Cox, 1967). Acutally this is evidence for the presence of a centromere.

I. *The different patterns of clonal evolution: summary*

The seven main patterns of clonal evolution are summarized in Table II. In fact not all examples of clonal evolution fit into one of these patterns: 2 or more of them may have been used, simultaneously or consecutively. This "composite" pattern explains some deeply rearranged karyotypes with pronounced aneuploidy, markers and minutes. An already mentioned "composite" type of evolution is hypo- or hypertetraploidy. Hypotetraploidy for instance can be achieved by doubling of the chromosomes set, followed by the "loss" of some chromosomes, or by the doubling of already hypodiploid cells (Hsu *et al.*, 1961).

Some particular modes of evolution, such as centromere fission or formation of isochromosomes, have not been discussed separately because they do not seem very frequent. However, they do exist (Hsu *et al.*, 1961) and may participate in the elaboration of a "new" karyotype. The models proposed in this section definitely support the fact that "although apparently haphazard, the chromosome variation in cancer has been demonstrated in

TABLE II. *The different patterns of clonal evolution*

Type	Number of chromosomes	Remarks
eudiploidy	$2n$	karyotype identical to the original
pseudodiploidy	$2n$	presence of marker chromosomes
centromeric fusion	$<2n$	NF constant
hyperdiploidy	$>2n$	supernumerary chromosomes with "normal morphology"
hypodiploidy	$<2n$	"missing" chromosomes
triploidy or tetraploidy	$3n$ or $4n$	very often near triplo- or near tetraploidy
minute chromosomes	$\pm 2n$	

all carefully analysed instances to be governed by strict rules" (Levan, 1969) and this applies also to tissue cultures.

III. MODELS OF KARYOTYPE EVOLUTION IN MAMMALS

A. *Introduction*

In the first section of this chapter only cell lines of mammalian origin were considered; the different models of clonal evolution were based solely on morphological grounds. In this second section, the same general scheme will be followed. Thus, only the morphological aspects of chromosome evolution, linked with speciation in mammals, will be discussed. However, exceptionally, reference to nuclei DNA content will be made when necessary.

On the other hand, it is known that "hundreds of species definitions have been proposed" (Mayr, 1969). Indeed, it is certainly not the purpose of this chapter to discuss the concept of species: thus taxonomy will be admitted here as it has been presented in the literature. Before analysing the data, some problems must be shortly commented upon and the advantages, or disadvantages, of studying species and not cell lines emphasized.

(a) *The stemline concept applied to living species*

The "stemline" in tumors, has been compared by Lejeune, as mentioned above, to the successful branches of evolution. Indeed, a living species is a successful branch of evolution, at least at the present moment. Thus, the number of "stemlines" with the same common origin may be very large: for instance, about 130 species are known in the superfamily Bovoidea. There is no example of 130 stemlines and sidelines reported in a single tumor or cell culture. Doubtless, the large number of species in some families is an advantage for the cytogeneticist, for numerous modes of karyotype variations with a common origin can be observed.

(b) *The unknown "originator"*

In tissue cultures, the original karyotype of the ancestral cell of an established cell line can easily be known. In tumors, the original karyotype is represented by the host's normal cells. On the contrary, when dealing with evolution of species, the karyotype of the "originator" of a given family is unknown. Nevertheless, hypothetical, "original" or "ancestral" karyotypes have been proposed for many families, using such concepts as the NF of Matthey, with satisfactory results. However, it must be emphasized that the study of karyotype evolution in mammals is not as accurate as clonal evolution in cells, because the beginning of the history is lacking. Only a few proofs that things happened as have been postulated exist: they concern the fertility of some hybrids.

(c) *Chromosome polymorphism inside species*

Chromosome mutations, with no specific signification, have been reported in many species. In fact, each species may present chromosome polymorphism. This particular form of chromosome variation shall not be considered here, although it seems likely that chromosome polymorphism may be an indicator of potential speciation; only those variations linked with speciation will be discussed. The order of presentation already used in the first section will be followed. Very often instead of the original publication reference will be made to "Comparative Mammalian Cytogenetics" (Benirschke, 1969), in which detailed articles, with complete references, appeared.

B. *Speciation without any detectable change of the karyotype*

This mode of speciation is very frequently observed in many families. Different species, sometimes with very different phenotypes, show identical karyotypes. The list is so long, that only a few examples will be given. In the Carnivora (Wurster, 1969) most bears show the same karyotype with $2n = 74$ chromosomes; many species of the genus *Canis* have a diploid number of $2n = 78$, the whole chromosome set being composed of telo- or acrocentrics except the X chromosome. Interestingly enough and this is not a unique case, *Lycaon pictus*, belonging to a different genus, has a karyotype identical to that of the dog. The aard wolf and the hyena, both present a similar karyotype with $2n = 40$ chromosomes. In the Insectivora (Borgaonkar, 1969) 3 species of 2 genera show an identical karyotype: *Hemicentetes semispinosus*, *H. nigriceps and Centetes ecaudatus*. Different species of *Cervinae* (Taylor, *et al.* 1969) possess a similar chromosome pattern, with $2n = 68$; Odocoileini also show the same diploid number of $2n = 70$ and the same NF = 74. Speciation in Camelidae happened without detectable modifications of the karyotype (Taylor *et al.*, 1968). Bovidae with such different phenotypes as *Bos taurus*, *Bison bonasus* and *Cephalophus grimmi* have the same karyotype (Koulischer *et al.*, 1967). In many families of primates such as Papinae, Hylobatinae etc. some or all species show the same karyotype (Chiarelli, 1962; Egozcue, 1969).

In fact, morphological studies using routine techniques are limited: small structural rearrangements escape detection. Complementary techniques such as the dosage of the nuclear DNA content, then appear very useful. This is the case for the Marsupialia (Hayman and Martin, 1969); many of the species with $2n = 14$ chromosomes belonging to different families have an apparently similar karyotype. However, DNA studies show that differences as high as 30% exist between some species: if the *Sminthopsis* nucleus has an arbitrary value of 100 for its DNA content, the DNA content of *Perameles* is 130. The similarity between karyotypes of these species is "hardly a matter of chance and could indicate that evolution of chromosome morphology is highly directed inside a family or an order, and must follow some patterns and not others." Another example of similar chromosome evolution in two different families is that of the Giraffidae and the Bovoidea. The karyotype of the Giraffe is almost identical to the karyotype of the Sitatunga (Fig. 1) and the karyotype of the Okapi (Ulrich and Schmitt, 1969) is apparently the same as *Anoa depressicornis* ($2n = 46$, 7 pairs of metacentrics or submetacentrics and NF = 60; see Fig. 2).

In conclusion, speciation is far from being linked with dramatic chromosome morphological changes.

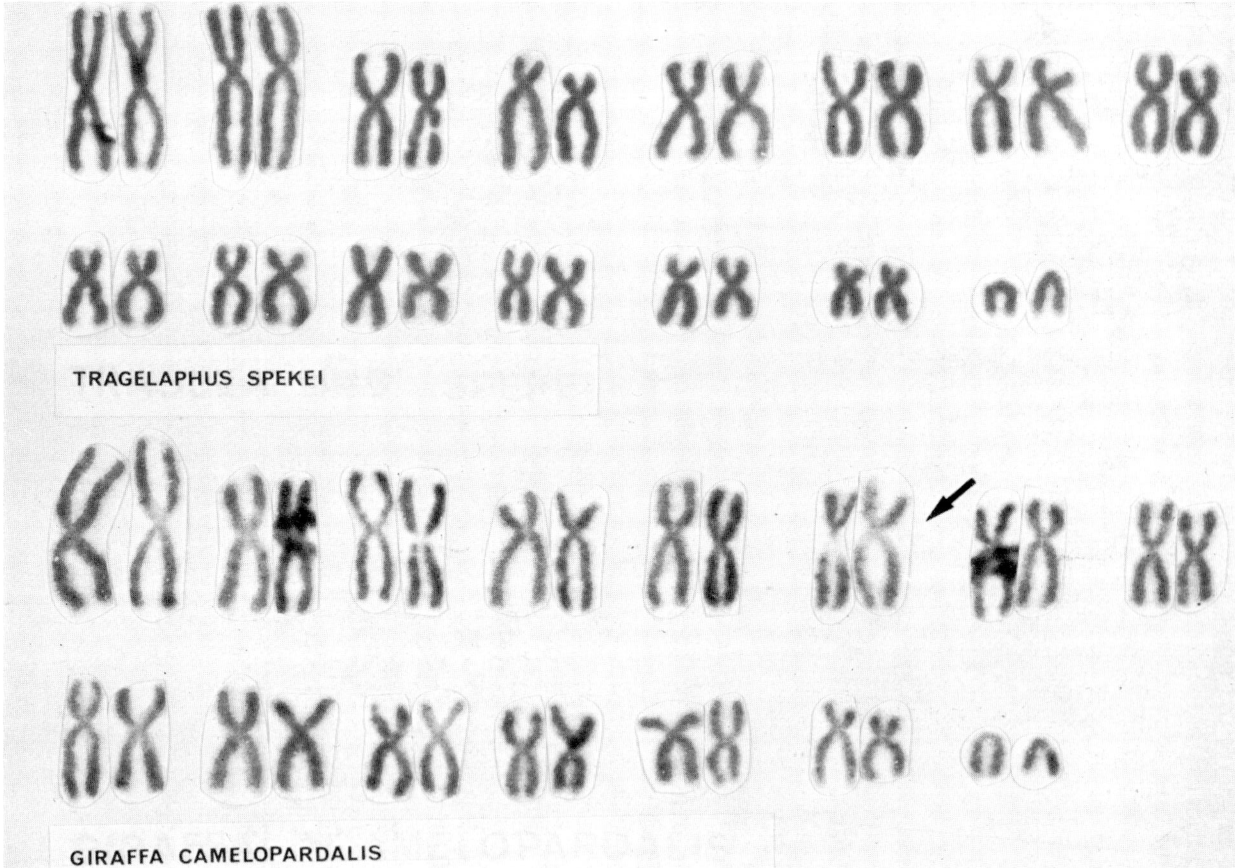

Fig. 1. Roughly similar karyotypes of a female Sitatunga (Tragelaphus spekei) and a female Giraffe (Giraffa camelopardalis). The diploid number is the same $2n = 30$; both species show a single pair of telocentrics. The sex chromosomes are represented by the largest pair in the Sitatunga and were not identified in the present specimen of Giraffe. The Giraffe shows a pair with a

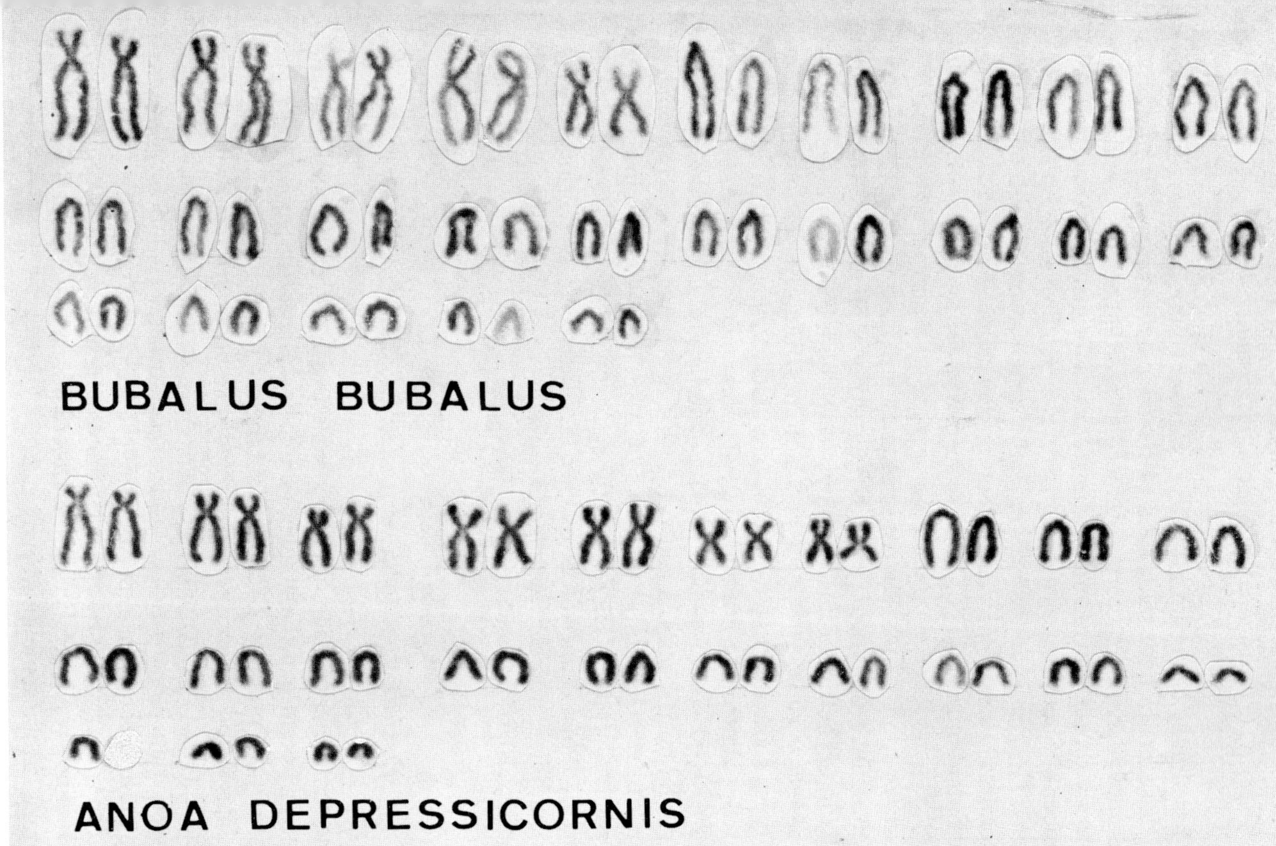

Fig. 2. Karyotypes of *Bubalus bubalus* ($2n = 50$) and *Anoa depressicornis* ($2n = 46$). The first five metacentric pairs of *Anoa* are apparently identical to the 5 pairs of metacentrics of *Bubalus*. The karyotype of *Bubalus* represents a "step" of karyotype evolution leading to *Anoa*.

C. *Speciation with the same diploid number but different karyotypes ("pseudodiploidy" of cellular clonal evolution)*

It has already been shown that for cellular evolution the original diploid number seems to have some kind of selective advantage. This number is maintained, despite structural chromosome rearrangements. Similar observations are reported in mammals. In Carnivora for instance (Wurster, 1969) a number of South American cats have $2n = 36$ chromosomes. But if *Felis pardalis*, *F. tigrina* and *F. wiedi* have the same karyotype with one pair of acrocentrics, *Felis geoffroyi* shows no pair of acrocentrics at all, with the same diploid number of 36. In those species of *Felis* with $2n = 38$ chromosomes, 3 pairs of acrocentrics are generally observed except in *Felis vagouaroundi*, in which all chromosomes are meta or submetacentric; pericentric inversions could explain the karyotype differences. In the genus *Erinaceus*, of the Erinaceidae (Gropp, 1969), both *Erinaceus europaeus* and *E. roumanicus* show $2n = 48$ chromosomes, but the obviously related karyotypes show important discrepancies. Borgaonkar reports that, in the subfamily Tenrecinae (Insectivora), *Echinops telfairi* and *Setifer setosus* have $2n = 40$ chromosomes, but the number of chromosome arms is different for both species. Again, pericentric inversions very likely occurred. In the Primates *Pongo*, *Gorilla* and *Pan* have $2n = 48$ chromosomes, but different karyotypes: *Pongo* shows 20 acrocentrics, *Gorilla* 16 and *Pan* 12 (Chiarelli, 1962; Hamerton *et al.*, 1963). Other species of Primates show common diploid numbers with more or less important karyotype differences (Chiarelli, 1968b; Egozcue, 1969).

D. *Centromeric or Robertsonian fusion of chromosomes*

The centromeric fusion of chromosomes, with a decrease of the diploid number, a decrease of the number of acro- or telocentric chromosomes, an increase in the number of meta- or submetacentrics and a constant NF is sometimes found as intraspecific chromosome polymorphism, without any specific signification. In many cases, however, Robertsonian fusion is linked with speciation. This particular mode of chromosome evolution is observed in a great number of mammalian families; the clearest example is perhaps that of the superfamily Bovoidea (Wurster and Benirschke, 1968). Indeed, the NF of nearly all members of this superfamily is close to 60. Species with the lowest chromosome numbers, i.e. $2n = 30$, show 28 or 30 metacentrics. Roughly, two different patterns of evolution are possible. First, speciation could occur without any apparent change of the original karyotype (the diploid number remaining $2n = 60$) or by fusion of telocentrics; concerning *all together* 2, 4, 6 or more pairs. Thus, species with diploid numbers lower than 60 could have emerged directly from an ancestral species with $2n = 60$ telocentrics. For each new chromosome number, further speciation could have happened without any detectable karyotype change, so two or more species with $2n = 58$, $2n = 56$, $2n = 54$ . . . $2n = 30$ chromosomes could exist (Fig. 3). The study of different Bovoidea karyotypes, however, suggests that more likely, things followed a second pattern (Fig. 4). Once a fusion occurred between 2 pairs of acrocentrics, it was transmitted to other species, each fusion appearing as one step of a progressive reduction of the diploid number. For instance, *Ammotragus lervia* has $2n = 58$ chromosomes, with a large pair of metacentrics, apparently resulting from the fusion of 2 large pairs of telocentrics. *Ovis ammon* (Schmitt and Ulrich, 1968) shows $2n = 56$ chromosomes; one pair of metacentrics is identical to the pair already observed in *Ammotragus lervia;* the second pair of metacentrics is also very large. In *Ovis aries* ($2n = 54$), the two pairs observed in *O. ammon*, are present, together with a third one (Koulischer, 1968).

The same progressive process may have happened for other species of the same superfamily, but cannot actually be observed because either the intermediate steps are fossils or have not been studied as yet. Thus, *Bubalus bubalus* has $2n = 50$ chromosomes, with 5 pairs of metacentrics; *Anoa depressicornis*, belonging to the same subfamily, has $2n = 46$ chromosomes (Fig. 2) with 7 pairs of metacentrics. The five first pairs of *Anoa* are identical to the 5 pairs of *Bubalus.*

A special comment is needed for some species with an apparently identical karyotype. Thus, *Antilocapra americana*, *Bibos frontalis* and *Oryx beisa*, belonging to different sub-families, do present $2n = 58$ chromosomes with a similar pair of submetacentric chromosomes (Fig. 5). This pair, however, could result from the fusion of different pairs of telocentrics and represent in each species the first step of a progressive evolution. This is why the 3 species mentioned here have not been presented in the previous section concerning speciation without apparent change of the karyotype (II, B). Doubtless fluorescence techniques will help to clarify this problem.

As the list of mammalian families showing centromeric fusions is very long only a few examples will be mentioned here.

In many rodents, speciation linked with centromeric fusion is observed (Nadler, 1969). *Mus poschiavinus* provides an example of extensive centromeric fusions with probable specific signification (Gropp,1969); its diploid number is $2n = 26$ with 14 metacentrics and 12 acrocentrics, the NF being of 40, like in *Mus musculus*. A Robertsonian mode is also observed in many families of the Carnivora (Wurster, 1969): for instance, the spectacle bear

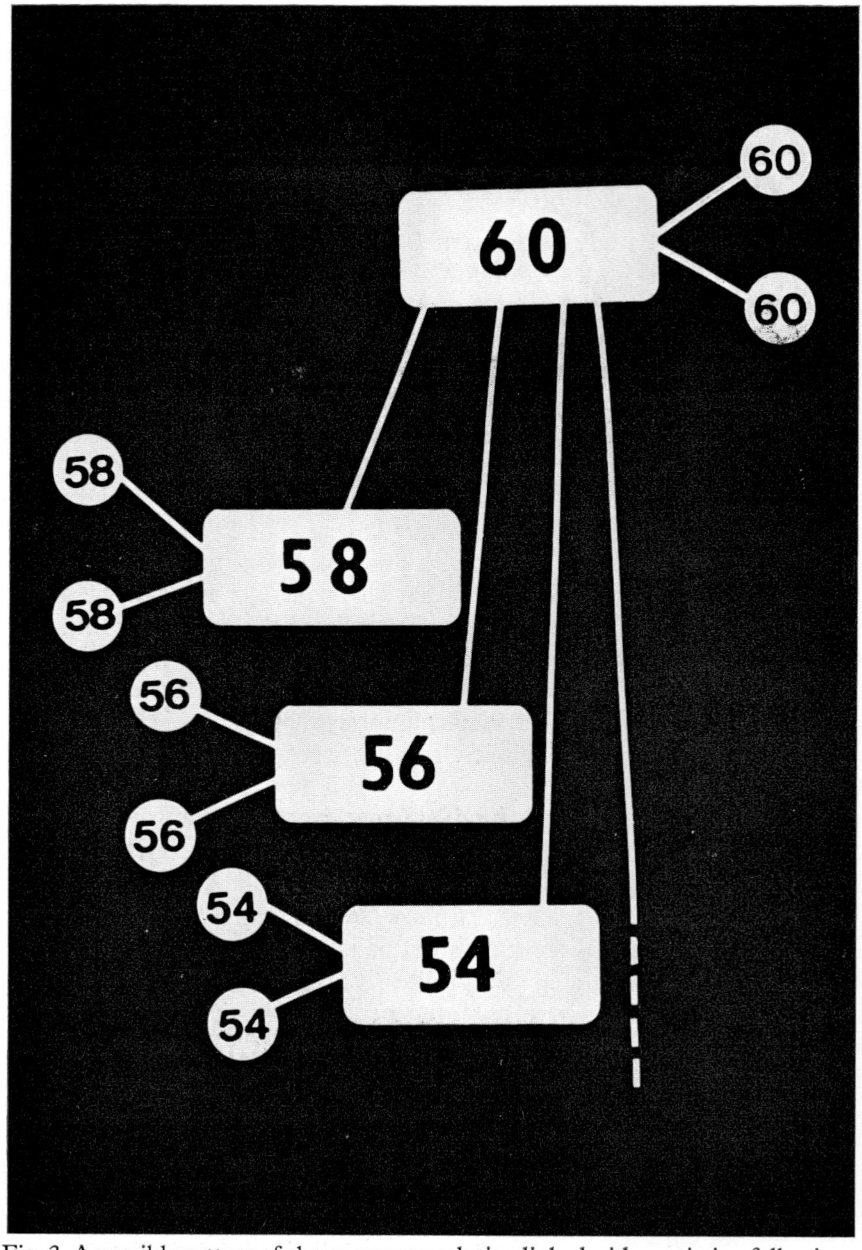

Fig. 3. A possible pattern of chromosome evolution linked with speciation following a Robertsonian mode, as in the Bovoidea. In this example all karyotypes originate directly from a single ancestral one with 60 telocentrics. For each chromosome number, different species with apparently identical karyotypes are observed.

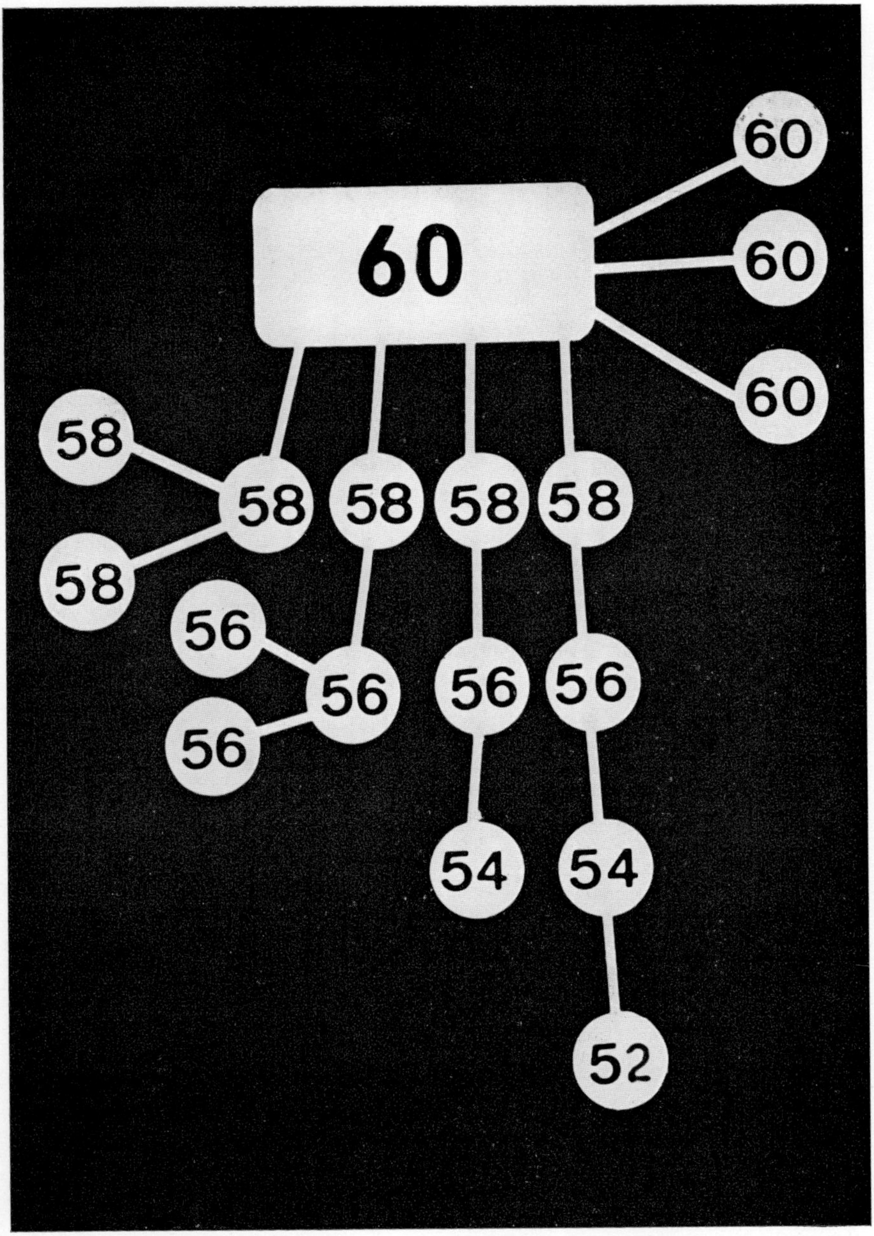

Fig. 4. Pattern of "step" chromosome evolution, which is actually observed in the Bovoidea (compare with Fig. 3). Karyotypes derive from one another and not directly from the ancestral one, except for those with $2n = 58$. Species with the same chromosome number may show the same karyotype or different karyotypes, depending upon the telocentric pairs involved in centromeric fusion.

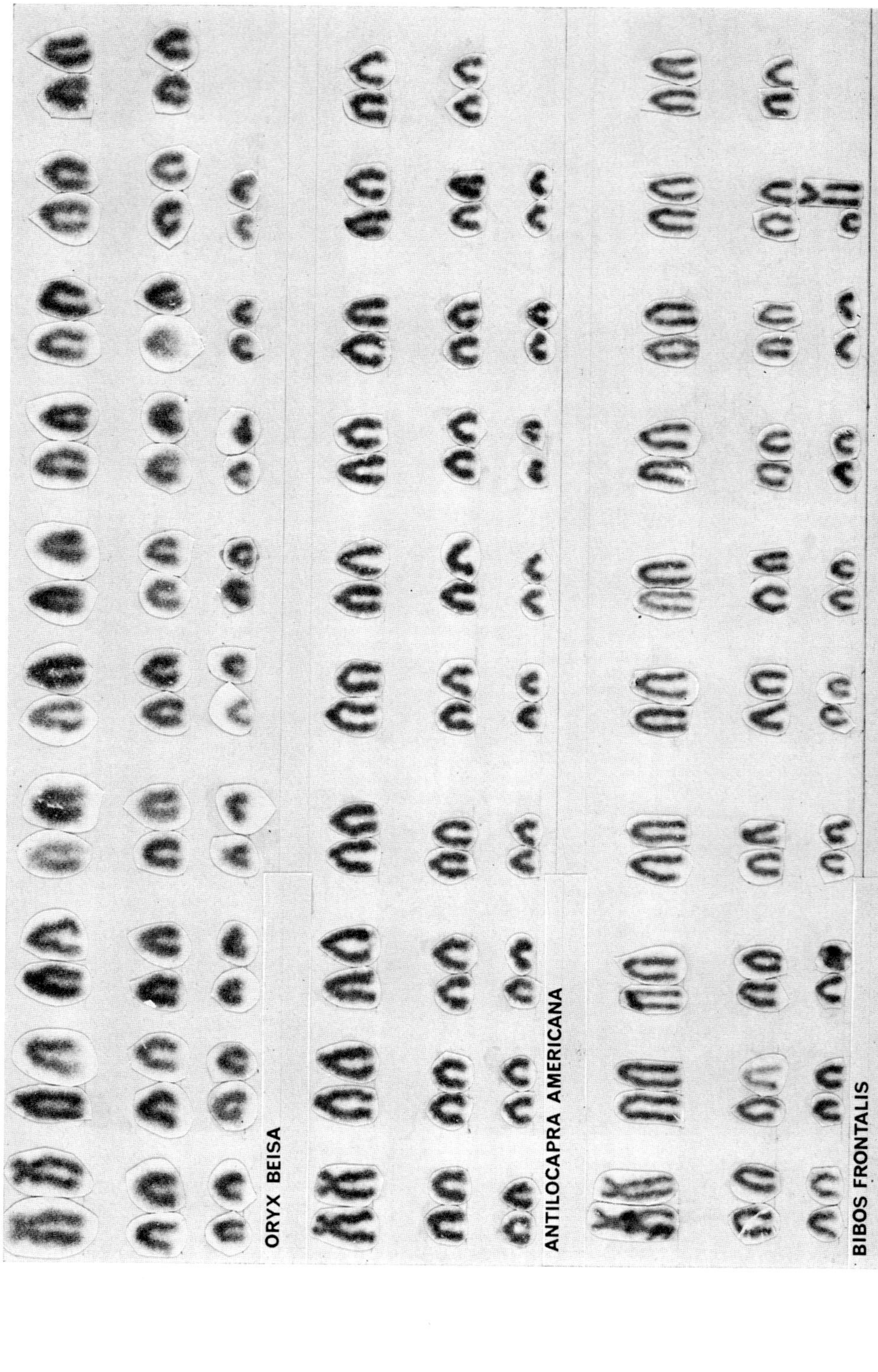

Fig. 5. Apparently identical karyotypes of different species belonging to different sub-families of the super-family Bovoidea.

has $2n = 52$ chromosomes with an NF $= 84$; the sun bear or the grizzly have $2n = 74$ chromosomes, the NF being also of 84.

Robertsonian fusion can be combined with other mechanisms of chromosome evolution, such as translocations hypo- or hyperdiploidy etc. On the other hand, Robertsonian fusion seems to have happened in almost the same manner in different families: already mentioned are the Bovoidea and the Giraffidae. To avoid taxonomic discussions this common chromosome evolution suggests at least two common ancestors with a similar karyotype. Perhaps *Homo sapiens* ($2n = 46$, 10 acrocentric pairs) should be mentioned in this paragraph. The closest karyotype is that of Pan ($2n = 48$, 12 acrocentric pairs); the main difference is a Robertsonian fusion, together with structural rearrangements of other pairs.

E. *Hyperdiploidy*

Families in which the most recent species have the highest diploid number are so infrequent that it has long been thought that evolution always proceeded in the sense of a reduction of the diploid number. Chromosome "fission" (the contrary of Robertsonian fusion) could be a possible mechanism, especially if the number of telocentrics or of isochromosomes is increased in recent species. However, in the best documented example of speciation apparently linked with progressive hyperdiploidy, the genus *Cercopithecus*, Chiarelli (1968) postulated a "reduplication of single and different autosomes". In this genus *Cercopithecus*, diploid numbers vary from $2n = 54$ to $2n = 72$, the most recent species showing the higher numbers. Chromosome measurements tend to confirm the general increase of the set; DNA measurements will be of interest to solve this problem. Thus Lampert (1968) has shown in cases of childhood leukemia, that cells with the highest chromosome numbers have also the highest DNA contents: similar results should be found in the *Cercopithecus*.

F. *Hypodiploidy*

Strict hypodiploidy, that is to say apparent "loss" of chromosomes involved in speciation, is not frequent. It has been suggested for *Equus prjewalskii* ($2n = 66$) and *E. caballus* ($2n = 64$): the only difference between the karyotypes of the 2 species is a "loss" of 2 telocentrics in *E. caballus* (Koulischer and Frechkop, 1966), although a Robertsonian mode has been postulated by others (Benirschke *et al.*, 1965). The South American bush dog *Speothos venaticus* has $2n = 74$ chromosomes, with 4 "missing" acrocentrics or telocentrics when compared to the karyotype of the dog with $2n = 78$ chromosomes (Wurster, 1969). Examples taken from different families of the same order may not be relevant, but it is interesting to point out that the lesser panda *Ailurus fulgens* "has a karyotype that is almost indistinguishable from that of the clouded leopard of Asia, except that it has one pair less in the F group" (Wurster, 1969). This hypodiploidy could signify a parallelism of evolution, similar to that of Giraffidae and Bovidea.

G. *Triploidy and tetraploidy*

Triploidy or tetraploidy are suggested, in tumors or tissue cultures, by the presence of chromosomes represented in 3 or 4 examples. Morphological data can be confirmed by the study of the DNA content. Polyploidization is so frequent in tumors or tissue cultures that chromosome counts alone suggest triploidy, tetraploidy etc.

In mammals chromosome counts are certainly not a useful criterion to detect eventual polyploidization inside an order or a family. Thus, in the order Marsupialia (Hayman and Martin, 1969), species with $2n = 12$ (*Dendrolagus ursinus*) or $2n = 24$ chromosomes (*Lagostrophus fasciatus*) are known, as one species with $2n = 16$ (different species of Macropus) and $2n = 32$ chromosomes (*Aepyprymnus rufescens*). Indeed, neither karyotype nor DNA studies suggest the possibility of a tetraploidization. In the family Cricetidae, *Cricetulus griseus*, the Chinese Hamster, has a diploid number of 22, while *Mesocricetus auratus*, the Golden hamster, has $2n = 44$ chromosomes. Karyotypes however are very different (Hsu and Benirschke, 1967) and certainly not suggestive of a "simple" tetraploidy. On the contrary, variations of the DNA content of mammalian nuclei show only small variations (Atkin *et al.*, 1965) far from reaching the range of triploidy. Polydiploidization is apparently linked with evolution, but in more remote periods and mammals are not concerned (for a discussion, see Ohno, 1970).

H. *Minute chromosomes*

Minute chromosomes are very seldom observed in mammals, while they are a very common finding in birds. Thus minute chromosomes, as triploidy or tetraploidy, do not seem to have been widely used in mammalian speciation.

I. *The different patterns of mammalian speciation at the chromosome level. Summary*

The main patterns of chromosome evolution linked with speciation in mammals are summarized in Table III. A comparison with cellular clonal evolution has been made (see Table II). Some common patterns appear: they will be discussed in the next section.

TABLE III. *Comparison between chromosome patterns of evolution in mammals and mammalian cells*

Modes of Chromosome evolution	Mammalian species	Mammalian cells
Eudiploidy	frequent	very frequent
Pseudodiploidy	less frequent	frequent
Centromeric fusion	very frequent	less frequent
Hyperdiploidy	less frequent	very frequent
Hypodiploidy	very infrequent	very infrequent
Triploidy or tetraploidy	never observed	very frequent
Minute chromosomes	exceptional	infrequent

As in clonal evolution of cells, structural rearrangements can be so important that chromosome evolution is impossible to reconstruct: this is the case in the Equidae (Benirschke and Malouf, 1967) with very different karyotypes, except for *Equus prjewalskii* and *E. caballus*.

IV DISCUSSION

Isolated mammalian cells cultivated *in vitro*, as well as mammalian tumoral cells, are evidently very different from normal cells, which form part of a whole mammalian organism. They are not involved in a general economy and are thus freer to show important variations, including chromosome changes, very likely linked with adaptation processes. Moreover, they are permanent in themselves. Each cell, *in vitro*, is a potential ancestor of a new stemline; chromosome evolution is easy, rapid and can be followed accurately with adequate techniques. A "whole organism", on the other hand, is permanent only through its germ cells, which are the only potential ancestors of a new species. To be transmitted, eventual chromosome changes must happen in the gonads and go through meiosis: complicated mechanisms such as pairing of homologue segments, chiasma formation and chromatid exchanges are needed for the transmission of the genome. This represents an important delay in evolution, implicating however a stability of the karyotype of each species, compared to the instability of the karyotype of isolated cells.

Despite fundamental differences, common patterns of chromosome evolution can be observed in both "systems". Does this analogy allow a better understanding of the origin of mammalian karyotypes ? The purpose of the present discussion is to try to answer to this question.

A. *The "highly condensed time scale" of clonal evolution*

As it has been pointed out by Levan (1967), it would take thousands (and even millions) of years to observe at the "whole organism" level the chromosome evolution reported in cells evoluting *per se*. A cellular generation time, 14 to 20 hours for mammalian cells *in vitro*, has often been compared to a generation time of a species, from some months to 20 years for Man. This comparison is perhaps not fully appropriate. In tissue cultures, dramatic karyotype changes may occur in one mother cell and be transmitted at once to the daughter cells, which may in turn form a new stemline. This is not conceivable in mammals: new chromosome rearrangements must overcome meiosis and the number or importance of chromosome changes at one time is indeed very limited. To form a new species fecundation must take place and a sufficient number of homozygotes must appear; this implies many generations. Even if it is admitted that karyotype evolution happened in a relatively short time, as in *Mus poschiavinus* (Gropp, 1969), this "short time" is still very long compared to clonal chromosome evolution. Indeed, the same fusion of telocentrics could happen in one generation *in vitro*. In cultures of bovine origin, chromosome fusions similar in morphology to those linked with speciation in the family Bovoidea have been observed after a few months *in vitro* (Nelson-Rees *et al.*, 1967).

Despite its high speed, chromosome evolution of isolated cell systems is not left to hazard: rules or "laws" apparently govern chromosome changes. Therefore cell lines maintained *in vitro*, or tumors *in vivo* or *in vitro*, could offer interesting models which could help to understand the formation of the karyotypes of the actual living species. This has been pointed out by Hughes (1968): "*in vitro* cell population provide us with a model system for the study of evolution on a highly condensed scale".

B. *The succession of events in chromosome evolution*

When the karyotype of the originator cells has been studied in a tissue culture or a tumor, the direction of chromosome evolution is easy to determine. The most recent karyotypes are known with certitude. This is not the case for mammalian evolution. It has long been proposed that karyotype evolution in mammals proceeded in the sense of a reduction of the total number of chromosomes, mainly by Robertsonian fusion. Thus, in a given family, the most recent species should be those with the lowest chromosome counts. Many exceptions however are known, some of which have already been referred to in the sections above. The genus *Cercopithecus* is one of these exceptions. It is interesting to note that both Lejeune (1965) for human leukemia and Chiarelli (1968a) for Cercopitheci, propose the same hypothesis to explain the origin of new karyotypes, namely duplication of single chromosomes. Needless to say the two "systems" are very different: nevertheless, chromosome evolution in a tumoral cell line and in a mammalian genus could have happened following common patterns. In this respect, it is very useful indeed to know beyond any doubt that in some mammalian cell systems the most recent cells are those with the highest number of chromosomes. Moreover, the "laws" of clonal evolution proposed by Lejeune for malignant tumors with increasing number of chromosomes are also suitable for evolution of species (Koulischer, 1968).

In view of these analogies, no *a priori* position should be taken as to the direction of chromosome evolution in species of the same family. This could raise some questions: for instance, at least theoretically, the karyotype of *Equus prjewalskii* ($2n = 66$) could be the result of selective duplication of chromosomes of *Equus caballus* ($2n = 64$); on the other hand, chromosome fission, instead of chromosome fusion, could have been used in speciation of other families in which the most recent species show the highest chromosome counts. Those problems replace cytogenetics in its general context and show that the full signification of chromosome studies, linked with speciation, appears only when related to other taxonomic features.

C. *Originality of some patterns of clonal evolution*

Although many similar trends are observed, cellular clonal evolution and chromosome evolution at the whole organism level show some important morphological differences. Thus Robertsonian fusion is much more frequent in mammalian speciation, while hyperdiploidy and polyploidy are more often observed in clonal evolution. Hyperdiploidy and polydiploidy are reminiscent of a remote type of evolution by gene duplication, which happened in amphibians and ceased at the reptilian stage (Ohno, 1970): isolated cells seem to tolerate, and even to find an advantage in, an excess of genes.

In mammalian evolution on the contrary, a general stability of the nuclear DNA content is observed; speciation seems mainly due to positional effects and a small number of genes are involved at the time Robertsonian fusion is used. On the other hand, an increase of the number of chromosomes could be linked with heterochromatinization: thus important chromosome morphological changes could have a small meaning as far as gene function is concerned. In general, whole organisms cannot tolerate wide variations of the genome. The example of medical cytogenetics demonstrates it very well: small chromosome aberrations, especially of the autosomes, have pronounced pathological effects. Isolated cells, in turn, are not affected by an increase or a reduction of the genome. Indeed, it is not known to what extent the active part of the cellular genome is affected. But it is striking, at least on morphological grounds, to observe that when mammalian cells are "free", they find almost immediately a remote mode of evolution, genome duplication, which could also be a future mode of evolution.

D. *Environment and chromosome evolution*

The part played by the environment in chromosome evolution can, of course, be studied much more easily in isolated cells *in vitro* than on whole organisms. Experimentally induced chromosome aberrations by means of chemical or physical agents have already been reviewed (Hirschhorn and Cohen, 1969). Thus, only a few examples shall be considered here. In one experiment the composition of the feeding medium of sub-cultures obtained from a single originator has been modified (Parshad and Sandford, 1968). Diploidy was retained for a longer time in mouse embryo cells C3Hf when fetal calf serum was used instead of horse serum. However this diploidy was independent of neoplastic transformation, which happened at the same time with calf serum, bovine serum and horse serum: chromosome morphology in these experiments was not related to neoplastic transformation. Similar observations were made by Evans *et al.* (1967). Fetuin, a seric proteic fraction found only in fetal calf serum, could be responsible for the longer conservation of a normal karyotype. Another comparable example has been reported by Hughes (1968); Burkitt's lymphoma cell line EBI changed from a bimodal distribution of chromosome numbers around $2n$ and $4n$ to a mostly hyperdiploid population after calf instead of human serum was added to the cultures. Somewhat different is the already mentioned observation of Nelson-Rees *et al.* (1967). A dog cell line maintained *in vitro* did not show Robertsonian fusions, as did bovine cell lines in the same conditions. On the other hand in neoplastic dog cell lines such as lymphoma or venereal sarcoma extensive centromeric fusions were reported.

In this case, malignancy is linked with chromosome changes similar to the most widely observed chromosome mechanism of evolution in mammalian speciation. Not only morphology, but also the speed of clonal evolution can be experimentally modified: polyploidization is accelerated when 5-bromodeoxyuridine is added to mouse tissue cultures (Hsu *et al.*, 1961).

When evolution is considered at the species level, karyotype changes linked with speciation must happen in the gonads, and meiotic chromosomes must be studied. This has been done in invertebrates. In mammals, meiotic chromosomes of rodents, especially of mice, have been extensively used. Besides their well known effect of punctiform mutations, physical (Leonard and Deknudt, 1966) as well as chemical agents (Sakkebaek *et al.*, 1968; Lavappa and Yerganian in the Armenian hamster, 1971) are able to induce chromosome changes affecting the germ cells. Indeed, this important field of research is more related to pathology than to physiology. It is hoped, however, that it will bring some information about karyotype evolution in mammals.

Two short remarks conclude this paragraph. First, it seems that environmental factors are linked with general morphological trends, but not with specific patterns of chromosome evolution. Secondly, as it has been pointed out, detectable chromosome rearrangements are not compulsory either for speciation or for the transformation of a cell strain into an "immortal" cell line. Thus, mutagenic agents other than those inducing chromosome aberrations could be important in evolution.

E. *Genetic determinism of chromosome evolution*

The part of genetic determinism in chromosome evolution is certainly difficult to ascertain. Indeed, a *specific* determinism seems improbable in clonal evolution. For instance, the numerous patterns of chromosome evolution of induced Rous sarcoma in mice, described by Mark (1970), are not in favour of a genetically specific effect. In the same way, the different karyotypes observed in human malignant tumors of the same histologic type do not suggest a specific reaction to the cancerous agent; human chronic myeloid leukemia and canine venereal sarcoma appear as exceptions. On the contrary, when the individual level and not the specific level is considered, the general impression is that the genome seems to be important in clonal chromosome evolution. Two different examples in favour of this view shall be given.

Sandberg *et al.* (1966) studied the chromosomes of dyzygotic twins presenting an acute myeloblastic leukemia. The onset of the disease happened at the same age, 7 months, and the clinical course was identical in both. Very likely, the 2 children had been submitted to the same environmental factors since fecundation. Nevertheless, one of them exhibited a leukemic stemline with 45 chromosomes while the other had a stemline with 52 chromosomes. Each child reacted to the same oncogenic factor(s) following its own genome and no common chromosome evolution could be observed.

The second example is very different and concerns tissue cultures of root voles (Ferno and Mandahl, 1970). After the 17th passage *in vitro*, the original culture was divided in two series: one continued to be grown, the other was frozen and kept in this state for 6 months. Thereafter, cells were thawed and grown again. Chromosome evolution was the same in both series, with elimination or rearrangement of the same chromosomes at the same passages. For the authors "it can be concluded from these similarities that the evolution of the cell line was not random but was predetermined by the genotype of the original cell".

In evolution of species, particular patterns seem to be used preferentially by some families, such as Robertsonian fusion in Bovoidea or selective duplication in *Cercopitheci*. However, this is far from being a general rule and exceptions are known in the two families mentioned. Nevertheless, chromosome evolution of species belonging to the same family is reminiscent of clonal evolution of cells with the same ancestor, and thus determined by the genome of the original cell. Indeed, more research is needed to understand clearly the genetic determinism of chromosome evolution, in cells as well as in species.

F. *Concluding remarks*

The aim of this chapter has been to outline some similarities between chromosome clonal evolution of mammalian cells and speciation in mammals. Indeed, the data discussed above were arbitrarily selected and condensed. On the other hand, comments concerning the environmental factors were strongly abbreviated and no discussion at all on the possible part of virus action on chromosome evolution was offered. However, it should be pointed out again that the intention was to consider mainly the morphology of chromosome evolution. Certainly, on morphological grounds, clonal evolution offers good models for chromosome evolution linked with speciation in mammals. Basically each family shows apparently the same kind of evolution that is observed *in vitro* in cells from the same originator. The importance of the individual genome appears to be predominant in both clonal evolution and mammalian speciation. An interesting point is that the direction of chromosome evolution is multiple and the most recent species, as the most recent cells, may exhibit identical, or higher, or lower chromosome numbers when compared to the originator.

In conclusion, it should be recalled that chromosome observations have a full meaning only when placed in the general context. Cytogenetics thus contributes to a better understanding of general problems of evolution, in conjunction with other related specialities.

Acknowledgements

The cytogenetic material used in the illustrations comes from the Royal Zoological Society of Antwerp, owner of the Antwerp Zoo. The help and assistance of Prof. J. Mortelmans and Mr. J. Tijskens are gratefully acknowledged. Technical assistance from Marie-Anne Cosme, Eliane and Christiane Canon and Franco Meraglia is also gratefuly acknowledged, as is the typing by Rose-Marie Fox and Chantal Porsont.

References

Atkin, N. B., Mattinson, G., Beçak, W., Ohno, S. (1965). The comparative DNA content of 19 species of placental mammals, reptiles and birds. *Chromosoma*, **17**, 1–10.

Barski, G. and Cassingena, R. (1963). Malignant transformation in vitro of cells from C 57 BL mouse normal pulmonary tissue. *J. Natn. Cancer Inst.* **30**, 865–883.

Barski, G. and Cornefert-Jensen, F. (1966). Cytogenetic study of Sticker venereal sarcoma in European dogs. *J. Natl. Cancer Inst.* **37**, 787–797.

Barski, G., Billardon, C., Jullien, P. M. and Carswell, E. (1966). Evolution in vitro et cancérisation des cellules pulmonaires de souris. *Int. J. Cancer*, **1**, 541–556.

Bayreuther, K. (1960). Chromosomes in primary neoplastic growth. *Nature*, (*Lond*) **186**, 6–9.

Benirschke, K. (1969). "Comparative Mammalian Cytogenetics". Springer-Verlag, Berlin, Heidelberg and New-York.

Benirschke, K. and Malouf, N. (1967). Chromosome studies of Equidae. *In* "Equus" Vol. I, pp. 253–284 (A. Dathe, ed.). Tierpark, Berlin.

Benirschke, K., Malouf, N., Low, R. J. and Heck, H. (1965). Chromosome complement: differences between Equus caballus and Equus prjewalskii, Poliakoff. *Science*, N.Y. **148**, 382.

Borgaonkar, D. S. (1969). Insectivora cytogenetics. *In* "Comparative Mammalian Cytogenetics" (K. Benirsche, ed.), 219–245. Springer-Verlag, Berlin, Heidelberg and New York.

Buttle, H. L. and Hancock, J. L. (1966). The chromosomes of goats, sheep and their hybrids. *Res. Vet. Sci.*, 7, 230–231.

Chiarelli, B. (1968a). Chromosome polymorphism in the species of the genus *Cercopithecus*. *Cytologia*, **33** 1–16.

Chiarelli, B. (1962). Comparative morphometric analysis of Primate chromosomes. II: the chromosomes of the genera *Macaca*, *Papio*, *Theropithecus* and *Cercocebus*. *Caryologia*, **15**, 401–419.

Chiarelli, B. (1968b). Caryological and hybridological data for the taxonomy and phylogeny of the Old World Primates. *In*: "Taxonomy and Phylogeny of Old World Primates with references to the origin of man." Rosenberg and Sellier, Torino.

Cox, D., Yunchen, C., Spriggs, A. L. (1965). Minute chromatin bodies in malignant tumors of childhood. *Lancet*, **2**, 55–58.

DeCarli, L. and Chiarelli, B. (1960). Variabilitá del numero di cromosomi tra cloni ed entro cloni in due ceppi stabilizzati di cellule umane coltivate in vitro. *Atti A.G.I.*, **5**, 255–264.

Egozcue, J. (1969). Primates. *In* "Comparative Mammalian Cytogenetics", pp. 357–389 (K. Benirschke, ed.) Springer-Verlag, Berlin, Heidelberg and New York.

Evans, V. J., Jackson, J. L., Andresen, W. F. and Mitchell, J. T. (1967). Chromosomal characteristics and neoplastic transformation of C3H mouse embryo cells in vitro in horse and fetal calf serum. *J. Natn. Cancer. Inst.*, **38**, 761–769.

Ferno, A. and Mandahl, N. (1970). In vitro chromosome studies in the root vole (Microtus oeconomus). *Hereditas (Lund)*, **66**, 153–164.

Gropp, A. (1969). Cytologic mechanisms of karyotype evolution in insectivores. *In* "Comparative Mammalian Cytogenetics", pp. 247–266 (K. Benirschke, ed.) Springer-Verlag, Berlin, Heidelberg and New York.

Grouchy, J. de, de Nava, C., Feingold, J., Bilski-Pasquier, G. and Bousser, J. (1968) Onze observations d'un modèle précis d'évolution caryotypique au cours de la leucémie myéloïde chronique. *Eur. J. Cancer*, **4**, 481–492.

Hamerton, J. L., Klinger, H. P., Mutton, D. E. and Lang, E. M. (1963). The somatic chromosomes of the Hominoidea. *Cytogenetics*, **2**, 240.

Hampel, K. E. and Levan, A. (1964). Breakage in human chromosomes induced by low temperature. *Hereditas*, **51**, 315–343.

Hayflick, L. (1965). Cell culture and the ageing phenomenon. *In* "Topics in the Biology of Ageing", pp. 83–100 (P. L. Krohn, ed.) Interscience Publ., New York and London.

Hayflick, L. and Moorhead, P. S. (1961). The serial cultivation of human diploid cell strains. *Exp. Cell Res.*, **25**, 585–621.

Hayman, D. L. and Martin, P. G. (1969). Cytogenetics of marsupials. *In* "Comparative Mammalian Cytogenetics", pp. 191–217 (K. Benirschke, ed.) Springer-Verlag, Berlin, Heidelberg and New York.

Hellström, K. E., Hellström, I. and Sjogren, H. O. (1963). Further studies on karyotypes of a variety of primary and transplanted mouse polyoma tumors. *J. Natn. Cancer Inst.* **31**, 1239–1253.

Hirschhorn, K. and Cohen, M. H. (1969). Induced chromosome observations with special reference to man. *In* "Comparative Mammalian Cytogenetics," pp. 49–67 (K. Benirschke, ed.), Springer-Verlag, Berlin, Heidelberg and New York.

Hsu, T. C., Billen, D. and Levan, A. (1961). Mammalian chromosomes in vitro. XV. Patterns of transformation. *J. Natn. Cancer. Inst.* **27**, 515–541.

Hsu, T. C. and Benirschke, K. (1967). "An Atlas of mammalian chromosomes". Springer Verlag, Berlin and New York.

Huang, C. C. and Moore, G. E. (1969). Chromosomes of 14 hematopoietic cell lines derived from peripheral blood of persons with and without chromosome anomalies. *J. Natn. Cancer Inst.*, **43**, 1119–1128.

Huang, C. C., Imamura, T. and Moore, G. E. (1969). Chromosomes and cloning efficiencies of hematopoietic cell lines derived from patients with leukemia, melanoma, myeloma and Burkitt lymphoma. *J. Natn. Cancer. Inst.*, **43**, 1129–1146.

Hughes, D. T. (1968). Cytogenetical polymorphism and evolution in mammalian somatic cell population *in vivo* and *in vitro*. *Nature* (*Lond*), **217**, 518–523.

Jacobs, P. A., Tough, I. M. and Wright, O. H. (1963). Cytogenetic studies in Burkitt's lymphoma. *Lancet*, **2**, 1144–1146.

Kohn, G., Mellman, W. J., Moorhead, P. S., Lofts, J. and Henle, G. (1967). Involvement of C group chromosomes in 5 Burkitt lymphoma cell lines. *J. Natn. Cancer Inst.*, **38**, 209–223.

Kondo, T., Muragishi, H. and Imaizumi, M. (1971). A cell line from a human salivary gland mixed tumor. *Cancer*, **27**, 403–410.

Koulischer, L. (1968). Concept of cellular clonal evolution of karyotypes applied to evolution of species. *Acta Zool. Path.* (*Antwerp*)., **48**, 21–28.

Koulischer, L. and Frechkop, S. (1966). Chromosome complement: a fertile hybrid between Equus prjewalskii and Equus caballus. *Science*, N.Y. **151**, 93.

Koulischer, L., Tijskens, J. and Mortelmans, J. (1967). Mammalian cytogenetics. I. The chromosomes of three species of Bovoidea: *Bos taurus*, *Bison bonasus* and *Cephalophus grimmi*. *Acta Zool. Path.* (*Antwerp*), **43**, 135–141.

Kuccheria, K. (1968). Double minute chromatin bodies in a subependymal glioma. *Brit. J. Cancer*, **22**, 696–697.

Lampert, F. (1968). Kerntrockengewicht, DNS-gehalt und chromosomen bei akuten Leukämien im Kindeslater. *Virchows Arch. B. Zellpath*, **1**, 31–48.

Lavappa, K. S. and Yerganian, G. (1971). Latent meiotic anomalies related to an ancestral exposure to a mutogenic agent. *Science*, N.Y. **172**, 171–173.

Lejeune, J. (1965). Leucémie et cancer. *In* "Les Chromosomes Humains", pp. 181–215. (R. Turpin and J. Lejeune, eds). Gauthier Villars, Paris.

Lejeune, J., Berger, R., Haines, M., Lafourcade, J., Vialatte, J., Satge, P. and Turpin, R. (1963) Constitution d'un clone à 54 chromosomes au cours d'une leucoblastose chez une enfant mongolienne. *C. R. Hebd. Séanc. Acad. Sci. Paris*, **256**, 1195–1197.

Leonard, A. (1964). Analyse chromosomique d'une tumeur d'ascite de souris après 200 transplantations. *C. R. Séanc. Soc. Biol.*, **158**, 1958.

Leonard, A. and Deknudt, G. (1966). Meiotic chromosome rearrangements induced in mice by irradiation of spermatogonial stages. *Can. J. Genet. Cytol.*, **8**, 520–527.

Levan, A. (1967). Some current problems of cancer cytogenetics. *Hereditas*, **57**, 343–355.

Levan, A. (1969). Chromosome abnormalities and carcinogenesis. *In* "Handbook of Molecular Cytology" (A. Lima-de-Faria, ed.) North-Holland Publishing Company, Amsterdam and London.

Levan, A., Manolov, G. and Clifford, J. (1968). Chromosomes of human neuroblastoma: a new case with multiple minute chromosomes. *J. Natn. Cancer Inst.*, **41**, 1377–1387.

Levan, G. (1970). Contributions to the chromosomal characterization of the PTKI rat kangaroo cell line. *Hereditas*, **64**, 85–96.

Lithner, F. and Ponten, J. (1966). Bovine fibroblasts in long term tissue culture: chromosome studies. *Int. J. Cancer*, **1**, 579–588.

Lubs, H. A., Salmon, J. H. and Flanigan, S. (1966). Studies of a glial tumor with multiple minute chromosomes. *Cancer*, **19**, 591–599.

Macallister, R. M., Gardner, M. B., Greene, A. E., Bradt, C., Nichols, W. W. and Landing, B. H. (1971). Cultivation in vitro of cells derived from a human osteosarcoma. *Cancer*, **27**, 397–402.

Makino, S. (1957). The chromosome cytology of the ascites tumors of rats, with special reference to the concept of stemline cell. *Int. Rev. Cytol.* **6**, 26–85.

Makino, S. (1963). Some epidemiologic aspects of venereal tumors of dogs as revealed by chromosome and DNA studies. *Ann. N.Y. Acad. Sci.* **108**, 1106–1122.

Mark, J. (1967). Chromosomal analysis of ninety one primary Rous sarcomas in the mouse. *Hereditas*, **57**, 23–82.

Mark, J. (1970). Rous sarcomas in mice: the chromosomal progression during early in vivo transplantation. *Hereditas*, **65**, 59–82.

Matthey, R. (1949). "Les Chromosomes des Vertébrés". Librairie de l'Université F. Rouge, Lausanne.

Mayr, E. (1969). Species, speciation and chromosomes. *In* "Comparative Mammalian Cytogenetics", pp. 6–7 (K. Benirschke, ed.) Springer-Verlag, Berlin, Heidelberg and New York.

Miles, C. P. and O'Neill, F. (1967). Chromosome studies in 8 in vitro lines of Burkitt's lymphoma. *Cancer Res.* **27**, 392–402.

Miles, C. P., O'Neill, F. O., Armstrong, P., Clarkson, B. and Keane, J. (1968). Chromosome patterns of human leukocyte established cell-lines. *Cancer Res.*, **28**, 481–490.

Moore, G. E. and Sandberg, A. A. (1964). Studies of a human tumor cell line with a diploid karyotype. *Cancer*, **17**, 170–175.

Moore, G. E., Porter, I. H. and Huang, C. C. (1969). Lymphocytoïd lines from persons with sex chromosome anomalies. *Science*, N.Y. **163**, 1453–1454.

Moorhead, P. S. (1965). Human tumor cell line with a quasi diploid karyotype (RPMI 2650). *Exp. Cell Res.*, **39**, 190–196.

Moorhead, P. S. and Saksela, E. (1965). The sequence of chromosome aberrations during SV 40 transformation of a human diploid strain. *Hereditas*, **52**, 271–284.

Muldal, S., Elejalder, R. and Harvey, P. W. (1971). Specific chromosome anomaly associated with autonomous and cancerous development in man. *Nature (Lond)*, **299**, 48–49.

Nadler, C. F. (1969). Chromosomal evolution in rodents. *In* "Comparative Mammalian Cytogenetics", p. 277–309 (K. Benirschke, ed.) Springer-Verlag, Berlin, Heidelberg and New York.

Nava, C. de (1970). "Les Anomalies Chromosomiques au Cours des Hémopathies Malignes et Non Malignes". Vol 1. Monographies des Annales de Génétique.

Nelson-Rees, W. A., Kniazeff, A. S. and Darby, N. B. (1967). Début and acumulation of centric fusion products: an index to age of certain cell lines. *Cytogenetics*, **6**, 436–450.

Nowell, P. C. and Hungerford, D. A. (1960). A minute chromosome in human chronic granulocytic leukemia. *Science*, N.Y. **132**, 1497.

Nuzzo, F., DeCarli, L. and Chiarelli, B. (1961). Analisi morfologica degli assetii cromosomici in un ceppo di cellule umane stabilizzate in vitro. *Atti A. G. I*, **6**, 111–124.

Ohno, S. (1970). "Evolution by Gene Duplication". Vol. 1. Springer Verlag, Berlin, Heidelberg, and New York.

Parshad, R. and Sanford, K. K. (1968). Effect of horse serum, fetal calf serum, calf serum, bovine serum and fetuin on neoplastic conversion and chromosomes of mouse embryo cells *in vitro*. *J. Natn. Cancer Inst.*, **41**, 767–779.

Piccotti, F. and Chiarelli, B. (1967). Caratterizzazione cariologica del tumore ascite di Ehrlich (Linea Torino). *Boll. Soc. It. Biol. Sper.* **63**, 1317–1319.

Pierre, R. V., Hoagland, C. H. and Linman, J. W. (1971). Microchromosomes in human preleukemia and leukemia. *Cancer*, **27**, 160–175.

Pollack, R., Wolman, S. and Vogel, A. (1970). Reversion of virus transformed cell-lines: hyperploidy accompanies retention of viral genes. *Nature (Lond)* **228**, 938–970.

Ponten, J. and Lithner, F. (1966). Absence of specific chromosome alterations in bovine lung fibroblasts exposed to Rous sarcoma virus. *Int. J. Cancer*, **1**, 589–598.

Rieger, R., Michaelis, A. and Green, M. M. (1968). "A Glossary of Genetics and Cytogenetics". Vol 1. Springer Verlag, Berlin, Heidelberg and New York.

Sakkebaek, N. E., Philip, J. and Rafaelson, O. J. (1968). L S D in mice abnormalities in meiotic chromosomes. *Science*, **160**, 1246.

Sandberg, A. A. (1966). The chromosomes and causation of human cancer and leukemia. *Cancer Res.*, **26**, 2064–2081.

Sandberg, A. A., Cortner, J., Takagi, N., Moghadan, M. A. and Crosswhite, L. H. (1966). Differences in chromosomes constitution of twins with acute leukemia. *New Eng. J. Med.*, **275**, 809–812.

Schmitt, J. and Ulrich, F. (1968). Die chromosomen der wilden Caprini Simpson, 1945. *Z Säugetierk.* **33**, 180.

Sinha, A. K. (1967). Spontaneous occurrence of tetraploidy and near haploidy in mammalian peripheral blood. *Exp. Cell Res.*, **47**, 443–448.

Sonnenschein, C. and Smilov, S. B. (1966). Karyotypic analysis of an epithelial like cell line derived from normal adult human gingiva. *Exp. Cell Res.*, **43**, 633.

Spriggs, A. I., Boddington, M. M. and Clarke, C. M. (1962). Chromosomes of human cancer cells. *Brit. Med. J.*, **2**, 1431–1435.

Taylor, K. M., Hungerford, D. A., Snyder, R. L. and Ulmer, F. A. (1968). Uniformity of karyotypes in the Camelidae. *Cytogenetics*, **7**, 8–15.

Taylor, K. M., Hungerford, D. A. and Snyder, R. L. (1969). Artiodactyl mammals: their chromosome cytology in relation to patterns of evolution. *In* "Comparative Mammalian Cytogenetics", p. 346–356. (K. Benirschke, ed.) Springer Verlag, Berlin, Heidelberg and New York.

Tjio, J. H. and Levan, A. (1955). See Leonard (1964).

Ulrich, F. and Schmitt, J. (1969). Die chromosomen von Okapia Johnstoni (Sclater, 1901). *Acta Zool. Path. (Antwerp)*. **49**, 123–124.

Weber, W. T., Nowell, P. C. and Hare, W. C. D. (1965). Chromosome studies of a transplanted and a primary canine venereal sarcoma. *J. Natn. Cancer Inst.*, **35**, 537–547.

White, L. and Cox, D. (1967). Chromosome changes in a rhabdomyosarcoma during recurrence and in cell culture. *Brit. J. Cancer*, **21**, 684–693.

Wurster, D. H. (1969). Cytogenetic and phylogenetic studies in Carnivora. *In* "Comparative Mammalian Cytogenetics", pp. 310–329 (K. Benirschke, ed.) Springer Verlag, Berlin, Heidelberg and New York.

Wurster, D. H. and Benirschke, K. (1968). Chromosome studies in the superfamily Bovoidea. *Chromosoma (Berl.)*, **25**, 152–186.

Yamada, K., Takagi, N. and Sandberg, A. Z. (1966). Chromosomes and causation of human cancer and leukemia. II. Karyotypes of solid tumors. *Cancer*, **19**, 1879–1890.

Yerganian, G. and Leonard, M. J. (1961). Maintenance of normal *in situ* chromosomal features in long terme tissue cultures. *Science*, N.Y. **133**, 1600–1601.

Zang, K. D. and Singer, H. (1967). Chromosomal constitution of meningiomas. *Nature (Lond)* **216**, 84–85.

Zur Hausen, H. (1967). Chromosomal changes of similar nature in seven established cell lines derived from the peripheral blood of patients with leukemia. *J. Natn. Cancer. Inst.*, **38**, 683–696.

Section IV

PHYLOGENETIC AND TAXONOMIC INTERPRETATION OF THE KARYOLOGICAL DATA

8

The Chromosomes of Lower Chordates

DARIO COLOMBERA
Institute of Animal Biology, University of Padua, Italy

I. INTRODUCTION

If we consider that the tadpole larva is believed to be an ancient Ascidian creation (Van Name, 1921; Garstang, 1921; Berrill, 1950) and we accept the hypothesis that Vertebrates originated from an ancient tadpole larva by becoming neotenic (Garstang, 1928; Berrill, 1955), an investigation of Ascidian chromosomes may be a good start for an analysis of the obscure origin of Chordates and Vertebrates.

Therefore in 1968, a research on Ascidian chromosomes was started in our laboratory and studies on the chromosomes of pelagic Tunicates and Cephalochordates are actually under way.

In this paper all the available data on lower Chordate chromosomes are briefly reviewed and their possible taxonomical significance considered.

II. CHROMOSOME NUMBERS

There seemed to be very high variability in chromosome numbers within the class Ascidiacea (Minganti, 1956); diploid numbers ranging from the exceptionally low value of $2n = 4$ for *Dendrodoa grossularia* up to $2n = 32$ for some Stolidobranchiata. Differences were reported in the chromosome numbers of different geographical populations of the same species too, so that suggestions were advanced about the presence of geographical races possessing different chromosome numbers (Minganti, 1956; Colombera, 1969). But, when the doubtful cases were checked (Colombera, 1970a, b), the same results were obtained from all the populations of the same species.

Our present data indicate that the chromosome numbers of Ascidians (Table I) are all near the diploid values 12, 18, 22, 32 and 40, and in consequence the hypothesis that tetraploidy played a role in Ascidian evolution

TABLE I. *Chromosome numbers of Ascidians*

	Species	2*n*	Reference
Ord. Enterogona Subord. *Aplousobranchiata*			
Fam. Clavelinidae	*Distaplia occidentalis*	24	Bancroft, 1899
	Distaplia occidentalis	>24	Minganti, 1967b
	Clavelina lepadiformis	18	Colombera, 1968
Fam. Polyclinidae	*Aplidium proliferum*	28	Zwillenberg and Zwillenberg, 1954
	Aplidium nordmanni	40	Colombera
	Sydnium elegans	18	Zwillenberg and Zwillenberg, 1954
	Sydnium turbinata	40[a]	Colombera
Fam. Didemnidae	*Trididemnum tenerum*	8	Salensky, 1895
	Trididemnum tenerum	30	Colombera
	Didemnum candidum	30	Colombera
	Diplosoma listerianum	30	Colombera and Sala, in preparation
Subord. *Phlebobranchiata*			
Fam. Cionidae	*Ciona intestinalis*	18	Boveri, 1890
	Ciona intestinalis	12	Morgan, 1942
	Ciona intestinalis	28	Taylor, 1967
Fam. Ascidiidae	*Ascidiella aspersa*	16	Minganti, 1952
	Ascidiella aspersa	18	Colombera, 1970b
	Ascidiella scabra	18	Colombera, 1968
	Ascidiella patula	18	Floderus, 1896
	Ascidia mentula	18	Boveri, 1890
	Ascidia paratropa	18	Minganti, 1967a
	Ascidia callosa	18	Minganti, 1967a
	Ascidia virginea	18	Colombera
	Ascidia malaca	16	Minganti, 1952
	Ascidia conchilega	16	Colombera
	Phallusia mammillata	16	Hill, 1895
Fam. Corellidae	*Corella willmeriana*	12	Minganti, 1967b
	Corella parallelogramma	12	Colombera
Ord. Pleurogona Subord. *Stolidobranchiata*			
Fam. Styelidae	*Styela partita*	16	Minganti, 1967b
	Styela gibsti	32	Minganti, 1967b
	Styela plicata	32	Minouchi, 1936
	Styela coriacea	32	Colombera
	Botryllus schlosseri	32	Colombera, 1963
	Botrylloides leachi	32	Colombera, 1967
	Dendrodoa grossularia	4	Julin, 1893

With some reserve.

	Species	2*n*	Reference
	Dendrodoa grossularia	40–42	Colombera
	Polycarpa pomaria	32	Colombera
Fam. Pyuridae	*Pyura microcosmus*	16	Zwillenberg and Zwillenberg, 1954
	Pyura haustor	22	Minganti, 1967b
	Pyura squamulosa	22	Colombera
	Boltenia villosa	32	Minganti, 1967b
Fam. Molgulidae	*Molgula manhattensis*	16	Crampton, 1898

(Minganti, 1956; Colombera, 1969) may be supported. But, if Ascidian phylogeny is considered (Fig. 1), there seem to be some tendencies toward

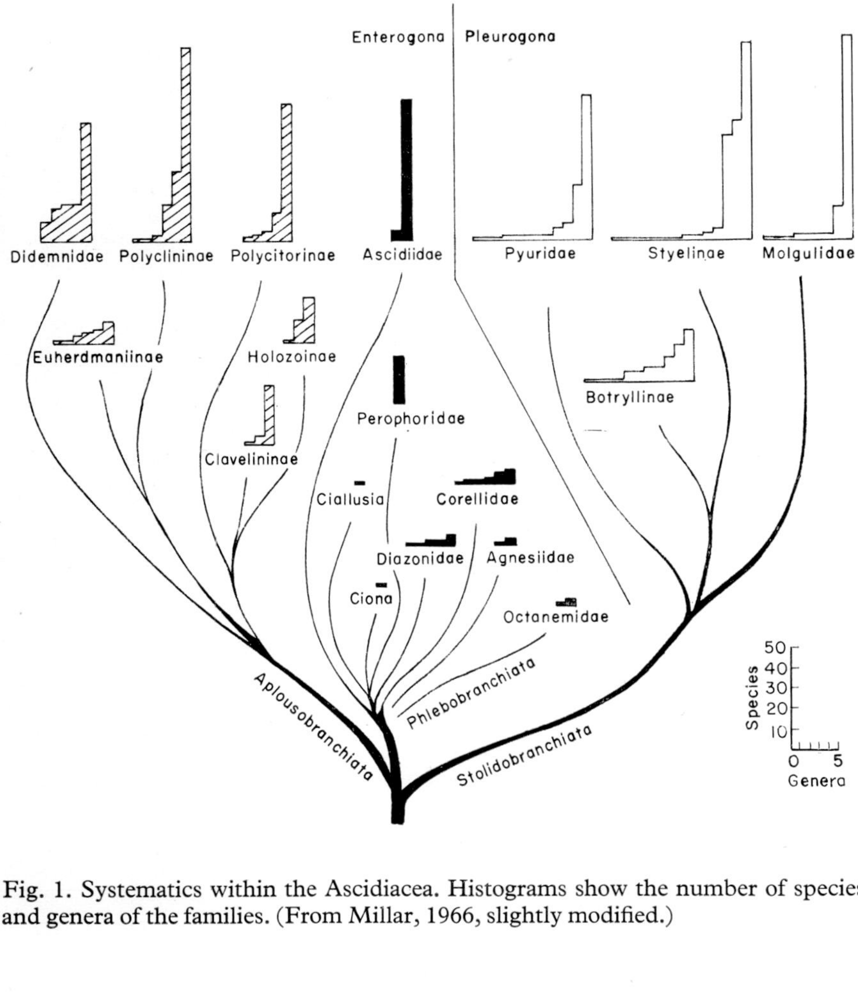

Fig. 1. Systematics within the Ascidiacea. Histograms show the number of species and genera of the families. (From Millar, 1966, slightly modified.)

reduction of chromosomes too: for instance, within Phlebobranchiata the diploid number $2n = 28$ is found in *Ciona intestinalis*, which, in some respects, is an ancestral form (Berrill, 1955), whereas the successful families of Ascidiidae and Corellidae possess $2n = 16$ or 18 and $2n = 12$ respectively. When evolution is accompanied by an increase in chromosome numbers, as it seems happened with Polyclinidae, a process of polyploidization is probably not involved. On the contrary, a polyploidization might have occurred within the family of Styelidae.

The chromosomes of pelagic Tunicates do not show the great variations found within Ascidiacea and, if the oldest reports (Table II) are ignored, as they seem to deserve, only the diploid values 8 and 16 for Larvacea and 22, 24 and 26 for Thaliacea are present.

TABLE II.

	Species	$2n$	Reference
Cephalochordata	*Amphioxus lanceolatus*	24	Cerfontaine, 1905
Thaliacea			
Ord. Pyrosomida	*Pirosoma giganteum*	8	Salensky, 1891
	Cyclosalpa pinnata	12	Todaro, 1886
	Cyclosalpa pinnata	8	Stier, 1938
Ord. Salpida	*Salpa maxima*	12	Todaro, 1886
	Salpa maxima	26	Colombera
	Salpa confoederata	8	Salensky, 1917
	Pegea (Salpa) confoederata	24[a]	Colombera
	Salpa fusiformis	10	Leloup, 1929
	Salpa fusiformis	24	Colombera
	Thalia democratica	24[a]	Colombera
Ord. Doliolida	*Doliolum denticulatum*	22	Colombera
Larvacea			
Ord. Copelata	*Oikopleura vanhoeffeni*	6	Salensky, 1903
	Oikopleura rufescens	7	Salensky, 1904
	Oikopleura dioica	6	Ihle, 1908 Koerner, 1952
	Oikopleura dioica	16[a]	Colombera and Fenaux
	Oikopleura albicans	16	Fenaux, 1963
	Oikopleura fusiformis	16	Colombera and Fenaux
	Oikopleura longicauda	16	Colombera and Fenaux
	Fritillaria pellucida	8	Colombera and Fenaux

[a] With some reserve.

If the chromosome numbers of Ascidiacea, Larvacea, Thaliacea and Cephalochordata are compared, Larvacea seem to be closer to Ascidiacea than to Thaliacea, whereas Cephalochordata and Thaliacea are unexpectedly akin. The "basic" number of lower Chordate chromosomes seems to be $2n = 24$ or close to 24.

III. THE TOTAL AMOUNT OF CHROMOSOMAL DNA

The few minute chromosomes of Larvacea (Colombera and Fenaux, in preparation) indicate a reduction of chromosomal DNA in the passage from the ascidian organization to the larvacean one (Van Name, 1921; Garstang, 1921; Berrill, 1955).

The analysis of the DNA content in two Ascidians, a Cephalochordata and two Ciclostomata (Taylor, 1967; Atkin and Ohno, 1967) indicates that the genome of the organism which served as ancestral stock of Vertebrates probably possessed a small amount of DNA, and that the passage from one kind of organization to a more complex one is accompanied by gene duplication both by regional duplication of chromosomal segments and by polyploidization (Atkin and Ohno, 1967).

IV. MITOTIC CHROMOSOMES

The mitotic chromosomes of Ascidians are characterized by the absence of heterochromosomes. The chromosome dimensions do not include the extremes found in Vertebrates, with the exception of *Clavelina lepadiformis* which possesses very minute chromosomes.

At present, data are available in the morphology of mitotic chromosomes of *Botryllus schlosseri*, *Ascidiella aspersa*, *Clavelina lepadiformis*, *Phallusia mammillata* (Colombera, 1970b) and of *Ciona intestinalis*, *Dendrodoa grossularia* and *Diplosoma listerianum* (Colombera, 1972). According to these data, good correlation is already evident between number and morphology of chromosomes: low chromosome numbers are correlated with metacentric chromosomes, high numbers of chromosomes are, on the contrary, correlated with both matacentric and subtelocentric and/or acrocentric chromosomes. Hence, reconsidering the possible trends of chromosome evolution within Ascidians, it would seem to be possible that there was a contemporaneous decrease in the number together with an increase in the symmetry of chromosomes, instead of an augmentation of asymmetry, as usually occurs when chromosome numbers lessen (Rieger *et al.*, 1968).

The morphology of mitotic chromosomes of Larvacea (Colombera and

Fenaux, in preparation) and Thaliacea (Colombera, in preparation) seem to be closely related to that of Ascidiacea. Considering the data actually at hand, Tunicate chromosomes seem to be very different from those of Cyclostomata, which are numerous and, with a single exception (Potter *et al.*, 1968), mainly with terminal or subterminal centromere (Howell and Duckett, 1971; Potter and Rothwell, 1970; Howell and Denton, 1969; Sasaki and Hitotusmachi, 1967; Taylor, 1967; Nogusa, 1960) and also from those very numerous of the lowest pisces (Matthey, 1949).

V. MEIOTIC CHROMOSOMES

Three distinct types of oöcyte bivalents are distinguishable within Ascidians: the *Botryllus schlosseri* type, that of *Ciona intestinalis* and that of *Clavelina lepadiformis*.

Botryllus schlosseri bivalents (Fig. 3) consist usually of two closely associated well stained roundish bodies. Analogous bivalents have been found in some Styelidae (Minganti, 1967b).

Ciona intestinalis bivalents are slightly condensed, difficult to stain and regularly clumped together. Similar bivalents have been found only in the families Cionidae and Ascidiidae (Fig. 4) (Colombera, 1970b).

Clavelina lepadiformis bivalents (Fig. 5) are well contracted and stainable, and the number and position of chiasmata remain distinguishable.

The *Ciona intestinalis* type of bivalents are found only within the Suborder Phlebobranchiata, and as far as Ascidian evolution is concerned (Millar, 1966), they also correlated with the most ancient Ascidians. In the relatively more evolved family Ascidiidae, bivalents resemble *Ciona* bivalents, but are more condensed and less sticky (Fig. 4).

The spermatocyte bivalents, so far known for *Botryllus schlosseri* (Colombera, 1969) (Fig. 6), *Botrylloides leachi* (Colombera, 1970a), *Clavelina*

Fig. 2. Pachitene chromosomes in the testis of *Diplosoma listerianum*. (Colombera and Sala, a.)
Fig. 3. Oocyte bivalents of *Botryllus schlosseri*. (Colombera, 1969.)
Fig. 4. Oocyte bivalents of *Ascidia mentula*.
Fig. 5. Oocyte bivalents of *Clavelina lepadiformis*.
Fig. 6. Spermatocyte bivalents of *Botryllus schlosseri*.
Fig. 7. Oocyte bivalents of *Ascidiella scabra*.
Fig. 8. Spermatocyte bivalents of *Clavelina lepadiformis*.
Fig. 9. Spermatocyte bivalents of *Diplosoma listerianum*. (Colombera and Sala, a.)
Fig. 10. Spermatocyte bivalents of *Ciona intestinalis*. (Colombera and Sala, b.)
Fig. 11. Metaphase chromosomes in dividing eggs of *Botryllus schlosseri*. Daughter kinetochores are well separated from prometaphase.

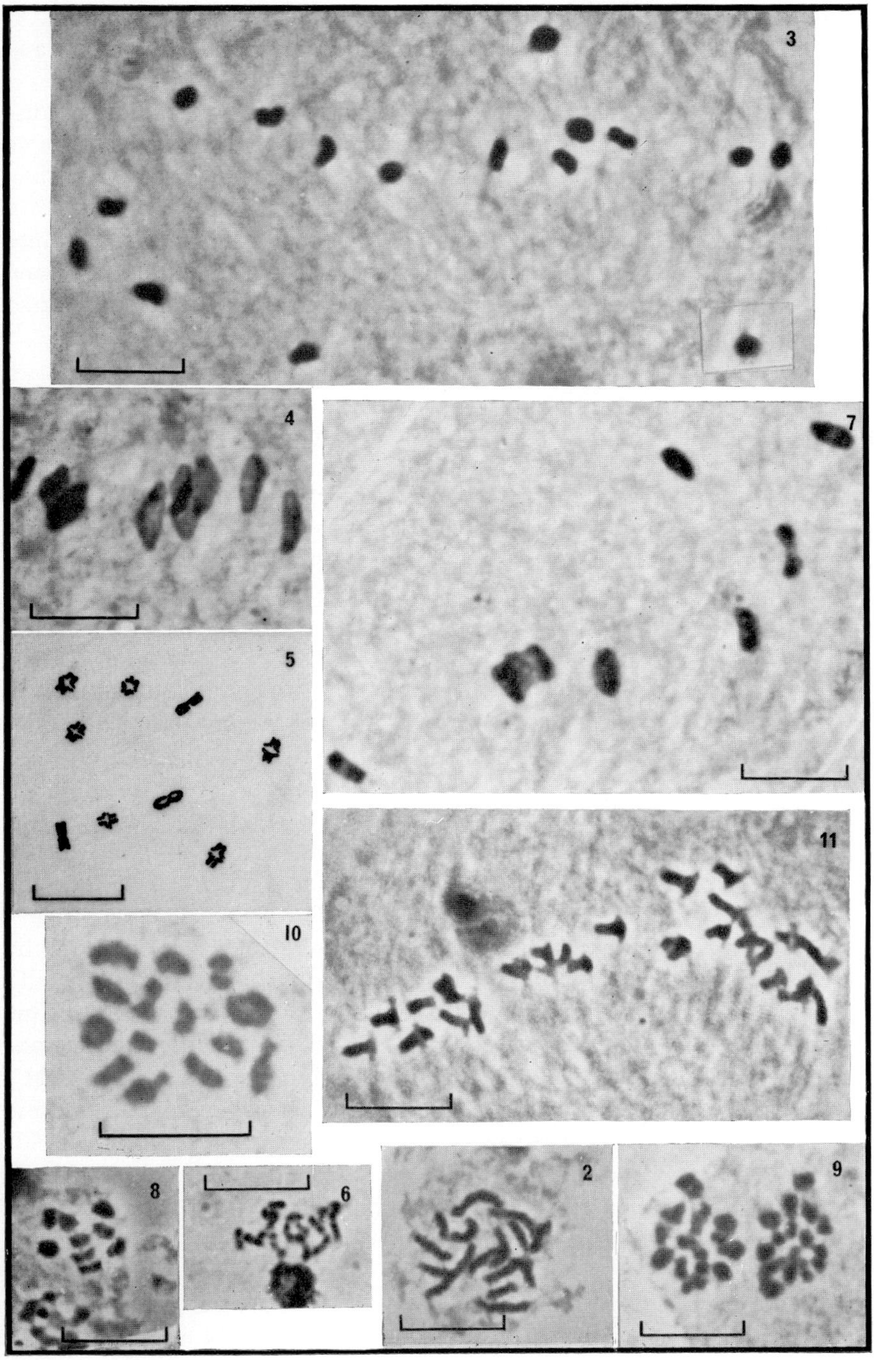
3
4
7
5
11
10
8
6
2
9

lepadiformis (Fig. 8) (Colombera, 1971a), *Diplosoma listerianum* (Fig. 9) (Colombera and Sala, 1972a) and *Ciona intestinalis* (Fig. 10) (Colombera and Sala, 1972b) are far more condensed than oöcyte bivalents.

At present two different types of spermatocyte bivalents can be distinguished: those of *Botryllus* and *Botrylloides* where the bivalents are so condensed that eventual chiasmata are not detectable and those of *Clavelina, Ciona and Diplosoma* where chiasmata, often terminalized, are evident.

The striking differences in the morphology of meiotic chromosomes found in the ambit of the class Ascidiacea are, in my opinion, the expression of a long evolutionary history which sometimes cannot be fully appreciated at a higher morphological level.

VI. THE EMBRYONIC MITOSES

The time spent by dividing eggs and embryos in the mitotic cycle is strictly correlated to condition of viviparity or oviparity. Viviparous species, mainly colonial forms, have eggs that divide very slowly, whereas oviparous species, all of which are solitary forms, have external fertilization and the eggs divide very fast (Colombera, 1970b).

If Ascidian evolution proceeds from colonial forms toward solitary individuals (Millar, 1966), the time necessary for the development of the swimming larva from the fecundated egg can be considered as a measure of this evolutionary trend.

VII. THE PRECOCIOUS SEPARATION OF DAUGHTER KINETOCHORES

Another very interesting characteristic of Ascidian chromosomes is the precocious separation of daughter kinetochores during prometaphase and metaphase, found so far without exception (Colombera, 1970b). This separation of daughter kinetochores is clearly visible because the chromosome arms are forcibly stretched in the kinetochore regions (Fig. 11). Although precocious centromeric separation has been reported for mitotic chromosomes of other organisms as well (Bajer, 1958; Davies, 1956; Cooper, 1951), the persistence of a repulsion between daughter kinetochores, which does not lessen at metaphase and is regularly present only in dividing eggs and young embryos makes this character very peculiar (Colombera, 1971b, c, d).

Considering the hypothesis (Matthey, 1945) that precocious centromeric activity may indicate a predisposition toward polyploidy and considering that morphologically differentiated sex chromosomes are absent among Ascidians, that in the genus *Styela* the two diploid numbers 16 and 32 are present and that tetraploid embryos have been experimentally obtained within the suborder Stolidobranchiata and remembering the increment of chromosomal DNA from Ascidians toward lower Vertebrates, there are many indications that polyploidy may have played a role in Ascidian as in lower Vertebrates evolution. The significance of the precocious repulsion between daughter kinetochores as a taxonomical parameter is still to be determined, but its constant presence in all the species investigated, representing 5 families and all the suborders of Ascidians, indicates its great antiquity. It is also reasonable to hope that precocious centromeric repulsion will be useful in comparison of Ascidians with other Chordates.

VIII. CONCLUSIONS

The present review of lower Chordate chromosomes shows how much Tunicates and Cephalochordates are ignored at a karyological level: in fact only the numbers of chromosomes of about 50 Tunicates and one Cephalochordata are actually known and behaviour and morphology of chromosomes are practically unknown outside the class Ascidiacea.

Within Ascidians, despite the low number of species investigated, about 2.5% of the known species, there are substantial indications that the morphology of mitotic and meiotic chromosomes, and the dynamics of embryonic mitotic divisions are all valuable taxonomical parameters. Speaking of the cytotaxonomical parameters of Ascidians, I would like to mention the experiments of interspecific hybridization (Minganti, 1950, 1952, 1955, 1967) which proved to be very effective in examining Ascidian relationships at the lowest levels.

In the light of a recent paper (Howell, W. M. & Boschung, H. T., Jr. (1971): Chromosomes of the lancelet *Branchiostoma floridae*, order Amphioxi. *Experientia* 27: 1495–1496.) and of the new results obtained in our laboratory on the *Branchiostoma lanceolatum* and about 20 species of Ascidians, the considerations on the phylogeny of lower Chordates presented in this paper should need to be reviewed.

REFERENCES

Atkin, N. B. and Ohno, S. (1967). *Chromosoma*, **23**, 10–13.
Bajer, A. (1958). *Exp. Cell Res.*, **14**, 245–256.

Bancroft, F. W. (1899). *Bull. Mus. Comp. Zool. Harv.*, **35**, 59–112.
Berrill, N. J. (1950). "The Tunicata" Ray Society, London.
Berrill, N. J. (1955). "The Origin of Vertebrates." Clarendon Press, Oxford.
Boveri, T. (1890). *Jena Z.*, **24**, 314.
Cerfontaine, L. (1905). *Arch. Biol.*, **22**, 229–418.
Colombera, D. (1963). *Ricerca. Scient.*, **33**, 443–448.
Colombera, D. (1967). *Boll. Zool.*, **34**, 105.
Colombera, D. (1968). *Boll. Zool.*, **35**, 446.
Colombera, D. (1969). *Caryologia*, **22**, 339–350.
Colombera, D. (1970a). *Caryologia*, **23**, 113–115.
Colombera, D. (1972). V°E.M.B.S., (ed. Piccin, Padova), Venice.
Colombera, D. (1971a). *Mar. Biol.*, **11**, 149–151.
Colombera, D. (1971b). *Caryologia*, **24**, 33–39.
Colombera, D. (1971c). *Caryologia*, **24**, 59–64.
Colombera, D. (1971d). *Caryologia*, **24**, 65–70.
Colombera, D. in preparation.
Colombera, D. and Fenaux, R. in preparation.
Colombera, D. and Sala, M. (1972a). *Caryologia*, 25.
Colombera, D. and Sala, M. (1972b). *Caryologia*, 25.
Cooper, K. W. (1951). *J. Morph.*, **89**, 37–59.
Crampton, H. E. (1898). *Am. Nat.*, **32**, 126.
Davies, D. R. (1956). *Chromosoma*, **8**, 221–228.
Fenaux, R. (1963). *Vie Milieu, suppl. Vol.* 16, **VIII**, 142 p.
Floderus, M. (1896). *Z. Wiss. Zool.*, **61**, 163–260.
Garstang, W. (1921). *J. Linn. Soc. Zool.*, **35**, 81–392.
Garstang, W. (1928). *Q. Jl. Micra. Sci.*, 72, 51–187.
Hill, B. A. (1895). *Q. Jl. Micra. Sci.*, **150**, 315–330.
Howell, W. M. and Denton, T. E. (1969). *Copeia*, **1969**, 393–395.
Howell, W. M. and Duckett, C. R. (1971). *Experientia*, **27**, 222–223.
Ihle, J. E. W. (1908). *Siboga-Exped.*, **56c**, 123 p.
Julin, C. (1893). *Bull. Scient. Fr. Belg.*, **25**, 93–154.
Koerner, W. F. (1952). *Z. Morph. Ökol. Tiere.*, **41**, 1–53.
Leloup, E. (1929). *Bull. Acad. R. Belg. Cl. Sci.*, **15**, 461–477.
Matthey, R. (1945). *Revue Suisse Zool.*, **52**, 1–109.
Matthey, R. (1949). "Les Chromosomes des Vertébrés." Rouge, Lausanne.
Millar, R. H. (1966). "Evolution in Ascidians", H. Barnes, London.
Minganti, A. (1950). *Pubbl. Staz. Zool. Napoli*, **22**, 293–308.
Minganti, A. (1952). *Rc. Acad. Naz. Lincei*, s. 8, **12**, 125–127.
Minganti, A. (1955). *Ricerca. Scient.*, **25**, 2625–2630.
Minganti, A. (1956). *Boll. Zool.*, **23**, 299–315.
Minganti, A. (1967a). *Boll. Zool.*, **34**, 142–143.
Minganti, A. (1967b). Personal communication.
Minouchi, O. (1936). *Z. Zellforsch. Mikrosk. Anat.*, **23**, 790.
Morgan, T. H. (1942). *J. Exp. Zool.*, **90**, 199–228.
Nogusa, S. (1960). *Mem. Hyogo Univ. Agric.*, **3**, 1–62.
Potter, I. C., Robinson, E. S. and Walton, S. M. (1968). *Experientia*, **24**, 966–967.

Rieger, R., Michaelis, A., Green, M. M. (1968). "A Glossary of Genetics and Cytogenetics", Springer Verlag, Berlin, Heidelberg and New York.
Salensky, W. (1891). *Zool. Jber. Abt. Anat.*, **5**, 1–98.
Salensky, W. (1895). *Mitt. Zool. Stn. Neapel*, **11**, 488–630.
Salensky, W. (1903). *Mém. Acad. Sci. Petersbourg*, s. 8, **13**, n. 7, 1–14.
Salensky, W. (1904). *Mém. Acad. Sci. Petersbourg*, s. 9, **15**, n. 1, 45–150.
Salensky, W. (1917). *Bull. Acad. Sci. Petrograd*, **11**, 219–240.
Sasaki, M. and Hitotsumachi, S. (1967). *Chromosome Inf. Serv.*, **8**, 22–24.
Stier, A. (1938). *Z. Morph. Ökol. Tiere*, **33**, 582–630.
Todaro, F. (1886). *Atti Acad. Naz. Lincei Rc.*, **4**, 641.
Taylor, K. M. (1967). *Chromosoma*, **21**, 181–188.
Van Name, W. G. (1921). *Bull. Am. Mus. Nat. Hist.*, **44**, 274–282.
Zwillenberg, L. O. and Zwillenberg, H. H. L. (1954). *Nature (Lond.)*, **173**, 876–877.

9

The Chromosomes of the Cyclostomes

I. C. POTTER
School of Biological Sciences, University of Bath, England
and
E. S. ROBINSON
School of Biological Sciences, Macquarie University, North Ryde, Australia

I. INTRODUCTION

The fossil record shows that the earliest vertebrates, the ostracoderms, lacked jaws. The jawed vertebrates (gnathostomes) evolved from this ostracoderm stock and the descendants of this mainstream of vertebrate evolution are the fishes, amphibians, reptiles, birds and mammals. The only other vertebrates to have survived until the present day are the lampreys and hagfishes, which have often been placed together in the Class Cyclostomata. Although there is some reason for regarding this as an unnatural taxonomic grouping (Berg, 1958; Adam and Strahan, 1963), it seems useful to retain the term cyclostome for these sole two extant agnathan groups (Hubbs and Potter, 1971).

As with the gnathostomes, problems of taxonomic and phylogenetic relationships exist within and among many cyclostome taxa. Morphological criteria will, of course, remain the most powerful taxonomic tool in the classification of fossils but, in the living Agnatha, other criteria, such as biochemical characters, may also be used to shed light on affinities. While the latter type of approach has been widely employed in some other vertebrates, notably the birds, there have been few comparable detailed studies on the cyclostomes. Recently, the restricted cytogenetic data on lampreys and hagfishes has however, been considered in conjunction with morphological criteria in an attempt to help clarify relationships (Robinson and Potter, 1969; Hubbs and Potter, 1971; Potter and Robinson, 1971).

In what ways and at what levels can cytotaxonomy be expected to contribute to a better understanding of cyclostome relationships? In a more restricted and pragmatic sense, cytotaxonomy might assist in taxonomic discrimination where morphological and other criteria have proved to be inadequate. Broader comprehensive cytotaxonomic studies on cyclostomes may also ultimately contribute, however, to a rational picture of genome evolution in the lower vertebrates. Included in such a picture should be not only data on changes in chromosome number and form but also information on genome size as a reflection of DNA content. However, any hypothesis of early genome evolution in the lower vertebrates must be formulated with considerable caution for the hazards are in some ways similar to those of morphology—namely, whether particular traits reflect primitive states or specialized conditions.

In this article, an attempt is first made to outline the currently accepted views on cyclostome taxonomy and phylogenetic relationships based largely on morphology. The meagre literature on the karyotypes of cyclostomes will then be reviewed and its bearing on taxonomy and phylogeny tentatively assessed.

II. RELATIONSHIPS OF FOSSIL FORMS

The first firm evidence of a relationship between the living cyclostomes and the ostracoderms of the Ordovician, Silurian and Devonian periods was provided by the classical palaeontological studies of Stensiö (1927, 1932). He holds the view that the lampreys evolved from cephalaspidomorph ostracoderms, a group containing the cephalaspids and anaspids. Although some authors (Denison, 1960, 1964; Heintz, 1963) contend that lampreys evolved specifically from anaspid stock, Stensiö (1968) considers, particularly on the basis of the number of prebranchial myomeres and the loss of anterior gills in this group, that they were too specialized to provide such an ancestry. There is thus broad agreement on the general stock from which lampreys evolved, but some dispute as to the precise ancestor and the time at which the group diverged.

In the case of hagfishes, there is a much more fundamental divergence of opinion. Stensiö (1932, 1968) maintains that the lampreys and hagfishes are diphyletic with the latter evolving independently from the Heterostraci, which belong to the other major division of agnathans, the Pteraspidomorphi. Several authors (Denison, 1960, 1964; Tarlo, 1961, 1962; Heintz, 1963) are, however, of the opinion that the hagfishes cannot be derived from a group exhibiting a diplorhinal condition and regard the hagfishes, like the lampreys, as having evolved from an anaspid-like vertebrate.

In the general context of cyclostome phylogeny, two recent works are of significance. The first concerns the discovery in the mid-Pennsylvanian of Illinois of the first fossil lamprey *Mayomyzon pieckoensis* (Bardack and

Zangerl, 1968, 1971). Unfortunately this exciting discovery does not help greatly in resolving the problems of relationships outlined above since its morphology is remarkably similar to living lampreys. This does, however, imply a conservative rate of evolution within the group during the subsequent 280 million years and indicates a considerably earlier separation of the hagfish lineage. A second and important recent work is the analysis, by Ritchie (1968), of the mid-Silurian ostracoderm *Jamoytius kerwoodi*. This anaspid-like fossil was shown to have several features that would fulfil the requirements of a basal stock from which the lampreys and hagfishes could have diverged. It would also place the time of divergence as rather earlier than may have been generally envisaged by the proponents of a cephalaspidomorph origin.

III. SYSTEMATICS OF LIVING CYCLOSTOMES

Lampreys are found in the temperate and cold temperate river systems of both the Northern and Southern Hemispheres (Lanzing, 1959; Fontaine, 1958; Potter and Strahan, 1968; Hubbs and Potter, 1971). Although it is often assumed that after metamorphosis, lampreys characteristically migrate to sea to feed primarily on teleost fishes, only 9 of the thirty-one species undergo an anadromous migration (Hubbs and Potter, 1971). Some of these species, such as *Petromyzon marinus* and *Lampetra fluviatilis*, have given rise to landlocked parasitic derivatives but there are several parasitic species which are entirely restricted to fresh water, such as those of the genera *Ichthyomyzon*, *Eudontomyzon* and *Tetrapleurodon*. In contrast to these parasitic lampreys are the nonparasitic species which do not feed after the onset of metamorphosis and reach sexual maturity some 6 to 9 months after transformation. In most examples, each of these nonparasitic species can be seen to be closely related to a particular parasitic lamprey similar to that from which it has probably been derived (Hubbs, 1925). The term "paired species" was applied by Zanandrea (1959) to each of these ancestral and derivative forms.

The various schemes for the taxonomic assignment of lamprey genera at the familial and sub-familial levels were reviewed by Potter and Strahan (1968) following studies on the Southern Hemisphere species which had previously received rather incomplete treatment. These authors recognized three major divisions: one comprising all holarctic species and the other two for *Geotria* and *Mordacia* (*Caragola*) respectively, the only two Southern Hemisphere genera currently recognized (de Buen, 1961; Potter and Strahan, 1968). This tripartite division was retained by Hubbs and Potter (1971), but each group was elevated to family status: Petromyzonidae,

Geotriidae and Mordaciidae. In this account of lamprey chromosomes, the genera and species and specific names are those listed by Hubbs and Potter (1971), a scheme differing at the generic level in only two aspects from that recently proposed for the Petromyzonidae by Vladykov and Follett (1967). Thus, the genera *Entosphenus*, *Lethenteron* and *Lampetra*, as recognised by Vladykov and Follett, are here regarded as subgenera of *Lampetra*, and *Okkelbergia* is listed as a monotypic genus containing a species which in general had been previously known as *Lampetra aepyptera*.

Ichthyomyzon was regarded by Hubbs and Potter (1971) as the most primitive of all groups on the basis of its simple isomerous dentition which could have provided the basis for the more differentiated patterns found in other holarctic genera and been similar to those that gave rise to the two Southern Hemisphere groups. While the 27 species, comprising the seven holarctic genera, obviously form a relatively compact and closely inter-related group, the distinctive patterns of dentition found in *Mordacia* and *Geotria* suggest a considerable divergence of the Southern Hemisphere genera. There is, however, a basic alate arrangement to the various tooth rows in all living lampreys, particularly well exhibited by the more primitive species belonging to *Ichthyomyzon* and *Petromyzon*. It thus seems possible that the differentiated patterns of dentition in Southern Hemisphere lampreys, especially that of *Mordacia*, may reflect their origin from a group similar to primitive holarctic species. Furthermore, the degree of differentiation between *Mordacia* and *Geotria* suggests that these two genera evolved from stocks that became isolated at different times (Hubbs and Potter, 1971).

Although the hagfishes, like lampreys, have an antitropical distribution, their present environment and life cycle is very different. This group is entirely restricted to a marine environment, where they feed on dead and dying fishes and on various bottom fauna, and they do not undergo a conspicuous metamorphosis.

The taxonomy of the hagfishes was reviewed by Adam and Strahan (1963), who followed Berg (1958) in assigning the various species to two separate families Myxinidae and Bdellostomidae (Eptatretidae). In an earlier review however, the small degree of variation found among the various species was considered by Holly (1933) to merit recognition of only the single family Myxinidae. This divergence of opinion is not critical and is mainly based upon the weight given to the significance of whether the different branchial ducts open by a common or by separate apertures but none of these authors would probably dispute the relative homogeneity of the group.

Adam and Strahan (1963) listed eighteen species but this is an under-estimate as there are several others awaiting description (C. L. Hubbs, personal communication). Since there is considerable taxonomic dispute as

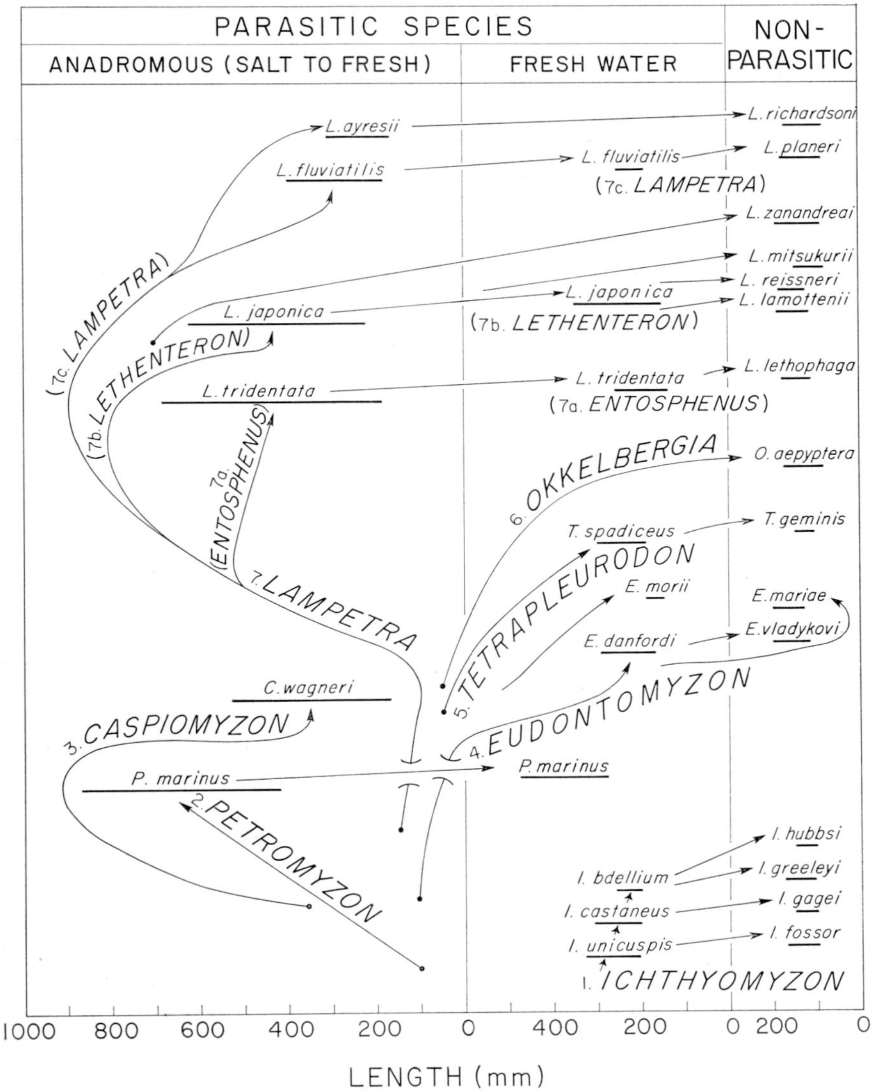

Fig. 1. Proposed phylogeny of the genera, subgenera and species of Northern Hemisphere lampreys (Petromyzonidae), with indication of trophic type, habitat and, very approximately, the usual size ranges attained just prior to spawning (from Hubbs and Potter, 1971).

The question of whether some nonparasitic species were derived from anadromous rather than from exclusively freshwater races is left open.

to which of the synonyms, *Bdellostoma* and *Eptatretus*, should be used for three species, *stoutii*, *burgeri* and *okinoseanum*, whose chromosomes have been investigated, we have elected to use the name *Eptatretus* as this appears to be the term most commonly in current use.

IV. LAMPREY KARYOTYPES

Our knowledge of lamprey karyotypes has come largely from studies carried out during the last decade (Table I). This is to a large extent due to the technical difficulties involved in preparing satisfactory spreads for counting large numbers of very small chromosomes (Potter and Robinson, 1971). With improved techniques, including the use of air-dried spreads and the application of colchicine for metaphase arrest, data have now been obtained for a number of different species. However, despite this advance in our knowledge, there is still a long way to go before we have a comprehensive picture, especially since there are a few apparently anomalous diploid numbers in the literature, and also because it is often very difficult to be absolutely certain of the location of the centromere. As can be seen from Table I, the material used in these studies has come from embryonic, larval and adult stages and from meiotic and mitotic tissues.

The earliest recorded lamprey diploid numbers for *L. fluviatilis* were those of Chubareva (1957a, b) who gave $2n \simeq 60$ but later Van Brink (in Lanzing, 1959) reported a value of "about 120" for this species. When these two counts are compared with the 156 obtained by Zanandrea and Capanna (1964), the difficulties in interpreting the results are only too apparent. *Lampetra reissneri* is the second species for which there is a similar large discrepancy. Nogusa (1960) estimated the diploid number of this species as 94–96, but Sasaki and Hitotsumachi (1967) later produced evidence for a higher value, with counts between 165 and 174.

Despite this apparent confusion, a clear pattern appears to be arising with respect to the diploid values of lampreys. Thus, recent investigations on *Okkelbergia aepyptera*, *Petromyzon marinus* (Fig. 2b) and *Ichthyomyzon gagei* (Fig. 2a) have all yielded mean diploid counts between 164 and 168 (Howell and Denton, 1969; Potter and Rothwell, 1970; Howell and Duckett, 1971). When taken in conjunction with those obtained by Zanandrea and Capanna (1964), not only for *L. fluviatilis* but also for *L. planeri* and *L. zanandreai*, and with those of Sasaki and Hitotsumachi (1967), all of which were supported by photographic documentation, there is strong evidence that counts of over 140 are characteristic of the group (Table I).

In the case of the Southern Hemisphere species of the genus *Mordacia* there is quite clearly a different situation (Table I; Potter *et al.*, 1968;

TABLE I. *A summary of current information on lamprey chromosomes (From Potter and Robinson, 1971)*

Genus	Species	Author and date	Material	Diploid number	Size Range (μm)	Centromere[a] position
Ichthymomyzon	*I. gagei*	Howell and Duckett, 1971	Gill	164	0.6–3.8	Subterminal to terminal
Petromyzon	*P. marinus*	Potter and Rothwell, 1970	Gill	168	0.7–2.7	Some metacentric, most appear acrocentric
Okkelbergia	*O. aepyptera*	Howell and Denton, 1969	Gill	164	0.8–3.0	Subterminal to terminal
Lampetra	*L. fluviatilis*	Zanandrea and Capanna, 1964	Embryo	156	0.5–1.0	—
	L. fluviatilis	Chubareva, 1957a, b,	Ovary	*c.* 60	—	—
	L. fluviatilis	Van Brink in Lanzing, 1959	Testis	*c.* 120	—	—
	L. planeri	Zanandrea and Capanna, 1964	Embryo	146	0.5–1.0	—
	L. zanandreai	Zanandrea and Capanna, 1964	Embryo	142	0.5–1.0	—
	L. reissneri	Nogusa, 1960	Testis	*c.* 94–96	—	Telomitic
	L. reissneri	Sasaki and Hitotsumachi, 1967	Kidney, gill and testis	165–174	—	Some mediocentrics
Mordacia	*M. mordax*	Potter *et al.*, 1968	Gut	76	0.8–3.5	Many metacentric or sub-metacentric
	M. praecox	Robinson and Potter, 1969	Testis	76	—	—

[a] The centromere position is as described by the individual authors.

Robinson and Potter, 1969). In the anadromous parasitic *Mordacia mordax*, metaphase spreads from gut tissue yielded a diploid value of 76 (Figs 3a, 4a, 4b), a count identical with that obtained from the testis of the non-parasitic derivative *Mordacia praecox* (Fig. 3b). Unfortunately we do not have extensive data on *Geotria australis*, the only species in the other Southern Hemisphere genus, but preliminary studies show that it has a very high diploid number resembling Northern Hemisphere lampreys (Potter and Robinson, 1971).

Several earlier workers commented on the small size and large number of the chromosomes of lampreys (Okkelberg, 1921; Montalenti, 1936), but the first photographs of metaphase spreads were not presented until the studies of Zanandrea and Capanna (1964) on three *Lampetra* species. A notable aspect of this work, which bore out the earlier observations, was that the largest chromosomes in the complement were only 1.0μm in length. These values are, however, rather smaller than the maximum lengths re-

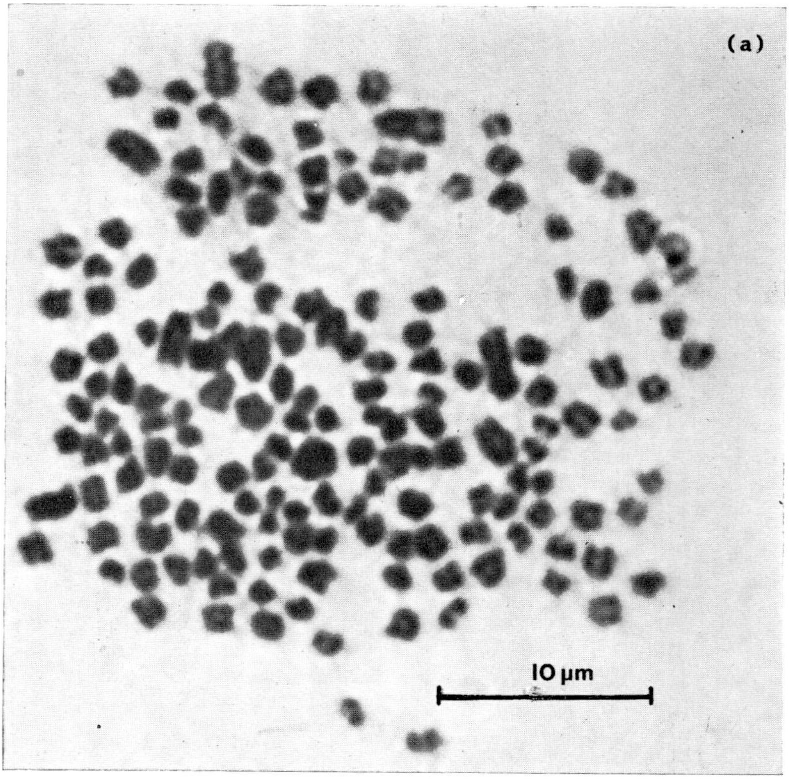

Fig. 2. Mitotic chromosomes from gill epithelium.
(a) *Ichthyomyzon gagei*, $2n = 164$ (from Howell and Duckett, 1971).

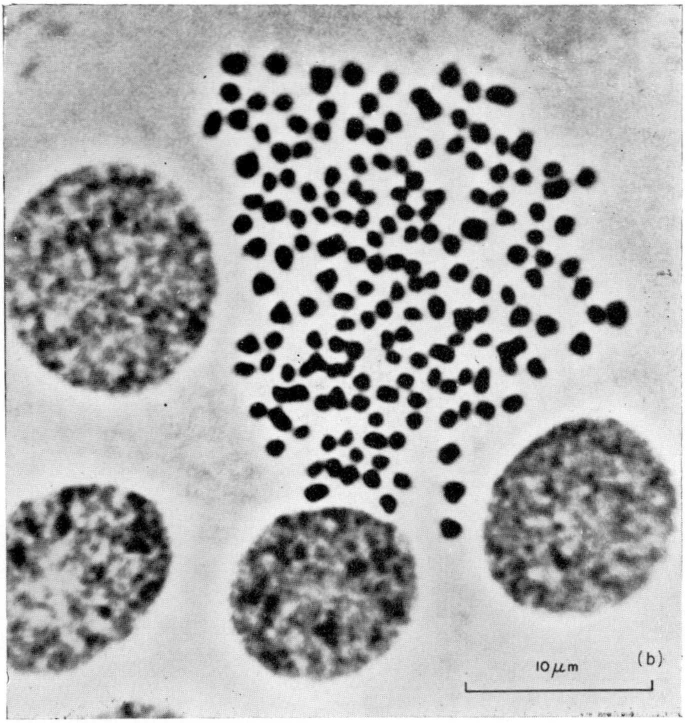

Fig. 2. (b) *Petromyzon marinus*, $2n = 168$ (from Potter and Rothwell, 1970).

corded in studies on other species, and this may either be attributable to the spreads being obtained from embryos or to the use of colchicine. However, that small chromosome size is characteristic of the group is borne out by the fact that the largest chromosome length so far recorded from mitotic metaphases is the 3.8μm in spreads from ammocoete gill tissue of *Ichthyomyzon gagei* (Howell and Duckett, 1971).

In discussing centromere position, we follow our previous review of lamprey karyotypes (Potter and Robinson, 1971) in grouping chromosomes with subterminal and terminal centromeres together under the term acrocentric since it appears generally impossible to distinguish the two categories in such small chromosomes. This feature highlights one of the major problems in reviewing lamprey karyotypes. The chromosomes are so small and numerous that often it is not only difficult to decide in which category the centromere position belongs but also to identify its presence. These problems may well provide the explanation for some of the apparent anomalies in the literature.

Nogusa (1960) in his studies on *L. reissneri* stated that the chromosomes

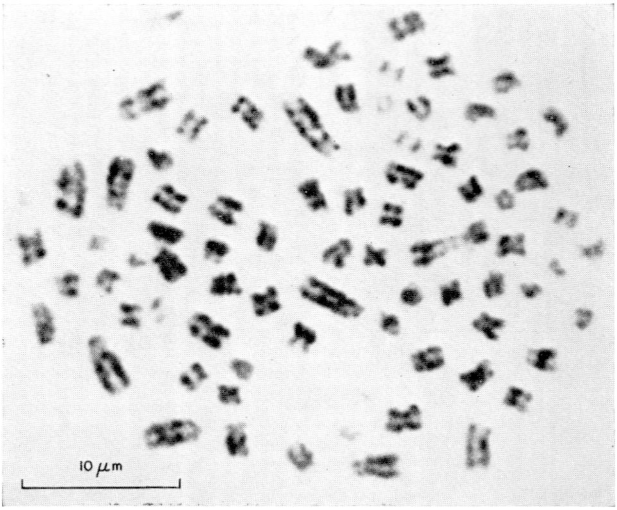

Fig. 3. (a) Mitotic metaphase chromosome spread from the gut of an ammocoete of *Mordacia mordax*, $2n = 76$, showing distinct chromatids and many showing the median position of the centromere (from Potter *et al.*, 1968).

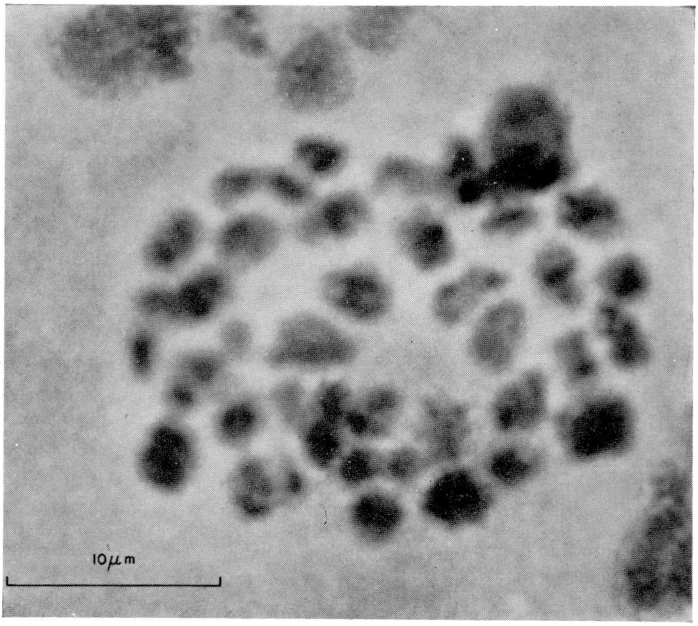

Fig. 3. (b) High power polar view of first meiotic metaphase from the testis of *Mordacia praecox* (from Robinson and Potter, 1969).

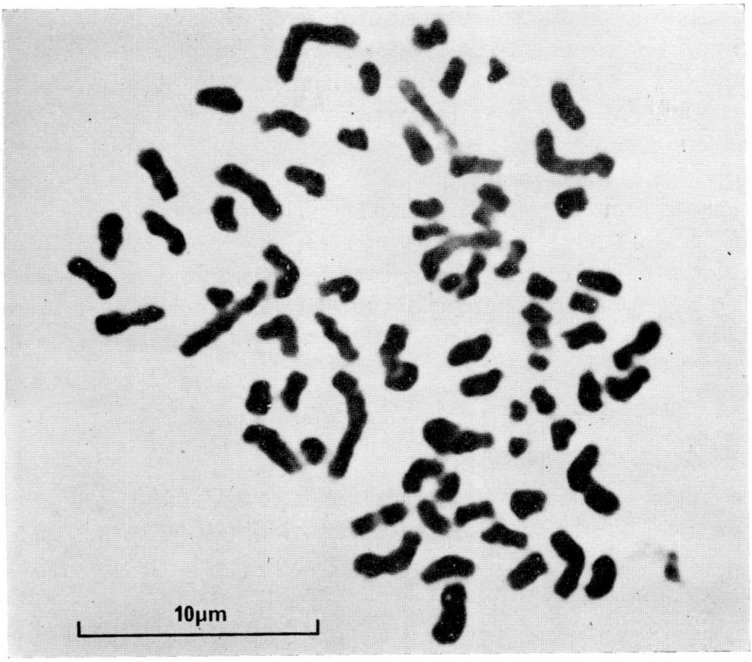

Fig. 4. Mitotic chromosomes from gut cell of *Mordacia mordax* (from Potter *et al.* 1968).
(a) Spread showing variation in length of chromosomes within a single complement.

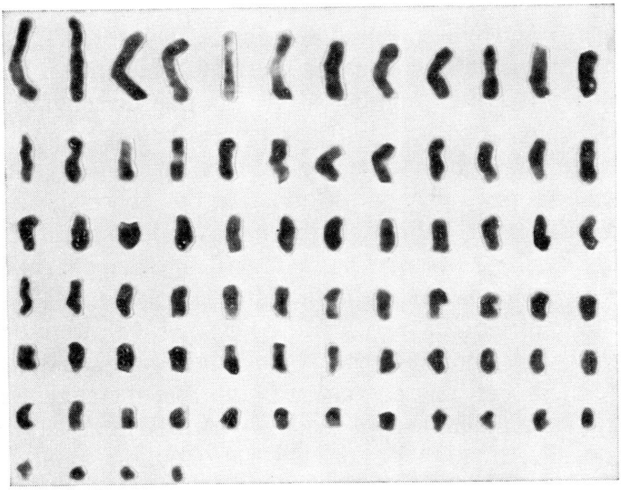

(b) Serial alignment of the chromosomes shown in 4(a).

were "telomitic", i.e. acrocentric, but in later work on the same species, Sasaki and Hitotsumachi (1967) recorded the presence of "some mediocentrics", i.e. metacentrics. The centromere position in *Okkelbergia aepyptera* and *Ichthyomyzon gagei* comes within our definition of acrocentric (Howell and Denton, 1969; Howell and Duckett, 1971) and the same is also probably true of the chromosomes of *Petromyzon marinus*, although in this latter species some meta- or submetacentric chromosomes could be identified (Potter and Rothwell, 1970).

The only estimate of the DNA content in lampreys has yielded in *L. planeri* a value of approximately 38% of that found in mammals (Atkin and Ohno, 1967).

V. HAGFISH KARYOTYPES

The initial studies on the chromosomes of hagfishes (Retzius, 1890; Schreiner and Schreiner, 1904a, b) were carried out far earlier than comparable investigations on lampreys. This is probably due to their much larger size and smaller number which might make the preparation of countable spreads more feasible. The techniques of analysing hagfish chromosomes involved the use of meiotic and mitotic material using squashes, spreads and sectioned material.

The early investigations on diploid number provided evidence that hagfishes might be characterized by the possession of approximately 50 chromosomes. Thus, Schreiner and Schreiner (1908) gave a count of 48 for *Eptatretus burgeri* and values of 50 and 52 were provided for *Myxine glutinosa* by Retzius (1890) and Schreiner and Schreiner (1904a, b). Nogusa (1960) re-investigated *Eptatretus burgeri* using spermatogonial metaphase material and confirmed the diploid number of 48. He also investigated *Eptatretus okinoseanum*, using primary and secondary spermatocytes, and found that this species had the slightly lower diploid number of 46. A fourth species *Eptatretus stoutii* formed part of a study made by Taylor (1967) on a number of lower chordates. Although the basic diploid number was 48, a varying number (1–4) of what were regarded as small supernumerary chromosomes were found in some of the spreads (Fig. 5c).

The chromosomes of *E. burgeri* and *E. okinoseanum* were described by Nogusa (1960) as isomorphic, each being short and rod-shaped with a terminally located centromere. Nogusa found that the slight variations in length amongst each complement made it virtually impossible to identify homologous pairs. Although Taylor (1967) also found a similar centromere position in *E. stoutii*, he showed that there was considerable variation in the size of the chromosomes (Figs 5, 6) with the majority of the complement being far larger than those of the larger lamprey chromosomes. Measurements of the chromosomal area showed that in *E. stoutii* this value approached that of placental mammals.

Although the majority of the chromosomes at Meiosis I were of typical

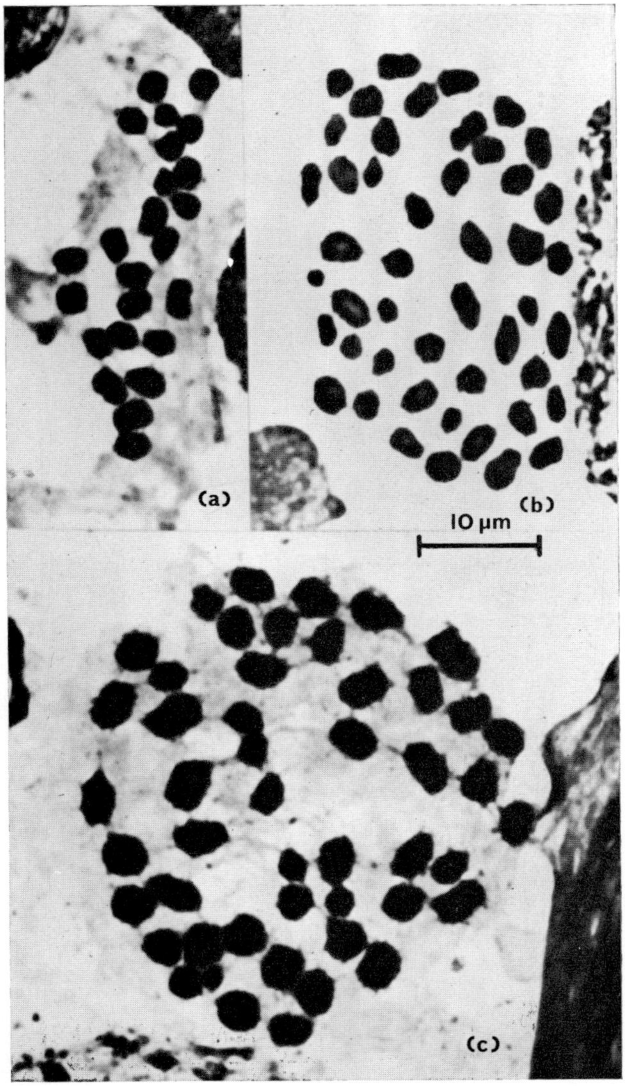

Fig. 5. Chromosomes of *Eptatretus stoutii* (from Taylor, 1967).
(a) First meiotic metaphase of a spermatocyte, $n = 24$ bivalents.
(b) Mitotic metaphase in a spermatogonium, $2n = 48$ chromosomes.
(c) Mitotic metaphase in a spermatogonium, $2n + 1 =$ chromosomes.

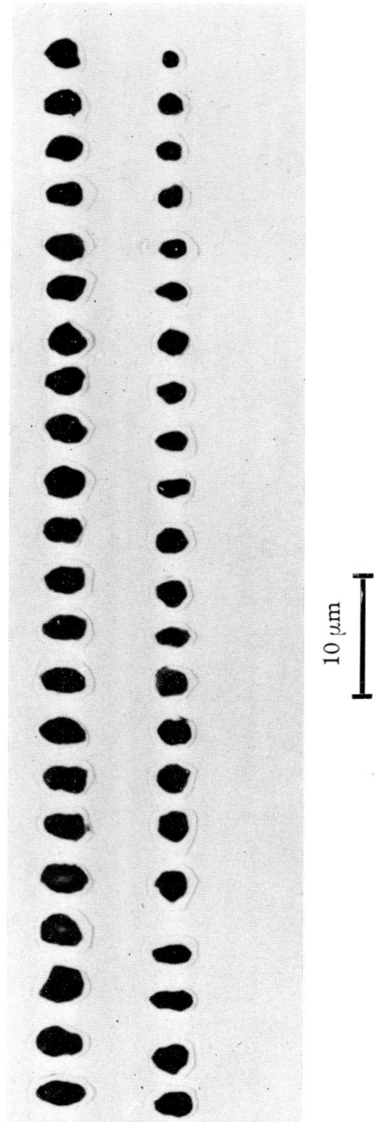

Fig. 6. Karyotype made from chromosomes shown in Fig. 5(b) (from Taylor 1967).

bivalent structure, Nogusa observed some cases in which 2–6 bivalents, characteristically dumb-bell shaped and with a transverse suture in their mid-region, came together to form a secondary association. The association of five such bivalents was most prevalent in *E. burgeri* whereas four was more frequent in *E. okinoseanum*. Although associations of this nature are found in many plants, this, according to Nogusa (1960), was only the second example found in the Animal Kingdom.

The only studies on the DNA content of hagfishes are those by Atkin and Ohno (1967) for *E. stoutii* which yielded values equivalent to 78% of that found in mammals.

VI. DISCUSSION

The major taxonomic separation of living cyclostomes into lampreys and hagfishes was based on anatomical, developmental and fossil evidence. The admittedly meagre cytotaxonomic information on living cyclostomes is entirely in accord with this subdivision. Probably in no other vertebrate group is there such a clear cut separation into two subgroups based on karyotypic differences. The hagfishes with their chromosomes of relatively large size and uniformity of number contrast dramatically with the lampreys, whose chromosomes are small in size and vary in number between families. This contrast lends further support to those systematists who regard the Class Cyclostomata, containing only the lampreys and hagfishes, as an unnatural grouping.

Amongst the hagfishes, cytology has in general failed to provide much additional evidence for an improvement of the current taxonomic schemes because of the similarity in chromosome complements of species so far examined, although it should be noted that the recorded counts for *Myxine* are slightly higher than those for *Eptatretus* (Table II). Within the lampreys, on the other hand, cytology provides help in the resolution of some taxonomic problems. For example, it is of interest that the diploid number of the monotypic genus *Geotria* should apparently differ so greatly from that of *Mordacia* since this is in line with their separation into separate families largely on the basis of their morphology (Hubbs and Potter, 1971). Additional support for this taxonomic distinction is provided by the complete lack of correspondence in the haemoglobin electrophoretic patterns of the two genera (Potter and Nicol, 1968). Although *Geotria* and *Mordacia* have some common morphological characters not present in Northern Hemisphere genera, the degree of divergence of the two Southern genera and the obvious affinities of *Geotria* with holarctic lampreys in terms of their morphology and karyotypes, suggests that *Mordacia* is the most widely

TABLE II. *A summary of current information on hagfish chromosomes*

Genus	Species	Author and date	Material	Diploid number	Centromere position
Myxine	*M. glutinosa*	Retzius, 1890	Testis	*c.* 50	—
	M. glutinosa	Schreiner and Schreiner, 1904a, b	Testis	*c.* 52	—
Eptatretus	*E. burgeri*	Schreiner and Schreiner, 1908	Testis	48	—
	E. burgeri	Nogusa, 1960	Testis	48	Acrocentric
	E. okinoseanum	Nogusa, 1960	Testis	46	Acrocentric
	E. stoutii	Taylor, 1967	Testis	48	Acrocentric

diverged of all genera. Again, if one can accept the view that the more recent diploid counts for the *Petromyzonidae* are typical of this group, one cannot help but be struck by the uniformity in the more recent studies. For example, counts for *I. gagei*, the most primitive species so far investigated and for *P. marinus*, *O. aepyptera* and *L. reissneri*, have all yielded mean values in the region of 164–170, and the counts for three *Lampetra* species are only slightly lower. These results add further support to the view that the Petromyzonidae may be regarded as a very compact group.

A study of chromosomes might also be used to shed light on another taxonomic problem at quite another level, namely the specific status of the so-called paired species. Unfortunately, data are available for only two such groups. In *Mordacia mordax* and *M. praecox* the diploid value is apparently identical in both species but in *Lampetra fluviatilis* and *L. planeri* this value differs by ten. This difference may be sufficient to prevent the successful production of fertile offspring even though hybrids which were still surviving at the end of two years have been produced by artificial fertilization (Hardisty, personal communication).

The importance of cyclostome cytotaxonomy, however, lies not only in the rigour it may bring to classification within the hagfish and lamprey groups but also in the information it may provide about karyotypic changes in the early evolution of the lower chordates. The studies of phylogenetic relationships based on genome characteristics have taken a new twist in recent years with interest in the quantitative changes in nuclear DNA that have occurred during vertebrate evolution. There is clearly no single trend or straightforward correlation in a vertebrate between the amount of DNA per nucleus and phylogenetic position, but attempts to explain known values for DNA in the cells of a variety of vertebrates (Goin and Goin, 1968; Ohno, 1970) suggest that the phylogenetic importance of such information cannot be overlooked. Can we then, using quantitative DNA values and karyotype analyses, construct a plausible sequence of genome changes involved in the origin and evolution of the cyclostomes? Assuming this is possible, other lower chordate groups must be considered and thus in the following discussion, both the protochordates and the teleosts are included.

Ohno *et al.* (1968) used, as a base for their hypothetical pattern of karyotype evolution in the lower vertebrates, the type of chromosomal complement and DNA content found within the protochordates. Thus, the urochordate *Ciona intestinalis* and the cephalochordate *Amphioxus lanceolatus* both have low DNA values (6% and 17% respectively of that found in mammals) and possess a relatively small number of chromosomes which, at least in the latter species, are acrocentric. The tunicates, probably the nearest living protochordate group to that from which the ancestral vertebrate evolved (Berrill, 1955; Whitear, 1957; Bone, 1960), have also been shown by other studies to have a diploid number of 32 or under (Minouchi, 1936; Zwillenberg and Zwillenberg, 1954). From these data, it is concluded that the chordate group ancestral to the vertebrates probably had a small number of gene loci. Within the protochordates, the evolution of an *Amphioxus*-like form from a tunicate-like larva was thought by Ohno (1970) to have involved a two- to threefold increase in genome size but the mechanism involved cannot at present be resolved. The size of the *Amphioxus* genome is, according to Ohno, very nearly the same as that found in several different teleost species exhibiting the lowest values found in their group.

In considering the teleosts, Ohno *et al.* (1968) pointed out that species from several diverse orders have an identical diploid complement of 48 acrocentric chromosomes (Nogusa, 1960; Post, 1960; Ohno and Atkin, 1966). However, it was notable that these particular species often had widely differing DNA values which were taken as implying that increase in DNA had taken place by unequal exchange, unequal crossing over and regional redundant duplication of DNA molecules. Contrasting with this particular situation, Wolf *et al.* (1969) showed that various species in the teleost family Cyprinidae could be classified into two categories; one with 50 chromosomes and a DNA content of 20 to 38% of that found in mammals and the other with about 100 chromosomes and a DNA value of 50% of that found in mammals. These data illustrate a second way in which DNA can be increased, namely by polyploidy although in this case there is of course a marked change in the diploid number. Diploid values may also be altered through the formation of Robertsonian centric fusions between acrocentric chromosomes coupled with splitting of a few original acrocentrics to two smaller acrocentric chromosomes through pericentric inversions. These processes, in combination with polyploidy, have been invoked by Ohno *et al.* (1968) as the means through which several salmonids evolved karyotypes with 60–80 chromosomes, many of which are metacentric and possess DNA contents of approximately 80% of that found in mammals.

The occurrence of differing amounts of gene duplication and polyploidy in teleosts may explain the marked differences between groups that are not too widely separated. For this reason, caution must be exercised in drawing conclusions from karyotypic data especially when groups of great antiquity are involved. Despite this, however, the very great differences in many aspects of the chromosomal constitutions, including DNA content, would seem to argue for an early separation of the lineages that led to the two groups of living cyclostome. This need not necessarily imply support for an heterostracan ancestry for the hagfishes but may equally reflect an early divergence of the two groups from an anaspid or anaspid-like stock.

The recurrence in the teleosts, as well as in the very widely separated group comprising the hagfishes, of chromosomal complements with 48 acrocentrics, is taken by Ohno *et al.* (1968) as indicating that this was probably the ancestral vertebrate karyotype derived from its chordate ancestor by tetraploidization. These authors also conclude that the ancestral DNA content was 20% of that found in mammals. Since hagfishes have a DNA content of 78% of that found in mammals (Atkin and Ohno, 1967) and a diploid complement of 48 acrocentric chromosomes, the increase in DNA content was probably increased by a factor of four in the same way as described above for some teleosts, i.e. by unequal exchange, unequal crossing over and regional redundant duplication of DNA molecules, without altering the shape of the chromosomes, although of course their size is increased. This hypothesis is supported by the fact that the monomeric haemoglobin polypeptides of this group are coded for by four independent gene loci (Ohno and Morrison, 1966).

If one accepts this hypothesis for the evolution of the hagfish karyotype, quite a different explanation must be sought to account for the evolution of the lamprey karyotype. Any discussion of the possible ways in which this could have occurred is hampered by the anomalies in the results reported for some lamprey species. However, we have taken the view that all recent work on the chromosomes of holarctic lampreys, which have been supported by photographs, provide strong evidence that this group is characterized by diploid counts of approximately 140 to 170 (Potter and Robinson, 1971). Yet, the possibility that the quite remarkable differences in the diploid numbers reported for *L. fluviatilis* and *L. reissneri* are a reflection of a polyploidic series cannot be ruled out until further studies have been made on specimens from different geographical areas. Endopolyploidy is unlikely to provide the explanation of such anomalies since Sasaki and Hitotsumachi (1967) apparently obtained spreads for *L. reissneri* from both larval and adult and somatic and gonadal material all corresponding to diploid values of 165–174. The third possibility is that they reflect differences in technique, as the chromosomes are so small that some may be missed particularly in sectional material.

Although there is conflict over the position of the centromere in Northern Hemisphere lampreys, it would appear from the most recent studies (Howell and Denton, 1969; Potter and Rothwell, 1970; Howell and Duckett, 1971) that this group is probably characterized by the possession of acrocentric chromosomes. Yet, a few meta- or submetacentric chromosomes have been identified in *L. reissneri* and *P. marinus* (Sasaki and Hitotsumachi, 1967; Potter and Rothwell, 1970).

As has been previously pointed out, there is a striking contrast between the typical Northern Hemisphere situation, in which the complement appears to comprise an extremely high number of mainly acrocentric chromosomes, and that found in the Southern Hemisphere genus *Mordacia*.

Thus, the diploid number of *Mordacia mordax* and *M. praecox* is much lower ($2n = 76$) and the chromosomes are very largely meta- or submetacentric, the centromere being clearly identifiable in many of the metaphase spreads.

This raises the question of how such differences arose in the two groups which, although apparently meriting separate familial status on the basis of their morphology, are at the same time obviously members of a relatively homogeneous group. In view of the high diploid numbers, quite exceptional for vertebrates in the case of the Petromyzonidae, it seems reasonable to invoke polyploidy, evidence for which is now accumulating, not only for the teleosts but also for a group of amphibians (Beçak *et al.*, 1966, 1967a, b). Although White (1954, 1957, 1959, 1965) has stressed the problems involved with respect to the resulting imbalance in the sex determining mechanism by implicating polyploidy in vertebrate karyotype evolution, it may be that the presence of a multifactorial and undifferentiated mechanism for sex differentiation, as suggested by the protracted intersexual condition in this group (Dodd, 1960; Hardisty, 1965a, b), provides a reasonably flexible template for polyploidic changes.

Ohno *et al.* (1968) used the DNA value of 38% of mammals found in *L. planeri* (Atkin and Ohno, 1967), and the 94–96 small acrocentric chromosomes described by Nogusa (1960) for *L. reissneri*, as the basis for suggesting an evolutionary pattern within the lamprey group. However, it should be noted that there is doubt as to the validity of Nogusa's count in view of the entirely different karyotype later described for the same species by Sasaki and Hitotsumachi (1967). Yet, this should not be taken as invalidating the concept of polyploidy for lamprey karyotype evolution: indeed this is supported by the very large values in the Petromyzonidae and the possibility in *M. praecox* of secondary associations which are common in polyploid plant species with small chromosomes.

If one assumes that 48 acrocentric chromosomes are representative of the "diploid" karyotype of the earliest vertebrates, it seems possible that *P. marinus* may represent an octaploid condition with some reduction in chromosomal number occurring as a result of centric fusions (Potter and Rothwell, 1970). Unfortunately, the small size and large number of the chromosomes makes it impossible to judge how many chromosomes are metacentric in this species and thus assess the validity of such a view. Future studies should elucidate the exact relationship between the number of acrocentric and meta- or submetacentric chromosomes in species belonging to the Petromyzonidae. In the case of *Mordacia*, karyotype evolution may have been partly the result of extensive centric fusions preceding or more probably following polyploidization of the genome. Centric fusions may have contributed to the greatly reduced diploid number in the Mordaciidae compared with that found in the Petromyzonidae, particularly since *Mordacia* has 76 chromosomes, of which the majority are meta- or sub-metacentric, compared with the 104 acrocentrics of *Ichthyomyzon gagei*, a species belonging to the most primitive genus. However, it seems likely that other factors, such as one of those listed by Morescalchi (1970), may have led to either a loss of chromosomes in the Mordaciidae or alternatively an increase in holarctic species. These are questions that can only be resolved when more data are available on diploid numbers, centromere position, length of chromosome arms and DNA ontent.

The preliminary investigations on *Geotria*, which show that it had a similar karyotype to Northern Hemisphere lampreys, suggest that this character was retained from an ancestral form that became isolated and differentiated at a later date than *Mordacia*, a view suggested by Hubbs and Potter (1971) on the basis of morphological criteria.

This review serves to emphasize the shortage of detailed information on the chromosomes of both lampreys and hagfishes, both with respect to diploid number, centromere position and DNA content. Despite this, however, a pattern is emerging and there is little doubt that future studies, especially if combined with biochemical investigations having a taxonomic basis, can help further to elucidate interrelationships within both cyclostome groups. Such studies will also contribute to our knowledge of karyotype evolution in the lower chordates and may reveal that in the lampreys polyploidy has played a more striking role than in any other vertebrate.

Acknowledgements

We are indebted to Dr. K. Taylor for providing prints of Figs 5 and 6 and to Dr. W. M. Howell for Fig. 2a.

References

Adam, H. and Strahan, R. (1963). Systematics and geographical distribution of Myxinoids 1–8. *In* "The Biology of Myxine". (A. Brodal and R. Fänge, eds.) Universitetforlaget, Oslo.

Atkin, N. B. and Ohno, S. (1967). DNA values of four primitive chordates. *Chromosoma* **23**, 10–13.

Bardack, D. and Zangerl, R. (1968). First fossil lamprey: A record from the Pennsylvanian of Illinois. *Science*, N.Y. **162**, 1265–1267.

Bardack, D. and Zangerl, R. (1971). Lampreys in the fossil record. *In* "The

Biology of Lampreys" Vol. 1, 67–84, (M. W. Hardisty and I. C. Potter, eds.) Academic Press, London and New York.

Beçak, M. L., Beçak, W. and Rabello, M. N. (1966). Cytological evidence of constant tetraploidy in the bisexual South American frog *Odontophyrynus americanus*. *Chromosoma* **19**, 188–193.

Beçak, M. L., Beçak, W. and Rabello, M. N. (1967a). Further studies on polyploid amphibians (Ceratophrydidae): I Mitotic and meiotic aspects. *Chromosoma* **22**, 192–201.

Beçak, W., Beçak, M. L., Lavalle, D. and Schreiber, G. (1967b). Further studies on polyploid amphibians (Ceratophrydidae). II. DNA content and nuclear volume. *Chromosoma* **23**, 14–23.

Berg, L. S. (1958). "System der Rezenten und Fossilen Fisehartigen und Fische." VEB Deutscher Verlag der Wissenschaften, Berlin.

Berrill, N. J. (1955). "The Origin of Vertebrates." Oxford University Press.

Bone, Q. (1960). The origin of the chordates. *J. Linn. Soc. Zool.* **44**, 252–259.

Chubareva, L. A. (1957a). Cytological and cytochemical study of the development, maturation and fertilisation of the eggs of the freshwater lamprey, *Lampetra fluviatilis*. *Vest. Leningr. Gos. Univ.* **9**, 93–98 (In Russian).

Chubareva, L. A. (1957b). Cytological and cytochemical studies of the process of fertilisation in the river lamprey (*Lampetra fluviatilis*). *Dokl. Akad. Nauk. S.S.S.R.* **112**, 945–948 (In Russian).

de Buen, F. (1961). Las lampreas (Marsipobranchii o Ciclostomi) en aguas de Chile. *Investnes Zool. Chil.* **7**, 101–124.

Denison, R. H. (1960). Fishes of the Devonian Holland Quarry Shale of Ohio. *Fieldiana: Geol.* **4** (10), 555–613.

Denison, R. H. (1964). The Cyathaspididae. A family of Silurian and Devonian jawless vertebrates. *Fieldiana: Geol.* **13**, 309–473.

Dodd, J. M. (1960). Genetic and environmental aspects of sex differentiation in cold-blooded vertebrates. *Mem. Soc. Endocr.* **No.** 7, 17–44.

Fontaine, M. (1958). Classe des Cyclostomes. Formes actuelles super-ordres des Petromyzonoidea et des Myxinoidea. *In* "Traité de Zoologie" **13**, (1): 13–172 (P. P. Grassé, ed.) Masson and Co, Paris.

Goin, O. B. and Goin, C. J. (1968). DNA and the evolution of the vertebrates. *Am. Midl. Nat.* **80**, 289–298.

Hardisty, M. W. (1965a). Sex differentiation in lampreys (Part I). *J. Zool.* (*Lond*) **146**, 305–345.

Hardisty, M. W. (1965b). Sex differentiation in lampreys (Part II). *J. Zool.* (*Lond*) **146**, 346–387.

Heintz, A. (1963). Phylogenetic aspects of myxinoids. *In* "The Biology of Myxine". 9–21 (A. Broadal and R. Fänge, eds.). Universitetsforlaget, Oslo.

Holly, M. (1933). Cyclostomata. *In* "Das Tierreich, Lfg." (F. E. Schulze, W. Kükenthal, K. Heider and R. Hesse, eds.) Walter de Gruyter, Berlin and Leipzig.

Howell, W. M. and Denton, T. E. (1969). Chromosomes of ammocoetes of the Ohio Brook Lamprey, *Lampetra aepyptera*. *Copeia.* **1969**, 393–395.

Howell, W. N. and Duckett, C. R. (1971). Somatic chromosomes of the lamprey

Ichthyomyzon gagei. (Agnatha: Petromyzonidae). *Experientia* **27**, 222–223.
Hubbs, C. L. (1925). The life cycle and growth of lampreys. *Pap. Mich. Acad. Sci.* **4**, 587–603.
Hubbs, C. L. and Potter, I. C. (1971). Distribution, taxonomy and phylogeny. *In*' 'The Biology of Lampreys". **1**, 1–65. (M. W. Hardisty and I. C. Potter, eds.). Academic Press, London and New York.
Lanzing, W. J. R. (1959). "Studies on the river lamprey, Lampretra fluviatilis, during its anadromous migration". Neerlandia, Utrecht.
Minouchi, O. (1936). Notiz über die Chromosomen von *Tethyum plicatum* Les. (Ascidacea). *Z. Zellforsch.* **23**, 790–794.
Montalenti, G. (1936). Analisi citologica della fecondazione e della attivazione artificiale della uova di Lampreda. *Memorie R. Accad. Ital.* **7**, 191-213
Morescalchi, A. (1970). Karyology and vertebrate phylogeny. *Boll. Zool.* **37**, 1–28.
Nogusa, S. (1960). A comparative study of the chromosomes in fishes with particular considerations on taxonomy and evolution. *Mem. Hyogo Univ. Agric.* **3**, 1–62.
Nygren, A., and Janke, M. (1972). Cytological studies in *Myxine glutinosa* (Cyclostomata) from the Gullmaren fjord in Sweden. *Swedish J. Agric. Res.* **2**: 83–88.
Ohno, S. (1970). "Evolution by Gene Duplication." George Allen and Unwin, London.
Ohno, S. and Atkin, N. B. (1966). Comparative DNA values and chromosome complements of eight species of fishes. *Chromosoma* **18**, 455–456.
Ohno, S. and Morrison, M. (1966). Multiple gene loci for the monomeric haemoglobin of the hagfish (*Eptatretus stoutii*). *Science* **154**, 1034–1035.
Ohno, S., Wolf, U. and Atkin, N. B. (1968). Evolution from fish to mammals by gene duplication. *Hereditas* **59**, 169–187.
Okkelberg, P. (1921). The early history of the germ cells in the brook lamprey, *Entosphenus wilderi* (Gage) up to and including the period of sex differentiation. *J. Morph.* **35**, 1–152.
Post, A. (1965). Verleichende Untersuchungen der Chromosomenzahlen bei Süsswasser-Teleosteern. *Z. Zool. Syst. Evol. Forsch.* **3**, 47–93.
Potter, I. C. and Nicol, P. I. (1968). Electrophoretic studies on the haemoglobins of Australian lampreys. *Aust. J. Exp. Biol. Med. Sci.* **46**, 639–641.
Potter, I. C. and Robinson, E. S. (1971). The chromosomes. *In* "The Biology of Lampreys" **1**, 279–293. (M. W. Hardisty and I. C. Potter, eds.). Academic Press, London and New York.
Potter, I. C. and Rothwell, B. (1970). The mitotic chromosomes of the lamprey, *Petromyzon marinus* L. *Experientia* **26**, 429–430.
Potter, I. C. and Strahan, R. (1968). The taxonomy of the lampreys *Geotria* and *Mordacia* and their distribution in Australia. *Proc. Linn. Soc.* (*Lond*). **179**, 229–240.
Potter, I. C., Robinson, E. S. and Walton, S. M. (1968). The mitotic chromosomes of the lamprey *Mordacia mordax* (Agnatha: Petromyzonidae). *Experientia* **24**, 966–967.
Retzius, G. (1890). Ueber Zellenteilung bei *Myxine glutinosa*. *Biol. Foren. Stockholm, Forhandl.* **11** No. 8: 80–91.

Ritchie, A. (1968). New evidence on *Jaymoytius kerwoodi* White, an important ostracoderm from the Silurian of Lanarkshire, Scotland. *Paleontology* **11**, 21–39.

Robinson, E. S. and Potter, I. C. (1969). Meiotic chromosomes of *Mordacia praecox* and a discussion of chromosome numbers in lampreys. *Copeia* **1969**, 824–828.

Sasaki, M. and Hitotsumachi, S. (1967). Notes on the chromosomes of a freshwater lamprey, *Entosphenus reissneri* (Cyclostomata). *Chromosome Inf. Serv.* (**8**), 22–24.

Schreiner, A. and Schreiner, K. E. (1904a). Die Reifensteilungen bei den Wirbeltieren. Ein Beitrag zur Frage nach der Chromatinreduktion. *Anat. Anz.* **24**, 561–578.

Schreiner, A. and Schreiner, K. E. (1904b). Ueber die Entwicklung der mannlichen Geschlechtozellen von *Myxine glutinosa* (L.) Vermehrungsperiode, Reifungsperiode und Reifungsteilungen. *Arch. Biol.* **21**, 183–355.

Schreiner, A. and Schreiner, K. E. (1908). Zur spermienbildung der Myxinoiden. Ueber die Entwicklung der Mannlichen Geschlechtszellen von *Myxine gluttinosa* (L.) III. *Arch. Zool.* **1**, 152.

Stensiö, E. A. (1927). The Downtonian and Devonian Vertebrates of Spitzbergen. Pt. 1. Family Cephalaspidae. *Skr. Svalbard Ishavet* **12**, 1–391.

Stensiö, E. A. (1932). "The Cephalspids of Great Britain." British Museum (Nat. Hist.), London.

Stensiö, E. A. (1958). Les cyclostomes fossiles ou ostracoderms. *In* "Traite de *Zoologie*" (P. P. Grassé, ed.). Paris: Masson and Co, Paris.

Stensiö, E. A. (1968). The cyclostomes with special reference to the diphyletic origin of the Petromyzontida and Myxinoidea. *In* "Current problems of lower vertebrate phylogeny" (T. Orvig, ed.). Interscience, New York, London, Sydney.

Tarlo, L. B. (1961). *Rhinopteraspis cornubica* (McCoy) with notes on the classification and evolution of pteraspids. *Acta Paleont. Pol.* **6**, 362–400.

Tarlo, L. B. (1962). The classification and evolution of the Heterostraci. *Acta Paleont. Pol.* **8**, 249–290.

Taylor, K. M. (1967). The chromosomes of some lower chordates. *Chromosoma* **21**, 181–188.

Vladykov, V. D. and Follett, W. I. (1967). The teeth of lampreys (Petromyzonidae): their terminology and use in a key to holarctic genera. *J. Fish. Res. Bd. Can.* **18**, 1125–1143.

White, M. J. D. (1954). "Animal Cytology and Evolution." 2nd Ed. Cambridge University Press.

White, M. J. D. (1957). Some general problems of chromosomal evolution and speciation in animals. *Surv. Biol. Prog.* **3**, 109–147.

White, M. J. D. (1959). Speciation in animals. *Aust. J. Sci.* **22**, 32–39.

White, M. J. D. (1965). Principles of karyotype evolution in animals. *In* "Genetics Today" **2**, 391–397. (S. J. Geerts, ed.). Pergamon Press, London and New York.

Whitear, M. (1957). Some remarks on the ascidian affinities of vertebrates. *Ann. Mag. Nat. Hist.* (12) **10**, 338–347.

Wolf, U., Ritter, H., Atkin, N. B. and Ohno, S. (1969). Polyploidization in the fish family Cyprinidae, Order Cypriniformes. 1. DNA content and chromosome sets in various species of Cyprinidae. *Hum. Gen.* 7, 240–244.

Zanandrea, G. (1959). Recenti ricerche sulle forme "appaite" di lamprede dell' Italia e del Danubio. *Boll. Zool.* **26**, 545–554.

Zanandrea, G. and Capanna, E. (1964). Contributo alla cariologia del genere. *Lampetra. Boll. Zool.* **31**, 669–677.

Zwillenberg, L. O. and Zwillenberg, H. H. L. (1954). Chromosome counts in Tunicates and a modified method for making permanent squash preparations. *Nature* (*Lond*) **173**, 876–877.

ADDENDUM

Since writing the above review we have received data which contrasts with the close correspondence in the diploid values ($2n$ = c. 50) of the various species of hagfishes recorded in this article. For example, Nygren and Jahnke (1972) have reported finding a great deal of intraspecific variation in *Myxine glutinosa* with diploid numbers ranging from 17 to 116, a feature they attribute to "misdivision in connection with a formation of isochromosomes." In a recent letter, Dr. K. M. Taylor reports that although he found no evidence for similar variation in recent preliminary studies on *Eptatretus stoutii* and *E. deani*, the counts for these two species were low ($2n = 34$). He also reports that recent work by Dr. C. L. Hubbs on the taxonomy and vertical distribution of myxinoids from the Pacific coast of North America indicate that the hagfish whose chromosomes he investigated previously (Taylor 1968) was probably a species other than *Eptatretus stoutii*. In view of the different diploid values obtained by the above workers, and their divergence from the previously homogenous pattern of chromosome numbers in hagfishes, it is important that future studies should establish the basis of this variation.

10

Checklist of Fish Chromosomes

EDITORS REVIEW*

I. INTRODUCTION

Various techniques devised in recent years for chromosome investigation have furthered our knowledge of chromosome numbers and their morphology in many mammalian species and other groups of higher vertebrates. A taxonomic revision has thereby been promoted leading to a thorough interpretation of the phylogenetic interrelationships between the various groups.

Beyond doubt, the Superclass *Pisces* with its key position in the evolution of vertebrates is a determinant for the identification of some evolutionary lines which have given origin to the higher vertebrates. For this reason, further insight into the chromosomes in this group is anxiously awaited. It can be foreseen that karyological information may lead to the solution of a number of the complex problems besetting both the Systematics and phylogeny of this group.

In the Superclass *Pisces*, in fact, Systematics and phylogenetic interrelations are indeed puzzling and still a matter of debate, owing to the extreme heterogeneity of the extant forms, which is the result of the differentiation that came into play over the millenia of their evolutionary history. In addition, this situation is complicated by the existence of complex speciation mechanisms depending on both the peculiarity of the aquatic environment and the reproduction pattern.

Present-day knowledge about the chromosome number and their morphology in Pisces is however very meagre. Only recently have some moderately satisfactory karyotypes been obtained either by *in vitro* culture methods (Chiarelli *et al.*, 1968; Ferrantelli *et al.*, 1968) or more simply by

* The Editors acknowledge the valuable cooperation of Mrs. L. Lamberti Montagnoni (Institute of Anthropology, University of Turin) and Dr. A. C. Rossi (Institute of Zoology, University of Parma), in preparing this reference list.

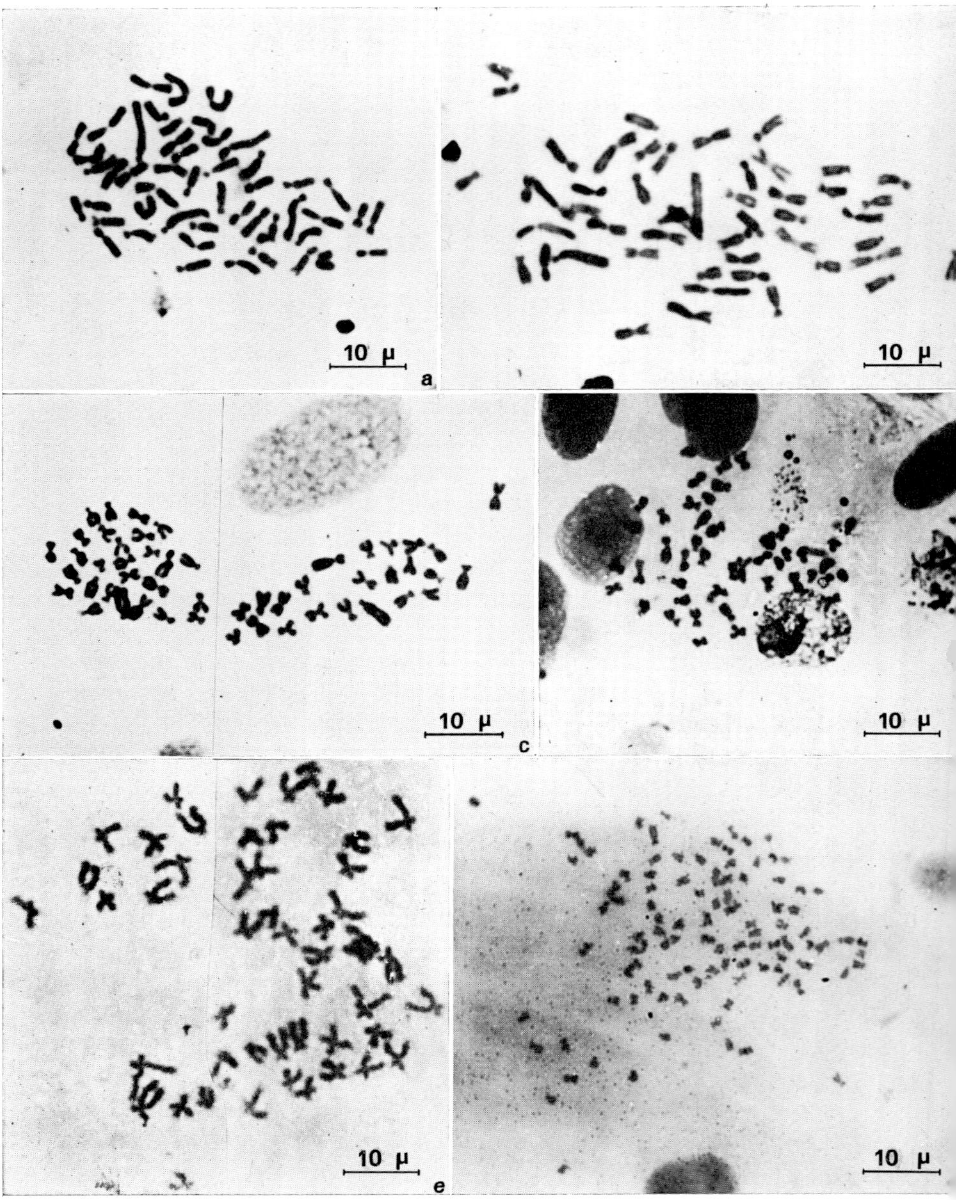

Fig. 1. Some metaphasic plates of fish chromosomes: (a) *Alburnus albidus alborella*; (b) *Leuciscus cephalus*; (c) *Leuciscus aula*; (d) *Tinc tinca*; (e) *Danio malabaricus*; (f) *Carassius auratus*, obtained from the Laboratory of Tissue Cultures of the Institute of Anthropology, University of Turin.

using the inner epithelium of the fish scales (Denton-Howell, 1969). Nowadays, a great many investigations are being carried out world wide and fresh data are appearing in the international literature.

Hence, in our opinion, any attempt whatever towards a synthesis, even solely about the chromosome number is not practicable today.

The purpose of the present work is to provide as full a survey as possible of the available information up to 1970, in order to make it easier for our colleagues both to collect further data and to revise critically older and incomplete information.

Besides our knowledge of the chromosome number and morphology in this vertebrate group, particular attention should be given to the chromosome size and the DNA nuclear content. Variations in these parameters have in fact been repeatedly observed in the Superclass *Pisces* by many workers, and this type of information is necessary for any approach to either Systematics or phylogenetic organization.

II. CHROMOSOME TABLES

	n	*2n*	References
CHONDROICHTHYES			
Order: Chimerae			
Family: Chimaeridae			
Hydrolagus colliei		58±	Ohno *et al.*, 1969.
Order: Selachii			
Family: Scylliorhinidae			
Pristiurus sp.		60–72	Ruckert, 1892a, b.
Scylliorhinus catula		62	Matthey, 1937.
Scyllyum canicula		24	Moore, 1895.
		28–40	Rawitz, 1899.
		60–100	Kastschenko, 1890.
Family: Squalidae			
Spinax niger		60–70	Schreiner and Schreiner, 1904.
Squalus acanthias		62	Nygren *et al.*, 1971d.
S. suckleyi		62	Makino, 1937a.
Order: Batoideea			
Family: Raijdae			
Raja batis		98	Nygren *et al.*, 1971d.
R. clavata		98	Nygren *et al.*, 1971d.
R. meerdevoortii		104	Makino, 1937a.
R. radiata		98	Nygren *et al.*, 1971d.
Family: Torpedinidae			
Torpedo ocellata		50	Kastschenko, 1890.

OSTEICHTHYES

ACTINOPTERYGII

Order: Olostei		
Family: Amiidae		
Amia calva	46±	Ohno *et al.*, 1969.
Family: Lepisosteidae		
Lepisosteus productus	69±	Ohno *et al.*, 1969.
Order: Chondrostei		
Family:		
Scarphirhynchus platorhyncus	112±	Ohno *et al.*, 1969.
Order: Anguilliformes		
Family: Anguillidae		
Anguilla anguilla	38	Sick *et al.*, 1962; Chiarelli *et al.*, 1969. Cucchi *et al.*, 1969.
A. rostrata	38	Sick *et al.*, 1962.
A. japonica	38	Sick *et al.*, 1962.
Family: Murenidae		
Murena pardalis	40	Nogusa, 1960.
Gymnothorax kidako	42	Nogusa, 1960.
Family: Echelidae		
Echelus uropterus	50	Nogusa, 1960.
Order: Clupeiformes		
Family: Engraulidae		
Engraulis mordax	48	Ohno *et al.*, 1967.
E. japonicus	48	Nogusa, 1960.
Family: Clupeidae		
Alosa pseudoharengus	48	Mayers and Roberts, 1969.
Order: Salmoniformes		
Family: Salmonidae (Coregonidae e Thymallidae)		
Thymallus thymallus	102	Svärdson, 1945.
Salmo carpio	96	Pomini, 1939.

	n	$2n$	References
Coregonus (fera) nasus		48	Karbe, 1963.
Salmo clarcki lewisi		64	Simon and Dollar, 1963.
S. trutta		70	Pomini, 1939.
		80	Svärdson, 1945; Rees, 1964; Nygren *et al.*, 1971c; Capanna *et al.*, 1971.
		96	Ojima *et al.*, 1963.
S. gairdneri irideus		60	Bungenberg de Jong, 1955; Wright, 1955; Lieder, 1956.
		58–64	Ohno *et al.*, 1965; Ohno and Atkin, 1966; Ohno, 1969.
		104	Nogusa, 1960; Ojima *et al.*, 1963.
S. trutta lacustris		70	Pomini, 1939.
S. marmoratus		84	Pomini, 1939.
S. alpinus		80	Svärdson, 1945; Nygren *et al.*, 1971c.
S. fario		84	Prokofieva, 1934; Pomini, 1939.
S. salar		60	Prokofieva, 1934; Svärdson, 1941, 1945; Rees, 1964, 1967.
		56	Boothroyd, 1959
		58	Rees, 1967; Nygren *et al.*, 1968b.
S. salar × S. trutta		69	Rees, 1967.
Oncorhynchus nerka	50		Nogusa, 1960.
	54		Ojima *et al.*, 1963.
		56	Simon, 1963.
		58	Sasaki *et al.*, 1968.
O. masu	50		Nogusa, 1960; Ojima *et al.*, 1963.
O. keta		74	Makino, 1937b; Simon, 1963; Sasaki *et al.*, 1968.
	50		Nogusa, 1960; Ojima *et al.*, 1963.
O. kisutch		58	Ohno, 1969.
		60	Simon, 1963.

O. gorbuscha		52	Simon, 1963.
O. rhodurus	50		Nogusa, 1960; Ojima *et al.*, 1963.
Salvelinus fontinalis		80	Prokofieva, 1934.
		84	Svärdson, 1945; Wright, 1955; Capanna *et al.*, 1971.
		100	Nogusa, 1960.
		50	Ojima *et al.*, 1963.
S. fontinalis × Salmo trutta		82	Capanna *et al.*, 1971.
S. namaycush		80	Wahl, 1960.
Coregonus albula		80	Svärdson, 1945; Nygren *et al.*, 1971a.
C. lavaretus		80	Svärdson, 1945; Nygren *et al.*, 1971b.
C. lavaretus baeri		80	Prokofieva, 1934.
C. lavaretus pidschian		80	Nygren *et al.*, 1971b.
C. schinzii helveticus		72	Kupka, 1948.
	48		Karbe, 1963.
C. schinzii duplex		72	Kupka, 1948.
C. wartmanni		36	Kupka, 1948.
	48	96	Karbe, 1963.
C. wartmanni coeruleus		70±2	Kupka, 1948.
C. wartmanni nobilis		48	Karbe, 1963.
C. (macrophthalmus) oxyrhyncus	48	96	Karbe, 1963.
C. acronius	48	96	Karbe, 1963.
Family: Osmeridae			
Osmerus eperlanus		54	Nygren *et al.*, 1971e.
		58	Svärdson, 1945.
Hypomesus pretiosus		52	Ohno *et al.*, 1967.
Family: Bathylagidae			
Bathylagus ochotensis		54	Chen, 1969.
B. stilbius		64	Chen, 1969.
B. milleri		60	Chen, 1969.
B. wesethi		36	Chen, 1969.

	n	$2n$	References
Family: Salangidae			
Salangichthys microdon		28	Nogusa, 1960.
Family: Esocidae			
Esox lucius		18	Svärdson and Wickbom, 1939.
		46	Lieder, 1953.
		48	Prakken, 1955.
		50	Nygren *et al.*, 1968a.
Family: Umbridae			
Umbra limi		20±2	Foley, 1926.
Family: Myctophidae			
Lampanyctus ritteri		47♂; 48♀	Chen, 1969.
Parvilux ingens		49♂	Chen, 1969.
Order: Cypriniformes			
Family: Characidae			
Megalamphodus megalopterus	24		Post, 1965.
Hyphessobrycon timari	24	48 ?	Post, 1965.
H. serpae	24		Post, 1965.
H. pulchripinnis	24		Post, 1965.
H. haraldschultzi	24		Post, 1965.
H. griemi	24		Post, 1965.
H. gracilis	24	48	Post, 1965.
H. flammeus	24		Post, 1965.
Hyphessobrycon stictus		52	Scheel, 1970.
Hemigrammus pulcher	24		Post, 1965.
Hemigrammus erythrozonus		48	Leuken and Foster, 1969.
H. aff. schmardae		52	Scheel, 1970.
H. ocellifer	24	48 ?	Post, 1965.
H. caudovittatus		50	Fontana *et al.*, 1970.
Hasemannia marginata	34		Post, 1965.
Gymnocorhymbus termetzi	24		Post, 1965.

Thaveria obliquua	24		Post, 1965.
Pristella riddlei	24		Post, 1965.
Micralestes interruptus	24		Post, 1965.
Moenkhausia sanctaefilomenae	24		Post, 1965.
M. pittieri	24		Post, 1965.
Astyanax mexicanus	24		Post, 1965.
Cheirodon axelrodi	24		Post, 1965.
Chalceus macrolepidotus	?	54	Muramoto *et al.*, 1968.
Paracheirodon innesi		36	Leuken and Foster, 1969.
		32	Rees, 1964.
Serrasalmus hollandi		64	Muramato *et al.*, 1968.
Anoptichthys jordani	24		Post, 1965.
Alestes longipinnis	24		Post, 1965.
Family: Cyprinidae			
Abbottina rivularis	25	50	Nogusa, 1955a.
Abramis brama	26	52	Lieder, 1954a; Post, 1965.
		50	Wolf *et al.*, 1969.
Acheilognathus cyanostygma	22	44	Nogusa, 1960.
A. rhombea	22–24	44–48	Nogusa, 1960.
A. lanceolatum	25	50	Makino, 1939.
A. limbata	22	44	Nogusa, 1960.
Alburnus albidus alborella		50	Fontana *et al.*, 1970.
Notemigonus crysoleucas		50	Lieppman and Hubb, 1969.
Labeo chrysophekadion		50	Muramoto *et al.*, 1968.
Hemigramnocypris rasborella	24	48	Nogusa, 1960.
Rhodeus ocellatus		44	Nogusa, 1960.
Pungtungia herzi		50	Nogusa, 1960.
P. hilgendorfi		50	Nogusa. 1960.
Moroco percnurus	27		Nogusa, 1960.
M. steindachneri	27	54	Nogusa, 1960.
Gnathopogon elongatus		50	Nogusa, 1960.
G. coerulescens		50	Nogusa, 1960.

	n	$2n$	References
Pelteobagrus nudiceps		56	Nogusa, 1960.
Garra rufa	22–26		Post, 1965.
Ishikauria steenackeri		48	Nogusa, 1960.
Hemibarbus longi rostris	25	50	Nogusa, 1960.
Pseudogobio esocinus	26	52	Nogusa, 1960.
Sarcocheilichthys variegatus	25	50	Nogusa, 1960.
Tribolodon hakuensis		50	Makino, 1939.
Notropis callistius		50	Denton and Howell, 1969.
N. stilbius		50	Denton and Howell, 1969.
Puntius conchonius	24		Post, 1965.
P. filamentosus	24 ?		Post, 1965.
Capoeta oligolepis	24		Post, 1965.
Barbus semifasciolatus.	26	52	Nogusa, 1960.
Danio malabaricus		50	Fontana *et al.*, 1970.
Brachydanio (Danio) frankei	24		Post, 1965.
Rutilus rutilus × Gobio fluviatilis		52	Lieder, 1954a.
Leuciscus souffia muticellus		50	Fontana *et al.*, 1970.
Gobio gobio	24		Post, 1965.
		52	Lieder, 1954a.
Leuciscus cephalus		50	Wolf *et al.*, 1969; Fontana *et al.*, 1970.
Typhlogarra widdowsoni	24		Post, 1965.
Tinca tinca		46–48	Chiarelli *et al.*, 1969.
	48		Post, 1965.
		48	Wolf *et al.*, 1969; Fontana *et al.*, 1970.
Scardinius erythrophthalmus		54	Lieder, 1954a.
		52	Post, 1965.
		48	Chiarelli *et al.*, 1969; Fontana *et al.*, 1970.
Rasbora trilineata	24		Post, 1965.
R. heteromorpha	24		Post, 1965.
Barbus barbus		100	Wolf *et al.*, 1969.

Barbus fasciatus		52	Nogusa, 1960; Ohno *et al.*, 1967; Vialli, 1970.
Barbus (Barbodes) pentazona	24		Post, 1965.
Barbus "schuberti"	24		Post, 1965.
Barbus (Capoeta) tetrazona	24		Post, 1965.
		50	Ohno *et al.*, 1967; Vialli, 1970.
Barbus (Capoeta) titteya	24		Post, 1965.
Barbus (Barbodes) viviparus	24		Post, 1965.
Danio devario		50	Fontana *et al.*, 1970.
Danio rerio		50	Endo and Ingalls, 1968; Fontana *et al.*, 1970.
	24		Post, 1965.
Danio albolineatus	24		Post, 1965.
		50	Fontana *et al.*, 1970.
Ctenopharyngodon idella		48	Nogusa, 1960.
Ospariichthyx uncirostris		66	Nogusa, 1960.
Cyprinus carpio		104	Makino, 1939.
		100	Ojima and Hitotsumachi, 1967; Labat, 1967.
		104 ?	Ohno *et al.*, 1967.
Carassius carassius auratus		94	Makino, 1939.
		104	Ohno and Atkin, 1966.
		100	Fontana *et al.*, 1970.
Carassius auratus		94	Makino, 1941b.
			Ojima *et al.*, 1966.
		100	Ojima and Hitotsumachi, 1967.
		102–104	Ohno and Atkin, 1966; Ohno *et al.*, 1967.
		104	Chiarelli *et al.*, 1969.
Rutilus rutilus × Scardinius erytrophithalmus		52	Lieder, 1954a.
Rutilus rutilus × Abramis brama		52	Lieder, 1954a.
Leuciscus aula		50	Fontana *et al.*, 1970.
Rutilus rutilus		50	Wolf *et al.*, 1969.
		52	Lieder, 1954a.
Phoxinus percnurus		27	Nogusa, 1960.
Pseudorasbora parva	25	50	Makino, 1939.

	n	2*n*	References
Family: Cobitidae			
Misgurnus fossilis	24		Post, 1965.
Misgurnus anguillicaudatus		50	Hitotsumachi *et al.* 1969.
Cobitis biwae	27	54	Post, 1965.
		96	Hitotsumachi, *et al.* 1969.
C. delicata		50	Hitotsumachi *et al.*, 1969.
Barbatula oreas	24	48	Makino, 1941a.
B. toni		50	Hitotsumachi *et al.*, 1969.
Lefua nikkonis		50	Hitotsumachi *et al.*, 1969.
L. echigonia	25	50	Nogusa, 1960.
Botia macracanthus		98	Muramoto *et al.*, 1968.
Acanthophthalmus kuhli		50	Muramoto *et al.*, 1968.
Order: Siluriformes			
Family: Bagridae			
Mystus mystus		54	Muramoto *et al.*, 1968.
M. vittatus		50	Srivastava and Das, 1969.
M. seenghala		50	Srivastava and Das, 1969.
Family: Siluridae			
Parasilurus asotus	29	58	Nogusa, 1960; Muramoto, 1969.
Ictalurus punctatus		56	Muramoto *et al.*, 1968.
Family: Loricariidae			
Loricaria parva	24		Post, 1965.
Family: Doradidae			
Hypostomus plecostomus		54	Muramoto *et al.*, 1968.
Order: Atheriniformes			
Family: Hemiramphidae (Exocoetidae)			
Dermogenys pusillus	24		Post, 1965.
Family: Cyprinodontidae			
Aphanius sophiae	24		Karbe, 1961.
A. iberus	24		Karbe, 1961.

A. fasciatus	22	44	Otzan, 1954.
	24	48	Karbe, 1961.
A. dispar	24	48	Otzan, 1954; Karbe, 1961.
Aphanius (cypris) mento	22	44	Otzan, 1954.
	24	48	Karbe, 1961.
Fundulus heteroclitus		36	Moenkhaus, 1904.
		45	Pinney, 1918.
	24		Post, 1965.
		48	Chen and Ruddle, 1970.
F. diaphanus		48	Chen and Ruddle, 1970.
F. parvipinnis		48	Chen and Ruddle, 1970.
F. majalis		48	Chen and Ruddle, 1970.
F. notti		48	Denton and Howell, 1969.
F. stellifer		48	Denton and Howell, 1969.
Epiplatys duboisi	24		Scheel, 1966b.
E. dageti	25		Scheel, 1966b.
E. spilargyreius	17	34	Scheel, 1966b, 1968.
E. chaperi	24		Post, 1965.
	25	50	Scheel, 1966b, 1968.
E. fasciolatus	19	38	Scheel, 1966b, 1968.
E. grahami	23	46	Scheel, 1968.
E. esekanus	21	42	Scheel, 1968.
E. bifasciatus	20	40	Scheel, 1966b, 1968.
E. barmoiensis	17	?	Scheel, 1968.
E. annulatus	25	50	Scheel, 1966b, 1968.
E. sexfasciatus	24		Post, 1965; Scheel, 1966b.
Aplocheilus blocki	24		Scheel, 1966b.
A. dayi	24		Scheel, 1966b.
Aplocheilichthys spilauchena	24		Post, 1965.
Nothobranchius guentheri	19		Post, 1965; Scheel, 1966b, 1968.
N. mayeri	19		Post, 1965.
Oryzias latipes		48	Iriki, 1932b; Katayame, 1937.

	n	$2n$	References
Pachypanohax playfairi	24		Post, 1965; Scheel, 1966b.
Cynolebias belotti	24		Post, 1965.
C. nigripinnis	24		Post, 1965.
C. whitei	24		Post, 1965.
	23		Scheel, 1970.
Cynopoecilus ladigesi	24		Post, 1965.
Cynopoecilus melanotaenia	24		Post, 1965.
Pterolebias longipinnis	10		Post, 1965.
Rivulus cylindraceus	24		Post, 1965.
R. santensis		48	Wickbom, 1943.
Jordanella floridae	24		Nogusa, 1955c.
Kosswigichthys asquamatus	24		Otzan, 1954; Karbe, 1961.
Aphyosemion sjörestedi	20	40	Post, 1965; Scheel, 1968.
A. filamentosum	18		Scheel, 1966a.
A. lujae	20	40	Scheel, 1968.
A. lonessense	10	20	Scheel, 1966a, 1968.
A. labarrei	14	28	Scheel, 1968.
A. gardneri Nigeria	18	36	Scheel, 1968.
A. gardneri Cameroons	20	40	Scheel, 1968.
A. franzwerneri	11	22	Scheel, 1970.
A. exiguum	18	36	Scheel, 1968.
A. cinnamomeum	20	40	Scheel, 1968.
A. christyi	9–15	18–30	Scheel, 1968.
A. celiae	10	20	Scheel, 1970.
A. cameronense	17	34	Scheel, 1968.
A. calliurum	16	32	Scheel, 1966a, 1968.
A. calliurum ahli	20	40	Post, 1965.
	16		Scheel, 1966a.
A. bualanum	18	36	Scheel, 1968.
A. arnoldi	20		Post, 1965.

	18		Scheel, 1966a.
A. bivittatum S.W. Nigeria	20	40	Scheel, 1966a, 1968.
A. bivittatum Cameroons	19	38	Scheel, 1968.
A. bivittatum E. Cameroons	17	34	Scheel, 1968.
A. bivittatum Rio Muni	13	26	Scheel, 1968.
A. bivittatum hollyi	20		Post, 1965.
A. australe	15		Scheel, 1966a.
A. bertholdi	21		Scheel, 1966a.
A. walkeri	18	36	Scheel, 1968.
A. santaisabellae	20	40	Scheel, 1968.
A. obscurum	17	34	Scheel, 1968.
A. vidianum	20	40	Scheel, 1968.
A. gulare	16	32	Scheel, 1968.
A. guineense	19		Scheel, 1966a.
A. spurelli	18		Scheel, 1966a.
A. nigerianum	18		Scheel, 1966a.
A. roloff	21		Scheel, 1966a.
Orestias luteus	24		Lueken, 1962.
O. agassizi	24		Lueken, 1962.
Pterolebias peruensis		20 ?	Post, 1965.
	27		Scheel, 1970.
Cyprinodon macularius	24		Post, 1965.
Nothobranchius palmquisti	19		Post, 1965.
	18		Scheel, 1968.
N. melanospilus	19		Post, 1965.
N. rachowi	8	16	Post, 1965.
	9		Scheel, 1966b, 1968.
Aplocheilichthys flavipinnis	24		Post, 1965.
A. katangae	24		Post, 1965.
Procatopus nototaenia	24		Post, 1965.
P. roseipinnis	24		Scheel, 1970.
P. similis	24		Scheel, 1970.

	n	$2n$	References
P. nigromarginatus	24		Scheel, 1970.
Nothobranchius orthonotus	18		Scheel, 1968.
Epiplatys sangmelinensis	24		Scheel, 1968.
Austrofundulus dolichopterus	22	44	Scheel, 1968.
Aplocheilus panchax	18	36	Scheel, 1966b, 1968.
A. lineatus	25	50	Scheel, 1966b, 1968.
Aphyoplatys duboisi	24	48	Scheel, 1966b, 1968.
Roloffia bertholdi	21	42	Scheel, 1966b, 1968.
R. geryi	20	40	Scheel, 1968.
R. guineensis	19	38	Scheel, 1966b, 1968.
R. occidentalis	23	46	Scheel, 1968.
R. roloffi	21	42	Scheel, 1966b, 1968.
Family: Poeciliidae			
Heterandria formosa	24		Post, 1965.
		46	Wickbom, 1943.
Gambusia holbrookii	24		Post, 1965.
		36	Geiser, 1924.
G. affinis		48	Chen and Ebeling, 1968.
Xiphophorus montezumae	24		Friedman and Gordon, 1934.
X. helleri	24	48	Friedman and Gordon, 1934; Wickbom, 1943; Karbe, 1958; Ohno and Atkin, 1966; Freye and Zernahle, 1967.
X. sp. X. jalapa	24		Ralston, 1934.
X. conchianus	24 ?		Friedman and Gordon, 1934; Ralston, 1934.
X. xiphidium	24		Friedman and Gordon, 1934; Lueken, 1962.
X. maculatus	24	48	Friedman and Gordon, 1934; Ralston, 1934; Wickbom, 1943; Freye and Zernahle, 1967.
Limia vittata	23	46	Wickbom, 1943.
L. melanogaster	24		Post, 1965.
L. caudofasciata reticolor		46	Wickbom, 1943.

Phalloceros caudomaculatus		46	Wickbom, 1943.
Phallichthys pittieri		46	Wickbom, 1943.
P. amates	24		Post, 1965.
Belonesox behzanus	24		Post, 1965.
Mollienisia sphenops	24		Post, 1965.
	19	38	Wickbom, 1943.
M. latipinna	24		Post, 1965.
M. sphenops	24		Post, 1965.
	23	46	Wickbom, 1943.
		36	Meyer, 1938.
M. sphenops × M. sphenops	24		Post, 1965.
	21	42	Wickbom, 1943.
M. velifera	23	46	Wickbom, 1943; Post, 1965.
Poecilia (Lebistes) reticulata	23	46	Winge, 1922; Iriki, 1932b; Wickbom, 1943; Post, 1965.
P. mexicana	23	46	Prehn and Rasch, 1969.
P. latipunctata	23		Prehn and Rasch, 1969.
P. vittata	23	46	Wickbom, 1943.
P. velifera	23	46	Wickbom, 1943; Post, 1965; Prehn and Rasch, 1969.
P. sphenops		36	Meyer, 1938.
	23	46	Wickbom, 1943; Prehn and Rasch, 1969.
	24		Post, 1965.
P. melanogaster	24		Post, 1965.
P. latipinna		36	Prehn and Rasch, 1969.
	23	46	Drewry, 1964.
	24		Post, 1965.
P. formosa	23	46	Wickbom, 1943; Drewry, 1964; Schultz, 1967; Prehn and Rasch, 1969.
P. caudofasciata	23	46	Wickbom, 1943.
Platypoecilus maculatus	24	48	Friedman and Gordon, 1934; Ralston, 1934; Wieckbom, 1943; Karbe, 1958; Freye and Zernahle, 1967.

	n	$2n$	References
P. variatus	24		Wikbom, 1943; Karbe, 1958.
	25		Friedman and Gordon, 1934.
P. xiphidium	24		Friedman and Gordon, 1934.
P. conchianus	24	48	Friedman and Gordon, 1934; Ralston, 1934; Wickbom, 1943; Karbe, 1958.
Poeciliopsis lucida		48	Schultz, 1967.
P. latidens		48	Schultz, 1967.
Family: Atherinidae			
Menidia notata		36	Moenkhaus, 1904.
Melanotaenia macculochi	24		Post, 1965.
Order: Beryciformes			
Family: Anoplogasteridae			
Scopelogadus microlepis bispinosus		46	Chen, 1969.
Order: Gasterosteiformes			
Family: Gasterosteidae			
Apeltes quadracus		46	Chen and Reisman, 1970.
Gasterosteus aculeatus		42	Swarup, 1959; Post, 1965; Chen and Reisman, 1970.
G. wheatlandi		42	Chen and Reisman, 1970.
Culaea inconstans		46	Chen and Reisman, 1970.
Pungitius pungitius		42	Makino, 1934a; Post, 1965; Chen and Reisman, 1970.
P. tymensis		42	Makino, 1934a.
Order: Scorpaeniformes			
Family: Synancejidae			
Inimicus japonicus		50	Nogusa, 1960.
Family: Hexagrammidae			
Hexagrammos octogrammus	24		Makino, 1937b.
H. otakii		48	Nogusa, 1960.

Family: Cottidae			
Cottus gobio	24		Post, 1965.
C. pollux	24	48	Nogusa, 1967a.
C. bairdii	18	36–38	Hann, 1927.
Order: Perciformes			
Family: Centropomidae			
Chanda ranga	22		Post, 1965.
Family: Serranidae			
Coreoperca kawametari	24	48	Nogusa, 1960.
Family: Centrarchidae			
Ambloplites rupestris		48	Roberts, 1964.
Notropis lutrensis	50		Lieppman and Hubb, 1969.
Acantharus pomotis		48	Roberts, 1964.
Pomoxis annularis		48	Roberts, 1964.
Ennecanthus chaetodon		48	Roberts, 1964.
Lepomis humilis		46	Roberts, 1964.
L. macrochirus		48	Roberts, 1964.
L. microlophus		48	Roberts, 1964.
L. marginatus		48	Roberts, 1964.
L. auritus		48	Roberts, 1964.
Chaenobryttus gulosus		48	Roberts, 1964.
Lepomis megalotis		48	Roberts, 1964.
L. gibbosus		48	Roberts, 1964.
		46	Fontana *et al.*, 1970.
L. cyanellus		46–48	Roberts, 1964; Ohno and Atkin, 1966.
Pomoxis nigromaculatus		48	Roberts, 1964.
Micropterus salmoides		46	Roberts, 1964.
M. dolomieu		46	Roberts, 1964.
Elassoma evergladei	24		Post, 1965.
E. zonatum	24	48	Roberts, 1964; Post, 1965.
Enneacanthus gloriosus		48	Roberts, 1964.
E. obesus		48	Roberts, 1964.

	n	$2n$	References
Centrarchus macropterus		48	Roberts, 1964.
Family: Percidae			
Perca fluviatilis		28	Svärdson and Wickbom, 1939.
		48	Nygren *et al.*, 1968a.
Lucioperca lucioperca		24	Svardson and Wickbom, 1939.
		46	Lieder, 1959.
		48	Nygren *et al.*, 1968a.
Acerina cernua	24	48	Lieder, 1963; Post, 1965; Nygren *et al.*, 1968a.
Family: Latilidae			
Sillago sihama		48	Nogusa, 1960.
Family: Monodactylidae			
Monodactylus argenteus	24		Post, 1965.
Family: Nandidae			
Nandus nandus		48	Kaur and Srivastava, 1965.
Family: Badidae			
Badis badis	24		Post, 1965.
Family: Cichlidae			
Homichromis bimaculatus	22		Post, 1965.
Haplochromis multicolor	22		Post, 1965.
Symphysodon aequifasciata		60	Ohno and Atkin, 1966
Pterophyllum eimekei	24		Post, 1965.
Nannacara anomala	24		Post, 1965.
Pelmatochromis kribensis	24		Post, 1965.
Apistogramma ramirezi	24		Post, 1965.
A. pertense	24		Post, 1965.
Cichlasoma severum	24		Post, 1965.
Lamprologus leleupi	24		Post, 1965.
Tilapia grahami	24	48 ?	Post, 1965.
Family: Labridae			
Family: Eleotridae			Nogusa, 1955b.

Pseudolabrus japonicus		46	Nogusa, 1960.
Ctenolabrus adspersus		38–48	Pinney, 1918.
Family: Eleotridae			
Mogrunda obscura	31	62	Nogusa, 1960.
Family: Gobiidae			
Acanthogobius flavimanus		44	Nogusa, 1957b.
Glossogobius giuris		46	Kaur and Srivastava, 1965.
Brachygobius nunus	24		Post, 1965.
Boleophthalmus pectinirostris	23	46	Nogusa, 1957b, 1960.
Gobius abei		52	Lieder, 1964.
Tridentiger obscurus		44	Nogusa, 1957b.
Chaenogobius isaza		46	Nogusa, 1957b.
Chaenogobius urotaenia		44	Nogusa, 1957b.
Family: Anabantidae			
Anabas testudineus		48	Kaur and Srivastava, 1956.
Macropodus opercularis	21		Post, 1965.
Betta splendens	21	42	Post, 1965.
Colisa fasciata		48	Kaur and Srivastava, 1965.
Ctenopoma ansorgei	24	?	Post, 1965.
Trichogaster trichopterus	23–24 ?		Post, 1965
Family: Mastacembelidae			
Rhynchobdella aculeata		48	Nogusa, 1957b.
Family: Mugilidae			
Mugil saliens		48	Cataudella and Capanna, 1971.
Family: Periophthalmidae			
Periophthalmus cantonensis	23	46	Nogusa, 1960.
Family: Callionymidae			
Callionymus richardsonii	19	38	Nogusa, 1960.
Family: Pholadidae			
Pholis pictus		46	Makino, 1937b.
Order: Pleuronectiformes			
Family: Bothidae			
Paralichthys olivaceus		46	Nogusa, 1960.

	n	$2n$	References
Xystreurys biolepis		48	Ohno and Atkin, 1966.
Family: Pleuronectidae			
Karenis bicoloratus		48	Nogusa, 1960.
Pleuronichthys verticalis		48	Ohno and Atkin, 1966.
Order: Tetraodontiformes			
Family: Balistidae			
Novodon modestus	20	40	Nogusa, 1960.
Stephanolepis cirrhifer		34	Nogusa, 1960.
CROSSOPTERYGII			
Order: Dipnoi			
Family: Lepidosirenidae			
Lepidosiren paradoxa		38	Agar, 1913; Makino, 1951; Ohno and Atkin, 1966.
Family: Protopteridae			
Protopterus annectens	—	34	Wickbom, 1945.
BRANCHIOPTERYGII			
Order: Polypteriformes			
Family: Polypteridae			
Calamichthys calabaricus	—	36	Capanna and Cataudella

References

Agar, W. E. (1913). Transverse Segmentation and Internal Differentiation of Chromosomes (*Lepidosiren*). *Q. Jl. Microsc. Sci.* **230**, 285–298.

Barigozzi, C. (1937). La gametogenesi e la sessualità di Cyprinus carpio var. specularis. *Atti. Soc. Ital. Sci. Nat.* **76**, 88–104.

Boothroyd, E. R. (1959). Chromosome studies on three Canadian populations of Atlantic salmon. *Salmo salar L. Can. J. Genet. Cytol.* **1** (2), 161–172.

Bungenberg de Jong, C. M. (1955). Cytological studies on *Salmo irideus*. *Genetica* **27**, 472–483.

Capanna, E., Cataudella, S. and Volpe, R. (1971). *Boll. Pesca Piscic. Idrobiol.* In press.

Capanna, E. and Cataudella, S. (1972). The Chromosomes of *Calamoichthys calabaricus* (*Pisces, Polyoterifones*). *Experientia* in press.

Cataudella, S. and Capanna, E. (1971). *Boll. Zool.* In press.

Chen, T. R. (1969). Karyological heterogamety of deep-sea fishes. *Postilla* **130**, 1–29.

Chen, T. R. and Ebeling, A. W. (1968). Karyological evidence of female heterogamety in the mosquito fish *Gambusia affinis*. *Copeia* **17**, 70–75.

Chen, T. R. and Ruddle, F. M. (1970). A Chromosome Study of four Species and a Hybrid of the Kill fish Genus *Fundulus* (Cyprinodontidae). *Chromosoma* **29**, 255–267.

Chen, T. R. and Reisman, H. M. (1970). A comparative chromosome study of the North American species of sticklebacks (Teleostei: Gasterosteidae). *Cytogenetics* **9**, 321–332.

Chiarelli, B., Ferrantelli, O. and Cucchi, C. (1969). The caryotype of some teleostean fish obtained by tissue culture in vitro. *Experientia* **25**, 426–427.

Cucchi, C. and Morittu, G. (1969). Il cariotipo di *Anguilla anguilla* delle acque delle valli di Comacchio. *Annali. Univ. Ferrara* **2**, 151–155.

Denton, T. E. and Howell, W. M. (1969). A Technique for obtaining chromosomes from the scale epitelinum of teleost fishes. *Copeia* **2**, 392–393.

Drewry, G. E. (1964). App. I Chromosome Number. *Bull. Tex. Mem. Mus.* **8**, 67–69.

Endo, A. and Ingalls, T. M. (1968). Chromosomes of the zebra fish. A model for cytogenetic, embriologic, and ecologic study. *J. Hered.* **59**, 382–384.

Ferrantelli, O., Chiarelli, B. and Cucchi, C. (1968). Colture "in vitro" di Tessuti di Pesce: risultati preliminari. *Boll. Soc. Ital. Biol. Sper.* **44**, 1165–1166.

Foley, J. O. (1926). The spermatogenesis of *Umbra limi* with special reference to the behavior of the spermatogonial chromosomes and the first maturation division. *Biol. Bull. Woods Hole* **50**.

Fontana, F. and Lamberti, L. (1970). Un metodo rapido per ottenere cromosomi somatici dall 'epitelio delle scaglie di pesci. *Boll. Biol. Sper.* **46**, 876–878.

Fontana, F., Chiarelli, B. and Rossi, A. (1970). Il cariotipo di alcune specie di Cyprinidae, Centrarchidae, Characidae studiate mediante colture in vitro. *Caryologia* **23 (4)**, 549–564.

Freye, H. A. and Zernahle, K. (1967). The chromosomes of *Xiphophorus helleri*, *Platypoecilus maculatus* and hybrids. *Biol. Zbl.* **86** (suppl.), 267–275.

Friedman, B. and Gordon, M. (1934). Chromosome numbers in Xiphophorin fishes. *Am Nat.* **68**, 446–455.

Gas, M. (1970). Revue bibliographique sur la caryologie des téléostéens. Etude critique des méthodes employées et des résultats obtenus. *Biol. Med.* **59**, 54–81.

Geiser, S. W. (1924). Sex ratios and spermatogenesis in the top-minnow, *Gambusia holbrooki* Grd. *Biol. Bull. Woods Hole* **47**, 175–212.

Hann, H. (1927). The history of the germ cells of *Cottus bairdii* Girard. *J. Morph. Philad.* **43**, 427–480.

Hitotsumachi, S., Sasaki, M. and Ojima, Y. (1969). A comparative Karyotype study in several species of Japanese loahces (Pisces, Cobitidae). *Jap. J. Genet.* **44 (3)**, 157–161.

Iriki, S. H. (1932b). Preliminary notes on the Chromosomes of Pisces. *Apeocheilus latipes* and *Lebistes reticulatus. Proc. Imp. Acad. Japan* **8**, 262–263.

Karbe, L. (1958). Chromosomenuntersuchungen an Fischen. Unveroeffentlichte Diplom. Arbeit.

Karbe, L. (1961). Cytologische Untersuchungen der Sterilitätsercheinungen bei anatolischen Zahnkarpfen, ein Beitrag zum Speziationsproblem. *Mitt. Hamb. Zool. Mus. Inst.* **59**, 73–104.

Karbe, L. (1963). Die Chromosomenverhältnisse bei den Coregonen des Bodensees und einiger weiterer voralpiner Seen, ein Betrag zum Problem der Speziation in der Gettung Coregonus. *Z. Zool. Syst. Evol.Forsch.* **2**, 18–40.

Kastschenko, W. (1890). *Z. Wiss. Zool.* **50**, 428–442.

Katayame, M. (1937). On the spermatogenesis of the Teleost *Oryzias latipes. Bull. Jap. Soc. Scient. Fish* **5**, 277–278.

Kaur, D. and Srivastava, M. D. L. (1965). The structure and behaviour of chromosomes in five fresh-water teleosts. *Caryologia* **18** (2), 181–191.

Kupka, E. (1948). Chromosomale Verschiedenheiten bei schweizerischen Coregonen. *Revue Suisse Zool.* **55**, 285–293.

Labat, R. (1967). Technique de culture de leucocytes de *Cyprinus carpio. L.C.R. Hebd. Séanc. Acad. Sci. Paris* **264**, 2473–2475.

Lieder, U. (1953). Chromosomenstudien an Knochenfischen. I. Die Chromosomen des Hechtes (Esox lucius L.) in den embryonalen Mitosen. *Z. Fisch.* **2**, 417–418.

Lieder, U. (1954a). Chromosomenstudien an Knochenfischen. II. Uber die chromosomenzahl und morphologie der Plötze (*Leuciscus rutilus* L.) und einiger ihrer Bastarde mit anderen Cypriniden. *Z. Fisch.* **3**, 375–382.

Lieder, U. (1956). Chromosomenstudien an Knochenfischen. IV. Die chromosomenverhältnisse bei der Regenbogen und Bachforelle und ihren Bastarden. *Z. Fisch.* **4**, 589–594.

Lieder, U. (1959). Die Chromosomenhältnisse von *Esox lucius* L., *Perca fluviatilis* L. und *Lucioperca Lucioperca L.* (Vertebrata Pisces). *Arch. Zool.* 2 Ser. 12, 123–126.

Lieder, U. (1963). On presumptive sex chromosomes in *Perca*, *Acerina* and *Anguilla. Biol. Zbl.* **82** (3): 296–302.

Lieder, U. (1964). Über den gegenwärtigen Stand und die Methodik der Chromosomen untersuchungen bei Fischen. *Z. Fisch.* **11**, 673–684.

Lieppman, M. and Hubb, C. (1969). A karyological Analysis of two cyprinid fishes, *Notemigonus crysoleucas* and *Notropis lutrensis. Tex. Rep. Biol. Med.* (2), 427–435.

Lueken, W. (1962). Chromosomenzahl bei *Orestias* (Pisces Cyprinodontidae). *Mitt. Hamb. Zool. Mus. Inst.* **60**, 195–198.

Lueken, W. and Foster, W. (1969). Chromosome studies on fish, using a simplified cell culture technique. *Zool. Anz.* **183**, 168–176.

Makino, S. (1934a). The chromosomes of the sticklebacks, *Pungitius tymensis* and *P. pungitius. Cytologia* **5**, 155–168.

Makino, S. (1934b). Notes on the chromosomes of some fresh-water teleosts. *Jap. J. Genet.* **9** (2), 100–103.

Makino, S. (1937a). The chromosomes of two Elasmobranch fishes. *Cyt. Fujii Jubilee* **867**.

Makino, S. (1937b). Notes on the chromosomes of some teleost fishes. *Z. M. (Japan)* **49**.

Makino, S. (1939). The chromosomes of the carp (*Cyprinus carpio*), including those of some related species of Cyprinidae for comparison. *Cytologia (Tokyo)* **9**, 430–440.

Makino, S. (1941a). The chromosomal relation between the two allied species of the loach (Cobitidae-Pisces). *Cytologia* **12**, 79–82.

Makino, S. (1941b). A Karyological study of goldfish of Japan. *Cytologia (Tokyo)* **12**, 96–111.

Makino, S. (1951). "An Atlas of the Chromosome Numbers in Animals". Iowa State College Press.

Matthey, R. (1937). La formule cromosomiale du Sélacien Scylliorhynus catula. *C. R. Séanc. Soc. Biol.* **126**, (27), 388–389.

Mayers, L. J. and Roberts, F. L. (1969). Chromosomal Homogeneity of Five Populations of Alewives *Alosa pseudoharengus. Copeia* 313–317.

Meyer, H. (1938). Investigations concerning the Reproductive Behaviour of Mollienesia "formosa". *J. Genet. Lond.* **36** (3), 329–366.

Moenkhaus, W. J. (1904). The development of the hybrid between Fundulus heteroclitus and *Menidia notata* with special reference to the behavior of the maternal and paternal chromatin. *Am. J. Anat.* **3**, 29–65.

Moore, J. E. S. (1895). *Q. Rev. Microsc. Sci.* **38**, 275–313.

Muramoto, J. (1969). On the chromosomes of Parasilurus asotus. *Chrom. Inform. Ser.* **10**, 18–19.

Muramoto, J., Ohno, S. and Atkin, N. B. (1968). On the Diploid State of the Fish Order Ostariophysi. *Chromosoma* **24**, 59–66.

Nayyar, R. P. (1966). Karyotype studies in thirteen species of fishes. *Genetica* **37**, 78–92.

Nogusa, S. (1955a). Note on the chromosomes of *Abbotina vulgaris* (Cyprinidae). *Jap. J. Genet.* **30** (1): 10–11.

Nogusa, S. (1955b). Chromosome studies in Pisces. IV. The chromosomes of

Mogrunda obscura (Gobiidae with evidence of male heterogamety). *Cytologia (Tokyo)* **20**, 11–18.

Nogusa, S. (1955c). Chromosome studies in Pisces. V. Variation of the chromosome number in *Acheilognathus rhombea* due to multiple chromosome formation. *Anat. Zool. Japan* **28**, 249–255.

Nogusa, S. (1957a). Chromosome studies in Pisces. VI. The X–Y chromosomes found in *Cottus pollux* Günther (Cottidae). *J. Fac. Sci. Hokkaido Univ.* **13**, 289–292.

Nogusa, S. (1957b). Chromosome studies in fishes. VII. A comparative study of the chromosomes in six species of Gobiidae. *Jap. J. Ichthyol.* **6**, 141–146.

Nogusa, S. (1960). A comparative study of the chromosomes of fishes with particular considerations on taxonomy and evolution. *Mem. Hyogo Univ. Agric.* **3** (1), 1–62.

Nygren, A., Edlund, P., Hirsch, W. and Ahsgre, L. (1968a). Cytological studies in perch (*Perca flaviatilis* L.) pike (*Esox lucius*), pike-perch (*Lucioperca lucioperca* L.) and ruft (*Acerina cernua* L.). *Hereditas* **59**, 518–524.

Nygren, A., Nilsson, B. and Jahnke, M. (1968b). Cytological studies in Atlantic Salmon (*Salmo salar*). *Ann. Acad. Reg. Sci. Ups.* **12**.

Nygren, A., Nilsson, B. and Jahnke, M. (1971a). Cytological studies in *Thymallus thymallus* and *Coregonus albula*. *Hereditas* **67**, 269–274.

Nygren, A., Nilsson, B. and Jahnke, M. (1971b). Cytological studies in Coregonus from Sweden. *Ann. Acad. Reg. Scien. Ups.* **15**, 5–20.

Nygren, A., Nilsson, B. and Jahnke, M. (1971c). Cytological studies in *Salmo trutta* and *Salmo alpinus*. *Hereditas* **67**, 259–268.

Nygren, A., Nilsson, B. and Jahnke, M. (1971d). Cytological study in Hypotremata and Pleurotremata (Pisces). *Hereditas* **67**, 275–282.

Nygren, A., Nilsson, B. and Jahnke, M. (1971e). Cytological studies in the smelt *Osmerus eperlanus* L. *Hereditas* **67**, 283–286.

Ohno, S. (1969). Diploid-tetraploid relationship in Clupeoid and Salmonid fish. Heredity 22. Suppl. Vol. *Chrom. Today* **2**, 139–147.

Ohno, S. and Atkin, N. B. (1966). Comparative DNA values and chromosome complements of eight species of fishes. *Chromosoma* **18**, 455–466.

Ohno, S., Stenius, C., Faisst, E. and Zemzes, M. T. (1965). Post-zygotic chromosomal rearrangements in rainbow trout (*Salmo irideus* Gibbons). *Cytogenetics* **4**, 117–129.

Ohno, S., Muramoto, J., Christiian, L. and Atkin, N. B. (1967). Diploid-Tetraploid Relationship among Old-World Members of the fish Family Cyprinidae. *Chromosoma* **23**, 1–9.

Ohno, S., Muramoto, J., Stenius, C., Christian, L., Kittrel, W. A. and Atkin, N. B. (1969). Microchromosomes in Holocephalian, Condrostean and Holostean Fishes. *Chromosoma* **26**, 35–40.

Ojima, Y. and Hitotsumachi, S. (1967). Cytogenetic studies in power vertebrates. IV. A note on the chromosomes of the carp (*Cyprinus carpio*) in comparison with those of the funa and the goldfish *Jap. J. Genet.* **42** (3), 163–167.

Ojima, Y., Maeki, K., Takayama, S. and Nogusa, S. (1963). A cytotaxonomic study on the Salmonidae. *Nucleus* **6**, 91–98.

Ojima, Y., Hitotsumachi, S. and Makino, S. (1966). Cytogenetic studies in lower vertebrates. I. A preliminary report on the chromosomes of the funa (*Carassius auratus*) and the gold fish. *Proc. Jap. Acad.* **24**, 62–66.

Otzan, N. (1954). Cytological investigation of the sexual differentiation in the hybrids of Anatolin Cyprinodontids. *Rev. Fac. Sci. Univ. Istanbul* **19**, 245–280.

Pinney, E. (1918). A study of the relation of the behaviour of the chromatin to development and heredity in teleost hybrids. *J. Morph.* **31**, 225.

Pomini, F. P. (1939). Ricerche sullo sviluppo delle Trote Italiane. 3° Stadi larvali e primi stadi post-larvali delle Trote di ruscello (*Salmo fario* L.). *Arch. Zool. Ital.* **27**, 383–404.

Post, A. (1965). Vergleichende Untersuchungen der Chromosomenzahlen bei Süsswasser-Teleosteen. *Z. Zool. Syst. Evol.Forsch.* **3** (1/2), 47–93.

Prakken, R. (1955). The chromosomes of *Esox lucius* L. *Genetica* **27**, 484–488.

Prehn, L. M. and Rasch, M. (1969). Cytogenetic studies of Poecilia (Pisces). I. Chromosome number of naturally occurring poeciliid species and their hybrids from eastern Mexico. *Can. J. Genet. Cytol.* **11**, 880–895.

Prokofieva, A. (1934). On the chromosome morphology of certain pisces. *Cytologia* **5**, 498–506.

Ralston, E. M. (1934). A Study of the chromosomes of *Xiphophorus*, *Platypoecilus* and Xiphophorus × Platypoecilus hybrids during spermatogenesis. *J. Morph.* **56**, 423–432.

Rawitz, B. (1899). *Arch. Mikrosk. Anat.* **53**, 481–513.

Rees, H. (1964). The question of polyploidy in the Salmonidae. *Chromosoma* **15**, 275–279.

Rees, H. (1967). The chromosomes of *Salmo salar*. *Chromosoma* **21**, 472–474.

Roberts, F. L. (1964). A chromosome study of twenty species of Centrarchidae. *J. Morph.* **115**, 401–417.

Rodolico, A. (1933). Differenziamento sessuale ed ovospermatogenesi nell'Anguilla. *Monitore. Zool. Ital.* **43** suppl. 212.

Ruckert, J. (1892a). *Anat. Anz.* **7**, 107–158.

Ruckert, J. (1892b). *Anat. Anz.* **8**, 44–52.

Sasaki, M., Hitotsumachi, S., Makino, S. and Terao, T. (1968). A comparative study of the chromosomes in the chum salmon, the kokanee salmon and their hybrids. *Caryologia* **21** (4), 389–394.

Scheel, J. J. (1966a). Notes on phenotypy, distribution and systematics of *Aphyosemion bivittatum* (Loenberg), with remarks on the chromosome number in the rivulinae. Ichthyologica. *Acquar. J.* **3**, 261–278.

Scheel, J. J. (1966b). Taxonomic studies of African and Asian toothcarps (Rivulinae) based on chromosome numbers, haemoglobin patterns, some morphological traits and crossing experiments. *Vidensk. Meddr. Dansk. Naturh. Foren.* **129**, 123–148.

Scheel, J. J. (1968). "Rivulins of the Old World". TFH Publ. USA.

Scheel, J. J. (1970). "Freshwater Fishes of the World." Vol. II pp. 358–444. G. Sterbe.

Schreiner, A. and Schreiner, K. E. (1904). Die Reifungsteilung bei den Wirbeltieren. Ein Beitrag zur Frage nach der Chromatin-reduktion. *Anat. Anz.* **24**, 561–578.

Shultz, R. J. (1967). Gynogenesis and Triploidy in the Viviparous Fish Poeciliopsis. *Science N.Y.* **157** (3796), 1564–1567.

Sick, K., Westergaard, M. and Frydenberg, O. (1962). Haemoglobin pattern and chromosome number of American, European, and Japanese eels (*Anguilla*). *Nature, (Lond.)* **193**, 1001–1002.

Simon, R. C. (1963). Chromosome morphology and species evolution in the five North American species of Pacific Salmon (*Oncorhynchus*). *J. Morph.* **112**, 77–95.

Simon, R. C. and Dollar, A. M. (1963). Cytological aspects of speciation in two North American teleosts, *Salmo gairdneri* and *Salmo Clarki Lewisi*. *Can. J. Genet. Cytol.* **5**, 43–49.

Srivastava, M. D. L. and Das, B. (1969). Somatic chromosomes of teleostean Fish. *J. Hered.* **60**, 57–58.

Svärdson, G. (1941). Somatic pairings in Salmo and Coregonus and its hypothetical explanation. *Arch. Zool.* **33B** 9, 1–6.

Svärdson, G. (1945). Chromosome studies in Salmonidae. *Rep. Short Pap. Inst. Freshwat. Res. Drottningholm* **23**, 1–151.

Svärdson, G. and Wickbom, J. (1939). Notes on the chromosomes of some Teleosts, *Esox lucius* L., *Lucioperca lucioperca* L. and *Perca fluviatilis* L. *Hereditas* **25**, 472.

Swarup, H. (1959). Production of triploidy in *Gasterosteus aculeatus*. *J. Genet.* **56**, 129–142.

Vaupel, J. (1928). The spermatogenesis of *Lebistes reticulatus*. *J. Morph.* **47**, 555–587.

Vialli, M. (1970). Contributo dei dati sul numero cromosomico alla soluzione negativa del discusso rapporto di specie diploide e tetraploide tra tinca e carpa. *Rc. Ist. Lomb. Sci. Lett.* **103**, 47–51.

Wahl, R. W. (1960). Chromosome morphology in laketrout *Salvelinus namaycush*. *Copeia* 1960 **(1)**, 16–19.

Wickbom, T. (1943). Cytological studies on the family Cyprinodontidae. *Hereditas* **29**, 1–24.

Wickbom, T. (1945). Cytological studies on Dipnoi, Urodela, Anura, and Emys. *Hereditas* **31**, 241.

Winge, O. (1929). The experimental alternation of sex chromosomes into autosomes and vice versa as illustrated by Lebistes. *C. R. Lab. Carlsberg Serphysiol.* **21**, 1–49.

Wright, J. E. (1955). Chromosome numbers in Trout. *Progve. Fish Cult.* **17**, 172–176.

Wolf, U., Ritter, H., Atkin, N. B. and Ohno, S. (1969). Polyploidization in the fish Cyprinidae order Cypriniformes. I. DNA content and chromosome sets in various species of Cyprinidae. *Human genetics* 7, 240–244.

11

Amphibia

ALESSANDRO MORESCALCHI
Institute of Histology and Embryology, University of Naples, Italy

I. INTRODUCTION

A. *Phylogenetic Problems on the Amphibia*

Among the Vertebrates, the living Amphibia—subdivided into three orders of Apoda (or Gymnophiona), Urodela (or Caudata) and Anura (or Salientia); although this last is often considered a super-order)—are perhaps the most problematical of all the tetrapods from the point of view of establishing a phylogeny; this is due, on the one hand, to the degree of specialization attained by the members of each of the three orders and, on the other hand, to the fact that between them and the Paleozoic group of amphibians is a broad evolutionary gap, not filled by fossil material (Romer, 1966).

The origins and relationships of the Recent Amphibia are uncertain and have been the subject of a great variety of theories; I will mention here the three that have proved the most acceptable, at various times.

According to the first theory, the Anura (with the Labyrinthodontia and the Amniota) are descended from the Crossopterygii, while the Urodela would seem to derive from a stock of Dipneusti (Holmgren, 1952). This hypothesis, based on theories on the evolution of the chiridium that have turned out to be unfounded, is now considered somewhat improbable, also because the Dipnoi appear too old and too "conservative" to enable them to have given rise to certain groups of modern Amphibia (cf. Schmalhausen, 1968); I have considered it opportune to mention it since beginning with Wickbom (1945, 1950a), numerous karyologists have considered it to be valid.

Also the second theory, advanced by Jarvik (1942) but taken up by various later authors (Huene, 1956; Bertmar, 1969), favours a different origin of the living Amphibia: according to it, the Anura (with the ancestors of the Amniota) are derived from a group of Rhipidistia, the Osteolepiformes, while Urodela and Apoda are derived from a second group of these Crossopterygii, the Porolepiformes. The subdivision of the Rhipidistia into two groups (orders or super-families), chiefly based on the snout morphology, is now considered to be rather unfounded, mainly on the basis of the discovery of new fossils revealing various stages of passage between the two groups in the morphology of the nasal cavities; consequently, the Rhipidistia would constitute a homogeneous taxon (Thomson, 1964).

According to the third theory, which today finds a great many supporters, the Recent Amphibia are derived from a relatively restricted group of Paleozoic Amphibia (though there exist various uncertainties regarding their identification) and therefore constitute a monophyletic group, the sub-class (or infra-class) Lissamphibia *sensu* Gadow (Parsons and Williams, 1963). According to certain Paleontologists, evidence of the relationships of the Lissamphibia to the Paleozoic Amphibia is unsatisfactory, and it is then preferable to recognize three independent subclasses: Labyrinthodontia, Lepospondyli and Lissamphibia (Baird, 1965; Panchen, 1967); recent findings, however, indicate the presence of labyrinthodont forms, possibly on the line of the Lissamphibia, in the Lower Permian of the United States (Bolt, 1969); this fact would strengthen the theories that the Lissamphibia originate from the Dissorophidae (Estes, 1965).

From the anatomical and biological point of view, the three orders of living Amphibia reveal various differences though, in the opinion of various authors, these are, on the whole, less important than the reciprocal resemblances, interpreted as true homologies (Szarski, 1962).

The Apoda, often considered as mono-familial, have recently been subdivided by Taylor (1968, 1969) into three or four families; they are worm-like limbless organisms, whose affinities with the remaining Amphibia are the most problematical (Eaton, 1959; Reig, 1964). No certain fossils of the order are known, and many of the morphological and biological characteristics shown by the present representatives require more thorough study, which is now being carried out (Wake, 1968). The Apoda, specialized for burrowing life, have reduced or lost various anatomical characteristics; however, there remain others, especially cranial and cutaneous, that are absent in the other present-day Amphibia but present in those of the Paleozoic period (Noble, 1931).

We shall see later that the karyological data seem to point to a greater affinity of these Amphibia with the Anura.

According to Noble's "classical" systematics, the present Urodela may be divided into 5 distinct suborders, among which that of the Cryptobranchoidea (families Hynobiidae and Cryptobranchidae) includes the more primitive forms, similar to the ancestral forms of the other suborders. These others are: the Ambystomatoidea (family Ambystomatidae), the Salamandroidea (families Salamandridae, Amphiumidae and Plethodontidae), the Proteida (family Proteidae) and the Meantes (family Sirenidae). The last two, each comprising few pedomorphic species, are the most problematical from the phylogenetic point of view, and their relationships are still debated today; Ambystomatoidea and Salamandroidea (the so-called "advanced" or "higher" Urodela) seem, however, to be of clear hynobiid derivation (especially the former). Nevertheless, the composition of the two suborders into families, as assumed by Noble, has aroused various criticisms, and today it seems more probable that the Plethodontidae (and perhaps also the Amphiumidae) are related to the Ambystomatidae (and hence included among the Ambystomatoidea) rather than to the Salamandridae (cf. Laurent, 1947; Regal, 1966; Wake, 1966; Salthe, 1967). The recent increase in paleontological studies on the Urodela has led to a knowledge of Mesozoic fossils of considerable systematic interest; progress in this type of research will probably lead, in the near future, to a better definition of the uncertain systematics of these Amphibia (cf. Estes, 1965, 1969a; Nevo and Estes, 1969).

The Urodela appear, morphologically, as the forms that are relatively the least differentiated from certain small Paleozoic Amphibia (Reig, 1964); also their characteristics of development indicate that they may be considered as the group of Lissamphibia that has remained closest to the ancestral stock of the subclass (cf. Romer, 1966). As we shall see, the karyological data, in addition to being in agreement in various points with the present theories on the phylogeny of the order, suggest some new facts about the position of certain pedomorphic families.

Although, in certain characteristics, Apoda and Urodela probably resemble their Paleozoic ancestors, the Anura reveal entirely new characteristics: "As an order, frogs are so specialized that a tremendous morphological gap separates them from all other amphibians" (Inger, 1967). Unfortunately, this gap is not filled by probative fossil findings: only the problematical *Triadobatrachus* (= *Protobatrachus*) of the Triassic might be a transitional form (Proanura) towards the living Anura (Euanura) (cf. discussion in Hecht, 1962); the scanty fossils of Jurassic and Cretaceous periods belong to forms already highly specialized, such that they can be included in families that are either still existing today or are closely related to them (Reig, 1958, 1961; Hecht, 1963; Nevo, 1968). Also the living Anura can be divided, on anatomical bases that have been confirmed by the paleontological findings, into two distinct levels of organization, with "primitive" and "higher" families; however, the phylogenetic relationships between and within these two levels are extremely difficult to clarify. This is due, in fact, to the high degree of specialization, with the consequent simplification of various osteological structures and the development along parallel lines of other structures, which makes this group of Amphibia very homogeneous, morphologically; as Inger states: ". . . the variation within the order, despite adaptations to fossorial, aquatic, arboreal or terrestrial life, is slight." The presence of a great number of evolutional parallelisms and convergences in various anatomical and biological characteristics of the Anura creates an uninterrupted series of problems of interpretation: it may be said that nearly every author that has been concerned with the phylogeny of the order has proposed modifications, in varying degrees, to the pre-existing systematic theories, attributing a different value or meaning to these characteristics (cf. Griffiths, 1963; Hecht, 1963; Tihen, 1965; Špinar, 1967; Laurent, 1967); the problem does not seem to have been simplified by modern attempts at a solution, made with the introduction of methods that are possibly objective in the taxonomy of the Anura (cf. Inger, 1967; Kluge and Farris, 1969).

I shall return to certain details of systematics in the sections devoted to the karyology of the various families of the order; in general, for the present, it may be said that various authors accept the theory of an early dichotomy among the Euanura, which would have given rise, on the one hand, to the Pipids and to related groups now extinct and, on the other hand, to the rest of the families; the original stock of the latter would be a group of amphicoelous forms, of which the Ascaphidae have survived. Apart from certain families judged to be "transitional" (such as Pelobatidae and Discoglossidae), the great branch of the higher Anura (Neobatrachia, acc. Reig, 1958) would have dichotomized early into a group of families with a prevalently arciferous pectoral girdle (Leptodactylidae, Hylidae and Bufonidae, approximately corresponding to Noble's Procoela) and another group with a firmisternous girdle (Ranidae, Rhacophoridae and, with various doubts, Microhylidae, corresponding approximately to Noble's Diplasiocoela) (Noble, 1931; Reig, 1958, 1961; Griffiths, 1963; Casamiquela, 1965; Hecht, 1969).

However, we are still far from having reached agreement in the various ideas on the relationships between the different families of Anura; hence the attempts—which have now been extensively developed—to approach the phylogenetic problem from other points of view than a strictly anatomical one. Among these studies, those of a cytotaxonomic nature seem capable —thanks to the researches of various authors of different continents—of making an important contribution to the solution of various phylogenetic questions; in particular, this type of study would appear to have a value that is probably decisive today for interpreting the speciation mechanisms in certain groups of Anura.

B. *Criteria for a Cytotaxonomic Study*

Starting from the assumption—by now generally unquestioned—that the morphology of the karyotype is one of the characteristics of the phenotype of a given organism, and that it is species-specific and evolves along its own lines, which are largely independent of those of the genotype (cf. White, 1954; Swanson *et al.*, 1967; John and Lewis, 1968), I have applied to the karyotype of Amphibia the same logical principles adopted by Inger (1967) for studying the evolution of the morphological and biological characteristics of the Anura. The central problem is to establish what has been the sequence of events that have characterized the karyological evolution of the Amphibia; it may be formulated under two aspects, which may be approached separately: (i) a matter of determining which of the characteristics of the karyotype of the Amphibia are primitive and which are derived and (ii) a matter of determining which of these characteristics may have arisen only once in the phylogeny of the Amphibia and which may have arisen several times, independently of one another.

As regards (i), I maintain that it is possible to give a fairly exhaustive answer to the problem. It must first be stated that of the specific morphological characteristics of the karyotype—number, shape and size of the chromosomes—only the first two can be utilized in problems of macrosystematics of the Amphibia; the chromosome sizes, which depend on the quantity of nuclear DNA (Ullerich, 1966; Griffin *et al.*, 1970; cf. Morescalchi, 1970), seem, like the latter, to assume a role of an adaptive nature in these organisms, varying at inter-specific level independently of the phylogenetic position (Mirsky and Ris, 1951; Stebbins, 1966; Goin *et al.*, 1968): these problems will be examined more extensively in Section II.F. It therefore remains to determine which characteristics in the number and shape of the chromosomes are primitive in the phylogeny of the Amphibia; this may be done according to two criteria.

The first consists in seeking a karyotype model that is common to the more generalized forms of all three Recent orders, because, as is the case in the criteria for determining the anatomical characteristics, a characteristic common to the less specialized members of related taxa must probably be considered primitive. As we shall see in the various sections on the karyotype of Apoda, Urodela and Anura (Sections II.A, B and C), there exists a karyological formula common to the more generalized families of the three orders, consisting of a high number of chromosomes, many of which are acrocentric and microchromosomes: in all probability, this formula must be considered the primitive one in the Amphibia.

The second criterion consists in considering as primitive a karyotype that may have given rise to a second one, from which it cannot be derived. As we shall see in the sections mentioned, the numerically reduced karyotype, lacking in microchromosomes and—with only a few exceptions—in acrocentrics, which is typical of the higher Amphibia, may be made to derive from one similar to that described here as primitive, by postulating a series of equal chromosome rearrangements in the three orders (cf. Section II.E). This theory may be corroborated by the fact that, as in the reptilian stock, which probably evolved parallel with that of the Lissamphibia, there is a numerical reduction in the microchromosomes and often also in the other chromosomes; however, for the present, it is not proved that a karyotype similar to that of the higher Amphibia cannot give rise to one of the type presumed primitive. Actually, certain phenomena of karyological evolution might even occur cyclically, as is assumed by certain authors (cf. Todd, 1967, 1970); in this case, however, the systematically highest position of the species involved in a secondary evolutional cycle may be determinant also for an understanding of its cytotaxonomic position (cf. the case of the Eleutherodactylini, Section II.C).

These last considerations have brought us to the second aspect of the problem (ii), regarding the determination of the uniqueness or multiplicity of a given characteristic of the karyotype arising. Just as in their anatomical evolution, so also in their karyological differentiation the Amphibia seem subject to considerable parallelisms and convergences, because the presence of almost identical karyotypes is noted in groups that are relatively far removed from one another; generally, it is the higher families that possess very similar karyotypes, whereas in the other families the inter-generic karyological variability may be high and may at times be useful for assessments of a systematic nature.

Consequently, the cytotaxonomic study of the higher families is found to be immediately useful only at very general levels, except in certain cases of broad inter-specific differentiation, which may become interesting for the problems of speciation.

This limitation is not exclusive, as is well known, of the karyological characteristics, especially as regards the Amphibia; consequently, the integration of these data with anatomical or other types of data (which I have always tried to bear in mind in this work) may always provide useful indications for a precise definition of the evolutional lines followed by these Vertebrates.

Regarding the nomenclature used in this work for the mitotic chromosomes, I have adopted, like various other authors, that which was proposed by Levan *et al.* (1964), excluding, for practical reasons, the two "extreme" classes of chromosomes (*M* chromosomes, with a perfectly median centromere, and *T* chromosomes, with a perfectly terminal centromere), the precise definition of which is difficult to establish in the works of various authors, and in any case, is of little use in a work having the characteristics of this present one. Furthermore, since it is the most commonly used, I have adopted the "non-recommended" terminology of Levan *et al.*; the nomenclature used here is the following, the arm ratio being the ratio between the lengths of the long and the short arms of each chromosome:

Centromeric position	Arm ratio	Chromosome designation
Median region	1.0–1.7	*m* = metacentric
Submedian	1.7–3.0	*sm* = submetacentric
Subterminal	3.0–7.0	*st* = subtelocentric
Terminal region	7.0–∞	*t* = telocentric (acrocentric)

II. CYTOTAXONOMY OF THE AMPHIBIA

A. *Apoda*

The systematics of these Amphibia, the least studied among the three modern orders, is still uncertain; in general, they are placed together in a single family (Caeciliidae), though Taylor (1969) has recently proposed a subdivision of the Apoda into 4 families: Ichthyophidae, the most primitive, with 4 genera; Typhlonectidae, with 4 genera; Scolecomorphidae, with one genus only and Caeciliidae, with 25 genera, the most widespread and rich in species.

Fossils of Apoda are not well known; Taylor (1968) refers to a discovery made by Marcus regarding a presumed member of the order of the Carboniferous period; the diffusion of the Gymnophiona, mostly relicts in the Tropics of the Old and New World, together with certain anatomical data of the living species, actually suggests that this group of Amphibia must be very old (Eaton, 1959).

The Apoda have been little studied, karyologically. The chromosomes of 4 species are known, one of Ichthyophidae and three of Caeciliidae:

Ichthyophidae:	*Ichthyophis glutinosus*:	$2n = 42$	(Seshachar, 1937)
Caeciliidae:	*Uraeotyphlus narayani*:	$2n = 36$	(Seshachar, 1939)
	Geotrypetes seraphini:	$2n = 38$	(This paper)
	Caecilia occidentalis:	$2n = 24$	(Barrio and Rinaldi de Chieri, 1970c)

Ichthyophis ($2n = 42$) seems to possess 10 two-armed chromosomes and 32 one-armed chromosomes (NF = 52); from a study of the figures given by Seshachar, it seems to me probable, on the basis of my experience in comparing the karyological findings obtained with the old and the new techniques, that certain of the "acrocentric" chromosomes described by the Indian author may be *st*, so that the NF would be higher than 52; moreover, certain of the smaller chromosomes, described as "puntiform", are microchromosomes. The karyotype of this Apod, which might be the most primitive in the order (Taylor, 1968), may therefore resemble, as regards the number, shape and size of the chromosomes, that of certain primitive Anura, such as *Ascaphus* and *Alytes*; according to Seshachar (1938, 1939), the spermatocyte bivalents of *Ichthyophis* are of the "urodelan" type, like those of *Alytes* and *Bombina*.

Among the Caeciliidae, *Uraeotyphlus narayani* ($2n = 36$) of India would have the same NF as *Ichthyophis*, possessing 16 two-armed chromosomes and 20 one-armed chromosomes; the spermatocyte meiosis of this species would then be similar to that of *Ichthyophis* (Seshachar, 1939, 1941).

I have studied the karyology of an African species of the family, *Geotrypetes seraphini*, with $2n = 38$: there are 6 pairs of large homologues, *m* or *sm*; 6 pairs of medium-small chromosomes, all *st* except for one pair of *m*; 3 pairs of small homologues, *m*, and 4 pairs of *t* microchromosomes (Fig. 1): the diploid set of this species, probably similar to that of *Uraeotyphlus*, resembles that of the Discoglossid *Alytes* (Morescalchi, 1966). In the spermatocyte line, however, *Geotrypetes* frequently reveals bivalents with terminal chiasmata, as occurs in the higher Anura but sometimes also in certain Discoglossids (cf. Section D below).

Caecilia occidentalis, of South America, has $2n = 24$, with 6 pairs of large and 6 pairs of small homologues, all *m* or *sm* (NF = 48) (Barrio and Rinaldi de Chieri, 1970c); according to these authors, another South American species, *Chthonerpeton indistinctum* (Typhlonectidae), reveals a karyotype that is only slightly different from that of *Caecilia*; the chromosome set of these two Apoda therefore approaches that of certain higher Anura; no data exist on the spermactoyte bivalents of these two species.

The data so far collected on the Apoda are too scanty to enable any conclusions on the karyological evolution of the order, comparable to those that can be drawn on the Anura and Urodela, as we shall see. In the morphology of the mitotic chromosomes the 4 species studied approach certain Anura; to be precise, *Ichthyophis*, *Uraeotyphlus* and *Geotrypetes* resemble

Fig. 1. Intestinal metaphase plate of *Geotrypetes seraphini* ♂ ($2n = 38$).
Fig. 2. Spermatogonial metaphase plate of *Hynobius nebulosus* ($2n = 56$).
Fig. 3. Intestinal metaphase plate of *Hynobius dunni* ♂ ($2n = 56$).
Fig. 4. The diploid karyotype of *H. nebulosus* (from the plate in Fig. 2).
Fig. 5. Spermatogonial metaphase of *Megalobatrachus japonicus* ($2n = 64$) (from Iriki, 1932).

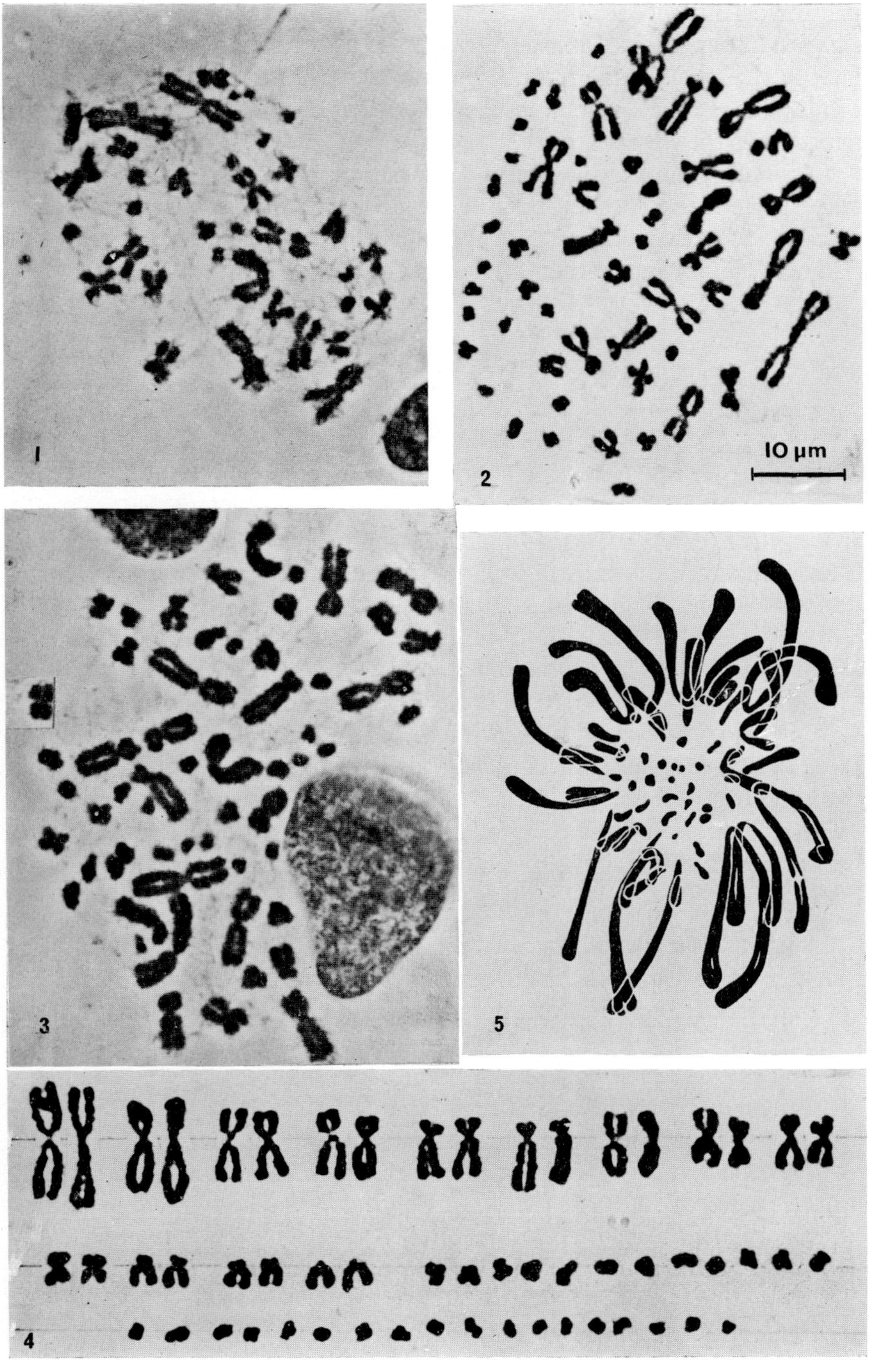
1
2
10 µm
3
5
4

primitive Anura (Ascaphids or, possibly, rather Discoglossids), whereas *Caecilia* (and perhaps *Chthonerpeton*) resemble higher Anura (Hylids etc. cf. later in the text). In the morphology of the spermatocyte bivalents, part of the Apoda studied resemble the primitive Anura, whereas *Geotrypetes* approaches the higher Anura.

Obviously, the karyological differentiation of the two orders may have taken place along parallel lines; however, in my opinion, the karyological resemblances between various Apoda and the more primitive families of the Anura are highly interesting (cf. the conclusions).

B. *Urodela*

(a) *Hynobiidae*

This family of Urodela, with complete metamorphosis, comprises the more generalized forms of the Order; it is of exclusively Asiatic diffusion and, of the 3 (or 4) genera that it includes, only *Hynobius*, the most primitive, is diffuse from Western Asia to Japan, while the others are localized forms, derived from *Hynobius* (Noble, 1931).

The genus *Hynobius* is often subdivided into 4 groups, the most primitive of which includes the continental species, while the other three comprise exclusively insular species (cf. Thorn, 1968). The karyology of species of the 4 groups is known, especially owing to the excellent work of two Japanese karyologists, Makino and Sato; personally, I have been able to study the chromosomes of three species of *Hynobius*.

The chromosome numbers of the Hynobiidae studied (with *Hynobius* subdivided into the above-mentioned groups) are shown in the following table:

1st group (continental species)
- *Hynobius* (*Salamandrella*) *keyserlingii*: $2n = 62$ (Makino, 1932).
- *H. leechii*: $2n = 56$ (Makino, 1934).

2nd group (*naevius*)
- *H. naevius*: $2n = 56$ (Makino, 1932).
- *H. kimurai* (from Kamitakara): $2n = 56$ (Makino, 1932).
  (from Tango) : $2n = 60$ (Sato, 1936a).
- *H. okiensis*: $2n = 56$ (Makino, 1951).
- *H. stejnigeri*: $2n = 56$ (Makino, 1951).

3rd group (*nebulosus*)
- *H. nebulosus*: $2n = 56$ (Makino, 1934; Sato, 1936a; this paper).
- *H. tsuensis*: $2n = 56$ (Sato, 1936a; this paper).
- *H. dunni*: $2n = 56$ (Sato, 1936a; this paper).

4th group (*lichenatus*)

H. lichenatus: $2n = 58$ (Makino, 1932).
H. nigrescens: $2n = 56$ (Makino, 1932).
H. retardatus: $2n = 40$ (Makino, 1932).
H. sadoensis: $2n = 56$ (Makino, 1951).

Finally, also *Pachypalaminus boulengeri* has $2n = 56$ (Sato, 1936a).

Because of the techniques used, the two Japanese authors have only been able to distinguish two types of chromosomes in the karyotype of the Hynobiidae that they studied: "V-shaped" chromosomes (*m* or *sm*, with two arms each, usually the largest in the set) and "rod-shaped" chromosomes (*t* or *st*, considered as having only one arm, usually the smallest in the set). From a calculation of the number of arms (NF) in the species studied, it is found that most of them have NF = 76; in one case (*H. keyserlingii*) the NF seems equal to 74 and in another (*H. retardatus*) to 62; as for the two populations of *H. kimurai* having different chromosome numbers, they would both have an NF = 76, because the one with $2n = 56$ has 20 "V" chromosomes and 36 "rod-shaped" ones, while the one with $2n = 60$ has 16 "V" and 44 "rod-shaped" chromosomes (Sato, 1936a).

With the modern cytological techniques, I have been able to study three species of *Hynobius* in the *nebulosus* group (*H. nebulosus, tsuensis* and *dunni*) (Figs 2–4); as regards the chromosome number ($2n = 56$), my results correspond with the data of the previous authors, whereas, regarding the morphology of the individual pairs of homologues, they differ from them to some extent. The chromosomes of each of the three species of *Hynobius* may, in fact, be subdivided into 4 groups (though these tend not to be clearly defined from one another): (1) a group of 9 pairs of large homologues, *m* or *sm*; (2) a group of 4 pairs of medium homologues, *m, st* or *t*; (3) a group of 7–8 pairs of small homologues, *m, sm* or *st* and (4) a group of 8–7 pairs of very small elements (less than 2 microns in length) possibly all acrocentric (*t*), similar to the "microchromosomes" typical of various lower Vertebrates and the Sauropsids (Fig. 4). The NF of the three species, which differ from one another only in the morphology of certain elements of the 2nd, 3rd and perhaps the 4th group, is more than 90, and is therefore much higher than that calculated by the Japanese authors.

The fact that most of the species of *Hynobius* show $2n = 56$ has led various authors to believe that a karyotype with these characteristics is the closest to the ancestral karyotype of the family, and that the karyotypes having numbers higher or lower than 56 are derived from one similar to it, through Robertsonian translocations (cf. Matthey, 1949); however, this theory should be re-examined in the light of more modern karyological findings, since the calculation of the NF, which is basic for postulating this type of chromosome rearrangement, is imprecise if performed by means of the old techniques.

(b) *Cryptobranchidae*

A small group of semilarval giant Urodela, derived directly from the Hynobiidae; today they comprise two genera: the Asiatic *Megalobatrachus* (*Andrias*) with the species *davidianus* and *japonicus*, and the American *Cryptobranchus*, with the single species *alleganiensis*.

M. japonicus has $2n = 64$, with 17 pairs of large chromosomes, "V" or rod-shaped, and 30 microchromosomes (Fig. 5); *C. alleganiensis* has $2n = 62$ and shows two pairs of large chromosomes less and two microchromosomes more than the other species (Iriki, 1932; Makino, 1935); according to Makino, the two species have the same NF, which Matthey calculates as equal to 74, like that of most of the Hynobiids. The differences between the karyotypes of these two species would therefore be of a Robertsonian nature.

The karyotype of the two Cryptobranchids is very similar to that of *H. keyserlingii*, which, significantly, is also one of the more primitive Hynobiids; this fact agrees with the systematic data on the two families, which indicate a close phyletic relationship between them, and it also supports the theory that the common ancestral form from which the present members of the two families derive must have possessed a high number of chromosomes and many microchromosomes; this form of karyotype is found in the more primitive groups of the lower Vertebrates and in the more primitive species of the other two orders of Amphibia (Morescalchi, 1970).

Hynobiids and Cryptobranchids are different from all the remaining Urodela in the high number of chromosomes and microchromosomes present in their karyotype: also on karyological bases, it therefore seems justified to group the two families in a suborder that is exclusive to them: the Crytopbranchoidea.

(c) *Ambystomatidae*

This family, of Nearctic diffusion, seems to be derived directly from the Hynobiids, with which it was previously included by systematicians; it differs from the latter in certain characteristics of greater morphological evolution (Noble, 1931; Regal, 1966). It comprises 4 genera: *Rhyacotriton* and *Dicamptodon*, monospecific and highly localized; *Rhyacosiredon*, little known and with 4 species and *Ambystoma*, rich in species and widespread throughout North America; this last is generally considered the most highly evolved genus in the family.

Rhyacosiredon does not yet seem to have been studied karyologically; according to Humphrey (1958), *Rhyacotriton olimpicus* has $2n = 26$, with 6 pairs of large and 7 pairs of smaller chromosomes, all *m* or *sm* (NF = 52); this author reports only the chromosome number of *Dicamptodon ensatus*, which has $2n = 28$, $n = 14$.

All the species of *Ambystoma* so far studied have $2n = 28$; the best known are the following:

A. mexicanum:	$2n = 28$ (Wickbom, 1945; Hauschka and Brunst, 1964; Callan, 1966)
A. maculatum:	$2n = 28$ (Wickbom, 1945)
A. tigrinum:	$2n = 28$ (Makino, 1962)
A. jeffersonianum:	$2n = 28$; $3n = 42$ (Uzzell, 1963)
A. gracile:	$2n = 28$ (Humphrey, 1958)
A. macrodactylum:	$2n = 28$ (Humphrey, 1958, this paper).

According to these authors, the chromosomes of the various species of *Ambystoma* are all *m*, *sm* or *st*; the NF of this genus is therefore equal to 56; the differences in the morphology of the various pairs of homologues, between the species of this genus, seem very small; in *A. jeffersonianum*, some populations consisting only of females are triploid.

In Fig. 6 I have shown the chromosome set of *Ambystoma macrodactylum*, of which, previously, only the chromosome number was known: in this species, it is possible to distinguish two groups of chromosomes of different size (though they differ only slightly from each other): in the first group, which includes the largest elements in the set, there are 6 pairs of homologues, all *m*; in the second group, the 7th and 8th pairs consist of *m* chromosomes, while the others consist of *sm* chromosomes, with the exception of the 11th and 12th pairs, which are *st*.

The derivation of the Ambystomatids from a hynobiid stock seems to be one of the few data that are not debated in the phylogeny of the Caudata; however, this family differs sharply from all the Cryptobranchoidea as regards the morphology and chromosome number: in fact, it possesses half (28) the number of chromosomes present in most of the Hynobiids (56), while its chromosomes are all relatively large, each with 2 arms (NF = 56). Assuming that the karyotype of the Ambystomatids is derived from one similar to those still present today in the Cryptobranchoidea, it must have differentiated through the loss of the microchromosomes (perhaps translocated onto other elements in the set) and of the large acrocentrics (perhaps the object of centric fusions); both phenomena lead, in fact, to a clear decrease in the number of the chromosomes and may result in the genesis of elements with two chromosome arms. These phenomena of karyological differentiation may have been in process also at the hynobiid level, seeing

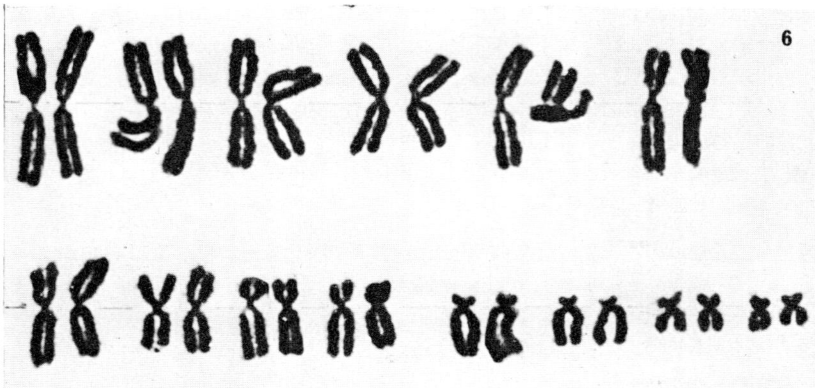

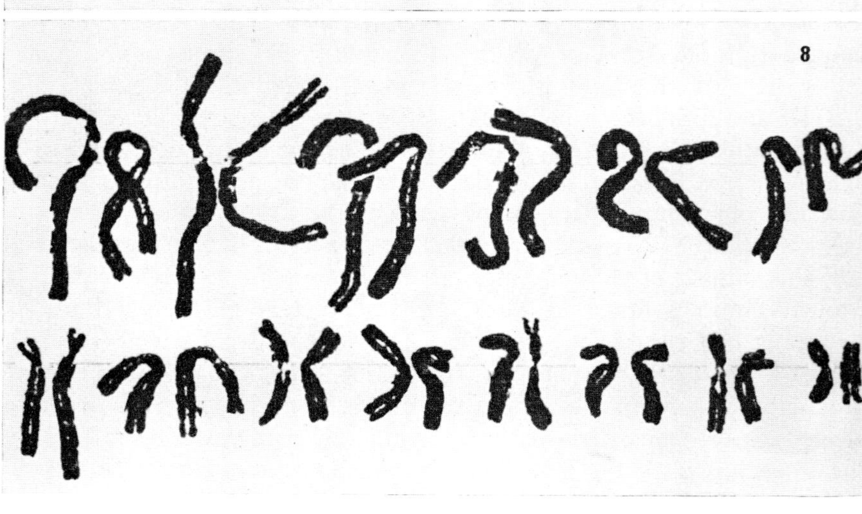

Fig. 6. Diploid chromosome set of *Ambystoma macrodactylum* ♂.
Fig. 7. Diploid chromosome set of *Batrachoseps wrighti* ♂ (Plethodontidae) (redrawn from Kezer, 1964).
Fig. 8. Diploid chromosome set of *Amphiuma means* ♂ (redrawn from Donnelly and Sparrow, 1965).

that in the present representatives of this family there are forms with a relatively low chromosome number (*H. retardatus*, with $2n = 40$).

It is interesting to note the difference in the chromosome number between *Rhyacotriton* on the one hand and the other Ambystomatids on the other: compared with *Ambystoma*, *Rhyacotriton* seems to possess two fewer chromosomes among the smaller ones (Humphrey, 1958); it is therefore probable that in this genus, which also anatomically is seen to be well separated from the other members of the family (Tihen, 1958), there has been a further step in the karyological evolution that has led from the Hynobiids to *Ambystoma*, with a further numerical reduction especially in the small elements: similar phenomena occur, as we shall see, also in the karyological differentiation of Plethodontids and Salamandrids.

(d) *Plethodontidae*

This is the most highly specialized family among the Urodela; it comprises lungless forms, adapted to various kinds of life, including arboreal. There is no agreement among authors regarding its subdivision into sub-families or tribes (cf. Wake, 1966); the classification followed here is that adopted by Regal (1966) and Brame (1967), who distinguish three sub-families: the Desmognathinae, with 2 or 3 genera, the Plethodontinae and the Bolitoglossinae, both rich in genera; all the Plethodontids are Nearctic, with the exception of certain Bolitoglossinae, which have European and South American species.

According to Noble (1931), the Plethodontids are derived from a stock of Salamandrids; however, present-day authors consider that this family has close relationships with the Ambystomatids, from the stock of which they would have originated (Laurent, 1947; Regal, 1966; Wake, 1966; Salthe, 1967; Monath, 1965; Estes, 1969*c*). With regard to the relationships between the three sub-families of Plethodontids, the Desmognathinae, which separated early from the rest of the family, comprise the genera having the greatest number of primitive characteristics; the Bolitoglossinae comprise the more highly specialized genera of the family.

The karyological data collected in the past on the family, with the exception of certain data (Fankhauser, 1939), are to be considered inaccurate (cf. Matthey, 1949); Kezer (1964) is largely responsible for a modern study of these Urodela, though many of his data have not yet been published (cf. Kezer and Macgregor, 1971). I give these data below, partly in the form of Kezer's personal communication and partly as reported by other authors (cf. Beatty, 1964; Mancino, 1965).

Kezer has studied species of most (20) of the 22 genera of living Plethodontids: in the Desmognathinae and in the Plethodontinae, the author has found a diploid number of chromosomes equal to 28, while in the Bolitoglossinae there are species with 28 and others with 26 chromosomes; the latter are species of *Batrachoseps* and the seven Latin American genera, also remarkable—and unique among Urodela—in possessing morphologically recognizable sex chromosomes in the male line of some genera and, in few cases, supernumerary chromosomes. According to Kezer, the Plethodontids with 26 chromosomes seem to lack the pair of smallest chromosomes present in the karyotype of the species with 28 chromosomes. All the chromosomes of the Plethodontids are two-armed (with the possible exception of the supernumerary ones) and the karyotype of most of the members of the family therefore seems very similar to that of the Ambystomatids (Fig. 7).

Hence, the karyological data support the theory of relationships between Plethodontids and Ambystomatids, which certain authors group together in the same sub-order (Ambystomatoidea); in both families, the karyological evolution seems to have continued in its process of reducing the diploid number, with karyotypes of 26 chromosomes in species that, at least as regards the Plethodontids, are also morphologically the most specialized in the family.

In my opinion, considerable interest, from a phyletic point of view, should be attached to the findings of Kezer (already reported) on the existence of morphologically recognizable sex chromosomes in species of the genera *Oedipina*, *Thorius* and *Chiropterotriton* (Bolitoglossinae), with male digamety of the XY type; there are indications of the presence of heterochromosomes also for certain Salamandrids (*Triturus*, *Pleurodeles*), which, however, would appear to have female digamety, with WZ chromosomes, distinguishable from one another only at the stage of lampbrush chromosomes (cf. Lacroix, 1970b).

Plethodontids and Salamandrids seem to be the Urodela that are most highly differentiated at present, and the karyological data generally support the anatomical and biological findings in this sense: as regards the sex chromosomes, which are generally absent in the lower Vertebrates and in most of the Amphibia, it must be assumed that the genesis of these peculiar elements occurred rather late in the phylogeny of the Urodela, which seem to lack them (at least at a level at which they can be shown cytologically) in most of the families.

(e) *Amphiumidae*

A mono-generic family, comprising only three species of semi-larval Urodela the best-known of which is *Amphiuma means*, the popular Conger eel of the south-eastern United States. According to Noble (1931) and other authors, *Amphiuma* belongs to the sub-order of the Salamandroidea and is derived from a salamandrid stock; however, more recent studies have cast doubts on these conclusions, revealing, on the contrary, that these Urodela have plethodontid or ambystomatid–plethodontid affinities (Regal, 1966; Salthe and Kaplan, 1966; Estes, 1969b).

According to Donnelly and Sparrow (1965), *Amphiuma means* has $2n = 28$, with large chromosomes, all *m* or *sm* (though some of them, in the terminology adopted in this work, may perhaps be considered *st*); consequently, the NF of this species is NF = 56 (Fig. 8).

In the number and form of the chromosomes, *Amphiuma* appears closer to the group of the Ambystomatoidea than to the Salamandrids, which generally have $2n = 24$; nevertheless, since the karyotype of the Salamandrids, in its process of differentiation, might also have passed from a possible ancestral set with a larger number of chromosomes (cf. later in the text) through a 28-chromosome stage, it is not possible, from a karyological aspect, to exclude entirely the theory of phyletic relationships between Amphiumids and Salamandrids; in comparison to the latter, however, the Amphiumids would appear less highly differentiated, from this point of view.

(f) *Salamandridae*

This great family of evolved Urodela is of uncertain origin; according to Noble (1931), it might have derived from a pro-hynobiid stock now extinct. To the present, this theory has not been replaced by any other that has been universally accepted, and the relationships between the salamandrid and the ambystomatid stock, of which last the hynobiid derivation appears more certain, are unclear; according to some authors, the two stocks have evolved independently of each other (Wake and Özeti, 1969).

Often, the numerous species of the family are subdivided into two large groups, to which is given the rank of sub-family: the first (Salamandrinae) comprises the species of the genera *Salamandra*, *Mertensiella* and *Chioglossa*, which are generally less highly specialized than the remaining members of the family; the second (Pleurodelinae) comprises the species of the genera *Tylototriton*, *Pleurodeles*, *Salamandrina*, *Euproctus*, *Triturus* and various genera closely related to *Triturus* and possibly derived from the same, which would comprise the more highly differentiated species of the family; however, the relationships between these genera of Salamandrids are much debated and are often the object of contradictory interpretations (cf. Wahlert, 1956, 1957; Salthe, 1967; Thorn, 1968; Wake and Özeti, 1969).

From a karyological point of view, the Salamandrids appear very homogeneous: except for the two Nearctic genera *Taricha* and *Notophtalmus*, which have $2n = 22$, all the others so far studied have $2n = 24$ (Figs. 9–12).

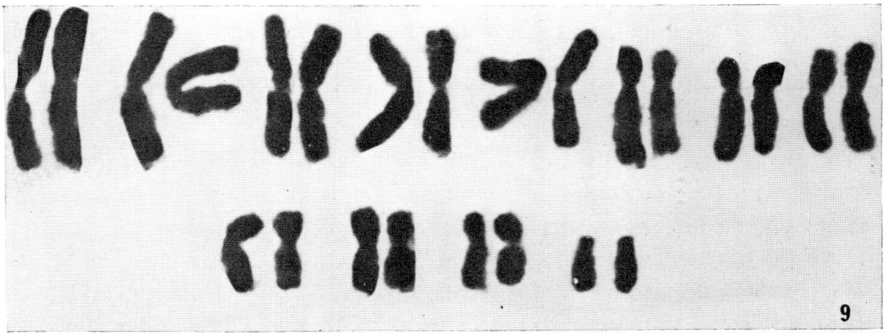

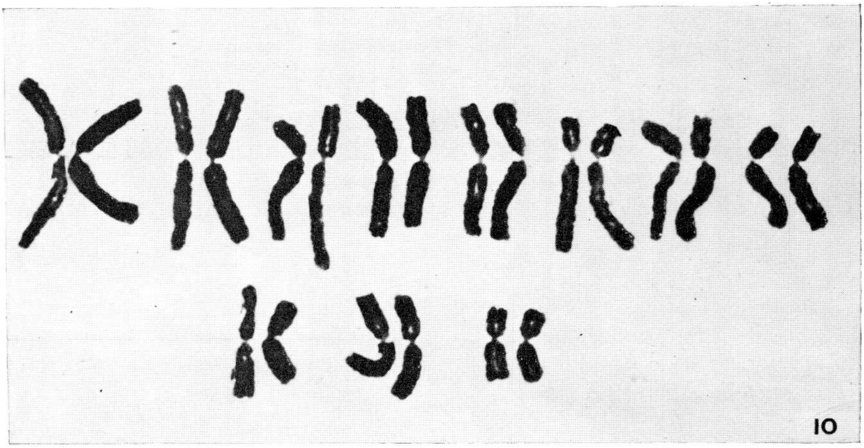

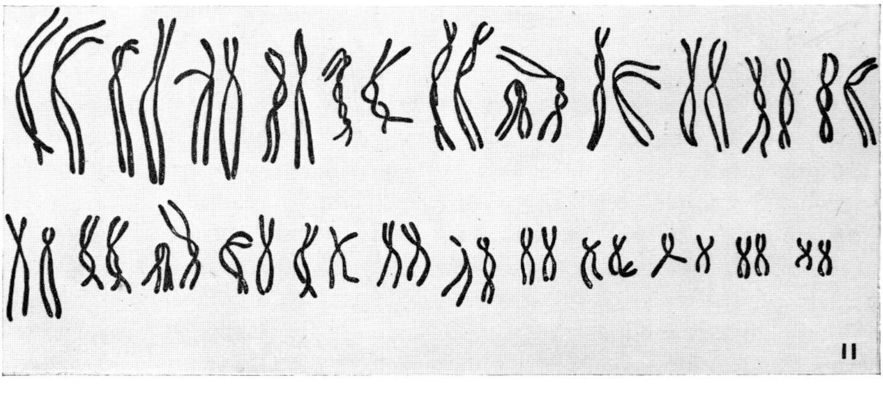

Fig. 9. Diploid chromosome set of *Tylototriton verrucosus* ♂.
Fig. 10. Diploid chromosome set of *Taricha granulosa* ♂.
Fig. 11. Diploid chromosome set of *Siren intermedia* ♂.

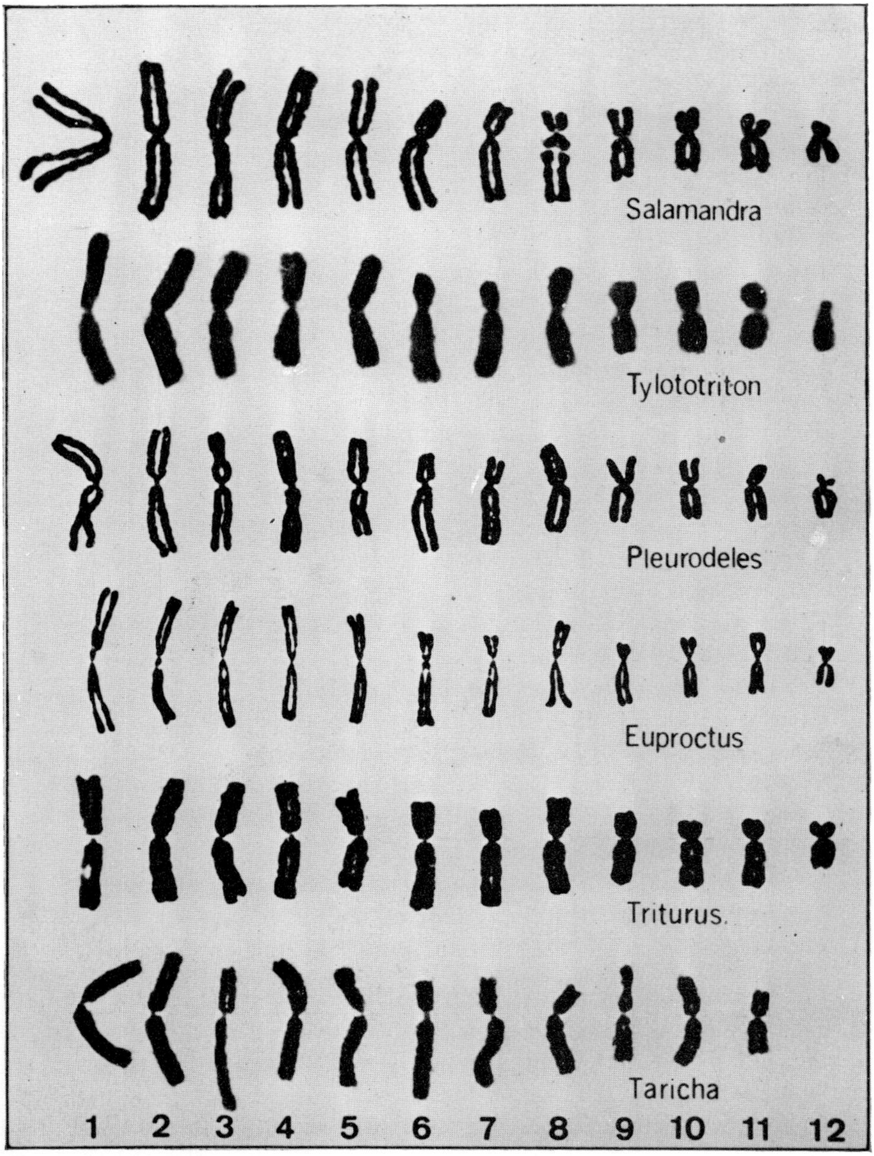

Fig. 12. The haploid karyotypes of species of some of the main Salamandrid genera to show the scarce morphological variability in the chromosome sets of this Urodelan family.

There follows a simplified list of the most recent karyological data on the various genera in the family.

Salamandrinae:

Salamandra salamandra: $2n = 24$ (Kessous *et al.*, 1968; Mancino *et al.*, 1969; Ullerich, 1970)

S. atra: $2n = 24$ (Wickbom, 1945; this paper).

Pleurodelinae:

Tylototriton andersoni: $2n = 24$ (Makino, 1962)

T. verrucosus: $2n = 24$ (This paper)

Pleurodeles waltlii: $2n = 24$ (Gallien *et al.*, 1965)

P. poireti: $2n = 24$ (Lacroix, 1970b)

Salamandrina perspicillata: $2n = 24$ (Mancino and Barsacchi, 1966a)

Euproctus asper: $2n = 24$ (Jaylet, 1966)

E. montanus: $2n = 24$ (Mancino and Barsacchi, 1966b)

E. platycephalus: $2n = 24$ (Mancino, 1965)

Triturus alpestris: $2n = 24$ (Fischberg, 1944; Wickbom, 1949; Selman, 1958; Mancino and Barsacchi, 1965; Ullerich, 1970)

T. cristatus: $2n = 24$ (Wickbom, 1945; Callan and Spurway, 1951; Lantz and Callan, 1954)

T. helveticus: $2n = 24$ (Fischberg, 1947; Selman, 1958; Mancino and Barsacchi, 1966c)

T. italicus: $2n = 24$ (Mancino and Barsacchi, 1969)

T. marmoratus: $2n = 24$ (Callan, 1955)

T. vulgaris: $2n = 24$ (Wickbom, 1945; Barsacchi *et al.*, 1970; Ullerich, 1970)

Cynops pyrrhogaster: $2n = 24$ (Fankhauser and Humphrey, 1942; Kawamura and Utsonomiya, 1957)

Neurergus crocatus: $2n = 24$ (Bhatnagar, personal communication)

Notophtalmus viridescens: $2n = 22$ (Fankhauser, 1938; Gall, 1954)

Taricha granulosa: $2n = 22$ (Fankhauser, 1932; this paper).

The 24-chromosome karyotype is very similar in all the species that have it (Fig. 12); in general, it is possible to distinguish 8 pairs of larger chromosomes, all *m* or *sm*, and 4 pairs of slightly smaller chromosomes, which are *m*, *sm* or *st* (the last often being represented by the smallest pair in the set); various authors further subdivide the 8 pairs of larger chromosomes into two groups of 4 pairs of different length, but this subdivision is not always clearly appreciable.

The inter-specific differences in the karyotype of the 24-chromosome forms seem to concern few chromosomes, which may be *m* or *sm*, and *sm* or *st* (in the 22-chromosome species, the two smallest chromosomes present in the karyotype of the 24-chromosome species seem to be missing (Fig. 10).

Since the Palearctic *Triturus* are generally less highly evolved than the related Nearctic forms and also possess the general diploid number of the family, it seems very probable that the 22-chromosome karyotype of *Notophtalmus* and *Taricha* is derived from one similar to that of the members of the genus *Triturus* (cf. Fankhauser, 1932).

The data on the somatic set, except for the Nearctic forms, therefore do not offer any new facts to be considered for interpreting the complex reciprocal relationships between the various genera of Salamandrids; as regards the relationships with the other families, it may be said that, on account of their chromosome number, which is the lowest found among the living Urodela, the Salamandrids may be considered, karyologically, as the most highly differentiated family in the order.

(g) *Proteidae*

This family of permanently larval Urodela comprises only two genera: *Proteus* (mono-specific), the blind salamander of the Carsic caves near Trieste, and *Necturus*, with about 5 species, the common "mud-puppy" of the Eastern United States. The relationships between this family and the other Urodela are problematical; Noble (1931) considers that it may have derived from a stock of Salamandrids.

Proteus anguinus and *Necturus maculosus* have $2n = 38$, with karyotypes that are very similar, consisting of 13 pairs of *m* and *sm* homologues, common to the two species, and 6 pairs of homologues that are all *t* in *Necturus* and *sm* or *t* in *Proteus* (one of these, the smallest, may perhaps be considered to consist of microchromosomes) (Kezer, 1962; Kezer *et al.*, 1965; Seto *et al.*, 1964).

The phyletic affinities between the two genera of living Proteids have been doubted by certain authors (Hecht, 1957; Salthe, 1967); however, the considerable karyological resemblance between *Proteus* and *Necturus* and the differences between the karyological formula of these two genera and that of the other Urodela seem to confirm the validity of a systematization that assigns them to the same family (Kezer *et al.*, 1965).

The chromosome number of the Proteids ($2n = 38$; NF $= 64$–68) is intermediate between that of many Hynobiids ($2n = 56$; NF $=$ over 90) and that of the more highly evolved families of Urodela ($2n = 28$–26–24; NF $= 56$–52–48) (Fig. 13); the Proteids differ from the latter also for the presence of true acrocentrics (*t*), which are missing in the higher Urodela. The main difference between the karyotype of the Proteids and that of the Hynobiids seems to consist in the total or almost total absence of microchromosomes in the set of the Proteids.

Karyologically, therefore, the Proteids constitute an interesting transitional family between the Cyptobranchoidea, the probable ancestors of most of the modern Urodela, and the more highly differentiated families in the order: they might have retained the karyological formula of the stock (or stocks) that has led from the Hynobiids to the Ambystomatids and/or the Salamandrids.

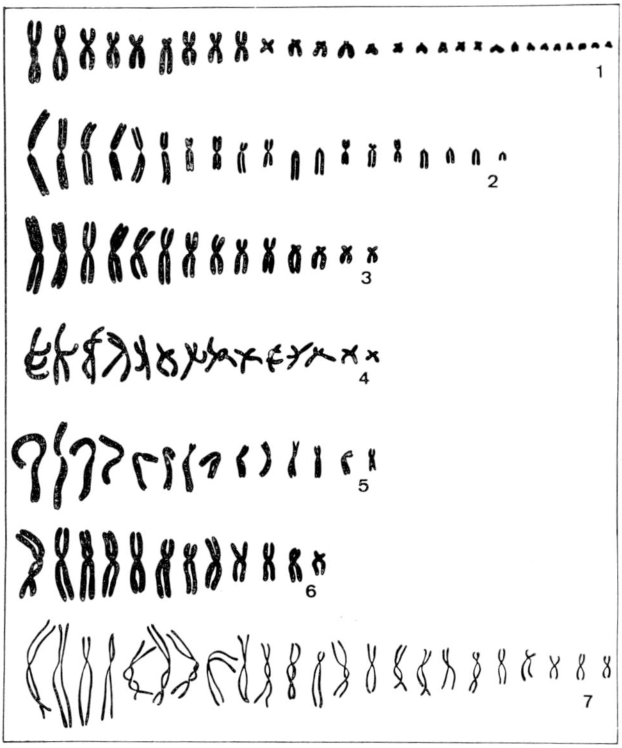

Fig. 13. The haploid karyotypes of species of the main Urodelan families. 1.—Hinobiidae (*H. nebulosus*); 2.—Proteidae (*Necturus maculosus*; redrawn from Kezer *et al.*, 1965); 3.—Ambystomatidae (*A. macrodactylum*); 4.—Plethodontidae (*Desmognathus fuscus*; redrawn from Kezer, 1964); 5.—Amphiumidae (*A. means*; redrawn from Donnelly and Sparrow, 1965); 6.—Salamandridae (*Pleurodeles waltlii*; redrawn from Gallien *et al.*, 1965); 7.—Sirenidae (*S. intermedia*).

(h) *Sirenidae*

A family of permanent larval forms, to which now belong the genera *Pseudobranchus* (mono-specific) and *Siren* (with two species); this is perhaps the most problematical family among Urodela, from a phylogenetic point of view. Noble (1931) considers that it probably derives from some hynobiid stock; other authors, on the contrary, consider the Sirenids as different from all the other Urodela, such that they classify them in an order apart (Trachystomata) (cf. Goin and Goin, 1962); however, these conclusions are not accepted by many authors (cf. Estes, 1965; Wake, 1966).

The only information that so far exists on the karyology of the Sirenids concerns *Siren intermedia*, a species in which Kezer and Bainbridge claim to have counted 42 chromosomes (though the authors offer this fact as provisional; cf. Kezer *et al.*, 1965). I myself have studied intestinal mitoses of two male specimens of this species, in which I constantly counted $2n = 46$, with all large chromosomes, *m* or *sm* (Fig. 11): I maintain that this number is the one typical of the species; unfortunately, the testicle of the two species contained no meiotic divisions. No karyological data exist on *Pseudobranchus*.

As regards the morphology of the chromosomes, all *m* or *sm*, and their relative size (the smallest are about 1/5 the size of the largest), *Siren* therefore recalls the higher Urodela, though it differs from them in the high chromosome number (46), which approaches that of the Cryptobranchoidea; however, it is not possible to see how a hynobiid karyotype can have given rise to one similar to that of *Siren*, because the high chromosome number in those primitive Urodela is a consequence of the presence of various microchromosomes, which are absent in *Siren*. Unless *Siren intermedia* is an allopolyploid species (a possibility that I cannot yet entirely exclude, on the basis of the few data available)—in which case it might very well be considered to belong to the higher Urodela—from a karyological point of view this species seems to be fairly isolated.

(i) *The evolution of the karyotype of the Urodela*

The following list shows the chromosome numbers characteristic of species of the eight families of modern Urodela:

Hynobiidae	: $2n = 62, 60, 58, 56$ (most species) and 40
Cryptobranchidae	: $2n = 64, 62$
Ambystomatidae	: $2n = 28$; *Rhyacotriton*: $2n = 26$
Plethodontidae	: $2n = 28$; many Bolitoglossinae: $2n = 26$
Amphiumidae	: $2n = 28$
Salamandridae	: $2n = 24$, Nearctic genera: $2n = 22$
Proteidae	: $2n = 38$
Sirenidae	: $2n$ (?) $= 46$

The morphological relationships between the chromosome formulae typical of the various families—excluding the Cryptobranchidae, on which no recent karyological data are available—are summarized in Fig. 13: the Figure clearly shows the tendency for the overall chromosome number to decrease and for the microchromosomes to disappear, a phenomenon that characterizes the phylogeny of the order (with the possible partial exception of the Sirenidae).

Hynobiidae and Cryptobranchidae (Cryptobranchoidea) comprise the more generalized forms of the order, and I have previously advanced the theory that their karyological formula, with a high number of chromosomes, acrocentrics and microchromosomes, may also be the most primitive among those revealed by the present-day Urodela (cf. also Matthey, 1949). This theory is essentially based on knowledge of the phylogenetic position of this sub-order, on the fact that all the more primitive groups of lower Vertebrates possess similar karyological formulae (cf. Morescalchi, 1970) and above all on the fact that also the more primitive Apoda and Anura possess high chromosome numbers and various microchromosomes, whereas the more highly differentiated forms of these two orders, like most of the advanced Urodela, possess low chromosome numbers and lack microchromosomes.

On the other hand, Wickbom (1945, 1949) has advanced the theory that the karyological formula of the Cryptobranchoidea may be derived by "fragmentation" from a formula with a high number of metacentric chromosomes, like that of the Dipnoi: however, the phyletic relationships between Dipnoi and Urodela seem far less close today than they were considered to be by the Swedish author (cf. Schmalhausen, 1968); moreover, the karyological data on which this theory is based—and which are essential to it—are partly mistaken.

Already at the hynobiid level, there seems an obvious evolutional tendency towards a reduction in the chromosome number, which is higher in the primitive continental forms; this tendency, albeit to a lesser extent, also appears evident in certain families of higher Urodela (Plethodontids, Salamandrids), where the species with lower chromosome numbers are among the most highly specialized of each (among the Ambystomatids, the case of *Rhyacotriton* is less clear; however, this genus seems to have long since separated from the rest of the family).

The difference in chromosome number and morphology that exists between most of the Cryptobranchoidea, on the one hand, and the higher Urodela on the other is partly filled by the peculiar karyological formula of the Proteids, which reveal intermediate conditions between those of these two groups of families: the Proteids lack microchromosomes (or possibly have one pair of them), but they retain various acrocentrics, which are lacking in the higher Urodela; their chromosome number is intermediate ($2n = 38$) between that of most of the Hynobiids ($2n = 56$) and that of the Ambystomatoidea ($2n = 28$), the largest group of higher Urodela.

Ambystomatids, Plethodontids and Amphiumids generally have $2n =$

28, with all two-armed chromosomes (NF = 56): the karyological homogeneity of the three families may constitute an argument in favour of the theories of close phyletic relationships between them, especially between the first two.

In the Salamandrids, the chromosome set is found to be further reduced ($2n = 24$); however, their karyological formula may easily be derived from a 26- or 28-chromosome formula by postulating the "loss" (possibly through unequal translocations) of some of the smallest chromosomes of the set; karyologically, these Urodela therefore appear highly differentiated forms in the order.

This simple scheme of chromosome evolution does not cover the Sirenids which, in the only species studied, show 46, all two-armed chromosomes that are relatively large: the set of *Siren intermedia* (if diploid) cannot be traced to that of any other Urodelan so far studied karyologically.

Obviously, this chromosome reduction may have been arrived at independently by certain large groups of higher Urodela (e.g. the Ambystomatoidea and the Salamandrids), starting from formulae similar to those of the present-day Hynobiids. However, it is a fascinating theory that the considerable karyological resemblances between the various families of higher Urodela (excluding, for the present, the Sirenids) and the karyological homogeneity attained by them may be interpreted as being favourable to a common origin from a Hynobiid stock, as various anatomical data seem to indicate.

In conclusion, by comparing the facts known today on the phylogeny of the Urodela with the karyological data described here, which point to a fairly precise and unique series of events, it is possible to make certain observations of general interest on the evolution of these Amphibians:

(1) the Cryptobranchoidea have retained a primitive karyotype model similar to that of various lower Vertebrates and the more primitive Amphibians of the other orders;

(2) the Proteids occupy an intermediate position between the Cryptobranchoidea and the advanced Urodela; this fact, even if it contrasts with certain theories regarding a derivation of this family from the Salamandrids, is in agreement with the paleontological data, which point to a considerable antiquity of these pedomorphic forms (cf. Estes, 1965; 1969a);

(3) Ambystomatids, Plethodontids and Amphiumids have reached the same level of karyological differentiation, with karyotypes very similar to one another; the Salamandrids seem to have surpassed this level of karyological differentiation, even if it appears likely that their ancestors have passed through 28- and 26-chromosome stages, similar to those of the other families;

(4) on the whole, it is possible to formulate the theory that the Urodela—

with the possible exception of the Sirenids—have pursued a single line of karyological evolution, some stages of which are still represented in living forms of the order.

Bearing in mind the morphological affinities between the various families, their present degree of specialization towards aquatic or terrestrial life (shown on the abscissae; cf. Noble, 1931) and what has been said about their karyological differentiation (shown on the ordinates), the relationships between the families of the order may be represented in the form of a diagram, as shown in Fig. 14.

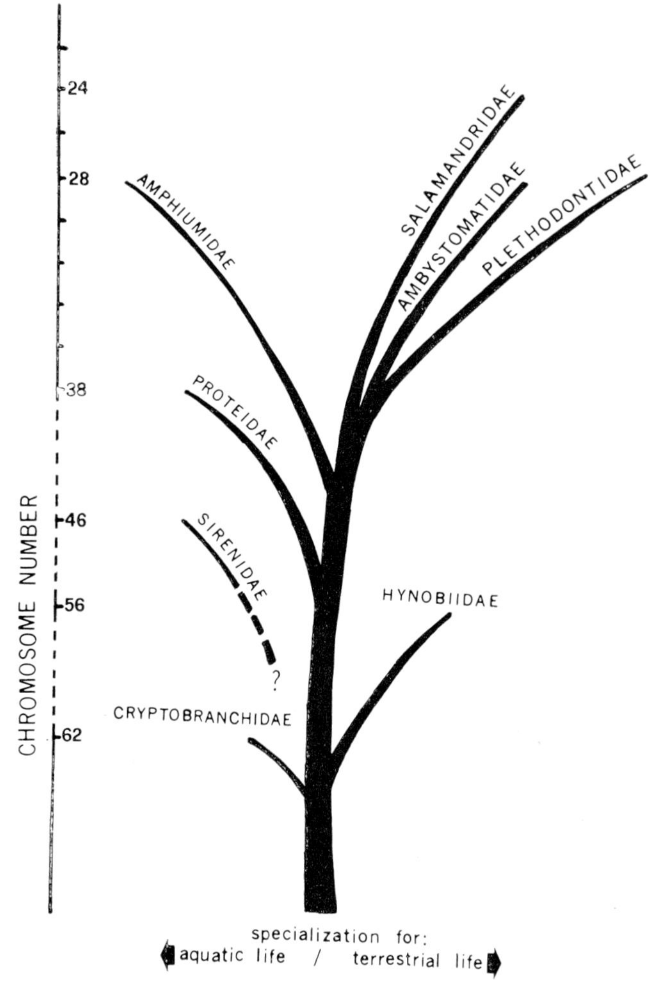

Fig. 14. Phyletic tree of the Urodela, built utilizing karyological and morphological data. Further explanations in the text.

C. *Anura*

(a) *Ascaphidae* (*Leiopelmatidae*)

This very old family, now reduced to a few remaining species, includes the most primitive Anura of the order; to these belong the North American *Ascaphus*, with the single species *truei*, and the New Zealand *Leiopelma*, with three species. The phyletic relationships between the two genera are generally considered to be very close by Systematicians; however, some of them prefer to consider *Ascaphus* and *Leiopelma* in two different families (cf. Gorham, 1966).

Ascaphus truei has been studied karyologically by Wickbom (1950a), by Morescalchi (1967a, 1968d) and by Bogart (1970). The first author counted $2n = 42$ chromosomes, with 6 pairs of large *m* and 15 pairs of small *t* (NF = 54). I myself, in various specimens of both sexes originating from two different localities in Oregon, found a diploid number of $2n = 44$, with 5 pairs of large homologues, *m* or *sm*, 4 pairs of medium-small homologues, three of them *t* and one *st*, and 13 pairs of very small *t* chromosomes (microchromosomes) (NF = 56); in the male germinal line I counted 22 spermatocyte bivalents, 5 of them large, with three or more chiasmata, often interstitial, and 18 other small ones with one or, more rarely, two chiasmata (Fig. 39) (I shall return to a fuller discussion of these findings in Section D). Bogart, in the specimens that he studied, counted $2n = 46$ chromosomes, 36 of which were *t*-chromosomes (NF = 56). Possible Robertsonian mechanisms could explain the differences in diploid number between Bogart's findings and my own, seeing that the NF is the same in both cases.

Regarding *Leiopelma*, the only karyological data so far collected are my own (Morescalchi, 1968b), on two female specimens of the species *hochstetteri* from the neighbourhood of Auckland. In the first specimen (here called ♀1), I counted $2n = 34$, with 5 pairs of large *m* or *sm* chromosomes, 6 pairs of medium-small *t* chromosomes and 6 pairs of microchromosomes, all *t* (NF = 44); this set differs from that of *Ascaphus* essentially on account of the lack of certain pairs of microchromosomes (Fig. 15). In the second specimen (♀2) only 23 chromosomes ($2n = 23$) were present, in intestinal and skin metaphases; compared with ♀1, as many as 11 of the 12 microchromosomes were lacking, while the rest of the set was identical (Figs 17–18). On account of the fact that ♀2 shows an odd number of chromosomes, I considered the karyotype of ♀1 as "normal": the karyotype of ♀2 may be derived from one similar to the former, through the loss of various microchromosomes. Seeing that ♀2 seemed morphologically quite regular and, on the basis of the anatomical data, was certainly a specimen of the same species as ♀1 (in any case, it came from the same population), the

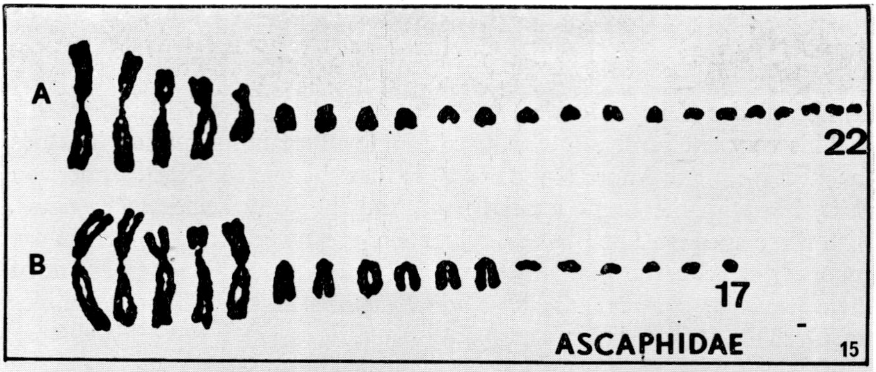

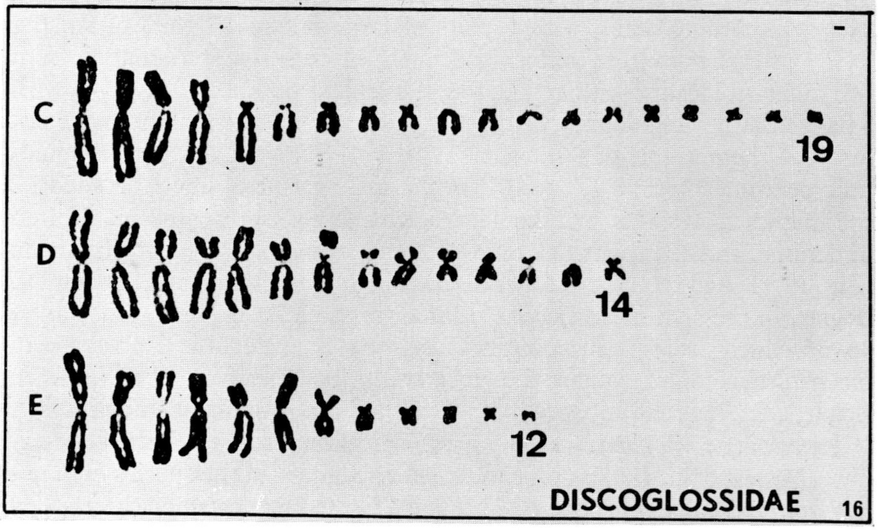

Fig. 15. The haploid karyotypes of *Ascaphus truei* (A) and *Leiopelma hochstetteri* ♀ 1 (B) (Ascaphidae).

Fig. 16. The haploid karyotypes of three Discoglossids: C: *Alytes obstetricans*; D: *Discoglossus pictus*; E: *Bombina variegata*.

Figs. 17–22. All at the same enlargement, as indicated in Fig. 17.

Fig. 17. Intestinal metaphase plate of *Leiopelma hochstetteri* ♀1 ($2n = 34$): as many as 12 microchromosomes can be found in this specimen.

Fig. 18. Intestinal metaphase plate of *L. hochstetteri* ♀2 ($2n = 23$): one microchromosome only is present in this specimen (arrow).

Fig. 19. Intestinal metaphase plate of *Discoglossus pictus* ♂: the two *t*-chromosomes are equal in length (heterochromosomes ZZ).

Figs. 20, 21, 22. Intestinal metaphase plates of *D. pictus* ♀: the two *t*-chromosomes of each plate have different length (heterochromosomes ZW).

Fig. 23. Intestinal metaphase plate of *Xenopus laevis* ♀, specimen coming from Kenya. No heterochromosomes can be found.

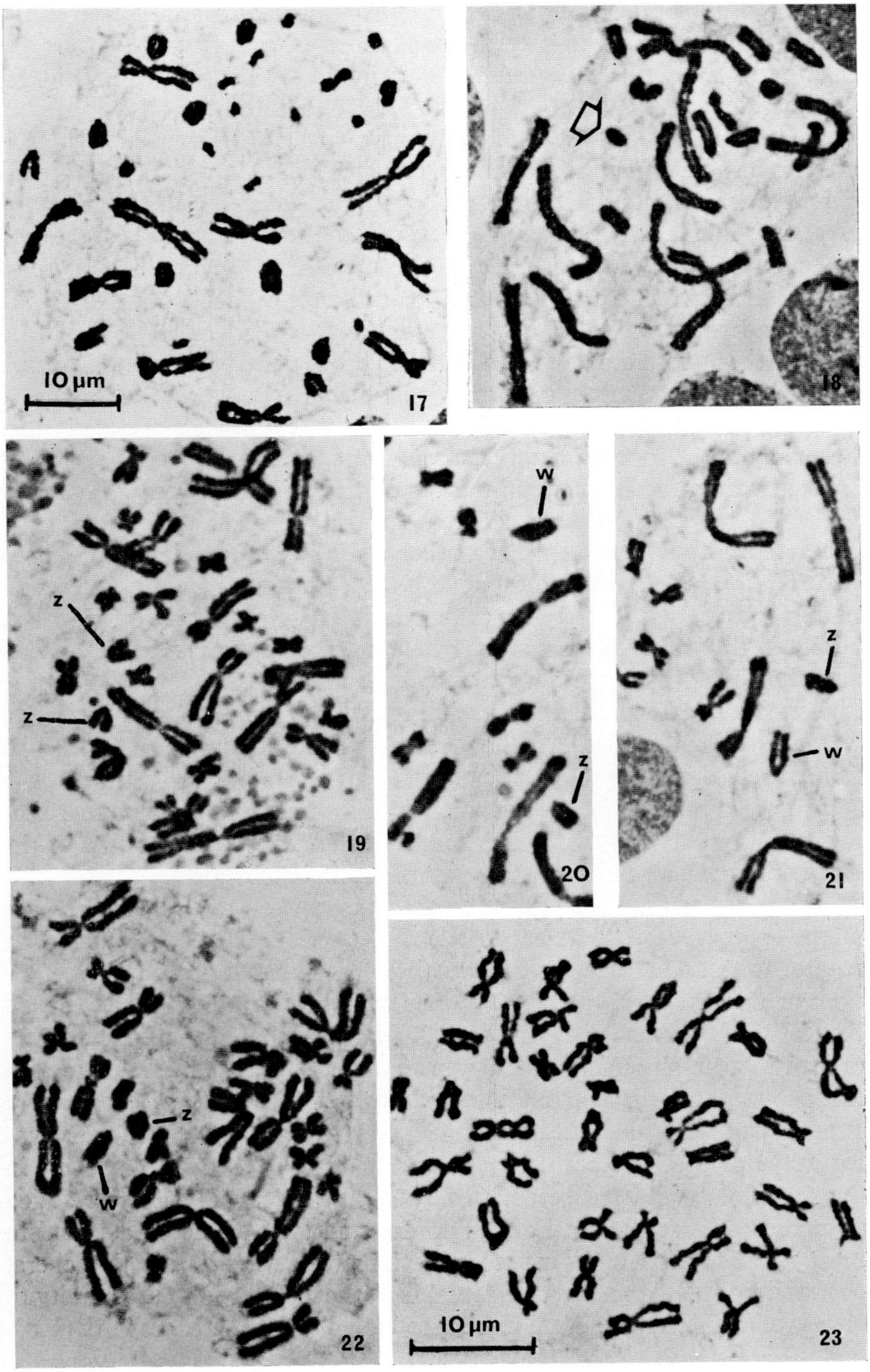
10 µm
17
18
z
z
19
w
z
20
z
w
21
z
w
22
10 µm
23

problem arises of determining how it is that such a great karyological difference has not expressed itself at the phenotype level: in other words, one may put forward the theory of a "supernumerary" nature of the microchromosomes of *Leiopelma* (Morescalchi, 1968b).

Today, a discovery of this kind is not new for the Amphibia: supernumerary chromosomes have been found in certain Plethodontid Urodela (Kezer, personal communication) and in other Anura (*Rana temporaria* and *Acris crepitans*; Ullerich, 1967; Nur and Nevo, 1969); however, in no case are they present in numbers comparable to those possibly present in *Leiopelma* (where there are at least 11 of them); on the other hand, in other Anura having microchromosomes (*Ascaphus* and *Alytes obstetricans*) I found no variability in the number of these elements in the populations that I studied.

Obviously, as regards *Leiopelma*, this problem can be solved only by studying new specimens, especially of the male sex, in view of the peculiar behaviour of the supernumerary chromosomes at meiosis.

Considering for the moment the karyotype of *Ascaphus* and that of only ♀1 of *Leiopelma*, it may be said that the karyotype of the Ascaphids is characterized by the presence of a relatively high number of chromosomes (that of *Ascaphus* is the highest among the Anura, not counting the polyploid species), by the presence of many "microchromosomes" (which are absent in the higher Anura) and by the presence of various medium-small acrocentrics (*t*) (these too are generally absent, except in a few cases, in the higher Anura). Also in the morphology of the spermatocyte bivalents, *Ascaphus* differs from the higher Anura, which have bivalents with two terminal chiasmata.

In general, therefore, the somatic set of the Ascaphids differs morphologically from that of the higher Anura to the same extent as that of the Cryptobranchoidea differs from the set of the higher Urodela.

On the basis of this analogy, it seems to me probable that the morphology of the karyotype of the Ascaphids is of the "primitive" type among the Anura; in this, the karyological data therefore faithfully reflect the conditions of anatomical primitiveness of the family that are generally recognized by nearly all systematicians; the data also appear favourable to the inclusion of the two genera in a single family, in view of the close similarities between the karyotype of *Ascaphus* and that of *Leiopelma* (at least the ♀1 specimen). It may be objected that *Ascaphus* and *Leiopelma* resemble each other karyologically only because both reflect a condition of general primitiveness in the morphology of their set; however, to the karyological data are added numerous other anatomical data, and the sum total of these data tend to make it unlikely that one is concerned with mere convergences in the general evolution of the two genera. In actual fact, the objections to the affinity of *Ascaphus* and *Leiopelma* seem to be based mainly on differences in the biology of the development between the two genera and on zoogeographical considerations: according to the opinion of some authors, it would seem that these objections can be easily overcome (cf. Stephenson, 1955; Reig, 1961).

Forms very similar to the recent Ascaphids have been found in the Middle and Lower Jurassic in Patagonia (Reig, 1957, 1961); the presence of members of this family in North America and in New Zealand suggests that this ancient group has radiated very extensively. As we shall see, the karyological data may be favourable to the theory that various modern families of Anura have originated from ascaphid forms (in a broad sense, in accordance with other authors, I have also included Reig's Notobatrachidae (1958) among them), which were spread throughout lands of Laurasian and Gondwanian origin.

In the phylogeny of the Anura, the Ascaphids may therefore represent what the Hynobiids represent for that of the Urodela: that is the forms nearest to the stem stock of most of the actual families of the order.

Finally, it is interesting to mention a surprising karyological discovery made by Macgregor and Kezer (1970) regarding the oogenesis of *Ascaphus truei*: in this species the oocytes appear as polynucleate (with 8 nuclei, each provided with a regular set of lampbrush chromosomes) up to advanced stages of the I meiotic division, when they eliminate 7 of the 8 nuclei; this discovery, at present unique among the Vertebrates, has been related by the authors mentioned above to peculiar methods of gene amplification.

The degree of "specialization" reached by the oogenesis of *Ascaphus*, and the fact that also *Leiopelma* possesses advanced methods of embryonal development, constitute biological peculiarities that are exclusive to Anura, which are otherwise primitive; they may have contributed to the survival of these bradytelic forms.

(b) *Discoglossidae*

These constitute one of the oldest families in the order and are considered by some authors transitional between the more primitive Anura (Ascaphids) and the higher ones; however, even if its affinities with the Ascaphids are generally admitted by present-day systematicians, there are still various doubts about its phyletic relationships with the higher Anura (cf. Reig, 1961; Tihen, 1965).

Today, the Discoglossids include 4 genera (*Alytes*, *Barbourula*, *Bombina* and *Discoglossus*), each with one or only a few species, spread throughout various Eurasiatic areas; however, certain paleontological findings show that the family is spread throughout all the Holarctic Region (Estes, 1969d).

A karyological study has been made of species of three genera; only the rare and monospecific *Barbourula*, confined to certain islands of the Philippines, has not been studied from this point of view. The list of the species studied is as follows:

Alytes obstetricans: $2n = 36$ (Wickbom, 1949)
A. o. obstetricans : $2n = 38$ (Morescalchi, 1966)
Bombina bombina : $2n = 24$ (Wickbom, 1945; Morescalchi, 1965b; Ullerich, 1970)
B. orientalis : $2n = 24$ (Sato, 1939)
B. variegata : $2n = 24$ (Wickbom, 1945; Morescalchi, 1965b; Ullerich, 1970)
Discoglossus pictus: $2n = 28$ (Wickbom, 1945, 1949; Morescalchi, 1964c).

The species of Discoglossidae having the greatest number of chromosomes is *Alytes obstetricans*; here I shall consider only my own data, since those of Wickbom are difficult to compare with those obtained by means of the most modern cytological techniques.

Alytes ($2n = 38$; NF = 64–72) has 5 pairs of large homologues, *m*, *sm* or *st* (the 5th pair), 6 pairs of medium-small *st* or *t* elements, and 8 pairs of small chromosomes, *st* or *t*, some of which are very similar to the microchromosomes of the Ascaphids. Also in morphology and the number of large and medium-small chromosomes, *Alytes* closely resembles the karyotype of the Ascaphids (Fig. 16). The morphology of the spermatocyte bivalents of *Alytes*, which have three or more chiasmata if large, or one only if small, is of the "ascaphid" model, as is also the case, though sometimes in a different way, with that of the other Discoglossids (cf. Section D).

Discoglossus pictus has $2n = 28$, NF = 54; it has 7 pairs of large homologues, *m* or *sm*, and 7 pairs of smaller homologues, *m*, *sm* or *st*, expect for the 13th pair, which is *t*: this last pair of chromosomes is peculiar, because the two homologues are found to be of different lengths in the ♀, whereas they are equal in the ♂, so that I have considered them as heterochromosomes (Morescalchi, 1964c; cf. Figs 19–22); this species, unique among the Anura so far studied, would therefore show cytologically recognizable sex chromosomes and female digamety.

The species of *Bombina* so far studied have $2n = 24$ and karyotypes practically equal to one another, consisting of 6 pairs of large homologues, one pair of medium-large chromosomes and 5 pairs of smaller homologues; all the chromosomes are *m* or *sm*, except those of the 8th and 12th pair, which are *st* (NF = 48) (the set thus described holds especially for the European species; *B. orientalis* has not been studied by modern techniques).

In *Discoglossus* and *Bombina* the spermatocyte bivalents have more than two chiasmata if large, and one or two if small; most of the chiasmata, as in the Aschaphids, are interstitial and not terminal as in the higher Anura and the Pipids; however, in *Discoglossus* there are also frequent meiotic figures with terminal chiasmata.

On anatomical bases it is difficult to determine which of these three genera of Discoglossids is the most primitive; for certain osteological characteristics, *Bombina* is the closest to the Ascaphids (Slabbert and Maree, 1945); from the karyological point of view, *Alytes* is the discoglossid closest to the Ascaphids and the most primitive in the family, while *Bombina*, at least as regards the somatic chromosomes, is the most highly differentiated, having a karyotype that resembles, in number and form of the chromosomes, that of various higher Anura (certain Microhylids, Bufonids etc.); *Discoglossus*, as regards the somatic set, occupies an intermediate position between *Alytes* and *Bombina*, whereas, as regards the spermatocyte chromosomes, it approaches the higher Anura (Fig. 16).

In the Discoglossids, the karyological evolution—which seems to have led from an "ascaphid" karyotype, rich in acrocentrics and with some microchromosomes, as that of *Alytes*, to one of a "higher" type, with few chromosomes, all with two arms, as that of *Bombina*—has possibly taken place with the apparent loss of the microchromosomes and the acrocentrics and with a reduction in the total number of the chromosomes.

We have seen how these methods of karyological evolution are those that are the most general in the Urodela; as we shall see in the higher Anura as well, there is a clear tendency towards a reduction in the number of the chromosomes: consequently, in the Discoglossids, the karyological differentiation seems to have repeated the main stages characteristic of the karyological evolution of Urodela and Anura. This family, derived at an early stage from an ascaphid stock, seems to have differentiated rapidly from the point of view of the somatic set, attaining the karyological formulae of the higher Anura; on the other hand, the karyological differentiation in the male germinal line does not seem to have proceeded *pari passu* with that of the somatic set, seeing that the Discoglossids, in general, maintain the type of bivalents characteristic of *Ascaphus* (and of nearly all the Urodela).

We shall see later on that the karyotype that is probably basic for the higher Anura is $2n = 26$, with all two-armed chromosomes ($NF = 52$); among the Discoglossids, this chromosome formula is not found; however, that of *Discoglossus* ($2n = 28$; $NF = 54$) does not seem very far from this, while that of *Bombina* resembles, as already stated, that of higher Anura that are karyologically already differentiated (Fig. 35).

Now even if, from an anatomical—and above all zoogeographical—point of view, various authors consider it improbable that the Discoglossids are on the same phyletic line as the higher Anura, from a karyological point of view, at least as regards the somatic set, they do in fact come close to the latter. Consequently, even if the Discoglossids may not be the direct ancestors of the higher Anura, their karyology shows that, very probably, they are derived from the same basal stock that gave rise to the latter; considering the karyology of *Alytes*, this stock may consist of forms very similar to the present Ascaphids.

(c) *Pipidae*

These are one of the oldest families of the Anura (Early Jurassic? cf. Hecht, 1963), comprising forms that were already highly specialized at the beginning of Cretaceous (Nevo, 1968). At present, the Pipids are spread throughout Africa and South America, with aglossan forms adapted to aquatic life; however, fossils of Pipids and of other families akin to them (Palaeobatrachidae; cf. Špinar, 1967) are evidence of a diffusion of this group of Anura also in holarctic regions. The origin of the family has been the subject of various conjectures; generally, it is considered to be derived from ascaphid or discoglossid forms (cf. Noble, 1931; Reig, 1958). However, according to recent authors, it might have a much older origin, perhaps being derived directly, like the ascaphid forms, from a stock of Proanura (Griffiths, 1963; Nevo, 1968).

Today, certain species of Pipids survive, subdivided into an American genus (*Pipa*: considered by some authors to be an ensemble of several genera) and two African genera (*Xenopus* and *Hymenochirus*; the problematical and very rare *Pseudohymenochirus merlini* seems rather to be identifiable as a species of *Hymenochirus*; cf. Lamotte, 1963).

In view of the high degree of specialization for aquatic life attained by the actual Pipids, it is difficult to determine which of them is the most primitive; in general, *Xenopus* seems the least specialized of the three, while *Pipa* would seem to be the most specialized. As regards certain aspects, *Hymenochirus* seems more akin to *Pipa* than to *Xenopus* (cf. Nevo, 1968).

Of the three genera of Pipids, the chromosomes of the following species are known:

Pipa pipa : $2n = 22$ (Wickbom, 1950b; Morescalchi *et al.*, 1970a)
P. (*Protopipa*) *parva* : $2n = 30$ (Morescalchi, 1968a)
Hymenochirus boettgeri: $2n = 24$ (Morescalchi, 1968a)
Xenopus laevis : $2n = 36$ (Wickbom, 1945; Weiler and Ohno, 1962; Morescalchi, 1963b; Mikamo and Witschi, 1966)
X. muelleri : $2n = 36$ (This paper)

Except for *P. parva*, which has procentric chiasmata, the other Pipids present spermatocyte bivalents with terminal chiasmata, like the higher

Anura (cf. Morescalchi, 1968a; Morescalchi, *et al.* 1970a). The karyotypes of the three genera of Pipids studied are seen to be very different from one another, as regards both the number and the morphology of the chromosomes (Fig. 24).

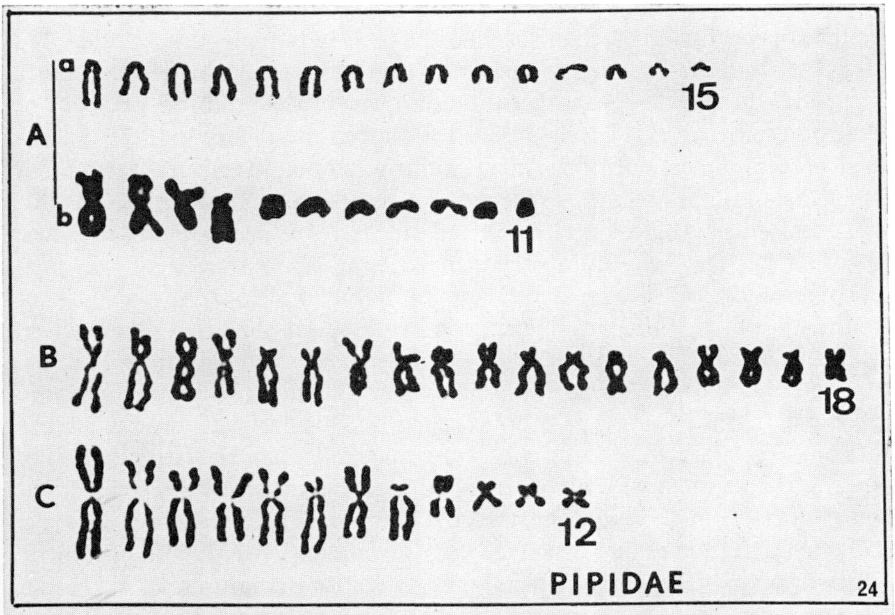

Fig. 24. The haploid karyotypes of four Pipids. A: a–*Pipa parva*, b–*P. pipa*; B: *Xenopus laevis*; C: *Hymenochirus boettgeri*.
Fig. 25. The haploid karyotype of *Rhinophrynus dorsalis* (Rhinophrynidae). (Courtesy of Dr J. Bogart.)

Pipa pipa ($2n = 22$; NF $= 30$) has 4 pairs of large homologues, *m* or *sm*, and 7 pairs of smaller homologues, all *t* (some resemble the microchromosomes of the Ascaphids); in various aspects, this karyological formula therefore resembles that of the ♀2 of *Leiopelma hochstetteri*; however, the bivalents of the male line of this species are of the ring type, like the higher Anura.

P. parva ($2n = 30$; NF $= 30$) has 15 pairs of homologues, all *t*, of regularly decreasing length; the smallest may perhaps be "microchromosomes" like those of the Ascaphids. The relationships between the two species of *Pipa*, which have the same NF, can easily be interpreted by postulating centric fusion/fission mechanisms in some pairs of homologues.

When I studied the karyotype of *P. parva* (Morescalchi, 1968a), I was struck by the discovery—at that time unique—of the exclusive presence of *t* chromosomes in this species; then I advanced the theory that this karyological formula was one of the most primitive among the Anura (seeing that the *t* chromosomes seemed exclusive to primitive forms among the Amphibia). Later karyological studies carried out on *P. pipa*, together with those of other authors on Anura of advanced families, in species of which *t*-chromosomes are also present, convinced me, however, of the lack of foundation of this theory. In fact, the karyological condition of *P. parva*, which is entirely peculiar among the Pipids, can more probably be interpreted as *secondary* than as *primitive*; it therefore seems to me more probable that the karyotype of this species may be derived, through centric fissions of the larger chromosomes, from that of *P. pipa*; the latter is also the species that is most widespread in South America, whereas *P. parva* might represent an old local endemism (cf. Dunn, 1948). In this case, also the peculiar condition of the spermatocyte bivalents of *P. parva*, with procentric chiasmata, may be derived from the generalized condition of the other Pipids, with terminal chiasmata (Fig. 42); in fact, it seems probable, that among the mechanisms of karyological differentiation here postulated for *P. parva* (centric fission and variation in the localization of the chiasmata), there are close causal relationships (cf. Stebbins, 1966).

On the basis of this hypothesis, I consider it possible that the karyotype of *P. pipa* is the most generalized among those of the two American Pipids studied; although it appears highly differentiated, on account of the considerable reduction in the number of chromosomes, it might nevertheless retain vestiges of primitivity—e.g. in the presence of microchromosomes (cf. Wickbom, 1950b)—which, as has been said, may bring it close to the set of certain Ascaphids.

Regarding the two genera of African Pipids, *Xenopus laevis* reveals 18 pairs of homologues of regularly decreasing length (the smallest are about one-fifth of the largest), all of which have two arms (*m*, *sm* or *st*) (NF $= 72$) (cf. Mikamo and Witschi, 1966). *Xenopus muelleri* has a chromosome set that is very similar to that of *X. laevis*.

X. laevis is one of the animals most widely employed in biological research laboratories, in view of the ease with which it can reproduce and be reared in captivity; in certain laboratory strains, heterochromosomes in the female line have been described (Weiler and Ohno, 1962; Morescalchi, 1963b; Ohno, 1967); however, Mikamo and Witschi (1966) have criticized these findings. In my own studies on animals coming direct from South Africa and on others collected in the wild state in Kenya and sent at once to the Laboratory, I was unable to detect any further indications of the presence of morphologically differentiable sex chromosomes: in the wild state, *X. laevis* (like *X. muelleri*) therefore seems to lack cytologically recognizable heterochromosomes (Fig. 23).

Hymenochirus boettgeri has $2n = 24$, with 8 pairs of large chromosomes, one intermediate and 3 small chromosomes; they are *m* or *sm*, except the 6th and 8th pairs, which are *st* (NF = 48): this karyological formula, as regards number and form of the chromosomes, approaches that typical of certain higher Anura (cf. later in text); among the primitive Anura, it is fairly similar to that of the discoglossid *Bombina*.

The three genera of recent Pipids are, in general, seen to be very different from one another, karyologically. Even if *Pipa* may retain certain characteristics of primitivity that bring it close to the Ascaphids (or perhaps also to the Proanura, seeing that the original stock of the Euanura, like various low Actynopterigi and Tetrapods, probably possessed microchromosomes; cf. Morescalchi, 1970), *Hymenochirus* seems to be specialized from this point of view and its karyological formula approaches that of certain higher Anura and of *Bombina*; on the other hand, *Xenopus* has a karyotype that, at present, appears unique among the present Anura and differs from the other Pipids to the same extent as, for example, the Sirenids differ from the higher Urodela. Obviously, the karyological heterogeneity may be related to the antiquity of the family; however, it might also be an indication of a polyphyletic origin of the present Pipids, seeing that in the other primitive families of Anura (Ascaphids, Discoglossids) the karyological formulae, even if different from one another, may be traced back to a single formula. On the other hand, it is very difficult to perceive morphological relationships between the karyotypes of the three modern Pipids, among which *Xenopus* in particular reveals a wholly peculiar karyological formula.

In general, therefore, the karyological findings emphasize the anatomical differences between the genera of modern Pipids; unfortunately, they do not seem to offer new facts for explaining the origin of this interesting family of Anura: even if a derivation of part of the Pipids from an ascaphid stock cannot be rejected from this point of view, neither is it possible to exclude other hypotheses, such as that of an early derivation of the Pipids from a stock of Proanura, in view of the fact that the present members of the family may have pursued different methods of karyological evolution from those of the rest of the Anura.

(d) *Rhinophrynidae*

Rhinophrynus dorsalis is a burrowing Anuran living in the tropical lowlands of Mexico, the morphological characteristics of which—a mixture of primitive and specialized ones—make it problematical to attribute it to one of the other families of the order (Walker, 1938); it is therefore considered preferable to classify it as a family on its own (Rhynophrynidae), which some consider to be related to the Pipids, especially on the basis of the morphology of the larval form; according to some authors, on the other hand, this family is related to the Discoglossids (regarding this problem see Hecht, 1963; Tihen, 1965; Meszoely, 1966; Nevo, 1968).

The karyotype of *Rhinophrynus* has been studied by J. Bogart, who has kindly provided me with a metaphase plate of this Anuran (Fig. 25): it has $2n = 22$, with 7 pairs of larger homologues and 4 pairs of smaller ones, which are *m*, *sm* and perhaps *st* (one pair only) (NF = 44).

According to Bogart, this karyological formula resembles that of certain American Microhylids (personal communication); in my opinion, it also resembles that of *Hymenochirus* (compared with which it has one less pair of large *sm*) or, perhaps to a greater extent, that of *Bombina* (compared with which it has one less pair of small chromosomes).

For the moment, therefore, the data on the chromosome set do not seem to be sufficient for providing a solution to the problem of the phyletic affinities of this peculiar Mexican Anuran, which is karyologically as close to certain Discoglossids and Pipids as it is possibly to certain higher Anura such as Microhylids. It would be interesting to study the spermatocyte bivalents of *Rhinophrynus*, in order to see whether they have interstitial chiasmata like those of *Bombina* or terminal ones like those of *Hymenochirus* and the higher Anura.

(e) *Pelobatidae*

This is a family of holarctic diffusion, of relatively ancient origins (probably prior to the Cretaceous period), often considered transitional between the Discoglossids and the group of the arciferal higher Anura (Noble, 1931); its origin from the Discoglossids is generally accepted by various contemporary authors (cf. Estes, 1970), whereas, as regards the role of the Pelobatids in the phylogeny of the higher families, there are almost diametrically opposed theories (cf. Griffiths, 1963; Tihen, 1965).

The living genera of the family are often grouped into three different subfamilies: the Megophryinae, Asiatic, comprising various genera and the most primitive forms of the family; the Pelobatinae, with the Eurasiatic *Pelobates* and the North American *Scaphiopus*, often subdivided into the two sub-genera *Scaphiopus* and *Spea* and the Pelodytinae, with the single genus Pelodytes, Eurasiatic, which is seen to be highly differentiated from the members of the other two sub-families, such that various authors consider it as a family apart, the Pelodytidae (Taylor, 1941, 1942; Zweifel, 1956). The Sooglossinae of the Seychelles are probably not Pelobatidae; Griffiths (1963) considers them specialized Ranids, while other authors have put forward the theory that they may be relict leptodactylid forms (Laurent, personal communication).

The karyological data so far collected on the Pelobatids are as follows:

Megophryinae:
Leptobrachium hasselti: $2n = 26$ (This paper)
L. nigrops: $2n = 26$ (This paper)
Megophrys monticola: $2n = 26$ (This paper)
M. nasuta: $2n = 26$ (This paper)
Pelobatinae:
Pelobates cultripes: $2n = 26$ (Morescalchi, 1964b)
P. fuscus: $2n = 26$ (Wickbom, 1945; this paper)
Scaphiopus (Scaphiopus) couchi: $2n = 26$ (Wasserman and Bogart, 1968)
S. (S.) holbrooki: $2n = 26$ (Wasserman and Bogart, 1968)
Scaphiopus (Spea) bombifrons: $2n = 26$ (Hughes, 1962; Duellman, 1967).
S. (S.) hammondi: $2n = 26$ (Hughes, 1962; Morescalchi, 1968d)
Pelodytinae:
Pelodytes punctatus: $2n = 24$ (Jaylet, 1964; Morescalchi, 1964a)

With the exception of *Pelodytes*, the Pelobatids studied all have $2n = 26$, with *m*, *sm* or *st* chromosomes (NF = 52); the karyotype is morphologically scarcely variable in the different species, appearing in two main formulae: the first, present in *Leptobrachium*, *Pelobates* and *Scaphiopus* of the subgenus *Scaphiopus* (significantly, the most primitive genera of the two subfamilies of "true" Pelobatids), consists of 6 pairs of large and 7 pairs of small chromosomes; the second, present in *Megophrys* and in the *Scaphiopus* of the sub-genus *Spea*, consists of 5 pairs of large and 8 pairs of small chromosomes (Fig. 26). The relationships between the two formulae are easy to interpret, seeing that the differences are in one pair of homologues, the 6th pair, which have a longer arm in the first group of Pelobatids and a shorter arm in the second: since the first group comprises the most primitive forms, it is possible to consider that the second karyological formula is derived from the first through the mere deletion of a fragment of the long arm of the 6th pair of homologues.

Pelodytes has 7 pairs of larger homologues and 5 pairs of smaller chromosomes, all *m*, *sm* or *st* (NF = 48); it differs karyologically from *Pelobates* in having one more pair of large homologues and two less small chromosomes: in theory, the latter, through an unequal translocation, might have given

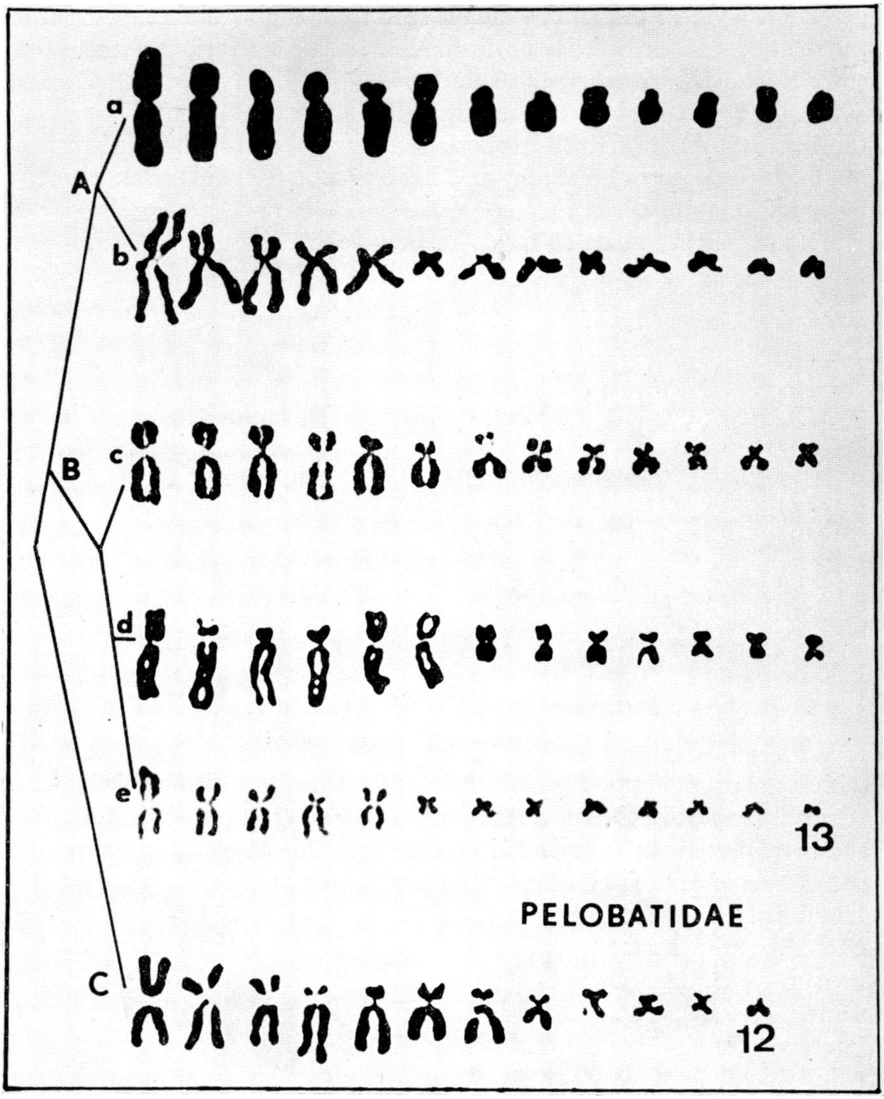

Fig. 26. The haploid karyotypes of some Pelobatids: A: Megophryinae, a–*Leptobrachium hasselti*, b–*Megophrys nasuta*; B: Pelobatinae, c–*Pelobates cultripes*, d–*Scaphiopus* (*Scaphiopus*) *holbrooki*, e–*Scaphiopus* (*Spea*) *hammondi*; C: Pelodytinae, *Pelodytes punctatus* (d: from Wasserman and Bogart, 1968).

rise to the large *st* typical of *Pelodytes* (cf. Morescalchi, 1964b).

In agreement with the systematic data and also from a karyological point of view, *Pelodytes* is thus seen to be well differentiated from the rest of the

Pelobatids, compared with which it appears to be more highly specialized.

Leptobrachium, *Pelobates*, *Scaphiopus* and *Pelodytes* have spermatocyte bivalents with two terminal chiasmata, as in the higher Anura; moreover, in the basic families of the latter there are also 26-chromosome karyotypes almost equal to those of the typical Pelobatids: on the basis of other data, Reig (1961) does not exclude the possibility that the Pelobatids may belong to the Neobatrachia, a sub-order in which this author includes all the higher Anura; from a karyological point of view, this possibility would seem to be fully supported.

The problem that arises here—though to a greater extent—is the same as that already posed by the karyology of the Discoglossids: even if, from a zoogeographical point of view, it appears improbable that the Pelobatids, typically holarctic, are on the same phyletic line as the higher Anura (the more generalized families of which are probably of Gondwanian origin), from a karyological point of view the Pelobatids seem to belong to this group of forms.

At this point, two hypotheses may be advanced: the Pelobatids, derived in holarctic regions from discoglossid ancestors, have had a karyological evolution that is closely parallel to that of the (leptodactylid) forms from which the higher Anura are descended; or else the Pelobatids are derived directly from the same stock, ubiquitous in the Middle Mesozoic, that has given rise to the leptodactylid radiation in the Gondwanian regions.

(f) *Leptodactylidae*

The Leptodactylids constitute one of the most extensive and interesting families of the higher Anura, of which they often represent generalized forms; they are spread throughout Australasia and the Neotropical Region; certain species have reached the Southern United States and a single genus, of problematical affinities, is present in southern Africa.

The systematics of these Anura are extremely complex and, according to some authors, they are variously subdivided into taxonomic entities of different degree: some maintain that the leptodactylids can also be grouped into about ten different families; however, in agreement with various authors, I have preferred a "unitary" classification of the leptodactylid forms, considering them all as Leptodactylidae and distributing them among various sub-families, which, in their turn, are often subdivided into minor groups; to these last groups I have ascribed—at times arbitrarily (as seems to be the case, however, to a varying extent, with every classification so far proposed for these Amphibia)—the role of tribes.

The sub-families recognized here are as follows:

(1) Myobatrachinae: these comprise all the Leptodactylids of Australasia

(Noble's Criniinae), subdivided by Parker (1940) into two sub-families, albeit closely related to each other, which have here been given the role of tribes, seeing that various authors are not in agreement over the dividing up of the Myobatrachinae: the Myobatrachini and the Cycloranini. As we shall see, the karyological data also testify to a considerable homogeneity in these two groups of leptodactylids.

(2) Heleophryninae: this sub-family includes the only African genus, *Heleophryne*, with few species, one of which has been studied karyologically and findings seem to confirm that these problematic forms belong to the family.

(3) Ceratophryinae: this includes certain genera of South American Leptodactylids, probably very old, morphologically and ehtologically quite distinct from the other forms of the family, such that various authors consider it a family in itself. In this sub-family, two groups of forms can easily be recognized (Reig and Limeses, 1963): one, with the genera *Chacophrys*, *Ceratophrys* and *Lepidobatrachus* (here considered as "Ceratophryini"), comprises the more typical forms of the sub-family, possessing curious well-developed dermic ossifications; the other, with the genera *Odontophrynus*, *Macrogenioglottus* and *Stombus* (here indicated as "Odontophrynini"), comprises forms that are morphologically closer to the typical members of the Leptodactylinae.

(4) Leptodactylinae: includes a very large number of species, all American, constituting the most representative group of the family. They may be further divided into a certain number of minor groups; the more important among those that comprise forms studied karyologically are perhaps the following (as already stated, there is no definite classification for these Amphibia, at least not one that is generally accepted): (i) Telmatobiini and possibly related forms (such as *Calyptocephalella* and *Cycloramphus*), which are among the oldest Leptodactylinae; (ii) Leptodactylini and possibly related forms: a vast group, with the most typical forms of the sub-family; (iii) Eleutherodactylini and possibly related forms, which constitute a group of Leptodactylids biologically highly specialized and in a stage of radiation in part of Central and South America and (iv) Batrachylini (Alsodini) and possibly related forms, perhaps strictly related to the Eleutherodactylini, but with some genera that seem very old (*Eupsophus*).

(5) Elosiinae: a small group of Brazilian Leptodactylids, long since geographically isolated from the Leptodactylinae and possibly having some affinities with the Ceratophryinae or Telmatobiini.

The following list shows the chromosome numbers so far obtained, especially by American research workers, for the Leptodactylidae, subdivided according to the methods now described. Some species are polyploid and are indicated by the sign (+); others have diploid and polyploid populations (perhaps cryptic species) and are indicated by the sign (++); in many cases I have reported only the most recent and better documented bibliographical data and have omitted the others, which are, however, easily obtainable by reference to the works listed.

(1) Myobatrachinae.

(i) Myobatrachini.

Crinia signifera: $2n = 24$ (Morescalchi, 1968e; Morescalchi *et al.*, 1968)

C. sloanei: $2n = 24$ (Morescalchi, 1968e; Morescalchi *et al.*, 1968)

Pseudophryne bibronii: $2n = 24$ (Morescalchi, 1968e; Morescalchi *et al.*, 1968)

Uperoleia marmorata: $2n = 24$ (Morescalchi, 1968e; Morescalchi *et al.* 1968).

(ii) Cycloranini.

Limnodynastes peronii: $2n = 24$ (Robinson and Stephenson, 1967; Morescalchi, 1968e; Morescalchi *et al.*, 1968)

L. tasmaniensis: $2n = 24$ (Ullerich, 1967)

L. fletcheri: $2n = 24$ (Morescalchi, 1968e; Morescalchi *et al.*, 1968)

(2) Heleophryninae.

Heleophryne purcelli: $2n = 26$ (This paper)

(3) Ceratophryinae.

(i) Ceratophryini.

Ceratophrys calcarata: $2n = 26$ (Morescalchi, 1967c)

(+) *C. dorsata*: $8n = 104$ (Beçak, M. L. *et al.*, 1967)

(++) *C. ornata*: $2n = 26$, $8n = 104$ (Bogart, 1967; Barrio and Rinaldi de Chieri, 1970a)

Chacophrys pierottii: $2n = 26$ (Bogart, 1967; Barrio and Rinaldi de Chieri, 1970a)

Lepidobatrachus asper: $2n = 26$ (Bogart, 1967; Barrio and Rinaldi de Chieri, 1970a)

L. laevis: $2n = 26$ (Bogart, 1967; Barrio and Rinaldi de Chieri, 1970a)

L. llanensis: $2n = 26$ (Bogart, 1967; Barrio and Rinaldi de Chieri, 1970a)

(ii) Odontophrynini.

Macrogenioglottus alipioi: $2n = 22$ (Beçak M. L. *et al.*, 1970)

(++) *Odontophrynus americanus*: $2n = 22$; $4n = 44$ (Beçak, M. L. *et al.*, 1966, 1967, 1970; Bogart, 1967)

O. carvalhoi: 2n = 22 (Beçak M. L. *et al.*, 1970)
O. cultripes: 2n = 22 (Beçak M. L. *et al.*, 1970)
O. occidentalis: 2n = 22 (Saez and Brum-Zorrila, 1966; Beçak M. L. *et al.*, 1970)
Stombus appendiculatus: 2n = 22 (Beçak M. L. *et al.*, 1970)
S. boiei: 2n = 22 (Beçak M. L. *et al.*, 1970)

(4) Leptodactylinae.

(i) Telmatobiini.

Telmatobius barrioi: 2n = 26 (This paper)
T. ceiorum: 2n = 26 (This paper)
T. schreiteri: 2n = 26 (Barbieri, 1950)
T. stephani: 2n = 26 (This paper)
T. (species of *marmoratus* group): 2n = 22 (Brum-Zorrilla and Saez, 1968).
Calyptocephalella gayi: 2n = 26 (Brum-Zorrilla and Saez, 1968; this paper)
Cycloramphus asper: 2n = 26 (Beçak M. L. *et al.*, 1970)
C. dubius: 2n = 26 (Beçak M. L. *et al.*, 1970)
C. fuliginosus: 2n = 26 (Bogart, 1970)

(ii) Leptodactylini.

Eupemphix nattereri: 2n = 22 (Beçak, 1968)
E. pustulosus: 2n = 22 (Morescalchi and Gargiulo, 1968)
Leptodactylus bufonius: 2n = 22 (Bogart, 1967)
L. chaquensis: 2n = 22 (Brum-Zorrilla and Saez, 1968)
L. fuscus: 2n = 22 (Beçak M. L. *et al.*, 1970)
L. melanonotus: 2n = 22 (Bogart, 1967)
L. ocellatus: 2n = 22 (Bianchi and Molina, 1967; Beçak, M. L. 1968)
L. pentadactylus: 2n = 22 (Morescalchi and Gargiulo, 1968)
L. prognathus: 2n = 22 (Brum-Zorrilla and Saez, 1968)
Physalaemus biligonigerus: 2n = 22 (Brum-Zorrilla and Saez, 1968)
P. cuvieri: 2n = 22 (Beçak M. L. *et al.* 1970)
P. fuscomaculatus: 2n = 22 (Beçak M. L. *et al.*, 1970)
P. gracilis: 2n = 22 (Brum-Zorrilla and Saez, 1968)
Pleurodema bibronii: 2n = 22 (Barrio and Rinaldi de Chieri, 1970b)
P. bufonina: 2n = 22 (Barrio and Rinaldi de Chieri, 1970b)
P. cinerea: 2n = 22 (Barrio and Rinaldi de Chieri, 1970b)
(+) *P. darwinii*: 4n = 44 (Barrio and Rinaldi de Chieri, 1970b)
P. guayape: 2n = 22 (Barrio and Rinaldi de Chieri, 1970b)
(+) *Pleurodema kriegi*: 4n = 44 (Barrio and Rinaldi de Chieri, 1970b)

P. tucumana: $2n = 22$ (Barrio and Rinaldi de Chieri, 1970b)
Pseudopaludicola ameghini: $2n = 20$ (Beçak, M. L. 1968)
P. falcipes: $2n = 18$ (Beçak, M. L. 1968)

(iii) Eleutherodactylini
Eleutherodactylus augusti: $2n = 22$ (Duellman, 1967)
E. discoidalis: $2n = 22$ (Brum-Zorrilla and Saez, 1968)
E. guentheri: $2n = 22$ (Beçak, M. L. 1968)
E. ranoides: $2n = 22$ (Brum-Zorrilla and Saez, 1968)
E. altamazonicus: 2n = 34 (Bogart, 1970)
E. diastema: $2n = 18$ (Bogart, 1970)
E. brandsfordi: $2n = 20$ (Bogart, 1970)
E. fleishmanni: $2n = 20$ (Bogart, 1970)
E. melanostictus: $2n = 22$ (Bogart, 1970)
E. podiciferus: $2n = 18$ (Bogart, 1970)
E. variabilis: $2n = 34$ (Bogart, 1970)
E. ventrimarmoratus: $2n = 36$ (Bogart, 1970)
Lithodytes lineatus: $2n = 18$ (Bogart, 1970)
Syrrophus marnocki: $2n = 30$ (Bogart, 1970)
Tomodactylus nitidus: $2n = 22$ (Duellman, 1967)

(iv) Batrachylini (Alsodini).
Batrachyla leptopus: $2n = 26$ (Bogart, 1970)
Eupsophus nodosus: $2n = 22$ (Bogart, 1970)
E. coppingeri: $2n = 26$ (Bogart, 1970)
E. roseus: $2n = 30$ (Bogart, 1970)
E. taeniatus: $2n = 26$ (Bogart, 1970)
E. vertebralis: $2n = 28$ (Bogart, 1970)
Thoropa miliaris: $2n = 26$ (Brum-Zorrilla and Saez, 1968; Bogart, 1970)

(5) Elosiinae.
Crossodactylus gaudichaudii: $2n = 26$ (Bogart, 1970)
C. grandis: $2n = 26$ (Beçak, M. L. 1968)
Elosia aspera: $2n = 26$ (Brum-Zorrilla and Saez, 1968)
E. lateristrigata: $2n = 26$ (Brum-Zorrilla and Saez, 1968)
Zachaenus parvulus (*Oocormus microps*): $2n = 26$ (Beçak M. L. *et al.*, 1970)

(1) Myobatrachinae: The Australian Leptodactylids are seen to be karyologically very homogeneous and well defined in comparison to the other members of the family: they all have $2n = 24$ (NF = 48), with 6 pairs of larger and 6 pairs of smaller chromosomes; the former are all *m* or *sm*, except for one pair of *st* in *Limnodynastes* and in *Pseudophryne* and two pairs of *st* in the other Myobatrachini; the latter are all *m* or *sm* and are very similar to

one another in all the species studied, except for the 12th pair, which in *Pseudophryne bibronii* is considerably smaller than in the remaining species; *Pseudophryne* also has relatively larger chromosomes than the other species.

Other karyological differences among the Myobatrachinae concern, as usual, the localization of the heterochromatic tracts; these are particularly abundant in *Pseudophryne*, possibly sensitive to the effects of colchicine (cf. Morescalchi and Gargiulo, 1967).

Excluding the peculiar *Pseudophryne*, the most important differences between the karyotype of the Cycloranini (*Limnodynastes*) and of the Myobatrachini (*Crinia* and *Uperoleia*) are essentially confined to those between the chromosomes of the 6th pair of the two groups, which are *m* in the first, and *st* in the second (Fig. 27): these differences can easily be interpreted by postulating a single chromosome rearrangement (of the *shift* type, or a pericentric inversion) that has led from one chromosome formula to the other (Morescalchi, 1968e).

The primitive morphology of the Australian Leptodactylids does not seem to be verified to any great extent in the karyological data, since these Anura reveal karyotypes that are relatively differentiated as regards morphology and number of chromosomes (however, more primitive karyotypes might perhaps be found in other genera of the subfamily, such as the generalized *Cyclorana*, *Mixophyes* etc. not yet studied). This karyological formula may perhaps be derived from one with 26 chromosomes and more generalized in various aspects (see later), as is found in various American Leptodactylids and also in *Heleophryne*: as sometimes occurs the Myobatrachinae, though morphologically primitive, are seen to be karyologically more differentiated than other members of the same family.

The fact that the 12th pair of *Pseudophryne* is much smaller than that of the other species may perhaps indicate a tendency towards their loss in some members of the sub-family, without a pair of small homologues, the karyotype of the Myobatrachinae (especially the Cycloranini) become very similar to that of the Bufonidae ($2n = 22$).

(2) Heleophryninae: The so-called Ghost-Frogs (genus *Heleophryne*) of South Africa have been variously attributed to the families of the Ranidae, the Bufonidae or the Leptodactylidae; this last theory, prevalent today, seems to find support in the karyological data: in fact, *H. purcelli* has $2n = 26$ (NF $= 52$), with 7 pairs of larger and 6 pairs of smaller chromosomes, which are in order similar to the 7 larger chromosomes of various Leptodactylinae having 22 chromosomes (with 7 pairs of large and 4 pairs of small homologues) and to the smaller chromosomes of various Ceratophryini. From a karyological point of view, *Heleophryne* therefore present a mosaic of conditions intermediate between those of primitive and of differentiated

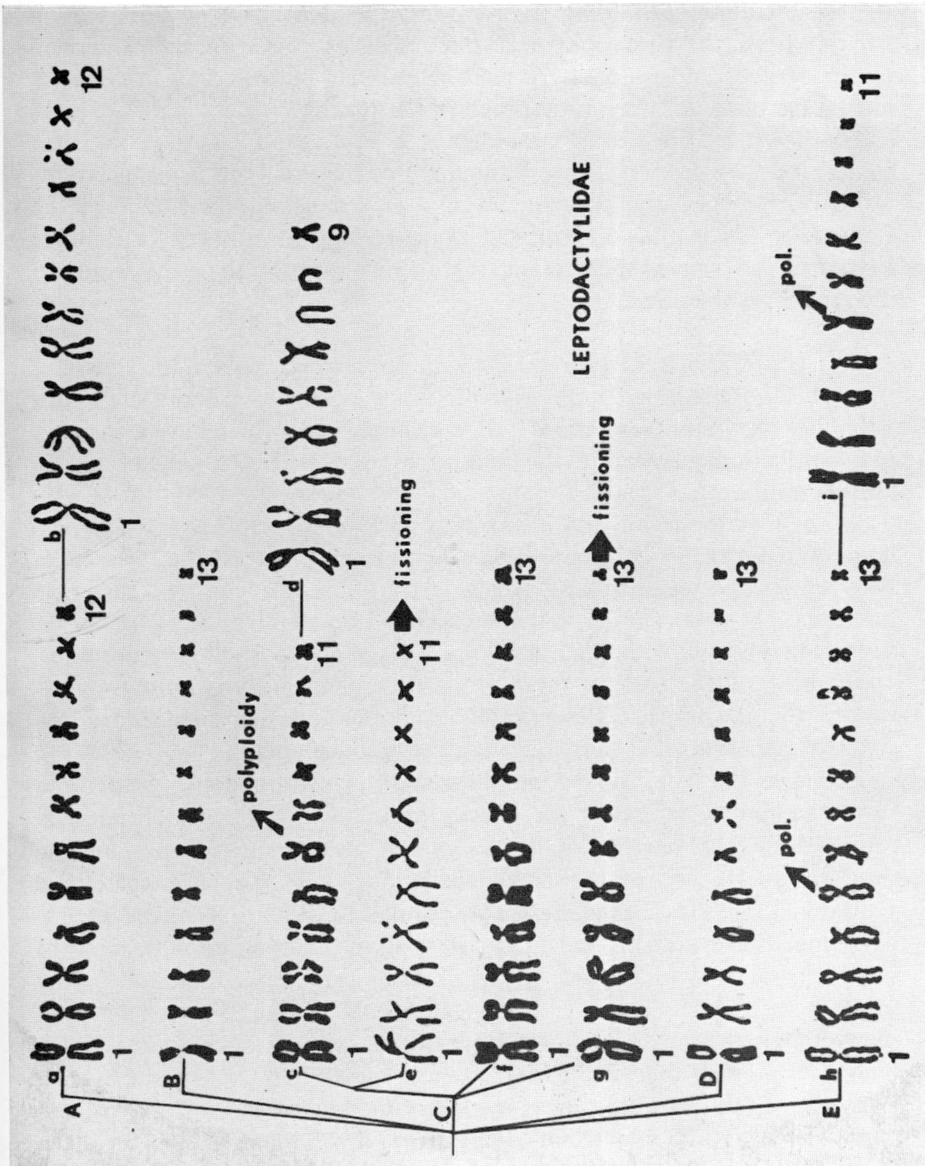

Fig. 27. The haploid karyotypes of some Leptodactylids.
A: Myobatrachinae, a–*Crinia signifera* (Myobatrachini), b–*Limnodynastes peronii* (Cycloranini);
B: Heleophryninae (*Heleophryne purcelli*);
C: Leptodactylinae, c–*Eupemphix pustulosus* and d–*Pseudopaludicola falcipes* (Leptodactylini), e–*Eleutherodactylus guentheri* (Eleutherodactylini); f–*Calyptocephalella gayi* (Telmatobiini), g–*Eupsophus coppingeri* (Batrachylini);
D: Elosiinae (*Crossodactylus grandis*);
E: Ceratophryinae, h–*Ceratophrys calcarata* (Ceratophryini), i–*Odontophrynus carvalhoi* (Odontophrynini). A part of the haploid karyotpes are redrawn from figures in the papers cited in the list of chromosomes (pp. 275–277).

Leptodactylids; this finding may suggest the derivation of this African Anuran from the same ancestral stock that gave rise to the American Leptodactylids and probably to the Australasian ones as well. This stock must have been far more widespread in the past than it is today, as seems to be shown by the fossils, probably of a leptodactylid nature, in India (*Indobatrachus*); from it, the spreading out of the family in lands of Gondwanian origin and the subsequent total or partial disappearance of the Leptodactylids in the territories of radiation of the competitive Ranidae have probably caused the present fragmentary dislocation of the family in the Austral hemisphere.

(3) Ceratophryinae: In agreement with the systematic data, the genera of this sub-family subdivide, also karyologically, into two groups (both with polyploid species or populations): the Ceratophryini, which have a basic set of $n = 13$ chromosomes, and the Odontophrynini, with a basic set of $n = 11$ chromosomes.

In the Ceratophryini, the haploid set of 13 chromosomes consists of 6 larger and 7 smaller chromosomes, which are *m*, *sm* or *st*; the differences between the species seem fairly small. In the Odontophrynini, on the other hand, the set is very similar to that of various Leptodactylinae with $n = 11$: in fact, they show, with few variations, 7 large and 4 small chromosomes, *m*, *sm* or *st*: on the basis of these similarities, some authors have even advanced the theory that the Odontophrynini are to be included in the group of Leptodactylids here indicated as Leptodactylinae (Bogart, 1967).

However, the set of the Odontophrynini ($n = 11$) might even be derived from one similar to that of the Ceratophryini ($n = 13$) through unequal translocations that have led to the genesis of a "new" large chromosome and the "disappearance" of two small chromosomes in the basic set; these evolutional methods may therefore be parallel to those that, in the Leptodactylinae, have possibly led to the genesis of a karyotype with $n = 11$ starting from one with $n = 13$, typical of the oldest species of the sub-family (cf. later in text). Reig and Limeses (1963) have advanced the theory that the Odontophrynini are the forms closest to the original stock of the Ceratophryinae (a stock that, according to these authors, is among the typical Leptodactylinae), whereas the Ceratophryini are the more differentiated (apomorphic) forms of the sub-family: from a karyological point of view, the contrary view seems more probable, in view of the fact that the Odontophrynini seem, karyologically, the most differentiated forms of this sub-family. On the other hand, the Ceratophryini possess the same karyological formula as most of the more generalized forms of various families of higher Anura ($n = 13$), which undoubtedly constitutes a proof of the primitive nature of this formula; furthermore, it is interesting to point out that the karyotype of these Leptodactylids is very similar to that of some Pelobatidae (Morescalchi, 1967c).

(4) Leptodactylinae: A general list of the chromosome numbers of these Anura may suggest irregular and heterogeneous lines of karyological evolution (cf. the list in Rabello, 1970); however their subdivision into minor groups that probably contain species related to one another suggests the existence of few lines of karyological evolution, which fit well into the more general evolution of the family.

(i) The ancient but specialized Telmatobiini (with the forms possibly related to them, and also specialized for aquatic life) have so far been studied little karyologically. All the species of *Telmatobius* that I examined show $n = 13$ in the male germinal line; these data correspond to others reported by Barbieri (1950) but are in disagreement with those of Brum-Zorrilla and Saez (1968), who found $n = 11$ in an unidentified species of the group *marmoratus*.

A diploid set of $2n = 26$ is found in *Calyptocephalella gayi*: this interesting relict form, unique in its genus, has remained almost unchanged since the beginning of Tertiary (Schaeffer, 1949); it shows 6 pairs of larger and 7 pairs of smaller homologues, with a morphology similar to those of the Ceratophryini. Also the problematical *Cycloramphus*, in the species studied, shows $2n = 26$, with 5 pairs of large and 8 pairs of smaller homologues (cf. the bibliographical list): the relations between the formula of *Cycloramphus* and that of *Calyptocephalella* seem to be of the same kind as those already seen, among the Pelobatidae, in various genera of the family, even within the genus *Scaphiopus*.

The basic 13-chromosome karyotype therefore seems the one typical of this very old group of Leptodactylinae, as it also is of other sub-families of Leptodactylidae.

(ii) The karyotype of the Leptodactylini is considerably homogeneous, these Anura showing $2n = 22$ (NF $= 44$), often with 7 pairs of larger and 4 pairs of smaller homologues; certain species (or populations) are polyploid. The species of *Pseudopaludicola* ($2n = 20$, 18) constitute exceptions, being also morphologically well differentiated from the other Leptodactylini (Barrio, 1945).

The basic set of the Leptodactylini is therefore $n = 11$ as in the Odontophrynini; parallel with the latter, it may be derived, by unequal translocations, from one with $n = 13$, as is found, for example, among the Telmatobiini (and, we shall see, in other very old Leptodactylinae).

Only *Pseudopaludicola* seems to have passed the stage of karyological differentiation reached by most of the Leptodactylini, with the further reduction in the number of the chromosomes: a fact that places the genus, from a karyological point of view, at the top of the evolution of the Leptodactylini so far studied.

(iii) The Eleutherodactylini possess chromosome numbers that are very different in the various species, which show diploid sets ranging between 18 and 36 (NF = 44–36); since various species have *t*-chromosomes, the number of which increases with that of the diploid set, Bogart (1970) has advanced the theory that the interspecific karyological differences may depend on phenomena of centric fusion, considering the original karyotype as one having 36 chromosomes, all *t* (as is found, for example, in *Eleutherodactylus ventrimarmoratus*); this theory is based particularly on the fact that, as already seen, the *t*-chromosomes are characteristic of the more primitive Anura (Ascaphidae etc.).

However, there are no proofs that the Eleutherodactylini are primitive forms among the Leptodactylidae; indeed, the species with many *t*-chromosomes belong to groups of Central American forms, and forms from the northern part of South America, where these frogs are at present in a stage of "amazing" radiation (Cei, 1968). On the other hand, the more southern *Eleutherodactylus* (*guentheri*, *discoidalis*), have $2n = 22$ (NF = 44), with a karyotype more or less equal to that typical of the Leptodactylini (Fig. 27c–e).

I therefore suggest that in the Eleutherodactylini the mechanisms of actual radiation and the interspecific (and perhaps also *intraspecific*) karyological variability are phenomena closely connected to each other: in the nordic and subequatorial forms of these Anura, processes of *centric fission* may have caused variations in the number of the chromosomes (an increase in them) and, with other less extensive chromosome rearrangements, may have caused the genesis of cytological barriers within the populations, favouring the phenomena of speciation (on this problem see White, 1969).

It therefore seems to me probable that the karyotypes rich in *t*-chromosomes may be derived by centric fission from others, which might be the 22- (or 26-)chromosome karyotypes found among the southern species of the group. On the basis of this hypothesis, the ancestral karyological formula of the Eleutherodactylini would therefore seem similar to that of the Leptodactylini ($2n = 22$, NF = 44); parallel with species of the latter (of the genus *Pseudopaludicola*) also in the Eleutherodactylini some species would have further reduced the chromosome number to 20–18 elements, possibly by unequal translocations; later, in the areas of "explosive" speciation, phenomena of centric fission would have created a whole series of karyotypes having a number intermediate between the initial number of 18–22 chromosomes (NF = 44–36) and the terminal number of 36 chromosomes, all *t* (NF = 36).

From a karyological point of view, and in greater agreement with the systematic data, the Eleutherodactylini, possibly derived from a stock of Leptodactylini, must perhaps be considered among the more specialized Leptodactylinae.

(iv) As regards the Batrachylini, *Batrachyla* and *Thoropa* have the type of karyological formula considered "primitive" among the Leptodactylidae, with $2n = 26$, NF $= 52$; of interest are the resemblances between the set of *Thoropa* and that of a probably old genus, *Cycloramphus* (Bogart, 1970): this fact may perhaps strengthen the concept of an origin of all the Leptodactylinae from ancestral 26-chromosome forms.

As regards *Eupsophus* ($2n = 22, 26, 28, 30$), some species seem to preserve the "primitive" karyological formula with 26 chromosomes, while others show different chromosome numbers and sets that can hardly be referred to a single form of karyotype (cf. Bogart, 1970). It is possible that this genus (perhaps a super-genus), which seems to have repeated, in other areas and in older periods (Schaeffer, 1949), processes of adaptive radiation of the type of those actually seen in the Eleutherodactylini, may preserve only some of the karyological formulae derived from previous phenomena of centric fission and further modified by other chromosome rearrangements; owing to this fact, comparison between karyotypes of different species is complex. The original formula for these processes might have been the 26-chromosome type, to be found in some species of *Eupsophus* and in the other Batrachylini.

(5) Elosiinae: The sub-family of the Elosiinae, which includes forms that long ago separated from the rest of the Leptodactylidae and is of uncertain relationship with the Ceratophryinae and the Leptodactylinae, also shows, in the species studied, the 26-chromosome (NF $= 52$) diploid set typical of the oldest forms of the family; evidently, the Elosiinae are karyologically conservative forms.

The relationships between the more common karyotypes in the various groupings of Leptodactylidae are shown in Fig. 27, which indicates the haploid sets of each of the groups.

In conclusion, among the 5 sub-families of present Leptodactylids, 4 show the formula with $n = 13$ ($2n = 26$) as basic karyotype in the oldest forms of each sub-family. There exist some versions (only slightly differentiated from one another) of this basic karyotype, though they can easily be related to one another by postulating few chromosome rearrangements.

Genera that probably occupy a key-position in the phylogeny of the family (such as *Heleophryne*, *Ceratophrys*, *Telmatobius*, *Calyptocephalella*, *Cycloramphus*, *Batrachyla*, *Thoropa*, *Crossodactylus*, *Elosia* etc.) all have this

26-chromosome set, which must therefore have been present also in the ancestral stock from which this vast family originated (this may be corroborated by the presence of this same set in the primitive Pelobatidae, and above all in the Hylidae, which have clear phyletic affinities with the Leptodactylidae).

In contrast with this theory there is the presence of $2n = 24$ to be found in the primitive Myobatrachinae (at least in the studied species); however, the karyotype of these Australian Leptodactylids may easily be made to derive from a 26-chromosome karyotype through simple rearrangements in few pairs of homologues: this seems to have occurred, as we shall see, also in the Hylidae.

Among the Leptodactylinae and among the Ceratophryinae, various groups of species seem to have further reduced their chromosome number, with the acquisition of 22-chromosome formulae (in the diploid species),

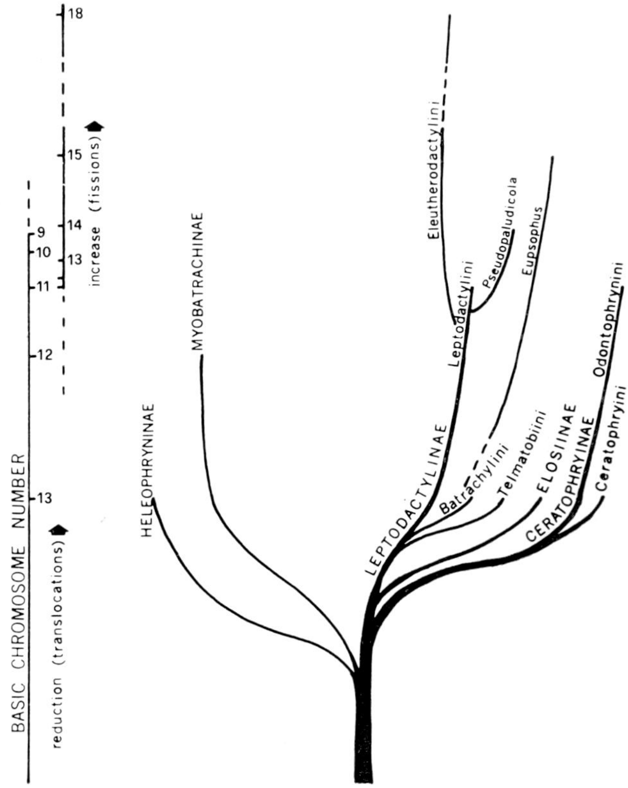

Fig. 28. A phyletic tree of the Leptodactylidae, built utilizing especially karyological data (on the ordinates); explanations on the intrafamilial relationships and on the modes of karyotype evolution are to be found in the text.

which, especially in the former sub-family, seem to have been "stabilized". Since, as has been said, the oldest species of each of the sub-families have $2n = 26$, it seems probable that, in order to acquire a 22-chromosome set, various groups of species of these Leptodactylids have reached intermediate conditions with 24-chromosome karyotypes, perhaps similar to those of the present Myobatrachinae.

Again among the Leptodactylinae, various groups of species may have differentiated their karyotype further, reducing it to 18–20 chromosomes, as occurs in *Pseudopaludicola* and in certain Eleutherodactylini; in the latter (and perhaps also in *Eupsophus*), successive mechanisms of karyotype fissioning (cf. Todd, 1967, 1970) and other minor rearrangements seem to be correlated with processes of "explosive" radiation.

On the basis of these data, the cytotaxonomic relationships among the Leptodactylids have been represented by a diagram (Fig. 28) in which I have shown, on the ordinates, the basic chromosome number (which tends to be reduced, possibly through unequal translocations, except for possible subsequent increases through centric fissions). The numerical variations in the chromosomes have been expressed arbitrarily on a logarithmic scale, the better to express the presumed importance that a decrease or increase in the chromosome number may have had in restricting—or increasing, respectively—the genetic potentiality of each formula (to a great extent, I defer further reference to these concepts till later in the text; see Section E).

(g) *Dendrobatidae*

These constitute a family of Neotropical diffusion, consisting of only 4 genera but rich in species (Savage, 1968); its phyletic relationships are uncertain: some authors consider them among the Ranidae (Griffiths, 1959, 1963), while most authors maintain that they are derived from the Leptodactylidae.

Only Duellman (1967) has so far studied the karyology of species of this family; *Dendrobates pumilio* has $2n = 20$, with 6 pairs of larger and 4 of smaller homologues, all with 2 arms (NF = 40), while *Prostherapis inguinalis* has $2n = 24$, and the morphology of the various chromosomes is not known. From the only mitotic figure presented by this author (on *D. pumilio*), it seems to me that the chromosomes of these Anura are of the "leptodactylid" type, and different from those of the Ranidae.

As we shall see, 20-chromosome sets are found in the Centroleninae, which are specialized Hylids, probably derived from Hylinae with 24 and 22 chromosomes: the Dendrobatidae might constitute a stock of specialized Leptodactylids (or, less probably, Hylids), the karyological evolution of which has occurred parallel with that of the Centroleninae (we shall see that cases of parallelism between Leptodactylids and Hylids are common in the karyological evolution).

(h) *Hylidae*

A large family of higher Anura, widespread throughout the area occupied bv the order, with the exception of Africa; *Hyla* is also present in almost the whole area of diffusion of the family. The greatest differentiation of the Hylids, which seem partly to be still flourishing today, is reached in the Neotropics; outside this area few genera are found, except for *Hyla*, which, however, probably constitutes a non-natural ensemble of species (Goin and Goin, 1962); it may even be that the family is polyphyletic and that the Australian and Papuan Hylids (Nyctimystinae) have originated from a different leptodactylid stock from that from which the other Hylids are derived (cf. Griffiths, 1963; Tyler, 1968; Laurent, personal communication).

However, the Hylids essentially differ from the Leptodactylids in the presence of intercalary phalanges, the genesis of which may have occurred at various times, as part of a general adaptation to arboreal life.

Most of the Hylids are included in the sub-family of the Hylinae, especially widespread in the Neotropics; it is from forms of this sub-family that the Centroleninae are probably derived. These are specialized tree-frogs, which some authors raise to the rank of a family, confined to Central and South American areas; however, there is still uncertainty regarding the relationships of the Neotropical Pseudinae (which are also sometimes raised to the rank of a family but are more commonly considered a sub-family of the Hylids, see Griffiths, 1963), which might even be derived directly from a leptodactylid stock. As already stated, the Australo-Papuan members of the family are included in the Nyctimystinae; among these, various species are commonly assigned to *Hyla*; however, the name *Litoria* is probably more adequate for the so-called *Hyla* of this geographic area (Tyler, 1971).

In the following chromosome list I have given the karyological data so far acquired by various authors on the family, subdivided into the above-mentioned sub-families; the Hylinae have been further subdivided into groups of genera (tribes), which, in the opinion of competent authors, may have differentiated very early from the central stock of the sub-family (cf. Goin and Goin, 1962; Duellman, 1968): the Hemiphractini, with the marsupial hylids, and the Phyllomedusini, group of presumably old tree-frogs, often considered in a subfamily (Phyllomedusinae); the remaining Hylinae have been considered as Hylini.

It is difficult to fit into place the numerous karyological data collected on the family (many of which are confined solely to the number of chromosomes) and this is partly due to the numerous uncertainties that exist today on the systematics of these Anura; I therefore consider it probable that, perhaps even in the near future, there may be modifications in the conclusions that I am about to draw.

(1) Nyctimystinae (Australasian Hylids)

Hyla (*Litoria*) *arfakiana* $n = 13$ ($2n = 26$) (Duellman, 1967)
H. (*L.*) *becki* $n = 13$ ($2n = 26$) (Duellman, 1967)
H. (*L.*) *darlingtoni* $n = 13$ ($2n = 26$) (Duellman, 1967)
H. (*L.*) *angiana* $n = 15$ ($2n = 30$) (Duellman, 1967)
H. (*L.*) *bicolor* $2n = 26$ (Morescalchi, 1968e)
H. (*L.*) *caerulea* $2n = 26$ (Stephenson and Stephenson, 1970)
H. (*L.*) *phyllochroa* $2n = 26$ (Stephenson and Stephenson, 1970)
Nyctimystes foricula $n = 13$ ($2n = 26$) (Duellman, 1967)
N. kubori $n = 13$ ($2n = 26$) (Duellman, 1967)
N. semipalmata $n = 13$ ($2n = 26$) (Duellman, 1967)

(2) Hylinae

(i) Hylini.

Eurasiatic Hylini

Hyla arborea $2n = 24$ (Seto, 1964; Morescalchi, 1965a; Ullerich, 1970)

American Hylini

Hyla albomarginata $2n = 24$ (Beçak, 1968)
H. albopunctata $2n = 22$ (Beçak, 1968)
H. andersoni $n = 12$ ($2n = 24$) (Duellman and Cole, 1965; Wasserman, 1970).
H. arenicolor $n = 12$ ($2n = 24$) (Duellman and Cole, 1965; Wasserman, 1970)
H. avivoca $2n = 24$ (Bushnell *et al.*, 1939)
H. bistincta $n = 12$ ($2n = 24$) (Duellman and Cole, 1965)
H. californiae $2n = 24$ (Maxson and Jameson, 1968)
H. cinerea $2n = 24$ (Bushness *et al.*, 1939)
H. crepitans $2n = 24$ (Rabello, 1970)
H. crucifer $n = 12$ ($2n = 24$) (Duellman and Cole, 1965)
H. euphorbiacea $n = 12$ ($2n = 24$) (Duellman and Cole, 1965)
H. exhimia $n = 12$ ($2n = 24$) (Duellman and Cole, 1965)
H. faber $2n = 24$ (Beçak, 1968)
H. fuscomarginata $2n = 24$ (Rabello, 1970)
H. hayi $2n = 24$ (Beçak, 1968)
H. lafrentzi $2n = 24$ (Duellman and Cole, 1965)
H. hazelae $n = 12$ ($2n = 24$) (Duellman and Cole, 1965)
H. multilineata $2n = 24$ (Beçak, 1968)

Fritziana (Beçak, 1968), belonging to a group of forms that have long since separated geographically from the rest of the family; however, *Fritziana*

H. pentheter $n = 12$ ($2n = 24$) (Duellman, 1967)
H. polytaenia $2n = 24$ (Rabello, 1970)
H. pulchella $2n = 24$ (Beçak, 1968)

possesses some pairs of acrocentrics, which may signify that the karyotype of this hylid has undergone processes of centric fission and may be derived from the 24-chromosome type generally present in the Hylini.

Further unequal translocations may have reduced the karyotype with $2n = 24$, giving rise to one with $2n = 22$, in the few Hylini that show this last chromosome number (this appears clear especially for *Acris*, derived from *Pseudacris*, which has $2n = 24$; Duellman, 1967).

Certain species of *Hyla* of Amazonic, Brasilian and Central American genocentres, have $2n = 30$, and in the karyotype of most of them there are various pairs of *t*-chromosomes, a fact that has led to the postulation of Robertsonian phenomena in their karyological differentiation. This seems all the more probable since these species are among the more highly evolved Hylini (cf. Duellman and Cole, 1965; Beçak, 1968; Rabello, 1970). This group of 30-chromosome species might therefore be derived from others, with the 24-chromosome set typical of the Hylini, through centric fission phenomena having the same evolutional significance as those already described in the Eleutherodactylini and in other Leptodactylids: in fact, also in this case, the numerical increase in the chromosomes (and the *t*-chromosomes) may be related to stages of adaptive radiation of these Hylini, rich in species and predominant in their environment (cf. Cei, 1968).

A further interesting fact is the discovery of polyploid species or populations in the Hylinae, which obviously emphasizes the parallelisms between the Hylidae and the Leptodactylidae in the ways in which their karyology has evolved.

(3) The Hylids of the sub-family of the Centroleninae are known karyologically only by their chromosome number, which is $n = 10$, $2n = 20$; this can easily be made to derive from a 24- (or 22-)chromosome formula, which is to be found in the South and Central American Hylini, from which the Centroleninae seem phyletically derived. Following the above-mentioned general theory on the karyological evolution of these Anura, this may have occurred through further unequal translocations in the small chromosomes, which have led to a reduction in the number of the set. This theory would therefore place the Centroleninae at the top of the karyological evolution reached by the Hylidae, in accordance with what is known about the morphology of the members of this sub-family.

(4) The Pseudinae have $2n = 24$ (NF $= 48$), with karyological formulae similar to those of various 24-chromosome Hylinae (Barrio and Pistol de Rubel, 1970); this may favour the theory that these Anura are derived from the Hylidae. However, the alternative theory that they are derived from the Leptodactylids cannot be rejected, since, although there are no 24-chromosome species of Leptodactylids among those so far studied in South America,

such species exist in other geographical areas (in Australia, with the Myobatrachinae) and it is not unlikely that, in their karyological evolution, various Leptodactylinae with 22 chromosomes have also passed through a 24-chromosome stage.

In conclusion, the vast family of the Hylids presents different chromosome numbers, with a range of variability that closely resembles that of the related Leptodactylids; in fact, the methods of karyological evolution of this family may be interpreted relatively simply (Figs. 29–30), by assuming for

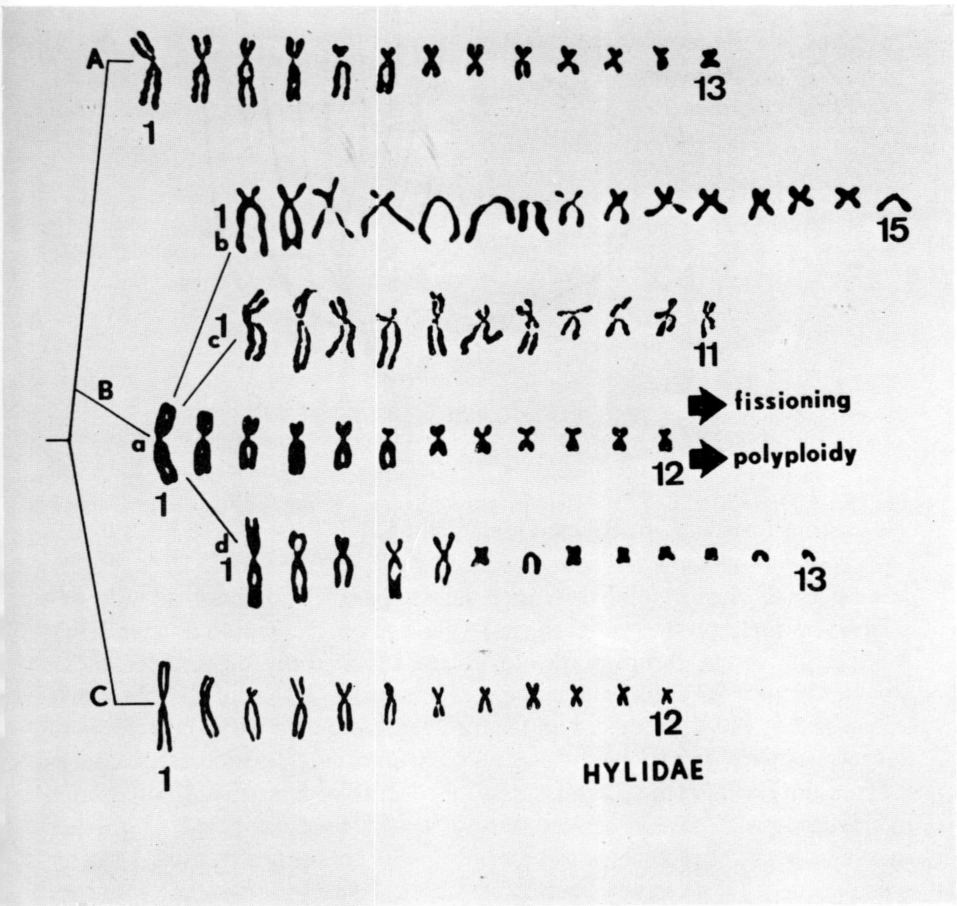

Fig. 29. The haploid karyotypes of some Hylids. A: Nyctimystinae (*Hyla* (*L.*) *bicolor*);
B: Hylinae, a–*Hyla regilla*, b–*H. microps*, c–*Acris crepitans*, d–*Fritziana goeldi*;
C: Pseudinae (*Pseudis paradoxus*). A part of the haploid karyotypes are redrawn from figures in the papers cited in the list of chromosomes (pp. 287–289).

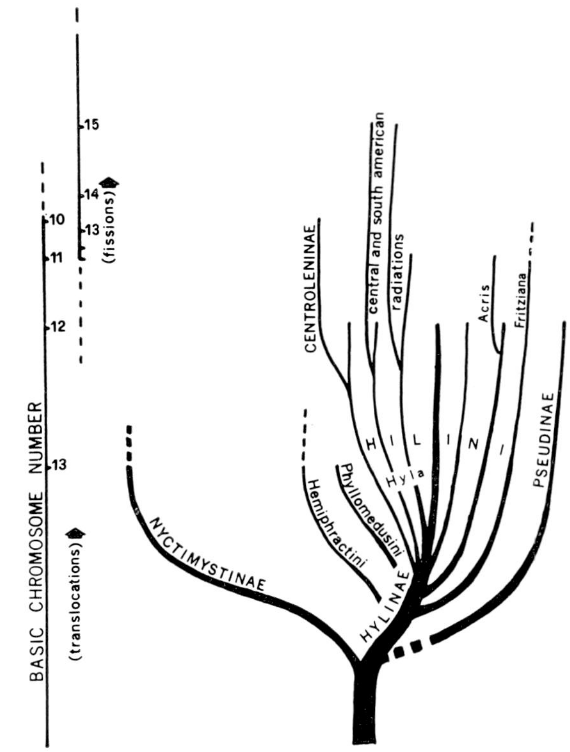

Fig. 30. A phyletic tree of the Hylidae treated in this section, built utilizing especially karyological data. Further explanations in the text.

it a series of chromosome rearrangements, parallel to those already described in the Leptodactylids: the most general process would be a tendency for the number of chromosomes to decrease (especially the smaller ones), starting from a 26-chromosome karyotype similar to that of the Nyctimystinae and of Phyllomedusini and some Hemiphractini with the genesis of karyotypes with 24 and later 22 and 20 chromosomes, which are found in most of the Hylinae, the Pseudinae and the Centroleninae; in certain groups of Hylini, processes of centric fission would then have led to further increases in the chromosome number ($2n = 30$; theoretically, even a greater number of chromosomes seems possible). Again in parallel with the Leptodactylids, also certain Hylids then differentiated by polyploidy.

Obviously these methods of karyological evolution, which start from and lead to chromosome formulae that are almost equal in the two families considered, do not contribute towards solving, karyologically, the problem of the mono- or poly-phyletic origin of the Hylids from the Leptodactylids.

(i) *Bufonidae*

This is one of the most widespread of the families of Anura, only in Madagascar and Australasia is it not found. The family derives from the Leptodactylids, from which it is differentiated mainly by the presence of a Bidder's organ (Davis, 1936). The most important genus is *Bufo*, various species of which are spread throughout the whole area of the family; other genera, far less widespread, are found especially in Africa and, to a lesser extent, in South America: among the former are *Nectophryne* and the old but biologically specialized *Nectophrynoides* (Tihen, 1960); among the latter are *Melanophryniscus* (= *Dendrophryniscus*), initially included in the South American family of the Atelopodidae (= Brachycephalidae), now subdivided by Griffiths (1959) into forms attributed to the Bufonidae and into others that are the true Atelopodidae.

Bufo is one of the genera of Anura that has been most studied from the karyological point of view; of the other Bufonids, the karyotypes known are those of species of the above-mentioned *Nectophryne*, *Nectophrynoides* and *Melanophryniscus*.

Bufo. Eurasian species:

andersonii $2n = 22$ (Singh *et al.*, 1970)
bufo $2n = 22$ (Morescalchi, 1964a; Ullerich, 1966)
calamita $2n = 22$ (Ullerich, 1966; Morescalchi and Gargiulo, 1968; Griffin *et al.*, 1970)
melanostictus $2n = 22$ (Manna and Bhunya, 1966; Morescalchi and Gargiulo, 1968)
raddei $2n = 22$ (Sato, 1936b)
stomatictus $2n = 22$ (Sharma *et al.*, 1965)
viridis $2n = 22$ (Morescalchi, 1963a; Ullerich, 1966)

American species:

alvarius $2n = 22$ (Cole *et al.*, 1968; Doyle and Beckert, 1970)
americanus $2n = 22$ (Doyle and Beckert, 1970)
arenarum $2n = 22$ (Bianchi and Laguens, 1964; Morescalchi and Gargiulo, 1968)
boreas $2n = 22$ (Sanders and Cross, 1964)
canorus $2n = 22$ (Witschi, 1933)
cognatus $2n = 22$ (Cole *et al.*, 1968; Morescalchi and Gargiulo, 1968)
compactilis $2n = 22$ (Hayes, 1933)
crucifer $2n = 22$ (Beçak, 1968)
fowleri $2n = 22$ (Witschi, 1933)
granulosus $2n = 22$ (Saez and Brum, 1960; Beçak M. L. *et al.*, 1970)

hemiophrys 2*n* = 22 (Morescalchi and Gargiulo, 1968)
houstonensis 2*n* = 22 (Bogart, 1968)
ictericus 2*n* = 22 (Beçak, 1968)
lentiginosus 2*n* = 22 (Witschi, 1933)
lonnbergi 2*n* = 22 (Bogart, 1968)
marinus 2*n* = 22 (Beckert and Doyle, 1967; Ullerich, 1967; Cole *et al.*, 1968; Morescalchi and Gargiulo, 1968; Volpe and Gebhardt, 1968; Doyle and Beckert, 1970; Griffin *et al.*, 1970)
marmoreus 2*n* = 22 (Duellman, 1967)
mazatlanensis 2*n* = 22 (Doyle and Beckert, 1970)
microscaphus 2*n* = 22 (Cole *et al.*, 1968)
paracnemis 2*n* = 22 (Beçak, 1968; Morescalchi and Gargiulo, 1968)
perplexus 2*n* = 22 (Duellman, 1967)
punctatus 2*n* = 22 (Cole *et al.*, 1968)
quercicus 2*n* = 22 (Witschi, 1933)
retiformis 2*n* = 22 (Cole *et al.*, 1968; Doyle and Beckert, 1970)
speciosus 2*n* = 22 (Hayes, 1933)
spinulosus 2*n* = 22 (Morescalchi and Gargiulo, 1968)
terrestris 2*n* = 22 (Doyle and Beckert, 1970)
valliceps 2*n* = 22 (Cole *et al.*, 1968; Doyle and Beckert, 1970)
woodhousi 2*n* = 22 (Cole *et al.*, 1968)

African species:

brauni 2*n* = 20 (Bogart, 1968)
carens 2*n* = 22 (Bogart 1968; Morescalchi and Gargiulo, 1968)
gariepensis 2*n* = 22 (Bogart, 1968)
garmani 2*n* = 20 (Bogart, 1968; Morescalchi and Gargiulo, 1968)
gutturalis 2*n* = 20 (Bogart, 1968)
latifrons 2*n* = 20 (Bogart, 1968)
mauritanicus 2*n* = 22 (Bogart, 1968; Morescalchi and Gargiulo, 1968)
perreti 2*n* = 20 (Bogart, 1968)
rangeri 2*n* = 20 (Bogart, 1968; Morescalchi and Gargiulo, 1968)
regularis 2*n* = 20 (Bogart, 1968; Doyle and Beckert, 1968; Morescalchi and Gargiulo, 1968)
regularis (*pardalis*) 2*n* = 22 (Griffin *et al.*, 1970)
rosei 2*n* = 22 (Bogart, 1968)

Other Bufonids:

Melanophryniscus moreirae 2*n* = 22 (Beçak M. L. *et al.*, 1970)
Melanophryniscus stelzneri 2*n* = 22 (Morescalchi and Gargiulo, 1968)

Nectophryne afra $2n = 22$ (Scheel, 1970)
Nectophrynoides occidentalis $2n = 22$ (This paper).

Only two types of chromosome formula seem to be present in the Bufonids so far studied: one with 22 and the other with 20 chromosomes; the latter seems to be confined to African species of *Bufo* of the *regularis* group and possibly to others related to it. In view of the ubiquity of the 22-chromosome formula in *Bufo*, this is probably the older of the two (Bogart, 1968; Morescalchi and Gargiulo, 1968); the discovery of 22 chromosomes in the old *Nectophrynoides*, which I report here for the first time, confirms this theory.

The 20-chromosome karyotype differs from the generalized karyotype of the family in the absence of a pair of small homologues, possibly translocated into large ones, and at any rate apparently "lost" by some South African *Bufo*; in any case, the recent discovery of a sub-species of *B. regularis* (*B.r. pardalis*) with 22 chromosomes (Griffin *et al.*, 1970) may suggest that this event must have been relatively recent and possibly still in process in the form of this vast group of *Bufo* (cf. Blair, 1963).

In the 22-chromosome *Bufo*, certain morphological differences are to be found in the karyotype of the different species, which generally present 6 pairs of large and 5 pairs of small homologues, all *m* or *sm*, with the exception of one or two pairs of *st* among the small chromosomes (NF = 44); the same chromosome formula is present in *Nectophryne*, *Nectophrynoides* and *Melanophryniscus* (Fig. 31); as regards this last genus, the finding may constitute a confirmation of the systematics proposed for these American forms by Baldauf (1959) and Griffiths (1959).

The above-mentioned differences in species in the genus *Bufo* mainly concern the relative sizes of the chromosomes of the 6th pair of homologues, which, in the South American species and in *B. marinus*, are smaller than in the other *Bufo*, and the morphology of the last pair of homologues, which vary between *sm* and *st* in different species; other differences concern the localization of the heterochromatic zones (the nucleolar organizers; cf. Miller and Brown, 1969), which may differ according to the species (cf. Morescalchi and Gargiulo, 1968).

In any case, groups of related species present smaller karyological differences than species far removed from one another: for example, the South American species *paracnemis*, *arenarum*, *ictericus*, *crucifer*, *granulosus* and *spinulosus*, like *marinus* from which they are probably derived, all possess a heterochromatic tract, sometimes very extensive, on the shorter arms of the 7th pair of homologues.

The discovery of chromosome sets that are numerically equal and morphologically almost equal to one another in most of the species of *Bufo*,

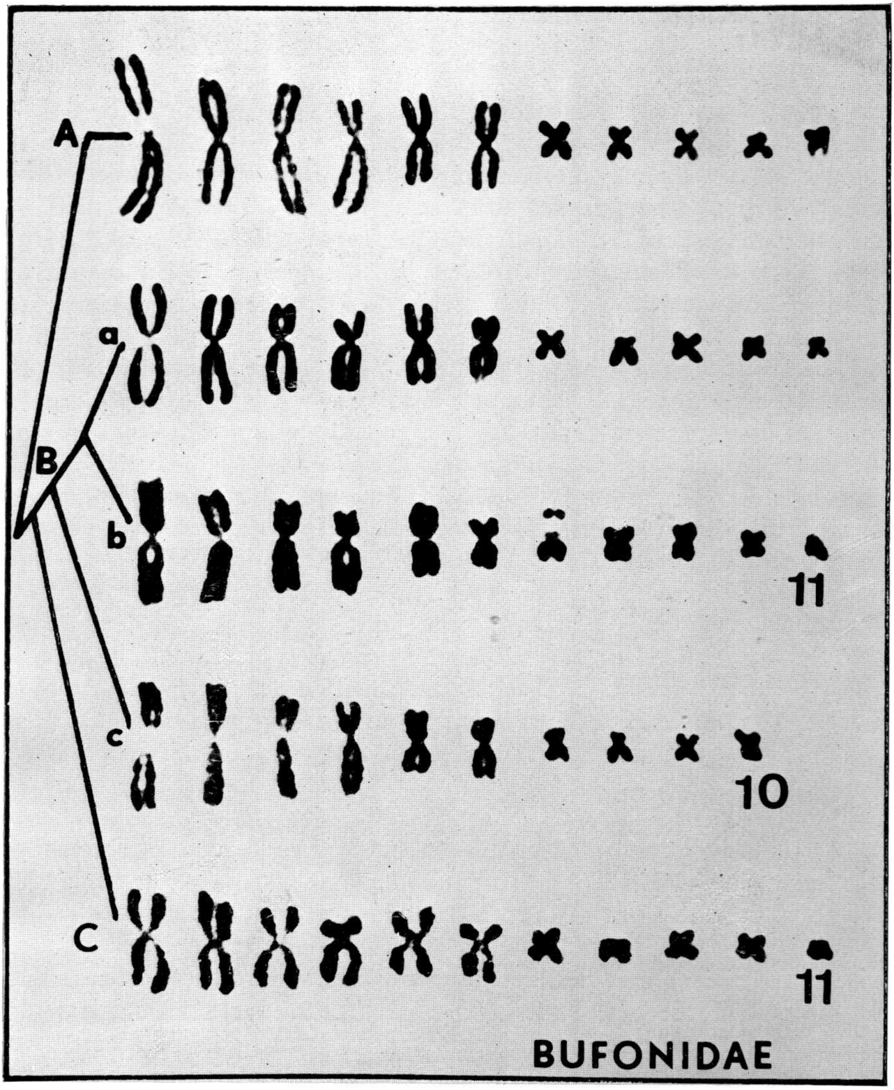

Fig. 31. The haploid karyotypes of some Bufonids. *A*: *Nectophrynoides occidentalis*; B: *Bufo*, a–*bufo*, b–*arenarum*, c–*regularis*; C: *Melanophryniscus stelzneri*.

spread throughout various continents, and in the genera *Nectophryne*, *Nectophrynoides* and *Melanophryniscus*, which are phyletically far removed from one another, suggests that the 22-chromosome formula, typical of the present Bufonidae, was acquired very early from the ancestral stock of the family and has then been preserved for long periods with fairly restricted modifications.

The origin of this family from the Leptodactylidae is generally not doubted by systematicians; also in these latter Anura, 22-chromosome formulae are common (e.g. in the Leptodactylinae). However, as regards the morphology of the chromosomes, the group of Leptodactylids nearest to the Bufonids is not that of the Leptodactylinae with 22 chromosomes (which possess various *st* elements among the large and small chromosomes); as already mentioned, it seems rather that of the Myobatrachinae ($2n = 24$), especially the Cycloranini, which differ from the Bufonids only in having one more pair of small homologues: these differences are therefore equivalent to those to be found, in the genus *Bufo*, between most of the species, with 22 chromosomes, and those of the *regularis* group, with 20 chromosomes. At present, as I have already said, the Myobatrachinae are confined to Australasia (where the Bufonids are absent); however, it is probable that long before the beginning of the Cenozoic, forms kariologically similar to them were spread throughout a large part of Gondwanaland (and perhaps also in other areas of the Oriental Region): it is from this morphologically primitive leptodactylid stock that Bufonidae might have originated (cf. Reig, 1961).

Regarding the "true" Atelopodidae, from a karyological point of view it is merely known that *Atelopus varius* has $n = 11$ ($2n = 22$) (Duellman, 1967); this fact may corroborate the theories about the origin of these South American Anura from a bufonid stock. However, since various forms of Leptodactylidae having the same chromosome number are present in this continent, studies must be carried out to establish the morphology of the individual pairs of homologues of the set of Atelopodidae before it is possible to say whether it resembles the set of Bufonidae more than that of other families of Anura.

(f) *Microhylidae*

Systematically, this is the most problematic family among the firmisternous Anura (*sensu* Griffiths, 1963): on the basis of the peculiarity of its larval form (type II of Orton, 1953, 1957), it is differentiated from all the other higher Anura; however, it is uncertain whether this larval form is really of primitive type or whether it may be derived from that common to the latter (type IV of Orton) (cf. Kluge and Farris, 1969).

Its diffusion tends to be relict in the Tropics in both Old and New Worlds which may show that it is of old origin (Laurent, 1942) (or, according to other authors, polyphyletic); however, there exist no fossils of Microhylids earlier than the Miocene and these Anura do not appear morphologically primitive, but seem to have many similarities with the Ranids (Noble, 1931).

On the basis of these data, which apparently disagree, two theories have been advanced regarding the phylogeny of the Microhylids, and the opinion of systematicians is still divided. According to one theory, this family is one of the oldest among the living Anura and has no phyletic relationships with the Ranids (Hecht, 1963; Inger, 1967); according to the other theory, the Microhylids are "true" higher Anura, with various relationships with the Ranids, being derived from the same ancestral stock as the latter (cf. Parker, 1934; Griffiths, 1963; Kluge and Farris, 1969).

Unfortunately, the karyological data so far collected on this family are still very scanty; they concern species of five sub-families (the Phrynomerinae, provided with intercalary phalanges, are often considered a family apart, though differentiated from "typical" Microhylids); among the Microhylinae, the most widespread of the sub-families, Asiatic and American species have been studied.

Microhylinae:	*Kaloula borealis* $2n = 28$ (Sato, 1936b)
	Kaloula pulchra $2n = 28$ (Morescalchi, 1968c)
	Dermatonotus muelleri $2n = 22$ (Rabello, 1970)
	Gastrophryne carolinensis $2n = 22$ (Morescalchi, 1968d)
	Microhyla rubra $2n = 26$ (Bai, 1956)
	Ramanella variegata $2n = 26$ (Bai, 1956)
	Uperodon systoma $2n = 26$ (Natarajan, 1953)
Sphenophryninae:	*Cophixalus riparius* $2n = 26$ (Cole and Zweifel, 1971)
	C. cryptotympanum $2n = 26$ (Cole and Zweifel, 1971)
Asterophryinae:	*Asterophrys sp.* $2n = 26$ (Cole and Zweifel, 1971)
Brevicipitinae:	*Breviceps gibbosus* $2n = 24$ (Morescalchi, 1968c)
Phrynomerinae:	*Phrynomerus bifasciatus* $2n = 26$ (Morescalchi, 1968c)

The haploid karyotypes of the species that I studied are shown in Fig. 32; *Dermatonotus* seems to differ from *Gastrophryne* only in the morphology of two pairs of homologues; *Cophixalus* is very similar to *Phrynomerus*.

The chromosome numbers found among the Microhylidae are therefore $2n = 28$, 26, 24 and 22; all seem to be present in the Microhylinae (Bogart, unpublished), which is the only sub-family having Eastern and American representatives.

The 24- and 22-chromosome karyotypes, consisting almost entirely of two-armed chromosomes (one pair per species, among the small ones, is of *t*-chromosomes) (NF = 46–42), appear morphologically differentiated and approach those of *Hemisus* and other Ranids; the 26-chromosome karyotype of *Phrynomerus* and *Cophixalus* is almost the same as that of the Ranid *Mantella* (NF = 52), while the 28-chromosome karyotype of *Kaloula*, consisting of *m* or *sm* or *st* chromosomes, is the only one that may have certain characteristics of primitivity in the family (Morescalchi, 1968c).

As regards the morphology and the number of the chromosomes, and the morphology of the spermatocyte bivalents (which are of the ring type), the Microhylidae so far studied, with certain reservations for *Kaloula*, therefore seem "typical" higher Anura and are karyologically more similar to the Ranidae than to other families.

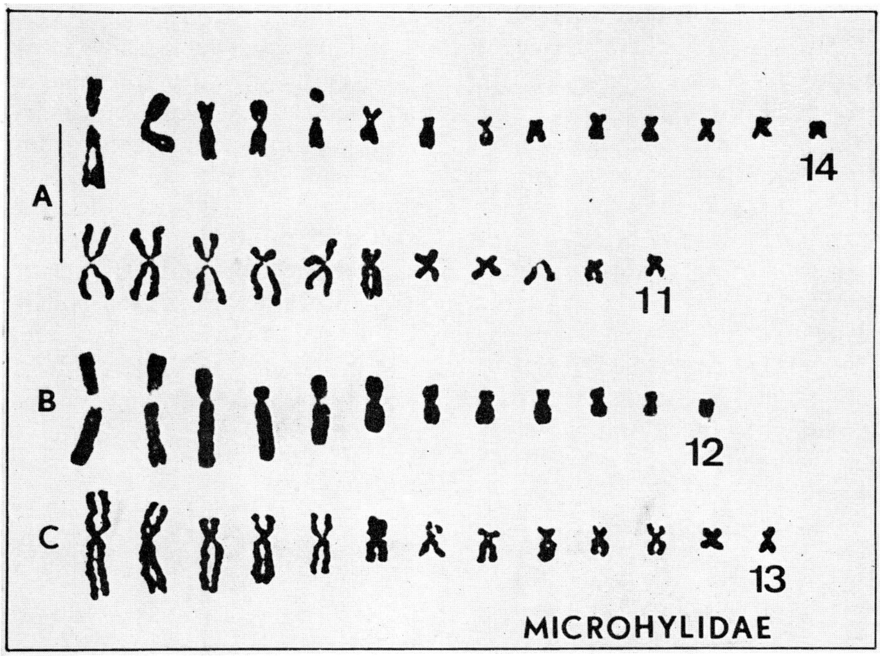

Fig. 32. The haploid karyotypes of some Microhylids. A: Microhylinae, *Kaloula pulchra* ($n = 14$) and *Gastrophryne carolinensis* ($n = 11$);
B: Brevicipitinae, *Breviceps gibbosus* ($n=12$);
C: Phrynomerinae, *Phrynomerus bifasciatus* ($n = 13$).

It remains to be seen whether other species show different chromosome formulae from those so far found in the family; for the moment, the karyological data appear to favour the theory according to which the Microhylids are derived from a pre-ranid stock. The affinities between the karyotype of certain Microhylids and that of Ranids of different sub-families may be interpreted either as proof of mechanisms of parallel karyological evolution between the two families or as proof of a polycentric evolution of the Microhylids themselves (cf. Poynton, 1964).

(k) *Ranidae*

This is one of the families of higher Anura that has asserted itself most strongly; it is absent only in parts of South America and Australia (continents where *Rana* seems in process of penetration). The Ranids are probably derived from certain stocks of leptodactylids, in Africa or in the Oriental Region; at present, the centre of their greatest diffusion seems to be the Ethiopian Region (cf. Laurent, 1951; Griffiths, 1963).

The Ranids comprise various sub-families; the most widespread is that of the Raninae, where the genus *Rana* (perhaps a super-genus) covers nearly all the area over which the family is spread; the other sub-families are spread throughout Africa and in the Oriental Region: unfortunately, most of the karyological findings at present concern species of the genus *Rana* and a few other African species, belonging to the sub-families of Mantellinae, Arthroleptinae and Hemisiinae; it is sometimes debated whether the last two belong to the Ranids (cf. Laurent, 1951).

Karyological data are therefore lacking for the more interesting sub-families, such as the Astylosterninae and the Cornuferinae, for an understanding of the karyological evolution of the family; obviously, therefore, the conclusions reached in this section are provisional and susceptible of modification as soon as the gaps begin to be filled in the karyology of this important family of Anura.

Raninae

Rana: adspersa $2n = 26$ (This paper)
arvalis $2n = 24$ (Kobayashi, 1962; Morescalchi, 1967b; Ullerich, 1967)
catesbeiana $2n = 26$ (Seto, 1965)
chensinensis $2n = 24$ (Kobayashi, 1962; Seto, 1965)
clamitans $2n = 26$ (Rafferty, 1969)
cyanophlyctis $2n = 26$ (Natarajan, 1958a)
dalmatina $2n = 26$ (Guillemin, 1967)
fuscigula $2n = 26$ (This paper)
esculenta $2n = 26$ (Morescalchi, 1962, 1967b; Ullerich, 1967; Gunther, 1970)
graeca $2n = 26$ (Morescalchi, 1967)
grisea $2n = 26$ (Duellman, 1967)
hexadactyla $2n = 26$ (Natarajan, 1958a)
japonica $2n = 26$ (Kobayashi, 1962; Seto, 1965)
limnocharis $2n = 26$ (Sato, 1934)
nigromaculata $2n = 26$ (Seto, 1965)
ornativentris $2n = 24$ (Kobayashi, 1962; Seto, 1965)

palustris $2n = 26$ (cf. Goin and Goin, 1962)
papua $2n = 26$ (Duellman, 1967)
pipiens $2n = 26$ (Di Berardino, 1962; Bogart *et al.*, 1969; Volpe *et al.*, 1970)
plancyi $2n = 26$ (Ting, 1939)
ridibunda $2n = 26$ (Morescalchi, 1967b; Günther, 1970)
rugosa $2n = 26$ (Seto, 1965)
sylvatica $2n = 26$ (Hennen, 1964; Rafferty, 1969)
temporaria $2n = 26$ (Guillemin, 1967; Morescalchi, 1967b; Ullerich, 1967)
tigrina $2n = 26$ (Singh *et al.*, 1970)

Mantellinae
Mantella aurantiaca $2n = 26$ (Morescalchi, 1967b)
Arthroleptinae
Arthroleptis stenodactylus $2n = 14$ (This paper)
Hemisiinae
Hemisus marmoratus $2n = 24$ (Morescalchi *et al.*, 1970a).

From the chromosome list, it may be seen that most species of the genus *Rana*, belonging to various continents, have $2n = 26$; those for which the karyotype has been studied in greater detail reveal pairs of chromosomes that differ only slightly from one species to another. As usual, the variations concern the arrangement of the heterochromatic areas on the chromosomes and at times the position of the centromere in some pairs of small elements. The 26 chromosomes can be divided into 5 pairs of large homologues, *m* or *sm*, and 8 pairs of smaller homologues, *m*, *sm*, or *st* (NF = 52). Hence the karyotype of *Rana* approaches, morphologically, that of certain members of the Pelobatidae and of certain arciferous higher Anura, such as various Leptodactylidae and Hylidae; however, it differs from all these in the lack of *st* chromosomes among the large elements and the morphology and relative size (greater in the Ranidae) of the small chromosomes. Even if the Raninae, perhaps the oldest stock among the living Ranids, are derived from some leptodactylid stock (as seems possible also from a karyological point of view) they must have been derived from forms having karyotypes morphologically different from those of the living Leptodactylids with 26 chromosomes (mostly South American). As we shall see, *Hemisus* resembles the Myobatrachini, karyologically; it might therefore be thought that this stock was the same that has given rise, in Australasia, to the Leptodactylids and Hylids now present there, and in Africa to *Heleophryne*. In the future, it will be interesting to compare the karyotype of the Papuan Ranids with that of the Nyctimystinae, which are perhaps the stock of arciferous Anura that most closely approach the Ranids from the karyological point of view.

Certain species of *Rana*, often grouped together among the so-called "grass-frogs", because they are prevalently terrestrial compared with the more primitive "water-frogs", have $2n = 24$, with one more pair of large homologues and two fewer pairs of small homologues compared with the 26-chromosome species: this reduction in the number of the chromosomes may have occurred through unequal translocations (Morescalchi, 1967b). This karyotype is found in the European *R. arvalis* and certain Japanese *Rana*, considered in the past as sub-species of *temporaria* but today considered more probably as true species (*R. chensinensis* and *R. ornativentris*) (Kawamura, 1952; Kobayashi, 1962; Seto, 1965), perhaps related to *R. arvalis* (Witschi *et al.*, 1958).

Mantella aurantiaca (Mantellinae) has $2n = 26$, with a karyotype that differs from that typical of the Raninae in a few morphological characteristics in the small chromosomes, which are all *m* or *sm* (and therefore lacking the *st* characteristics of *Rana*); as we shall see, this karyotype approaches that of the Rhacophoridae.

A spectacular reduction is found in the chromosomes of *Arthroleptis stenodactylus* (Arthroleptinae), which shows only $2n = 14$, with 5 pairs of large homologues similar to those of the Raninae (and the Rhacophorids) and two pairs of medium-large homologues, *m*, which are typical of this species (NF = 28) (Fig. 33). I have encountered this karyological formula in few intestinal metaphases of two female specimens and so I give these data here with some reservations. I have decided to do so because a reference to certain still unpublished data of Scheel (in Bogart, 1970) would be in agreement with my observations.

The genesis of this "reduced" set from a 26- (or 24-)chromosome set typical of various Ranids (and also of the Rhacophorids) would require a great number of chromosome rearrangements, which places *Arthroleptis* among the more highly evolved forms of the family, from a karyological point of view. I consider it probable that karyotypes indicating a transition towards that of *Arthroleptis* will be found in African Anura of this family, or perhaps also among the Rhacophorids, seeing that medium-large chromosomes are characteristic of specialized forms of this second family (e.g. *Kassina*); in fact, according to Laurent (1951), *Arthroleptis* might be derived from forms that are here included among the Rhacophorids.

Hemisus marmoratus (Hemisiinae) has $2n = 24$, like various specialised Ranids; however, in the morphology of the chromosomes, it differs from them (and from the Rhacophorids) as much as it differs, for example, from certain African Microhylids (cf. Morescalchi *et al.*, 1970a): in fact, this species has 6 pairs of large homologues, in which the 6th pair consists of *st*

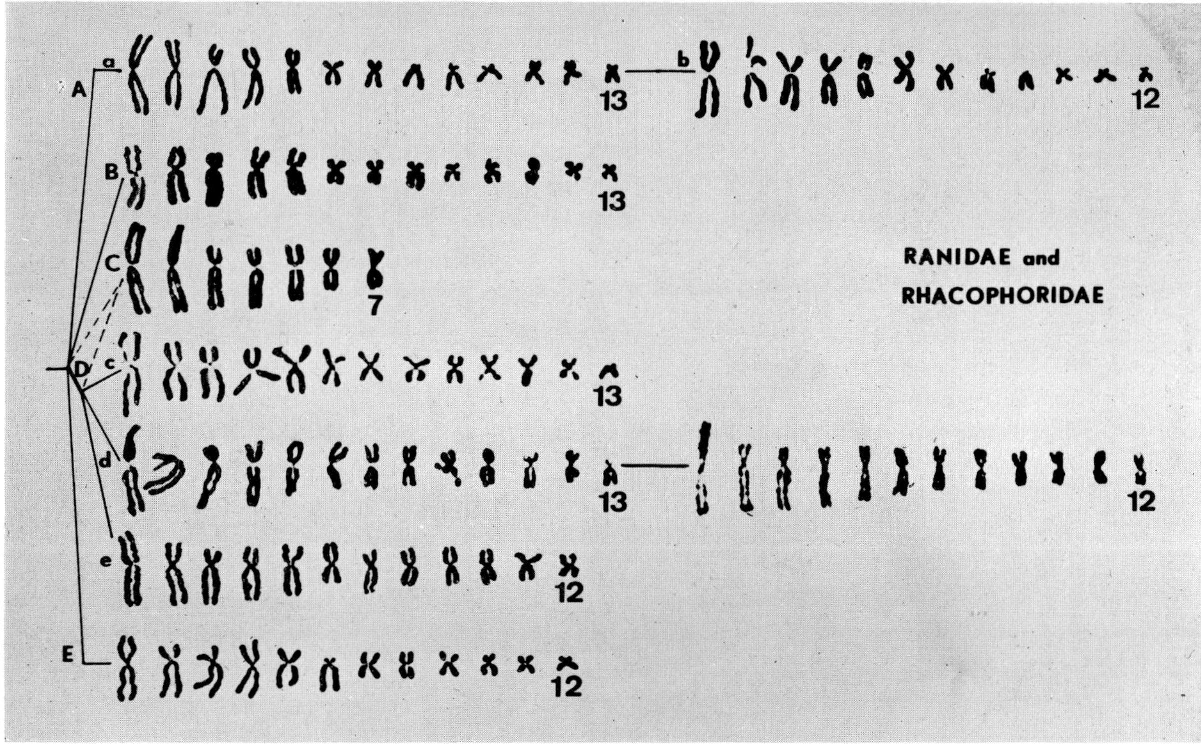

Fig. 33. The haploid karyotypes of some Ranids and Rhacophorids. A: Raninae, a–*Rana esculenta*, b–*R. arvalis*; B: Mantellinae (*Mantella aurantiaca*); C: Arthroleptinae (*A. stenodactylus*); D: Rhacophoridae, c–*Rhacophorus leucomystax*, d–*Hyperolius cinctiventris* ($n = 13$) and *H. sp.* ($n = 12$), e–*Kassina senegalensis*; E: Hemisiinae, *Hemisus marmoratus*.

elements that are lacking in the other Ranids; the smaller chromosomes of *Hemisus* are *m* or *sm*, except for the 12th pair, in which they are *st*. It is interesting to note how *Hemisus* is one of the Anura that has aroused a considerable amount of discussion as regards its systematic attribution; in fact, the adult has the general morphological characteristics of the Microhylids (such that it has often been included in this family; cf. Noble, 1931), whereas its larval form is that typical of the Ranids: it is on the basis of this last characteristic that it is nowadays preferred to consider *Hemisus* among the Ranids (cf. Parker, 1934; Griffiths, 1963).

If *Hemisus* has the anatomical characteristics of Ranidae and Microhylidae, from the karyological point of view it holds in store an interesting surprise; its karyotype is almost identical to that of some Myobatrachini (*Crinia*). Obviously, this may be the result of convergences in the karyological evolution of these Anura; however, the discovery seems to me to be of interest if considered in the light of the problems that are still open as regards the phylogeny of the higher firmisternous families.

In fact, at least part of these families seem to have originated from a leptodactylid stock; among them, the Ranids, from various aspects, seem to have remained closest, morphologically, to the original stock (Parker, 1934). *Hemisus* might have retained certain karyological characteristics from the stock of leptodactylid origin; moreover (in view of its morphological, and also karyological, affinities with the Microhylidae), it might have remained similar to those forms that, early in the phylogeny of the Ranids, gave rise to the radiation (or radiations) of the Microhylids (cf. Griffiths, 1963).

For confirmation of these theories, which might shed new light on the origin of a vast group of African Anura, it would be necessary to carry out new karyological studies on a great number of species of Ranids and Microhylids of this continent.

(l) *Rhacophoridae*

These constitute a group of Anura specialized for arboreal life; they differ from the Ranids essentially in the development of the intercalary phalanges (Ahl, 1931). They are spread throughout the Tropics of the Old World, and their origin (perhaps diphyletic) is commonly sought among the Ranids.

While various authors consider the Rhacophorids a group of natural forms (Noble, 1931), Laurent (1951) is of the opinion that the Eastern forms of the so-called Rhacophorids merely constitute a sub-family (Rhacophorinae) of the Ranids, whereas the Ethiopian forms (Hyperoliinae) and other sub-families (included among the Randis by various authors) constitute a family apart (Hyperoliidae), almost as old as the Ranidae and, like them, directly derived from a Leptodactylid stock.

Liem (1970), in his turn, raises to the rank of families (separated from the Ranids and derived from them) both the Rhacophorinae and the Hyperoliinae, and once more includes Astylosterninae, Arthroleptinae and Hemisiinae (included by Laurent among the Hyperoliidae) among the Ranidae.

In agreement with other authors, I have preferred to consider this family as Ahl does, though recognizing the strong probability of a different origin of the Oriental and Ethiopian forms of the Rhacophorids (here treated as sub-families), because the karyological data seem to show a considerable homogeneity in the species so far studied.

The chromosome sets of 10 species are known, belonging to 4 genera of Rhacophorids, 2 among the Rhacophorinae and 2 among the Hyperoliinae.

Rhacophorinae:
- *Chiromantis petersi*: $2n = 26$ (This paper)
- *Rhacophorus eiffingeri*: $2n = 26$ (Sato, 1934)
- *Rhacophorus leucomistax*: $2n = 26$ (Morescalchi *et al.*, 1970a)
- *Rhacophorus maculatus*: $2n = 26$ (Natarajan, 1958b; Singh *et al.*, 1970)
- *Rhacophorus schlegelii*: $2n = 26$ (Makino, 1951)

Hyperoliinae:
- *Hyperolius argentovittis*: $2n = 26$ (Morescalchi, 1968c)
- *Hyperolius cinctiventris* (?): $2n = 26$ (this paper)
- *Hyperolius madagascariensis* (?): $2n = 24$ (this paper)
- *Kassina senegalensis*: $2n = 24$ (Morescalchi, 1968c)
- *Kassina wealii*: $2n = 24$ (Morescalchi *et al.*, 1970a)

In the Rhacophorids so far studied, 26- and 24-chromosome karyotypes are found, as in the Ranids (Fig. 33). *Rhacophorus leucomystax* and some *Hyperolius* have 26 chromosomes, with 5 pairs of large homologues, similar to the large chromosomes of *Rana* and *Mantella*, and 8 pairs of smaller homologues, resembling especially those of *Mantella*; the karyotypes of species of these two genera are practically identical to one another. *Chiromantis* ($2n = 26$) also has a karyotype that is almost the same as that of *Rhacophorus*. Even if, on the one hand, these data are favourable to Laurent's theory, according to which the Rhacophorinae are derived from the Mantellinae, on the other hand, in view of the karyological identity between *Rhacophorus* and *Hyperolius*, they appear to be in agreement with Ahl's and Noble's classification.

Certain Madagascan *Hyperolius* (the specific determination of which is at present doubtful) and two species of *Kassina* have $2n = 24$, all with *m* or *sm* elements, consisting of 6 pairs of large homologues and 6 small ones (though the differences between the two are very small): the genesis of this karyological formula from one with 26 chromosomes, which is perhaps the basic formula of the family, may have occurred by the same methods as those that, in the Ranids, have led to 24-chromosome formulae, starting from other 26-chromosome formulae.

The greatest difference between the karyotypes of various Raninae and those of the Rhacophoridae lies in the small degree of differentiation, in the latter, between the size of the large and small chromosomes, which is far more accentuated in the Raninae; moreover, the small chromosomes of the Rhacophorids generally have the centromere in a more medial position than in the Ranids. From this point of view, *Arthroleptis*, as I have already said, approaches certain specialized Rhacophorids, such as *Kassina*.

In conclusion, the scanty karyological data so far collected on this family seem to corroborate the theories that it originates from the Ranids, in which there are specialized forms (such as *Mantella*) that are very similar to *Rhacophorus* and *Hyperolius* (though, according to Liem, 1970, *Mantella* also belongs to the Rhacophoridae); furthermore, there are only slight karyological differences between the genera of the families that have been studied so far.

Obviously, as has already been stated for Ranids and Microhylids, further studies are necessary, especially on the sub-families about which there are various doubts regarding their attribution to a certain family (e.g. the *Astylosterninae*), in order to explain the mechanisms of karyological evolution of the Rhacophorids. For the present, it appears that Ranids and Rhacophorids can be closely linked, in view of the morphology of the chromosome set, and the cytotaxonomic relationships between the two families might be those illustrated in Fig. 36, constructed on the basis of the (provisional) observations already expressed regarding the possible monophyletic origin of all the firmisternous higher Anura.

(m) *Karyology and phylogeny of the Anura*

On the basis of the karyological data here described, the various families of Anura may be divided into three groups:

(1) Families with spermatocyte bivalents with interstitial chiasmata (cf. Section D), very variable chromosome numbers at the inter-generic level (which are often high) and, in some cases, rich in acrocentrics and microchromosomes: Ascaphids and Discoglossids;

(2) Families with spermatocyte bivalents with terminal chiasmata, very variable chromosome numbers at the inter-generic level and, in some cases, karyotypes having acrocentrics and microchromosomes: Pipids;

(3) Families with spermatocyte bivalents with terminal chiasmata, chromosome numbers that are only slightly variable at the inter-generic level and lacking in acrocentrics and microchromosomes (with some exceptions, possibly capable of being related to this scheme; cf. Section E): all the remaining families (with the possible exception of the Rhinophrynids, of which the morphology of the spermatocyte bivalents is not known) (Fig. 34).

This sub-division on karyological bases is interesting because it corroborates similar taxonomic sub-divisions based on findings of another nature.

The phyletic affinity between Ascaphids and Discoglossids is admitted by various authors and seems to rest on sound anatomical and paleontological bases (Orton, 1957; Reig, 1958, 1961; Tihen, 1965); the latter family is considered more recent than the former, and the karyological data support this theory: if certain Discoglossids have an "ascaphid" karyotype—that is, primitive, since it is similar to that of primitive Apoda and Urodela—others, on account of the morphology of the somatic and spermatocyte set, approach the typical condition of the families of group (3).

In the evolution from the Ascaphids to the Discoglossids, there appears a clear tendency towards a reduction in the number of the chromosomes (from $2n = 44$ to $2n = 24$), which mainly takes place through the apparent loss of acrocentrics and microchromosomes; then there are also modifications at the level of the spermatocyte bivalents, which tend to have terminalized chiasmata in some Discoglossids.

Regarding the Pipids, their condition of isolation among the modern Anura is revealed by the most recent works (Nevo, 1968), though other authors consider them at the basis of the higher families (Griffiths, 1963). The karyotype of these aglossan Anura generally appears less primitive than the ascaphid karyotype, while it may have evolved along different lines from those followed by the families of group (3) at least partly by the living Pipids (*Xenopus*, *Pipa*). If the family is not polyphyletic (as the systematic data would tend to exclude; cf. Dunn, 1948), it must be assumed that the forms from which it originated had a different karyological situation from that of the ancestors of the higher Anura (the Pipids might have derived at an early date from a stock of Proanura); however, it seems possible that also this family may be derived from primitive ascaphid forms (cf. Reig, 1958).

The families of group (3) comprise the Pelobatids and the so-called higher Anura (Reig's Neobatrachia), which appear as a karyologically homogeneous ensemble, having the same chromosome set ($2n = 26$, NF $= 52$) in the basic forms of nearly all the families; within the group it is often possible, especially on the basis of differences in the morphology and relative size of the small chromosomes, to distinguish two sub-groups. To one of these belong the Pelobatids, Leptodactylids, Hylids, Bufonids and possibly other minor families (i.e. in addition to the Pelobatids, the Anura that are classified in the sub-order of the Procoela by Noble and considered as "arciferous" higher Anura by Griffiths); to the other belong the Microhylids, Ranids and Rhacophorids (i.e. the Diplasiocoela according to Noble, considered as "firmisternous" higher Anura by Griffiths).

Generally, the more highly evolved species or families of this group have reduced chromosome sets compared with the more primitive species or families: the tendency towards the numerical reduction of the karyotype is therefore clear also in this group of families, even if it is less marked than that found in the families of group (1); in some families that are perhaps now in process of radiation, there exist possibly "secondary" tendencies towards an increase in the chromosome number through centric fission or polyploidy.

The karyological data that are perhaps of most general interest in this group of families are those that support the relationship of the Pelobatids with the arciferous higher Anura, and possibly that of the Microhylids with the group of the ranid forms; as already stated, these are two problems that have aroused innumerable discussions among systematicians.

On morphological bases, the Pelobatids are differentiated less than the higher Anura and various authors have placed them at the base of the group of the arciferous families (Noble, 1931); other authors, on the other hand, consider them a phyletically sterile branch, derived at an early date from the original stock of the higher Anura (Griffiths, 1963). The almost complete karyological identity between certain Pelobatids and Ceratophryinae (Morescalchi, 1967c) may make one incline towards the theory that Pelobatids and Leptodactylids have had the same ancestors and been diffused over different territories (karyologically, the primitive Megophryinae do not seem less differentiated than various Leptodactylids).

The radiation of the vast complex of the leptodactylid forms may have occurred before Cretaceous, prior to the final breaking up of the Gondwanian super-continent; among these Anura, the Australian representatives (Myobatrachinae) are morphologically the least differentiated in the group, whereas the Neotropical ones and *Heleophryne*, possess, in the more generalized species, the 26-chromosome karyotype characteristic of the Pelobatids and other arciferous families. The phyletic relationships between Myobatrachinae and the remaining Leptodactylids are somewhat doubted

Fig. 34. The haploid karyotypes of species of the main Anuran families: a: Ascaphidae (*A. truei*); b: Discoglossidae (*Alytes obstetricans*); c: Pelobatidae (*P. cultripes*); d: Leptodactylidae (*Ceratophrys calcarata*); e: Hylidae (*H. bicolor*); f: Bufonidae (*B. arenarum*); g: Ranidae (*R. esculenta*); h: Rhacophoridae (*Kassina senegalensis*); i: Microphylidae (*Gastrophryne carolinensis*); k: Pipidae (*Xenopus laevis*).

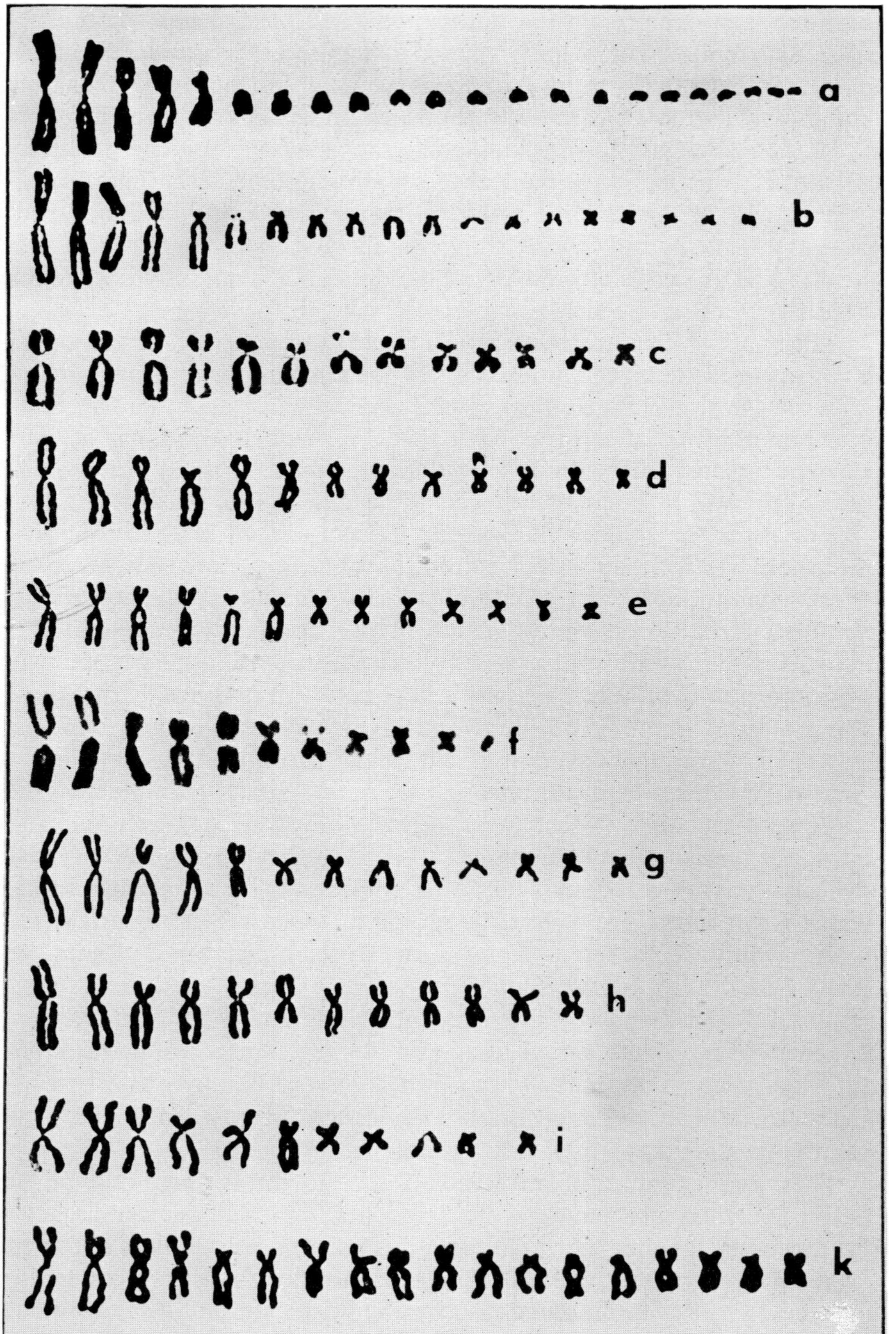
a
b
c
d
e
f
g
h
i
k

by more than one present-day author; on karyological bases, the studied Australian forms are more differentiated than the others: however, from this point of view, they might have derived from forms having the same chromosome set as *Heleophryne* and various American Leptodactylids.

In the Leptodactylids, in addition to the general evolutional tendency towards a reduction in the number of the chromosomes, other more localized methods of chromosome evolution have probably become established, such as polyploidy and centric fission; parallel to this, in the related Hylids we also find processes of karyological differentiation similar to those of the Leptodactylids (cf. Section E).

The relative "conservative nature" of the karyotype, which characterizes a large part of the higher Anura (possibly terminal forms of a long process of karyological evolution), thus displays considerable exceptions in these Neotropical forms; it seems possible that similar phenomena may later be encountered in Ranids, Bufonids and Rhacophorids in the tropics of the Old World, where these families seem to be in process of radiation.

As regards the Hylids, the Australo-Papuan forms seem karyologically more conservative than most Neotropical forms, generally possessing the 26-chromosome karyotype typical of the less differentiated higher Anura; however, in the morphology of the karyotype, there may exist other differences between Nyctimystinae and Hylinae, which will be more precisely determined in the future. *A priori*, it does not seem improbable that these two groups of tree-frogs may have closer phyletic relationships with the leptodactylid forms of the respective regions than between one another.

As regards the other great family of arciferous Anura, the Bufonids, the living representatives appear to be karyologically very homogeneous and more differentiated than the basic groups of Leptodactylids and Hylids; the karyotype of the Bufonids closely resembles that of the Myobatrachinae, which may confirm the theories according to which the family originated from primitive leptodactylid forms, formerly widespread in Africa and/or South America.

The second sub-group of group (3) comprises families on which there is, in general, very scanty karyological data. Ranids and Rhacophorids seem karyologically closely related to each other; all together, they may be considered as "Ranoidea", as is also the case if they are regarded from a comparative–anatomical point of view (cf. Laurent, 1967).

The basic karyotype of the Ranoidea is a 26-chromosome one, but it differs in certain morphological characteristics from the basic karyotype of the arciferous families; however, from this point of view, it does not seem far-fetched to assume that this group of Anura originates from leptodactylid forms, as is indicated also by the systematic findings (Laurent, 1951; Reig, 1961). The centre of origin of the Ranids, probably the oldest group among the Ranoidea, seems to be the Ethiopian Region (according to some authors, however, it might be the Oriental Region); in this case it must be assumed that the African leptodactylid forms that were the ancestors of the Ranids were later replaced by them.

Future karyological findings on African and Eastern Ranid forms may perhaps clarify some of the open problems regarding the phylogeny of this important group of Anura; similarly, other studies of this type seem necessary in order to define the position of the problematical Microhylids in the systematics of the order. As already mentioned, the karyological data point to a relationship between this last family and the Ranids. The existence of forms, such as *Hemisus* which morphologically and karyologically seem to be transitional between Ranids and Microhylids, corroborate this theory.

In conclusion, from a karyological point of view—and often from a morphological one as well—the families of group (3) may all have originated from a leptodactylid stock, which must have been ubiquitous throughout continents derived from Gondwanaland at the end of the Jurassic or just after it. In its turn, this stock may be derived from primitive forms of group (1), the fossils of which are found to date back as early as the Lower–Middle Jurassic (cf. Reig, 1958, 1961; Casamiquela, 1965). The fossil remains of these ancient Anura and the geographical distribution of the present-day Ascaphids are evidence of a generalized dispersal of this group of amphicoelous forms in the central part of Mesozoic; the Pelobatids, with the same karyotype as that of the primitive Leptodactylids, may constitute an early derivation, in Laurasian territories, of this stock, which in southern territories has given rise to the early leptodactylid forms.

From the karyological point of view, the transition between the ancestral ascaphid stock and the subsequent leptodactylid–pelobatid forms seems to have occurred with radical modifications in the number and morphology of the somatic set (Fig. 35) and also in the spermatocyte set. This stage of intense rearrangements may perhaps be correlated with the rapid cladogenesis of the families of higher Anura, which seems to have taken place in the early Cretaceous periods.

The Discoglossids, whose origin must probably date back to ascaphid forms of Jurassic, seem to "recapitulate", in their karyological situation, various stages of the chromosome differentiation that has led to the formulae typical of the higher Anura. Karyologically, therefore, this family appears to be transitional between the ascaphid and the leptodactylid–pelobatid stock; however, its strictly holarctic diffusion suggests that it may have close phyletic relationships with only the pelobatid stock. The Discoglossids might also be a third branch, derived, like the pelobatid and leptodactylid

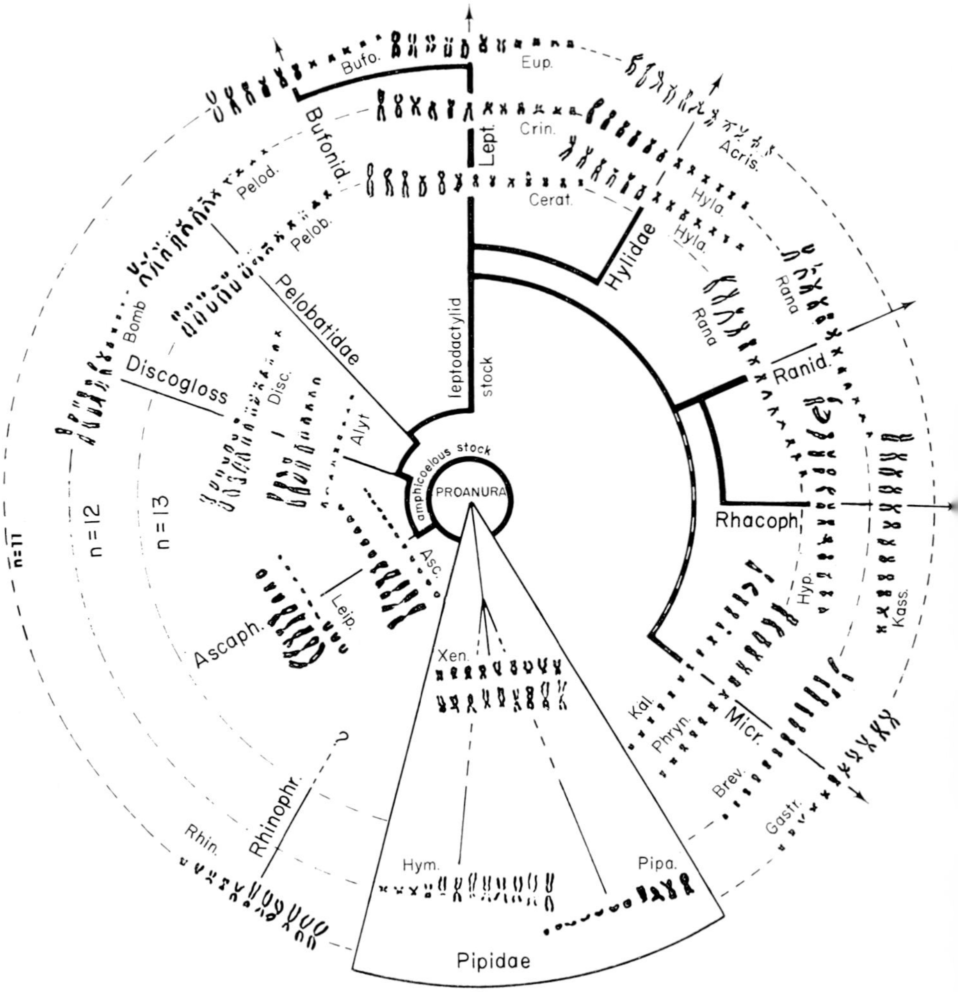

Fig. 35. Scheme of the possible karyological relationships of the main Anuran families. The basic chromosome number is reduced from the centre to the periphery of the scheme; many cases of parallel evolution in the karyotype morphology of the higher families can be seen along the same circumference.

branch, from the ascaphid stock, but morphologically and (partly) karyologically bradytelic.

There originated very early from the leptodactylid stock, perhaps diphyletically, the Australo-Papuan and the Neotropical Hylids and, in Africa or the Oriental Region, the Bufonids and the Ranids and later the Rhacophorids; from the same stock took rise, in Africa or South-America, the

Bufonids. The genesis of the Microhylids remains obscure; karyologically, the family might have originated, perhaps even polyphyletically, from African and Eastern ranid stocks.

Finally, the karyology of the families of Anura in group (3) shows that there are numerous parallelisms in the inter- and intra-family chromosome differentiation (cf. previous Sections), comparable with those known in the anatomical differentiation of these Amphibia (Hecht, 1963; Inger, 1967) and, like them, tending to mask the behaviour of the phylogeny (Fig. 35).

Taking account not only of the karyological data so far acquired but also of present knowledge on the anatomy and paleontology of the Anura, I have placed the conclusions of this Section on a diagram showing a graphical representation of the relationships between the various families, in accordance with the theories so far advanced (Fig. 36).

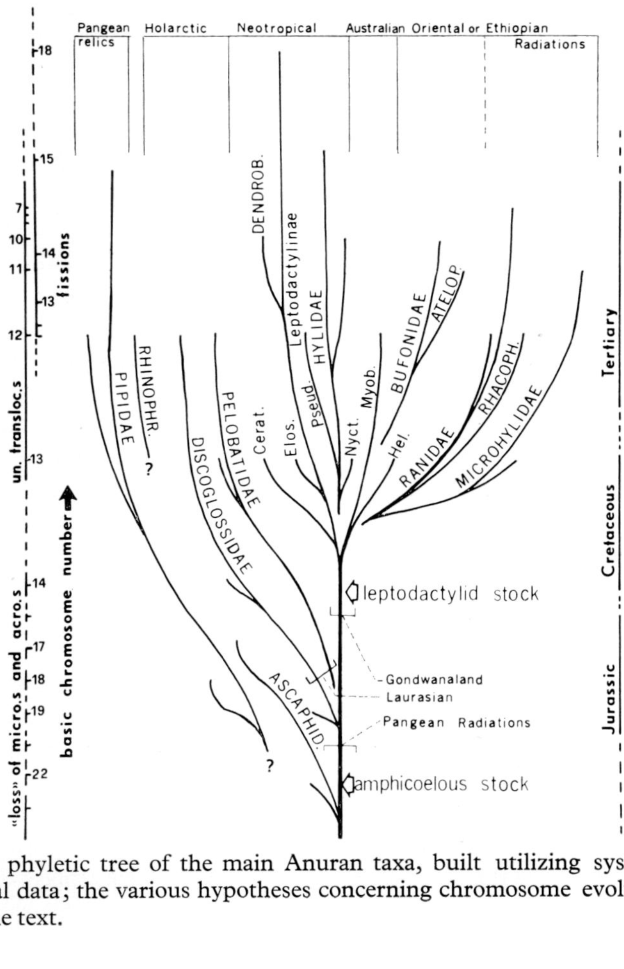

Fig. 36. A phyletic tree of the main Anuran taxa, built utilizing systematic and karyological data; the various hypotheses concerning chromosome evolution can be found in the text.

D. *Meiotic Chromosomes and their Taxonomic Interest*

A cytological study of the meiotic processes may result in findings that are useful from a systematic point of view; this seems true especially as regards the male line, where these processes can be studied more easily and in greater abundance (for a review, see John and Lewis, 1965); however, in the case of the Amphibia, also the meiosis of the female line has offered interesting contributions in this sense.

In the Urodela, numerous studies have been carried out on the meiosis of the two sexes. In the females, the morphological (and often metabolic) characteristics of the bivalents of various species are well known, at the stage of lampbrush chromosomes (diplotene). In these peculiar elements, the position and structure of the lateral loops along the chromosome axes vary according to the species; certain variations are even on an individual scale and a study of them may provide indications on the genetic constitution of a single organism, seeing that the structural differences in the loops are merely the expression of different genic activities (cf. Callan, 1963). Again in the female line of certain Urodela, interspecific differences have been found in the methods of nucleologenesis (Macgregor, 1963) and in the number and localization of the chiasmata on the lampbrush chromosomes or the metaphase bivalents (Callan and Lloyd, 1960; Watson and Callan, 1963; Mancino and Barsacchi, 1966a).

Intra- or inter-specific differences in the spiralization of the bivalents, in the localization of the chiasmata and in their terminalization have also been described in the male line of species of Salamandrids; in some cases differences exist in the frequency and localization of the chiasmata between the two sexes of the same species (Watson and Callan, 1963; Mancino and Poggi, 1963; Mancino, 1965).

However, a study of these differences at present appears to be of exclusively genetic or *microsystematic* interest, seeing that no correlations have been described between certain morphological characteristics of the meiosis and the phylogenetic position of the species that present them. Indeed, since considerable differences seem to exist at times (within one sex), between sub-species of the same species or between species of the same genus (or *Artenkreis*), it is doubtful whether, in the Urodela, this type of research can be used at a taxonomic level higher than the generic.

In general, the spermatocyte bivalents of the Urodela are seen to be relatively despiralized in diplotene—a stage which, on account of its length, is usually one of the most important in the meiosis of these Amphibia. As I have already said, the number and localization of the chiasmata, as well as their terminalization in metaphase, vary according to species. However, the most frequent condition is possibly that of bivalents with more than two chiasmata, many of them in an interstitial position (cf. Galgano, 1933; Callan and Spurway, 1951; Kezer, 1964); the bivalents of this form are often indicated, by the experts on the karyology of the Amphibia, as being of "urodelan" type (cf. Wickbom, 1945), to distinguish them from those typical of other Amphibia, such as the Anura, which in most cases have terminal chiasmata. This obviously does not exclude the possibility that also the spermatocyte bivalents of some species of Urodela have terminal chiasmata.

As regards the morphology of the meiosis of the Apoda, the problem is still to be studied in its broad outlines. In two Indian species, one of them *Ichthyophis glutinosus* being one of the most primitive of the order, Seshachar (1937, 1938) found spermatocyte bivalents of the "urodelan" type: that is, relatively despiralized and with various interstitial chiasmata retained even in metaphase I. In *Geotrypetes seraphini*, an African Apod that is more specialized than the Indian ones, I found, on the contrary, various spermatocyte bivalents with terminal ring-type chiasmata such as are found in the higher Anura.

For the present the findings are too scanty to allow any conclusion to be drawn on their possible taxonomic value; however, it is not improbable that in this order of Amphibia the meiotic processes of the male line may assume the same significance that we now see in the Anura.

In the Anura, the meiotic processes in the female line have been studied far less thoroughly than in the Urodela; information is available on the morphology of the lampbrush chromosomes especially of certain species of the genus *Rana* (for a review see Lacroix, 1968) and of species of *Bombina*, *Bufo* and *Hyla* (Morescalchi, 1965b; Ullerich, 1970).

In general, the morphology of these oocyte bivalents and the phenomena of nucleologenesis associated with them (cf. Morescalchi and Filosa, 1965; Gall, 1969; Macgregor and Kezer, 1970) do not appear different from the Urodela; for the Anura, however, there are no precise "maps" of the lampbrush chromosomes comparable with those for the various Tritons.

Certainly, also in the Anura there exist interspecific differences in the way in which the oocyte meiotic processes take place; the most striking is perhaps that which has already been described, discovered by Macgregor and Kezer (1970) in *Ascaphus truei*, the oocytes of which are seen to be polynucleate up to advanced stages of the meiotic prophase; so far this condition appears to be unique, not only in the Amphibia but also in all the Vertebrates.

As regards the number and localization of the chiasmata, few data exist on the female line of the Anura. In the species that I studied (*Bombina variegata*, *Hyla arborea*, *Bufo viridis* and *Rana esculenta*), the diplotene bivalents generally show chiasmata having no preferential localization, in numbers proportional to the length of the bivalents (Fig. 37a, b). These considerations seem to me possible, in general, also from an examination of the figures in the work by Ullerich (1970) on species that are partly different from those that I studied. This unrestricted chiasma distribution, which concerns species of Anura that are systematically far removed from one another, might represent a general condition in the female line of this order; these findings therefore seem to limit the taxonomic interest of the study of the morphology of the lampbrush chromosomes of the Anura. However, no recent data are available on the processes that take place in the bivalents in the stages later than the diplotene, and it remains to be shown whether the terminalization of the chiasmata in the oocyte metaphase of the Anura occurs by the same methods in all the species.

In the Anura, as in the Urodela, there are also differences in the morphology of the bivalents between the two sexes of one and the same species, though their extension differs according to the species: for example, they are considerable in *Rana esculenta* (Galgano, 1936; Galgano and Morescalchi, 1966) and probably in most of the higher Anura, whereas they are only slightly accentuated in *Bombina variegata* (Morescalchi, 1965b) and probably in the Ascaphids and in the other Discoglossids.

As regards the male line of the Anura, it has been known for some time that the spermatocyte bivalents of most of the species of the order normally have two terminal chiasmata each, which give them a typical ring form; the discovery of certain species with a different situation aroused the attention in the past of Galgano (1933), who emphasized the fact that the bivalents of these species had a morphology of the "urodelan" type; these observations have been taken up by other authors and have also extended to the bivalents of the Apoda (Seshachar, 1938; Sato, 1939). Wickbom (1945, 1949), in an attempt to show karyological differences between Anura and Urodela, which he considered to be of different origin, advanced the theory that the type of spermatocyte bivalents of the Discoglossids cannot be considered as a primitive characteristic; he based this theory on the fact that the Pipids, considered by Noble (1931) as being in the same sub-order as the Discoglossids, reveal spermatocyte bivalents of typical "anuran" shape, which, according to Wickbom, may suggest that the condition presented by the male meiosis of the Discoglossids is isolated. However, this objection is seriously diminished by the fact that the most recent studies on the systematics of these two families show that probably they have few (if any) phyletic relationships with each other (cf. Griffiths, 1963; Nevo, 1968); moreover, the karyological findings on the Ascaphids, the oldest family of living Anura, by showing that in their spermatocyte line there are bivalents of the same type as those of the Discoglossids, seem to strengthen the theory that the

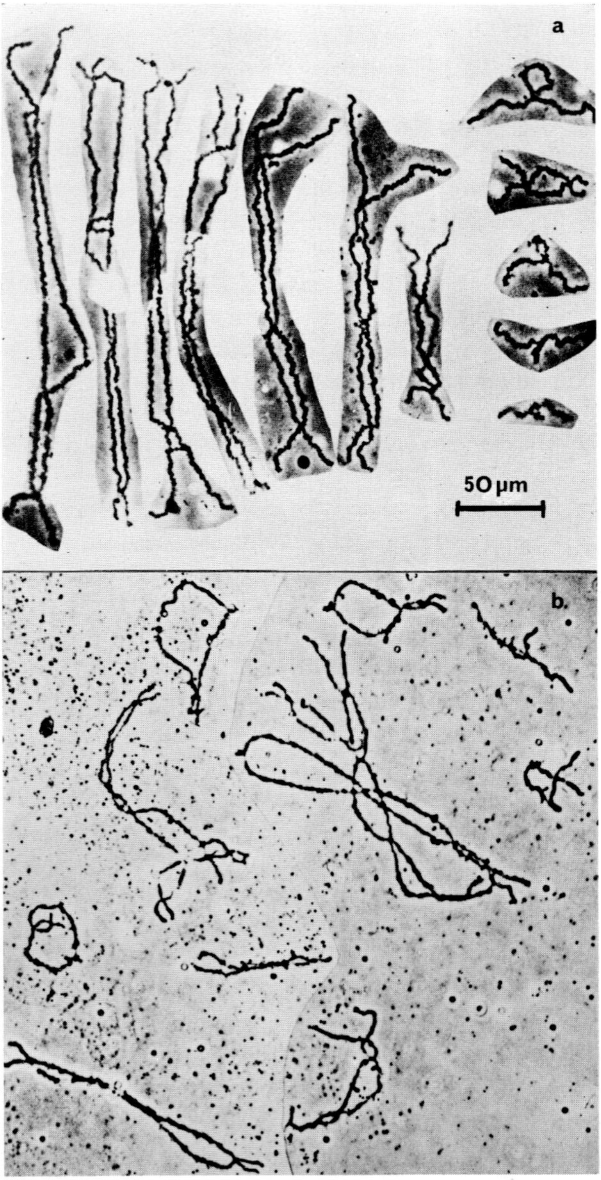

Fig. 37. Meiotic chromosomes in the female line of the Anura, to show the "random" distribution of chiasmata. In a, the set of lampbrush chromosomes in a primitive Anuran, *Bombina variegata*; in b, that of a higher Anuran, *Rana esculenta*.

bivalents of "urodelan" type are actually of primitive type in the Anura (Morescalchi, 1967a, 1968d, 1970).

In *Ascaphus truei* (Ascaphidae), *Discoglossus pictus*, *Bombina variegata*, *B. bombina* and *Alytes obstetricans* (Discoglossidae), the spermatocyte diplotene is a very long stage, during which the bivalents, relatively only slightly spiralized, have very sharp outlines and reveal chiasmata, mostly interstitial, without a precise localization. Their number is generally greater than two in the large bivalents and is one or two in the small ones. On the other hand, the diakinesis is a very short stage, during which the chromosome spiralization, the number and the terminalization of the chiasmata undergo relatively small changes compared with those undergone by the diakinetic chromosomes of the other Anura. At metaphase I, the large bivalents reveal 2 or 3 chiasmata and the small ones generally only one; only some of these chiasmata terminalize. The chromosome spiralization does not attain the high degrees reached by the metaphase bivalents of the other Anura (Figs. 38–41).

These conditions are revealed with certain differences in the 4 genera studied, and sometimes there are variations between the figures of the same specimen, though these are infrequent; in general, *Ascaphus* and *Bombina* adhere closely to the scheme, and frequently also *Alytes; Discoglossus*, and sometimes *Alytes*, reveals more highly spiralized bivalents and a lower frequency of chiasmata compared with the other genera; *Discoglossus* often reveals metaphase bivalents with terminal chiasmata.

In the other families of Anura (Pipidae, Pelobatidae, Leptodactylidae, Hylidae, Bufonidae, Microhylidae, Ranidae, Rhacophoridae) the diplotene of the male line is a very short stage, at the beginning of which the bivalents have an ill-defined morphology, appearing relatively spiralized but entangled and having blurred and spiky outlines. Towards the middle or late diplotene, when the homologues, continuing to spiralize and open out, assume a more sharply definable morphology, the chiasmata appear in a terminal position in practically all cases; their number, though not always well definable before the late diplotene, is 2 for each bivalent. The diakinesis is very long; at this stage the bivalents appear strongly spiralized and each is provided with two terminal chiasmata, assuming a ring form. This shape is retained also in metaphase, when there is a further increase in the spiralization (Figs. 43–45).

The essential characteristics of the spermatocyte meiotic processes in the two groups of families of Anura are summarized in Table I.

Among the families that possess the most widespread type of spermatocyte meiosis in the order, there are some exceptions.

Species of African Ranids and Rhacophorids (*Rana adspersa*, *Kassina wealii* and *senegalensis*, *Hyperolius sp.*) frequently reveal diakinetic bivalents

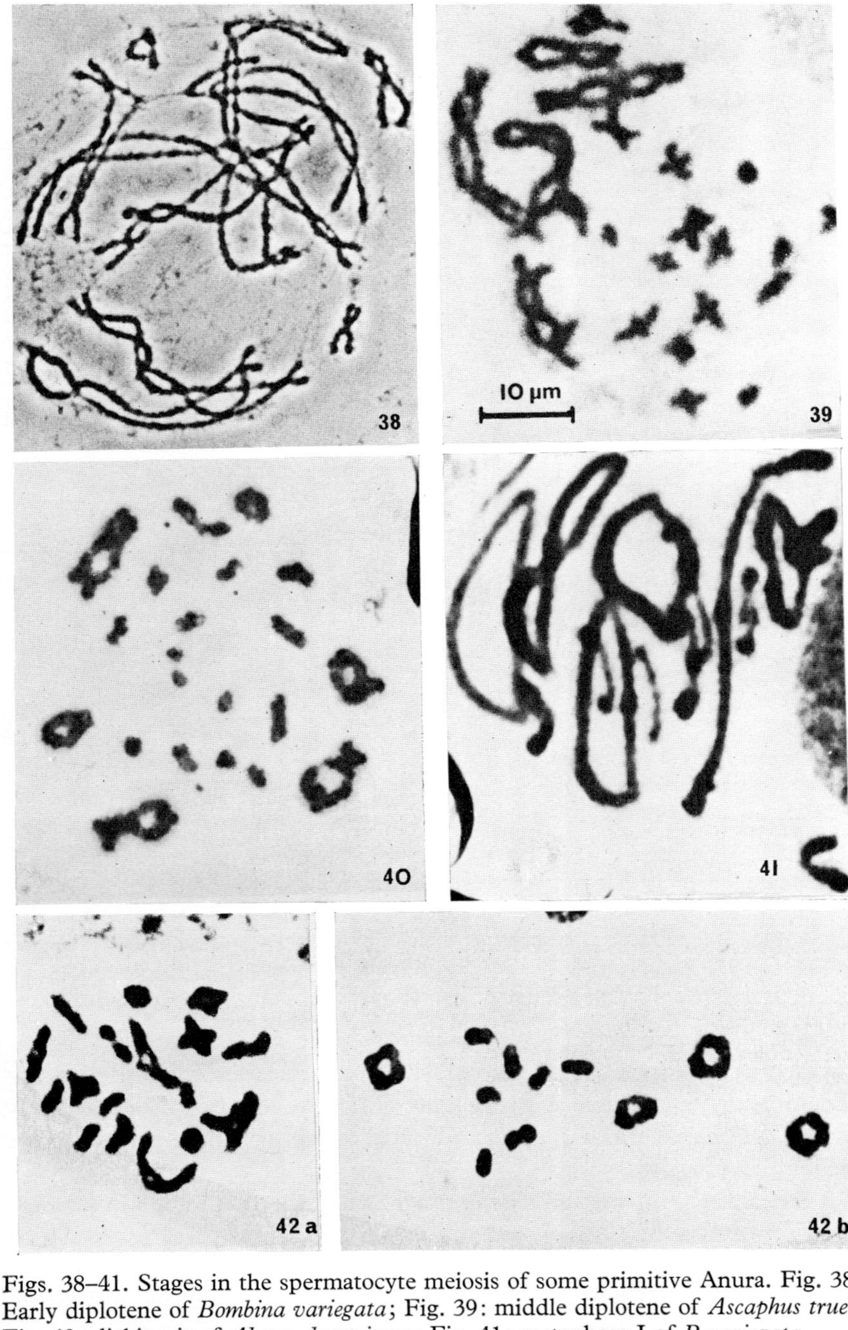

Figs. 38–41. Stages in the spermatocyte meiosis of some primitive Anura. Fig. 38: Early diplotene of *Bombina variegata*; Fig. 39: middle diplotene of *Ascaphus truei*; Fig. 40: diakinesis of *Alytes obstetricans*; Fig. 41: metaphase I of *B.variegata*.
Fig. 42. Spermatocyte metaphases I of two Pipids: a–*Pipa parva*, with many procentric chiasmata; b-*P. pipa*, with terminal chiasmata.

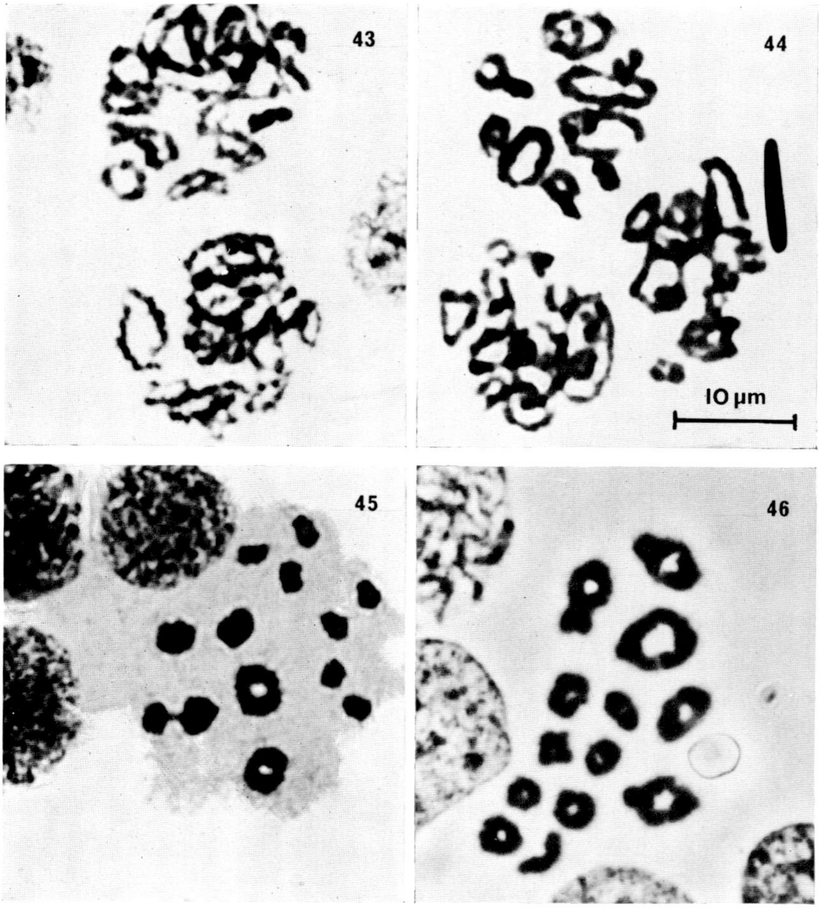

Figs. 43–45. Stages in the spermatocyte meiosis of a higher Anuran, *Rana esculenta*; Fig. 43: early diplotene; Fig. 44; middle and late diplotene; Fig. 45: diakinesis. Fig. 46. Diakinesis of *Hyperolius argentovittis*, showing some interstitial chiasmata.

provided with proterminal chiasmata and, in a few cases, interstitial chiasmata in the large bivalents; with metaphase I, however, the bivalents mostly assume the characteristic ring form, typical of most of the members of the two families (Fig. 46).

A particularly interesting case seems to me to be that of the Pipid *Pipa parva* (Morescalchi, 1968a) and of certain Neotropical Hylids, such as *Hyla microps* and *H. rubicundula* ($2n = 30$) (cf. figures in the works by Beçak, 1968 and Rabello, 1970), species which, in the spermatocyte bivalents, often reveal—albeit to a different extent in each—interstitial chiasmata,

TABLE I.

Characteristics of the Meiosis of the ♂	Ascaphids and Discoglossids	Other families of Anura
Relative length of stages of late prophase I.	Long diplotene, short diakinesis.	Short diplotene, long diakinesis.
Degree of spiralization of the bivalents in late prophase I.	Bivalents relatively little spiralized.	Bivalents highly spiralized.
Mean number of chiasmata in late diplotene and in diakinesis.	More than two chiasmata in the large bivalents; one only in the small ones.	Two chiasmata in both large and small bivalents.[a, b]
Position of chiasmata in the diplotene and in diakinesis.	Interstitial chiasmata in the majority.	Terminal chiasmata.[a, b]
Chiasma terminalization at metaphase I.	Normally never total.	Normally total.[a]

[a] Except for *Pipa parva*, some Neotropical Hylids and possibly other higher Anura, with bivalents mostly provided with interstitial (often procentric) chiasmata.
[b] Except for some African Ranids and Rhacophorids, with bivalents provided with proterminal, sometimes interstitial chiasmata at late diplotene and diakinesis.

frequently procentric; the bivalents are therefore highly spiralized, as in the higher Anura. Since the other species of the two families have the "normal" ring-type bivalents of the majority of the Anura, it seems to me probable that the peculiar condition revealed by *P. parva* and the two species of *Hyla* may be considered to be derived, and not primitive, in these families (Fig. 42).

These last species are peculiar also because they possess *t*-chromosomes in the diploid set, a relatively rare condition among the non-Ascaphids. According to the theory that I stated previously, these *t*-chromosomes might have derived, except in the case of the most primitive Anura, from processes of centric fission of *m*, *sm* or *st* chromosomes, present in the somatic set of *Pipa pipa* and of most of the species of *Hyla*. Thus there seems to exist a correlation between the presence of interstitial chiasmata in the spermatocyte set and the presence of *t*-chromosomes in the somatic set; on the basis of this correlation, it is possible that other cases of spermatocyte meiosis of this type will be found in Hylids and Leptodactylids with *t*-chromosomes.

It is difficult to interpret the possible evolutional significance of this correlation, since there is no information, about eventual modifications of the localization of the chiasmata in the female meiosis of the described species, which has not been studied. On the contrary, I consider it interesting to return to the observation that the morphology of the meiotic processes in the male line of the Anura seems to make for a dichotomic division of the families of the order: on the one hand, Ascaphids and Discoglossids, with bivalents of "urodelan" type; on the other hand, with some exceptions, the remaining families, with ring-type bivalents.

The presence, though not constant, of ring-type spermatocyte in *Discoglossus* (and, less frequently, in *Alytes*) suggests, that, between the two conditions described, the second may be derived from the first. This is possible because the bivalents of "urodelan" type are found, as already stated, in the most primitive families of the three orders (Hynobiidae, besides other families, among the Urodela; Ichthyophidae among the Apoda; Ascaphidae among the Anura); moreover, some of the characteristics revealed by the meiosis of the type of the higher Anurans (the short diplotene, the rapid spiralization of the bivalents, the terminal localization of the chiasmata) must probably be considered as indications of karyological "specialization".

The generalized presence of bivalents with terminal chiasmata in the spermatocyte meiosis of the Anura suggests that the acquisition of this type of meiosis may have had a certain importance in the evolution of the order; however, it is difficult to judge its significance.

In the female line, we have seen that the Anura generally seem to show an unrestricted distribution of chiasmata; in the male line, on the other hand, with the exception of Ascaphids and Discoglossids, chiasmata are restricted to terminal regions of the bivalents. It is therefore probable that the two sexes of most of the Anura possess different systems of gene recombination.

This fact would invalidate any purely genetic interpretation of the significance of the localization of the chiasmata in the spermatocyte line, because the non-recombined groups of genes in the male meiosis may be recombined in the female meiosis (see discussion in Watson and Callan, 1963).

As a hypothesis, the significance of this phenomenon might be of an adaptive nature: it is well known that the spermatogenesis of the Anura is highly thermo-sensitive (cf. synthesis in Galgano, 1952); the acquisition of a type of meiosis in which there is a considerable shortening in some of the stages that are most critical for the morpho-functionality of the bivalents might have favoured the Anura with short diplotene and rapid chromosome spiralization by removing them from the harmful effects of temperature changes on the spermatogenesis; obviously, other selective factors might also have given rise to this peculiar type of spermatocyte meiosis.

The fact that the old Pipids also reveal the type of bivalents of the higher Anura suggests that this type of spermatogenesis appeared very early in the phylogeny of the order; the Discoglossids, from this point of view, appear as transitional forms between the Ascaphids and the remaining families: as already stated in a previous section, these conclusions seem probable also on the basis of the morphology of the somatic set of the Discoglossids.

E. *Mechanisms of Karyological Evolution in the Amphibia*

In the previous Sections I have mentioned the possible mechanisms of karyological differentiation that seem to be of general interest in the Amphibia.

Among these mechanisms, in the past, great importance was attached to the processes generally known as "centric fusion", which lead (by methods about whose details there is still discussion today; see White, 1954, 1969; John and Hewitt, 1968) to the formation of two-armed chromosomes (*m*, *sm* or *st*) starting from pairs of non-homologous one-armed chromosomes (*t*), in such a way as to leave the NF of the set unaltered; the chief supporter of the importance of these processes (sometimes known as Robertsonian processes) for the Amphibia was Wickbom (1945, 1949), who has been followed in this by certain others, for instance Goin and Goin (1962). However, the schemes drawn up by the Swedish author are to be considered inaccurate: in fact, for technical reasons, the interpretation of the karyological formulae offered by Wickbom for certain species of Discoglossids, Pipids and Ascaphids—an interpretation necessary for supporting this theory—has been found to be wrong (Morescalchi, 1966).

According to Beçak (1968), the mechanisms of centric fusion may be valid, in the Anura, only in order to interpret the evolution at intra-family level, not at inter-family level, which seems to involve other types of chromosome rearrangement, such as translocations of various forms, duplications and inversions (see also Rabello, 1970). For my part, I agree with these theories, though it must be borne in mind that various cases of presumed "centric fusion", described by these last authors in certain Neotropical Anura, can perhaps be interpreted better—as I shall mention later—as cases of "centric fission".

Moreover, some authors tend to attach considerable importance, in the karyological evolution of the Amphibia, to phenomena of polyploidy (cf. Beçak, M. L. *et al.*, 1970); however, it must be remembered that, for the present, cases of polyploidy seem to be confined to a few Urodela and to

Anura of certain families, such as Hylidae and Leptodactylidae; according to Bogart (unpublished), certain African Ranids and Bufonids are also polyploid. Among the Urodela, populations composed of polyploid specimens are known in *Ambystoma jeffersonianum* and consist only of females, triploid and gynogenetic (Uzzell, 1963); sporadic cases of polyploid specimens, often only slightly fertile or infertile, are recorded in other Ambystomatids and in certain Salamandrids (Fankhauser and Humphrey, 1942; Kawamura, 1951). In the Anura, where the polyploid species so far studied seem amphimictic, this phenomenon may have had a considerable evolutional value for certain genera; however, for the present, it is difficult to attribute similar importance to the part played by this phenomenon at a higher taxonomic level. Polyploidy is viewed, by the Beçak school, as one of the possible methods of gene duplication; in fact, according to various authors, these methods have played a fundamental role in the evolution of the genetic set of the Amphibia and, in general, of the Vertebrates (Ohno, 1967; Beçak, 1969).

Schematically, it seems to me that, in the karyological evolution of the three orders of Amphibia (though few studies have been carried out on the Apoda in this sense), two phases of general importance can be identified; in some groups of Anura, there also seems to exist a third phase of karyological differentiation.

In the first phase, which characterizes the transition between the more primitive and the "higher" forms of the Amphibia, the chromosome rearrangements lead to the apparent loss (perhaps through unequal translocations) of the acrocentrics and the microchromosomes, with a consequent large reduction in the chromosome number. The "loss" of the acrocentrics may be attributed to mechanisms of centric fusion, the evolutional significance of which may easily be guessed at by postulating the genesis of new gene associations of general utility. The "loss" of the microchromosomes may have other significances—in addition to the one just mentioned—that appear to be associated with the peculiar nature of these elements, which, at least in the Vertebrates, behave differently from the other chromosomes in certain characteristics (cf. White, 1954). In the birds, for example, the microchromosomes are the nucleolar organizers and are sometimes heterochromatic and contain particular fractions of repetitive DNA (Comings and Mattoccia, 1970). The microchromosomes of the Amphibia are generally larger, in an absolute sense, then those typical of the Sauropsids and other low Vertebrates; however, their relative size compared with the "normal" chromosomes in the set is the same in Amphibia and Sauropsids; furthermore, the microchromosomes of certain Ascaphids and Caeciliids, the length of which is less than one micron, approach very close to the dot-chromosomes of the Reptiles: it therefore seems to me that it is not possible to exclude the theory that these elements may have the same functions in these two groups of Tetrapods (as, in fact, in other Vertebrates also). An explanation must also be given of the relationships between the microchromosomes and the supernumerary chromosomes (also known as accessory or B-chromosomes), the latter often being dot-like in appearance and not always numerically variable (Puteyevsky and Zohary, 1970), so that I consider that they are sometimes hardly distinguishable from the former. In fact, supernumerary chromosomes are present in certain Anura and Urodela (*Rana temporaria*, *Acris crepitans*, possibly *Hyla trachitorax* and certain Bolitoglossinae; Kezer, 1964, personal communication; Ullerich, 1967; Nur and Nevo, 1969; Rabello, 1970), and it is known that the B-chromosomes possess functional and genetic characteristics that differentiate them from the other chromosomes (cf. Swanson, 1957; Battaglia, 1964).

The early disappearance of the microchromosomes in the phylogeny of the Amphibia, which in general differentiates these organisms from the reptilian stock, may perhaps be related to the assumption, by the normal chromosomes of the set, of particular functions, some of them "metabolic" (see Section F), that have probably been transmitted to the microchromosomes in the lower Vertebrates, the Sauropsids and the more primitive Amphibia. In other words, the theory is advanced that the reductional process in the microchromosomes may constitute a further characteristic of karyological specialization in organisms that are already considerably differentiated from this point of view (cf. Morescalchi, 1970).

In the second phase, which characterizes the inter- and intra-family differentiation of the higher Amphibia, the karyotype seems to undergo only small further reductions in the chromosome number, especially in the small elements; this process may depend on unequal translocations. Other types of chromosome rearrangement may also explain inter-specific differences of smaller degree.

In general, therefore, this second phase, contrary to the first, seems to be characterized by a considerable "stability" in the chromosome morphology and number (the diploid set generally consists of two-armed elements, with numbers ranging from 28 to 22), which has often been considered a general characteristic of all the Amphibia, whereas it is only a characteristic of most of the higher families of the order.

Obviously, this karyological stability is only an *apparent* characteristic, the result of observations at a relatively superficial level on chromosome numbers and morphology. In fact, there is evidence that, in nature, chromosome rearrangements are not rare in the higher Amphibia. Personally (Morescalchi, 1963c, 1964a), I have identified a probable case of duplication of the chromosome section containing a heterochromatic area (possibly the nucleolar organizer) in only one of the homologues of the 10th pair in a young female of *Rana esculenta* (Fig. 47) and a possible case of simple translocation in a female of *Pelodytes punctatus*; on closer examination, even specimens that appear to have a normal somatic karyotype reveal, on an analysis of the loops of the oocyte lampbrush chromosomes, the presence of extensive chromosome rearrangements. This is the case with certain Salamandrids, described in detail by Lacroix (1970a) and Mancino *et al.* (1970).

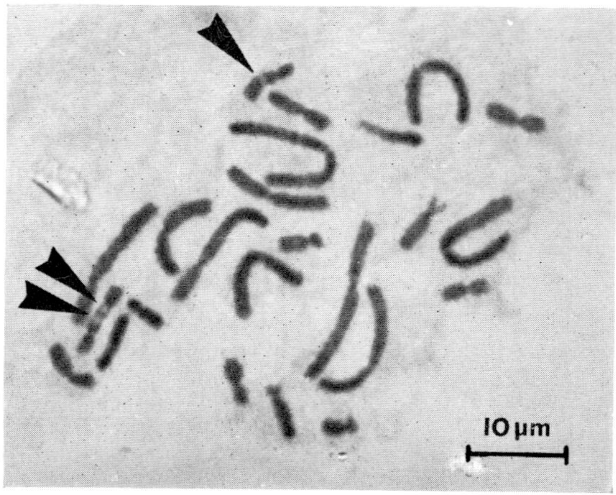

Fig. 47. Intestinal metaphase plate of a female specimen of *Rana esculenta* showing a rearrangement (possibly a duplication of the nucleolar organizer) in one homologous chromosome of the 10th pair, which has two heterochromatic areas (two arrows) and is longer than its "normal" partner, which has one heterochromatic tract (single arrow).

The "apparent" conservative nature of the karyotype of the higher Amphibia therefore does not necessarily signify the absence of structural rearrangements. These may be rigidly controlled and canalized by various endonuclear factors, in such a way that they result in a karyotype that is morphologically hardly variable on cytological examination. For example, the considerable quantitative differences in the nuclear DNA of interrelated species of Amphibia (such as species of the genus *Rana* or *Bufo*), possibly due to duplications (or deletions), seem to be "adjusted" in such a way as to modify only the absolute dimensions of certain chromosomes but to produce very little change in the form, and none at all in the number, of the elements of the set, so that the karyotypes remain similar to one another within each genus (cf. Ullerich, 1966, 1967; Griffin *et al.*, 1970); in these phenomena, it is possible that the repetitive DNA assumes an important role, as we shall see in the following Section.

That an almost identical karyological formula, in related species or subspecies, does not signify that the genetic heritage is standardized, is also shown by the fine chromosome maps obtained from an analysis of the oocyte lampbrush chromosomes of various Urodela of the genera *Triturus* and *Pleurodeles*, which reveal differences in the main marker loops, probably due to allelic changes in single genetic loci (cf. White, 1969).

An identical chromosome "phenotype" may therefore reflect very different conditions in the genetic constitution; in any case, the karyological formulae typical of the higher Amphibia must be particularly well balanced as regards the various selective factors that act on the chromosome phenotype (cf. John and Lewis, 1968), to be preserved in such a stable manner in this group of Vertebrates.

The significance of the conservative nature of the morphology of the karyotype of these Amphibia may perhaps be interpreted by what has been suggested by Stebbins (1966) for the higher plants: "*In a sexually reproducing, outcrossing species, the most efficient method of reducing genetic variability is the tightening of genetic linkages. One is by reduction in frequency and localization of crossovers, so that large segments of chromosomes are virtually devoid of genetic recombination; and the other is by reduction of the number of chromosomes or linkage groups*" (p. 1466). The former condition seems to be attained in the male line, even if not in the female line, of the higher Anura and perhaps of various Apoda (see Section D); the latter condition, on the other hand, seems generelly valid for the higher Amphibia (with few exceptions) which possess chromosome numbers that, on the average, are among the lowest in the present-day Vertebrates. This fact, with the consequent reduction in genetic variability, would be useful for these organisms (again following Stebbins' theory), because it would hinder the segregation and recombination of the advantageous linkage-groups gained in the previous stage of considerable numerical reduction; this seems even more probable if it is remembered that the Amphibia are organisms that are particularly dependent on the external environment.

As regards the morphology of the karyotype, most of the higher Amphibia therefore seem to have reached a terminal point in the karyological differentiation. Among the species sc far studied, only certain American Leptodactylids and Hylids (though it is probable that other higher families reveal similar phenomena) seem to have been capable of emerging from that kind of evolutional cul-de-sac in which the karyotype of most of the Amphibia is confined; significantly, these two groups of Anura have explored the same means for overcoming this *impasse*: the first seems to be polyploidy, useful because it doubles the chromosome number and possibly on account of the lack of chromosome mechanisms of sex determination in these Amphibia (Beçak M. L. *et al.*, 1970); the second possibly consists in the establishing of extensive phenomena of centric fission, with the genesis of a number of *t*-chromosomes that is potentially double that of the initial set, consisting of two-armed elements in the higher Amphibia.

The authors that have studied the karyotype of the Neotropical species of Anura with *t*-chromosomes (various Eleutherodactylini and Hylini) have noted that the number of these particular chromosomes is proportional to the total number of the elements of the set (variable in these species) and seem inclined to interpret these observations as favourable to the existence of phenomena of centric fusion (Beçak, 1968; Rabello, 1970; Bogart, 1970). For my part, I attach greater probability to the theory that the *t*-chromosomes originated secondarily through fission phenomena which amount to doubling the number of linkage-groups of these Anura; in fact, the latter are differentiated karyologically from the basic species of the subfamilies or families to which they belong (which have two-armed chromosomes) and therefore can hardly be considered primitive (see Section C). In my opinion, another significant fact is that the *Eleutherodactylus* (and perhaps also the species of *Hyla*) with *t*-chromosomes, are localized in areas where the genus is represented by hundreds of species: this may suggest that there are correlations between the three phenomena: interspecific variability in the chromosome number, the presence of *t*-chromosomes, and the actual stage of adaptive radiation on the part of these tree-frogs (it would also be interesting, in these species, to check the existence of phenomena of chromosome polymorphism, to be expected in this evolutional stage).

In conclusion, the phenomena of centric fission might result advantageously in an increase in the "genetic potentiality", parallel with the increase in the chromosome number (cf. Todd, 1967, 1970), or else, more simply, they might be useful in creating cytological barriers between different populations, by favouring or provoking processes of speciation in the adaptive radiation stage (cf. White, 1968). The fact that some of the species with *t*-chromosomes possess spermatocyte bivalents with interstitial chiasmata (see Section D) suggests that other chromosome rearrangements, in addition to centric fission, may have contributed towards modifying the localization of the crossover in these Anura in the evolutional stage.

In the present state of affairs, it is very difficult to prove these theories on cytological bases; on the other hand, some of the cytogenetic studies so far carried out on the Amphibia have had aims other than those of cytotaxonomy (see synthesis in Beatty, 1964). The discovery in nature of structural rearrangements in certain Amphibia, as mentioned before, is too sporadic to have any new significance in this problem; other lines of study are necessary, some of which are probably capable of producing interesting findings in this respect. For example, the treatment of live specimens of

Bombina variegata with prolonged cold (2°C ± 1 for 4 days) induces modifications in the mitotic and spermatocyte chromosomes, such as: fragmentations at the level of the centromere, with the genesis of apparently telocentric chromosomes; extensive phenomena of centromere heterochromatization; reduction in the frequency of the chiasmata, a good part of which become terminal, as is normally the case in the higher Anura (but

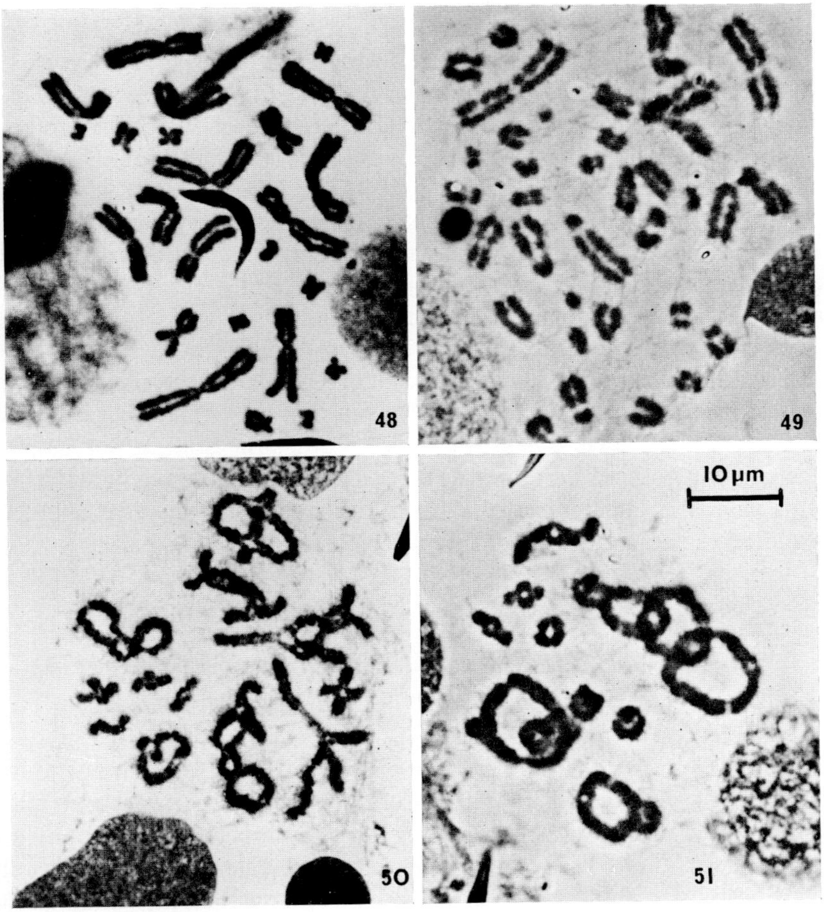

Figs. 48, 49. Spermatogonial chromosomes of an untreated specimen of *Bombina variegata* (all two-armed) (Fig. 48), and of a cold-treated specimen of the same species, showing some *t*-chromosomes (Fig. 49).
Figs. 50, 51. Spermatocyte bivalents of an untreated (Fig. 50) and of a cold-treated specimen of *B. variegata* (Fig. 51); in the last case, many chiasmata are terminalized.

not in *Bombina*) (Figs 48–51) (Morescalchi *et al.*, unpublished). In view of the fact that the Amphibia are particularly thermo-sensitive, these findings may theoretically assume a certain value, showing that certain rearrangements typical of the karyological differentiation of these Vertebrates are reproducible at an individual level with the intervention of factors that are probably not rare in nature.

In any case, the greater number of contributions to the solution of the above-mentioned problems may spring from studies of a quantitative and qualitative nature on the nuclear DNA of the Amphibia. The former studies have already made it possible to reveal, at inter-specific level, the existence of extensive local variations in the quantity of chromosome DNA (which also leads to variations in the absolute size of the chromosomes), which have been attributed to processes of interstitial duplication and deletion (Ullerich, 1966, 1967; Beçak, 1969). The latter studies, which have only just been started, for the moment point to the theory that these inter-specific differences are due especially to variations in a particular fraction of the repetitive DNA, having little or no genetic significance but a probable functional or adaptive role (see following Section).

It is clear that if this last theory is confirmed and—as seems possible—future studies point to a "heterochromatic" nature of the microchromosomes of the Amphibia (which would mean that they consist especially of repetitive DNA), then the significance of the processes of karyological differentiation in these Vertebrates will be viewed in a different light. In fact, in this case, the phenomena of a functional or adaptive nature might be considered as pre-eminent over the purely genetic phenomena.

F. *Role of the Variations in Nuclear DNA on the Karyological Differentiation*

It is well known that the Amphibia (probably with the Dipnoi) differ from the rest of the Vertebrates in certain quantitative aspects in their nuclear DNA. In fact, whereas in most of the members of the subphylum the DNA content is lower than 10 picograms per nucleus (pg/*n*) and is only very slightly variable at inter-specific level, among the Amphibia all the Urodela so far studied and some of the Anura and Apoda reveal nuclear DNA contents that are higher—in some cases, considerably so—than the above quantity, with an inter-specific variability that is often very great (cf. Ohno, 1967; Goin and Goin, 1968; Beçak W. *et al.*, 1970; Morescalchi. *et al.*, 1970b).

The results reported by the authors that have studied the DNA in the

Amphibia, often using techniques and relative units of measurement that differ from one another, do not always appear readily comparable; however, with a certain approximation due to the necessity of adopting a single unit of measurement for heterogeneous data, I have shown in Table II, in pg/N, the maximum and minimum values of nuclear DNA encountered in the different families studied (the figure in brackets denotes the number of species studied in each family) (see Table II). In addition to the data

TABLE II. *Minimum and maximum amounts of nuclear DNA in the Amphibia.*

	DNA (pg/N)
APODA	
Caeciliidae (3)	7.4–27.9
URODELA	
Hynobiidae (5)	33.0–40.9
Ambystomatidae (5)	43.7–105.0
Plethodontidae (7)	20.0–62.2
Amphiumidae (1–2)	128.3–189.0
Salamandridae (9)	28.5–98.0
Proteidae (4–5)	48.4–189.0
Sirenidae (1)	107.7
ANURA	
Ascaphidae (2)	6.0–12.7
Discoglossidae (4)	10.5–21.1
Pipidae (3)	4.9–8.2
Rhinophrynidae (1)	7.5
Pelobatidae (6)	1.6–8.9
Leptodactylidae (32)	3.0–12.7
Hylidae (31)	2.3–11.1
Bufonidae (16)	5.4–14.6
Atelopodidae (1)	5.5
Microhylidae (2)	3.1–7.5
Ranidae (18)	6.6–15.8
Rhacophoridae (1)	8.8
CHOANICHTHYES	
Latimeria (1)	5.6
Dipnoi (2)	100.0–177.8
(Other Lower Vertebrates)	0.4–6.7
AMNIOTA	1.7–10.5

The number in brackets denotes the number of species studied.

reported in the above-mentioned works, I have used those of Goin *et al.* (1968), Bachmann (1970), Griffin *et al.* (1970) and Olmo *et al.* (1970); other data have been supplied to me, in the form of personal communications, with permission to report them, by J. G. Gall and, in considerable quantities, by E. Sexsmiths; further data have been collected in this Laboratory by E. Olmo and are still unpublished.

To make it easier to compare the data given in Table II, I have shown them in the form of a diagram (Fig. 52); along the abscissae I have indicated the various families of Amphibia and along the ordinates the minimum

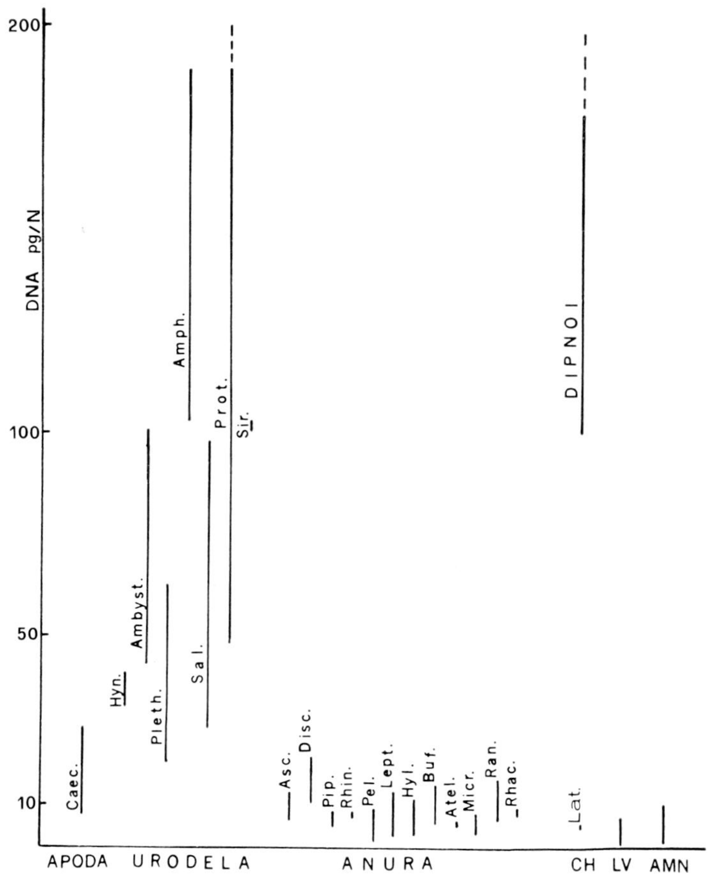

Fig. 52. Minimum and maximum amounts of nuclear DNA in Amphibia and in other Vertebrates: along the abscissae, the quantity of DNA (in picograms per nucleus); along the ordinates, the various taxa here considered. CH = Choanichthyes; LV = other Lower Vertebrates; AMN = Amniotes; Lat = *Latimeria*; other abbreviations are referred to the Amphibian families considered in Table II.

and maximum values of nuclear DNA found in each of them; so as to provide a comparison with other Vertebrates, Table II and the diagram show the DNA values of the main groups of Vertebrates, with particular reference to the Dipnoi. The broken lines on the diagram refer to values of DNA, found in certain families of Urodela and in the Dipnoi, on which there is no agreement among the various authors (it must be mentioned that, with the present techniques, this kind of research is one that still gives only approximate results).

From the Table and from Fig. 52, it is clearly seen that the Urodela possess, on average, far more nuclear DNA than the Anura, while the three species of Apoda so far studied possess values of DNA intermediate between those of species of the other two orders.

Among the Urodela, the maximum values of DNA are reached by species of Amphiumids, Proteids and Sirenids, which, on account of this characteristic, resemble the Dipnoi and, with them, possess the highest values encountered among the Vertebrates (and perhaps in all the eukaryotes). On the other hand, the minimum values are reached by species of Plethodontids and Salamandrids, the most highly differentiated families of the order; the species of Hynobiids so far studied have relatively low DNA values. Of the three families with the maximum nuclear DNA contents, none can be considered basic in the phylogeny of the Urodela, while the reciprocal phyletic relationships (if they exist) appear, as previously stated, highly problematical; the only characteristic revealed in common by these Urodela seems to be the possession of a larval or semi-larval *habitus*, which has induced certain authors to consider that there may be a close correlation between this characteristic and the high DNA contents (Commoner, 1964); Goin *et al.* (1968) consider it probable that the quantity of DNA in all the Amphibians is directly correlated with the length of the larval period.

The resemblances between these three families in the quantity of DNA may therefore assume an "adaptive" rather than a phyletic significance; this same conclusion is also suggested by the observation of the great intra-family variability that characterizes the nuclear DNA of all the Urodela.

It is interesting to point out that the other families, some of which are central in the phylogeny of this order of Amphibia, reveal certain species with relatively low DNA values; among these, the minimum values seem to be reached by the forms best adapted to terrestrial life, such as the Plethodontids. This fact, even if it may confirm the theories of an "adaptive" nature of the variations in the quantity of DNA, nevertheless suggests the possibility that the ancestors of the living Urodela had less DNA than some of the present-day species, the pedomorphic habitus of which is probably of secondary origin; on the basis of these data, the resemblances between the present Dipnoi and these last Urodela in the quantity of the DNA may easily be interpreted as mere convergence phenomena (Morescalchi, 1970; Morescalchi *et al.*, 1970b).

As regards the Anura, the species with the greatest quantity of nuclear DNA are to be found among the primitive—but perhaps phyletically isolated—Discoglossids (*Bombina variegata* has about 20 pg/*n* of DNA); these amounts equal the minimum quantities found among the Urodela; the minimum among the Anura (less than 2 pg/*n* of DNA) are revealed by certain Pelobatids having a rapid larval development (Goin *et al.*, 1968). The intra-family and often the inter-specific variability in the Anura is generally high and in various cases higher than the inter-family variability; in any case, the average values found in this order are about 7 pg/*n* and resemble those revealed by many Amniota. This may be seen by plotting on a histogram the DNA values found in the various species of Anura (which cannot yet be done for Urodela and Apoda, in which few species have been studied): the shape of the resulting graph tends to be Gaussian (Fig. 53). This fact suggests the idea that the DNA values that depart from those revealed by most of the species (between 4 and 8 pg/*n*) may have been

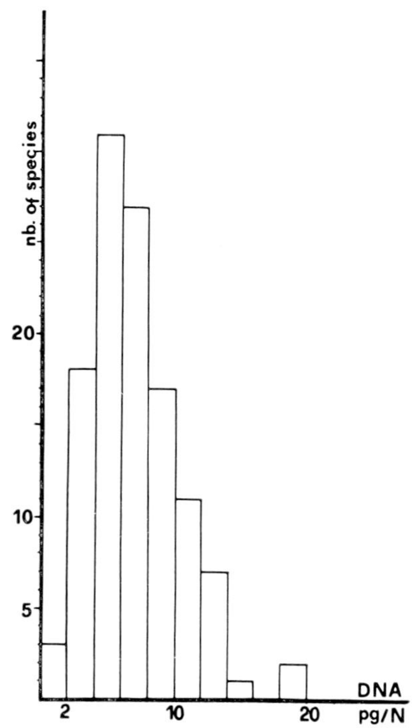

Fig. 53. Histogram showing the distribution of the amount of nuclear DNA (along the abscissae, in picograms/nucleus) in the studied species of Anura (number along the ordinates).

acquired "secondarily" through a sort of karyological differentiation at inter-specific level.

The considerable inter-specific variability in the DNA of the Amphibia thus seems to invalidate the theories that attribute a great phylogenetic importance to this characteristic, at least as regards these Vertebrates; the only fact that might assume this importance consists in the considerable average differences between the DNA values of Urodela and Anura (cf. Ohno, 1967; Ohno *et al.*, 1968; Beçak W. *et al.*, 1970). However, considering the theories, already partly stated, of a probable adaptive nature of the variations in nuclear DNA in the Amphibia, these present differences might depend on the long process of specialization towards differing conditions that has been undergone by the two orders of Amphibia. In fact, the more terrestrial Urodela have similar quantities of DNA to those of certain Anura, while the Apoda, adapted to other conditions of life, have species with quantities of DNA similar to those of species of Anura, and other species with quantities similar to those of species of Urodela.

Since all these theories are based on a particular interpretation of the biological role of the variability of nuclear DNA in the Amphibia, I wish to dwell on this interesting problem.

According to Szarski (1970), there is in the Vertebrates a close correlation between the evolution of the nuclear DNA and that of the cell size, both being dependent upon one and the same cause, adaptation to different metabolic conditions: the increase in metabolism makes small cells more "economical", and vice-versa. Szarski's theory is interesting also because it makes it possible to revise others, which attributed a role of genetic importance to the variations in DNA.

In order to understand how these adaptive processes on the part of the nuclear DNA are possible, we must briefly consider the results of modern studies on the existence, in the genome of the eukaryotes, of nucleotide sequences repeated millions of times, which, in some species (such as certain Urodela), form over 90% of the nuclear DNA (Britten and Kohne, 1969). This "repetitive" DNA contains different fractions (besides that which codifies the ribosome RNA; Brown and Dawid, 1968), sometimes separable by technical means from the rest of the DNA, in this case known as "satellite" (McCarthy, 1969). Cytologically, the satellite DNA can often be localized in the centromere and telomere heterochromatin, through certain observations on the polytene chromosomes of *Drosophila* make probable the theory that part of it is spread over various portions of the genome (Hennig *et al.*, 1970; Eckhardt and Gall, 1971). The association between repetitive DNA and "constitutive" heterochromatin (Brown, 1966) or α-heterochromatin (Heitz, 1934) at present appears to be a general phenomenon in the studied species of Amphibia (Barsacchi, 1970; Macgregor and Kezer, 1971). This type of heterochromatin is genetically inert; it is not certain whether the repetitive DNA is as well (Pardue and Gall, 1970), but most of the data accumulated, including the possibility of obtaining fertile hybrids by crossing specimens of species having very different contents of satellite DNA, seem to support this theory (Hennig *et al.*, 1970).

Once the possibility of a strictly genetic role of the repetitive DNA has been discarded (though not completely at present), most authors—especially in consideration of its pre-eminent localization in chromosome areas of kinetic interest—seem inclined towards the theory of a "functional" role of this DNA in the complex and precise phenomena that occur in the chromosome phenotype, particularly during mitosis and meiosis (the chromosomal "housekeeping" of Walker *et al.*, 1969; cf. also Jones and Robertson, 1970; Hennig *et al.*, 1970; Macgregor and Kezer, 1971).

The information that has been accumulating for some time on the possible biological significance of the α-heterochromatin would tend, however, to broaden this theory, by ascribing to it (and hence also the repetitive DNA) roles by which it would favour (and perhaps canalize) the evolution of the chromosome phenotype and consequently the speciation processes dependent upon this evolution (cf. Darlington, 1937; Swanson, 1957; Natarajan and Ahnström, 1969).

Experimental studies showing a positive correlation between the quantity of nuclear DNA and the frequency of the chromosome rearrangements in the Amphibia (Griffin *et al.*, 1970) seem to make these last observations valid; actually the great differences between the nuclear DNA contents in related species probably depend, at least in part, on variations in the quantity of the repetitive DNA (Hennig *et al.*, 1970; Ris and Kubai, 1970).

The theories of an adaptive nature of the differences in the nuclear DNA in the Amphibia are therefore in agreement both with the present known facts on the biological role of the repetitive DNA (which seems the main factor responsible for these differences) and with all that was previously known about the distribution and behaviour of α-heterochromatin, which, in the Amphibia, is variable at inter-specific level and seems very sensitive to various physical and chemical agents (Callan, 1942; Morescalchi and Gargiulo, 1967), thus revealing "metabolic" characteristics that differentiate it from the so-called "euchromatin". These last phenomena are significantly revealed also by other large-celled organisms, such as some Monocotyledons, which, like the Amphibia, show high inter-specific variability in the nuclear DNA (Dyer, 1963, 1964; Stebbins, 1966).

In conclusion, the great differences in nuclear DNA do not seem to depend only on variations in the number of the structural genes but especially on variations in the number of the sequences of the repetitive DNA, whose coding properties (if any) are peculiar. In the Amphibia, these variations seem to be subject to the selective pressure of various factors, among which metabolic and developmental ones are probably important (Goin *et al.*, 1968); hence there is no correlation between the total quantity of DNA of the genome of a species and its phylogenetic position. For this reason, various doubts exist on a phyletic importance of the differences in nuclear DNA between Anura and Urodela.

Referring back to Szarski's theory, the variations in the repetitive DNA might also have the function of adjusting—without very great modifications in the genetic pool—the chromosome sizes to those of the nucleus during the stages of cell adaptation to the various environments and ways of life which, according to the author, have been explored by the Vertebrates in the course of their evolution. However, only the Amphibia (and perhaps the Dipnoi) seem capable of great inter-specific variations in nuclear DNA content: it is therefore probable that the "plasticity" in the DNA of the Amphibia may be considered as a form of karyological specialization, particularly useful for these organisms, which are the Vertebrates most closely associated with the external environment and sensitive to modifications in it. In the other members of the subphylum, these adaptation mechanisms are perhaps replaced by others, of a more general nature, which act on the organism as a whole and not only at endo-cellular level.

III. CONCLUSIONS

Karyologists have interpreted the data collected on the Amphibia as favouring, to varying degrees, the theory of a polyphyletism of the living orders; Anura and Reptiles have often been associated with each other, and contrasted with Dipnoi and Urodela (and sometimes also Apoda), on the basis of karyological resemblances or differences, respectively. However, various criticisms may be raised on these conclusions.

First and foremost, a study of the more generalized forms of the three orders of Amphibia reveals the presence of karyological formulae having the same characteristics: a high chromosome number, with various acrocentrics and microchromosomes. These formulae are also typical of various lower Vertebrates and of the Sauropsids; in these they are generally preserved, whereas in the Amphibia there is a rapid karyological differentiation which, early in the phylogeny, leads to numerically reduced formulae, lacking in microchromosomes and (with some exceptions) in acrocentrics. It is therefore probable that the ancestral amphibian stock of the Recen Amphibia and of the Reptiles possessed a karyological formula similar to that of the more primitive representatives of Apoda, Urodela and Anura: diverging lines of karyological differentiation would then have led to the differences that characterize the chromosome set of most of the living species of the two groups of tetrapods (cf. Morescalchi, 1970).

In the Amphibia, the considerable inter-specific variability in the quantity of the nuclear DNA may indicate that components of a metabolic nature have been of great importance in this differentiation; in the Reptiles, where the karyological differentiation is apparently far slower, there has perhaps been a prevalence of point mutations (not detectable cytologically) over chromosome rearrangements (save some flourishing group of Lacertilians), as also seems to be indicated by the very low inter-specific variability in the DNA of these organisms.

The "plasticity" in the nuclear DNA content shown by the Amphibia at all taxonomic levels is a cytological characteristic that differentiates them sharply from the rest of the Vertebrates (with the possible exception of the Dipnoi) and may therefore be interpreted as a kind of endonuclear specialization, typical of the living forms of the class; its biological role is not yet clear, though there are various indications suggesting that it is of an adaptive nature.

The presumed karyological resemblances (also in the high DNA contents) between Urodela and Dipnoi are actually limited—as regards these Amphibia—to certain pedomorphic forms (especially to *Siren*) belonging to families that do not seem basic in the phylogeny of the order: they may therefore be the result of evolutional parallelisms. The fact that *Latimeria* possesses low quantities of DNA and that the most primitive bony fishes have karyological formulae rich in chromosomes and microchromosomes (whereas the present Dipnoi have relatively simplified formulae, with 34 to 38 large two-armed chromosomes) suggests that the lines of karyological development followed by the Dipnoi may have been to a certain extent parallel to those followed by some families of larval Urodela; the facts known about the adaptive role that seems to be played, in these organisms, by the extensive variations in the cell dimensions and in the DNA (probably in the repetitive fraction) fit in well with this theory.

It is also interesting to note that the karyological differentiation in the three orders of Recent Amphibia seems to have passed through similar stages and employed the same types of chromosome change, ending in formulae that are only slightly different from one another in the higher families.

Hence, none of the karyological proofs so far obtained seems contrary to the theory of a monophyletic nature of the three orders of living Amphibia. Indeed, the known facts that are accumulating appear increasingly to be in contrast with the polyphyletic theories; because, on the one hand, they confirm the karyological homogeneity of the basic forms of the three orders and, on the other hand, they accentuate the differences in the methods of karyological evolution and in the type of chromosome specialization between these Amphibians and the other groups of Vertebrates.

Of particular interest are the close karyological resemblances between the few Apoda so far studied and certain species of Anura, which suggest that there are good reasons for the problematical Apoda, from a karyological point of view, to be considered among the Lissamphibia.

Regarding the Urodela, it appears that about 34% of the known species have been studied karyologically. The data acquired confirm the existence in the order of two different levels of differentiation: one, typical of forms that possess the karyotype of the primitive model among the Amphibia (Hynobiidae and Cryptobranchidae); the other, typical of the "higher" families, with numerically reduced formulae. The Proteidae have karyotypes transitional between the two levels, while within the second level it is possible to distinguish a group of families that have reached an almost identical karyological formula (Ambystomatidae, Plethodontidae and Amphiumidae) and a family (Salamandridae) with the most highly specialized forms, karyologically, in the order. The Sirenidae, with a single species studied, seem to be outside the scheme described above.

Regarding the Anura, the chromosomes of over 280 species are known, corresponding to about 15% of the known species. In the order, one family (Pipidae) seems karyologically separated from the rest of the Anura; the Ascaphidae and some of the Discoglossidae possess the formula of the primitive type, while the remaining families (the basic species of which often have 26 chromosomes) are seen to be differentiated from the karyological point of view. The data collected on the cytology of the spermatogenesis confirm this subdivision into primitive and higher families, revealing, in the former, the presence of bivalents with "urodelan" characteristics.

The general karyological conservativeness of the higher Amphibia, considered as terminal forms of a long process of chromosome evolution, finds certain exceptions at the level of groups of tropical species of Anura, which at present seem to be in a stage of intense karyological differentiation (through phenomena of polyploidization or centric fission), probably correlated with speciation mechanisms.

Acknowledgements

Many persons have provided assistance and encouragement in the long period of karyological research that has resulted in the present work. I wish to express my gratitude for this, first to my wife and also for both scientific and personal assistance to Profs. M. Galgano, G. Ghiara, R. F. Laurent, O. A. Reig, F. A. Saez, W. F. Blair, R. F. Inger, J. Kezer and B. Lanza, and to Mr. R. Thorn.

For useful exchanges of information and permission to publish data—in some cases hitherto unpublished—I am particularly grateful to Drs M. L. and W. Beçak, J. P. Bogart, E. Sexsmiths, J. G. Gall, and E. Olmo.

I also thank the numerous persons that often and disinterestedly sent animals, indispensable for my researches. Finally, I thank Mr G. Falcone for the technical help he has given me.

The research has been financed by a contribution from the C.N.R.

References

Ahl, E. (1931). *In* "Das Tierreich" Lfg. **55**. (W. de Gruyter and Co. eds.), Berlin and Leipzig.

Bachmann, K. (1970). *Chromosoma, Berlin* **29**, 365–374.

Bai, A. R. K. (1956). *Proc. Natn. Inst. Sci. India* **22**, 1–5.

Baird, D. (1965). *Am. Zool. Utica* **5**, 287–294.

Baldauf, R. J. (1959). *J. Morph.* **104**, 527–560.

Barbieri, F. D. (1950). *Acta Zool. Lilloana* **9**, 455–462.

Barrio, A. (1945). *Revta. Argent. Zoogeogr.* **5**, 37–43.

Barrio, A. and Pistol de Rubel, D. (1970). *Physis* **29**, 505–510.

Barrio, A. and Rinaldi de Chieri, P. (1970a). *Physis* **30**, 321–329.

Barrio, A. and Rinaldi de Chieri, P. (1970b). *Physis* **30**, 309–319.

Barrio, A. and Rinaldi de Chieri, P. (1970c). *Physis* **30**, 305–307.

Barsacchi, G. (1970). *Boll. Zool.* **37**, 474–475.

Barsacchi, G., Bussotti, L. and Mancino, G. (1970). *Chromosoma, Berlin* **31**, 255–279.

Battaglia, E. (1964). *Caryologia* **17**, 245–299.

Beatty, R. A. (1964). *In* "Intersexuality in Vertebrates Including Man" (C. N. Armstrong and A. J. Marshall, eds.) pp. 17–143. Academic Press, New York and London.

Beçak, M. L. (1968). *Caryologia* **21**, 191–208.

Beçak, M. L., Beçak, W. and Rabello, M. N. (1966). *Chromosoma, Berlin* **19**, 188–193.

Beçak, M. L., Beçak, W. and Rabello, M. N. (1967). *Chromosoma, Berlin* **22**, 192–201.

Beçak, M. L., Denaro, L. and Beçak, W. (1970). *Cytogenetics* **9**, 225–238.

Beçak, W. (1969). *Genet. Suppl.* **61**, 183–290.

Beçak, W., Beçak, M. L., Schreiber, G., Lavalle, D. and Amorim, F. O. (1970). *Experientia* **26**, 204–206.

Beckert, W. H. and Doyle, W. (1967). *Can. J. Genet. Cytol.* **9**, 297–301.

Bertmar, G. (1969). *Evolution* **23**, 131–152.

Bianchi, N. O. and Laguens, R. (1964). *Cytologia* **29**, 151–154.

Bianchi, N. O. and Molina, O. J. (1967). *Chromosoma, Berlin* **22**, 391–400.

Blair, W. F. (1963). *Evolution* **17**, 1–16.

Bogart, J. P. (1967). *Can. J. Genet. Cytol.* **9**, 531–542.

Bogart, J. P. (1968). *Evolution* **22**, 42–45.

Bogart, J. P. (1970). *Cytogenetics* **9**, 369–383.
Bogart, J. P., Blair, W. F. and Hubbs, C. (1969). *Cytogenetics* **8**, 30–34.
Bolt, J. R. (1969). *Science, N.Y.* **151**, 417–425.
Brame, A. H., Jr (1967). *Herpeton (J. Southw. Herpetol. Soc.)* **2**, 1–26.
Britten, R. J. and Kohne, D. E. (1969). *In* "Handbook of Molecular Cytology" (A. Lima de Faria, ed.) pp. 21–36. North Holland Publ. Co., Amsterdam and London.
Brown, D. D. and Dawid, I. B. (1968). *Science, N.Y.* **160**, 272–280.
Brown, S. W. (1966). *Science, N.Y.* **151**, 417–425.
Brum-Zorrilla, N. and Saez, F. A. (1968). *Experientia* **24**, 969.
Bushnell, R. J., Bushnell, S. P. and Parker, M. V. (1939). *J. Tenn. Acad. Sci.* **14**, 209–215.
Callan, H. G. (1942). *Proc. R. Soc. Lond. B* **130**, 324–335.
Callan, H. G. (1955). *In* "Symposium on Fine Structure of Cells" pp. 89–109. Interscience Publ. Inc., New York.
Callan, H. G. (1963). *Int. Rev. Cytol.* **15**, 1–34.
Callan, H. G. (1966). *J. Cell Sci.* **1**, 85–108.
Callan, H. G. and Lloyd, L. (1960). *Phil. Trans. R. Soc. Lond. B* **243**, 135–219.
Callan, H. G. and Spurway, H. (1951). *J. Genet.* **50**, 235–249.
Casamiquela, R. (1965). *Revta. Mus. La Plata* **27**, 265–317.
Cei, J. M. (1968). *Publ. Cons. Nat. Recherches, Paris. Biol. Am. Aust.* **4**, 199–214.
Cole, C. J. (1966). *Copeia* **1966**, 578–580.
Cole, C. J. and Zweifel, R. G. (1971). *Herpetol. Rev.* **3**, 15–16.
Cole, C. J., Lowe, C. H. and Wright, J. W. (1968). *Copeia* **1968**, 96–100.
Comings, D. E. and Mattoccia, E. (1970). *Chromosoma, Berlin* **30**, 202–214.
Commoner, B. (1964). *Am. Sci.* **52**, 365–388.
Darlington, C. D. (1937). "Recent Advances in Cytology", 2nd ed., J. A. Churchill, London.
Davis, D. D. (1936). *Zool. Ser. Field Mus. Nat. Hist.* **20**, 115–121.
Di Berardino, M. A. (1962). *Dev. Biol.* **5**, 101–126.
Donnelly, M. G. and Sparrow, A. H. (1965). *J. Hered.* **56**, 90–98.
Doyle, W. and Beckert, W. H. (1970). *Caryologia* **23**, 143–154.
Duellman, W. E. (1967). *Syst. Zool.* **16**, 38–43.
Duellman, W. E. (1968). *Univ. Kansas Publ., Mus. Nat. Hist.* **18**, 1–10.
Duellman, W. E. and Cole, C. J. (1965). *Syst. Zool.* **14**, 139–143.
Dunn, E. R. (1948). *Am. Mus. Novitates* **1384**, 1–13.
Dyer, A. F. (1963). *Chromosoma, Berlin* **13**, 548–576.
Dyer, A. F. (1964). *Cytologia* **29**, 171–190.
Eaton, T. H. Jr. (1959). *Univ. Kansas Publ., Mus. Nat. Hist.* **12**, 155–180.
Eckhardt, R. A. and Gall, J. G. (1971). *Chromosoma, Berlin* **22**, 407–427.
Estes, R. (1965). *Amer. Zool.* **5**, 319–334.
Estes, R. (1969a). *Nature (Lond)* **224**, 87–88.
Estes, R. (1969b). *Breviora Mus. Comp. Zool.* **322**, 1–11.
Estes, R. (1969c). *Copeia* **1969**, 225–234.
Estes, R. (1969d). *Breviora Mus. Comp. Zool.* **328**, 1–7.
Estes, R. (1970). *Bull. Mus. Comp. Zool.* **139**, 293–340.

Fankhauser, G. (1932). *Anat. Rec.* **54**, 73.
Fankhauser, G. (1938). *Proc. Amer. Phil. Soc.* **79**, 715–739.
Fankhauser, G. (1939). *J. Hered.* **30**, 379–388.
Fankhauser, G. and Humphrey, R. R. (1942). *Biol. Bull. Woods Hole* **83**, 367–374.
Fischberg, M. (1944). *Revue Suisse Zool.* **51**, 430–436.
Fischberg, M. (1947). *Arch. Julius Klaus-Stift. Vererbforsch.* **22**, 331–336.
Galgano, M. (1933). *Arch. Ital. Anat. Embr.* **32**, 171–200.
Galgano, M. (1936). *Arch. Ital. Anat. Embr.* **36**, 1–51.
Galgano, M. (1952). *Arch. Zool. Ital.* **37**, 193–230.
Galgano, M. and Morescalchi, A. (1966). *Rend. Acc. Sci. Fis. Mat. (Naples)* **33**, 79–87.
Gall, J. G. (1954). *J. Morph.* **94**, 283–353.
Gall, J. G. (1969). *Genet. Suppl.* **61**, 121–132.
Gallien, L., Labrousse, M., Picheral, B. and Lacroix, J. C. (1965). *Revue Suisse Zool.* **72**, 59–85.
Goin, C. J. and Goin, O. B. (1962). "Introduction to Herpetology." W. H. Freeman and Co., San Francisco and London.
Goin, C. J. and Goin, O. B. (1968). *Am. Midl. Nat.* **80**, 289–298.
Goin, O. B., Goin, C. J. and Bachmann, K. (1968). *Copeia* **1968**, 532–540.
Gorham, S. W. (1966). *In* "Das Tierreich" Lfg. **85**. (W. de Gruyter and Co. eds.), Berlin.
Griffin, C. S., Scott, D. and Papworth, D. G. (1970). *Chromosoma, Berlin* **30**, 228–249.
Griffiths, I. (1959). *Proc. Zool. Soc. (Lond)* **32**, 457–487.
Griffiths, I. (1963). *Biol. Rev.* **38**, 241–292.
Guillemin, C. (1967). *Chromosoma* **21**, 189–197.
Günther, R. (1970). *Biol. Zbl.* **89**, 327–342.
Hauschka, T. S. and Brunst, V. V. (1964). *Hereditas* **52**, 345–356.
Hayes, E. S. (1933). Cited in Bogart, 1970.
Hecht, M. (1957). Proc. Zool. Soc. Calcutta, Mookerijee Mem. Vol., 283–292.
Hecht, M. (1962). *Syst. Zool.* **11**, 39–44.
Hecht, M. (1963). *Syst. Zool.* **12**, 20–35.
Hecht, M. (1969). *Ann. N.Y. Acad. Sci.* **167**, 74–79.
Heitz, E. (1934). *Biol. Zbl.* **54**, 588–609.
Hennen, S. (1964). *J. Hered.* **55**, 124–128.
Hennig, W., Hennig, I. and Stein, H. (1970). *Chromosoma, Berlin* **32**, 31–63.
Holmgren, N. (1952). *Acta Zool. (Stockholm)* **33**, 1–115.
Huene, F. von (1956). "Palaentologie und Phylogenie der Niederen Tetrapoden." G. Fischer, Jena.
Hughes, N. (1962). *Texas J. Sci.* **14**, 255–258.
Humphrey, D. G. (1958). *Science N.Y.* **128**, 304.
Inger, R. F. (1967). *Evolution* **21**, 369–384.
Iriki, S. (1932). *Scient. Rep. Tokyo Bunrika Daigaku B* **1**, 91–103.
Jarvik, E. (1942). *Zool. Bidr. Upps.* **21**, 235–675.
Jaylet, A. (1964). *C. R. Hebd. Séanc. Acad. Sci. Paris* **258**, 5523–5526.
Jaylet, A. (1966). *Chromosoma, Berlin* **18**, 79–87.

John, B. and Hewitt, G. M. (1968). *Chromosoma, Berlin* **25**, 40–72.
John, B. and Lewis, K. R. (1965). *Protoplasmatol.* **6 F1**, 1–335.
John, B. and Lewis, K. R. (1968). *Protoplasmatol.* **6 A**, 1–206.
Jones, K. W. and Robertson, F. R. (1970). *Chromosoma, Berlin* **31**, 331–345.
Kawamura, T. (1951). *J. Sci. Hiroshima Univ. B* **12**, 1–10.
Kawamura, T. (1952). *J. Sci. Hiroshima Univ. B* **13**, 139–149.
Kawamura, T. and Utsunomiya, Y. (1957). *J. Sci. Hiroshima Univ. B* **17**, 1–12.
Kessous, A., Caussinus, H., Jaylet, A. and Beetschen, C. (1968). *Chromosoma, Berlin* **23**, 324–332.
Kezer, J. (1962). *Biol. Věst.* **10**, 45–48 (Reprinted from Author).
Kezer, J. (1964). *In* "The Mechanics of Inheritance" (F. W. Stahl, ed.) pp. 101–114. Prentice-Hall Inc., New Jersey.
Kezer, J. and Macgregor, H. C. (1971). *Chromosoma, Berlin* **33**, 146–166.
Kezer, J., Seto, T. and Pomerat, C. M. (1965). *Am. Nat.* **99**, 153–158.
Kluge, A. G. and Farris, J. S. (1969). *Syst. Zool.* **18**, 1–32.
Kobayashi, M. (1962). *J. Sci. Hiroshima Univ. B* **20**, 157–180.
Lacroix, J. C. (1968). *Ann. Embr. Morph.* **1**, 179–202.
Lacroix, J. C. (1970a). *Ann. Embr. Morph.* **3**, 205–248.
Lacroix, J. C. (1970b). *C. R. Hebd. Séanc Acad. Sci. Paris* **271**, 102–104.
Lamotte, M. (1963). *Bull. I.F.A.N.* **25**, 944–952.
Lantz, L. A. and Callan, H. G. (1954). *J. Genet.* **52**, 165–185.
Laurent, R. (1942). *Bull. Mus. R. Hist. Nat. Belg.* **18**, 1–20.
Laurent, R. (1947). *Bull. Mus. R. Hist. Nat. Belg.* **23**, 1–4.
Laurent, R. (1951). *Rev. Zool. Bot. Afr.* **45**, 116–122.
Laurent, R. (1967). *Acta Zool. Lilloana* **22**, 207–102.
Levan, A., Fredga, K. and Sandberg, A. A. (1964). *Hereditas* **52**, 201–220.
Liem, S. S. (1970). *Fieldiana (Zool)* **57**, 1–145.
Macgregor, H. C. (1963). *Q. Jl. Microsc. Sci.* **106**, 215–228.
Macgregor, H. C. and Kezer, J. (1970). *Chromosoma* **29**, 189–206.
Macgregor, H. C. and Kezer, J. (1971). *Chromosoma* **33**, 167–182.
Makino, S. (1932). *J. Fac. Sci. Hokkaido Univ.* **2**, 97–108.
Makino, S. (1934). *Trans. Sapporo. Nat. Hist. Soc.* **13**, 351–354.
Makino, S. (1935). *J. Morph.* **58**, 573–583.
Makino, S. (1951). "An Atlas of the Chromosome Numbers in Animals", 2nd ed. Iowa State Coll. Press, Ames.
Makino, S. (1962). *In* "Growth", pp. 1–7. Feder. Amer. Soc. Exper. Biol., Washington.
Mancino, G. (1965). *Rend. Accad. Naz. Lincei* **39**, 540–548.
Mancino, G. and Barsacchi, G. (1965). *Caryologia* **18**, 637–665.
Mancino, G. and Barsacchi, G. (1966a). *Boll. Zool.* **33**, 252–267.
Mancino, G. and Barsacchi, G. (1966b). *Rend. Accad. Naz. Lincei* **41**, 396–401.
Mancino, G. and Barsacchi, G. (1966c). *Riv. Biol. (Perugia)* **59**, 311–351.
Mancino, G. and Barsacchi, G. (1969). *Ann. Embr. Morph.* **2**, 355–377.
Mancino, G. and Poggi, M. (1963). *Rend. Accad. Naz. Lincei* **35**, 591–597.
Mancino, G., Barsacchi, G. and Nardi, I. (1969). *Chromosoma, Berlin* **26**, 365–387.
Mancino, G., Nardi, I. and Barsacchi, G. (1970). *Cytogenetics* **9**, 260–271.

Manna, G. K. and Bhunya, S. P. (1966). *Caryologia* **19**, 403–411.
Matthey, R. (1949). "Les Chromosomes des Vertébrés." F. Rouge, Lausanne.
Maxson, L. R. and Jameson, D. L. (1968). *Copeia* **1968**, 704–707.
McCarthy, J. B. (1969). *In* "Handbook of Molecular Cytology" (A. Lima de Faria, ed.) pp. 3–20. North Holland Publ. Co., Amsterdam and London.
Mexzoely, C. A. M. (1966). Abstr. 46th Ann. Meeting Amer. Soc. Icht. Herpetol., Miami, Florida, June 1966.
Mikamo, K. and Witschi, E. (1966). *Cytogenetics* **5**, 1–19.
Miller, L. and Brown, D. L. (1969). *Chromosoma, Berlin* **28**, 430–444.
Mirsky, A. E. and Ris, H. (1951). *J. Gen. Physiol.* **34**, 451–462.
Monath, T. (1965). *J. Morph.* **116**, 149–170.
Morescalchi, A. (1962). *Boll. Zool.* **29**, 601–609.
Morescalchi, A. (1963a). *Rend. Accad. Naz. Sci. Fis. Mat. (Naples)* **30**, 236–240.
Morescalchi, A. (1963b). *Rend. Accad. Naz. Sci. Fis. Nat. Mat. (Naples)* **30**, 310–314.
Morescalchi, A. (1963c). *Rend. Accad. Naz. Lincei* **34**, 696–699.
Morescalchi, A. (1964a). *Rend. Accad. Naz. Sci. Fis. Mat. (Naples)* **31**, 238–245.
Morescalchi, A. (1964b). *Rend. Accad. Naz. Sci. Fis. Mat. (Naples)* **31**, 326–332.
Morescalchi, A. (1964c). *Caryologia* **17**, 327–345.
Morescalchi, A. (1965a). *Caryologia* **18**, 193–206.
Morescalchi, A. (1965b). *Boll. Zool.* **32**, 207–218.
Morescalchi, A. (1966). *Riv. Biol. (Perugia)* **19**, 3–40.
Morescalchi, A. (1967a). *Atti. Soc. Pelorit. Sci. Fis. Mat. Nat.* **13**, 23–30.
Morescalchi, A. (1967b). *Caryologia* **20**, 65–85.
Morescalchi, A. (1967c). *Experientia* **23**, 1071–1072.
Morescalchi, A. (1968a). *Experientia* **24**, 81–82.
Morescalchi, A. (1968b). *Caryologia* **21**, 37–46.
Morescalchi, A. (1968c). *Experientia* **24**, 280–283.
Morescalchi, A. (1968d). *Experientia* **24**, 964–966.
Morescalchi, A. (1968e). Act. 4th Congr. Latinoamer. Zool., Caracas, November 1968 (published 1973).
Morescalchi, A. (1970). *Boll. Zool.* **37**, 1–28.
Morescalchi, A. and Filosa, S. (1965). *Atti. Soc. Pelorit Sci. Fis. Mat. Nat.* **11**, 211–219.
Morescalchi, A. and Gargiulo, G. (1967). *Rend. Accad. Sci. Fis. Mat. (Naples)* **34**, 3–9.
Morescalchi, A. and Gargiulo, G. (1968). *Rend. Accad. Sci. Fis. Mat. (Naples)* **35**, 117–120.
Morescalchi, A., Gargiulo, G. and Olmo, E. (1968). *Boll. Zool.* **35**, 333–334.
Morescalchi, A., Gargiulo, G. and Olmo, E. (1970a). *J. Herpetol.* **4**, 77–79.
Morescalchi, A., Olmo, E. and Gargiulo, G. (1970b). *Boll. Zool.* **37**, 504–505.
Natarajan, R. (1953). Proc. 14th Ind. Sci. Congr. Lucknow, 180–181.
Natarajan, R. (1958a). *J. Zool. Soc. (India)* **9**, 114–120.
Natarajan, R. (1958b). *J. Zool. Soc. (India)* **9**, 16–21.
Natarajan, A. T. and Ahnström, G. (1969). *Chromosoma, Berlin* **28**, 48–61.
Nevo, E. (1968). *Bull. Mus. Comp. Zool.* **136**, 255–318.

Nevo, E. and Estes, R. (1969). *Copeia* **1969**, 540–547.
Noble, G. K. (1931). "The Biology of the Amphibia". McGraw-Hill Book Co., New York.
Nur, U. and Nevo, E. (1969). *Caryologia* **22**, 97–102.
Ohno, S. (1967). "Sex Chromosomes and Sex-linked Genes." Monogr. Endocr., Springer Verlag, Berlin, Heidelberg and New York.
Ohno, S., Wolf, U. and Atkin, N. B. (1968). *Hereditas* **59**, 169–187.
Olmo, E., Gargiulo, G. and Morescalchi, A. (1970). *Boll. Zool.* **37**, 513–514.
Orton, G. L. (1953). *Syst. Zool.* **2**, 63–75.
Orton, G. L. (1957). *Syst. Zool.* **6**, 79–86.
Pardue, M. L. and Gall, J. G. (1970). *Science, N.Y.* **168**, 1356–1358.
Parker, H. W. (1934). "A Monograph of the Frogs of the Family Microhylidae." British Mus. (Nat. Hist.), London.
Parker, H. W. (1940). *Novit. Zool.* **42**, 1–106.
Panchen, A. L. (1967). *In* "The Fossil Record", pp. 685–694. Geological Soc., London.
Parsons, T. and Williams, E. E. (1963). *Q. Rev. Biol.* **38**, 26–53.
Poynton, J. C. (1964). *Ann. Natal Mus.* **17**, 1–334.
Puteyevsky, E. and Zohary, D. (1970). *Chromosoma, Berlin* **32**, 135–141.
Rabello, M. N. (1970). *Caryologia* **23**, 45–59.
Rafferty, K. A. Jr. (1969). *In* "Biology of Amphibian Tumours" (M. Minzell, ed.) pp. 52–81. Springer Verlag, Berlin, Heidelberg and New York.
Regal, P. J. (1966). *Evolution* **20**, 392–407.
Reig, O. A. (1957). *Acta Geol. Lilloana* **1**, 231–297.
Reig, O. A. (1958). *Physis* **21**, 109–118.
Reig, O. A. (1961). *Act. Trans. Prim. Congr. Latam. Zool.* **1**, 271–278.
Reig, O. A. (1964). *Ameghiniana* **3**, 191–211.
Reig, O. A. and Limeses, C. E. (1963). *Physis* **24**, 113–128.
Ris, H. and Kubai, D. F. (1970). *Ann. Rev. Genet.* **4**, 263–294.
Robinson, E. S. and Stephenson, E. M. (1967). *Cytologia* **32**, 200–207.
Romer, A. S. (1966). "Vertebrate Paleontology." 3rd ed., Univ. Chicago Press.
Saez, F. A. and Brum, N. (1960). *Nature (Lond)* **285**, 945.
Saez, F. A. and Brum-Zorrilla, N. (1966). *Caryologia* **19**, 55–63.
Salthe, S. N. (1967). *Copeia* **1967**, 100–117.
Salthe, S. N. and Kaplan, N. O. (1966). *Evolution* **20**, 603–616.
Sanders, O. and Cross, J. C. (1964). *Herpetologica* **19**, 248–255.
Sato, I. (1934). *Zool. Mag.* **46**, 98–104.
Sato, I. (1936a). *J. Sci. Hiroshima Univ. B* **4**, 143–154.
Sato, I. (1936b). *Zool. Mag.* **48**, 958–960.
Sato, I. (1939). *J. Sci. Hiroshima Univ. B* **6**, 53–70.
Savage, J. M. (1968). *Copeia* **1968**, 745–776.
Schaeffer, B. (1949). *Bull. Am. Mus. Nat. Hist.* **93**, 41–68.
Scheel, J. J. (1970). *Rev. Zool. Bot. Afr.* **81**, 225–336.
Schmalhausen, I. I. (1968). "The Origin of Terrestrial Vertebrates." Academic Press, New York and London.
Selman, G. G. (1958). *J. Embr. Exp. Morph.* **6**, 634–637.

Seshachar, B. L. (1937). *Z. Zell. Forsch. Mikrosk. Anat.* **27**, 133–158.
Seshachar, B. L. (1938). *Nature (Lond)* **142**, 757.
Seshachar, B. L. (1939). *Cellule* **48**, 96.
Seshachar, B. L. (1941). *Curr. Sci.* **10**, 281–284.
Seto, T. (1964). *J. Fac. Sci. Hokkaido Univ. Zool.* **15**, 366–373.
Seto, T. (1965). *Cytologia* **30**, 437–446.
Seto, T., Pomerat, C. M. and Kezer, J. (1964). *Am. Nat.* **98**, 71–78.
Sharma, G. P., Parshad, R. and Sethi, A. (1965). Proc. Combd. 51st and 52nd Ind. Sci, Congr., Pt. III (Abstract), 439.
Singh, L., Sharma, T. and Ray-Chauduri, S. P. (1970). *Mamm. Chrom. Newslett.* **11**, 91–94.
Slabbert, G. K. and Maree, W. A. (1945). *Annale Univ. Stellenbosch* **23**, 91–97.
Špinar, Z. V. (1967). *Sb. Ústřed. Úst. Geol.* **19**, 457–488.
Stebbins, G. L. (1966). *Science, N.Y.* **152**, 1463–1469.
Stephenson, E. M. and Stephenson, N. G. (1970). *Chromosoma, Berlin* **30**, 38–50.
Stephenson, N. G. (1955). *Proc. Zool. Soc. (Lond)* **124**, 785–795.
Swanson, C. P. (1957). "Cytology and Cytogenetics." Prentice-Hall Inc., New Jersey.
Swanson, C. P., Merz, T. and Young, W. J. (1967). "Cytogenetics." Prentice-Hall Inc., New Jersey.
Szarski, H. (1962). *Q. Rev. Biol.* **37**, 189–241.
Szarski, H. (1970). *Nature (Lond)* **226**, 651–652.
Taylor, E. H. (1941). *Kans. Univ. Sci. Bull.* **27**, 61–69.
Taylor, E. H. (1942). *Kans. Univ. Sci. Bull.* **28**, 199–235.
Taylor, E. H. (1968). "The Caecilians of the World." Univ. of Kansas Press, Lawrence.
Taylor, E. H. (1969). *Kans. Univ. Sci. Bull.* **48**, 585–687.
Thomson, K. (1964). *Sci. Progr.* **57**, 451–459.
Thorn, R. (1968). "Les Salamandres d'Europe d'Asie et d'Afrique du Nord." Lechevalier, Paris.
Tihen, J. A. (1958). *Bull. Fla St. Mus. Biol. Sci.* **3**, 1–50.
Tihen, J. A. (1960). *Copeia* **1960**, 225–233.
Tihen, J. A. (1965). *Am. Zool.* **5**, 308–318.
Ting (1939). Cited in Rabello, 1970.
Todd, N. B. (1967). *Mamm. Chrom. Newslett.* **8**, 268–269.
Todd, N. B. (1970). *J. Theor. Biol.* **26**, 445–480.
Tyler, M. J. (1968). *Zool. Verh. (Leiden)* **96**, 1–203.
Tyler, M. J. (1971). *Univ. Kansas Publ. Mus. Nat. Hist.* **19**, 319–360.
Ullerich, F. (1966). *Chromosoma* **18**, 316–342.
Ullerich, F. (1967). *Chromosoma* **21**, 345–368.
Ullerich, F. (1970). *Chromosoma* **30**, 1–37.
Uzzell, T. M. (1963). *Science N.Y.* **139**, 113–115.
Volpe, E. P., Duplantier, D. and Earley, E. M. (1970). *Cytogenetics* **9**, 161–172.
Volpe, E. P. and Gebhardt, B. M. (1968). *Copeia* **1968**, 570–576.
Wake, D. B. (1966). *Mem. Sth Calif. Acad. Sci.* **4**, 1–111.
Wake, D. B. and Özeti, N. (1969). *Copeia* **1969**, 126–137.

Wake, M. H. (1968). *J. Morph.* **126**, 291–346.
Wahlert, G. von (1956). *Zool. Anz. Suppl.* **19**, 274–279.
Wahlert, G. von (1957). *Zool. Jber. Syst.* **85**, 253–282.
Walker, C. F. (1938). *Occ. Pap. Mus. Zool. Univ. Mich.* **372**, 1–11.
Walker, P. M. B., Flamm, W. G. and McLaren, A. (1969). "Handbook of Molecular Cytology" (A. Lima de Faria, ed.) pp. 52–55. North Holland Publ. Co., Amsterdam and London.
Wasserman, A. O. (1970). *Science, N.Y.* **167**, 385–386.
Wasserman, A. O. and Bogart, J. P. (1968). *Copeia* **1968**, 303–306.
Watson, I. D. and Callan, H. G. (1963). *Q. Jl. Microsc. Sci.* **104**, 281–295.
Weiler, C. and Ohno, S. (1962). *Cytogenetics* **1**, 217–232.
White, M. J. D. (1954). "Animal Cytology and Evolution." 2nd ed., Cambridge University Press, London and New York.
White, M. J. D. (1968). *Science, N.Y.* **159**, 1065–1070.
White, M. J. D. (1969). *Ann. Rev. Genet.* **3**, 75–98.
Wickbom, T. (1945). *Hereditas* **31**, 241–346.
Wickbom, T. (1949). *Hereditas* **35**, 33–48.
Wickbom, T. (1950a). *Hereditas* **36**, 363–368.
Wickbom, T. (1950b). *Hereditas* **36**, 406–418.
Witschi, E. (1933). *Cytologia* **4**, 174–181.
Witschi, E., Kodani, M. and Mikamo, K. (1958). *Anat. Rec.* **131**, 610.
Zweifel, R. G. (1956). *Am. Mus. Novit.* **1762**, 1–45.

ADDENDUM

Many interesting contributions on amphibian cytotaxonomy have been produced since this paper was submitted for publication.

Barrio *et al.* (1971a, 1972) have studied the chromosomes of two South American Apoda, *Siphonops paulensis* (Caeciliidae), with $2n = 24$, and *Chtonerpteton indistinctum* (Typhlonectidae), with $2n = 20$.

In the author's laboratory, new data on the karyology of some Urodela have been collected: *Megalobatrachus japonicus* (Cryptobranchidae) has $2n = 60$ (of which 10 pairs are *t*-elements and 14 pairs microchromosomes); new studies on some Ambystomatidae (*Ambystoma opacum* and *A. talpoideum*) and Plethodontidae (*Aneides aeneus*, *Desmognathus fuscus*, *Ensatina erschscholtzi*, *Eurycea longicauda* and *Pseudotriton ruber*), all with $2n = 28$ and large bi-armed chromosomes, corroborate the hypotheses on a close phyletic affinity between the two families; then, we found 52 chromosomes in the somatic cells of four juvenile specimens of *Siren lacertina*, which are all bi-armed except 4 *t*-elements; the chromosomes of *Siren* may be arranged in 13 groups of four homologues each ($4n = 52$). The last results, if confirmed, provide some new facts on the problem of the phyletic relationships of Sirenidae, given that *Siren lacertina* (probably polyploid) has the same basic chromosome set ($n = 13$) of some Ambystomatoidea and approaches that of most of the Salamandridae ($n = 12$).

About the Anura, new contributions have been supported by many authors especially on American and African species. Leon (1970) has determined the chromosome numbers of 21 Costa Rican species of seven families; Cole (1971) has described the karyotype of *Rhinophrynus dorsalis* ($2n = 22$) and of *Phyllomedusa dacnicolor* ($2n = 26$); Karyological data on 14 species of Patagonian Leptodactylidae (here included in Telmatobiini and Batrachylini), all with $2n = 26$, have been provided by Barrio (1971 *et al.*): of special interest seems the report that the systematically problematical *Rhinoderma darwinii* has the same chromosome set of these Leptodactylidae ($2n = 26$).

An excellent systematic revision of Leptodactylidae, with new theories on their *intra*- and interfamilial relationships, has been written by Lynch (1971); some of the cytotaxonomic conclusions gained in the present paper seem to be in agreement with Lynch's hypotheses.

Scheel (1971) has found $2n = 14$ in the African Ranid *Arthroleptis poecilonotus*; numerically reduced karyotypes have been found by J. Bogart in African Ranidae of the subfamilies Petropedetinae and Arthroleptinae (personal communication); Olmo (1972) has compared the karyotype of *Hemisus marmoratus* with that of other (presumably) related species.

Wasserman (1970) has studied the chromosomes of some Pelobatidae of the *Spea* and *Scaphiopus* subgenera; Bogart and Wasserman (1972) have provided new data on polyploid Anura and on the evolutionary mechanisms of polyploidization in the order.

References

Barrio, A., Saez, F. A. and Rinaldi de Chieri, P. (1971a). *Caryologia* **24**, 435–445.

Barrio, A. and Rinaldi de Chieri, P. (1971b). *Physis* **30**, 667–685.

Barrio, A. and Rinaldi de Chieri, P. (1972). *Physis* **31**, 273–276.

Bogart, J. P. (1970a). *Act. IV Congr. Latin. Zool.* **1**, 65–78.

Bogart, J. P. (1970b). *Act. IV Congr. Latin. Zool.* **1**, 79–86.

Bogart, J. P. (1972). *In* "Evolution in the genus Bufo" (Blair, W. F. ed.), University of Texas Press, Austin.

Bogart, J. P. and Wasserman, A. O. (1972). *Cytogenetics* **11**, 7–24.

Cole, C. J. (1971). *Herpstol. Rev.* **3**, 37–38 and 55–56.

Leon, P. (1970). *Rev. Biol. Trop.* **17**, 119–124.

Lynch, J. D. (1971). Univ. Kansas Mus. Nat. Hist., Misc. Publ. no 53.

Olmo, E. (1972). *Caryologia* **25**, 33–42.

Scheel, J. J. (1971). *Hereditas* **67**, 287–290.

Špinar, Z. V. (1972). "Tertiary Frogs from Central Europe". Dr. W. Junk, N.V. Publ., The Hague.

Wasserman, A. O. (1970). *Southwest. Natural.* **15**, 235–248.

12

The Chromosomes of the Reptilia, a Cytotaxonomic Interpretation

GEORGE C. GORMAN

Museum of Vertebrate Zoology, University of California, Berkeley, California, U.S.A.

I. INTRODUCTION

A. *The Living Reptiles*

Living reptiles comprise a composite class of vertebrates grouped together as much by what they lack as by what they have in common. They may be defined as non-homiothermic amniotes. Modern forms can be conveniently divided into three orders and a fourth superorder which have descended from three ancient lineages. Turtles and tortoises comprise the order Chelonia, and are divided into two suborders that have been distinct since the Jurassic, a period of at least 200 million years. The order itself last shared common ancestry with other modern reptiles no more recently than the Pennsylvanian. The order Crocodilia is derived from an Archosaurian stock, the same ancestral stock that gave rise to the birds. In fact, many zoologists agree that crocodilians are more closely related to birds than they are to any other surviving reptilian stock.

The other extant orders are derived from a Lepidosaurian stock and are thus relatively closely related to one another. The most ancient order, Rhyncocephalia, consists of but a single relict species, *Sphenodon punctatum*, the tuatara, which lives on New Zealand islets.

The remaining reptiles, snakes and lizards, comprise the superorder Squamata. Recent workers (e.g. Underwood, 1967; Miller, 1966, 1968) have demonstrated that snake and lizard lineages must have had an ancient divergence time. Thus, some workers consider snakes (Ophidia) and lizards (Sauria) as distinct orders. Gans (1960) considered the amphisbaenians to be as distinct from lizards as snakes are, and he raised this taxon to ordinal rank. The limits of higher taxa are not universally agreed upon (many consider Squamata an order), so for convenience of presentation I follow Underwood (1957) and consider the amphisbaenians as a lizard family, and I accord ordinal rank to snakes and lizards.

For a basic paleontological history and classification of extant reptiles the reader is referred to Romer (1966).

B. *Reptilian Cytogenetics: Historical Overview*

Reptilian chromosome studies have a history of about 50 years (Dalcq, 1920; Painter, 1921). In the 1930s and 1940s growth of knowledge was gradual. A major credit must be given to workers in Japan (e.g. Nakamura) and to Prof. R. Matthey of Switzerland for organizing the data in a systematic manner. This early growth period culminated with a major review published by Matthey in 1949. Most chromosome studies to that date were made by examination of histological sections, making the analysis of chromosome morphology relatively difficult. Through 1949 the available chromosome data permitted only the broadest comparisons. Representatives of many families had been studied, but there were no specific attempts to analyse chromosome variation within a genus, nor to examine representatives of many genera in large families.

This began to change in the 1950s, again with credit to Professor Matthey, who, in a series of papers, analysed the karyotypes of some two dozen species of chamaeleons, and discussed the systematic and zoogeographic implications of the data (e.g. Matthey and van Brink, 1956; Matthey, 1957). The 1950s also saw the publication of a surprisingly simple yet important advance in technique for the study of chromosome morphology (Ford and Hamerton, 1956). This involves the use of colchicine to arrest mitotic divisions at metaphase, and hypotonic pretreatment to swell the cells and spread the chromosomes. From this time on, resolution of chromosome morphology was greatly improved, and different authors could directly compare their results.

In the 1960s reptilian cytotaxonomy moved into an explosive phase. Part of this resulted directly from the work of Moorhead *et al.* (1960) which outlined a simple technique for the culture of peripheral blood cells. The indirect effect of leukocyte culture was perhaps most important for the new spurt of interest. As mammalian chromosome studies proliferated, herpetologists interested specifically in questions about reptilian relationships commenced to examine karyotype systematically. To cite but a few examples, Dr. M. M. Cohen and associates studied the karyotypes of all 21 species of crocodilians, Dr. J. Wright and co-workers have studied about 30 species in the single lizard genus *Cnemidophorus*, and I have been involved in the systematic examination of karyotypes in the lizard genus *Anolis*. We now have chromosome data on more species in this one genus than the total number of lizard species listed in Matthey's 1949 review.

The 1960s was also the time when sex chromosome heteromorphism was definitively established for certain reptilian groups, and when chromosome information was utilized for the analysis of interesting biological problems related to the origin and proliferation of parthenogenetic species. Within the next decade, general surveys of the karyotypes of most important reptile groups will probably be available, and reptilian cytotaxonomy may evolve to a finer level, intraspecific variation and the analysis of population dynamics.

In this Chapter, I have undertaken the review of all chromosome data that are taxonomically relevant. Because the writing of a review paper is laborious and infrequent (this is the first comprehensive tabulation since Matthey (1949) and Makino (1951)) I have attempted to make it as up-to-date as possible by including information still unpublished that has resulted from work in my own laboratory and that has been communicated to me by numerous colleagues.

C. *Terminology*

Chromosomes of many reptile species fall into two distinct size groups, conveniently termed Macrochromosomes and Microchromosomes. I have not attempted formal definition of these categories—but the terminology is used if the break is marked. The abbreviations "M" (for macro) and "m" (for micro) are used throughout.

I also use descriptive abbreviations for the morphology of the chromosomes. The abbreviation "V" is used for a metacentric chromosome—a two-armed element with the centromere central or virtually central; and the abbreviation "I" is used for an acrocentric chromosome, one with a terminal or virtually terminal centromere. The abbreviation "s" represents the prefix sub- and is used in combination, "sV" for submetacentric; and "sI" for subacrocentric. The former is for a chromosome with long but subequal arms, the latter is used when there is a distinct short arm. I emphasize that these descriptions are qualitative, and it is sometimes an arbitrary decision whether a chromosome is termed "V" or "sV"; or "sV" or "sI". A more formal, quantitative, terminology has been proposed by Levan *et al.* (1964). However, it is often difficult to apply this terminology, especially to older works, and I feel that there has not been a great loss of information content with my old-fashioned terminology.

The term *Nombre fondamental* ("N.F.") has been used extensively by Matthey (1949) and other workers to represent the total number of arms in the karyotype. "V" and "sV" have a value of 2, "sI" and "I" have a value of 1, and all microchromosomes, regardless of morphology, have a value of 1. Often, within a given reptilian taxon there is variation in diploid number, but conservativeness of the "N.F." Apparently centric fusion or centric fission of whole arms has been involved in the evolution of the karyotype of this group. Such variation is termed Robertsonian. In this work I have not emphasized the "N.F." but I frequently point out the Robertsonian relationships between taxa.

II. SYSTEMATIC REVIEW

A. *Order Rhyncocephalia*

An accurate description of the karyotype of the tuatara has recently been published by Wylie *et al.* (1968). They found that both sexes have 18 pairs of chromosomes none of which are heteromorphic (Fig. 1). They divided

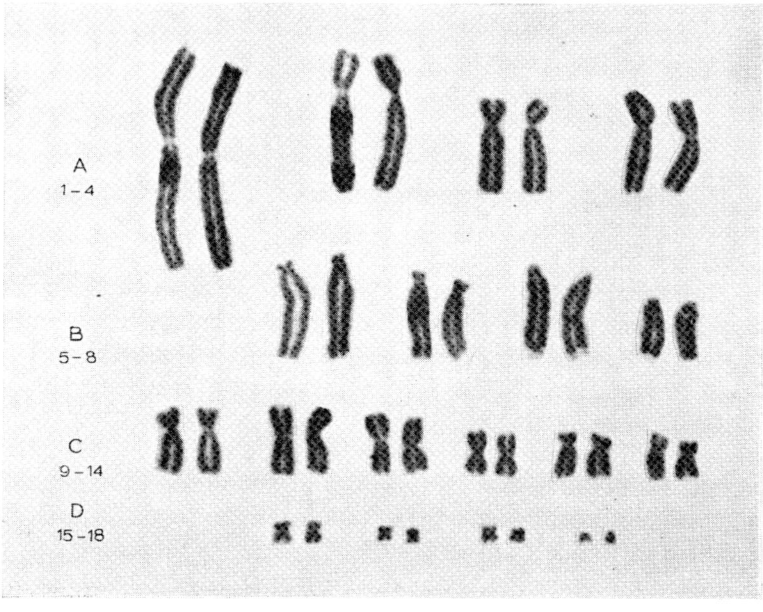

Fig. 1. Karyotype of a female tuatara, *Sphenodon punctatus*. $2n = 36$. Largest chromosomes are about 20 μ long. Reprinted from Wylie *et al.*, 1968 by permission of Dr. A. P. Wylie.

the karyotype into four groups:

(A) Four pairs of large meta- and submetacentric chromosomes:
(B) Four pairs of medium sized acrocentrics;
(C) Six pairs of smaller chromosomes with median to almost subterminal centromeres;
(D) Four pairs of very small chromosomes.

Each chromosome pair is specifically identifiable by size or centromere position. In many squamate reptiles the karyotype divides easily into macro- and microchromosomes. Although the group (D) chromosomes were small, they do not break sharply in size from the smallest of group (C), and Wylie *et al.* do not consider them microchromosomes. As will be reported, a diploid number of 36 is common in snakes and lizards, but the details of morphology make the tuatara karyotype unique.

The karyotype of *Sphenodon* was first reported by Keenan (1932), who also found a diploid number of 36, but techniques then available did not permit so accurate a description of chromosome morphology.

B. *Order Sauria*

The families of lizards are well defined (Camp, 1923), and the boundaries have been stable for the past half-century with but few exceptions. The most important involves the Gekkonidae which Underwood (1954) divided into three families but which Kluge (1967) reunited into a single family with four subfamilies. In this account, we shall review available chromosome data on a family-by-family basis, accepting Kluge's definition of the Gekkonidae, but otherwise following Underwood's (1957) classification (Table I).

(a) *Family Gekkonidae* (Table II)

The geckoes comprise a huge family of world-wide distribution. The cytotaxonomy of the group is still in its infancy, for no worker has attempted a broad comparison of genera, or species within a genus. Although data have been accumulating for the past 40 years (Matthey, 1931, 1933; Nakamura, 1932), the most recent review (Kluge and Eckardt, 1969) showed no data available for one of the subfamilies (Spherodactylinae), and data for only 19 species in 12 genera. Their list is supplemented by unpublished observations on 5 additional species representing genera not previously sampled.

The range of diploid numbers is 32–46. The typical gekkonid karyotype consists primarily of a graded series of acrocentric chromosomes, without a distinct break between macrochromosomes and microchromosomes (Fig.

TABLE I. Classification of Lizards, Order Sauria (adapted from Underwood, 1957).[a]

Order Sauria	Approx. no. genera/species in family	Approx. no. genera/species with karyotype data
Suborder Ascalabota		
Infraorder Gekkota		
Superfamily Gekkonoidea		
Family Gekkonidae	82/650	17/24
Superfamily Pygopodoidea		
Family Pygopodidae	8/15	2/2
Infraorder ?		
Family Xantusiidae	3/15	2/10
Infraorder Iguania		
Superfamily Iguanoidea		
Family Iguanidae	50/600	24/135
Family Agamidae	33/300	7/13
Infraorder Rhiptoglossa		
Family Chamaeleontidae	2/84	2/25
Suborder Autarchoglossa		
Infraorder Scincomorpha		
Superfamily Scincoidea		
Family Scinidae	50/800	7/19
Superfamily Lacertoidea		
Family Gerrhosauridae	10/50	1/1
Family Lacertidae	20/150	7/32
Family Teiidae	40/200	11/44
Infraorder Anguinomorpha		
Superfamily Diploglossa		
Family Anguidae	7/67	4/10
Family Anniellidae	1/2	1/1
Superfamily Platynota		
Family Helodermatidae	1/2	1/1
Family Varanidae	1/30	1/4
Infraorder Amphisbaenia		
Family Amphisbaenidae	23/140	11/19

[a] Families not represented in the chromosome literature are omitted. Underwood considered the Gekkonoidea to be comprised of 3 families. I follow Kluge (1967) in reducing these to a single family.

2). Large metacentric elements are not frequent and the majority of two-armed chromosomes, when they occur, appear to have subterminal centromeres.

The above description is almost sufficient to serve as a cytological definition of the Gekkota since it overlaps with few other families. Confusion can occur with lacertids which typically have 38 acrocentric chromosomes (but

Species	2n	Description	Reference
Eublepharinae			
Coleonyx variegatus ♂	32	Rods graded to dots	Gorman, unpublished; Matthey, 1933
Eublepharis macularius ♂	38	Rods graded to dots	Gorman, unpublished
Diplodactylinae			
Diplodactylus elderi ♂	38	Rods graded to dots	Gorman, unpublished
Hoplodactylus pacificus ♂	38	Rods graded to dots	Gorman, unpublished
Lucasium dameum ♂	38	Rods graded to dots	Gorman, unpublished
Phyllurus milii ♂ (*Gymnodactylus milliusi*)	38	Rods graded to dots	Matthey, 1933
Oedura lesuerii ♂	36	"more than 6 J-shaped pairs, and two of the smallest pairs are V-shaped, the remainder rod shaped"	Werner, 1956
Gekkoninae			
Bunopus tuberculatus ♂ (*Alsophylax tuberculatus*)	42[a]	?	Werner, 1956
Cyrtodactylus kotschyi ♂	42[a]	?	Werner, 1956
Gehyra variegata ♀	63[b]	Rods graded to dots	Makino and Momma, 1949
Gekko gekko ♂♀	38	6 large sM, 4M, 2sI, 26I in graded series	Cohen *et al.*, 1967
Gekko japonicus ♂	38	4M, 34I in graded series	Nakamura, 1932
Hemidactylus bowringii ♂	46	Rods graded to dots	Nakamura, 1932
Hemidactylus brookii ♂	40	4V, 36 rods graded to dots	Bhatnagar, 1962
Hemidactylus flaviviridis ♂	46	Rods graded to dots	Makino and Momma, 1949
Hemidactylus flaviviridis ♂♀	40	8V, 6sV, 6sI, 20I	Singh *et al.*, 1970a; Asana and Mahabale, 1941
Hemidactylus frenatus ♂	46	Rods graded to dots	Makino and Momma, 1949
Hemidactylus garnotti ♀	70[b]	2V, 3sV, 1sI, 64I rods graded to dots	Kluge and Eckardt, 1969
Hemidactylus turcicus ♂	44	Rods graded to dots	Werner, 1956
Lepidodactylus lugubris ♀	44[c]	All acrocentrics	Cuellar and Kluge, in preparation
Ptyodactylus hasselquistii ♂	40	Rods graded to dots	Werner, 1956
Stenodactylus doriae ♂ (*Ceramodactylus doriae*)	38[a]	?	Werner, 1956
Stenodactylus sthenodactylus ♂	38	10sI, 28I rods graded to dots	Werner, 1956
Tarentola mauritanica ♂	42	Rods graded to dots	Matthey, 1931
Tropiocalotes steudneri ♂	38	Rods graded to dots	Werner, 1956

[a] 2n number is based on doubling of first meiotic metaphase. [b] Probably triploid, parthenogenetic. [c] Probably parthenogenetic.

a distinct pair of microchromosomes can usually be defined), and with some teiids which have all acrocentric chromosomes. However, teiids with fully acrocentric karyotypes have diploid numbers in excess of 46 (see below).

Within the family, chromosomes do not appear promising for analysing subfamilial relationships. Representatives of the three subfamilies sampled can be found with 38 chromosomes, all acrocentric. This may be taken as

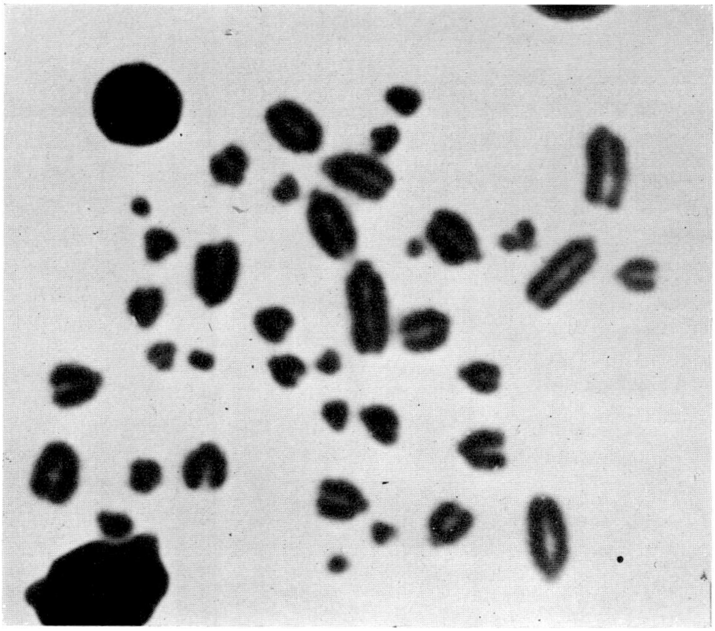

Fig. 2. Karyotype of the eublepharine gekkonid, *Eublepharis macularius* male. $2n = 38$, all chromosomes acrocentric, no break between macro- and microchromosomes.

supporting evidence for Kluge's (1967) uniting of the subfamilies into a single family.

In many lizard groups (see below) there is a distinct conservativeness of the number of arms in the karyotype, and evolution can be interpreted as having been accompanied by centric fusions or fissions. This is not true for geckoes. For example, *Coleonyx* with the lowest reported diploid number (32) has all acrocentrics, whereas several species with diploid numbers of 38 or 40 have several two-armed chromosomes.

With the exception of a few recent works (Cohen *et al.*, 1967; Kluge and Eckardt, 1969; Gorman, unpublished), all data are based on preparations made without colchicine–hypotonic pretreatment. Thus details of the morphological descriptions are questionable. For example, we cannot be certain that the chromosome morphology of *Gekko japonicus*, described by

Nakamura in 1932, really differs from *Gekko gecko* described by Cohen *et al.* (1967).

The only genus for which there are chromosome data for more than two species is *Hemidactylus*. Data have been published for 6 species, and there appears to be enough variation to make this genus of interest to the cytotaxonomist if pursued further. Three species, *H. bowringii*, *H. frenatus* and *H. flaviviridis*, supposedly have $2n = 46$, all chromosomes acrocentric, grading in size from "rods" to "dots" (Nakamura, 1932; Makino and Momma, 1949; Asana and Mahabale, 1941). However, Singh *et al.* (1970a) reported $2n = 40$ for *H. flaviviridis* with 8 metacentrics and 6 submetacentrics in the karyotype. We simply have no way of knowing if we are dealing with intraspecific karyotype variation or misidentification, but suspect the latter. *H. turcicus* reportedly has $2n = 44$, all acrocentrics. That is, it has "lost" or incorporated a pair of small chromosomes into its karyotype (Werner, 1956). Bhatnagar (1962) reported $2n = 40$ for *H. brooki*. The karyotype consisted of 4 metacentrics, and 36 acrocentrics. Parthenogenesis involving triploidy has been documented for *H. garnotii* (Kluge and Eckardt, 1969). They found that $3n = 70$. This is a modified triploid condition, presumably derived from $n = 23$ ancestor or ancestors. Interspecific hybridization has been implicated in the origin of parthenogenetic triploids in some lizards (Lowe and Wright, 1966). For *H. garnotii* there are no obvious "ancestors" because all the $n = 23$ *Hemidactylus* that are known have completely acrocentric karyotypes, whereas *H. garnotii* has several metacentric and submetacentric elements.

Hall (1970) pointed out another probable case of parthenogenetic triploidy in the family. The karyotype of *Gehyra variegata* was described as $2n$ (*sic*) $= 63$ (Makino and Momma, 1949). Only females were obtained for study. It is probable that this form was derived from an $n = 21$ ancestor. O. Cuellar and A. Kluge (personal communication) have found an apparent case of diploid parthenogenesis in *Lepidodactylus lugubris*, $2n = 44$, all acrocentric.

(b) *Family Pygopodidae*

This is a small family of limbless lizards, restricted in distribution to the Australia–New Guinea region. They are closely related to and probably derived from the gekkonids (Underwood, 1957). Chromosomes of only two species have been examined.

Observations on meiotic chromosomes of single male *Delma fraseri* revealed that $n = 19$, all chromosomes acrocentric, no obvious sex chromosome bivalent (Gorman, unpublished). This is a karyotype frequently found in the Gekkonidae.

Lialis burtonis has a more specialized chromosome complement. There is an obvious trivalent in the male meiosis. Male $2n = 33$ (Fig. 3), female $= 34$. This is an excellent example of an X_1X_2Y sex chromosome system (Gorman and Gress, 1970a). The Y is a small metacentric, X_1 and X_2 are acrocentrics. Sex chromosomes have not been reported for other members of the superfamily Gekkonoidea. Except for the sex chromosomes, the

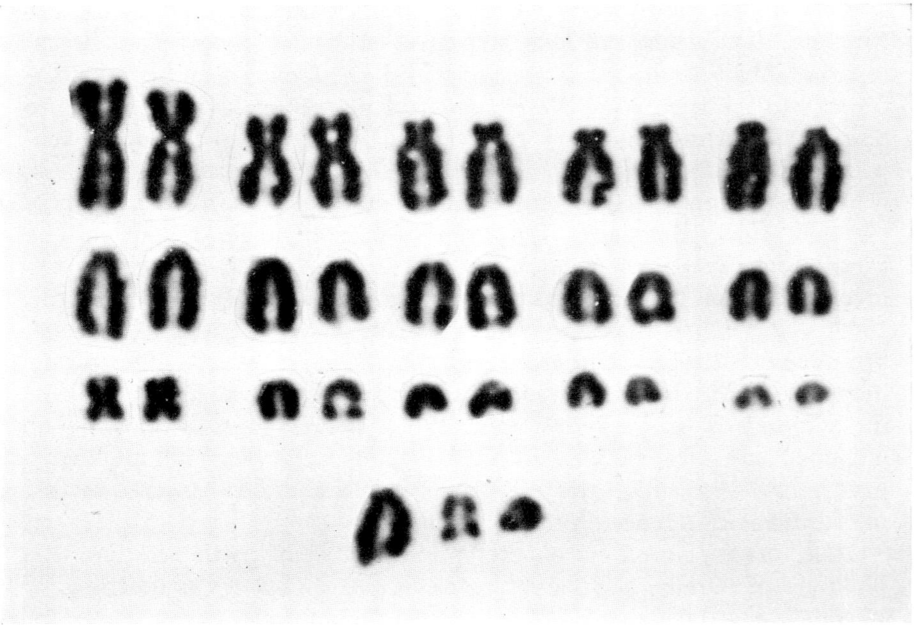

Fig. 3. Karyotype of the pygopodid lizard *Lialis burtonis*, male. $2n = 33$. There are three unpaired sex chromosomes (bottom row), the metacentric Y is between the X_1 and X_2. Reprinted from Gorman and Gress, 1970a with permission.

karyotype of *Lialis* is not unlike that of gekkonids. There is no obvious size break between M and m, the majority of chromosomes are acrocentric, and the diploid number is within the gekkonid range.

(c) *Family Xantusiidae*

This small family of lizards is restricted to the western United States, Mexico, Central America and Cuba. The relationships of the family are not at all clear. Note that Underwood (1957) placed it in an uncertain infraorder (see Table I). The karyotype of *Xantusia henshawi* was described by Matthey (1931). The diploid number is 42 consisting of $6V + 12I$ macrochromosomes, and 24 microchromosomes. The division of the karyotype into 24 macrochromosomal arms and 24 microchromosomal arms is found in lizards of many families. It is not characteristic of the Gekkonidea, to which Xantusiidae is sometimes thought related.

Dr. R. L. Bezy is involved in a cytotaxonomic study of the entire family (see Addendum).

(d) *Family Iguanidae*

Iguanid lizards are distributed throughout the Western Hemisphere, from Canada to Argentina, with numerous species on the West Indian Islands and several on the Galapagos Islands. In addition, there is a monotypic genus endemic to the region of Fiji and Tonga, and two endemic genera on Madagascar—providing a choice puzzle to zoogeographers.

There are more available data on the karyotypes of iguanid lizards than on any comparable reptilian group. For convenience, this section will be subdivided. Formal subfamilies have not been proposed for the Iguanidae, but systematists familiar with the group often refer to "sceloporines", "anolines", "iguanines", etc. (see Etheridge, 1964). There are extensive cytotaxonomic data for sceloporines and anolines, and these shall be treated separately.

(1) Anoline lizards (Table III): The most species-rich genus of all reptiles is *Anolis*, with at least 200 valid species. Certain anatomical specializations have led to the recognition of four minor genera (1–5 species each) that are grouped with *Anolis* as "anolines." The basic classification of *Anolis* was formulated by Etheridge (1960) who divided the large genus into major groups (alpha and beta) and defined series of species within these groups.

The most commonly observed karyotype among the alpha anoles is $2n = 36$, with 12 *ca.* metacentrics, 24 microchromosomes and no obvious sex chromosome heteromorphism (Fig. 4a). In several species-series the only departure from this karyotype is the apparent loss of microchromosomes (Gorman and Atkins, 1967, 1968b; Gorman *et al.*, 1967; Williams *et al.*, 1970). The karyotype of 12V + 24m is seen in many other iguanid lizards, particularly those that appear osteologically primitive, and it is a karyotype that is found in numerous lizard families. Elsewhere (Gorman *et al.*, 1967, 1969a) we have argued that for many groups this is most likely the primitive condition, and karyotype evolution has been *away* from this configuration. Among the species groups characterized by 12V + 24m, only a single species has been found that has an obviously derived karyotype. *Anolis monticola* has 24I + 24m (Preston Webster, personal communication). This

TABLE III. Chromosomes of anoline lizards

Series	Karyotype description	Species (reference)
α *Anolis*		
latifrons	$2n = 36, 12V + 24m$	*bonairensis* ♂ (Gorman and Atkins, 1967)
		luciae ♂♀ (Gorman and Atkins, 1967)
		trinitatis ♂♀ (Gorman and Atkins, 1967)
		richardi ♂♀ (Gorman and Atkins, 1967)
	$2n = 34, 12V + 22m$	*aeneus* ♂♀ (Gorman and Atkins, 1967)
		roquet ♂♀ (Gorman and Atkins, 1967)
	$2n = 32, 12V + 20m$	*jacare* ♂♀ (Williams *et al.*, 1970)
equestris	$2n = 36, 12V + 24m$	*equestris* ♂ (Gorman *et al.*, 1967)
coelestinus	$2n = 36, 12V + 24m$	*coelestinus* ♂ (Gorman *et al.*, 1967)
carolinensis	$2n = 36, 12V + 24m$	*carolinensis* ♂ (Matthey, 1931)
		allisoni ♂ (Gorman *et al.*, 1967)
		argenteolus ♂ (Gorman and Atkins, 1968b)
		bartschi ♂ (Gorman and Atkins, 1968b)
		chlorocyanus ♂ (Gorman *et al.*, 1967)
		hendersoni ♂ (Gorman *et al.*, 1967)
		lucius ♂ (Gorman and Atkins, 1968b)
		maynardi ♂ (Gorman and Atkins, 1968b)
		olssoni ♂ (Gorman *et al.*, 1967)
		porcatus ♂ (Gorman and Atkins, 1968b)
		semilineatus ♂ (Gorman *et al.*, 1967)
	$2n = 34, 12V + 22m$	*vermiculatus* ♂ (Gorman and Atkins, 1968b)
	$2n = 48, 24I + 24m$	*monticola* ♂♀ (T. P. Webster, unpublished)
ricordi	$2n = 36, 12V + 24m$	*ricordi* ♂ (Gorman *et al.*, 1967)
cuvieri	$2n = 36, 12V + 24m$	*cuvieri* ♀ (Gorman and Atkins, 1969)
incertae sedis	$2n = 36, 12V + 24m$	*occultus* ♂ (Gorman and Atkins, 1969)
bimaculatus/[a] *acutus*	$2n$ (♂) $= 31, 12V + 2$ smaller $V + 14m + X_1X_2Y$	*acutus* (Gorman and Atkins, 1969)

	$2n$ (♂) = 33, 10V + 4 smaller V + 16m + X_1X_2Y (♀) = 34, same but $X_1X_1X_2X_2$	*distichus* (Gorman and Atkins, 1969)
	$2n$ (♂) = 29, 12V + 2 smaller V + 12m + X_1X_2Y	*stratulus* (Gorman and Atkins, 1969)
	$2n$ ♂♀ = 26, 12V + 2 smaller V + 10m + XY/XX	*evermanni* (Gorman and Atkins, 1969)
bimaculatus/[a] *wattsi*	$2n$ ♂ = 29, 8V + 6sI + 4 smaller V + 8m, + X_1X_2Y	*wattsi* (Gorman and Atkins, 1969)
bimaculatus/[a] *bimaculatus*	$2n$ (♂) = 29, 8V + 6sI + 4 smaller (♀) = 30, V + 8m + X_1X_2Y or $X_1X_1X_2X_2$	*bimaculatus, leachi, gingivinus, ferreus, marmoratus, lividus, sabanus* (Gorman and Atkins, 1969)
	$2n$ (♂) = 31 (♀) = 32 6V + 6sI + 4 smaller V + 4I + 8m + X_1X_2Y or $X_1X_1X_2X_2$	*oculatus* (Gorman and Atkins, 1969)
cristatellus[a]	$2n$ = 36, 12V + 24m, no sex chromosome heteromorphism	*cybotes* (Gorman and Atkins, 1966)
	$2n$ (♂) = 29, 12V + 4 smaller V + 10m + X_1X_2Y	*cooki* (Gorman *et al.*, 1968) *krugi* (Gorman and Atkins, 1969) *pulchellus* (Gorman and Atkins, 1969) *poncensis* (Gorman and Atkins, 1969) *gundlachi* (Gorman and Atkins, 1969)
	$2n$ (♂) = 27, 12V + 4 smaller V + 8m + X_1X_2Y	*cristatellus* (Gorman *et al.*, 1968) *scriptus* (Gorman *et al.*, 1968)
β *Anolis*		
grahami	$2n$ = 30, 14V + 16m, no obvious heteromorphism	*lineatopus* ♂♀ (Gorman, 1969b) *garmani* ♀ (Gorman and Atkins, 1968b)
	$2n$ = 30, 14V + 16m, possible X–Y heteromorphism, pair 7	*opalinus* ♂ (Gorman, 1969b)
	$2n$ = 30, 14V + 16m ♀, 13V + 1I + 16m ♂, distinct heteromorphism	*conspersus* ♂♀ (Gorman and Atkins, 1968a)
	$2n$ = 32, 12V + 4I + 16m, no obvious heteromorphism	*grahami* ♂♀ (Gorman, unpublished)
valencienni	$2n$ = 30, 14V + 16m no obvious heteromorphism	*valencienni* ♂ (Gorman and Atkins, 1968b)

Series	Karotype description	Species (reference)
sagrei	$2n = 28$, 14V + 14m no obvious heteromorphism	*sagrei* ♂, *luteosignifer* ♂, *homolechis* ♂, *mestrei* ♂, *quadriocellifer* ♀, *rubribarbus* ♂ (Gorman and Atkins, 1968b)
petersi	$2n$ (♂) = 29, 12V + 14m + $X_1 X_2 Y$	
	(♀) = 30, 12V + 14m + $X_1 X_1 X_2 X_2$	*biporcatus* (Gorman and Atkins, 1966)
	$2n = 40$[b], 24M + 16m[c]	*capito* (Gorman, unpublished)
chrysolepis	$2n = 30$, 14V + 16m no obvious heteromorphism	*chrysolepis* (Gorman and Atkins, 1967)
		lineatus (Gorman and Atkins, 1967)
		auratus (Gorman *et al.*, 1967)
	$2n = 40$, 24M + 16m[c]	*cupreus* (Gorman, 1967)
		humilis (Gorman, 1967)
		tropidolepis (Gorman, unpublished
		tropidonotus (Gorman, unpublished)
	$2n + 40$, 20I + 4V + 16m	*tropidogaster*[d] (Gorman, 1967)
fuscoauratus	$2n = 40$, 24M + 16m[c]	*fuscoauratus* (Gorman, unpublished)
		lionotus (Gorman, unpublished)
		limifrons (Gorman, 1967)
		polylepis (Gorman, unpublished)
	$2n = 36$, 8V + 12I + 16m	*gracilipes* (Gorman, unpublished)
nebuloides	$2n$ (♂) = 30, 13V + 17m	*nebulosus* (Gorman, unpublished)
anoline genera		
α affinity	$2n = 36$, 12V + 24m	*Phenacosaurus heterodermus* (Gorman *et al.*, 1969b)
	$2n = 36$? 12V + > 20m	*Chamaeleolis porcus* (Gorman *et al.*, 1969b)
β affinity	$2n = 30$, 14V + 16m possible X–Y heteromorphism Pair 7	*Tripidodactylus onca* (Gorman, 1969b)

[a] In those species with multiple sex chromosomes, Y is a small metacentric, X_1 and X_2 are acrocentric and differ in size.
[b] Obtained by doubling haploid complement.
[c] Morphology of macrochromosomes has not been analysed.
[d] This species appeared osteologically intermediate between the *chrysolepis* or *fuscoauratus* series.
[e] Y is a large metacentric, X_1 and X_2 are acrocentric and differ in size.

is a specialized form in an advanced species-series. Hence it appears to be a clear-cut case of establishment of centric fission of the macrochromosomal complement (see Table III).

Two osteologically defined species-series of alpha anoles are particularly interesting for their highly specialized and derived karyotypes. The *bimaculatus* series of Etheridge (1960) was subdivided into the *acutus*, *wattsi* and *bimaculatus sensu stricto* groups on the basis of cytological and biochemical characters (Gorman and Atkins, 1969). The *acutus* group is comprised of species that all have diploid numbers below 36. They are characterized by 12V, a reduced number of m and one or two pairs of chromosomes intermediate in size between M and m. They all have distinctive sex chromosomes. The species with the lowest diploid number (*A. evermanni*, $2n = 26$) has X–Y, the others have X_1X_2Y. The true *bimaculatus* and *wattsi* groups have karyotypes that are even more derived. The ancestral 12V has been modified. The autosomes consist of 8V, 6 sV, 4 intermediate sized V, and 10 microchromosomes. The Y chromosome is a small metacentric, X_1 and X_2 are acrocentrics that differ in size (Fig. 4b). This karyotype characterizes all but one of the species, which has incorporated an additional centric fission.

The chromosome evolution in the *bimaculatus* series aided in an understanding of the biogeography of these insular lizards. There are anoles in the Greater Antillean islands of Puerto Rico and Hispaniola with primitive karyotypes (12V + 24m). We postulated that from one of these ancestors, the *acutus* group evolved by reduction of a number of microchromosomes and establishment of the X_1X_2Y sex chromosome system. The *acutus* group is to be found in the Greater Antilles and on St. Croix, just east of the Puerto Rico bank. The *wattsi* and *bimaculatus* groups are found further south and east on the small islands of the Lesser Antilles. Thus increasing specialization in karyotype correlated with colonization towards the south and east (Gorman and Atkins, 1969).

The second species series with specialized karyotype is the *cristatellus* series, found primarily on the Puerto Rican bank. Osteologically, Etheridge (1960) found that this series was closely related to the *bimaculatus* series, and in karyotype we found that there is a strong resemblance to the *acutus* group. The *cristatellus* series also has 12V plus intermediate sized metacentrics, reduced number of microchromosomes, and an X_1X_2Y sex chromosome system (Gorman and Atkins, 1969; Gorman *et al.*, 1968). There was only a single species of alpha anole that did not have a karyotype in concordance with its osteological characters. Etheridge (1960) found that *A. cybotes* of Hispaniola was a perfectly ordinary member of the *cristatellus* series, however, in karyotype it is a perfectly ordinary primitive alpha anole (12V + 24m, no sex chromosome heteromorphism).

The beta anoles can also be placed in distinct groups on the basis of karyotype. Some of these groups agree closely with the osteological classification, others seem to cross species-series lines. The "typical" beta karyotype consists of 14V + 16m without obvious sex chromosome heteromorphism. This karyotype is found in some members of the *chrysolepis* series and most members of the *grahami* series (Fig. 4c). The only exception in the *grahami* series is *A. grahami* itself, which seems to have established a karyotype involving centric fission of one of the pairs of metacentrics (12V + 4I + 16m). In *A. conspersus*, a close relative of *A. grahami*, there is a clear case of X–Y heteromorphism, with a metacentric X and acrocentric Y. This occurs on chromosome pair 7. Other members of the *grahami* series may have heteromorphism in pair 7, but it involves relative size and centromere position of two-armed chromosomes (Gorman and Atkins, 1967, 1968a, b; Gorman, 1969b). Six species of the *sagrei* series all have 14V + 14m. Thus all members of this series have the typical beta karyotype, except that it is reduced by a pair of microchromosomes (Gorman and Atkins, 1968b).

The typical beta karyotype appears to be a derived condition, not unlike that of the *bimaculatus* alpha anoles, but the similarities in karyotype almost certainly reflect convergent patterns of chromosome evolution. The increase in number of macrochromosomes (14V instead of 12V) is accompanied by a reduced diploid number. The "loss" of microchromosomes may be accounted for, at least partially, by their incorporation in the "new" macrochromosomes.

Another frequently observed chromosome complement among beta anoles is $n = 20$ (Fig. 4d). Again there are 8 pairs of microchromosomes, but 12 pairs of macrochromosomes. Morphological details of the karyotype have not yet been studied for all species, but it appears that most of the macrochromosomal elements are acrocentric, and it is probable that the 12 pairs of M chromosomes are homologous to the seven pairs of V of the "typical" betas. The $n = 20$ complement is common in anoles placed in the

Fig. 4a. Karyotype of a male *Anolis cybotes*, $2n = 36$, 12V + 24m, no obvious sex chromosome heteromorphism. This pattern is widespread among many species of α anoles, as well as among many genera of iguanid lizards.
b. Karyotype of a male *Anolis ferreus*, a highly modified α anole complement. $2n = 29$, there are three unpaired sex chromosomes (far right of second row), the Y is metacentric, X_1 and X_2 are acrocentric and differ in size.
c. Karyotype of *Anolis chrysolepis*, female. $2n = 30$, 14V + 16m. This pattern is common to many β anoles.
d. Metaphase II of *Anolis cupreus*, male. $n = 20$. There are 12 macrochromosomes and 8 microchromosomes. Only two of the macrochromosomes are metacentric, the remainder are acrocentric or subacrocentric. This pattern is found among several groups of β anoles.

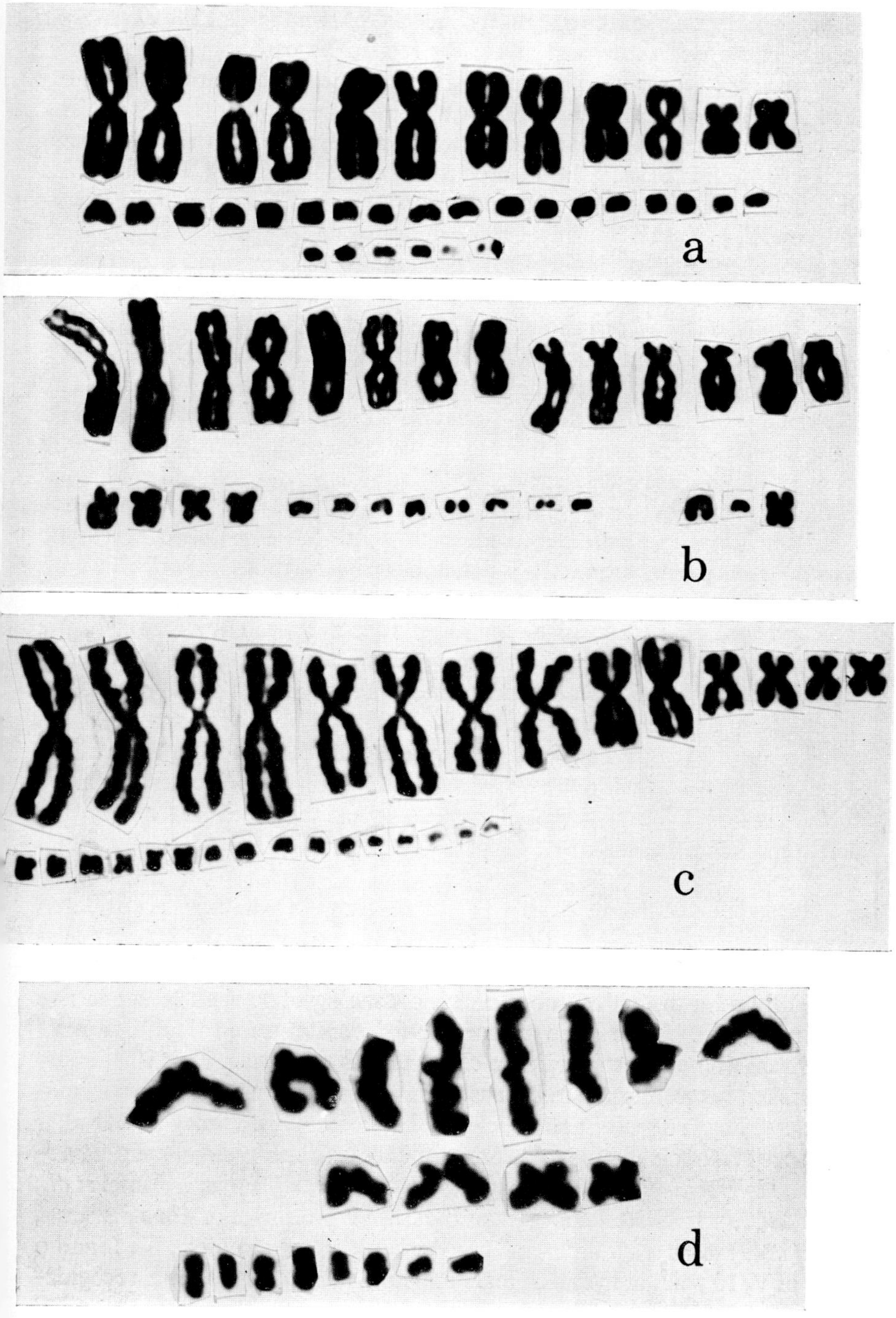
a
b
c
d

fuscoauratus series, and is also found in several members of the *chrysolepis* series. These two series appear closely related by osteological characters. Perhaps $n = 20$ was established only a single time, and chromosomes reflect the true boundaries between the species-series. *Anolis gracilipes* of the *fuscoauratus* series has an intermediate condition, $n = 18$, with eight pairs of microchromosomes. The macrochromosomes consist of 12I and 8V; thus there are 28 macrochromosomal arms, which is identical to the "typical" condition of 14V (Gorman, 1967, in preparation).

Two additional species series of beta anoles present some problems of concordance between osteological and cytological evidence. *A. capito* and *A. biporcatus* were both placed in the *petersi* series. The former has $n = 20$ distributed in the usual fashion, the latter has a modified $n = 15$, complicated by X_1X_2Y male heteromorphism (♂ $2n = 29$, ♀ $2n = 30$). The Y chromosome is metacentric, X_1 is an acrocentric macrochromosome, and X_2 is a microchromosome. Either there has been multiple origin of $n = 20$ from $n = 15$ (or the reverse), or *A. capito* and *A. biporcatus* should not belong to the same species series (Gorman and Atkins, 1966; Gorman, in preparation). *A. nebulosus* of the nebuloides series also has a modified karyotype (Gorman, in preparation). Only males have been studied. $2n = 30$, with 13V and 17m. Presumably this is a clear example of X–Y heteromorphism.

Karyotype data are available for three species representing three of the anoline genera. *Phenacosaurus* and *Chamaeleolis* are osteologically similar to alpha anoles. The karyotype of the former is precisely that of a typical alpha anole, with 12V and 24m; the exact number of microchromosomes for *Chamaeleolis porcus* has not been determined, but there are 12V and at least 20m (Gorman *et al.*, 1969b). *Tropidodactylus onca* is osteologically similar to the beta anoles, and karyotype agrees. It has the typical $2n = 30$, 14V and 16m (Gorman, 1969b).

(2) Sceloporine lizards (Table IV): This group of iguanids is considered osteologically advanced by Etheridge (1964). Except for the genus *Sceloporus* itself, the group is remarkably uniform in karyotype. Pennock *et al.* (1968) found a diploid number of 34, consisting of 12V and 22m in five species of *Uta*. These same authors (1969) demonstrated that X–Y male sex chromosome heteromorphism characterizes the genus. This is an easily overlooked heteromorphism because it involves the smallest microchromosomes. They suggest that such a heteromorphism may be rather widespread among iguanids. Chromosomes of *Callisaurus*, *Urosaurus*, *Petrosaurus*, *Holbrookia*, and *Uma* all are 12V + 22m (Hall, 1965; Gorman *et al.*, 1967, 1969a). In a detailed study of meiosis of two males of *Uma inornata*, Král (1969) reported a different type of male heteromorphism. He found 6 pairs of V, 10 pairs of m and a single unpaired microchromosome, recogniz-

TABLE IVA. Chromosomes of sceloporine lizards (exclusive of *Sceloporus*) (References in parentheses)

$2n = 34, 12V + 22m$
Callisaurus draconoides[a] ♂ (Gorman *et al.*, 1969a), *Petrosaurus mearnsi*[a] ♂ (Gorman *et al.*, 1969a), *Phrynosoma cornutum* (Robinson, personal communication), *P. coronatum*[a] (Gorman *et al.*, 1969a), *P. douglassi* (Robinson, personal communication), *P. m'calli*[a] (Gorman *et al.*, 1969a), *P. modestum* (Robinson, personal communication), *P. platyrhinos*[a] ♂ (Gorman *et al.*, 1969a), *P. solare* (Robinson, personal communication), *Uma inornata*[a] ♂ (Gorman *et al.*, 1969a), *Uma notata* ♂ (Gorman *et al.*, 1967), *Uma scoparia*[a] ♂ (Gorman *et al.*, 1969a), *Urosaurus graciosus*[a] ♂ (Hall, 1965), *Urosaurus microscutatus*[a] ♂ (Gorman *et al.*, 1969a), *Uta antiquus*[b] ♂ (Pennock *et al.*, 1968, 1969), *Uta nolascensis*[b] ♂ ♀ (Pennock *et al.*, 1968, 1969), *Uta palmeri*[b] ♂ ♀ (Pennock *et al.*, 1968, 1969), *Uta stansburiana*[b] ♂ ♀ (Pennock *et al.*, 1968, 1969), *Uta squamata*[b] ♂ ♀ (Pennock *et al.*, 1968, 1969).
$2n = 36, 12V + 24m$
Phrynosoma cornutum ♂ (Cavazos, 1951)
$2n = 33, 12V + 20m + 1$ unpaired large microchromosome
Uma inornata (Král, 1969)

[a] Diploid complement obtained by doubling observations of first meiotic metaphase.
[b] X–Y heteromorphism demonstrated for smallest pair of microchromosomes.

TABLE IVB. Chromosomes of sceloporine lizards: genus *Sceloporus*

Species group	Species/Sex	2*n*	Description[a]	Reference
undulatus	*undulatus*	22	12V + 10m	Lowe *et al.*, 1966
	occidentalis ♂ ♀	22	12V + 10m	Jackson and Hunsaker, 1970; Cole *et al.*, 1967
	virgatus ♂ ♀	22	12V + 10m	Cole and Lowe, 1968
graciosus	*graciosus* ♂ ♀	30	12V + 18m	Jackson and Hunsaker, 1970; Cole, 1971a
poinsetti	*poinsetti* ♂	31	12V + 18m + small V Y-chromosome	Cole *et al.*, 1967
	♀	32	12V + 20m ($X_1X_1X_2X_2$ all m)	
	jarovi ♂	31	As *poinsetti*	Cole *et al.*, 1967
	♀	32		
scalaris	*scalaris*	24	12V + 12m	Lowe *et al.*, 1966
chrysostictus	*chrysostictus* ♂ ♀	34	12V + 22m X–Y heteromorphism among m	Cole, 1971a
utiformis	*utiformis* ♂ ♀	34	12V + 22m X–Y heteromorphism among m	Cole, 1971a
maculosus	*maculosus* ♂	33	12V + 20m + intermediate I Y-chromosome	Cole, 1971a
	♀	34	12V + 22m ($X_1X_1X_2X_2$ all m)	
merriami	*merriami* ♂ ♀	46	24I (actually sI) + 22m X–Y heteromorphism, smallest m	Cole, 1971a
spinosus	*clarki* ♂ ♀	40[b]	4V + 16I + 20m	Cole, 1970; Lowe *et al.*, 1967
	melanorhinus ♂ ♀	40[b]	4V + 16I + 20m	Cole, 1970
	orcutti ♂ ♀	34	12V + 22m	Cole, 1970

	magister ♂ ♀	26	12V + 14m	Cole, 1970; Lowe *et al.*, 1967
	lundelli ♂ ♀	22	12V + 10m, largest m heteromorphic X = I, Y = sI	Cole, 1970
	edwardtaylori ♂ ♀	22	12V + 10m	Cole, 1970
	olivaceus ♂ ♀	22[b]	12V + 10m	Cole, 1970
	spinosus ♂ ♀	22[b]	12V + 10m	Cole, 1970
	horridus ♂ ♀	22[b]	12V + 10m	Cole, 1970
pyrocephalus	*pyrocephalus* ♂ ♀	34	2sI + 10V + 22m X–Y heteromorphism among m	Cole, 1971b
	gadoviae ♂ ♀	34	12V + 22m X–Y heteromorphism among m	Cole, 1971b
	nelsoni ♂ ♀	34	12V + 22m X–Y heteromorphism among m	Cole, 1971b

[a] Abbreviated terminology is retained for descriptive convenience. In species with low $2n$, the m chromosomes are relatively large with distinct morphology.
[b] See original reference for description of intraspecific karyotype variation.

able because it was the largest of its group. This was not observed by Gorman *et al.* (1969a), but our study was less detailed. Král considered this a case of X_1X_2Y heteromorphism, but such a judgment is premature in the absence of data on females. It could also be interpreted as X–O male, X–X female. The minuteness of the Y in the related genus *Uta* (Pennock *et al.*, 1969) is suggestive of such an interpretation. Král found a univalent, not a trivalent in diakinesis.

The karyotype of the specialized horned lizards *Phrynosoma* is of interest in terms of taxonomic interpretation. There has been debate concerning their relationships to sceloporines. Etheridge (1964) concluded that they were not true sceloporines, but subsequent osteological work (Presch, 1969) demonstrated that this was a true sceloporine genus. Cavazos (1951) published the karyotype of *Phrynosoma cornutum*, and claimed that the species was characterized by 12V and 24m. Gorman *et al.* (1967) used this information to support Etheridge's (1964) argument that *Phrynosoma* was not a true sceloporine. We accepted the proposition that the ancestral stock of horned lizards diverged from the true sceloporine ancestors, before establishment of the reduced diploid number. However, subsequent investigation of the chromosomes of three other species of *Phrynosoma* showed that all were characterized by 12M + 22m (Gorman *et al.*, 1969a). In the light of reinterpretation of the osteological data, it is of note that preliminary studies by M. Robinson (personal communication) on the karyotype of *P. cornutum* and several other members of the genus showed that $2n = 34$, with 12V and 22m.

The genus *Sceloporus* with about 60 species is second only to *Anolis* in species diversity within the family, and like *Anolis* shows a great deal of karyotypic variation and a variety of sex chromosome mechanisms. Lowe *et al.* (1966) provided basic karyotype data on nine species; Cole *et al.* (1967) presented expanded information on the sex chromosomes of two of these species; Lowe *et al.* (1967) discussed karyotype evolution in two (of the originally described 9) species. Cole (1970, 1971a, b) presented detailed chromosome information on 17 species, and Jackson and Hunsaker (1970) described karyotypes of two species, confirming the observations of Cole and co-workers.

Diploid numbers include 22 (*undulatus* groups, *spinosus* group, in part); 24 (*scalaris* group); 26 (*spinosus* group, in part); 30 (*graciosus* group); ♂ 31, ♀ 32 (*torquatus* group); ♂ 33, ♀ 34 (*maculosus* group); ♂ and ♀ 34 (*spinosus* group, in part; *chrysostictus* group; *utiformis* group; *variabilis* group; *siniferus* group; *pyrocephalus* group); 40 (*spinosus* group, in part) and 46 (*merriami* group).

There is relatively little variation in the macrochromosomal complement. All species with $2n = 34$ or less have 12 *ca.* metacentric macrochromosomes. The only variation described in their macrochromosomal complements involves location (or absence) of secondary constrictions, and in one case an apparent pericentric inversion in pair 1, such that it is sI (Cole, 1971b). The three species with elevated diploid numbers ($2n = 40, 46$) essentially have 24 macrochromosomal arms. Thus whole-arm translocation (Robertsonian variation) accounts for virtually all of the macrochromosomal variability.

The evolutionarily labile part of the karyotype involves the microchromosomes. The species with greatly reduced diploid numbers ($2n = 22$) have obviously large "microchromosomes" that were probably derived by centric fusion. The sex chromosomes (X–Y and X_1X_2Y systems occur within the genus) all are among the microchromosomes.

The detailed study of the *spinosus* group (Cole, 1970) provides a wealth of information on intra- and interspecific variation. Among the species in this group with $2n = 22$, Cole found interesting variation in chromosome pair 7, the largest of the microchromosomes. It is heteromorphic in *S. lundelli*, with the X a small metacentric, the Y a small acrocentric. In the related species *S. edwardtaylori*, there is no sex chromosome heteromorphism, pair 7 in both sexes is metacentric. In *S. spinosus* pair 7 is metacentric in some populations, subtelocentric in others. *S. horridus* shows a similar type of interpopulation variation. In *S. olivaceus* two individuals were found with pair 7 heteromorphic, but this was not sex correlated.

Sceloporus clarki has three observed karyotypes, all with $2n = 40$. The commonest is 4V + 16I + 20m. The second most frequently observed has the same formula but one pair of m shows size heteromorphism that was not sex correlated. The third karyotype shows heteromorphism in macrochromosomal pair 5. This is normally an acrocentric pair, but individuals have been found that apparently have established an inversion, because there is a distinct small arm on one of the two chromosomes. Again, this heteromorphism was not sex correlated. Homozygotes for the rare chromosomal conditions or heterozygotes for two rare conditions have not been found. But the frequency of these conditions is such that homozygotes would not have been expected within the sample sizes examined. Thus, there is no evidence for selection against any of these atypical karyotypes.

Cole (1970) proposed a phylogeny of the *spinosus* group based upon karyotypic evidence combined with morphological and ecological data. He considered the $2n = 40$ karyotype primitive, with chromosome evolution proceeding strictly by centric fusion. In a subsequent work (Cole, 1971a) he considered the karyotype of *S. merriami* (*merriami* group, $2n = 46$) as representing the primitive condition. Such an hypothesis discounts the evidence that all sceloporine genera with the exception of *Sceloporus* are virtually restricted to a karyotype of 12V + 22m.

Sceloporines are an osteologically advanced group of iguanids. Most primitive iguanids have 12V + 24m. Furthermore, many species groups of *Sceloporus* have 12V + 22m. On the principle of parsimony, invoking the fewest convergent events, it appears logical that karyotype evolution in the sceloporine lizards has been away from a 12V + 22m condition (Gorman *et al.*, 1969a). Cole (1970) claimed that acceptance of our hypothesis would mean that *S. orcutti*, a highly advanced species would have to be ancestral to the *spinosus* group. This is not true. Cole's own proposed phylogeny shows that *S. orcutti* is not particularly closely related to members of the *spinosus* group with high diploid numbers. And with very minor perturbation of his phylogenetic diagram (Fig. 16, in Cole, 1970) one could claim that this highly advanced species has retained the ancestral karyotype.

(3) Other iguanid lizards (Table V): The typical iguanid karyotype is 12V + 24m. This has been found in *Ctenosaura*, *Cyclura*, and *Sauromalus* (iguanines); *Basiliscus* (basiliscine); *Anisolepis* which may be of anole affinity; *Tropidurus* and *Liocephalus* (tropidurines); *Oplurus*, one of the Madagascar genera; *Cupriguanus* from southern South America and the collared and leopard lizard group, *Crotaphytus* (Cohen *et al.*, 1967; Montanucci, 1970). In *Iguana iguana*, we reported 12V + 24m (Gorman *et al.*, 1967) but a more detailed study (Cohen *et al.*, 1967) revealed only 22m.

Departure from the 12V + 24m complement is minor among most iguanids. Thus, two tropidurines that differ from the typical condition illustrate a case of centric fusion and a case of centric fission. In *Liolaemus lutzae* the complement may be described as 12V + 22m; but the largest pair of microchromosomes is distinctly larger than all the others, and probably was derived through centric fusion. In *Plica plica* there are 8V, 8I, 24m. As this is a specialized species, and typical tropidurines have 12V, the simplest interpretation is that the karyotype was derived by stabilization of centric fissions (Gorman *et al.*, 1967).

In *Cupriguanus achalensis* with 12V + 24m, the largest pair of microchromosomes appeared heteromorphic (Gorman *et al.*, 1967). The data were based on a single specimen and it would be premature to consider this sex chromosome heteromorphism. It is not unlike some of the variation found as non-sex correlated in *Sceloporus* (Cole, 1970). In fact, Pennock *et al.* (1969) point out that from our figure, it appears that the smallest pair of microchromosomes is heteromorphic; this could be similar to the case of the minute Y in the genus *Uta*.

The genus *Polychrus* is outstanding among the Iguanidae for having a karyotype that bears no obvious relationship to any other lizards (Gorman *et al.*, 1967, 1969b) (Fig. 5).

TABLE V. Chromosomes of the Iguanidae[a]

Species	2*n*	Description	Reference
Iguana iguana	36	12V + 24m	Gorman *et al.*, 1967
Iguana iguana ♂ ♀	34	12V + 22m	Cohen *et al.*, 1967
Ctenosaura pectinata	36	12V + 24m	Gorman *et al.*, 1967
Cyclura cornuta	36	12V + 24m	Gorman *et al.*, 1967
Sauromalus obesus ♀	36	12V + 24m	Gorman *et al.*, 1967
Basiliscus vittatus ♂	36	12V + 24m	Gorman *et al.*, 1967
Anisolepis grilli ♂	36	12V + 24m	Gorman *et al.*, 1967
Tropidurus torquatus ♂	36	12V + 24m	Gorman *et al.*, 1967
Liocephalus schreibersi ♂	36	12V + 24m	Gorman *et al.*, 1967
Liolaemus lutzae ♂	34	12V + 22m	Gorman *et al.*, 1967
Plica plica ♂	40	8V + 8I + 24m	Gorman *et al.*, 1967
Oplurus sebae ♀	36	12V + 24m	Gorman *et al.*, 1967
Cupriguanus achalensis ♂	36	12V + 24m	Gorman *et al.*, 1967
Crotaphytus collaris ♂ ♀	36	12V + 24m	Cohen *et al.*, 1967; Montanucci, 1960
Crotaphytus wislizenii ♂	36	12V + 24m	Montanucci, 1970
Crotaphytus silus ♂	36	12V + 24m	Montanucci, 1970
Polychrus marmoratus ♂	29	1V (=Y) + 19I + 9m	
♀	30	20I + 10m	Gorman *et al.*, 1967
Polychrus femoralis ♀	26	2V + 20I + 4m	Gorman *et al.*, 1969b
Polychrus peruvianus ♀	28	20I + 8m	Gorman *et al.*, 1969b

[a] See also Tables III and IV.

Data are available for three species. In *P. marmoratus* the female has 20I + 10m (2n = 30) (Fig. 5). The male diploid number is 29, with a metacentric Y and two unpaired X's. In *P. peruvianus* the female has 28 chromosomes, all acrocentric; and *P. femoralis* differs only slightly, with 1 submetacentric pair of macrochromosomes, and fewer microchromosomes.

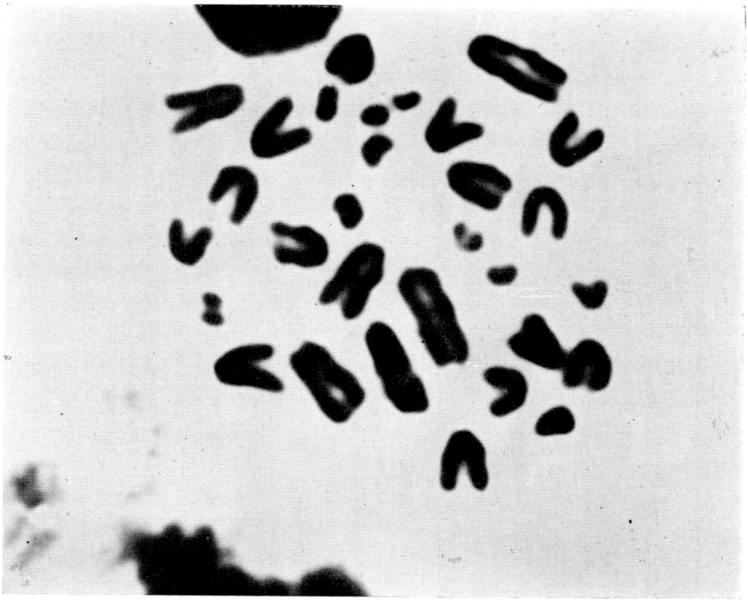

Fig. 5. Mitotic metaphase of *Polychrus marmoratus*, female. $2n = 30$. This is the most atypical iguanid karyotype, unusual in the complete absence of metacentric elements coupled with a low number of arms in the karyotype.

There is no "Robertsonian" relationship between the karyotypes of *Polychrus* and other iguanids. In general, 24 macrochromosomal arms is very conservative in the family. The only known exceptions occur in the genus *Anolis* (beta, and *bimaculatus* and *cristatellus* alpha) and in these cases there is an *increase* in the number of macrochromosomal arms. Presumably neo-M chromosomes arose through centric fusion of micro elements.

In morphological characters *Polychrus* is also a puzzle. In some features it is distinct among iguanids; in other features it resembles anoles. But it is not clear that the resemblance is not convergence for characters involved in arboreal specialization (Etheridge, 1960). Karyotype shows no obvious anoline affinity. (See Addendum.)

TABLE VI. Chromosomes of the Agamidae

Species	2*n*	Description	Reference
Agama caucasica	34	12V + 22m	Arronet, 1965
Agama pallida	46	24I + 22m	Gorman, unpublished
Agama savignii	46	24I + 22m	Gorman, unpublished
Agama sinaita	44	22I + 22m	Gorman, unpublished
Agama stellio	36	12V + 24m	Matthey, 1931
Agama tuberculata	34	12V + 22m	Dutt, 1969
Calotes versicolor	34	12V + 22m	Makino and Asana, 1948; Singh *et al.*, 1970a
Japalura swinhonis	46	24I + 22m	Nakamura, 1935b
Japalura polygonata	46	24I + 22m	Makino and Momma, 1949
Leiolepis belliana	36	12V + 24m	Hall, 1970
Leiolepis belliana (parthenogenetic)	54	18V + 36m	Hall, 1970
Phrynocephalus helioscopus	44	2V + 18I + 24m	Arronet, 1965
Sitana ponticeriana	46	24I + 22m	Makino and Asana, 1948
Uromastyx hardwicki	36	12V + 24m	Matthey, 1931

(e) *Family Agamidae* (Table VI)

This family is often considered the Old World counterpart of the iguanids. However, it has not received the same attention by cytotaxonomists. Chromosome data have been published for only 11 species, and I have added unpublished observations on several additional members of the genus *Agama*.

In general, there is a conservativeness for total number of arms in the karyotype. All diploid species have 22 or 24m, and all but two have 24 macrochromosomal arms, 12V, or 24I (Fig. 6). The exceptions are *Phrynocephalus helioscopus* with 2V + 18I (22 macrochromosomal arms) and *Agama sinaita* with 22I (Arronet, 1965; Gorman, unpublished).

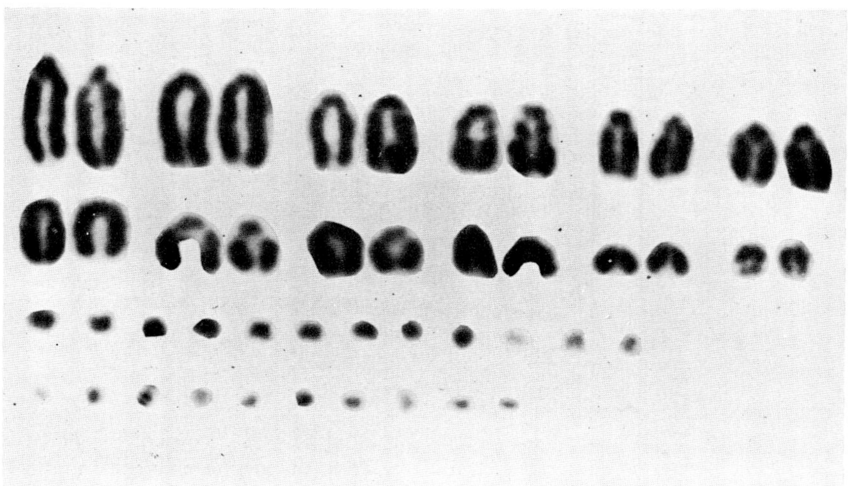

Fig. 6. Karyotype of *Agama pallida*, male. $2n = 46$, all chromosomes acrocentric.

Hall (1970) has described a clear case of parthenogenetic triploidy in *Leiolepis belliana*. He found that some samples contained both males and females, and these had 12V + 24m, whereas other samples were comprised only of females with 18V + 36m ($2n = 54$).

Makino and Asana (1948) claimed X–O female heterogamety in the genera *Calotes* and *Sitana*—but this has never been confirmed, and has recently been specifically denied (Singh *et al.*, 1970a).

(f) *Family Chamaeleontidae* (Table VII)

Karyotypes are known for about 25 species in this family of highly

			Mainland	Madagascar
I.	Sharp separation between 12 *ca.* metacentric macrochromosomes and a series of microchromosomes.			
	(A)	12V + 24m	*Chamaeleo fischeri*	*Brookesia stumpfi*
			C. cristatus	
			C. owenii	
			C. quadricornis	
			C. wiedersheimi	
	(B)	12V + 22m	*C. pumilis*	*C. gallus*
	(C)	12V + 12m	*C. chamaeleon*	
			C. dilepis	
			C. senegalensis	
			C. calcaratus[b]	
	(D)	12V + 14m		*C. campmani*
II.	Less sharp separation between M and m, M 12.			
	(E)	14M + 14m		*C. willsi*
	(F)	16M + 18m		*C. nasutus*
	(G)	18M + 14m		*C. brevicornis*
III.	Separation between M and m indistinct, only 4 chromosomes, at most, could be considered m.			
	(H)	2n = 20, 2m	*Brookesia spectrum* (*Rhampholeon*)	
		2n = 22, 2m		*C. lateralis*
		2n = 22, 2m		*C. pardalis*
		2n = 22, 4m		*C. oustaleti*
		2n = 22, 4m		*C. monoceras*
		2n = 24, 2m		*C. voeltzkowi*
		2n = 28, 4m		*C. cephalolepis* (Grande Comore Island)
IV.	No sharp separation between M and m. The M tend toward terminalized centromeres.			
	(I)	$2n = 24$, 4m	*C. jacksoni*	
			C. bitaeniatus	

[a] Adapted from Matthey and van Brink, 1960 and Matthey, 1961. [b] *C. calcaratus* = *C. chamaeleon* according to Hillenius, 1963.

specialized lizards, well known for their colour changing abilities, their independently movable eyes and their elongate protrusible tongue which is used in prey capture. All cytotaxonomic studies on this group have been done by Matthey (1957, 1961; Matthey and van Brink, 1960). Chromosome variation is reminiscent of that found in *Sceloporus* (see above) because the first six pairs of chromosomes are very conservative in morphology, being similar or identical in all species, but the smaller elements show a great amount of variation. Diploid numbers range from 20 (the lowest diploid number of any known reptile) to 36. Sex chromosome heteromorphism is not known.

Matthey divided the chamaeleons into four major chromosome groups and found a correlation between karyotype and distribution (Table VII), some groups predominating in Africa, others on Madagascar. Hillenius (1963) compared cytological groupings with his classification based upon external morphology and, in general, found excellent concordance. There were some exceptions. For example, Matthey found that *Chamaeleo monoceras* and *C. voeltzkowi* differ in the number of microchromosomes, and in relative centromere position of several of the macrochromosomes. Yet Hillenius was not convinced that the two were distinct species. Until more is known about the extent of intraspecific variation in karyotype (all data are based on small sample sizes), and more detailed studies are completed on the systematics of the family, such problems cannot be resolved.

I consider it significant that 12V + 24m is found both in Africa and Madagascar, and both in *Chamaeleo* and *Brookesia*. This may well be the ancestral condition with karyotype evolution involving a reduction of microchromosomes, in part, at least, by centric fissions which reduces the break between M and m. The genus *Rhampholeon* (mainland) is usually synonomized with *Brookesia* (Madagascar). Hall (1970) suggests that recognition of two genera is in order. *B. stumpffi* has the typical karyotype of 12V + 24m, whereas *B.* (*Rhampholeon*) *spectrum* has the highly derived karyotype of $2n = 20$. (See addendum).

(g) *Family Scincidae* (Table VIII)

Cytotaxonomic data are woefully sparse for this, the lizard family most rich in species. Recent reviews (Dallai and Talluri, 1969; Dutt, 1969) listed data for but 13 species in 7 genera. Subsequently, Deweese and Wright (1970) provided data for six additional species in one of these genera.

Because of superficial external uniformity, definition of genera and relationships between scincid genera have been problematical. Greer (1970a) has made a major contribution to the study of higher categories of skinks. Chromosome data are not adequate to compare with his proposals.

TABLE VIII. Chromosomes of the Scincidae

Species	2*n*	Description		Reference
Chalcides chalcides (*C. tridactylus*)	28	8V + 20m		Matthey, 1931
C. chalcides	28	10V + 18m, Pair 5 intermediate between M and m		Dallai and Talluri, 1969
C. ocellatus	28	10V + 18m		Dallai and Talluri, 1969
C. ocellatus	26	8V + 18m		Mendietta, 1961
Cryptobrephalus boutonii	28	10V + 18m		Makino and Momma, 1949
Eumeces brevilineatus	26	12V + 14m		Deweese and Wright, 1970
E. copei	26	12V + 14m		Deweese and Wright, 1970
E. dugesii	26	12V + 14m	all m two -armed	Deweese and Wright, 1970
E. obsoletus	26	12V + 14m		Deweese and Wright, 1970
E. gilberti	26	12V + 14m		Deweese and Wright, 1970
E. skiltonianus	26	12V + 14m		Deweese and Wright, 1970
E. latiscutatus	26	12V + 14m		Nakamura, 1931b
E. elegans	26	12V + 14m	all m one-armed	Makino and Momma, 1949
E. marginatus	26	12V + 14m		Makino and Momma, 1949
E. schneideri	32	No distinction between M and m		Gorman, unpublished
Mabuya carinata	30	12 large V, 4 smaller V, 14m		Dutt, 1969
Mabuya carinata	32	14V, 2sV, 2I, 14m		Singh *et al.*, 1970a
Mabuya macularia	26	10V, 6I, 10m		Asana and Mahabale, 1941
Mabuya macularia	32	No description		Makino, 1951
Mabuya striata	28	8 large V, 2 intermediate V, 18m		Dallai and Talluri, 1969
Riopa punctata	24	12V + 12m		Bhatnagar, 1962
Scincus officinalis	32	4V + 8I + 20m		Matthey, 1931
Sphenomorphus indicus	28	10V + 18m		Makino and Momma, 1949

As a group, diploid numbers are relatively low, ranging from 24 to 32. In general, there is a series of large *ca.* matacentric chromosomes and a series of distinctly smaller chromosomes. Because these smaller chromosomes have distinctive morphology and they are perhaps not so small as in lizards such as iguanids or agamids, some authors make a point of not calling the smallest elements microchromosomes (e.g. Dallai and Talluri, 1969). There have not been any descriptions of sex chromosome heteromorphism in this family.

Practically half the available chromosome data are for the genus *Eumeces*. Deweese and Wright (1970) described the karyotypes of 6 North American species, all with $2n = 26$ with 12V + 14m, all the m clearly metacentric. The macrochromosomes of three Asian species were similarly described (Makino and Momma, 1949; Nakamura, 1931b), but the microchromosomes were described as acrocentrics—rods grading to dots. Whether this reflects a real difference, or merely a difference in technique, is an open question. I recently studied the chromosomes of *E. schneideri* from Israel (Gorman, unpublished) and found a distinctly different karyotype; $2n = 32$, no break between M and m, most elements two-armed (Fig. 7).

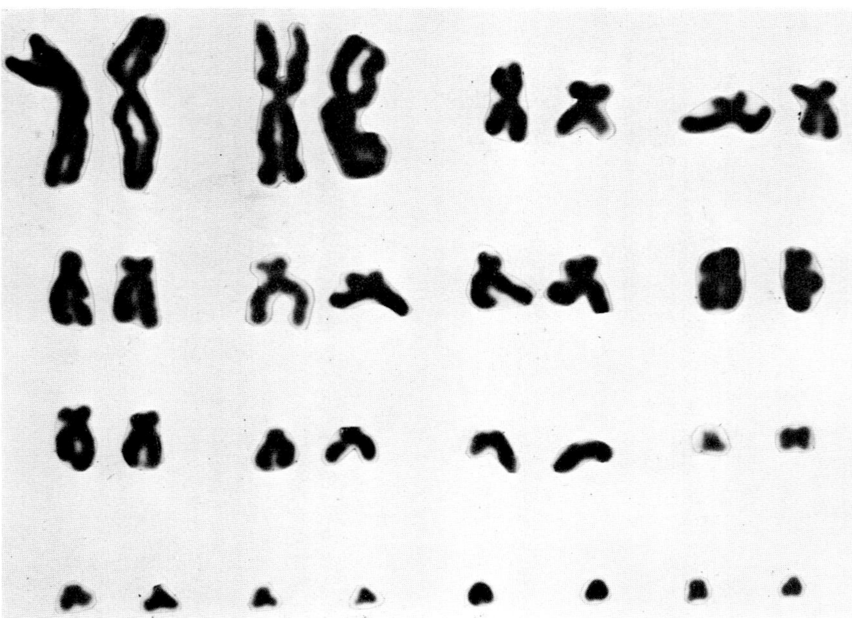

Fig. 7. Karyotype of the scincid lizard *Eumeces schneideri*, male. $2n = 32$, no distinct break between M and m. Nine other species in this genus have $2n = 26$, with 12V and a distinctly smaller set of m chromosomes.

Chromosome literature on the genus *Mabuya* is confusing. Dutt (1969) described the karyotype of *M. carinata* based on 8 specimens as 12 large V, 4 smaller V, and 14m ($2n = 30$). Singh *et al.* (1970a) described the karyotype supposedly of the same species as 14V, 2sV, 2I and 14M ($2n = 32$). Dallai and Talluri (1969) found $2n = 28$ in *Mabuya striata*, 8 large V, 2 intermediate V and 18 smaller chromosomes, mostly two-armed. Makino (1951) listed conflicting reports of $2n = 26$ and $2n = 32$ for *Mabuya macularia*. This genus, widely distributed throughout the world, deserves better treatment!

(h) *Family Gerrhosauridae*

This is a relatively small family of plated or spiny lizards restricted to Africa and Madagascar. Only a single species has been studied, *Gerrhosaurus flavigularis* (Matthey, 1933). It has the frequently observed saurian karyotype of 12V + 24m.

(i) *Family Lacertidae* (Table IX)

This is a moderately large family that has been relatively well studied for karyotype data. It is remarkable for chromosomal uniformity. 32 species representing 7 genera have been examined, and 29 of the species have virtually identical karyotypes; $2n = 38$, all chromosomes acrocentric, only the smallest chromosome pair is considered microchromosomal.

The three species that differ from this pattern, all in the genus *Lacerta*, also differ from each other. *L. ocellata* has $2n = 36$ with one pair of metacentric macrochromosomes (Matthey, 1938; Martin and Burgos, 1957). The number of arms in the karyotype is again 38. Such a karyotype probably arose through establishment of centric fusion. *L. parva* also has 38 chromosomal arms, but a diploid number of only 24 (Gorman, 1969a). There are seven large pairs of metacentric macrochromosomes, a distinct break in size, and five smaller pairs of chromosomes. Multiple centric fusions would account for this karyotype. *Lacerta vivipara* has $2n = 36$, it lacks the typical pair of microchromosomes. Oguma (1934) claimed that this species had female heterogamety, with male $2n = 36$, female $2n = 35$. Margot (1946) specifically denied this claiming that both sexes have 36 acrocentric chromosomes. Chevalier (1969) returned to this problem. She found distinct female heterogamety (Fig. 8). The male has 36 acrocentric chromosomes, the female has 35 chromosomes, one clearly metacentric, the remainder acrocentric. It is possible that none of the workers were "wrong" but that this widespread species has intraspecific variation. Female heterogamety has also been suggested by Gorman (1969a) for *L. trilineata*

TABLE IX. Chromosomes of the Lacertidae

Species	2*n*	Description	Reference
Acanthodactylus boskianus ♂	38	All I, 2m	Gorman, 1969a
A. pardalis ♂	38	All I, 2m	Gorman, 1969a
A. schreiberi ♂	38	All I, 2m	Gorman, 1969a
A. scutellatus ♀	38	All I, 2m	Gorman, 1969a
Eremias olivieri ♀	38	All I, possible heteromorphism 3m ?	Gorman, 1969a
E. guttulata ♂	38	All I, 2m	Gorman, 1969a
E. arguta ♂	38	All I, 2m	Gorman, unpublished
E. stauchi ♀	38	All I, 2m	Kupriyanova, 1968
E. velox ♂	38	All I, 2m	Kupriyanova and Arronet, 1969
Ophisops elegans ♂♀	38	All I, 2m	Gorman, 1969a; Arronet, 1968a
Psammodromus hispanicus ♂	38	All I, 2m	Matthey, 1939
Takydromus tachydromoides ♂	38	All I, 2m	Nakamura, 1935a
T. formosanus ♂	38	All I, 2m	Nakamura, 1935a
T. septentrionalis ♂	38	All I, 2m	Nakamura, 1935a
T. smaragdinus	38	All I, 2m	Nogusa, 1953
Tropidosaurus algirus	38	All I, 2m	Matthey, 1931
Lacerta ocellata ♂	36	2V, 34I of which 2m	Matthey, 1939
L. oxycephala ♂	38	All I, 2m	Gorman *et al.*, 1971
L. mellisellensis ♂	38	All I, 2m	Gorman *et al.*, 1971
L. sicula ♂♀	38	All I, 2m	Dallai and Urbani, 1967
L. leavis ♂	38	All I, 2m	Gorman, 1969a
L. trilineata ♀	38	All I, possible heteromorphism 3m ?	Gorman, 1969a
L. trilineata ♂ (also referred to as *L. media*)	38	All I, 2m	Kupriyanova, 1968
L. viridis	38	All I, 2m	Matthey, 1949
L. muralis	38	All I, 2m	Matthey, 1931
L. agilis	38	All I, 2m	Matthey, 1949
L. strigata ♂	38	All I, 2m	Kupriyanova, 1968

L. saxicola ♂♀	38	All I, 2m	Kupriyanova, 1969
L. unisexualis ♀ (parthenogen)	38	All I, 2m	Kupriyanova, 1969
L. dahli ♀ (parthenogen)	38	All I, 2m	Kupriyanova, 1969
L. armeniaca ♀ (parthenogen)	38	All I, 2m	Kupriyanova, 1969
L. vivipara ♂♀	36♂	All I, 2m	Oguma, 1934
	35♀	♂ lacking medium sized I	
L. vivipara ♂♀	36	All I, no m	
		No heteromorphism	Margot, 1946
L. vivipara ♂♀	36♂	♂ all I, no m	
	35♀	♂ 1V, 34I, no m	Chevalier, 1969
L. parva ♂	24	14V, 24I	Gorman, 1969a

and *Eremias olivieri*. Mitotic spreads were few for both of these species, but the females seemed to have a heteromorphic pair of acrocentric chromosomes, one a macrochromosome and one a microchromosome (3 microchromosomes in all).

Fig. 8. Karyotype of *Lacerta vivipara*, female. $2n = 35$. There is a single unpaired metacentric. Males have $2n = 36$, all acrocentric. Reprinted from Chevalier, 1969.

Parthenogenetic species of *Lacerta* normally appear to be diploid and to have the typical karyotype of 38 chromosomes (Darevsky, 1966; Kupriyanova, 1968). However, natural triploids are known. They are derived from matings between parthenogenetic females with normal bisexual males (Darevsky and Kulikova, 1964; Darevsky and Danielyan, 1968).

The lacertid karyotype, typical for the family, is distinctly different from all other lizards.

(j) *Family Teiidae* (Table X)

This is a large New World family with about 40 genera and 175 species. It has been a particularly rewarding group for cytotaxonomic study. Karyotypes seem to point out natural groupings of genera; relationships between seemingly close genera; and intrageneric relationships suggestive of species groups. In addition, chromosome data have been extremely useful in assessing the origin and relationships of parthenogenetic species.

The family is usually informally divided into macroteiids and microteiids (e.g. Vanzolini and Valencia, 1965). The macroteiids, with 9 genera, are clearly divided into two groups on the basis of karyotype (Gorman, 1970) and these two groups agree completely with the morphological findings of Vanzolini and Valencia. The first, termed the *Dracaena*-group, is comprised of 4 genera all characterized by 5–6 pairs of metacentric macrochromosomes and a distinctly smaller series of microchromosomes. The karyotype of each genus could have been derived with minor modification from a 12V + 24m ancestry (Fig. 9). Although no teiid was found with that exact karyotype,

Fig. 9. Half karyotypes of representatives of the *Dracaena*-group of macroteiids. Bars represent 5 μ. a. *Callopistes maculatus*. b. *Callopistes flavipunctatus*. c. *Dracaena guianensis*. d. *Tupinambis nigropunctatus*. e. *Crocodilurus lacertinus*. Reprinted from Gorman, 1970.

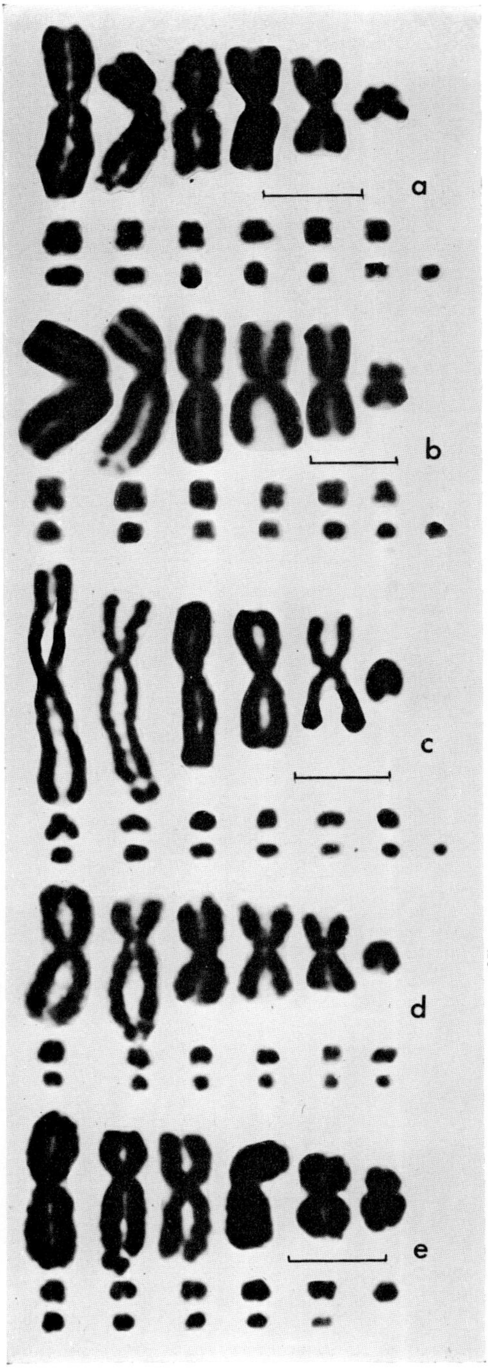
a
b
c
d
e

TABLE XA. Chromosomes of the Teiidae, excluding *Cnemidophorus*[a]

Species	2*n*	Description
Microteiids		
Anadia bitaeniata ♂	46	4V + 16sV and sI + 26m
Pholidibolus montium ♀	46 ?	4V + 16sV and sI + 26 ?m
Macroteiids		
Dracaena-group		
Crocodilurus lacertinus ♂	34	12V + 22m
Dracaena guianensis ♀	38	10V + 2I + 26m
Tupinambis nigropunctatus ♂	36	10V + 2I + 24m
Callopistes flavipunctatus ♀	38	12V + 26m
Callopistes maculatus ♀	38	12V + 26m
Ameiva-group		
Ameiva ameiva ♂	50	26I + 24m
A. exsul ♂	50	26I + 24m
A. dorsalis ♂	50	4sI + 22I + 24m
A. maynardi ♀	50	4sI + 22I + 24m
A. chrysolaema	50	6sI + 20I + 24m
Kentropyx striatus ♂	50	26I + 24m
Dicrodon guttulatum ♀	56	32I + 24m
Teius teyou ♀	54	2V + 6sV + 22I + 24m

[a] Adapted from Gorman, 1970.

I. Sexually reproducing species

Species group	2n	Description of karyotype	Species examined
deppei	52	28I + 24m	*deppei, guttatus, lineatissimus, hyperythrus*
lemniscatus[b]	50	2sI + 24I + 24m	*lemniscatus, murinus*
tigris[c]	46	6 large V + 16 smaller sI + 24m	*tigris*
sexlineatus	46	2 large V + 4sI + 20I + 20m (distinction between sI and I not always firm)	*angusticeps, burti, caldipes, communis, costatus, gularis, inornatus, mexicanus, motaguae, parvisocius, sacki, septemvittatus, sexlineatus*

II. Parthenogenetic species

Species group	Species	Ploidy	Comments
deppei[d]	*cozumelus*	diploid	modified diploid. 2n = 49 and 2n = 47 have been illustrated
sexlineatus	*exsanguis*	all triploid	
	flagellicaudus	all triploid	
	opatae	all triploid	
	sonorae	all triploid	
	uniparens	all triploid	
	velox	all triploid	
tesselatus	*neomexicanus*	diploid	derived presumably by *tigris-inornatus* hybridization; has *tigris*-group and *sexlineatus*-group haploid complements
	tesselatus	diploid and triploid	when diploid resembles *neomexicanus*; when triploid has 1 *tigris*-group and 2 *sexlineatus* group complements.

[a] Adapted from Lowe *et al.*, 1970a except where indicated.
[b] Gorman, 1970 was in agreement.
[c] Cole *et al.*, 1969 described X–Y heteromorphism for centromere position in chromosome pair 3.
[d] Fritts, 1969 found $2n = 47$ at least in some individuals of *C. cozumelus*. He reported $2n = 44$ for *C. angusticeps* and $2n = 50$ *C. deppei*. In all cases he records one chromosome pair fewer than Lowe *et al.*, 1970a.

one species, *Crocodilurus lacertinus* was found with 12V + 22m. As has been shown with several examples within the genus *Anolis*, loss of a pair of microchromosomes is a relatively frequent evolutionary event. The second group, the *Ameiva*-group is characterized by high diploid numbers (46–56), few two-armed chromosomes, and an indistinct break between M and m (Fig. 10). I postulated that the karyotypes in this group may have been derived by establishment of a completely fissioned karyotype from a

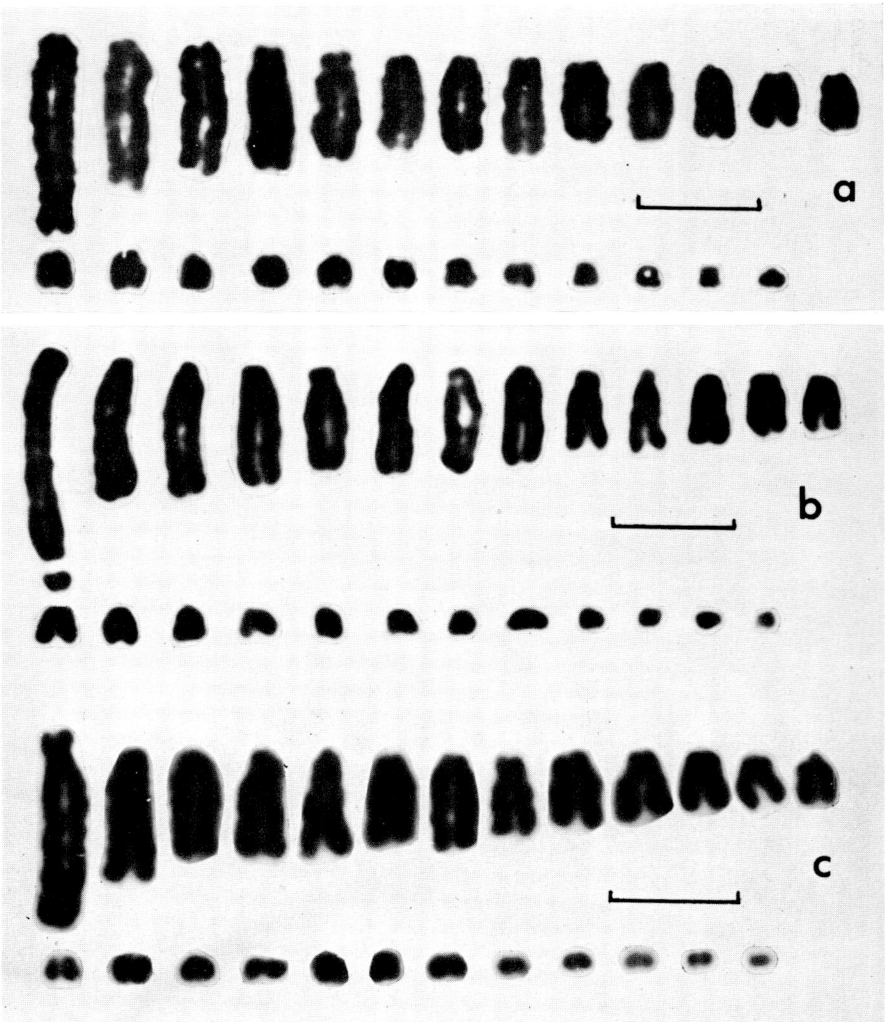

Fig. 10. Half karyotypes of representatives of the *Ameiva*-group of macroteiids. a. *Ameiva exsul*. b. *Kentropyx striatus*. c. *Cnemidophorus murinus*. (From Gorman, 1970.)

Dracaena-group ancestor (Gorman, 1970). The microteiids probably comprise several natural groups. Only two species in two closely related genera were examined and they were identical to each other and distinct from all of the macroteiids.

The large genus *Cnemidophorus* (*Ameiva*-group of macroteiids) has been extensively studied by Lowe and co-workers, who most recently published a review of the karyotype data for 29 species representing all recognized species groups (Lowe *et al.*, 1970a). They found that each species group had a distinctive karyotype, with virtually no chromosome variation *within* species groups.

Parthenogenetic triploidy within the genus *Cnemidophorus* was first described by Pennock (1965). Lowe and Wright (1966) and Lowe *et al.* (1970a, b) followed up this initial finding with extensive cytological studies on parthenogens. They found diploid as well as triploid species and have even reported tetraploid males that resulted from mating of a triploid parthenogen with a diploid male of another species. Chromosome data imply that *Cnemidophorus* parthenogens originated through interspecific hybridization. This was clearly the case when the presumed parental forms were of different species groups, because chromosome markers could be found representing both species groups. Biochemical genetic studies by Neaves and Gerald (1968) and Neaves (1969) convincingly support the hypothesis of hybrid origin. Ecological studies by Wright and Lowe (1968) show that there is a common denominator to the habitats of virtually all the parthenospecies. They live in marginal, perpetually disturbed areas.

Sex chromosomes have been demonstrated for *Cnemidophorus tigris*. X–Y heterogamety was found, involving centromere position (Cole *et al.*, 1969). Heterogamety is clearly absent in the other species groups of *Cnemidophorus*, and was not evident among the other species of the family that have been studied.

(k) *Families Anguidae* and *Anniellidae* (Table XI)

Anguids are a relatively small family with a large amount of karyotypic variability. *Diploglossus costatus* has $2n = 36$, with 12V and 24m (Bury *et al.*, 1969). McDowell and Bogert (1954) considered the diploglossines the most primitive of the anguids. Once again the "typical" saurian karyotype is found in the morphologically most primitive representative of the taxon, and karyotype evolution can be interpreted as away from this condition.

The limbless anguids *Anguis fragilis* and *Ophisaurus apus* have $2n = 44$. They both are reported to have 24m, and 24 macrochromosomal arms, 4V + 16I (Matthey, 1931; Margot, 1946). Thus there is an obvious Robertsonian relationship between these species and the primitive form. Another

TABLE XI. Chromosomes of the Anguinomorpha

Family	Species	2*n*	Description	Reference
Anguidae				
	Diploglossus costatus	36	12V + 24m	Bury *et al.*, 1969
	Anguis fragilis	44	4V + 16I + 24m	Margot, 1946
	Ophisaurus apus (*Pseudopus*)	44	4V + 16I + 24m	Matthey, 1931
	Ophisaurus ventralis	30	4V + 16I + 10m	Matthey, 1931
	Gerrhonotus multicarinatus	48	22I + 26m	Bury *et al.*, 1969
	G. multicarinatus	47	1V + 20I + 26m	Ohno, 1967
	G. paucicarinauts	46	2V + 18I + 26m	Bury *et al.*, 1969
	G. coeruleus	38	10V + 2I + 26m	Bury *et al.*, 1969
	G. monticolus	30	4V + 14I + 12m	Bury *et al.*, 1969
	G. kingi	45	1V + 20I + 24m	See Matthey, 1949, p. 193
Anniellidae				
	Anniela pulchra	22	8 large V + 8 smaller M (2sV + 2sI + 4I) + 6m	Gorman, unpublished
Helodermatidae				
	Heloderma suspectum	38	10V + 4I + 24m	Matthey, 1931
Varanidae				
	Varanus gouldi	40	8V + 8I + 24m	Matthey, 1931
	V. monitor	40	8V + 8I + 24m	Dutt, 1968
	V. monitor	40	12V + 2sV + 2I + 24m	Singh *et al.*, 1970a
	V. flavescens	40	10V + 4sV + 2I + 24m	
	V. rudicollis	40	8V + 6sI + 2I + 24m	Gorman and Gress, 1970b

limbless anguid, *Ophisaurus ventralis* has the same macrochromosomal complement as the above two species but a greatly reduced number of microchromosomes ($2n = 30$). It may be of some taxonomic significance that *Anguis* and *Ophisaurus apus*, both Old World species are more similar to each other in karyotype than are *O. apus* and the New World species *O. ventralis*.

The genus *Gerrhonotus* exhibits both intra- and interspecific karyotype variation. Bury *et al.* (1969) found that two *G. multicarinatus* had 48 chromosomes, all acrocentric. The break between M and m was not sharp. They considered the karyotype to be 22I + 26m, but it is not unlikely that such a karyotype could be derived from establishment of complete fissioning of a primitive 12V + 24m ancestor. Ohno (1967) published a figure of a mitotic spread of *G. multicarinatus* with $2n = 47$, one of the macrochromosomes was metacentric. It is thus likely that individuals will be found with $2n = 46$, with one pair of V. *G. paucicarinatus* is considered very closely related to *G. multicarinatus*. The single individual studied by Bury *et al.* in fact had $2n = 46$ with a single pair of V chromosomes.

G. coeruleus has $2n = 38$. There are 10V + 2sI + 26m. Again there are 48 arms in the karyotype. *G. monticolus* has the most distinctly different karyotype. There are 4V + 14I + 12m. The largest pair of m is clearly metacentric. The macrochromosomal complement has the same or similar number of arms as in the other members of the genus, but the number of microchromosomes is greatly reduced.

Matthey (1931) reported chromosome variation in *G. scincicauda* (=*G. multicarinatus scincicauda*) with $2n = 44$, one pair of V; and $2n = 45$ with a single unpaired V. We considered it likely that he failed to resolve the smallest pair of microchromosomes, and that the diploid numbers were 46 and 47. For *G. kingi* Matthey (1931) also reported $2n = 45$, with a single metacentric element.

Anniellidae is a monogeneric family (two species) closely related to the Anguidae. The karyotype of *Anniella pulchra* is highly derived and distinct from the anguids. Matthey (1931) described the karyotype as $2n = 24$, 8V + 8I + 8 short and globose m. He assumed that these "globose" microchromosomes were homologous to the typical 24m. Careful re-examination of the karyotype (unpublished) showed a slight discrepancy with Matthey's description. The diploid number is only 22. The 4 largest pairs of chromosomes are V, pair 5 is sI, pairs 6 and 7 are I, but pair 8 is clearly two-armed (sV). There are only 6 microchromosomes, and they are not large enough to warrant the assumption that they are directly homologous to the typical condition of 24m (Fig. 11). (See Addendum.)

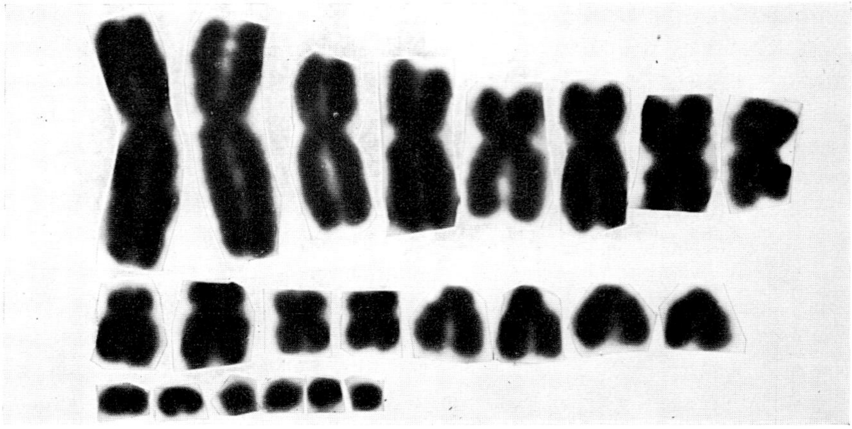

Fig. 11. Karyotype of *Anniella pulchra*, male, $2n = 22$.

(1) *Families Helodermatidae and Varanidae* (Table XI)

There are only two species of venomous lizards in the world, both in the genus *Heloderma*. Matthey (1931) described the karyotype of *H. suspectum* as $2n = 38$, 10V + 4I + 24m. There has been no further examination of this or the closely related species *H. horridum*.

The family Varanidae, which includes the world's largest lizards, is monogeneric. There are about 30 species of *Varanus*, with chromosome information available for four: *V. gouldi* (Matthey, 1931); *V. monitor* (Dutt, 1968; Singh *et al.*, 1970a); *V. flavescens* (Singh *et al.*, 1970a) and *V. rudicollis* (Gorman and Gress, 1970b). All have $2n = 40$ with 16M and 24m. There may be minor differences in the morphology of the macrochromosomes. Matthey described the M chromosomes of *V. gouldi* as 8 large V and 8 smaller I. Dutt's description of *V. monitor* was the same. We found a similar condition in *V. rudicollis* except that 6 of the 8 so-called I chromosomes had distinct short arms and thus are subtelocentric. Singh *et al.* described the macrochromosomal complement of *V. monitor* slightly differently from Dutt, claiming 12V + 2sV + 2I. For *V. flavescens* these authors state that there are 10V + 4sV + 2I.

Varanid lizards are sometimes considered as close relatives of primitive snakes (McDowell and Bogert, 1954). We felt that the karyotype does not support this contention (Gorman and Gress, 1970b). Many saurian groups typically have 24 m chromosomes, an unusual condition among snakes. Furthermore the macrochromosomal complement of varanids appears to be a relatively simple derivative of the common saurian 12V condition.

(m) *Family Amphisbaenidae* (Table XII)

Matthey (1932, 1933) published the first reports of amphisbaenid karyotypes, but more accurate descriptions of the chromosomes of the three species that he studied were published by Huang *et al.* (1967) in a paper describing the karyotypes of 13 species of amphisbaenids. These data were further discussed by Gans *et al.* (1967). Subsequently, Huang and Gans (1970) issued a short note on the karyotypes of 6 additional species.

Of the 19 species that have been studied, 5, representing 4 genera and both subfamilies, have the typical saurian karyotype of 12V + 24m. Another 9 species have the 12V, but a reduced number of microchromosomes. Because the only other reptiles that have 12V + 24m are true lizards, this argues for strong saurian affinity even if one chooses to raise the Amphisbaenia to ordinal level.

Karyotype details have been published for only three of the five species that lack the typical 12V (Huang *et al.*, 1967). In *Amphisbaena camura* there are 24 macrochromosomes. Twenty of these are clearly I, and 4 are sI. Thus, the macrochromosomal complement could have been derived by Robertsonian fission of all the M chromosomes, with subsequent pericentric inversions established in two pairs. In *Rhineura floridana* the largest pair of chromosomes is V, the next ten pairs are all virtually I. The break between M and m is not sharp. From the figure published by Huang *et al.* (1967) I would place the break between pairs 11 and 12. This would leave 11 pairs of microchromosomes and 24 macrochromosomal arms among the 11 pairs of macrochromosomes. Again there is a Robertsonian relationship between the M complement of *Rhineura* and the more typical amphisbaenid. This was not the interpretation of Huang *et al.* (1967). They claimed that there were 30 macrochromosomal arms in the karyotype. This was because they arbitrarily placed pair 12 with the macrochromosomes, and because pairs 2 and 3 have slightly subterminal centromeres. However, such subtelocentrics probably evolved by inversion following centric fission. Thus, only one species remains that differs significantly in macrochromosomal complement. This is *Bipes biporus* with 20 two-armed macrochromosomes and 20m ($2n = 40$). Huang *et al.* (1967) emphasize that classical anatomical studies indicate that *Bipes* is far removed from the main line of amphisbaenian evolution. Even in *Bipes* most of the two-armed macrochromosomes are sI, and therefore might be considered derived by inversion following centric fission.

(n) *Summary of lizard karyotypes*

Chromosome data are available for approximately 300 species (about

TABLE XII. Chromosomes of the Amphisbaenidae

Species	2*n*	Description	Reference
Trogonophinae			
Diplometopon zarudnyi	36	12V + 24m	Huang *et al.*, 1967
Trogonophis wiegmanni	36	12V + 24m	Huang *et al.*, 1967
Amphisbaeninae			
Amphisbaena caeca	36	12V + 24	Huang *et al.*, 1967
A. manni	36	12V + 24m	Huang and Gans, 1970
A. xera	36	12V + 24m	Huang *et al.*, 1967
A. angustifrons	30	12V + 18m	Huang *et al.*, 1967
A. darwini	30	12V + 18m	Huang *et al.*, 1967
A. alba	38	22M (morphology not described) + 16m	Huang and Gans, 1970
A. camura	44	20I + 4sI + 20m	Huang *et al.*, 1967
A. innocens	50	22M (morphology not described) + 28m	Huang and Gans, 1970
Anops kingi	26	12V + 14m	Huang and Gans, 1970
Bipes biporus	40	6V + 14sI + 20m	Huang *et al.*, 1967
Blanus cinereus	32	12V + 20m	Huang *et al.*, 1967
B. stauchi	32	12V + 20m	Huang *et al.*, 1967
Chirindia leucura	30–32	12V + 18 ? 20 ? m	Huang and Gans, 1970
Leptosternon microcephalum	32 ?		
	34 ?	12V + 20 ? 22 ? m	Huang *et al.*, 1967
Monopeltis capensis	34	12V + 22m	Huang and Gans, 1970
Rhineura floridana	44	2V + 20I + 22m	Huang *et al.*, 1967
Zygaspis quadriforns	36	12V + 24m	Huang *et al.*, 1967

10% of the extant forms). Diploid numbers range from 20 to 56. Male chromosome heteromorphism has been confirmed in representatives of three families. Female heteromorphism has been confirmed in one species of a fourth family. In general, sex chromosome heterogamety is not obvious. Parthenogenesis has been confirmed in 4 families, and in three of these families at least some of the parthenogenetic species are triploid. Interspecific hybridization has been implicated as the causative agent of at least some of the parthenogens. There is no *a priori* method to ascertain at what level karyotype will be taxonomically relevant. For example, the Lacertidae are remarkably uniform. Ninety per cent of the species studied have virtually identical karyotypes. The supposedly closely related family Teiidae is just the reverse. It is rare for two genera to have the same karyotype, and usually species groups differ. The single most commonly observed karyotype among lizards is 12V + 24m. Often the anatomically primitive representatives of a family have this karyotype. Although one could argue for multiple convergent events leading to such a karyotype, I prefer to consider this a primitive condition with evolution away from 12V + 24m. The following families have representatives with 12V + 24M: Iguanidae, Agamidae, Chamaeleontidae, Gerrhosauridae, Anguidae and Amphisbaenidae. In addition, representatives of the Scincidae and Teiidae can be found with a slight modification of this karyotype, a reduced number of microchromosomes; and the Helodermatidae and Varanidae are characterized by another sort of minor rearrangement, apparent centric fission of one or two pairs of macrochromosomes. Our knowledge is greatest for the New World families, Iguanidae and Teiidae, and the European Lacertidae. Our ignorance is maximum in the paleotropics and Australia. The latter situation could be easily ameliorated, because there is a relatively high density of excellent cytologists and herpetologists.

C. *Order Ophidia* (Table XIII)

The snakes rival lizards for numbers of species. They are more difficult to classify because of basic similarity of bodily proportions. There is far from uniformity of opinion about what constitutes a family, nor do systematists agree about relationships within and between supposed subfamilies. I have chosen to present the data within the framework of the classification of Underwood (1967, personal communication). It is undoubtedly controversial, but it is one way of ordering the chromosome data, and karyotype may point out strengths and weaknesses of the proposed scheme.

TABLE XIII. Chromosomes of Snakes

Taxon[a]	$2n$	Description[b]	Z–W[c]	Reference
Typhlopidae				
Typhlops simoni ♂	32	16M + 16m (10V + 6I)	?	Werner, 1959
Leptotyphlopidae				
Leptotyphlops phillipsi ♂	36	16M + 20m (10V + 6I)	?	Werner, 1959
Xenopeltidae				
Xenopeltis unicolor ♂	36	16M + 20m (8V + 8I)	?	Cole and Dowling, 1970
Boidae				
Boa constrictor ♂ ♀	36	16M + 20m (8V + 8I)	No	Beçak *et al.*, 1962
Charina bottae ♀	36	16M + 20m (8V + 8I)		Gorman and Gress, 1970b
Corallus caninus ♂ ♀	44	24M + 20m (24I)	No	Beçak, 1965
Corallus enhydris ♀	40	20M + 20m (4V + 16I)	No	Gorman and Gress, 1970b
Epicrates cenchria ♂ ♀	36.	6M + 20m (8V + 8I)	No	Beçak, 1965
Epicrates striatus ♂	36	16M + 20m (8V + 8I)	?	Gorman and Gress, 1970b
Eryx jaculus ♂	34	16M + 18m (8V + 8I)	?	Werner, 1959
Eryx johni ♂ ♀	34	16M + 18m (8V + 8I)	No	Singh *et al.*, 1968a, 1970a

Eunectes murinus ♂	36	16M + 20m		
♀		(8V + 8I)	No	Beçak, 1965
Lichanura roseofusca ♂	36	16M + 20m		
		(8V + 8I)	?	Gorman and Gress, 1970b
Python molurus ♀	36	16M + 20m		
		(8V + 8I)	No	Singh *et al.*, 1968a
'Colubrines"				
Arizona elegans ♂	36	16M + 20m	?	Bury *et al.*, 1970
Chilomeniscus stramineus ♂	36	16M + 20m	?	Bury *et al.*, 1970
Chionactis occipitalis ♂	36	16M + 20m	?	Bury *et al.*, 1970
Chironius bicarinatus ♂ ♀	36	16M + 20m	V, sV	Beçak, 1965
Contia tenuis ♂	36	16M + 20m	?	Bury *et al.*, 1970
Coluber constrictor ♂	36	16M + 20m	?	Bury *et al.*, 1970
Coluber gemonensis ♂ (*Zamensis gemonensis*)	36	16M + 20m	?	Matthey, 1931
Coluber jugularis ♂	36	16M + 20m	?	Gorman, unpublished
Coluber rogersi ♂	36	16M + 20m	?	Gorman, unpublished
Coluber viridiflavus ♂	36	16M + 20m	?	Kobel, 1967
Coronella austriaca ♂	36	16M + 20m	?	Kobel, 1967
Diadophis punctatus ♂	36	16M + 20m	?	Bury *et al.*, 1970
Dryadophis bifossatus ♂ ♀	36	16M + 20m	V, I	Beçak, 1965
Drymarchon corais ♂ ♀	36	16M + 20m	V, I	Beçak, 1965
Elaphe carinata ?	36	16M + 20m	?	Fischman *et al.*, 1968
Elaphe climacophora ♂	36	16M + 20m	?	Nakamura, 1935c
Elaphe obsoleta ♂	36	16M + 20m	?	Fischman *et al.*, 1968
Elaphe quadrivirgata ♂	36	16M + 20m	?	Nakamura, 1935c
Lampropeltis getulus ♂	36	16M + 20m	?	Bury *et al.*, 1970
Malpolon monspessulanus ♂	42	20M + 22m		
(*Coelopeltis lacertina*)		M, m indistinct	?	Matthey, 1931

Taxon[a]	2n	Description[b]	Z–W[c]	Reference
Masticophis flagellum ♂	36	16M + 20m	?	Bury *et al.*, 1970
Pituophis melanoleucus ♂	36	16M + 20m	?	Bury *et al.*, 1970
Phyllorhynchus decurtatus ♂	36	16M + 20m	?	Bury *et al.*, 1970
Pseustes sulphureus ♂	38	18M + 20m (6V + 8sV + 4I)	No	Beçak and Beçak, 1969
Pytas mucosus ♂ ♀	34	16M + 18m	No	Singh *et al.*, 1970a; Bhatnagar, 1960a
Rhinocheilus lecontei ♂	36	16M + 20m	?	Bury *et al.*, 1970
Spilotes pullatus ♂ ♀	36	16M + 20m	V sI	Beçak, 1965
Zaocys nigromarginatus ♂	36	16M + 20m	?	Nakamura, 1935c
"Xenodontines"				
Erythrolamprus aesculapii ♂	28	M, m indistinct	?	Beçak and Beçak, 1969
Heterodon platyrhinos ♂	36	16M + 20m	?	Addendum
Hydrodynastes bicinctus ♂	24	M, m indistinct	?	Beçak and Beçak, 1969
Hydrodynastes gigas ♂	24	M, m indistinct	?	Beçak and Beçak, 1969
Liophis miliaris ♂ ♀	28	M, m indistinct	V, I	Beçak and Beçak, 1969
Oligodon arnensis ♂	46	24M + 22m (24I)	?	Bhatnagar, 1959
Oligodon formosanus ♂ (*Holarchus formosanus*)	36	16M + 20m	?	Nakamura, 1935c
Philodryas aestivus ♂	36	16M + 20m	?	Beçak and Beçak, 1969
Philodryas olfersii ♂	36	16M + 20m	V, I	Beçak, 1965
Philodryas serra ♂ ♀	28	M, m indistinct	V, I	Beçak and Beçak, 1969
Thamnodynastes pallidus ♂ ♀	34	M, m indistinct	sV, I size	Beçak and Beçak, 1969
Thamnodynastes strigatus ♂	32	18M + 14m		
♀		(6V + 8sV + 4I)	sV, I size	Beçak and Beçak, 1969

Tomodon dorsatus ♂ ♀	32	18M + 14m (6V + 8sV + 4I)	sV, V size	Beçak and Beçak, 1969
Trimorphodon vandenburghi ♂	38	16M + 22m (8V + 8I) M m, indistinct	?	Bury *et al.*, 1970
Xenodon merremii ♂ ♀	30	16M + 14m (10V + 4sV + 2I)	V, v size	Beçak, 1965
Xenodon neuwiedii ♀	30	16M + 14m (10V + 4sV + 2I)	V, v size	Beçak and Beçak, 1969
"Lycodontines"				
Clelia occipitolutea ♂ ♀	50	14M + 36m (2V + 12I)	V, I size	Beçak, 1965
Dinodon rufuzonatus ♂	46	16M + 30m (2V + 14I)	?	Nakamura, 1935c
Lycodon aulicus ♂ ♀	36	16M + 20m	Yes, no details	Nakamura, 1935c
Oxyrophus petolarius ♂ ♀	46	16M + 30m (2V + 4sV + 10I)	sV, I size	Beçak and Beçak, 1969
"Natricines"				
Amphiesma piscator ♂ (*Natrix piscator*) ♀	40	10M + 30m (8V + 2sI)	V, I size	Singh *et al.*, 1968b
Amphiesma piscator ♂ ♀	42	18M + 24m (8V + 2sI + 8I)	Yes, no details	Singh *et al.*, 1970a
Macropisthodon rudis ♂	46	16M + 30m (2V + 14I)	?	Nakamura, 1935c
Natrix maura ♂ ♀	34	16M + 18m	V, I size	Kobel, 1967

Taxon[a]	2*n*	Description[b]	Z–W[c]	Reference
Natrix natrix ♂	34	16M + 18m	V, I	Kobel, 1967
♀			size	
Natrix tesselata ♂	34	16M + 18m	?	Kobel, 1967
Natrix rhombifera ♂	36	M, m indistinct		
♀		10 largest sI		
		most others V	No	van Brink, 1959
Rhabdophis tigrina ♂	40	16M + 24m	?	Nakamura, 1928
(*Natrix tigrina*)				
Xenochrophis stolata ♂	36	16M + 20m		
(*Natrix stolata*		(10V + 6I)	?	Bhatnagar, 1960a
Xenochrophis stolata ♂	36	14M + 22m	Yes,	
♀		(12V + 2sI)	no details	Singh *et al.*, 1970a
"Homalopsines"				
Cerberus rhynchops ♂	36	16M + 20m	Yes, no	Singh *et al.*, 1970a
♀			details	
Cerberus rhynchops ♂	38	?	?	Dutt, 1966 cited in Beçak and Beçak, 1969
Gerardia prevostiana ♀	36	16M + 20m	Yes, no	Singh *et al.*, 1970a
			details	
"Boigines"				
Imantodes cenchoa ♂	36	No details	?	Beçak and Beçak, 1969
Telescopus fallax ♂	36	16M + 20m	?	Matthey, 1931
(*Tarbophis fallax*)				
Viperidae				
Crotalinae				
Agkistrodon acutus ♂	36	16M + 20m	?	Nakamura, 1935c
Agkistrodon halys ♂	36	16M + 20m	?	Nakamura, 1935c
Bothrops alternatus ♀	36	16M + 20m	V, I	Beçak, 1965
			size	
Bothrops insularis[d]	36	16M + 20m	V, V	Beçak, 1965
			size	

Bothrops jararaca ♂	36	16M + 20m	V, sV	Beçak, 1965
♀				
Bothrops jararacussu ♂	36	16M + 20m	V, sV	Beçak, 1965
♀			size	
Bothrops moojeni ♂	36	16M + 20m	V, I	Beçak, 1965
(*Bothrops atrox*) ♀			size	
Bothrops pradoi ♂	36	16M + 20m	V, sI	Beçak, 1965
♀			size	
Crotalus cerastes ♀	36	16M + 20m	V, sI	Ohno, 1967
			size	
Crotalus durissus ♂	36	16M + 20m	V, sI	Beçak, 1965
♀			size	
Crotalus horridus	36	16M + 20m	?	William Johnson, unpublished
Crotalus mitchelli	36	16M + 20m	?	William Johnson, unpublished
Crotalus molossus	36	16M + 20m	?	William Johnson, unpublished
Crotalus ruber	36	16M + 20m	?	William Johnson, unpublished
Crotalus scutulatus	36	16M + 20m	?	William Johnson, unpublished
Crotalus tigris	36	16M + 20m	?	William Johnson, unpublished
Crotalus viridis ♂	36	16M + 20m	?	Monroe, 1962; William Johnson, unpublished
Lachesis muta ♂	36	16M + 20m	V, I	Beçak and Beçak, 1969
♀			size	
Trimeresurus flavoviridis ♂	36	16M + 20m	?	Makino and Momma, 1949
Trimeresurus gramineus ♂	36	16M + 20m	?	Nakamura, 1935c
Trimeresurus mucrosquamatus ♂	36	16M + 20m	?	Nakamura, 1935c
Trimeresurus okinavensis ♂	36	16M + 20m	?	Makino and Momma, 1949
Viperinae				
Echis carinatus ♀	36	16M + 20m	Yes, no details	Singh *et al.*, 1970a
Vipera aspis ♂	42	22M + 20m	V, sI	Kobel, 1967; Kobel, 1963; Matthey, 1931
♀		2V + 14sI + 61	size	
Vipera berus ♂	36	16M + 20m	V, sV	Kobel, 1967; Kobel, 1962;
♀			size	Makino and Momma, 1949

Taxon[a]	2n	Description[b]	Z–W[c]	Reference
Vipera ursinii ♀	36	16M + 20m	V, sV size	Kobel, 1967
Elapidae				
Elapinae				
Bungarus caeruleus ♂ ♀	44/43	24M + 20m (2sV + 22I)	Z_1 sV Z_2 I W V	Singh *et al.*, 1970b
Bungarus multicinctus ♂	36	22M + 14m (4V + 18I)	?	Nakamura, 1935c
Micrurus lemniscatus ♂	42	22M + 20m (2sV + 20I)	?	Beçak and Beçak, 1969
Naja naja ♂	38	14M + 24m (6V + 4sV + 4I)	?	Nakamura, 1935c
Naja naja ♂ ♀	38	10M + 6 intermediate + 22m (4V + 4sV + 2sI; + 2V + 4sV)	No	Singh *et al.*, 1970a
Hydrophiinae				
Laticauda semifasciata ♂	38	14M + 24m (8V + 6I)	?	Nakamura, 1935c

[a] *Taxon.* I have followed traditional familial classification except for the Colubridae, *sensu lato*, which is divided into groupings proposed by Underwood. Within each family or group, snakes are listed alphabetically, not by presumed taxonomic affinity. Names in brackets are those given in original papers.

[b] *Description.* M, m indistinct—no clear break in size between macrochromosomes and microchromosomes. In brackets the macrochromosomal morphology is abbreviated where relevant and available. This has not been done for colubrines or viperids, except where the karyotype is unusual. Older literature descriptions of macrochromosomal morphology are often erroneous. The widespread report that the "typical" colubrid or viperid macrochromosomal complement is 10V + 6I is erroneous. This karyotype has not been found in any species studied since 1965. The most "typical" karyotype of these higher snakes might be abbreviated 10V + 4sV + 2I. The abbreviated description is for males (homogametic sex).

[c] *Z–W.* All known cases of sex chromosome heteromorphism involve females. Chromosome morphology is described for the Z, then the W. The word "size" means that Z and W differ not only in the centromere position, but also in length. "Yes, no details" appears when authors claim heteromorphism without other data.

[d] An intersex was studied. It had the normal female karyotype.

(a) *The lower snakes*

The infraorder Scolecophidia consists of two families, Typhlopidae and Leptotyphlopidae. Both are world-wide in distribution and rich in species. Few systematists doubt that these represent primitive families. Chromosome data are available for only one species in each family (Werner, 1959). The karyotype of *Leptotyphlops phillipsi* was described as $2n = 36$, 10V + 6I + 20m. This karyotype had been described for many of the advanced snakes, but not in recent works. I am suggesting that unless the karyotype of *Leptotyphlops* is atypical, some of the macrochromosomes described as "I" probably have distinct short arms, which would give this primitive burrower a commonly observed karyotype. *Typhlops simoni*, $2n = 32$ was described as having the same macrochromosomal complement, but only 16 microchromosomes.

The infraorder Henophidia includes Xenopeltidae and Boidae. Cole and Dowling (1970) described the karyotype of *Xenopeltis unicolor* as $2n = 36$, 8V, 8I, 20m.

The condition 8V, 8I also seems characteristic of the Boidae. The very complement described for *Xenopeltis* has been found in the following boids: *Python molurus*, subfamily Pythoninae (Singh *et al.*, 1968a); *Boa constrictor*, *Epicrates cenchria Eunectes murinus* (Beçak, 1965); *Charina bottae* and *Lichanura roseofusca* (Gorman and Gress, 1970b) (Fig. 12). All of the foregoing are in subfamily Boinae, tribe Boini. Within the tribe Boini the only variation from the above described complement has been reported for the genus *Corallus*, *C. enhydris* has 4V, 16I, 20m (Gorman and Gress, 1970b) and *C. caninus* has 24I, 20m (Beçak, 1965). It can be seen that *Corallus* has the identical number of macrochromosomal arms as the other Boini. The karyotypes found in that genus were presumably derived by centric fission of metacentric macrochromosomes.

Only in the genus *Eryx*, tribe Erycini, is there any described deviation from the formula 24 macrochromosomal arms, 20 microchromosomal arms. Each of the three species of *Eryx* studied (Werner, 1959; Singh *et al.*, 1968a, 1970a) has $2n = 34$ with 18m. Thus there is a reduction in the number of microchromosomes. The macrochromosomes are similar to those of the other boids.

Sex chromosome heteromorphism has not been found in any member of the family Boidae. Only males have been studied in representatives of the other families of lower snakes.

(b) *The "colubrid" snakes*

Snakes lacking hollow fangs for the conduction of venom in the front of the mouth, and that are not clearly "lower" snakes are often grouped in this

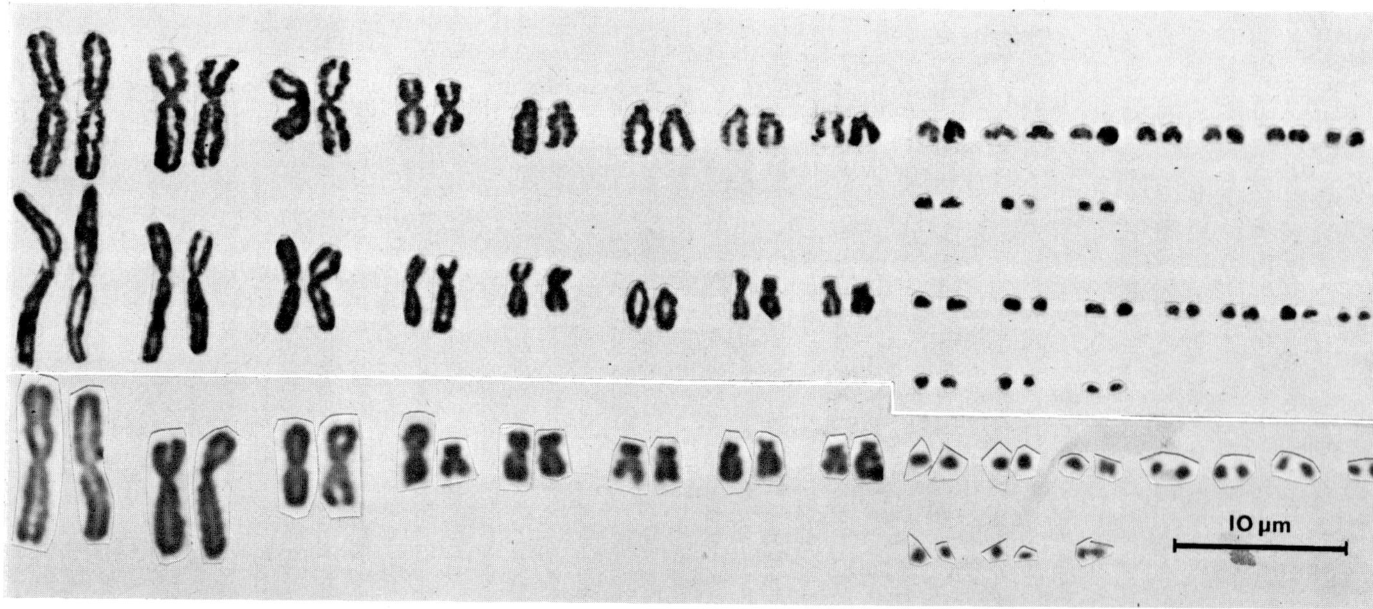

Fig. 12. Karyotypes of female snakes in diverse families with "typical" complements of $2n = 36$, 16M + 20m. Z–W sex chromosome heteromorphism variation between families is also well illustrated. Top row: *Epicrates cenchria*, Boidae. 8V + 8I + 16m, no sex chromosome heteromorphism. The typical boid karyotype. Middle row: *Drymarchon corais*, "colubrine". Except for pair 6, all macro-chromosomes are two-armed. The fourth largest are the sex chromosomes. Z and W are equal in size, differ slightly in centromere position. *Bottom row: Crotalus cerastes*, Viperidae. Marked heteromorphism in size in chromosome pair 4, W much smaller. From Ohno, 1967 with permission.

heterogeneous "family". As defined (or ill-defined) this includes about 70% of all snake species. The classification of this heterogeneous group is much in dispute, and I am not in a position to "take sides" on the basis of chromosome data. The data are presented following suggestions made by Underwood (1967, personal communication) with the express hope that certain taxonomic hypotheses can be tested with the further accumulation of data. I do not follow Underwood's formal familial and subfamilial classification of snakes typically considered "colubrid", but place each clear taxonomic unit in an "informal" subfamily.

(1) The colubrine snakes: Chromosome data are available for about 30 species of colubrines. Of these, the great majority have $2n = 36$ with the karyotype clearly divided into 16 macrochromosomes and 20 microchromosomes. From descriptions and figures in the literature of the 1930's (e.g. Matthey, 1931; Nakamura, 1935c) the macrochromosomal complement of the typical colubrine was considered to be 10V + 6I. More recent studies (Beçak, 1965; Kobel, 1967; Bury *et al.*, 1970) show that such a description is in error. The five largest chromosome pairs are most commonly metacentric, pair six is often with a terminal centromere, but the smaller pairs seven and eight are often submetacentric or subtelocentric. In cases where both males and females have been studied, the fourth chromosome pair (by size) is usually heteromorphic in females (Z–W) (Fig. 12).

Within this large group of snakes with basically similar karyotypes, detailed analysis of relative lengths of chromosome pairs and centromere position may be helpful in pointing out certain natural groups. For example Bury *et al.* (1970) found that chromosome pair four of *Chionactis* and *Chilomeniscus* has a subterminal centromere, supporting the idea that these two genera are closely related to each other. Fischman *et al.* (1968) described differences in centromere position of chromosome pairs six and eight within the genus *Elaphe*, and interpreted this as probable incorporation of pericentric inversions.

Malpolon, *Pseustes*, and *Ptyas* are the only colubrines that depart from the typical $2n = 36$ karyotype. Matthey (1931) described the karyotype of *Malpolon monspessulanus* as $2n = 42$, with about 22 macrochromosomes and 20 microchromosomes, the break between the categories not being well marked. This might be a derived condition, resulting from centric fission of some of the macrochromosomes. The karyotype of a female *Pseustes sulphureus* (Beçak and Beçak, 1969) has 38 chromosomes, 18M and 20m. Since there are two pairs of acrocentric macrochromosomes in this species, the karyotype may have been derived by centric fission of a single pair of metacentrics. This species is also interesting because there are no heteromorphic chromosome pairs, an exception among the higher snakes. The karyotype of *Ptyas mucosus* has been described twice (Bhatnagar, 1960a; Singh *et al.*, 1970a). The authors agree that $2n = 34$ with 16M and 18m, although they do not agree fully on their description of the macrochromosomal morphology. Such a karyotype is probably derived by loss or incorporation of a pair of microchromosomes.

(2) Xenodontine snakes: Underwood (1967) placed the xenodontines as a subfamily of the Dipsadidae. Chromosomes of 17 species have been studied. Only four of the species, *Oligodon formosanus* and three *Philodryas* have the common ophidian karyotype of 16M + 20m (Nakamura, 1935c; Beçak, 1965; Beçak and Beçak, 1969). Within the genus *Philodryas*, Beçak and Beçak found a distinctly different karyotype. *P. serra* has $2n = 28$, with no marked distinction between macro- and microchromosomes. Among the ten largest chromosome pairs, two were I, the others all V or sV. Sex chromosome heteromorphism was on the fourth largest pair. Similarly, $2n = 28$ with the absence of M,m distinction was reported for *Erythrolamprus* and *Liophis*; and $2n = 24$, again without the marked M,m distinction characterized two species of *Hydrodynastes* (Beçak and Beçak, 1969). The karyotype of *Philodryas serra* suggests re-examining the generic allocation of this species.

Several other distinctive karyotypes were described for this group. Two species of *Xenodon* have $2n = 30$, with a weak break between 16M and 14m (Fig. 13). *Tomodon dorsatus* and *Thamnodynastes strigatus* both have $2n = 32$, 18M, 14m. A second species of *Thamnodynastes*, *T. pallidus*, has $2n = 34$, no clear delineation between M and m (Beçak and Beçak, 1969). The authors suggest a Robertsonian relationship between the two species of *Thamnodynastes*, with acrocentric chromosome pairs 9 and 10 of *T. pallidus* homologous to submetacentric pair 9 of *T. strigatus*. Such a relationship would also account for the M,m size break seen in *strigatus* but not in *pallidus*. *Trimorphodon* with $2n = 38$ lacks the clear division into macro- and microchromosomes (Bury *et al.*, 1970); the karyotype of *Heterodon* has been incompletely described (Edgren, 1953) as $n = 20$–22. Finally, *Oligodon arnensis* has $2n = 46$ with 24 all acrocentric M and 22m (Bhatnagar, 1959).

(3) Lycodontine snakes: This is another group which Underwood (1967) placed in the family Dipsadidae. Chromosome data of four species are available. One seems "typical". *Lycodon aulicus* has $2n = 36$, 16M, 20m, the M chromosomes consisting of 12V and 4I (Bhatnagar, 1961). The other three species have high diploid numbers. Nakamura (1935c) reported $2n = 46$ with 16M + 30m for *Dinodon rufuzonatus*. *Oxyrophus petolarius* also has $2n = 46$ with 16M + 30m (Beçak and Beçak, 1969). *Clelia occipitolutea* with $2n = 50$ has the highest diploid number among the snakes (Beçak, 1965). The karyotype was considered to be comprised of 14I + 36m. The

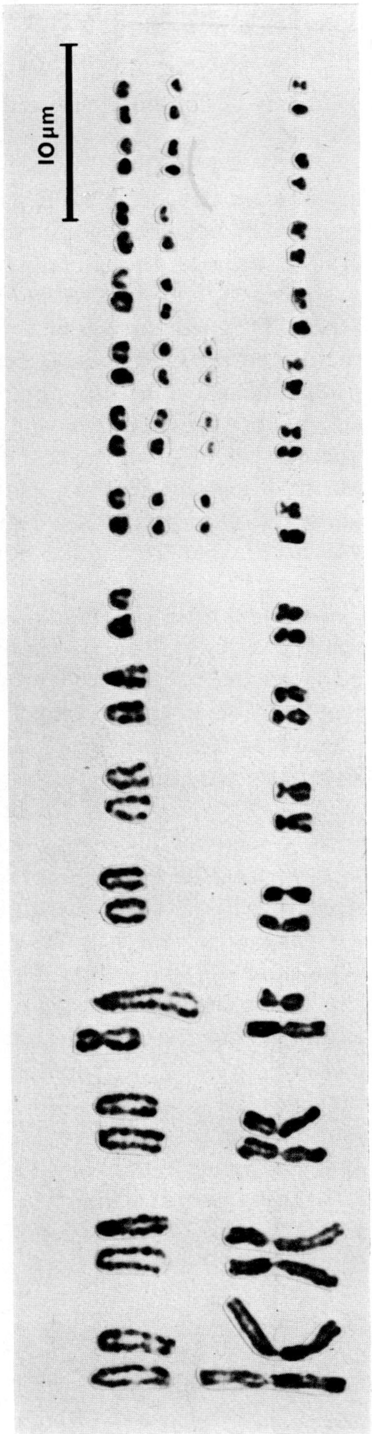

Fig. 13. Karyotypes of non-colubrine female "colubrid" snakes with highly derived chromosome complements. *Top row: Clelia occipitolutea*, "lycodontine". $2n = 50$. The W chromosome is nearly twice the size of the metacentric Z. *Bottom row: Xenodon merremii*, "xenodontine". $2n = 30$. The W chromosome is distinctly smaller than the Z. A break between 16M and 14m can be seen, but it is not clearly marked. From Ohno, 1967.

sex chromosomes are unusual, the Z is metacentric, the W acrocentric and markedly elongate (Fig. 13).

Anatomical studies subsequent to his 1967 publication suggest to Underwood that *Clelia* and *Oxyrophus* may not be closely related to true lycodontines (personal communication).

(4) Natricine snakes: Chromosome data have been published for only eight species in this large group which Underwood (1967) placed in the family Natricidae. Kobel (1967) studied three closely related European species of *Natrix*, *N. natrix*, *N. maura*, and *N. tessellata* and found $2n =$ 34 with 16M + 18m for all three. Detailed analysis of the karyotype (relative length of chromosome pairs, centromere position) permitted not only minor distinctions between the species, but also pointed out that the macrochromosomes of these *Natrix* are similar *inter se* and distinct from the basic macrochromosomal complement of other snakes. Kobel found sex chromosome heteromorphism in European *Natrix*. Matthey (1931) had reported $2n = 36$ for *N. maura* and *N. natrix*, but the high resolution of Kobel's preparations clearly establish that the diploid number is reduced by a pair of m chromosomes.

The North American *N. rhombifera* differs markedly in karyotype (van Brink, 1959). The diploid number is 36, but there is no distinct break between M and m, and there is apparently no sex chromosome heteromorphism. Dr. D. Rossman and Mr. G. Eberle (personal communication) inform me that other North American natricines have $2n = 36$ without the M,m distinction. This is a potentially extremely useful systematic character for it may imply that New World *Natrix* are more closely related to other New World genera than they are to their supposed congeners in Europe.

Chromosome data have been published for three species of Asian natricines that were considered to be *Natrix* by the authors, but which represent distinct genera. *Amphiesma stolata* has $2n = 36$. Bhatnagar (1960a) described the karyotype as 10V + 6I + 20m. Singh *et al.* (1970a) had a different interpretation. They claimed that there were only 14M (12V + 2sI) and 22m. Nakamura (1928) described the karyotype of *Xenochrophis piscator* as $2n = 40$, 10V + 6I + 24m. Singh *et al.* (1968b) reported $2n = 40$ in *Rhabdophis tigrina*. They found 10M + 30m. One pair of macrochromosomes was subtelocentric, the others clearly two-armed. Sex chromosomes were distinct, the Z was sV, the W a smaller I. In a subsequent report (Singh *et al.*, 1970a) these authors claimed $2n = 42$, with 8V + 2sI + 8I + 24m. They made no comment about the conflict in description, and this is all the more puzzling because the 1968 work was accompanied by excellent figures.

Another natricine with an elevated diploid number is *Macropisthodon rudis*, described by Nakamura (1935c) as $2n = 46$, 2V + 14I + 30m.

(5) Homalopsines and Boigines: The two groups considered here were placed by Underwood (1967) in the family Homalopsidae. Singh *et al.* (1970a) provided data on two homalopsines, *Gerardia prevostiana* ($2n = 36$, 16M + 20m, female heteromorphic) and *Cerberus rhynchops* ($2n = 38$, 16M + 22m, female heteromorphic). Chromosome data are available for only two boigines. *Telescopus fallax* has the typical $2n = 36$, 16M + 20m (Matthey, 1931), and *Imantodes cenchoa* also has $2n = 36$, although no details of karyotype have been published (Beçak and Beçak, 1969).

(c) *Vipers and pit-vipers* (*Viperidae*)

Vipers are venomous snakes with movable hollow fangs in the front of the mouth. The pit-vipers are further characterized by a facial pit with heat sensitive nerve endings that aid in the detection of prey. Sometimes placed in the family Crotalidae, we shall follow Underwood (1967) and consider the crotalines as a subfamily of the Viperidae.

The pit-vipers are extremely conservative in karyotype. Data have been published for two species of *Agkistrodon*, six *Bothrops*, four *Trimeresurus*, one *Lachesis*, and two rattlesnakes (*Crotalus*). In addition, Mr. William Johnson has studied, but not yet published the karyotypes of eight species of *Crotalus*. All crotalines have $2n = 36$ with 16M and 20m. The karyotypes in detail look similar to typical colubrines. The older literature (e.g. Nakamura, 1935c) characterizes the macrochromosomes as 10V + 6I. However, the more recent work shows that most of the macrochromosomes are two-armed. Z–W heteromorphism has been found when females have been studied (Fig. 12).

Among the true vipers there are published data for only three species of *Vipera* and one *Echis*. *Echis* and two *Vipera* have the typical $2n = 36$, 16M, 20m with most of the macrochromosomes two-armed. *V. aspis* has $2n = 42$ (Matthey, 1931; Kobel, 1963, 1967). This species has the typical microchromosomal complement of 20, and the difference in diploid number and chromosomal morphology seems to result from centric fissions accompanied by some pericentric inversions. Kobel (1963, 1967) was able to demonstrate probable homologies between the chromosomes of *V. berus* ($2n = 36$) and *V. aspis* ($2n = 42$).

(d) *Cobras, coral snakes and sea snakes* (*Elapidae*)

The family Elapidae is comprised of venomous snakes with fixed front fangs (cobras, kraits, coral snakes and sea snakes). The sea snakes, considered here as a subfamily (Hydrophiinae), are often raised to familial status.

Karyotype data have been published for only five species in this interesting and important family. Interestingly, none of them have the "typical" ophidian karyotype of 16M and 20m. Nakamura (1935c) described the karyotype of the cobra *Naja naja* as $2n = 38$ (10V + 4I + 24m); Singh *et al.* (1970a) divide the karyotype into 10 large and 6 smaller macrochromosomes and 22 microchromosomes. The karyotype of the krait *Bungarus multicinctus* was described as $2n = 36$; 4V, 18I 14m (Nakamura, 1935c). A second member of the genus, *B. caeruleus* proved to be cytologically interesting because of multiple sex chromosomes. Singh *et al.* (1970a, b) found $2n = 44$ in males, $2n = 43$ in females. The male has 2sV and 22I macrochromosomes and the female, 1sV, 1V, and 21I macrochromosomes. This is a Z_1Z_2W sex determining system, and is the only known case of compound sex chromosomes in a snake. Bhatnagar (1960b) described the karyotype of male *B. caeruleus* that was essentially in agreement with the subsequent study. The single coral snake studied, *Micrurus lemniscatus* has $2n = 42$ with 22 macrochromosomes, one pair submetacentric, the remainder acrocentric (Beçak and Beçak, 1969). The only description of a sea snake karyotype is for *Laticauda semifasciata* $2n = 38$, 8V + 6I + 24m (Nakamura, 1935c).

(e) *Summary of snake karyotypes*

Chromosome data are available for about 100 species (about 3% of the living forms). Diploid numbers range from 24 to 50, but $2n = 36$ with 16 macrochromosomes and 20 microchromosomes is by far the most frequent complement, and characterizes "lower snakes" (*Leptotyphlops*, *Xenopeltis*, Boidae), the majority of typical colubrines, the vipers and pit-vipers. It is a karyotype that is notably absent from the elapids, and uncommon among certain highly derived "colubrid" lines (xenodontines and lycodontines). Thus I interpret the general trend of karyotype evolution as away from the 16M, 20m condition. When a group shows departure from the typical condition, there is likely to be a great deal of interspecific (or intergeneric) karyotype variation, rather than stabilization of a new karyotype. It is in such groups that chromosome information will be taxonomically most useful. In general, there is a conservativeness of the total number of chromosome arms in the karyotype ("nombre fondamental" *sensu* Matthey, 1949), this number remaining near 50. Most of the forms with increased microchromosomal complements (> 20) have a reduced number of V and sV elements in the macrochromosomes.

Sex chromosome heteromorphism is unknown in the lower snakes, but is the rule rather than the exception in the higher snakes. Females are the heterogametic sex (Z–W). Among the colubrines, heteromorphism is expressed as a difference in centromere position, whereas among the natricines, xenodontines and viperids there is a distinct size heteromorphism as well. The W is generally smaller than the Z. The lycodontine *Clelia* is an exception, since it has a huge W chromosome. The elapid *Bungarus caeruleus* has multiple sex chromosomes and the W is larger than Z_1 or Z_2.

D. *Order Crocodilia* (Table XIV)

There are 21 species of living crocodilians, and the karyotypes of *all* have been studied and reviewed in a recent paper (Cohen and Gans, 1970). Crocodilians are divided into three subfamilies (sometimes considered families): the alligatorines (4 genera, 7 species); the gavaialines (monotypic) and the crocodilines (2 monotypic genera and 11 species of *Crocodylus*).

Diploid numbers range from 30 to 42, and do not clearly separate the subfamilies. $2n = 32$ is found in *Alligator* (both species), *Gavialis*, six species of *Crocodylus* and *Tomistoma*. The total number of arms in the karyotype is relatively conservative ranging from 56 (one species) to 62 (one species); 14 species have 58 arms.

Sex chromosome heteromorphism has not been found in the 14 species for which data are available for both sexes.

Cohen and Gans do not divide the karyotype into macrochromosomes and microchromosomes, because all the elements have distinct morphology, and many of the smallest elements are metacentric. However, they show (Fig. 5 and Table 11 of Cohen and Gans, 1970) that the karyotypes of all species can be divided into 24 large arms (ranging from 12V to 24I) and a series of distinctly smaller elements. Thus they suggest that there may be a Robertsonian relationship among the 24 large arms of all the species, and that the variation in centromere position among the smaller elements (all species have 18 of these smaller chromosomes) may be a result of independent establishment of inversions (Fig. 14).

Because there has probably been much convergence in the evolution of crocodilian karyotypes, only a few meaningful systematic statements can be made from karyotype. The new world caimans (*Caiman*, *Paleosuchus* and *Melanosuchus*) form a distinct group. They are the only species with $2n = 42$, all large chromosomes acrocentric; and the American and Chinese alligators have virtually identical karyotypes, distinguishable in details from all other species. *Crocodylus acutus* and *C. johnsoni* have identical karyotypes, but their distribution (New World and Australia, respectively)

TABLE XIV. Chromosomes of the Crocodilia

	$2n$	V	Group A: larger chromosomes I	Group B: smaller chromosomes
Paleosuchus trigonatus	42	—	24	18
Paleosuchus palpebrosus	42	—	24	18
Melanosuchus niger	42	—	24	18
Caiman latirostris	42	—	24	18
Caiman crocodilus	42	—	24	18
Osteolaemus tetraspis	38	4	16	18
Crocodylus siamensis	34	8	8	18
Crocodylus porosus	34	8	8	18
Crocodylus intermedius	32	10	4	18
Crocodylus niloticus	32	10	4	18
Crocodylus johnsoni	32	10	4	18
Crocodylus acutus	32	10	4	18
Crocodylus novaeguineae	32	10	4	18
Crocodylus moreletii	32	10	4	18
Alligator sinensis	32	10	4	18
Alligator mississipiensis	32	10	4	18
Tomistoma schlegelii	32	10	4	18
Gavialis gangeticus	32	10	4	18
Crocodylus cataphractus	30	12	—	18
Crocodylus rhombifer	30	12	—	18
Crocodylus palustris	30	12	—	18

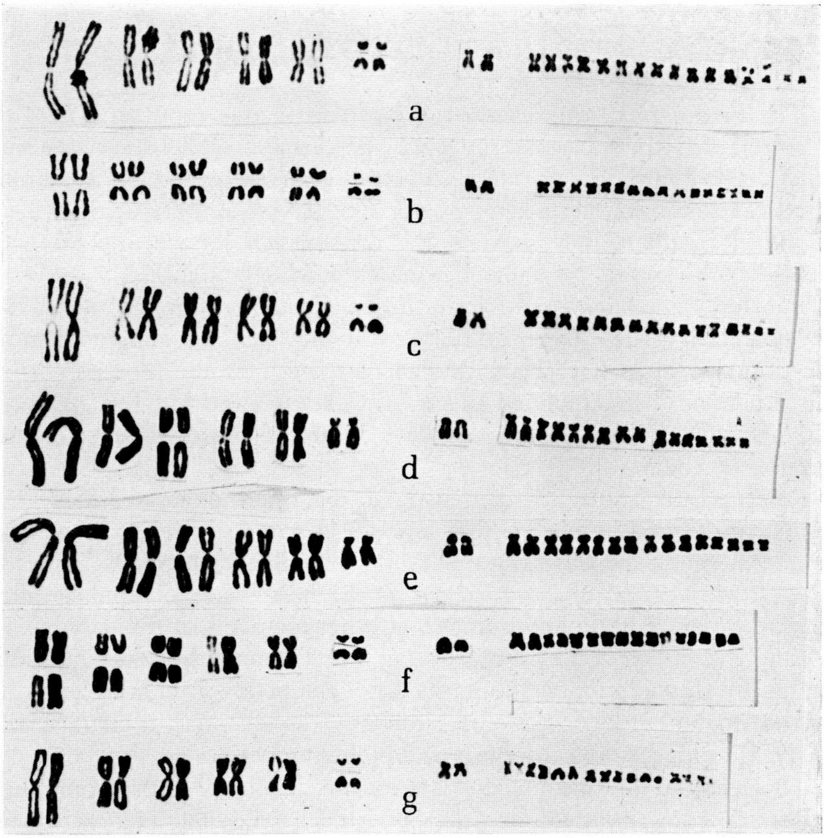

Fig. 14. Karyotypes of representative crocodilians. Arranged to show probable Robertsonian homologies among the large chromosomes. a. *Alligator mississipiensis*, $2n = 32$. b. *Melanosuchus niger*, $2n = 42$. c. *Gavialis gangeticus*, $2n = 32$. d. *Crocodylus intermedius*, $2n = 32$. e. *Crocodylus palustris*, $2n = 30$. f. *Osteolaemus tetraspis*, $2n = 38$. g. *Tomistoma schlegelii*, $2n = 32$. Adapted from Cohen and Gans, 1970 courtesy of Dr. M. M. Cohen.

argues against special affinity; likewise, *C. niloticus* (Africa) and *C. intermedius* (Orinoco region of South America) have identical karyotypes, but again this may be convergence, not phylogenetic affinity.

Greer (1970b) speculated that nesting behaviour may be useful in elucidating crocodilian phylogeny. He found that all the alligatorines and several of the crocodilines were mound nesters, whereas the majority of *Crocodylus* and the gavial were hole nesters (perhaps a behaviourally more primitive character). However, there is no emergent pattern when one compares nesting habits with karyotype. The mound nesting *Crocodylus* are not chromosomally distinct from the hole nesting group.

E. *Order Chelonia* (Table XV)

The turtles and tortoises are the most distinctive reptilian order, and have retained their basic form since the Triassic. There are two sub-orders. The Pleurodira, or side-neck turtles, are represented by two families restricted to the southern hemisphere. The Cryptodira are world-wide in distribution and include most of the well known groups, including the marine turtles and the land tortoises (about 8 living families).

For the cytotaxonomist, this is the most neglected group of reptiles. There are no recent review papers dealing with the Chelonia and, with minor exceptions, species have been studied randomly. At the time of writing, this situation is being rectified by Mr. A. R. Kiester and Mr. J. R. Childress of Harvard University. I am heavily indebted to these workers for their unpublished data.

Reference to Table XV demonstrates that there is a remarkable uniformity in diploid number within families and minor differences between families. Among the soft-shell turtles (Trionychidae), the recent report by Singh *et al.* (1970a) of $2n = 52$–54, is the first example of major intra-familial variation in diploid number, as other members of this family were distinctly characterized by $2n = 64$–66. The Pelomedusidae are outstanding for low diploid numbers, but this is not characteristic of the Pleurodira, for the second family of "side neck turtles", the Chelidae, is characterized by species with the typically elevated diploid numbers.

There are simply insufficient data on the details of chromosome morphology to know if karyotype will be useful for studying taxonomic problems within families. For example, chromosome morphology of *Chrysemys picta* and *Emys orbicularis*, both described by van Brink (1959) and both with $2n = 50$, can be directly compared. In both species the first two chromosome pairs are distinctly larger than all the others and metacentric. Pair three is subtelocentric in both species. In chromosome pairs 4–6 there is variation in centromere position between the species, but in pairs 7–25 there are no obvious differences. This may imply that fine analysis of karyotype will be helpful to the systematist. However, this specific example, comparing a North American endemic with a European form in another genus emphasizes how similar in karyotype are the distantly related forms. The best published comparative study is by Ayres *et al.* (1969) who examined five species of *Podocnemis*. They found that differences revealed in a detailed examination of chromosome pair 10 may have been species-specific markers. But most impressive was the similarity between the four species sharing $2n = 28$. Nakamura (1949) studied four species of emydid turtles and found $2n = 52$ in all cases. All had four large V, but details varied among the smaller elements, and he claimed that each species could be recognized as distinct.

There has been a long standing controversy about the existence of sex chromosome heteromorphism in the Chelonia. Oguma (1937) reported that female *Trionyx japonica* had one fewer chromosomes than the male. The "missing" element was a microchromosome. Nakamura (1949) made a similar suggestion for *Caretta*. Other reports of similar observations were published by Japanese workers (Suzuki, 1950; Makino, 1952). No other workers have been able to confirm this. van Brink (1959) argued that the techniques used by Oguma and others were not suitable for accurate resolution of all the microchromosomes. Ayres *et al.* (1969) found no evidence of female heteromorphism in their excellent preparations of *Podocnemis*. Although the book is not closed on the question of heterogamety, the burden of proof rests on those who claim to observe it. What are needed are photomicrographs of some good cells.

Figures 15 and 16 illustrate a "typical" and a reduced diploid karyotype.

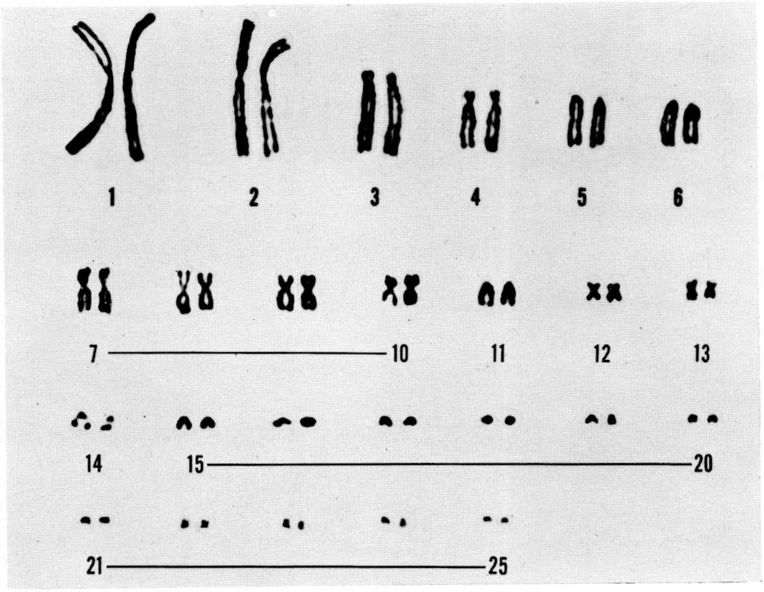

Fig. 15. Karyotype of the box turtle, *Terrapene carolina*, female. $2n = 50$. Fifteen species have been studied in this family, all with diploid numbers of 50 or 52. From Huang and Clark, 1967 courtesy of Dr. C. C. Huang.

TABLE XV. Chromosomes of the Chelonia

Taxon	2*n*	Reference
Cryptodira		
Chelydridae		
Chelydra serpentina	52	Kiester and Childress, personal communication
Chelydra osceola	52	Kiester and Childress, personal communication
Macroclemys temminckii	52	Kiester and Childress, personal communication
Kinosternidae		
Claudius angustatus	56	Kiester and Childress, personal communication
Kinosternon baurii	56	Kiester and Childress, personal communication
Kinosternon leucostomum	56	Kiester and Childress, personal communication
Kinosternon subrubrum	56	Kiester and Childress, personal communication
Staurotypus triporcatus	56	Kiester and Childress, personal communication
Sternotherus odoratus	56	Kiester and Childress, personal communication
Platysternidae		
Platysternon megacephalum	54	Kiester and Childress, personal communication
Emydidae		
Chrysemys picta	50	van Brink, 1959
Clemmys japonica	52	Nakamura, 1935d
Clemmys mutica	52	Nakamura, 1949; Kiester and Childress, personal communication
Cuora amboinensis	50	Kiester and Childress, personal communication
Cuora flavomarginata	52	Nakamura, 1949
Cyclemys dentata	50–52 ?	Kiester and Childress, personal communicaton
Emys orbicularis	50	van Brink, 1959
Geoemyda spengleri	52	Nakamura, 1949
Ocadia sinensis	52	Nakamura, 1949
Pseudemys concinna	50	Kiester and Childress, personal communication
Pseudemys rubriventris	50	Kiester and Childress, personal communication
Pseudemys decorata	50	Kiester and Childress, personal communication
Pseudemys nelsoni	50	Kiester and Childress, personal communication

Siebenbrockiella crassicollis	50	Kiester and Childress, personal communication
Terrapene carolina	50	Huang and Clark, 1967
Testudinidae		
Gopherus agassizi	52	Ohno, 1967; Kiester and Childress, personal communication
Gopherus berlandieri	52	Kiester and Childress, personal communication
Cheloniidae		
Caretta caretta	58	Nakamura, 1949
Chelonia japonica	56	Makino, 1952
Chelonia mydas	56	Waddell and Siegel, 1965
Trionychidae		
Lissemys punctata	66	Singh *et al.*, 1970a; Kiester and Childress, personal communication
Trionyx cartilagineus	66	Kiester and Childress, personal communication
Trionyx ferox	66	Ohno, 1967
Trionyx japonica	64	Oguma, 1937
Trionyx leithii	52–54 ?	Singh *et al.*, 1970a
Trionyx sinensis	66	Susuki, 1950
Pleurodira		
Pelomedusidae		
Pelomedusa subrufa	34	Kiester and Childress, personal communication
Pelusios subniger	34	Kiester and Childress, personal communication
Podocnemis cayennensis	28	Ayres *et al.*, 1969
Podocnemis dumeriliana	26	Ayres *et al.*, 1969
Podocnemis expansa	28	Ayres *et al.*, 1969, Huang and Clark, 1969
Podocnemis sextuberculata	28	Ayres, *et al.* 1969
Podocnemis unifilis	28	Ayres *et al.*, 1969; Huang and Clark, 1969
Chelidae		
Batrachemys nasuta	58	Kiester and Childress, personal communication
Platemys platycephala	68	Kiester and Childress, personal communication

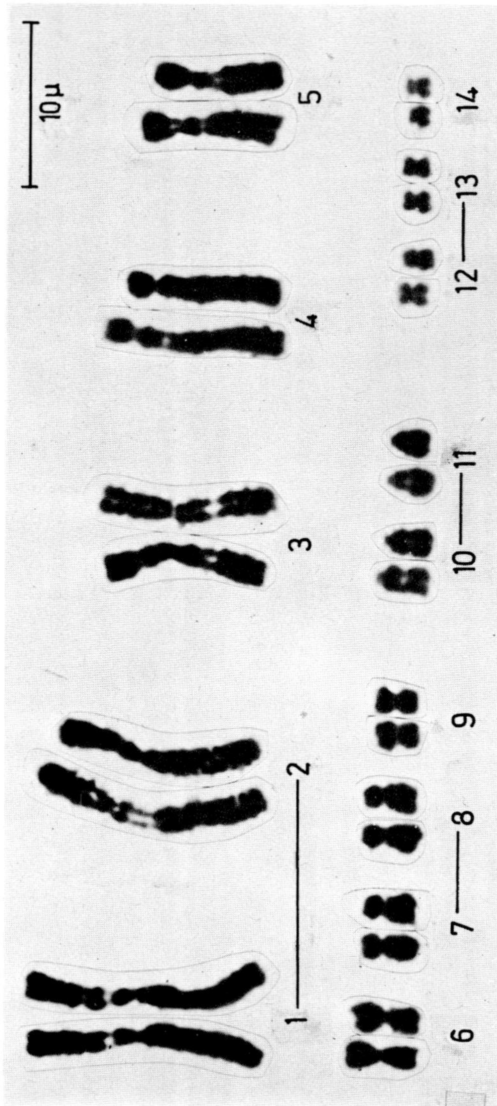

Fig. 16. Karyotype of the side-necked turtle *Podocnemis unifilis*. $2n = 28$. Only in the chelonian family Pelomedusidae are there diploid numbers below 50. From Huang and Clark, 1969 courtesy of Dr. C. C. Huang.

References

Arronet, V. N. (1965). *Tsitologia* 7, 237–239.

Arronet, V. N. (1968). *Tsitologia* **10**, 121–122.

Asana, J. J. and Mahabale, T. S. (1941). *Curr. Sci.* **10**, 377–379.

Ayres, M., Sampaio, M. M., Barros, R. M. S., Dias, L. B. and Cunha, O. R. (1969). *Cytogenetics* **8**, 401–409.

Beçak, W. (1965). *Mem. Inst. Butantan* **32**, 37–38.

Beçak, W. and Beçak, M. L. (1969). *Cytogenetics* **8**, 241–246.

Beçak, W., Beçak, M. L., and Nazareth, H. R. S. (1962). *Cytogenetics* **1**, 305–313.

Bhatnagar, A. N. (1959). *Cytologia* **24**, 459–465.

Bhatnagar, A. N. (1960a). *Caryologia* **12**, 349–361.

Bhatnagar, A. N. (1960b). *Cytologia* **25**, 173–176.

Bhatnagar, A. N. (1961). *Caryologia* **14**, 35–42.

Bhatnagar, A. N. (1962). *Caryologia* **15**, 335–349.

Brink, J. M. van (1959). *Chromosoma, Berlin* **10**, 1–72.

Bury, R. B., Gorman, G. C. and Lynch, J. F. (1969). *Experientia* **25**, 314–316.

Bury, R. B., Gress, F. and Gorman, G. C. (1970). *Herpetologica* **26**, 461–466.

Camp, C. L. (1923). *Bull. Am. Mus. Nat. Hist.* **48**, 289–481.

Canvazos, L. F. (1951). *Am. Nat.* **85**, 373–379.

Chevalier, M. (1969). *C. R. Hebd. Séanc. Acad. Sci. Paris, D* **268**, 2098–2100.

Cohen, M. M. and Gans, C. (1970). *Cytogenetics* **9**, 81–105.

Cohen, M. M., Huang, C. C. and Clark, H. F. (1967). *Experientia* **23**, 769–771.

Cole, C. J. (1970). *Am. Mus. Novit.* **2431**, 1–42.

Cole, C. J. (1971a). *Am. Mus. Novit.* **2450**, 1–17.

Cole, C. J. (1971b). *Herpetologica* **27**, 1–8.

Cole, C. J. and Dowling, H. G. (1970). *Herpetol. Rev.* **2**, 35–36.

Cole, C. J. and Lowe, C. H. (1968). *J. Ariz. Acad. Sci.* **5**, 128–130.

Cole, C. J. Lowe, C. H. and Wright, J. W. (1967). *Science, N.Y.* **155**, 1028–1029.

Cole, C. J., Lowe, C. H. and Wright, J. W. (1969). *Am. Mus. Novit.* **2395**, 1–14.

Dalcq, A. (1920). *C. R. Séanc Biol.* **83**, 995–997.

Dallai, R. and Talluri, M. V. (1969). *Chromosoma, Berlin* **27**, 86–94.

Dallai, R. and Urbani, C. B. (1967). *Caryologia* **20**, 347–353.

Darevsky, I. S. (1966). *Ohio Herpetol. Soc.* **5**, 115–152.

Darevsky, I. S. and Danielyan, F. D. (1968). *J. Herpetol.* **2**, 65–69.

Darevsky, I. S. and Kulikova, V. N. (1964). *Dokl. Akad. Nauk SSSR* **158**, 202–205.

Deweese, J. E. and Wright, J. (1970). *Mamm. Chrom. Newslett.* **11**, 95–96.

Dutt, K. (1966). Abstract in Proc. Symp. Anim. Venoms. Inst. Butantan, São Paulo, pp. 15–16.

Dutt, K. (1968). *Caryologia* **21**, 1–10.

Dutt, K. (1969). *Microscope* **17**, 213–218.

Edgren, R. A. (1953). *Evolution* 7, 88.

Etheridge, R. (1960). Ph.D. Thesis, Dept. of Zool., Univ. of Michigan, Ann Arbor.

Etheridge, R. (1964). *Copeia* **1964,** 610–631.
Fischman, H. K., Mitra, J. and Dowling, H. G. (1968). *Genetics* **60,** 177.
Ford, C. E. and Hamerton, J. L. (1956). *Stain Technol.* **31,** 247–251.
Fritts, T. H. (1969). *Copeia* **1969,** 519–535.
Gans, C. (1960). *Bull. Am. Mus. Nat. Hist.* **119,** 129, 204.
Gans, C., Huang, C. C. and Clark, H. F. (1967). *Copeia* **1967,** 485–487.
Gorman, G. C. (1967). Ph.D. Thesis, Dept. of Biol., Harvard Univ., Cambridge, Mass.
Gorman, G. C. (1969a). *J. Herpetol.* **3,** 49–54.
Gorman, G. C. (1969b). *Mamm. Chrom. Newslett.* **10,** 222–224.
Gorman, G. C. (1970). *Copeia* **1970,** 230–245.
Gorman, G. C. and Atkins, L. (1966). *Am. Nat.* **100,** 579–583.
Gorman, G. C. and Atkins, L. (1967). *Syst. Zool.* **16,** 137–143.
Gorman, G. C. and Atkins, L. (1968a). *Copeia* **1968,** 159–160.
Gorman, G. C. and Atkins, L. (1968b). *Herpetologica* **24,** 13–21.
Gorman, G. C. and Atkins, L. (1969). *Bull. Mus. Comp. Zool.* **138,** 53–80.
Gorman, G. C. and Gress, F. (1970a). *Experientia* **26,** 206–207.
Gorman, G. C. and Gress, F. (1970b). *Herpetologica* **26,** 308–317.
Gorman, G. C., Atkins, L. and Holzinger, T. (1967). *Cytogenetics* **6,** 286–299.
Gorman, G. C., Thomas, R. and Atkins, L. (1968). *Breviora* **293,** 1–13.
Gorman, G. C., Baptista, L. and Bury, R. B. (1969a). *Mamm. Chrom. Newslett.* **10,** 6–11.
Gorman, G. C., Huey, R. B. and Williams, E. E. (1969b). *Breviora* **316,** 1–17.
Gorman, G. C., Jovanovic, V., Nevo, E. and McCollum, F. (1971). *Genetika (Belgrade)* in press.
Greer, A. E. (1970a). *Bull. Mus. Comp. Zool.* **139,** 151–183.
Greer, A. E. (1970b). *Nature (Lond)* **227,** 523–524.
Hall, W. P. (1965). Abstract from Annual Meeting Amer. Soc. Ichthyologists and Herpetologists, Lawrence, Kansas.
Hall, W. P. (1970). *Experientia* **26,** 1271–1273.
Hillenius, D. (1963). *Beaufortia* **9,** 201–218.
Huang, C. C. and Clark, H. F. (1967). *Can. J. Genet. Cytol.* **9,** 449–461.
Huang, C. C. and Clark, H. F. (1969). *Chromosoma* **26,** 245–253.
Huang, C. C. and Gans, C. (1970). *Mamm. Chrom. Newslett.* **11,** 65
Huang, C. C., Clark, H. F. and Gans, C. (1967). *Chromosoma, Berlin* **22,** 1–15.
Jackson, L. and Hunsaker, D. (1970). *Experientia* **26,** 198–199.
Keenan, R. D. (1932). *J. Anat.* **67,** 1–17.
Kluge, A. G. (1967). *Bull. Am. Mus. Nat. Hist.* **135,** 1–59.
Kluge, A. G. and Eckardt, M. J. (1969). *Copeia* **1969,** 651–664.
Kobel, H. R. (1962). *Experientia* **18,** 173–174.
Kobel, H. R. (1963). *Arch. Julius Klaus-Stift. Vererb. Forsch.* **38,** 68–75.
Kobel, H. R. (1967). *Genetica* **38,** 1–31.
Král, B. (1969). *Zool. Listy* **18,** 185–194.
Kupriyanova, L. A. (1968). *Tsitologia* **10,** 892–895.
Kupriyanova, L. A. (1969). *Tsitologia* **11,** 803–814.
Kupriyanova, L. A. and Arronet, V. N. (1969). *Tsitologia* **11,** 157–160.

Levan, A., Fredga, K. and Sandberg, A. A. (1964). *Hereditas* **52**, 201–220.
Lowe, C. H. and Wright, J. W. (1966). *J. Ariz. Acad. Sci.* **4**, 81–87.
Lowe, C. H., Wright, J. W. and Cole, C. J. (1966). *Mamm. Chrom. Newslett.* **22**, 201–203.
Lowe, C. H., Cole, C. J. and Patton, J. L. (1967). *Syst. Zool.* **16**, 296–300.
Lowe, C. H., Wright, J. W., Cole, C. J. and Bezy, R. L. (1970a). *Syst. Zool.* **19**, 128–141.
Lowe, C. H., Wright, J. W., Cole, C. J. and Bezy, R. L. (1970b). *Syst. Zool.* **19**, 114–127.
Makino, S. (1951). "An Atlas of the Chromosome Numbers in Animals". Iowa State College Press, Ames, Iowa.
Makino, S. (1952). *Ann. Zool. Jap.* **25**, 250–257.
Makino, S. and Asana, J. J. (1948). *Chromosoma, Berlin* **3**, 208–219.
Makino, S. and Momma, E. (1949). *Cytologia* **15**, 96–108.
Margot, A. (1946). *Revue Suisse Zool.* **53**, 555–595.
Martin, G. G. and Burgos, J. A. (1957). *Genet. Iber.* **9**, 301–308.
Matthey, R. (1931). *Revue Suisse Zool.* **38**, 117–186.
Matthey, R. (1932). *Arch. Zool. Exp. Gen.* **74**, 193–203.
Matthey, R. (1933). *Revue Suisse Zool.* **40**, 281–316.
Matthey, R. (1939). *Cytologia* **10**, 32–39.
Matthey, R. (1949). "Les Chromosomes des Vertebres." F. Rouge, Lausanne.
Matthey, R. (1951). *Bull. Soc. Vaud. Sci. Nat.* **65**, 111–120.
Matthey, R. (1957). *Revue Suisse Zool.* **64**, 709.
Matthey, R. (1961). *Zool. Anz.* **166**, 153–159.
Matthey, R. and van Brink, J. M. (1956). *Revue Suisse Zool.* **63**, 241–246.
Matthey, R. and van Brink, J. M. (1960). *Bull. Soc. Vaud. Sci. Nat.* **67**, 333–348.
McDowell, S. B. and Bogert, C. M. (1954). *Bull. Am. Mus. Nat. Hist.* **105**, 1–142.
Mendietta, L. (1961). *Lincei Rend. Sci. Fis. Mat. Nat.* **30**, 90–93.
Miller, M. R. (1966). *Proc. Calif. Acad. Sci. 4th Ser.* **33**, 255–359.
Miller, M. R. (1968). *Proc. Calif. Acad. Sci. 4th Ser.* **35**, 425–476.
Monroe, J. E. (1962). *Herpetologica* **17**, 217–220.
Montanucci, R. R. (1970). *Copeia* **1970**, 104–123.
Moorhead, P. S., Nowell, P. C., Mellman, W. J., Battips, D. M. and Hungerford, D. A. (1960). *Exp. Cell Res.* **20**, 613–616.
Nakamura, K. (1928). *Mem. Coll. Sci. Kyoto Univ. Ser. B* **4**, 1–8.
Nakamura, K. (1929). *Zool. Mag. Tokyo* **41**, 398.
Nakamura, K. (1931a). *Proc. Imp. Acad. Tokyo* 7, 26.
Nakamura, K. (1931b). *Cytologia* **2**, 385–401.
Nakamura, K. (1932). *Cytologia* **3**, 156–158.
Nakamura, K. (1935a). *Mem. Coll. Sci. Kyoto Univ. Ser. B* **10**, 341–353.
Nakamura, K. (1935b). *Mem. Coll. Sci. Kyoto Univ. Ser. B* **10**, 355–360.
Nakamura, K. (1935c). *Mem. Coll. Sci. Kyoto Univ. Ser. B* **10**, 361–402.
Nakamura, K. (1935d). *Mem. Coll. Sci. Kyoto Univ. Ser. B* **10**, 403–406.
Nakamura, K. (1949). *Kromosomo* **5–6**, 205–213.
Neaves, W. B. (1969). *J. Exp. Zool.* **171**, 175–184.
Neaves, W. B. and Gerald, P. (1968). *Science, N.Y.* **160**, 1004–1005.

Nogusa, S. (1953). *Cytologia* **18**, 29–35.
Oguma, K. (1934). *Arch. Biol.* **45**, 27–46.
Oguma, K. (1937). *J. Genet.* **34**, 247–264.
Ohno, S. (1967). "Sex Chromosomes and Sex-linked Genes." Springer Verlag, Berlin, Heidelberg and New York.
Painter, T. S. (1921). *J. Exp. Zool.* **34**, 281–327.
Pennock, L. A. (1965). *Science, N.Y.* **149**, 539–540.
Pennock, L. A., Tinkle, D. W. and Shaw, M. W. (1969). *Chromosoma, Berlin* **24**, 467–476.
Pennock, L. A., Tinkle, D. W. and Shaw, M. W. (1969). *Cytogenetics* **8**, 9–19.
Presch, W. (1969). *Copeia* **1969**, 250–275.
Romer, A. S. (1966). "Vertebrate Paleontology." 3rd Edition. Univ. of Chicago Press.
Singh, L., Sharma, T. and Ray-Chaudhuri (1968a). *Cytogenetics* **7**, 161–168.
Singh, L., Sharma, T. and Ray-Chaudhuri (1968b). *Experientia* **24**, 79–80.
Singh, L., Sharma, T. and Ray-Chaudhuri (1970a). *Mamm. Chrom. Newslett.* **11**, 91–94.
Singh, L., Sharma, T. and Ray-Chaudhuri (1970b). *Chromosoma, Berlin* **31**, 386–391.
Susuki, K. (1950). *Jap. J. Genet.* **25**, 222.
Underwood, G. (1954). *Proc. Zool. Soc. (Lond)* **124**, 469–492.
Underwood, G. (1957). *J. Morph.* **100**, 207–268.
Underwood, G. (1967). "A Contribution to the Classification of Snakes". British Museum (National History), London.
Vanzolini, P. E. and Valencia, J. (1965). *Arq. Zool. Sao Paulo* **13**, 7–35.
Waddell, G. H. and Siegel, M. M. (1965). *Bact. Proc.* **1965**, 99.
Werner, Y. L. (1956). *Bull. Res. Coun. Israel,* **5B** section 8, 319.
Werner, Y. L. (1969). *Bull. Res. Coun. Israel,* **8B** section 8, 197–198.
Williams, E. E., Reig, O. A., Kiblisky, P. and Rivero-Blanco, C. (1970). *Breviora* **353**, 1–15.
Wright, J. W. and Lowe, C. H. (1968). *Copeia* **1968**, 128–138.
Wylie, A. P., Veale, A. M. O. and Sands, V. E. (1968). *Proc. Univ. Otago Med. Sch.* **46**, 22–23.

ADDENDUM

It is the inevitable fate of a review such as this to be out of date by the time of publication. Vertebrate cytotaxonomy is in its rapid growth phase. I cannot attempt to summarize all the literature that has appeared since this chapter was first sent to the editors, but they have kindly given me permission to incorporate data from several papers that I believe are important for their taxonomic or cytological implications.

Order Sauria

Family Xantusiidae

Karyotypes of 10 species of Xantusiids have been discussed and illustrated by Bezy (1972). Diploid numbers range from 36 to 40 with 16 to 18 macrochromosomes, 18–22 microchromosomes, 50 to 58 chromosome arms. Of the several groups of lizards that have been considered possibly related to xantusiids, Bezy finds that microteiids have the most similar karyotypes.

Family Iguanidae

Further peculiarities in the genus *Polychrus* were reported by Peccinini, Frota-Pessoa and Ferrari (1971). They described a "Pseudo-XO/XX" sex determining system in an undescribed species of *Polychrus*. The male diploid number is 19 (the lowest of any reptile) and female diploid number is 20. The karyotype consists primarily of two-armed macrochromosomes.

Family Chamaeleontidae

Bourgat has become actively involved in the cytotaxonomy of this family, and has produced several recent papers on the chamaeleons of Madagascar (1971).

The karyotype that I have consistently considered primitive (12V/24m) is exactly the karyotype that Bourgat concludes is primitive on the basis of his biogeographic analysis.

Family Anniellidae

Bezy and Wright (1971) made several interesting observations on *Anniella pulchra*. They found that *A. p. pulchra* had $2n = 20$, and the subspecies *A. p. nigra* had $2n = 22$. The *Anniella* that I had studied was *A. p. pulchra* from Contra Costa County in California, yet I found $2n = 22$. Thus there is within-subspecies variation. More important they point out similarities between the anniellid karyotype and certain scincid karyotypes. It is not clear whether these similarities reflect relationship or convergent evolutionary pathways but an important question about interfamilial relationships is raised.

Order Ophidia

Baker, Mengden, and Bull (1972) published a comprehensive paper on the karyotypes of 38 species of North American colubrids (*sensu lato*) and crotalines. Of particular taxonomic import is the great similarity *inter se* of the North American natricines representing the genera *Natrix*, *Storeria*, and *Thamnophis*, all of which lack the usual 16M 20m, but have instead a diploid number of 36 with few or no microchromosomes.

Additionally, this same paper re-examines the Xenodontine genus *Heterodon*. Unlike the previous report of Edgren the diploid number is 36 with 16M and 20m. Recall that this formula was the exception among xenodontines.

A tabulation of chromosome data on twelve species of snakes (Fischman, Mitra and Dowling, 1972) provides information on some interesting species. An additional typhlopid (only the second to be karyotyped) has $2n = 32$, with 16M/16m. The problematical Mexican dwarf python, *Loxocemus* as well as another species in the genus *Python* appear to be "typical" boids. Of cytological importance is the recent paper by Ray–Chaudhuri, Singh, and Sharma (1971) which demonstrates a heteropycnotic body in the interphase nuclei of females of a large number of snake species, including some in which the Z and W chromosomes are morphologically indistinguishable. Thus the Indian cobra *Naja naja*, belonging to an advanced group that typically shows morphologial sex chromosome heteromorphism, has indistinguishable Z and W chromosomes. But, differences in the replication pattern between Z and W, and W chromatin in the interphase nuclei can both be found. The more primitive boids, however, are isocyclic in DNA replication pattern and do not show the heteropycnotic body in the females.

Order Chelonia

The work cited as in preparation by Kiester and Childress has been discontinued. There are presently no plans for formal publication.

References

Baker, *R. J.*, Mengden, and Bull, (1972). *Copeia*, 257–265.

Bezy, R. L. (1972). *Contrib. Sci., Nat. Hist. Mus. Los Angeles County* No. **227**, 1–29.

Bezy, R. L. and Wright, J. W. (1971). *Herpetological Rev.* **3**, 71–72.

Bourgart, R. (1971). *C.r. Seanc. Soc. Biol.* **5**, 202–208.

Fischman, H. K., Mitra, J. and Dowling, H. G. (1972). *Mammalian Chromosome Newsletter*, **13**, 1, 7–9.

Peccinini, D., Frota-Pessoa, O. and Ferrari, J. (1971). *Caryologia* **24**, 129–139.

Ray-Chaudhuri, Sing, L., and Sharma, T. (1971). *Chromosoma, Berlin* **33**, 239–251.

13

Cytotaxonomy and Chromosome Evolution in Birds

RATNA RAY-CHAUDHURI
Cytogenetics Laboratory, Department of Zoology, Banaras Hindu University, Varanasi, India

I. INTRODUCTION

The karyotype of birds is characterized by having a high diploid number of chromosomes ranging between 60 and 80 which are usually grouped into two more or less clear cut size classes, the so called macro- and micro-chromosomes. Usually 6–8 pairs belong to the former group and the rest to the latter. The reptilian suborder serpentes shows a karyological kinship with birds (Beçak *et al.*, 1964) and possesses about 8 pairs of macro- and 10 pairs of microchromosomes. In spite of the difference in their chromosome numbers, both the groups have more or less similar DNA content per nucleus. A study of DNA content per nucleus estimated by microspectro-photometric measurements as well as by measurements of the total chromosomal area of a complement shows that reptiles can be divided into two distinct groups: (i) Serpentes, having similar DNA content as in birds and (ii) Crocodilia and Chelonia, having nearly 80% of DNA per nucleus found in mammals (Atkin *et al.*, 1965). These results show closer kinship between birds and snakes on the one hand and Crocodilia, Chelonia and mammals on the other. A further cytological closeness between the snakes and birds is in their sex determining mechanism. In both the groups, where sex chromosomes have been identified, the females are found to be the heterogametic sex.

Genetically the best known species of bird is *Gallus** *domesticus*, where only six linkage groups have been described corresponding to six pairs of macrochromosomes. This situation led Newcomer (1957) and Newcomer and Brant (1954) to doubt even the chromosomal status of the microchromosomes. This view is not shared by other authors (see Ray-Chaudhuri,

* *Gallus domesticus* is the domestic variety of *Gallus gallus*.

1967a). It is only in recent years that we have started getting some evidence about the function of the microchromosomes. Ohno *et al.* as early as in 1962 reported the association of nucleolus with at least 12 microchromosomes in chick and suggested that they are responsible for the organization of the nucleoli. Microchromosomes in Japanese quail fibroblasts were also shown to be nucleolus organizing (Comings and Mattoccia, 1970). Moreover, analytical centrifugation of quail DNA demonstrated a main band with a buoyant density of 1.701 g/cm^3 and a satellite constituting about 5% of the DNA with a buoyant density of 1.715 and the G–C content of the main and the satellite band was 42 and 55% respectively. The above authors postulate that microchromosomes presumably carry ribosomal DNA cistrons. Further confirmation for this conclusion should be obtained by *in situ* hybridization studies. The same authors, however, could not find any satellite band in the chick DNA although some heterogeneity was present. A different line of investigation has also shown results pertinent to the functional aspect of the microchromosomes and this is the study of the time and pattern of DNA replication of the chromosomes during cell cycle by means of autoradiography.

The replication pattern of the microchromosomes has been found to be different in different species of birds. Schmid (1962) and Donnelly and Newcomer (1963) found that DNA synthesis in *Gallus domesticus* terminates earlier in the smaller chromosomes than in the larger ones. A more critical study by Bianchi and Molina (1967) of the same species supported the above conclusions. On the other hand, Galton and Bredbury (1966) found in *Columba livia domestica** that at least some of the microchromosomes are late labelling. Comings and Mattoccia (1970) found that in the Japanese quail, the maximum intensity of silver grains in microchromosomes occurs in the late phase of the S period when the replication of macrochromosomes has ceased. If both late and early replication are considered as a property of heterochromatin, it may be presumed that microchromosomes largely contain heterochromatin and are therefore likely to possess relatively fewer Mendelian genes per unit length in the DNA than in the macrochromosomes.

It has been mentioned that female heterogamety operates in birds which is simple ZZ–ZW type and no multiple sex chromosomes have so far been reported from this group. In many species of birds the Z can be singled out at the metaphase stage with absolute certainty. In cases where the identification is clear cut, it appears that it has not undergone many structural changes during evolution because its relative length compared to the total haploid length of the genome excluding the microchromosomes has been found to vary within a relatively narrow range in a large number of species

* Domestic variety of *Columba livia*.

of birds so far studied. The mammalian X chromosome is also conservative in evolution in this sense (Ohno *et al.*, 1964a). There are however several differences between the mammalian X and the avian Z chromosomes which reflect different evolutionary origins of the chromosomal sex determining mechanisms in the two classes. For example, due to the operation of a dosage compensating mechanism in mammals, one of the two X chromosomes in the females becomes genetically inactive by becoming heterochromatic during early embryogenesis and consequently it forms a sex chromatin body in the interphase nuclei. This heterochromatic X is also late replicating and if similar dosage compensation mechanism were operative in both the groups one of the Z chromosomes in homogametic male birds is expected to behave like one of the X's in female mammals. Studies on the DNA replication pattern in birds, although limited to very few species have shown that the Z replicates its DNA synchronously with the larger autosomes. No nuclear sex chromatin has been detected in male birds (Ohno *et al.*, 1960; Hammar, 1964). This result is also consistent with the finding that there is absence of dosage compensation for certain Z linked plumage factors in male fowls, pigeons and canaries.

Although no sex chromatin has been reported in male birds, Kosin and Ishizaki (1959) and Ohno *et al.* (1960) have reported the presence of sex chromatin-like bodies, in the somatic nuclei of the females of *Gallus domesticus*. The former authors suggest that the presence or absence of sex chromatin may not be related to the sex chromosomes at all. The latter authors are, however, of the opinion that the sex chromatin body detected in the female represents the single Z in positive heteropycnosis. Ray-Chaudhuri *et al.* (1970, 1971) have demonstrated the presence of a sex chromatin like body in the somatic cells of the females in as many as 13 species of snakes which according to them is formed by the W chromosome and has been termed as W chromatin. It is detectable only in those species of snakes where the W is either morphologically distinguishable from the Z and/or when it is allocyclic in its DNA replication pattern. In view of the evolutionary closeness of reptiles and birds it is highly likely that the sex chromatin in female birds is also formed by the W chromosome rather than by the Z as proposed by Ohno *et al.* (1960). The W chromosome in birds is however not definitely identifiable in many species, and in them it is probably one of the microchromosomes.

II. COMPARATIVE KARYOLOGY AND TAXONOMY

Chromosome morphology, like other species-specific characters are expected to provide a means of assessing the phylogenetic relationship and evolutionary trends within and between different taxa. Cytotaxonomical studies in different groups of mammals among vertebrates have been carried out extensively with a good deal of success but such attempts have rarely been made in birds (see Ray-Chaudhuri *et al.*, 1969). In spite of the recent advances in the cytological techniques an exact determination of the chromosome number in birds still remains extremely difficult due to the presence of unusually large number of microchromosomes which can not be counted accurately (Table I). Fortunately, the cytological clarity of the macrochromosomes is comparable to the chromosomes of any other group. Moreover, the number of individually identifiable chromosomes or the so called marker chromosomes is more common in birds than in mammals (Ray-Chaudhuri *et al.*, 1969). This is a distinct advantage for comparison of chromosomes of groups of birds of different taxonomical categories in order to derive possible homologies between chromosomes of different species.

The purpose of this article is to review all the recent studies of avian chromosomes and evaluate them within an evolutionary context. With a view to understanding the nature of evolutionary changes in karyotypes, individual macrochromosomes of different species are measured and expressed as percentage of the total length of the haploid set of the species minus the microchromosomes. These measurements are used in drawing comparative idiograms, a study of which forms the basis of our cytotaxonomical conclusions. The classification and arrangements of the chromosomes in the comparative idiograms presented in the paper are made by following the system of nomenclature proposed by Levan *et al.* (1964) as far as practicable. We are compelled to omit the microchromosomes because they are not often represented in the karyotypes published by various authors. Even in publications where they are illustrated it is extremely difficult to measure them accurately. In order to arrive at an estimate of the error introduced in our calculations of relative lengths due to the omission of different numbers of microchromosomes in different species, we made a rough estimate of the percentage length of the microchromosome content of several species at random and the values thus obtained are found to vary between 30.4 % and 42.7 % of the total genome. The total genetic material in them is therefore very roughly constant. Thus the error introduced is less serious than appears from a superficial consideration of the method. The amount of error is further reduced when we consider individual groups of species in a particular idiogram. In most of these cases, the species compared have very similar numbers of macro-and microchromosomes and therefore from every species we have ignored more or less equal amounts of microchromosomal material thus making the idiograms of the macrochromosomes more or less accurate for the present purpose.

A. *Podicipiformes and Gaviiformes*

We know the chromosomes of only one species of each of the above orders. They are *Podiceps cristatus*, 2n = 90 ±, and *Gavia stellata*, *2n* = 72 ± (Table I). A superficial examination of their chromosomes (Fig. 1a and b) does not reveal any obvious similarity between the two, for instance,

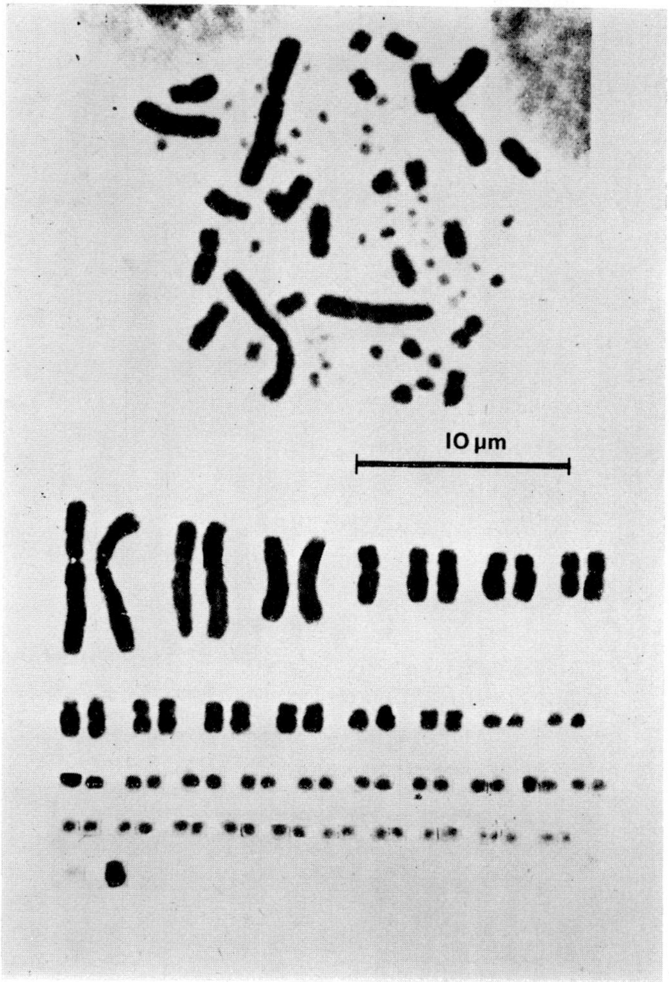

Fig. 1a. *Podiceps cristatus*, female metaphase chromosomes. (After Hammar, 1970).

TABLE I. Diploid Number and Sex Chromosome Constitution

	$2n$	Number of macro-chromosomes	Number of micro-chromosomes	Z	W	References
Order: Podicipiformes						
Family: Podicipitidae						
Podiceps cristatus	72±	22	50	sm	t	Hammar, 1970
Order: Gaviiformes						
Family: Gaviidae						
Gavia stellata	90±	40	50	sm	m	Hammar, 1970
Order: Falconiformes						
Family: Falconidae						
Subfamily: Falconinae						
Falco tinnunculus	52	—	—	t	t	Renzoni and Vegni-Talluri, 1966
Family: Accipitiridae						
Subfamily: Buteoninae						
Buteo buteo	68	—	—	m	t	Renzoni and Vegni-Talluri, 1966
Order: Ciconiiformes						
Family: Ardeidae						
Subfamily: Ardeinae						
Ardea cinerea	53–64, 68	22	31–42, 46	m	sm	Hammar, 1970; Itoh *et al.*, 1969
Ardeola grayii	68	22	46	m	t	Ray-Chaudhuri, 1967a
Order: Anseriformes						
Family: Anatidae						
Subfamily: Anserinae						
Tribe: Anserini						
Cygnus olor	80	20	60	t	t	Hammar, 1970
Cygnopsis cygnoid	80	20	60	m	m	Bhatnagar, 1968; Hammar, 1966
Anser anser	80	20	60	m	t	Itoh *et al.*, 1969
Anser albifrons	72±	20	52±	m	t	Itoh *et al.*, 1969

Eulabeia indica	72±	20	52±	m	t	Itoh *et al.*, 1969
Branta canadensis	80	20	60	sm	m	Hammar, 1970
Subfamily: Anatinae						
Tribe: Tadornini						
Tadorna tadorna	82	20	62	t	t	Hammar, 1970
Tribe: Anatini						
Anas platyrhynchos	78	20	58	t	sm	Hammar, 1966; Mott *et al.*, 1968
Tribe: Somateriini						
Somateria mollissima	80	20	60	t	st	Hammar, 1970
Tribe: Aythyini						
Aythya ferina	80	20	60	t	sm	Hammar, 1970
Aythya fuligula	78	20	58	t	sm	Hammar, 1966
Tribe: Cairini						
Cairina moschata	78–80	20	58–60	t	t	Mott *et al.*, 1968
Tribe: Mergini						
Mergus merganser	82	20	62	t	sm	Hammar, 1970
Bucephala clangula	84	24	60	t	t	Hammar, 1970
Order: Galliformes						
Family: Phasianidae						
Subfamily: Phasianinae						
Coturnix coturnix japonica	78	18	60	m	t	Vegni-Talluri and Vegni, 1965
Gallus domesticus	78±	22	56±	m	m	Ohno, 1961; Stenius *et al.*, 1963; Owen, 1965; Krishan and Shoffner, 1966
Gallus gallus bankiva	78	18	60	m	?	Ohno *et al.*, 1964b
Phasianus colchicus	78	22	56	m	sm	Stenius *et al.*, 1963; Krishan and Shoffner, 1966
Syrmaticus reevesii	72–82	22	50–60	m	?	Itoh *et al.*, 1969
Pavo cristatus	66±	18	48±	m	?	Ray-Chaudhuri *et al.*, 1969
Family: Numididae						
Numida meleagris	78±	20	58±	m	t	Piccini and Stella, 1970

	2n	Number of macro-chromosomes	Number of micro-chromosomes	Z	W	References
Family: Meleagridae						
Meleagris gallopavo	78±	22	56±	m	sm	Stenius *et al.*, 1963; Krishan and Shoffner, 1966
Order: Gruiformes						
Family: Rallidae						
Subfamily: Rallinae						
Gallinula chloropus	78	20	58	sm	sm	Hammar, 1970
Porphyrio poliocephalus viridis	62–77	22	40–55	m	?	Itoh *et al.*, 1969
Subfamily: Fulicinae						
Fulica atra	92	20	72	sm	sm	Hammar, 1970
Order: Charadriiformes						
Superfamily: Charadroidae						
Family: Haematopodidae						
Haematopus ostralegus	66	20	46	m	t	Hammar, 1970
Family: Charadriidae						
Subfamily: Charadriinae						
Vanellus vanellus	76	20	56	sm	sm	Hammar, 1970
Charadrius hiaticula	76	20	56	m	sm	Hammar, 1970
Subfamily: Scolopacinae						
Numenius arquata	78	28	50	sm	m	Hammar, 1970
Tringa totanus	88	20	68	m	sm	Hammar, 1970
Gallinago gallinago	98	28	70	m	m	Hammar, 1970
Subfamily: Recurvirostrinae						
Recurvirostra avosetta	76	20	56	m	st	Hammar, 1970
Order: Lariformes						
Family: Laridae						
Subfamily: Larinae						
Larus ridibundus	66	28	38	m	m	Hammar, 1966

Larus canus	66	28	38	m	m	Hammar, 1966
Larus argentatus	56–72	28	28–44	m	m	Itoh *et al.*, 1969
Larus fuscus	70	28	42	m	sm	Hammar, 1970
Subfamily: Sterninae						
Sterna hirundo	68	28	40	m	m	Hammar, 1970
Sterna paradisaea	70	28	42	m	m	Hammar, 1970
Sterna albifrons	68	28	40	m	m	Piccini and Stella, 1970
Order: Cuculiformes						
Family: Cuculidae						
Subfamily: Cuculinae						
Eudynamys scolopacea scolopacea	78	12	66	m	m	Ray-Chaudhuri, 1967b
Columbiformes						
Columba livia domestica	80	18	62	m	m	Galton and Bredbury, 1966; Itoh, *et al.* 1969
Columba palumbus	78	18	60	m	sm	Hammar, 1966
Streptopelia orientalis	72	18	54	m	?	Ray-Chaudhuri *et al.*, 1969
Streptopelia risoria	77–81	16	61–66	m	t	Itoh *et al.*, 1969
Geopelia cuneata	72	16	56	sm	t	Itoh *et al.*, 1969
Order: Psittaciformes						
Family: Psittacidae						
Subfamily: Psittacinae						
Psittacula krameri borealis	70	16	54	sm	t	Ray-Chaudhuri *et al.*, 1969
Psittacula alexandri fasciata	68	16	52	m	t	Ray-Chaudhuri *et al.*, 1966
Psittacula cyanocephala bengalensis	66	14	52	m	t	Ray-Chaudhuri *et al.*, 1969
Loriculus vernalis	72	16	56	t	?	Ray-Chaudhuri *et al.*, 1969
Melopsittacus undulatus	58	?	?	?	?	Rothfels *et al.*, 1963
Order: Strigiformes						
Family: Strigidae						
Subfamily: Striginae						
Otus scops japonicus	70–78	18	52–60	sm	?	Itoh *et al.*, 1969
Bubo v. virginianus	82–84	16	66–68	m	m	Krishan *et al.*, 1965
Athene noctua	82	12	70	m	t	Renzoni and Vegni-Talluri, 1966

	2*n*	Number of macro-chromosomes	Number of micro-chromosomes	Z	W	References
Athene brama	82	12	70	m	?	Ray-Chaudhuri *et al.*, 1969
Strix aluco	82	20	62	m	t	Renzoni and Vegni-Talluri, 1966; Hammar, 1970
Subfamily: Tytoninae						
Tyto alba	92	—	—	m	m	Renzoni and Vegni-Talluri, 1970
Order: Piciformes						
Family: Picidae						
Picus viridis	94	20	74	t	t	Hammar, 1970
Order: Passeriformes						
Family: Motacillidae						
Motacilla flava	78	16	62	st	sm	Hammar, 1970
Family: Campephagidae						
Coracina melanoptera	72–74	16	56–58	t	m	Ray-Chaudhuri *et al.*, 1969
Family: Mimidae						
Toxostoma rufum	78	22	56	m	?	Jovanovic and Atkins, 1969
Family: Muscicapidae						
Subfamily: Turdinae						
Phoenicurus phoenicurus	76	24	52	sm	sm	Hammar, 1966
Turdus sibiricus davisoni	74–82	22	52–60	sm	?	Itoh *et al.*, 1969
Turdus merula	80	22	58	m	t	Hammar, 1970
Turdus migratorius	80	24	56	m	m	Jovanovic and Atkins, 1969
Oenanthe oenanthe	80	22	58	m	t	Hammar, 1970
Saxicola torquata	86	22	64	st	t	Piccini and Stella, 1970
Subfamily: Timaliinae						
Turdoides striatus striatus	68	22	46	sm	m	Ray-Chaudhuri *et al.*, 1966
Family: Paridae						
Parus palustris	78	20	58	st	sm	Castroviejo *et al.*, 1969; Hammar, 1970

Parus major	78	20	58	st	sm	Castroviejo *et al.*, 1969; Hammar, 1970
Family: Certhiidae						
Certhia familiaris	80	20	60	t	t	Hammar, 1970
Family: Fringillidae						
Serinus canarius	80±	18	62	m	m	Ohno *et al.*, 1964b
Fringilla coelebs	80	18	62	m	m	Piccini and Stella, 1970
Zonotrichia albicollis	82–84	16	66–68	sm	sm	Thorneycroft, 1966
Family: Ploceidae						
Passer domesticus	76–78	16	60–62	m	?	Castroviejo *et al.*, 1969
Passer montanus	76–78	16	60–62	m	m	Castroviejo *et al.*, 1969; Hammar, 1970
Family: Sturnidae						
Subfamily: Sturninae						
Sturnus contra	68	12	56	m	?	Ray-Chaudhuri *et al.*, 1969
Family: Oriolidae						
Oriolus oriolus	78	18	60	t	?	Ray-Chaudhuri *et al.*, 1969
Family: Corvidae						
Cyanocitta cristata	78	22	56	m	sm	Jovanovic and Atkins, 1969
Pica pica	76	22	54	sm	m	Hammar, 1966
Corvus brachyrhynchos	80	22	58	sm	sm	Jovanovic and Atkins, 1969

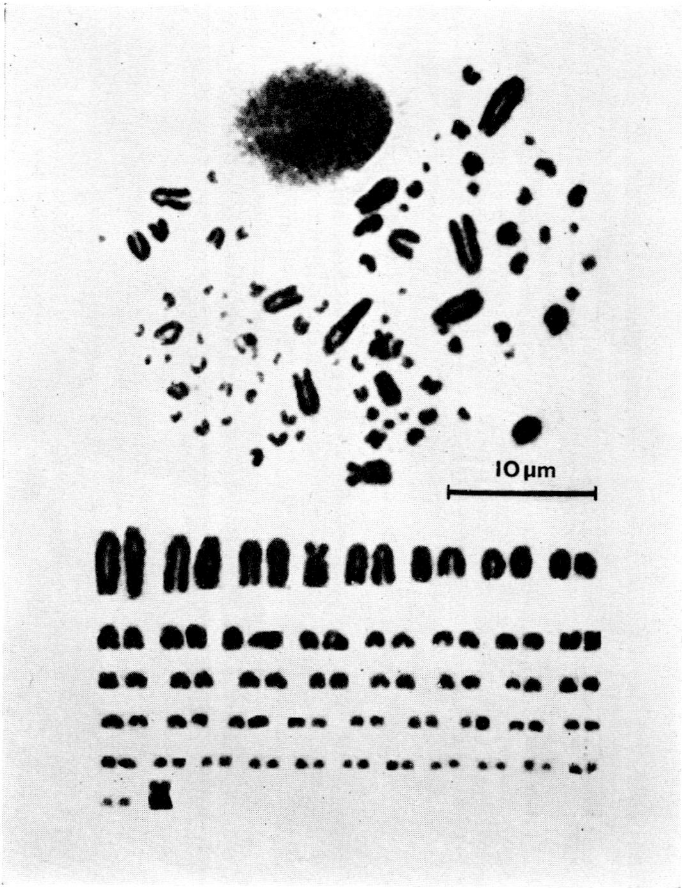

Fig. 1b. *Gavia stellata*, female metaphase chromosomes. (After Hammar, 1970.)

in *Gavia stellata* all the chromosomes are telocentrics except the Z and W which are submetacentrics, while in *Podiceps cristatus* they are metacentrics or submetacentrics except chromosome no. 3 and W which are telocentrics. But an examination of the comparative idiogram presented in Fig. 2 reveals a good deal of similarity between the chromosomes of the two species. The 9 pairs of metacentric chromosomes (chromosomes 1–2 and 4–10 in Fig. 2) in *Podiceps* can be considered to have been derived, during evolution, from 18 pairs of telocentrics of *Gavia* or *vice versa*. This conclusion finds further support from the data on relative lengths given in Table II where it can be seen that the combined relative lengths of the two acrocentrics are nearly equal to the corresponding metacentric chromosome which is supposed to have been derived from them. The rest of the chromosomes of the comple-

TABLE II. Podicipiformes and Gaviiformes

	Relative length %		Centromeric index %	
	Gavia stellata	*Podiceps cristatus*	*Gavia stellata*	*Podiceps cristatus*
(1)	10.1 — I	20.3	— I	38.4
	8.9 — II		— II	
(2)	7.5 — I		— I	
		16.6		37.0
	6.0 — II		— II	
(3)	12.3	12.1	—	—
(4)	5.9 — I		— I	
		8.1		32.2
	5.2 — II		— II	
(5)	4.0 — I		— I	
		7.2		35.7
	3.6 — II		— II	
(6)	3.7 — I		— I	
		7.0		45.7
	3.3 — II		— II	
(7)	3.3 — I		— I	
		5.8		23.2
	2.3 — II		— II	
(8)	3.0 — I		— I	
		5.5		45.4
	2.4 — II		— II	
(9)	2.4 — I		— I	
		4.3		45.4
	2.3 — II		— II	
(10)	2.4 — I		— I	
		4.3		28.5
	2.3 — II		— II	
Z	8.1	8.6	30.3	35.7
W	5.5	3.8	34.4	—

ments of both the species are microchromosomes and their number is 50 in each.

The Z chromosomes of the two species, as already indicated, are very similar both being submetacentric and having relative lengths 8.1 and 8.6. The W chromosome is bigger and submetacentric with a percentage length 5.5 in *Gavia* and smaller and telocentric with a percentage length 3.8 in *Podiceps*.

The two orders Podicipiformes and Gaviiformes were erected by breaking up the old order Colymbiformes which included both Podicipitidae and Gaviidae. It was pointed out by Stolpe (see Mayr and Amadon, 1951) that

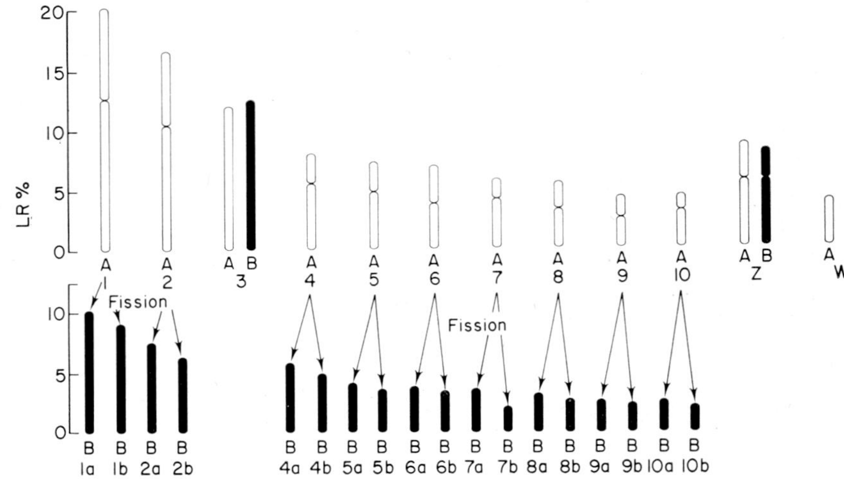

Fig. 2. Composite idiogram of Podicipiformes (A) and Gaviiformes (B).

the anatomical similarities cited as evidence for close relationship between grebes (Podicipiformes) and loons (Gaviiformes) are actually due to functional convergence and not due to closer taxonomic relationship. Mayr and Amadon (1951), however, pointed out that the grebes have been thought to be remote allies of petrels and since loons may be a specialized offshoot of the petrel stock, it is possible that the grebes and loons have some distant or indirect relationship.

Our comparative karyological data tend to show that the morphological similarities observed between these groups are probably due to their phylogenetic closeness rather than due to their functional convergence.

B. *Ciconiiformes*

Of the two species so far worked out in this order, *Ardeola grayii* has a diploid number 68, while *Ardea cinerea* has been reported to have a diploid number of 68 by Hammar (1970) but of 53–64 by Itoh *et al.* (1969). The morphology of the macrochromosomes described by them are, however, exactly similar. The two species belong to the family Ardeidae.

The comparative karyogram (Fig. 3) shows that among the bigger autosomes and the Z and W chromosomes bearing numbers 1, 3, 4, 5, 6, 11 and Z are more or less similar in the two species and therefore can be considered as "shared" chromosomes (Bianchi and Bianchi, 1969). Four other chromosomes with median and submedian centromeres (nos. 2, 7, 8 and 9) are present only in *Ardea* (A) and four telocentric chromosomes nos. 10, 12–14 are present only in *Ardeola* (B). If the relative lengths of the chromosomes, 2, 7, 8 and 9 on one hand and 10, 12–14 on the other are added up, they become 27.5% and 20.7% respectively (Table III). It may, therefore, be assumed that structural rearrangements leading to the divergence of the

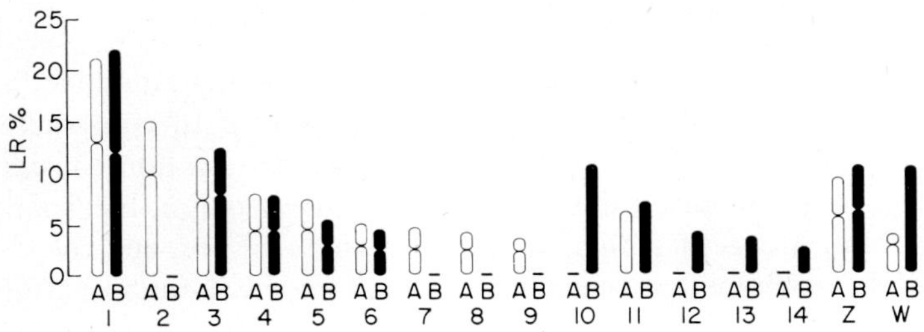

Fig. 3. Composite idiogram of Ciconiiformes.
A: *Ardea cinerea* B: *Ardeola grayii*

chromosome complements of the two species might have been restricted mainly to these chromosomes.

TABLE III. Ciconiiformes

	Relative length % *Ardea cinerea*	*Ardeola grayii*	Centromeric index % *Ardea cinerea*	*Ardeola grayii*
(1)	21.2	22.5	38.4	47.3
(2)	15.1	—	38.4	—
(3)	11.4	11.9	35.7	35.0
(4)	8.0	8.6	47.6	46.0
(5)	7.7	5.6	47.6	44.4
(6)	5.0	4.0	35.7	33.3
(7)	4.7	—	50.0	—
(8)	4.0	—	43.4	—
(9)	3.7	—	34.4	—
(10)	—	10.6	—	—
(11)	6.4	—	7.3	—
(12)	—	4.3	—	—
(13)	—	3.3	—	—
(14)	—	2.5	—	—
Z	8.7	10.6	43.4	47.1
W	3.7	10.5	33.3	—

The Z chromosomes in both the species are metacentrics and have relative lengths of 8.7% (A) and 10.6% (B). The most interesting difference in both size as well as centromeric index is observed in the W chromosome which is the smallest macrochromosome having a centromeric index of 33.3 in *Ardea* while in *Ardeola* it is a telocentric chromosome of about the same size as the Z (Table III).

C. *Anseriformes*

The order Anseriformes consists of a single family Anatidae, and we know the chromosomes of 14 species of the group. In constructing the composite idiogram presented in Fig. 4, we had to take into consideration only 13 of them because the karyotype of *Eulabeia indica*, incorporated in the idiogram, is exactly similar to that of *Anser albifrons*, and the chromosomes of the latter species have not been illustrated by the author (Itoh *et al.*, 1969).

The diploid number of chromosomes in Anatidae ranges between 72 and 84 although in the majority of the species the range is even narrower, viz. between 80 and 84 (Table I). An overall similarity of the karyotypes of all the species is evident from the comparative karyograms (Fig. 4, species A to M). The important differences worth pointing out are: (i) the two largest metacentric chromosomes widespread in the family, i.e. chromosome numbers 1 and 2 are absent in *Bucephala clangula* (M), being replaced by 4 telocentric chromosomes; (ii) chromosome no. 3 has terminal (t) centromeres in all the species except in *Cygnopsis cygnoid* (J), *Anser anser* (K) and *Eulabeia indica* (L) where it is subterminal; (iii) chromosome no. 4 is mediocentric (m) in the species J, K and L and in the rest it is telocentric and (iv) the Z chromosome is either median or submedian in the species J, K, L and also in C and terminal in the remaining species.

The four largest chromosomes of *B. clangula* (M) can be derived from centromeric fission of the two largest metacentrics common to the family thus making all its chromosomes telocentric. This view finds additional support from the measurements of the relative lengths of the original chromosomes and those derived from them (see Table IV). Thus, if we add up the relative lengths of the two largest chromosomes of *B. clangula* (M) we find that they correspond very closely to the relative length of the largest metacentric chromosome of the family (Table IV). The same is true for the next two acrocentrics whose total relative lengths are again nearly equal to the relative length of the second metacentric chromosome of the group. The rest of the chromosomes of the complement, viz. Nos. 3–9 and Z, W in *Bucephala* are similar in length and morphology to the corresponding chromosomes of the other species.

Apart from *B. clangula* (M), three more species deviate from the usual

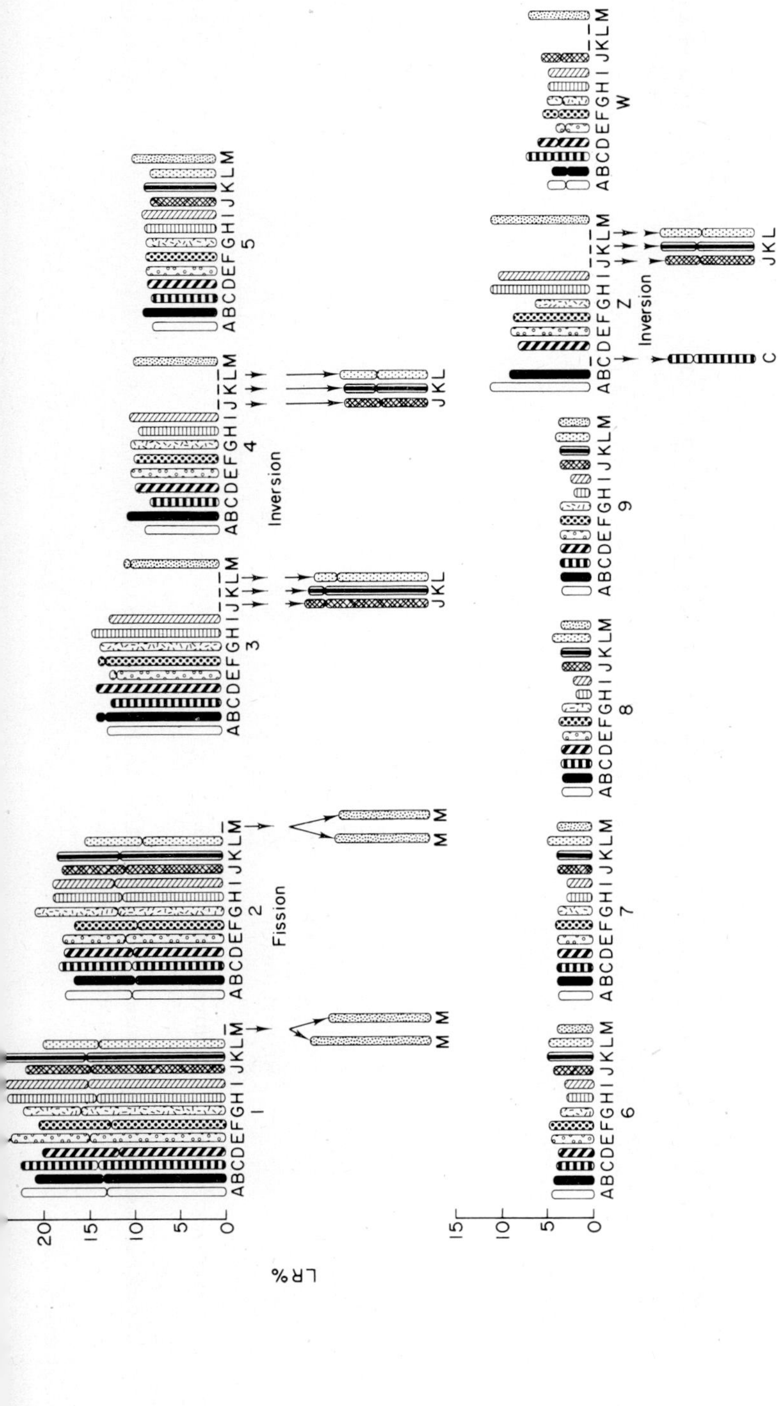

Fig. 4. Composite idiogram of Anseriformes.

A: *C. olor*
D: *Aythya ferina*
G: *Aythya fuligula*
J: *C. cygnoid*
M: *B. clangula*

B: *T. tadorna*
E: *S. mollissima*
H: *Anas platyrhynchos*
K: *Anser anser*

C: *B. canadensis*
F: *M. merganser*
I: *C. moschata*
L: *Eulabeia indica*

TABLE IV. Anseriformes

Species		Macrochromosome pair nos. 1	2	3	4	5	6	7	8	9	10	Sex chromosomes Z	W
Cygnus olor	Relative length	22.3	17.6	12.3	8.2	6.9	5.0	4.4	4.1	3.4		11.0	4.4
	Centromeric index	37.0	43.4	—	—	—	—	—	—	—		—	—
Tadorna tadorna	Relative length	21.2	16.4	13.6	10.4	8.0	5.2	4.4	4.0	3.6		8.8	4.0
	Centromeric index	38.4	43.3	9.2	—	—	—	—	—	—		—	—
Branta canadensis	Relative length	22.6	18.1	12.1	7.6	6.9	4.5	4.1	3.8	3.4		9.4	6.9
	Centromeric index	37.0	43.4	—	—	—	—	—	—	—		34.4	43.4
Aythya ferina	Relative length	21.7	17.6	13.5	9.0	7.7	5.3	4.0	3.6	3.6		8.2	5.3
	Centromeric index	40.0	43.4	—	—	—	—	—	—	—		—	37.0
Somateria mollissima	Relative length	23.5	17.4	12.2	9.6	7.7	4.8	4.5	3.8	3.5		9.0	3.5
	Centromeric index	35.7	40.0	8.1	—	—	—	—	—	—		—	23.8
Mergus merganser	Relative length	20.4	17.0	13.2	8.8	7.4	5.7	5.1	4.4	3.7		8.5	5.1
	Centromeric index	40.0	41.6	5.4	—	—	—	—	—	—		—	27.3
Aythya fuligula	Relative length	22.1	20.4	13.1	9.6	7.6	4.8	3.8	3.8	3.4		6.5	4.5
	Centromeric index	37.0	43.4	—	—	—	—	—	—	—		—	35.7
Anas platyrhynchos	Relative length	23.7	18.1	13.9	8.3	8.3	4.6	2.8	2.8	2.1	2.1	11.1	4.6
	Centromeric index	38.2	46.4	—	—	—	—	—	—	—	—	—	—
Cairina moschata	Relative length	24.1	18.4	12.0	8.0	8.0	4.5	3.2	2.3	2.3	2.3	9.8	4.5
	Centromeric index	35.4	37.5	—	—	—	—	—	—	—	—	—	—

Cygnopsis cygnoid	Relative length	22.1	17.5	13.4	8.6	6.9	4.5	4.3	4.1	3.5	3.5	9.5	5.2
	Centromeric index	37.2	40.8	14.5	50.0	—	—	—	—	—	—	36.4	41.6
Anser anser	Relative length	24.4	18.1	13.1	8.8	7.8	5.3	4.9	3.9	3.5	3.5	9.9	—
	Centromeric index	38.4	40.0	11.7	35.7	—	—	—	—	—	—	35.7	—
Eulabeia indica	Relative length	20.2	15.0	12.7	9.8	6.9	5.7	5.2	4.6	4.6	4.6	10.4	—
	Centromeric index	31.0	42.2	18.1	40.9	—	—	—	—	—	—	44.4	—
Bucephala clangula	Relative length	1a 13.4 1b 11.3	2a 10.5 2b 10.1	10.0	9.0	9.0	4.4	4.0	3.8	3.6	3.6	10.7	6.4
	Centromeric index	—	—	—	—	—	—	—	—	—	—	—	—

chromosome configuration of the family. They are *C. cygnoid* (J), *A. anser* (K) and *E. indica* (L) where chromosome nos. 3, 4 and Z have undergone pericentric inversions when compared to the respective chromosomes of other species of the order. A reference to the data in Table IV will show that chromosome No. 3 has a relative length varying between 12.0 and 13.9 whereas the same chromosome in the above three species has a relative length which varies between 12.7 and 13.4 (Table IV). A similar situation is also observable in chromosome No. 4. The size of the Z chromosome is however much more variable in the group but it is always telocentric except in species J, K and L as well as in *Branta canadensis* (C).

The percentage length of the Z chromosome in the family varies from 6.5 in *Aythya fuligula* to 11.1 in *Cygnus olor*. The W also is equally variable in their relative lengths varying between 3.5% and 6.6%.

On the basis of the study of the karyotype alone the 14 species of the family Anatidae can be divided into three groups. In the first group we can put the following species: (1) *C. cygnoid*; (2) *A. anser*; (3) *A. albifrons* and (4) *Eulabeia indica*. In the second group belongs a single species, i.e. *B. clangula*, while the rest of the species form a third group. Further, *B. clangula* should be more closely related to the third group rather than to the first group because the chromosome complement of the former can easily be derived from the latter group by two centromeric fissions only.

The chromosomal similarity of the first group can very easily be supported from usual taxonomic criteria. As a matter of fact the morphological similarities are so close among the different species of the first group that various taxonomists have doubted their distinct generic status (see Delacour and Mayr, 1945). Delacour and Mayr (1945) write that there is no ground for retaining the genera *Cygnopsis* and *Eulabeia* because all the species of these genera are actually the members of genus *Anser*.

The two other groups, classified according to the chromosome homology, cannot be justified from the taxonomist's point of view. Anatidae has been divided into two subfamilies, Anserinae and Anatinae. To the former subfamily belongs not only our species of group I but also two more cytologically known genera, viz. *Cygnus* and *Branta*, but the karyotypes of the latter resemble very closely those obtained in species belonging to Anatinae rather than Anserinae. Here we find a genuine contradiction between the cytological and taxonomical affinities, the latter being very strong between swans (*Cygnus*) and geese (*Anser*, *Branta*). According to Delacour and Mayr (1945) these two groups are so closely related that they cannot even be separated into different tribes. Here we have an example of cytologically dissimilar species being taxonomically similar. The reverse of this situation is found in the Anatinae subfamily. The members belonging to this group have been divided taxonomically into several well defined tribes but the chromosome picture is one of uniformity except that of *Bucephala clangual* which apparently shows a different chromosome pattern. But, as pointed out earlier, its chromosome pattern must have been derived from an ancestral species where chromosomes were similar to those of the Anatinae subfamily and therefore its inclusion in that subfamily is cytologically justified. Taxonomically *Bucephala* is closest to *Mergus* and they both belong to the tribe Mergini. Hybrid between these two genera have been found repeatedly in the wild state thus indicating their close affinity.

Hybrids from crosses of the same genus or even from crosses between related genera are often fertile in the family Anatidae (Delacour and Mayr, 1945). The uniformity of the karyotype of this group therefore is not only limited to the size and morphology of the chromosomes but also extends to the genic level. It should always be borne in mind, however, that chromosomal rearrangements can occur with or without very minute visible alterations of the morphology of the chromosomes; Mott *et al.* (1968) analysed the karyotypes of *Anas platyrhynchos* and *Cairina moschata* and hybrid between the two. Centromeric indices of both chromosome nos. 1 and 2 were slightly different in these two species but this difference was more than enough to make the hybrid sterile because of the failure of synapsis in meiosis.

D. *Galliformes*

From the point of view of chromosome number, structure and morphology, the different members of the order Galliformes appear to be fairly uniform. We know the chromosomes of eight species so far and in all of them the diploid number is 78$\pm$ except *Pavo cristatus* where it is only 66$\pm$ (Table I). A comparative karyogram of eight species is given in Fig. 5, six of them belong to the family Phasianidae and one each to Meleagridae and Numididae. The obvious differences between the karyotypes (Fig. 5) are as follows: Chromosome no. 2 is absent in *Syrmaticus reevesii* (E), *Phasianus colchicus* (F) and *Meleagris gallopavo* (G); Chromosome no. 3 is present only in *Pavo cristatus* (A) and *Numida meleagris* (H) and no. 4 is present only in *Pavo cristatus*.

When we compare the chromosomes of two taxonomically closest species viz. *Gallus domesticus* (B) and *Gallus gallus* (C) of Phasianidae we find all the chromosomes are very similar (compare B and C in 1–2, 5–10). The absence of chromosomes Nos. 11 and 12 in C and their presence in B is not to be taken seriously because the size of these chromosomes is such that they are likely to be considered as macrochromosomes in the latter and therefore included in the idiogram and microchromosomes in the former and hence

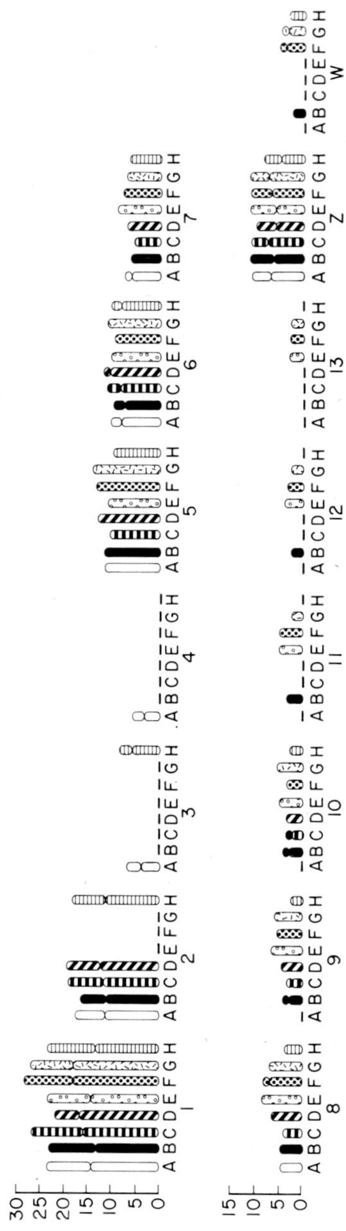

Fig. 5. Composite idiogram of Galliformes.

A: *P. cristatus* B: *G. domesticus*
C: *G. gallus* D: *C. c. japonica*
E: *S. reevesii* F: *Ph. colchicus*
G: *M. gallopavo* H: *N. meleagris*

TABLE V. Galliformes

Species		Macrochromosome pair nos. 1	2	3	4	5	6	7	8	9	10	11	12	13	Z	W
Pavo cristatus	Relative length	23.5	17.5	7.0	6.5	12.0	10.5	7.5	4.5	—	—	—	—	—	11.5	—
	Centromeric index	38.3	36.4	39.6	45.1	—	23.6	17.1	—	—	—	—	—	—	38.8	—
Gallus domesticus	Relative length	23.5	16.5	—	—	12.0	10.0	7.0	4.5	4.0	4.0	3.0	2.5	—	10.0	2.5
	Centromeric index	42.5	30.3	—	—	—	20.0	—	—	25.0	25.0	—	—	—	47.8	—
Gallus gallus	Relative length	27.0	19.0	—	—	11.0	11.5	6.0	4.0	3.0	3.0	—	—	—	11.0	—
	Centromeric index	40.7	36.5	—	—	—	21.7	—	—	16.6	16.6	—	—	—	31.8	—
Coturnix coturnix japonica	Relative length	22.0	19.5	—	—	13.5	12.0	7.5	6.5	4.5	3.5	—	—	—	10.0	—
	Centromeric index	23.0	36.5	—	—	—	8.3	—	—	—	—	—	—	—	40.0	—
Syrmaticus reevesii	Relative length	23.6	—	—	—	11.6	10.9	9.4	8.4	6.7	5.2	5.0	4.0	3.1	11.6	—
	Centromeric index	37.8	—	—	—	—	—	—	—	—	—	—	—	—	50.0	—
Phasianus colchicus	Relative length	28.5	—	—	—	14.0	10.0	8.5	8.0	6.5	5.0	5.0	3.5	2.5	11.0	—
	Centromeric index	36.8	—	—	—	—	—	—	—	—	—	—	—	—	36.3	—
Meleagris gallopavo	Relative length	27.0	—	—	—	14.5	11.5	7.5	7.0	6.0	5.5	2.5	2.5	2.5	11.5	—
	Centromeric index	33.3	—	—	—	—	—	—	—	—	—	—	—	—	—	—
Numida meleagris	Relative length	23.5	18.5	8.5	—	10.5	10.5	7.0	3.5	2.6	2.5	—	—	—	8.5	3.0
	Centromeric index	42.5	37.9	35.2	—	—	19.0	—	—	—	—	—	—	—	41.1	—

omitted. When the karyotypes of the five species of the family Phasianidae are compared among themselves, it is found that the two species of *Gallus*, on the one hand, and *Coturnix c. japonica* (D), on the other, differ in the following respects. Chromosome No. 1 has less median centromere in *Coturnix* than in *Gallus* (Table V). Chromosome No. 6 has a more terminal centromere, and Nos. 9 and 10 have no second arms in *Coturnix*. *P. cristatus* (A), another member of the family, has obvious similarities in its chromosomes with all the above species but its karyotype is somewhat distinct in having two mediocentric chromosomes viz. Nos. 3 and 4 which are not represented in any other members of the family Phasianidae. Moreover the telocentric chromosomes Nos. 9 and 10 are not present only in *Pavo cristatus*.

It is interesting to note that *Syramaticus reevesii* (E) and *Phasianus colchicus* (F) though belonging to the family Phasianidae, have karyotypes identical to that of *Meleagris gallopavo* (G) of the family Meleagridae (compare E, F and G in 1–13 and Z). The only difference is found in the structure of the W chromosome between the two families. The W chromosome could not however be identified in *Syrmaticus reevesii*. Stenius *et al.* (1963) pointed out the similarity of the karyotypes of *Phasianus colchicus* and *Meleagris gallopavo*. Since then we have come to know the chromosomes of *Syrmaticus reevesii* and this species also has chromosome affinities with Meleagridae. *Numida meleagris* (H) though belonging to the family Numididae has chromosomes similar to those of the members of Phasianidae except *Phasianus* and *Syrmticus* and its distinctiveness is only in the presence of metacentric chromosome No. 3 which is also present in *Pavo cristatus*.

From the point of view of chromosomal affinities as revealed through the study of comparative karyology, the order can be divided into three groups. In the first group we can put the two species of *Gallus* and *Coturnix c. japonica* and in the second *Syrmaticus reevesii*, *Phasianus colchicus* and *Meleagris gallopavo*. In a third group we put *Pavo cristatus* and *Numida meleagris*.

E. *Gruiformes*

The order Gruiformes is known chromosomally by only three species all belonging to the family Rallidae. An overall similarity of their karyotypes is clearly revealed in the composite idiogram (Fig. 6). If we compare the macrochromosomes alone, *Fulica atra* (A) and *Gallinula chloropus* (B) are almost identical (Compare A and B, 1–9 and Z, W) in spite of the difference in the number of microchromosomes between the two species, being 22 in the former and 58 in the latter (Table I). Since a comparison of relative lengths between species having an unequal number of microchromosomes is very risky, we measured the length of the microchromosomes of the two species to find out exactly the amount of material eliminated by us in calculating the relative lengths of the macrochromosomes. It has however been found that the total lengths of microchromosomes eliminated was 44.2% in A and 41.5% in B. The similarities between their chromosomes are therefore real. *Porphyrio poliocephalus viridis* (C) on the other hand has a

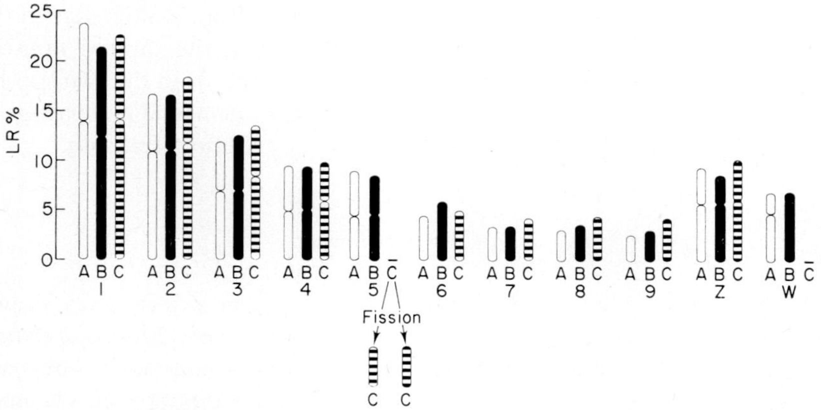

Fig. 6. Composite idiogram of Gruiformes.
A: *F. atra* B: *G. chloropus*
C: *P. poliocephalus viridis*

TABLE VI. Gruiformes

	Relative length			Centromeric index		
	Fulica atra	*Gallinula chloropus*	*Porphyrio poliocephalus viridis*	*Fulica atra*	*Gallinula chloropus*	*Porphyrio poliocephalus viridis*
(1)	23.4	20.8	22.5	41.6	41.6	39.0
(2)	16.6	16.5	19.5	32.2	32.2	40.0
(3)	11.9	12.4	14.6	40.0	47.6	40.0
(4)	9.7	9.6	10.1	50.0	47.6	42.7
			3.9 I			— I
(5)	9.0	8.4		50.0	47.6	
			3.9 II			— II
(6)	4.3	5.9	4.9	—	—	—
(7)	3.6	3.7	3.9	—	—	—
(8)	2.8	3.4	3.3	—	—	—
(9)	2.5	3.1	3.1	—	—	—
Z	6.8	7.1	?	35.7	33.3	—

slightly different karyotype from the previous two species. The metacentric chromosome No. 5 present in A and B is represented by two telocentrics in C whose combined relative length is more or less equal to the metacentric chromosome (Table VI) showing Robertsonian fission or fusion.

The morphology of the macrochromosomes is so similar in all the three species that any further subgrouping is cytologically unwarranted although the taxonomists have preferred to put *Fulica atra* in the subfamily Fulicinae and *Porphyrio poliocephalus viridis* and *Gallinula chloropus* in Rallinae. The only karyological support of this subdivision comes from a difference in the number of microchromosomes and consequently in the diploid number between Rallinae and Fulicinae being 78 and 72 (Table I) in the two species of Rallinae and 92 in the only species studied in Fulicinae. The cytological data are obviously too meagre in this group for any useful discussion.

F. *Charadriiformes*

A comparative study of the karyotypes of the seven cytologically known species of this order reveals curiously that there is very little correlation between their taxonomic subgroupings into families and subfamilies and the groups into which they can be classified from a study of their chromosome homology.

Out of the seven species, one, viz. *Haematopus ostralegus* (A) (Fig. 7a), belongs to Haematopodidae and the six others to Charadriidae. The latter family includes the chromosomes of three subfamilies: (i) Charadriinae, comprising *Vanellus vanellus* (B) and *Charadrius hiaticula* (C); (ii) Scolopacinae which includes *Numenius arquata* (D), *Tringa totanus* (E) and *Gallinago gallinago* (F) and (iii) Recurvirostrinae represented by a single species *Recurvirostra avosetta* (G).

In order to compare the chromosomes of this apparently heterogenous order we had to construct four separate karyograms. In Fig. 7a is included A, the only member of Haematopodidae, two of the subfamily Charadriinae B and C, and G of the subfamily Recurvirostrinae. These four species though distantly related taxonomically surprisingly show homology between their chromosomes (Table VIIA). The karyotype of A is almost identical to that of B. The only difference is in the centromeric position of the chromosome No. 8 which has a terminal (t) centromere in A and sub-terminal (st) in B. A single pericentric inversion is the simplest explanation of the derivation of one karyotype from the other. Again C and G have identical karyotypes though they belong to different subfamilies. The karyotypes of A and B on the one hand and C and G on the other are different in respect to the structure of two chromosomes only. The medium sized metacentric chromo-

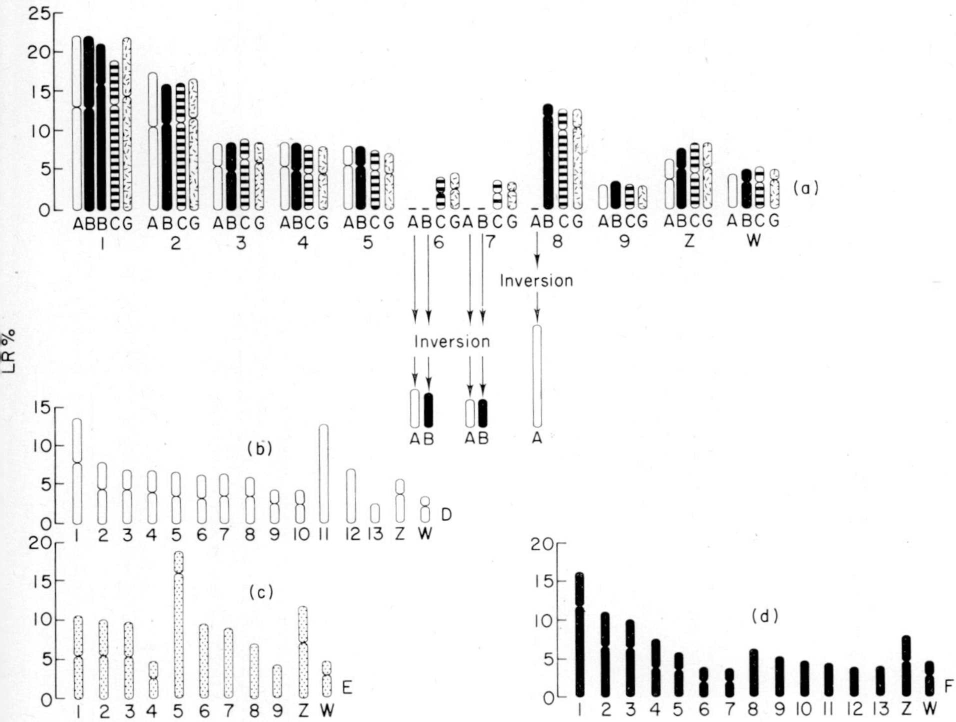

Fig. 7. Composite idiogram of Charadriiformes.

A: *H. ostralegus* B: *V. vanellus*
C: *C. hiaticula* D: *N. arquata*
E: *T. totanus* F: *G. gallinago*
G: *R. avosetta*

somes Nos. 6 and 7 in C and G are replaced by two similar sized (Table VIIA) telocentric chromosomes in A and B. Two pericentric inversions can account for the origin of one karyotype from the other. The diploid number is 66 in *Haematopus* and 76 in the other three.

The three members (D, E, F) belonging to the Scolopacinae subfamily of the family Charadriidae are so widely different both from the point of view of diploid number, having 78, 88 and 98 respectively (Table I), and chromosome morphology that the construction of a composite idiogram for comparing their karyotypes is useless. They are therefore plotted separately (Fig. 7b, c and d.). An examination of these diagrams shows certain obvious chromosomal differences between the species viz. (i) the biggest metacentric chromosome, i.e. chromosome No. 1, has different relative lengths in the different species; (ii) the number of mediocentric autosomes is 10 in D, 4

TABLE VIIA. Charadriiformes

	Relative length				Centromeric index			
	Hematopus ostralegus	*Vanellus vanellus*	*Charadrius hiaticula*	*Recurvirostra avosetta*	*Hematopus ostralegus*	*Vanellus vanellus*	*Charadrius hiaticula*	*Recurvirostra avosetta*
(1)	22.0	22.1 a 21.3 b	19.2	21.8	40.0	40.0 a 25.0 b	23.4	34.4
(2)	17.3	16.2	15.8	16.5	41.6	30.3	30.3	23.4
(3)	8.6	8.6	9.0	8.4	41.6	38.4	25.6	27.0
(4)	8.6	8.3	8.2	8.1	33.3	45.4	47.6	47.6
(5)	8.0	7.9	7.5	7.2	32.2	31. 2	34.4	37.0
(6)	4.9	4.3	4.1	4.4	—	—	45.4	50.0
(7)	3.4	3.9	3.4	2.5	—	—	47.6	41.6
(8)	13.0	13.0	12.6	12.6	—	11.4	19.2	15.8
(9)	13.1	13.5	12.9	12.3	—	—	—	—
Z	6.2	7.5	8.7	8.4	41.6	34.4	43.4	38.4
W	4.3	5.1	4.8	4.4	—	33.3	28.5	25.0

TABLE VIIB. Charadriiformes (Scolopacinae)

Species		Pairs of macrochromosomes 1	2	3	4	5	6	7	8	9	Z	W
Tringa totanus	Relative length	10.4	9.9	9.4	9.4	18.9	9.4	9.0	7.1	4.2	11.8	4.7
	Centromeric index	50.0	47.6	47.6	50.0	14.7	—	—	—	—	41.6	35.7

TABLE VIIC. Charadriiformes (Scolopacinae)

Species		Pairs of macrochromosomes 1	2	3	4	5	6	7	8	9	10	11	12	13	Z	W
Numenius arquata	Relative length	13.7	7.7	6.8	6.8	6.5	6.2	6.2	5.7	4.2	4.2	12.8	7.1	2.5	5.4	3.4
	Centromeric index	45.4	45.4	34.4	41.6	47.6	47.6	43.4	37.0	45.4	40.0	—	—	—	37.0	40.0

TABLE VIID. Charadriiformes (Scolopacinae)

Species		Pairs of macrochromosomes 1	2	3	4	5	6	7	8	9	10	11	12	13	Z	W
Gallinago gallinago	Relative length	16.2	11.0	10.0	7.6	5.8	3.8	3.8	6.2	5.2	4.8	4.4	4.1	4.1	7.9	4.4
	Centromeric index	25.6	40.0	37.0	38.4	40.0	50.0	47.6	—	—	—	—	—	—	47.6	38.4

in E and 6 in F; (iii) the number of chromosomes with terminal centromere is 3 in D and 4 in E and 6 in F and (iv) the very big subterminal (st) chromosome No. 5 of E is not represented in any other species. The morphometric data of the macrochromosomes of three species are given in Tables VII B, C and D.

G. *Lariformes*

The chromosome data of the order Lariformes are available from only two genera, viz. *Larus* and *Sterna*, belonging to the families Laridae and Sternidae respectively. The former genus is known through the study of four species and the latter from three. The available data therefore are in contrast with those of many other orders dealt with in this paper where the samples come, in many cases, from diverse families and subfamilies.

A study of the comparative karyogram (Fig. 8) of the seven species reveals, not unexpectedly, a good deal of similarity of karyotypes not only in the congeneric species but also between the species of the two genera. The diploid number of chromosomes is between 66 and 70 in all the species (Table I). When we compare the chromosomes of the various species of the same genus we find them to be almost identical (compare A–D and E–G). Chromosomal affinity between *Larus* and *Sterna* is evident from the relative lengths and structure of chromosomes Nos. 1–6 and 9–13 (Table VIII) and ZW. The main differences between them are (i) the presence of two small metacentric chromosomes Nos. 7 and 8 in all the three species of *Sterna* (E, F, G) and their absence in *Larus* and (ii) the presence of small telocentric chromosomes Nos. 14 and 15 in all the four species of *Larus* (A, B, C, D) which are not found in *Sterna*. The above differences may be explained by simple Robertsonian fission or fusion mechanism but it is not very obvious from the study of the idiogram presented. This is because of the fact that the fission or fusion, as the case may be, has involved chromosomes, which may either be classified as small macro- or large microchromosomes. It can be conceived that chromosomes Nos. 14 and 15 and the next two telocentric microchromosomes in order of size in *Larus* have undergone centromeric fusion to give rise to the metacentric chromosomes Nos. 7 and 8 of *Sterna* or vice versa.

Currently systematists consider that the birds of the order Lariformes should belong to the family Laridae under the order Charadriiformes. If some more genera of birds belonging to Lariformes could be shown to possess similar karyotypes as have been found so far, there will be much justification, at least from the karyological point of view, for keeping Lariformes a separate order rather than grouping it with Charadriiformes which is extremely heterogenous.

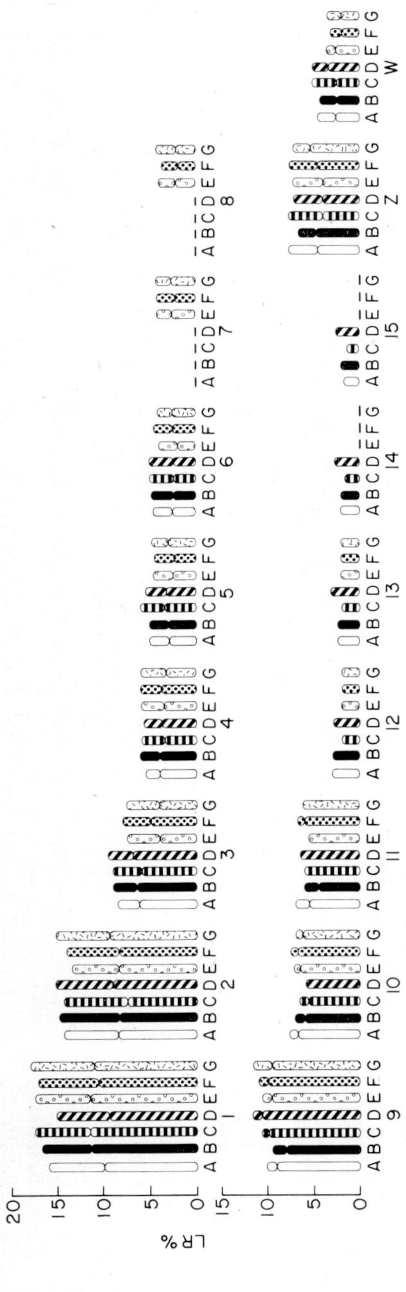

Fig. 8. Composite idiogram of Lariformes.
A: *L. canus* B: *L. ridibundus*
C: *L. argentatus* D: *S. hirundo*
E: *S. paradisaea* F: *S. albifrons*

TABLE VIII. Lariformes

Species		Macrochromosome pair nos.																
		1	2	3	4	5	6	7	8	9	10	11	12	13	14	15	Z	W
Larus canus	Relative length	16.2	13.8	8.4	5.4	5.1	4.5	—	—	10.2	7.5	6.9	3.0	2.4	2.4	2.1	7.5	4.5
	Centromeric index	37.0	41.6	27.7	27.0	38.4	45.4	—	—	9.1	10.0	23.2	—	—	—	—	41.6	47.6
Larus ridibundus	Relative length	16.8	14.6	9.0	6.1	5.5	4.8	—	—	9.7	6.8	5.8	2.9	2.6	1.9	1.9	6.8	4.2
	Centromeric index	34.4	43.4	27.7	29.4	38.4	47.6	—	—	14.2	14.2	23.2	—	—	—	—	45.4	47.6
Larus argentatus	Relative length	18.2	14.3	9.6	6.2	6.2	5.8	—	—	10.4	6.8	6.2	2.9	2.9	1.8	1.7	7.9	5.2
	Centromeric index	36.6	48.6	34.0	37.5	37.5	46.6	—	—	8.1	14.2	—	—	—	—	—	43.9	48.1
Larus fuscus	Relative length	16.6	15.0	10.1	5.7	5.3	4.9	—	—	12.0	7.1	6.4	3.4	3.4	3.0	3.0	7.5	5.6
	Centromeric index	38.4	40.0	29.4	32.2	45.4	5.0	—	—	9.1	—	—	—	—	—	—	47.6	31.2
Sterna hirundo	Relative length	17.8	13.9	7.7	6.0	4.6	4.3	4.1	3.8	10.9	7.1	6.5	1.9	1.9	—	—	7.4	3.5
	Centromeric index	35.7	40.0	45.4	47.6	47.6	47.6	47.6	45.4	12.5	12.3	—	—	—	—	—	45.4	47.6
Sterna paradisaea	Relative length	17.2	14.3	8.2	6.1	4.7	4.5	4.2	3.2	11.1	7.4	6.9	2.2	2.2	—	—	7.4	4.5
	Centromeric index	38.4	37.0	47.6	43.4	50.0	50.0	47.6	50.0	9.1	12.1	11.1	—	—	—	—	41.6	47.6
Sterna albifrons	Relative length	18.5	15.0	7.4	6.1	4.9	4.3	4.2	3.9	11.7	7.1	5.4	4.4	4.4	—	—	7.3	3.8
	Centromeric index	38.1	37.5	48.7	49.2	43.1	43.5	44.3	43.9	21.4	10.8	—	—	—	—	—	45.8	42.6

H. *Columbiformes*

Five species in this order are known cytologically and all of them belong to a single subfamily Columbinae, family Columbidae. The diploid number of chromosomes ranges from 72 to 80 (Table I) in this family. An examination of the idiogram presented in Fig. 9 reveals that the different species

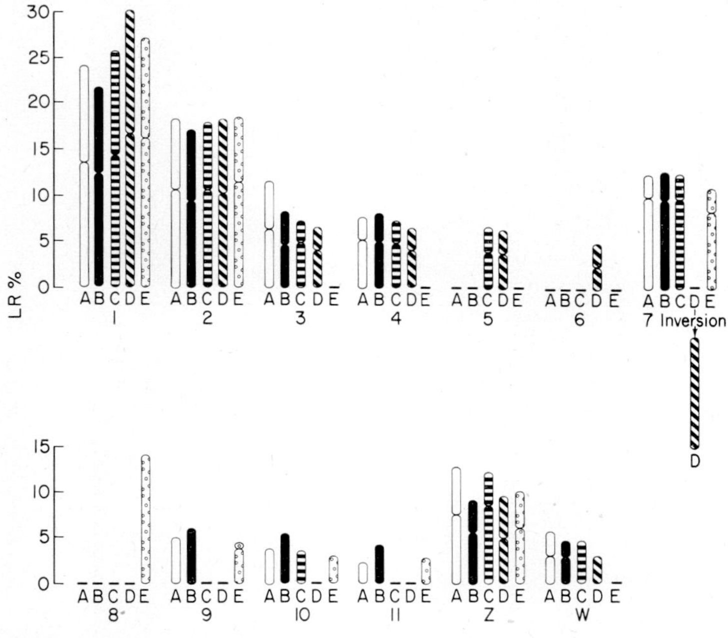

Fig. 9. Composite idiogram of Columbiformes.
A: *C. livia domestica* B: *C. palumbus*
C: *G. cuneata* D: *S. risoria*
E: *S. orientalis*

belonging to this group share among themselves four chromosomes viz. chromosomes Nos. 1, 2, 7 and Z. The other chromosomes are distributed amongst the species in such a manner that some chromosomes are common in some members while others are common to other members.

The two species of *Columba*, viz. *Columba livia domestica* (A) and *Columba palumbus* (B), have identical karyotypes. *Geopelia cuneata* (C) is the species which is closest to *Columba* having similarity with the latter in chromosomes Nos. 1, 2, 3, 4, 7, 11 and Z. The metacentric chromosome No. 5 in *Geopelia* and the telocentric chromosomes Nos. 9 and 10 in *Columba* have made the karyotype of these two genera different.

TABLE IX. Columbiformes

	Relative length					Centromeric index				
	Columba livia domestica	*Columba palumbus*	*Geopelia cuneata*	*Streptopelia risoria*	*Streptopelia orientalis*	*Columba livia domestica*	*Columba palumbus*	*Geopelia cuneata*	*Streptopelia risoria*	*Streptopelia orientalis*
(1)	24.2	21.3	25.8	30.3	27.2	43.2	43.6	43.9	44.4	40.0
(2)	18.4	17.2	18.2	18.5	18.8	42.8	44.6	41.7	45.2	38.1
(3)	11.8	8.6	7.6	6.7	—	44.4	44.4	33.3	37.5	—
(4)	7.9	8.2	7.6	6.7	—	33.3	36.4	33.3	37.5	—
(5)	—	—	6.7	6.6	—	—	—	40.0	37.5	—
(6)	—	—	—	5.0	—	—	—	—	50.0	—
(7)	12.4	12.9	12.6	12.4	11.3	21.1	26.0	20.0	—	23.6
(8)	—	—	—	—	14.2	—	—	—	—	—
(9)	5.2	6.2	—	—	4.7	—	—	—	—	10.5
(10)	3.9	5.7	3.8	—	3.3	—	—	—	—	11.8
(11)	2.6	4.7	—	—	3.1	—	—	—	—	—
Z	13.1	9.5	12.6	10.1	10.5	40.0	40.0	30.0	50.0	40.0
W	6.1	5.1	5.0	3.3	?	50.0	35.7	—	—	?

The two other congeneric species in this group are *Streptopelia risoria* (D) and *Streptopelia orientalis* (E) but they show more chromosomal divergence than is expected in two congeneric species. As a matter of fact only those chromosomes which are common throughout the family are also common in these two species. They are chromosomally related to *Columba* and *Geopelia* in the sense that some chromosomes of *Columba* and *Geopelia* are shared by *S. orientalis* while others are shared by *S. risoria*. For example metacentric chromosomes Nos. 3 and 4 are present in *Geopelia* (C), *Columba* (A and B) and *S. risoria* (D) while the telocentric chromosomes Nos. 9, 10 and 11 are common to *Columba* (A and B) and *S. orientalis* (E). In other words the two species of *Columba* have a karyotype pattern which is intermediate between those of *S. risoria* and *S. orientalis*.

The Z chromosome is conservative, its relative length varying from 9.5 to 10.5 in *Columba* and *Streptopelia* though in *Geopelia* it is slightly bigger being 12.5 (Table IX). It is submedian in *Geopelia* and median in the rest of the species. The W chromosome was identified in all except *S. orientalis* and its relative length varies from 3.3 to 6.5 and the centromeres are median in *Columba* and terminal in *Streptopelia* and *Geopelia*.

I. *Psittaciformes*

The order Psittaciformes is known cytologically from the study of five species belonging to a single family Psittacidae. Three out of these five worked out species are congeneric and the remaining two belong to two other genera. A study of their karyotypes (Fig. 10a and b) shows that only four species viz. *Psittacula krameri borealis* (A), *P. alexandri fasciata* (B), *P. cyanocephalo bengalensis* (C) and *Loriculus vernalis* (D) have karyotypes which are sufficiently similar for a meaningful comparison while the fifth one, *Melopsittacus undulatus* (E), is very different chromosomally (Fig. 10b).

An examination of the comparative karyogram (Fig. 10a) shows very convincingly that species A and B are very closely related because they have almost identical karyotypes. Species C, although belonging to the same genus, has a slightly different karyotype. In the latter species telocentric chromosomes Nos. 6 and 7 are absent and they are replaced by a metacentric chromosome which is equal to the combined relative lengths of these two chromosomes (Table X). We would like to interpret this change as a case of Robertsonian fusion because three out of four species studied herein have two telocentrics instead of one metacentric chromosome.

When we analyse the karyotype of *Loriculus* (D) we find that chromosomes Nos. 1, 6 and 7 are exactly similar to the three species of the genus *Psittacula*. Although the submetacentric chromosomes Nos. 2, 3, 4 and 5 of *Psittacula*

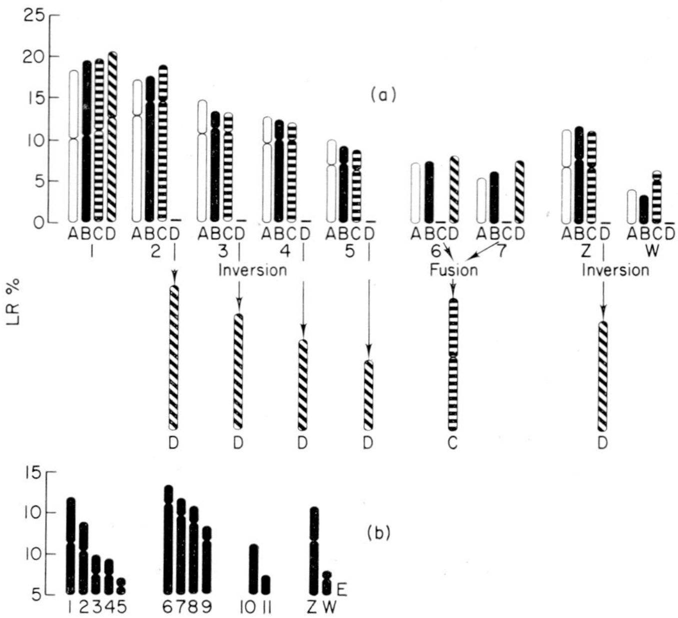

Fig. 10. Composite idiogram of Psittaciformes.
a. A: *P. krameri borealis* B: *P. alexandri fasciata*
C: *P. cyanocephala bengalensis* D: *L. vernalis*
b. E: *Melopsittacus undulatus*

are absent here, they are replaced by exactly similar sized chromosomes with centromeres in the terminal region (Table X). A series of pericentric inversions are the only explanation for the derivation of the karyotype of this species. The Z chromosome is also telocentric in *Loriculus* (D) while it is metacentric in all the three species of *Psittacula*. The chromosome number ranges from 66 to 72 in these four species.

Melopsittacus undulatus not only has a lower diploid number which is 58$\pm$, but its karyotype pattern is also different. The submetacentric chromosomes Nos. 6, 7, 8 and 9 of *Melopsittacus* resemble morphologically the chromosomes Nos. 2, 3, 4 and 5 of *Psittacula*. Their lengths differ, which may be due to the fact that a greater number of chromosomes have been measured in *Melopsittacus* than in *Psittacula* and therefore the chromosomes of the former have become relatively smaller. It seems these four chromosomes have not undergone much change during evolution of these species.

W was not identified in *Loriculus* as no female specimen could be studied. The length of the W in the family from 3.0 to 6.4 and morphology varies from median to subterminal and terminal.

TABLE X. Psittaciformes

	Relative length				Centromeric index			
	Psittacula kramari borealis	*Psittacula alexandri fasciata*	*Psittacula cyanocephala bengalensis*	*Loriculus vernalis*	*Psittacula krameri borealis*	*Psittacula alexandri fasciata*	*Psittacula cyanocephala bengalensis*	*Loriculus vernalis*
(1)	18.4	19.8	20.1	20.2	45.3	47.1	44.2	46.6
(2)	17.5	17.7	16.8	17.5	25.6	17.1	23.3	—
(3)	15.2	13.7	13.6	14.0	26.7	16.3	16.7	—
(4)	13.2	12.5	12.4	11.0	25.8	19.1	16.8	—
(5)	10.2	9.7	9.1	7.6	28.9	23.0	28.5	—
(6)	7.6	7.7	16.0	7.6	—	—	45.0	—
(7)	5.7	6.5		7.5	—	—		—
Z	11.6	12.1	11.6	12.5	38.2	35.3	38.4	—
W	4.4	3.8	6.4	?	—	—	14.4	?

J. *Strigiformes*

The chromosomes of six species of this order have been worked out and all of them belong to a single family, Strigidae; five belong to the subfamily Striginae and one to the subfamily Tytoninae.

Owing to the diversity in the chromosome constitution between different species of Strigiformes, three separate karyograms had to be constructed for a comparison of their karyotypes. In the first group belong the two congeneric species, *Athene noctua* and *A. brama* (Fig. 11a), in the second *Bubo virginianus virginianus*, *Strix aluco* and *Otus scops japonicus* (Fig. 11b) and

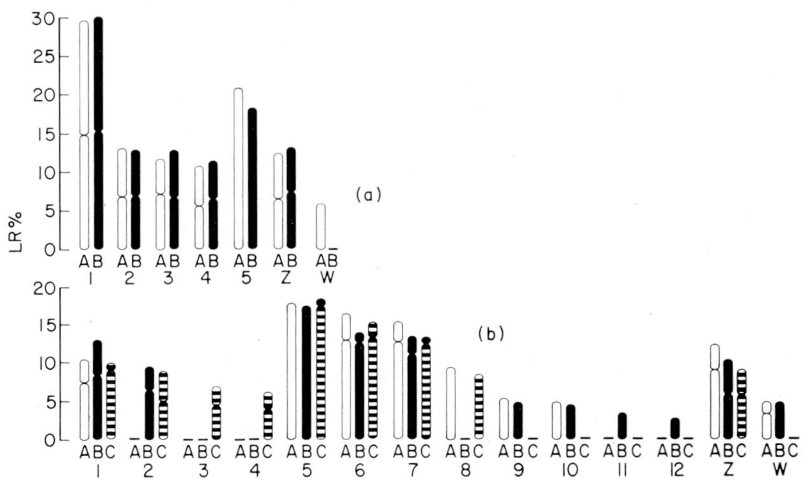

Fig. 11. Composite idiogram of Strigiformes.
a. A: *A. noctua* B: *A. brama*
b. A: *B. virginianus virginianus* B: *S. aluco*
C: *O. scops japonicus*

in the third a radically different karyotype of *Tyto alba* (Fig. 12a and b) which has a very high diploid number of 92 while all the other members of this order have chromosome numbers between 76 and 82.

In the karyotype of *Athene noctua* described by Renzoni and Vegni-Talluri (1966), 9 pairs of chromosomes including the Z are considered as macrochromosomes but an examination of the metaphase plate illustrated by those authors shows clearly only six pairs of such chromosomes. Moreover, the congeneric species *Athene brama* worked out by Ray-Chaudhuri *et al.* (1969) also shows only 6 pairs of macrochromosomes. The diploid number of chromosomes in both the species is 82±. While constructing the

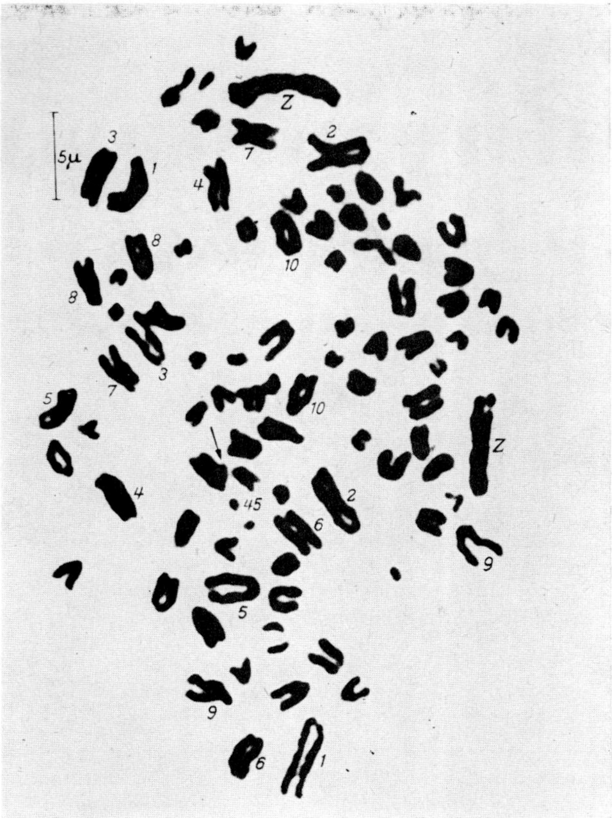

Fig. 12a. *Tyto alba* Metaphase plate from a female specimen. (After Renzoni and Vegni-Talluri, 1966.)

composite idiogram for comparison, we decided to measure only six pairs of macrochromosomes. The karyotypes of the same two species were compared by Ray-Chaudhuri *et al.* (1969) but we, according to Renzoni and Vegni-Talluri's designation, included three more pairs of chromosomes as macrochromosomes in *A. noctua* and therefore were not able to find out the close similarity which is now evident (see Fig. 11a). Chromosomes Nos. 1–5 and Z seem to be exactly alike (Table XIA).

While comparing the idiograms of three genera, *Bubo* (A), *Strix* (B) and *Otus* (C), we found that there are certain similarities in chromosomes Nos. 1, 5, 6, 7 and Z (Fig. 11b) with some minor differences. For example, chromosome No. 5 has a minute second arm in *Otus* and the second arm of chromosome No. 6 is slightly bigger in *Bubo* (Table XIB).

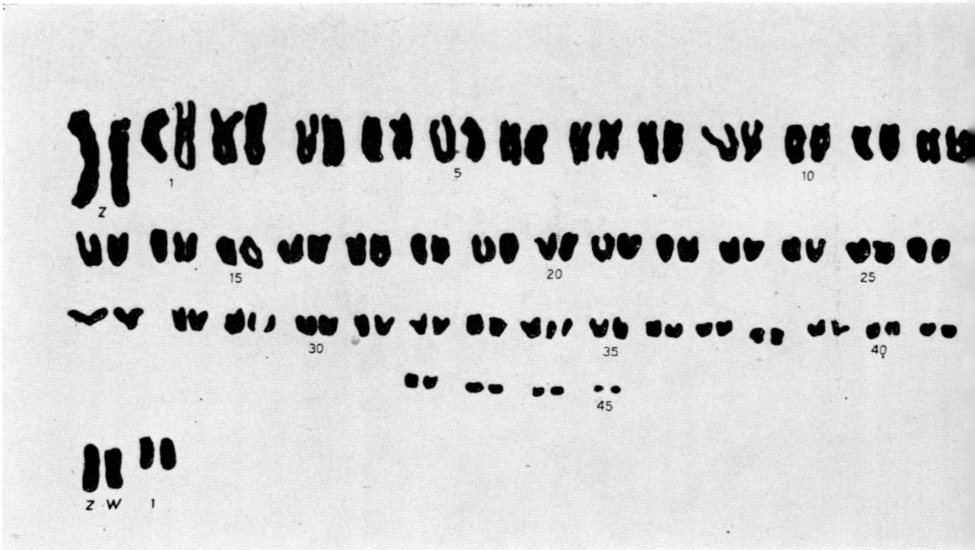

Fig. 12b. *Tyto alba* Chromosomes from the metaphase of Fig. 12a arranged in order of size. (After Renzoni and Vegni-Talluri, 1966.)

TABLE XIA. Strigiformes

	Relative length		Centromeric index	
	Athene noctua	*Athene brama*	*Athene noctua*	*Athene brama*
(1)	29.7	30.3	50.0	48.1
(2)	13.3	13.2	48.6	47.7
(3)	11.8	12.9	38.2	50.0
(4)	11.1	11.5	47.4	43.7
(5)	21.1	18.5	—	—
Z	12.6	13.2	48.2	42.8
W	5.9	—	—	—

The Barn owl, *Tyto alba* has a unique karyotype as far as birds are concerned. All the chromosomes of this species are acrocentric except the Z and W. Moreover, there is no distinction between the macro- and microchromosomes (Fig. 12a and b). This species seems to be clearly unrelated chromosomally to any other species of owl so far studied.

The length of Z chromosome varies from 9.2 to 12.8 and all are metacentrics or submetacentrics. The W could be identified in four species of this order and not in *A. brama* or *Otus scops japonicus*. The size and morphology

TABLE XIB. Strigiformes

	Relative length			Centromeric index		
	Bubo v. virginianus	*Strix aluco*	*Otus scops japonicus*	*Bubo v. virginianus*	*Strix aluco*	*Otus scops japonicus*
(1)	10.5	13.0	9.9	30.0	34.4	38.1
(2)	—	9.3	9.5	—	32.2	35.0
(3)	—	—	7.0	—	—	33.0
(4)	—	—	6.7	—	—	35.4
(5)	17.9	17.5	19.1	—	—	5.0
(6)	16.8	13.8	15.6	17.9	12.0	12.1
(7)	15.5	13.4	13.7	17.7	15.8	6.9
(8)	9.3	—	8.6	—	—	—
(9)	5.7	5.3	—	—	—	—
(10)	4.8	4.5	—	—	—	—
(11)	—	3.6	—	—	—	—
(12)	—	3.2	—	—	—	—
Z	12.5	10.9	9.6	42.0	47.6	35.0
W	6.5	5.2	—	44.0	—	—

of W vary quite a bit. The relative length varies from 4.3 in *A. noctua* to 11.0 in *Tyto alba* and it is telocentric in *Strix aluco* and *Athene noctua* but metacentric in *Bubo virginianus virginianus* and *Tyto alba*.

K. *Passeriformes*

Twenty three species are cytologically known in Passeriformes and this is the maximum number of species so far worked out in any particular order. This most highly evolved order in the class Aves also includes the maximum number of families, genera and species. Since twenty three worked out species belong to 12 different families, an opportunity is provided in this group to study the extent of interfamilial variation in the chromosome constitution within a single order.

The diploid number is very uniform throughout the order being 78 or 80 in most of the species with a few exceptions which have chromosome numbers between 66 and 76 (Table I).

An examination of the karyotypes of all the different species, will show that the various families in this order have more or less their own distinctive karyotypes. In other words, the different species so far worked out within a single family have karyotypes which are quite similar. There are, however, certain exceptions specially in certain families which unfortunately have only one cytologically worked out species. They possess karyotypes similar to species belonging to a different family. To facilitate comparison their idiograms have been included in the respective composite idiograms of those families to which they show chromosomal affinity.

(a) *Muscicapidae and Mimidae*

Only one species of the family Mimidae, viz. *Toxostoma rufum*, is chromosomally known and it has a karyotype which resembles that of the subfamily Turdinae of Muscicapidae, and therefore, for the purpose of chromosome comparison we have treated both the families together.

The family Muscicapidae has been divided into several subfamilies, of which we know the chromosomes of six species of Turdinae and only one species, *Turdoides striatus striatus* of Timaliinae.

Two composite idiograms had to be constructed to represent the homology of chromosomes properly. Species A–E (Turdinae) and H (Mimidae) are in one idiogram (Fig. 13a), one species of Turdinae (F) and one of

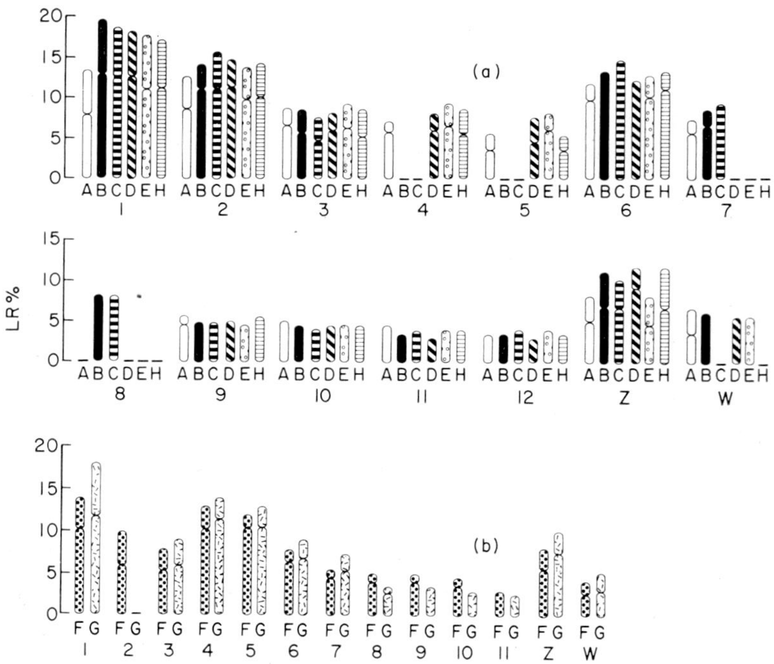

Fig. 13. Composite idiogram of Passeriformes.

a. Turdinae and Mimidae
A: *T. migratorius*
B: *T. merula*
C: *T. sibiricus davisoni*
D: *S. torquata*
E: *O. oenanthe*
H: *T. rufum*

b. Turdinae and Timaliinae
F: *P. phoenicurus*
G: *T. striatus striatus*

Timaliinae (G) in the other (Fig. 13b). An examination of the idiograms will justify the arrangement.

(b) *Turdinae*

The diploid number is uniformly 80 in all the species in the subfamily except in *Saxicola torquata* (D) where it is 86.

The overall similarity of all the chromosome complements of all these species is evident in chromosomes Nos. 1, 2, 3, 7, 10, 11, 12 and Z (compare A–E and H in Fig. 13a) except chromosome No. 1, which is distinctly smaller in *Turdus migratorius* (A).

When we compare the three species of *Turdus* (A, B, C) we find that besides the above mentioned similarities these three species have one more chromosome, i.e. No. 7 in common, which is absent in the other species although the size and morphology of this chromosome vary amongst the species A, B and C (Table XIIA).

Curiously enough, *Turdus migratorius* (A) is found to have more chromosomal similarity to species D and E belonging to two other different genera of the subfamily than to the two congeneric species B and C. For instance, the metacentric chromosomes Nos. 4 and 5 are present in A, D, E and even in H (Mimidae) but not in B and C. Again telocentric chromosome No. 8 is found in B and C but not in A (Fig. 13a).

The Z chromosomes in A and E are smaller having relative lengths 7.8 and 8.1 respectively while the others have their relative lengths between 10.0 and 11.5. The centromere is median in all the species except in D where it is subterminal with a centromeric index 21.4 (Table XIIA). The W chromosome could be identified in A, B, D and E which, except in A, are all acrocentrics of similar relative length.

(c) *Timaliinae*

We have already pointed out the chromosomal similarities of the only species of Timaliinae studied (G) with one species (F) of Turdinae. If we compare the idiograms (Fig. 13b) of these two species we find they are very similar. The metacentric chromosome No. 2 in *Phoenicurus* makes this species different from *Turdoides*. The other points of difference are a bigger No. 1 chromosome in *Turdoides* and absence of any second arm in chromosomes Nos. 9 and 10 in the same species (Table XIIB). They differ also in their diploid numbers, being 76 in F and 68 in G.

The chromosomal similarity of *Toxostoma rufum* belonging to Mimidae with some of the members of Turdinae and of Muscicapidae is not very surprising from the taxonomist's point of view since Mayr and Amadon

TABLE XIIA. Passeriformes (Muscicapidae and Mimidae)

Species		Macrochromosome pair nos. 1	2	3	4	5	6	7	8	9	10	11	12	Z	W
Turdus migratorius	Relative length	13.5	12.8	8.5	7.1	5.7	11.4	7.1	—	5.7	5.0	4.2	3.5	7.8	7.1
	Centromeric index	42.2	33.3	26.0	26.0	37.5	18.7	24.0	—	24.0	24.0	—	—	45.4	40.0
Turdus merula	Relative length	19.4	13.9	8.4	—	—	13.0	8.4	8.0	4.6	3.7	3.4	3.4	10.9	5.9
	Centromeric index	34.4	26.0	38.4	—	—	—	22.3	—	—	—	—	—	45.4	—
Turdus sibiricus davisoni	Relative length	18.6	15.3	7.3	—	—	14.6	9.3	8.0	4.6	4.0	4.0	4.0	10.0	—
	Centromeric index	32.1	26.1	45.0	—	—	—	—	—	—	—	—	—	33.0	—
Saxicola torquata	Relative length	17.9	14.4	8.2	7.9	7.3	11.8	—	—	4.7	4.3	3.0	2.9	11.5	5.5
	Centromeric index	32.2	25.6	25.6	26.7	38.4	—	—	—	—	—	—	—	21.4	—
Oenanthe oenanthe	Relative length	17.2	13.3	9.0	9.0	8.1	12.5	—	—	4.7	4.3	3.8	3.8	8.1	5.6
	Centromeric index	38.4	29.4	31.2	27.0	26.0	20.0	—	—	—	—	—	—	45.4	—
Toxostoma rufum	Relative length	17.0	14.0	8.5	8.5	5.1	13.0	—	—	5.5	4.5	4.0	3.5	11.5	—
	Centromeric index	35.3	28.5	41.1	40.9	33.3	15.3	—	—	—	—	—	—	45.6	—

TABLE XIIB. Passeriformes (Muscicapidae, Turdinae and Timaliinae)

	Relative length		Centromeric index	
	Phoenicurus phoenicurus	*Turdoides s. striatus*	*Phoenicurus phoenicurus*	*Turdoides s. striatus*
(1)	13.8	18.2	26.3	37.3
(2)	9.7	—	40.0	—
(3)	8.5	8.9	33.0	35.7
(4)	12.8	13.9	18.1	18.3
(5)	12.2	13.0	11.7	18.7
(6)	7.8	8.8	20.8	23.3
(7)	5.3	7.6	18.1	23.3
(8)	5.0	3.5	16.6	—
(9)	5.0	3.1	20.0	—
(10)	4.4	3.0	20.2	—
(11)	3.1	2.2	—	—
Z	7.8	10.2	28.6	20.6
W	4.0	5.0	31.2	41.9

(1951) point out that satisfactory association of many components of the Muscicapidae group is virtually impossible, and according to many authors Mimidae and several other families should be considered as subfamilies of Muscicapidae.

According to Ripley (1952) *Saxicola*, *Oenanthe* and *Turdus* are more recently evolved species than *Phoenicurus*. This can be taken as an explanation for the difference of the chromosome constitution of *Phoenicurus* from others of the same subfamily. There is however no support in the taxonomic literature for the chromosomal similarity of *Phoenicurus* with *Turdoides*. On the contrary, many taxonomists believe that the subfamilies Turdinae and Timaliinae are sufficiently dissimilar to deserve separate familial status. Moreover, *Turdoides* is purely an oriental species whereas *Phoenicurus* is palearctic. Their chromosomal similarities, therefore, remain unexplained.

(d) *Paridae*

Two species of this family, viz. *Parus major* (A) (Fig. 14) and *Parus palustris* (B), are cytologically known from the works of Castroviejo *et al.* (1969) and Hammar (1970). In both species the diploid number is 78±. The latter author has been able to identify the W chromosome in females and also provides chromosome measurements. The idiograms prepared by us are therefore constructed from his data. These two species have almost identical karyotypes (Table XIII). Castroviejo *et al.* (1969) point out and illustrate certain minute differences in chromosomes Nos. 4 and Z which

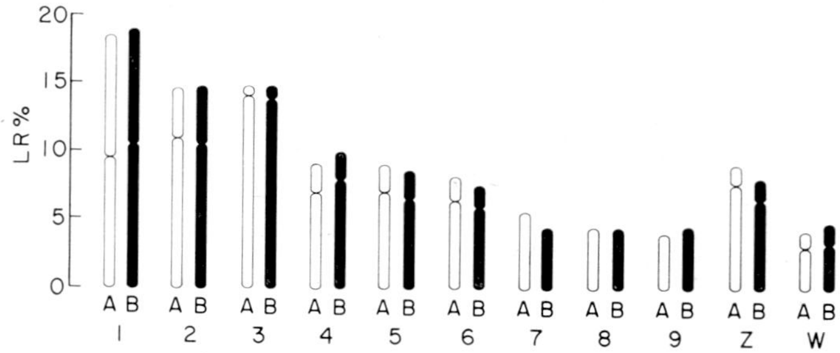

Fig. 14. Composite idiogram of Passeriformes (Paridae).
A: *P. major* B: *P. palustris*

TABLE XIII. Passeriformes (Paridae)

	Relative length		Centromeric index	
	Parus major	*Parus palustris*	*Parus major*	*Parus palustris*
(1)	18.4	18.8	45.4	45.4
(2)	14.2	14.5	25.6	27.7
(3)	14.2	14.5	4.3	4.0
(4)	9.2	10.0	20.8	20.0
(5)	8.8	8.5	24.3	24.3
(6)	7.9	7.3	19.2	22.7
(7)	5.4	4.7	—	—
(8)	4.6	4.3	—	—
(9)	3.7	3.8	—	—
Z	8.8	8.1	12.0	21.0
W	4.2	?	34.4	?

have longer second arms in *Parus palustris* than in *Parus major*. These differences have not been pointed out by Hammar (1970) nor are they obvious from his illustrations. The latter author, however, points out that the W chromosome is slightly larger in *Parus palustris*.

(e) *Fringillidae and Ploceidae*

Z. albicollis (Fringillidae) show chromosomal similarity with two species of Ploceidae viz *P. montanus* and *P. domesticus*. Their composite idiogram is drawn together (Fig. 15).

To facilitate comparison, 7 pairs of macrochromosomes and the Z and W are measured in all the species and plotted in the idiogram. The chromosomes are very similar (compare A, B, C, 1–7 and Z) in all the species. There are however slight differences, for example chromosome No. 3 in *Zono-*

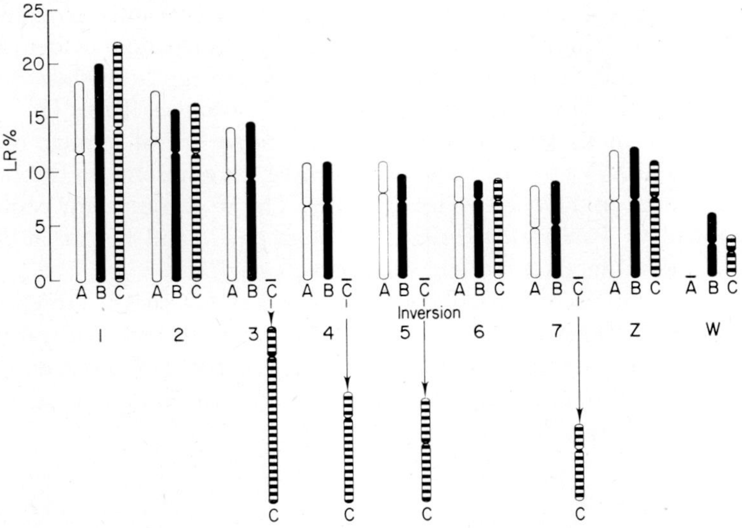

Fig. 15. Composite idiogram of Passeriformes (Fringillidae and Ploceidae)
A: *P. montanus* B: *P. domesticus*
C: *Z. albicollis*

trichia has a more terminal centromere while it is mediocentric in *Passer* (Table XIV). The change might have arisen due to an inversion.

TABLE XIV. Passeriformes (Ploceidae and Fringillidae)

	Relative length			Centromeric index		
	Passer domesticus	*Passer montanus*	*Zonotrichia albicollis*	*Passer domesticus*	*Passer montanus*	*Zonotrichia albicollis*
(1)	18.4	19.9	22.4	37.3	38.4	37.3
(2)	17.3	15.7	16.2	26.9	25.0	27.7
(3)	14.0	14.5	15.8	31.7	35.7	16.0
(4)	10.4	10.7	9.9	36.1	34.4	21.5
(5)	10.4	9.5	9.5	25.5	25.6	44.8
(6)	9.3	8.8	8.9	23.8	17.8	26.0
(7)	8.2	8.8	6.8	45.9	45.4	31.4
Z	11.5	11.8	10.1	38.4	40.0	32.0
W	?	5.7	3.7	?	47.6	42.1

It should be pointed out that the chromosome No. 8 (not shown in the idiogram) is metacentric and chromosomes Nos. 9 onwards are acrocentric chromosomes in both the species of *Passer*, but in *Zonotrichia albicollis* chromosomes Nos. 8, 9, 10 and 11 are distinctly metacentric. These are chromosomes of intermediate size between the macros and micros and the difference in the karyotypes between the two genera is only evident when we compare these chromosomes. They also differ slightly in their diploid numbers which are 78± in *Passer* and 82± in *Zonotrichia* (Table I).

The Z chromosomes in *Zonotrichia* and *Passer* are of similar relative lengths (Table XIV) but it is a submetacentric chromosome in the former and metacentric in the two species of *Passer*. The W chromosome could not be identified in *Passer domesticus*. It is metacentric and bigger in *Passer montanus* and submetacentric and smaller in *Zonotrichia*.

Two other species of Fringillidae, *S. canarius* and *F. coelebs* are chromosomally related. The diploid number is 80± in both and as regards the morphology, they are quite similar (compare A and B, 1–8 and Z and W in Fig. 16). The minute differences in their lengths and centromeric indices are shown in Table XV.

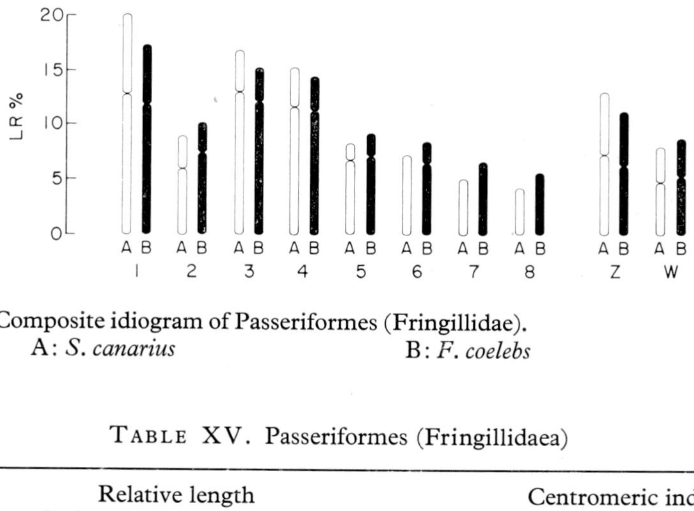

Fig. 16. Composite idiogram of Passeriformes (Fringillidae).
A: *S. canarius* B: *F. coelebs*

TABLE XV. Passeriformes (Fringillidaea)

	Relative length		Centromeric index	
	Serinus canarius	*Fringilla coelebs*	*Serinus canarius*	*Fringilla coelebs*
(1)	20.0	17.1	35.2	33.3
(2)	9.2	10.7	34.0	27.7
(3)	16.9	14.8	23.8	23.2
(4)	15.5	14.3	24.6	24.5
(5)	8.4	9.3	17.6	24.3
(6)	7.2	8.6	—	24.2
(7)	5.0	6.8	—	—
(8)	4.3	6.3	—	—
Z	13.2	11.1	45.4	46.9
W	8.0	8.9	42.5	39.1

(g) *Corvidae*

Three species, all belonging to three distinct genera are known cytologically in this family. They are *Pica pica* (A), *Cyanocitta cristata* (B) and *Corvus brachyrhynchos* (C).

All the species are strikingly similar both from the point of view of diploid number and chromosome morphology. The diploid number ranges from 76 to 80 (Table I) and the extent of karyotype similarity will be evident from Fig. 17. The difference of some significance is in the position of centromeres: e.g. chromosome No. 1 is a submetacentric chromosome in *Corvus* (C) and metacentric in the other two and Z is an m chromosome in *Cyanocitta* and sm in the other two species. Chromosomes Nos. 7 and 8 also vary to a certain extent in their centromeric indices in all the three species (Table XVI).

TABLE XVI. Passeriformes (Corvidae)

	Relative length			Centromeric index		
	Pica pica	*Cyanocitta cristata*	*Corvus brachyrhynchos*	*Pica pica*	*Cyanocitta cristata*	*Corvus brachyrhynchos*
(1)	17.2	16.9	16.4	40.0	39.6	25.0
(2)	13.5	12.8	13.8	10.5	24.3	15.6
(3)	13.2	11.6	13.1	18.1	15.0	19.6
(4)	8.4	8.4	8.6	16.6	27.6	25.0
(5)	8.4	8.1	8.6	22.1	21.4	19.0
(6)	7.4	7.3	7.9	16.6	19.8	21.6
(7)	4.4	5.2	5.3	28.5	3.0	19.8
(8)	4.4	4.6	5.1	24.3	3.0	16.6
(9)	3.7	4.6	4.3	—	—	—
(10)	3.4	3.5	3.0	—	—	—
Z	10.1	8.7	8.6	28.5	40.0	25.0
W	2.7	4.0	4.3	40.0	42.8	30.0

The relative length of Z varies from 8.6 to 10.1. Compared to Z, W is more variable in length but all are mediocentrics or sub-mediocentrics.

III. DISCUSSION

There is a rough correlation between the degree of taxonomic divergence and structural divergence in the chromosomes of birds. We analysed the karyotypes of twelve different genera where more than one species have been cytologically worked out. The amount of chromosomal variation between congeneric species has been found to be rather limited, since they

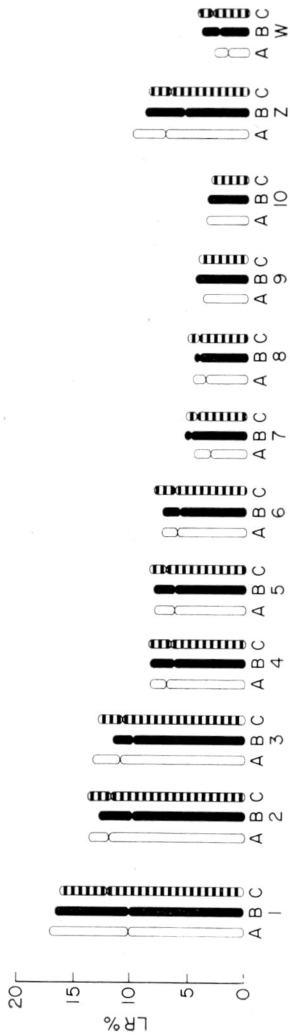

Fig. 17. Composite idiogram of Passeriformes (Corvidae).
A: *P. pica* B: *C. cristata*
C: *C. brachyrhynchos*

possess almost the same diploid number and the relative lengths and the position of the centromeres of the macrochromosomes do not vary beyond the limits of the possible error inherent in our method of study. The W chromosome has not been taken into consideration when similarities of these karyotypes are emphasized because being minute in size, in many cases, its identification is not possible. The following exceptions to the above general situation are however, found. Two species of the genus *Streptopelia*, viz. *S. orientalis* and *S. risoria*, have only three pairs of autosomes and the Z chromosome in common while the rest of the complements are dissimilar. No two congeneric species in birds have shown this degree of difference in their karyotypes. The situation which is found in the two species of the genus *Turdus* which differ among themselves in two pairs of chromosomes is not so serious. The differences in the karyotypes between three species of the genus *Psittacula* can be explained by assuming simple Robertsonian fusion.

The model of speciation, which ascribes a special role to chromosomal rearrangements in generating genetic isolating mechanisms, would not be plausible if closely related species possessed identical karyotypes. The congeneric species belonging to the dipterous genera *Drosophila, Chironomous, Anopheles* and *Sciara* apparently have similar karyotypes but they usually differ in banding sequence in the giant polytene chromosomes in ways which are clearly due to structural rearrangements of the chromosomes that have arisen in course of phylogeny. At present the study of the DNA replication and chromosome banding pattern by Giemsa staining and DNA binding flurochromes are the available means by which possible longitudinal differentiation in chromosomes of nondipterous species can be detected. Unfortunately, such studies are in their infancy in so far as birds are concerned. Hammar (1966) has, however, found that closely related avian species with identical karyotypes often have significant difference in the amount and distribution of interphase heterochromatin.

Evidences of speciation in action through structural rearrangements are quite common among mammals where interpopulational studies of various species have revealed extensive chromosomal polymorphism (Bianchi *et al.*, 1969; Baker *et al.*, 1970; Matthey, 1966). Indications of similar situations are also available in birds. In *Vanellus vanellus*, a member of the family Charadriidae, two types of chromosome No. 1 have been found (Hammar, 1970). In type (a) the chromosome No. 1 has a subterminal centromere and in type (b) the same chromosome is mediocentric (Fig. 7a). He has found the homozygotes of type (a) and heterozygotes between types (a) and (b) but not the homozygotes of type (b). Moreover, the homozygotes with type (a) chromosomes are more frequent thus indicating that this has been the original chromosome structure of the species and a single pericentric inversion has given rise to the type (b). The polymorphism has perhaps not yet been fixed in the population. A much more interesting case has been reported from another bird, *Zonotrichia albicollis* of the family Fringillidae (Thorneycroft, 1966). In a study of 35 sparrows, five distinct karyotypes are observed. A chromosomal polymorphism is present which involves at least two pairs of macrochromosomes, and rearrangements which involve translocations and pericentric inversions might have been the cause of observed polymorphism (Fig. 18). Evidences of selective breeding between morphs are also present in the data. Certain phenotypes are related to particular chromosome constitution and selective breeding appears to maintain heterozygosity within the population.

When we compare the chromosome structure between genera grouped under distinct families we find a mixed situation. Within the limits of the data available, the genera included under Anatidae, Rallidae and all the families of the order Passeriformes seem to be more conservative than those of the other known families. For example, the genera within the families Ardeidae, Charadriidae, Strigidae, Psittacidae and Columbidae are chromosomally much more variable.

Eight genera of the family Anatidae, from twelve worked out, have similar diploid numbers and very similar karyotypes. Of the remaining four, three differ from the general pattern by three pericentric inversions and the fourth by two centric fissions. According to Wurster and Benirschke (1968), not a single instance of verifiable fissioning in man or other mammals can be cited (see also Nadler, 1969), but here in the family Anatidae, we can say with reasonable certainty that the first four pairs of acrocentric chromosomes in *Bucephala* have arisen by the centromeric fissioning of the first two pairs of metacentric chromosomes common to the members of the Anatidae family. As 13 out of 14 species in this family have 2 pairs of metacentric chromosomes and only *Bucephala* has 4 pairs of acrocentric chromosomes it can safely be assumed that the karyotype of Bucephala has arisen through centromeric fission and not the other way round. A controversy exists as to the mechanism of centromeric fission and fusion. Robertson postulated that the ends of two acrocentric chromosomes may link together to form a metacentric chromosome and implicitly the reverse processes may occur also. This mechanism is accepted by many workers but White (1969) is of the opinion that reciprocal translocation rather than simple fission or fusion is the mechanism involved. Matthey (1966), however, argues that since centric fissions and fusions are common types of chromosome mutations and the reciprocal translocation theory requires the formation of two simultaneous breaks, both in the immediate vicinity of the centromere, the probability of the latter is very low. Moreover, formation of two acrocentrics from a metacentric needs one additional centromere which is difficult to obtain.

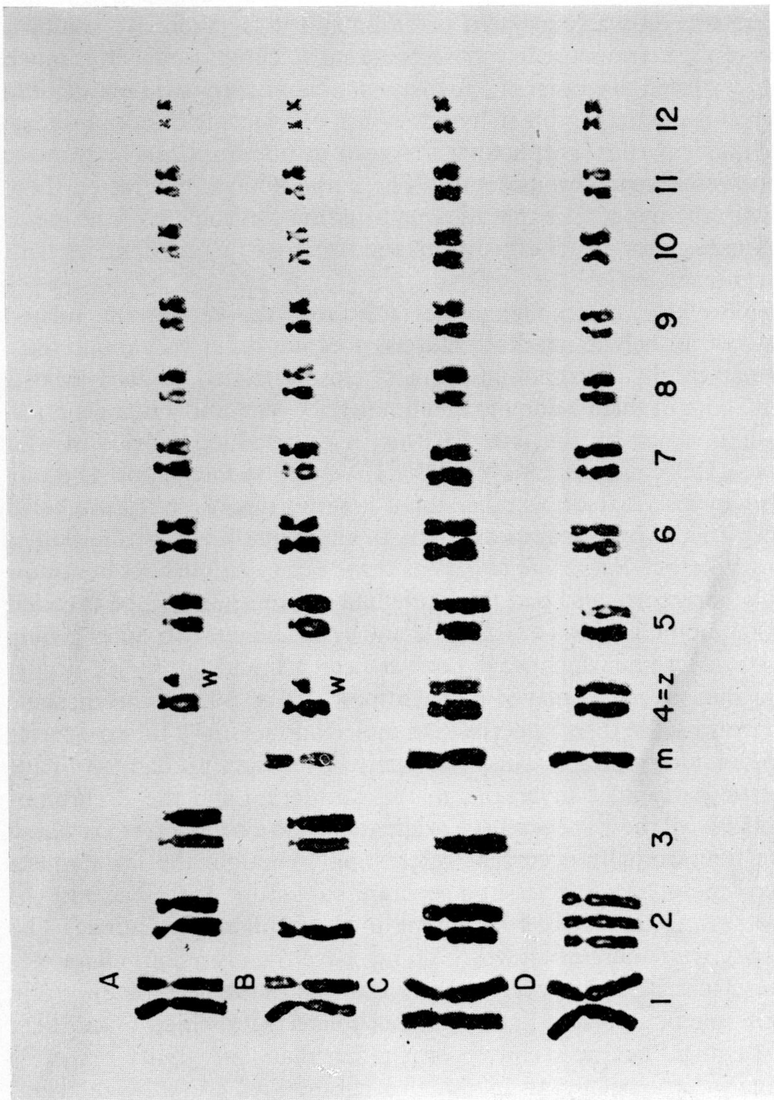

Fig. 18. The first 24 chromosomes in the karyotypes of four white-throated sparrows, *Zonotrichia albicollis*. (After Thorneycroft, 1966.)

Recently Comings and Okada (1970) by whole-mount electron microscopy of metaphase mouse L cells showed that in many cases the region of the centromere was located at one extreme end of the chromosome with no evidence for short arms of any significant size. In such telocentric chromosomes there was only a single area of chromatid association. By contrast, the region of the centromere in mouse metacentric chromosomes was much longer and was seen to consist of two distinct areas of chromatid association, each of which was comparable in length to that of telocentric chromosomes. A similar quadripartite morphology was seen in human, Chinese hamster and sheep metacentric chromosomes. The authors have interpreted these findings with the hypothesis that the quadripartite centromere of the metacentric chromosomes has resulted from the fusion of two telocentrics with bipartite centromeres.

Bucephala clangula, although it deviates from the rest of the related species by two Robertsonian dissociations, does not differ very much from them phenotypically. Taxonomically most closely related to *Bucephala* is *Mergus* and both of them belong to a common tribe Mergini. Their genetical closeness is revealed by the fact that they often produce hybrids in wild populations (Delacour, and Mayr 1945). It would be interesting to study the hybrid meiosis to see whether the 4 acrocentrics of *Bucephala* form trivalents with the homologous portions of the metacentric chromosomes of *Mergus*. Whether *Bucephala* alone has come out as an offshoot or simultaneously other genera also had this same line of evolution will be revealed from the karyotype analysis of *Clangula* and *Histrionious*, two other genera belonging to Mergini which have not yet been worked out cytologically. It appears that inversion played a much more active role in chromosome evolution in Anatidae than Robertsonian dissociation. Out of the six species so far worked out in the Anserinae subfamily of the family Anatidae 4 have characteristic pericentric inversions in two autosomes and the Z chromosome. Probably all these species have originated from a common stock where these inversions actually occurred. Interestingly enough, the taxonomists have placed these four species in a separate subfamily Anserinae and the species possessing the original karyotype in the subfamily Anatinae. The two genera *Cygnus* and *Branta* which do not have the inversions characteristic of Anserinae have, however, been retained in the same subfamily. We feel that the phenotypic and behavioural characteristics which place them in Anserinae might be due to convergence rather than true affinity.

All three genera belonging to the two subfamilies so far worked out in Ralidae are similar in their macrochromosome morphology, the deviation of the genus *Porphyrio* from the other two can be best explained by a single Robertsonian fission or fusion. But *Fulica atra* the only member of the Fulicinae subfamily worked out has 92 chromosomes in contrast to 78 and 62–77 in the other two genera. The difference in diploid number is obviously due to an increase in the number of microchromosomes in *Fulica* but the mechanism involved in this change in the number of microchromosomes remains unexplained. The identical number, relative length and morphology of the macrochromosomes of *Fulica* and *Gallinula* (Table VI) and almost the same total percentage length of the respective microchromosomes in spite of the difference in number apparently indicate that there was no transfer of material (in the form of translocations) from micro- to macrochromosomes or *vice versa* during the evolution of these species. Probably there was some ancestral stock common to both genera in which few of the bigger microchromosomes (at least 10 pairs) were metacentrics. In one direction of evolution all of these relatively bigger metacentric microchromosomes underwent centromeric fissions and this evolved into *Fulica* and in the other direction pericentric inversions of these chromosomes gave rise to *Gallinula*, and that primitive stock might have been extinct or might be present in some still cytologically unknown species.

The families so far discussed have relatively more uniform karyotype, but when we examine the chromosomes of different genera in families like Ardeidae, Charadriidae, Psittacidae and Strigidae, we find a different picture. Two genera worked out in Ardeidae are quite dissimilar. A number of translocations and inversions have to be postulated to account for this divergence.

Karyotypic comparisons are confusing in the genera included under different families of the order Charadriiformes. Four species belonging to two different families and three different subfamilies have identical karyotypes while three species belonging to one family and even to a single subfamily have karyotypes with widely different diploid numbers and chromosome morphology. A similar situation has been observed in mammals by Wurster and Benirschke (1968) where three species of the order Carnivora belonging to three different families have very similar, if not identical, karyotypes. These authors are of the opinion that these are cases of parallelism in karyotype evolution in completely independent groups. It seems to us quite unlikely that two unrelated species having dissimilar karyotypes could develop karyotypic similarity by parallelism, which is only conceivable for phenotypic similarities owing to functional convergence. It seems more logical to think that the karyotype pattern common to four species of Charadriiformes belonging to different families and subfamilies is the most primitive one from which all other patterns have evolved through a good deal of structural rearrangements.

In the family Psittacidae three genera, *Psittacula*, *Loriculus* and *Melopsittacus*, are known cytologically and all of them have different karyotypes. But the extent of dissimilarity of the last named genus with either of the two remaining genera is more pronounced compared to the dissimilarities between *Psittacula* and *Loriculus*. The chromosomal differences between *Psittacula* and *Loriculus* have arisen due to a series of pericentric inversions in all of the macrochromosomes including the Z except chromosome number 1. The number of inversions producing the difference in this case is the maximum so far found amongst birds.

The family Strigidae is known by five genera all of which are karyologically different. Four of these belong to the subfamily Striginae and the fifth one belongs to the subfamily Tytoninae. The karyotype of *Tyto alba* of the subfamily Tytoninae is totally different which renders it almost unique among Aves so far studied. It has a high diploid number, $2n = 92$, and does not have distinct microchromosomes so characteristic of the bird karyotype. It would be interesting to measure the DNA content per nucleus in this species which is likely to be higher than is usual for birds. The greater quantity of genetic material, if really present in *Tyto*, will be very interesting from the point of view of phylogenetic relationship. A parallel situation has been discovered in mammals. In aardvark, an eutherian mammal (*Orycteropus afer*) worked out by Benirschke *et al.* (1970), the DNA content of cultured cells was found to be 1.673 times that of human lymphocytes though the diploid number is as low as 20 with most of the chromosomes metacentric. These authors are of opinion that this is not readily explainable by contemporary knowledge of evolution.

Examples of orders in birds having more than one worked out family known cytologically are Galliformes, Charadriiformes, Lariformes and Passeriformes. The order Galliformes is really conservative because here three families which have very similar karyotypes, are worked out cytologically, and hence it can be inferred that very little change has taken place in the karyotype along with the marked morphological variation during evolution. The diploid number is uniformly $78\pm$ in all the species so far known in this order except in *Pavo cristatus*. A re-examination of some more individuals in this species is necessary to make any comment on this single variant. A more or less similar situation is also found in the order Lariformes where two families Laridae and Sternidae are known cytologically.

The karyotype heterogeneity of the order Charadriiformes has already been discussed and the same is true also of the order Passeriformes where the maximum number of families known chromosomally and where almost every family has its own distinctive karyotype and therefore families have both marked phenotypic, as well as karyotypic, alterations.

From the trend of chromosomal evolution discussed so far, it is expected that the different orders of the class Aves are not likely to be similar in their karyotypes. That is almost exactly what we observe. The only exception has been found between the orders Gaviiformes and Podicipiformes where the karyotypes, though distinct, still retain recognizable similarities. It appears that a large number of simple Robertsonian changes are involved in their evolutionary divergence.

Acknowledgements

The author is deeply indebted to Prof. S. P. Ray-Chaudhuri, Department of Zoology, Banaras Hindu University, Varanasi, India and Prof. E. Mayr, Museum of Comparative Zoology, Cambridge, Massachusetts, U.S.A. for their valuable suggestions and critically going through the manuscript. The assistance of Dr. B. Biswas, Zoological Survey of India for providing information about taxonomy of birds and the financial help of the Council of Scientific and Industrial Research, India, gratefully acknowledged.

References

Atkin, N. B., Mattinson, G., Beçak, W. and Ohno, S. (1965). The comparative DNA contents of 19 species of placental mammals, reptiles and birds. *Chromosoma, Berlin* **17**, 1–10.

Baker, R. J., Mascarello, J. T. and Jordan, R. G. (1970). Polymorphism in the somatic chromosomes of *Neotoma micropus* Baird, the plains woodrat. *Experientia* **26**, 426–428.

Beçak, W., Beçak, M. L., Nazareth, H. R. S. and Ohno, S. (1964). Close karyological kinship between the reptilian suborder Serpentes and the class Aves. *Chromosoma, Berlin* **15**, 606–617.

Benirschke, K., Wurster, D. H. and Low, R. J. (1970). The chromosome complement of the aardvark, *Orycteropus afer*. *Chromosoma, Berlin* **31**, 68–78.

Bhatnagar, M. K. (1968). Mitotic chromosomes of white Chinese geese. *J. Hered.* **59**, 191–195.

Bianchi, N. O. and de Bianchi, M. S. A. (1969). Origin of the pattern and chronology of chromosome replication in vertebrates. *Genet. Suppl.* **61**, 275–287.

Bianchi, N. O. and Molina, O. J. (1967). Chronology and pattern of replication in the bone narrow chromosomes of *Gallus domesticus*. *Chromosoma, Berlin* **21**, 387–397.

Bianchi, N. O., Contreras, J. and Dulout, F. N. (1969). Intraspecies autosomal polymorphism and chromosome replication in *Akodon molinae* (Rodentia: Cricetidae). *Can. J. Genet. Cytol.* **11**, 233–242.

Castroviejo, J., Christian, L. C. and Gropp, A. (1969). Karyotypes of four species of birds of the family Ploeceidae and Paridae. *J. Hered.* **60**, 134–136.

Comings, D. E. and Mattoccia, E. (1970). Studies of microchromosomes and a G-C rich DNA satellite in the quail. *Chromosoma, Berlin* **30**, 202–214.

Comings, D. E. and Okada, T. A. (1970). Whole mount electron microscopy of the centromere region of metacentric and telocentric mammalian chromosomes. *Cytogenetics* **9**, 436–449.

Delacour, J. and Mayr, E. (1945). The family Anatidae. *Wilson Bull.* **57**, 3–55.

Donnelly, G. M. and Newcomer, E. H. (1963). Autoradiographic patterns in cultured leukocytes of the domestic fowl. *Exp. Cell Res.* **30**, 363–368.

Galton, M. and Bredbury, P. R. (1966). DNA replication patterns of the sex chromosomes of the pigeon (*Columba livia domestica*). *Cytogenetics* **5**, 295–306.

Hammar, B. (1964). Absence of sex differences in avian chromocenters. *Hereditas* **51**, 191–197.

Hammar, B. (1966). The karyotypes of nine birds. *Hereditas* **55**, 367–385.

Hammar, B. (1967). Differences in heterochromatin between some avian species. *Hereditas* **57**, 209–216.

Hammar, B. (1970). The karyotypes of thirty one birds. *Hereditas.* **65**, 29–58.

Itoh, M., Ikeuchi, T., Shimba, H., Mori, M., Sasaki, M. and Makino, S. (1969). A comparative karyotype study in fourteen species of birds. *Jap. J. Genet.* **44**, 163–170.

Jovanovic, V. and Atkins, L. (1969). Karyotypes of four passerine birds belonging to the families Turdidae, Mimidae and Corvidae. *Chromosoma, Berlin,* **26**, 388–394.

Kosin, I. L. and Ishizaki, H. (1959). Incidence of sex chromatin in *Gallus domesticus. Science, N.Y.* **130**, 43–44.

Krishan, A. and Shoffner, R. N. (1966). Sex chromosomes in the domestic fowl (*Gallus domesticus*) turkey (*Meleagris gallopavo*) and the Chinese pheasant (*Phasianus colchicus*). *Cytogenetics* **5**, 53–63.

Krishan, A., Haiden, G. J. and Shoffner, R. N. (1965). Mitotic chromosomes and the W sex chromosome of the great horned owl (*Bubo v. virginianus*). *Chromosoma, Berlin* **17**, 258–263.

Levan, A., Fredga, K. and Sandberg, A. A. (1964). Nomenclature for centromeric position on chromosomes. *Hereditas* **52**, 201–220.

Matthey, R. (1966). "Cytogenetic Mechanisms and Speciation of Mammals" Amer. Tissue Cult. Assoc. Meeting, Miami. 1–11.

Mayr, E. and Amadon, D. (1951). "A Classification of Recent Birds." American Museum Novitates, City of New York, Number 1496.

Mott, C. L., Lockhart, L. H. and Rigdon, R. H. (1968). Chromosomes of the sterile hybrid duck. *Cytogenetics* 7, 403–412.

Nadler, C. F. (1969). Chromosomal evolution in rodents. *In* "Comparative Mammalian Cytogenetics" (K. Benirschke, ed.). Springer Verlag, Berlin, Heidelberg and New York.

Newcomer, E. H. (1957). The mitotic chromosomes in the domestic fowl. *J. Hered.* **48**, 227–234.

Newcomer, E. H. and Brant, J. W. A. (1954). Spermatogenesis in the domestic fowl. *J. Hered.* **45**, 79–87.

Ohno, S. (1961). Sex chromosomes and microchromosomes of *Gallus domesticus. Chromosoma, Berlin* **11**, 484–498.

Ohno, S., Kaplan, W. D. and Kinosita, R. (1960). On the sex chromatin of *Gallus domesticus. Exp. Cell Res.* **19**, 180–183.

Ohno, S., Christian, L. C. and Stenius, C. (1962). Nucleolus organizing microchromosomes of *Gallus domesticus*. *Exp. Cell Res.* **27**, 612–614.

Ohno, S., Beçak, W. and Beçak, M. L. (1964a). X-autosome ratio and behaviour pattern of individual X-chromosomes in placental mammals. *Chromosoma, Berlin* **15**, 14–30.

Ohno, S., Stenius, C., Christian, L. C., Beçak, W. and Beçak, M. L. (1964b). Chromosomal uniformity in the avian subclass Carinatae. *Chromosoma, Berlin* **15**, 280–288.

Owen, J. J. T. (1965). Karyotype studies of *Gallus domesticus*. *Chromosoma, Berlin* **16**, 601–608.

Piccini, E. and Stella, M. (1970). Some avian karyograms. *Caryologia* **232**, 189–202.

Ray-Chaudhuri, R. (1967a). Chromosomes of Indian birds. Ph.D. Thesis. Banaras Hindu University, Banaras.

Ray-Chaudhuri, R. (1967b). Mitotic and Meiotic chromosomes of the koel, *Eudynamys scolopacea scolopacea*. *Nucleus* **10**, 179–189.

Ray-Chaudhuri, R., Sharma, T. and Ray-Chaudhuri, S. P. (1969). A comparative study of the chromosomes of birds. *Chromosoma, Berlin* **26**, 148–168.

Ray-Chaudhuri, S. P., Ray Chaudhuri, R. and Sharma, T. (1966). The W chromosome in the females of two Indian species of birds. *Chromosoma, Berlin* **20**, 151–154.

Ray-Chaudhuri, S. P., Singh, L. and Sharma, T. (1970). Sexual dimorphism in somatic interphase nuclei of snakes. *Cytogenetics* **91** 410–423.

Ray-Chaudhuri, S. P., Singh, L. and Sharma, T. (1971). Evolution of sex-chromosomes and formation of W-chromatin in snakes. *Chromosoma, Berlin* **33**, 239–251.

Renzoni, A. and Vegni-Talluri, M. (1966). The karyograms of some Falconiformes and Strigiformes. *Chromosoma, Berlin* **20**, 133–150.

Ripley, S. D. (1952). "The Thrushes." *Postilla* **13**, 1–48.

Rothfels, K., Aspden, M. and Mollison, M. (1963). The W-chromosome of the budgerigar, *Melopsittacus undulatus*. *Chromosoma, Berlin* **14**, 459–467.

Sasaki, M., Ikeuchi, T. and Makino, S. (1968). A feather pulp culture technique for avian chromosomes, with notes on the chromosomes of the pea fowl and the ostrich. *Experientia* **24**, 1292–1293.

Schmid, W. (1962). DNA replication patterns of the heterochromosomes in *Gallus domesticus*. *Cytogenetics* **1**, 344–352.

Stenius, C., Christian, L. C. and Ohno, S. (1963). Comparative cytological study of *Phasianus colchicus*, *Meleagris gallopavo* and *Gallus domesticus*. *Chromosoma, Berlin* **13**, 515–520.

Thorneycroft, H. B. (1966). Chromosomal polymorphism in the white-throated sparrow, *Zonotrichia albicollis* (Gmelin). *Science, N.Y.* **154**, 1571–1572.

Vegni-Talluri, M. and Vegni, L. (1965). Fine resolution of the karyogram of the quail, *Coturnix coturnix japonica*. *Chromosoma, Berlin* **17**, 264–272.

White, M. J. D. (1969). Chromosomal rearrangements and speciation in animals. *Ann. Rev. Genet.* **3**, 75–98.

Wurster, D. H. and Benirschke, K. (1968). Comparative cytogenetic studies in the order Carnivora. *Chromosoma, Berlin* **24**, 336–382.

14

The Chromosomes of Non-Eutherian Mammals

G. B. SHARMAN
School of Biological Sciences, Macquarie University, North Ryde, Australia, 2113

I. INTRODUCTION

The chromosomes of about one hundred of the two hundred and thirty living species of marsupials and two, of a total of perhaps only three, extant monotremes have been described. The marsupials have diploid chromosome numbers of between 10 and 32 and the two monotremes have chromosome numbers of 54 and 64 (female). The lowest number recorded in a eutherian mammal is $2n = 6♀, 7♂$ (Wurster and Benirschke, 1970) and a few have numbers as low as $2n = 17$ but numbers as high as $2n = 88$ occur (George and Weir, 1972). The desoxyribose nucleic acid (DNA) contents of Australian marsupial and eutherian mammal nuclei are similar (Rendel and Kellerman, 1955) and two of the extant monotremes have 92% to 97% as much DNA as have humans and rats (Bick and Jackson, 1967a). Apparently both marsupial and monotreme non-eutherian mammals share with the eutherians extensive gene duplication, which presumably occurred early in mammalian evolution, while marsupials have fewer, and monotremes more, chromosomes than most eutherians.

The major classification used in this review follows Simpson (1945) in its primary division of living extant mammals into two subclasses—Prototheria containing the single order Monotremata (egg laying mammals) and Theria containing the infraclasses Metatheria and Eutheria. The Metatheria is regarded as consisting of the order Marsupialia (the marsupials) which is divided into four superfamilies all of which are extant. Simpson (1945) included a fifth superfamily of extinct marsupials, the Borhyaenoidea, but later (Simpson, 1970) deleted this group because it graded imperceptibly with the Didelphoidea.

The classification adopted assumes a primary dichotomy of the mammals into prototherian (monotreme) and therian (marsupial and eutherian) groups. Such a classification is not universally accepted. Gregory (1947) proposed a primary division of mammals into the subclasses Marsupionta, containing the marsupials and monotremes (the living non-eutherian mammals), and Monodelphia (placental (= eutherian) mammals). On the other hand MacIntyre (1967) proposed that the extinct ictidosaurs, multituberculates, docodonts, etc., and the living monotremes be grouped informally as "quasi-mammals" and classified with the therapsid reptiles. In view of the controversy about the taxonomic and phylogenetic position of the living monotremes cytological evidence of their reptilian, avian or mammalian relationships is particularly important. Sharman (1973) reviewed existing classifications of marsupials.

II. THE CHROMOSOMES OF THE MONOTREMATA

The first substantially accurate observations on the chromosomes of the monotremes were reported by Matthey and van Brink (1957) and van Brink (1959) who briefly summarized the earlier literature. These authors studied spermatogonial mitoses in a male echidna (*Tachyglossus aculeatus* (Shaw, 1792)), from an unstated part of its very wide range in Australia, and obtained diploid chromosome counts of 62, 63 and 64. The presence of a large, unpaired, acrocentric chromosome "superior by one-fourth of its length to the members of the largest pair of autosomes" indicated the occurrence of male heterogamety. The authors were not able to decide whether the male sex chromosome constitution was XO or XY.

Bick and Jackson (1967b, c), using improved techniques, studied the somatic and spermatogonial chromosomes of echidna and platypus (*Ornithorhynchus anatinus* (Shaw 1799)) from Tasmania. They obtained counts of 64 female, 63 male in the echidna and 54 female, 53 male in the platypus, the male of each species having an unpaired large metacentric or submetacentric chromosome, designated the X, which was represented twice in females. Bick and Jackson (1967b) concluded that "the chromosome numbers of both monotremes strongly suggest an XX–XO sex-determining mechanism—possibly the sex-determining system—is more complex, similar to the $X_1X_1X_2X_2/X_1X_2Y$ type found in two placental mammals and recently in an Australian marsupial".

Our observations (G. B. Sharman and P. G. Johnston, unpublished) on echidnas from Cootamundra (N.S.W.), Kangaroo Island (South Australia) and eastern Victoria are at variance with those previously published. The chromosome number is 64♀ (Fig. 1), 63♂ (Fig. 2) as stated by Bick and Jackson (1967b, c). Four chromosomes (1–4, Fig. 3) are represented twice in both sexes and, although not always represented by pairs of exactly the same size, they are obviously homologues. No suggestion to the contrary was found in the critical examination of over one hundred metaphase cells from two females and one male. The unpaired element of the male is therefore not the largest chromosome as stated by Matthey and van Brink (1957). The female has a pair of submetacentrics amongst the next largest set of chromosomes which appear to be replaced in the male by a heteromorphic pair (6a,

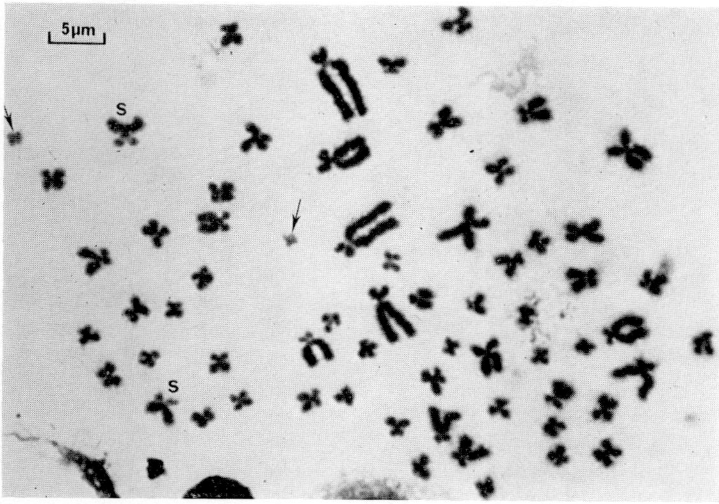

Fig. 1. Metaphase chromosomes from bone marrow of female echidna (*Tachyglossus aculeatus*) from Kangaroo Island, South Australia. Air dried preparation, fixed 2½ hours after colcemid injection. The smallest (metacentric) chromosome pair are indicated by arrows; S—satellited chromosomes.

b, Fig. 3). Since the male has one chromosome less than the female this indicates the probability of a sex determination mechanism of greater complexity than the XX♀/XO♂ system suggested by Bick and Jackson (1967b, c). As neither chromosome 5a or 6a has an obvious homologue in the male the pairing illustrated in Fig. 3 has no significance—the positions of 5b and 6b could be interchanged. Indeed it is possible that 5b finds a homologue amongst the remaining forty-five smaller chromosomes of the male, not illustrated in Fig. 3. Chromosome 5a, which frequently has a pronounced secondary constriction in its shorter arm in both male and female, seldom has a nearly equal sized potential homologue amongst the female complement. If this is a sex chromosome the seeming difference between it and its potential homologue in the female (5b, Fig. 3) may be the result of an extreme form of dose compensation (Section III E 2). Alternatively both female and male echidnas possess some form of permanent structural heterozygosity not connected with the sex determination mechanism.

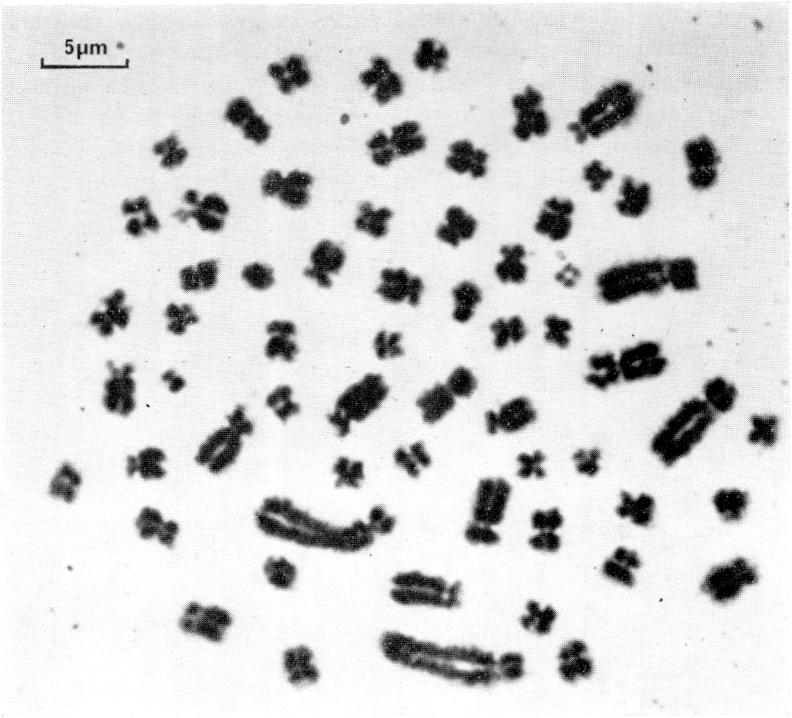

Fig. 2. Metaphase chromosomes from bone marrow of male echidna from Cootamundra, N.S.W. Air dried preparation, fixed 4 hours after colcemid injection.

III. THE CHROMOSOMES OF THE MARSUPIALIA

A. *Chromosome Number and Morphology*

1. *Superfamily Didelphoidea, Family Didelphidae* (a) Subfamily Didelphinae. Chromosome numbers of $2n = 22$, $2n = 18$ and $2n = 14$ are found and sex-determination mechanisms are of the XX♀ : XY♂ type in the eighteen (of a total of about sixty-five) species of the subfamily studied (Table I).

Extant species are confined to South and Central America with the single exception of the Virginian opossum (*Didelphis virginiana*) which apparently entered North America in Pliocene times and is not a surviving remnant of the extensive late Cretaceous North American marsupial radiation, which became extinct on that continent in early Miocene times (Clemens, 1968, 1971).

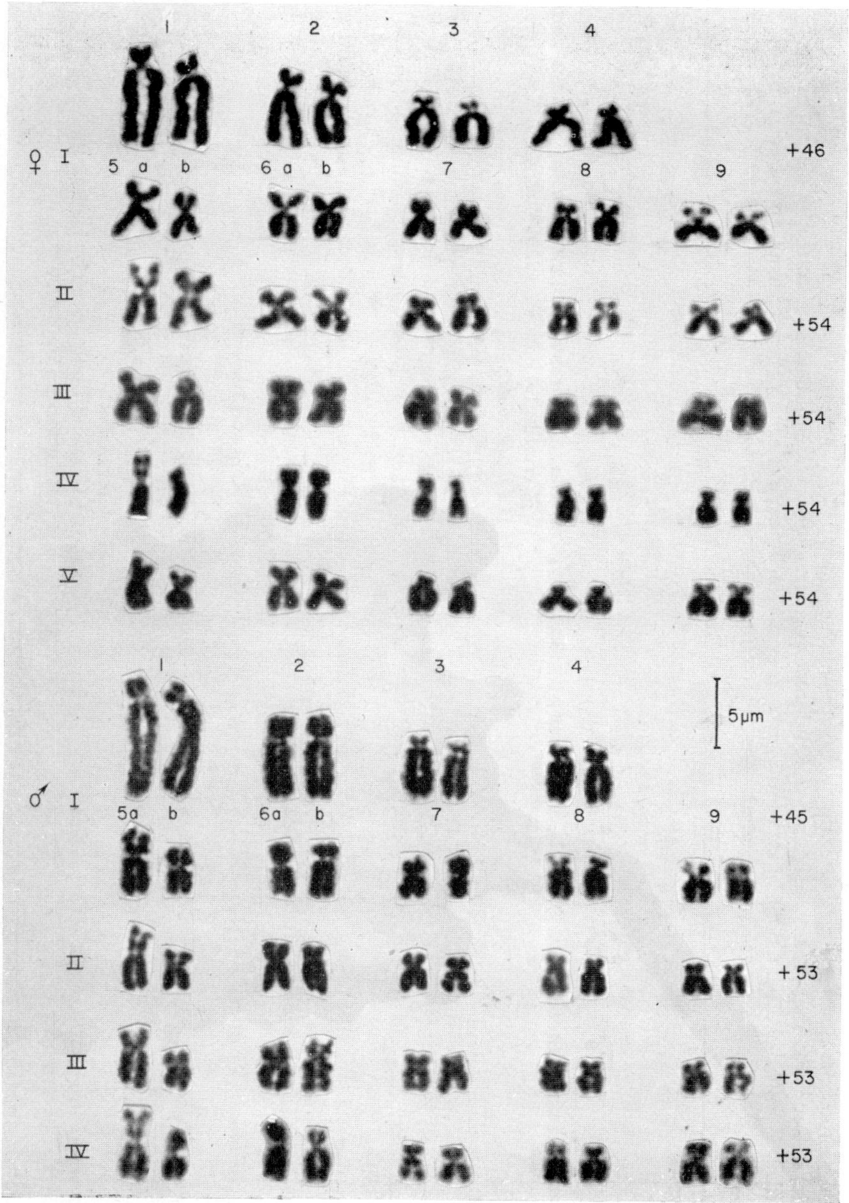

Fig. 3. Partial karyotypes of five cells from a female (above) and four cells from a male echidna (below). I (♀ and ♂)—The four largest autosome pairs and the next ten largest chromosomes. II–V (♀); II–IV (♂)—The ten largest chromosomes not including the four largest autosomal pairs. For further explanation see text.

TABLE I. Chromosome numbers, sex determination mechanisms and type of X chromosomes (A = acrocentric, M = metacentric) in American marsupials, subfamily Didelphinae (Didelphoidea, Didelphidae) and subfamily Caenolestinae (Caenolestidae, Caenolestoidea). Count of autosome arms does not include short arms of acrocentrics unless equal to, or greater in length than, shortest long arm (excluding Y)

Species	$2n$	X shape	No. of autosome arms	References, etc.
Didelphinae				
Didelphis azarae Temminck, 1825	20+XX/XY	A	20	Saez, 1931 (as *D. paraguayensis*); Perondini and Perondini, 1965; Reig and Lobig, 1970 (as *D. albiventris*).
Didelphis marsupialis Linne, 1758	20+XX/XY	A	20	Dreyfus and Campos, 1939 (as *Didelphys aurita*); Cestari and Perondini, 1964.
Didelphis virginiana Kerr, 1792	20+XX/XY	M	20	Shaver, 1962.
Lutreolina crassicaudata Desmarest, 1804	20+XX/XY	M	20	Saez, 1938, X telocentric according to Reig Bianchi, 1969.
Philander opossum (Linne, 1758)	20+XX/XY	A	20	Biggers, *et al.*, 1965; Perondini and Perondini, 1965.
Chironectes minimus (Zimmerman, 1780)	20+XX/XY	A	20	Yunis, *et al.*, 1972.
Monodelphis dimidiata (Wagner, 1847)	16+XX/XY	A	20	Reig and Bianchi, 1969.
Monodelphis brevicaudata (Erxleben, 1777)	16+XX/XY	A	20	Reig and Bianchi, 1969.
Marmosa robinsoni Bangs, 1898	12+XX/XY	M	20	Reig, 1968; Biggers, *et al.*, 1965 (as *M. mexicana*).
Marmosa murina (Linne, 1758)	12+XX/XY			Reig and Lobig, cited by Reig, *et al.*, 1972.
Marmosa cinerea (Temminck, 1824)	12+XX/XY			Reig and Lobig, 1970.

Marmosa elegans (Waterhouse, 1839)	12+XX/XY	M	20	Reig, *et al.*, 1972.
Marmosa fuscata Thomas, 1896	12+XX/XY	M	20	Reig and Sonnenschein, 1970.
Metachirus nudicaudatus (Geoffroy, 1803)	12+XX/XY			Reig and Lobig, 1970.
Caluromys derbianus (Waterhouse, 1841)	12+XX/XY	A	20	Biggers, *et al.*, 1965; Legator, *et al.*, 1966.
Caluromys lanatus (Illiger, 1815)	12+XX/XY		20	Yunis, *et al.*, 1972.
Caluromys philander (Linne, 1758)	12+XX/XY			Reig and Lobig, 1970.
Dromiciops australis (Phillipi, 1894)	12+XX/—	A	22	Reig, *et al.*, 1972.
Caenolestinae				
Caenolestes fuliginosus (Tomes, 1863)	12+XX/XY			Hayman, *et al.*, 1971.
Caenolestes obscurus Thomas, 1895	12+XX/XY	A	24	Hayman, *et al.*, 1971
Lestoros inca (Thomas, 1917)	12+XX/XY	A	24	Hayman, *et al.*, 1971.

References to the most recent literature on the chromosomes of the didelphine marsupials will be found in Table I. In general only those accounts which include a karyotype are referred to. Other references are given in the text. Table I (and other tables) also include a distinction, when accounts are available, between acrocentric (A) and metacentric or submetacentric (M) X chromosomes. When authors indicated a terminal centromere on the X chromosome (telocentric condition) the chromosome is regarded as acrocentric. While American marsupials are very uniform as regards number of recognizable autosome arms (as defined in Table I) there is considerable variation in X chromosome morphology. For this reason the count of chromosome arms does not take into account the X chromosomes so it differs from the "nombre fondamental" (NF) of Matthey (1958) and other papers.

(b) Other Didelphidae. The Microbiotheriinae (Simpson, 1929) are not recognized as being distinct, at the subfamily level, from the Didelphinae (following Simpson, 1970). The microbiotherines were regarded as entirely fossil (Simpson, 1945) but Reig (1955) referred the living *Dromiciops* to that subfamily and considered that *Caluromys*, *Glironia* and *Caluromysiops* should be regarded as specialized microbiotherines rather than didelphines. The last four species listed under Didelphinae in Table I would thus be placed by Reig in the Microbiotheriinae. All, except *Dromiciops australis*, show remarkable cytological affinities with the fourteen chromosome Didelphinae. *Dromiciops* alone has twenty-two autosomal arms all but one pair of its autosomes being metacentric or submetacentric (Fig. 4).

The remaining didelphid subfamilies are extinct as are the families Pediomyidae and Stagodontidae (see Clemens, 1966).

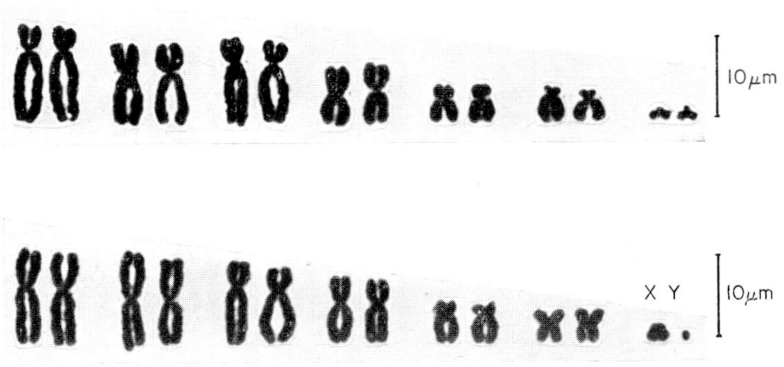

Fig. 4. Karyotypes of *Dromiciops australis* ♀ (Didelphoidea), above and *Antechinus stuartii* ♂ (Dasyuroidea) below. The fifth pair of autosomes in *Antechinus* have secondary constrictions (not obvious in illustration) in their short arms. X — X sex chromosome; Y — Y sex chromosome.

2. *Superfamily Caenolestoidea, Family Caenolestidae*
Subfamily Caenolestinae. There are six living species of caenolestoids, all placed in this subfamily, of which three have been studied cytologically (Table I). They are characterized by a uniform complement of $2n = 14$ and sex determination is of the XX female: XY male type.

Living caenolestoids are a remnant of a more extensive past radiation which included at least two families with a fossil history extending from Palaeocene times. They are exclusively of South American distribution.

3. *Superfamily Dasyuroidea, Family Dasyuridae*
(a) Subfamily Dasyurinae. The chromosome number of the fourteen (of a total of about forty-seven) species studied is exclusively $2n = 14$ (Table II). There are usually 22 autosomal arms (occasionally 24), the X chromosome is acrocentric and the Y minute (Fig. 4). In the classification adopted here subfamily Dasyurinae includes also subfamily Phascogalinae of other authors. Simpson (1945) placed the larger species (*Dasyurus*, *Sarcophilus*) in the Dasyurinae, the smaller species in the Phascogalinae and the marsupial "wolf" (*Thylacinus*) in a separate subfamily. Tate (1947) grouped some of the smaller species and *Thylacinus* as Phascogalinae while his Dasyurinae included both small and large species. *Thylacinus* has not been studied cytologically but the conflicting classifications of others and the extreme chromosomal uniformity thus far found do not appear to justify the recognition of Phascogalinae and Dasyurinae as separate subfamilies at present.

The Dasyurinae are exclusively of Australasian distribution where they extend from Tasmania (40°–44°S) and through continental Australia to the equatorial regions of New Guinea and adjacent islands.

(b) Subfamily Myrmecobiinae. The single representative of this subfamily *Myrmecobius fasciatus* (the banded anteater) is at present found only in southwestern Australia. Its many unique features justify its placing in a separate subfamily of the Dasyuridae but cytologically it is essentially similar to the dasyurine marsupials.

4. *Superfamily Perameloidea, Family Peramelidae*
(a) Subfamily Peramelinae. This subfamily contains about sixteen species half of which have been studied cytologically (Table III). All have $2n = 14$, the number of chromosome arms being the same as in the majority of the didelphine marsupials and the diploid number the same as that of the $2n = 14$ Didelphoidea and all Caenolestoidea and Dasyuroidea thus far examined. The perameline bandicoots differ at once from these superfamilies, however, in possessing larger X and Y chromosomes and in the elimination of

TABLE II. Chromosome numbers, sex determination mechanisms and type of X chromosomes (A = acrocentric, M = metacentric) in Australian marsupials, family Dasyuridae (Dasyuroidea). Count of autosome arms does not include short arms of acrocentrics unless equal to, or greater in length than, shortest long arm (excluding Y)

Species	2*n*	X shape	No. of autosome arms	References, etc.
Dasyurinae				
Antechinomys spenceri Thomas, 1906	12+XX/XY			Hayman and Martin, 1969.
Antechinus stuartii Macleay, 1841	12+XX/XY	A	22	Sharman, 1961 (mistaken ident. *A. flavipes*); this paper.
Dasycercus cristicauda (Krefft, 1867)	12+XX/XY			Martin and Hayman, 1967.
Dasyuroides byrnei Spencer, 1896	12+XX/XY			Hayman and Martin, 1969.
Dasyurus geoffroii Gould, 1841	12+XX/XY	A	24	Martin and Hayman, 1967 (as *Dasyurinus geoffroyi*).
Dasyurus hallucatus Gould, 1842	12+XX/XY	A	22	Sharman, 1961.
Dasyurus maculatus (Kerr, 1792)	12+XX/XY	A	22	Koller, 1936; Martin and Hayman, 1967 (as *Dasyurops maculatus*).
Dasyurus viverrinus (Shaw, 1800)	12+XX/XY	A	22	Drummond, 1938; Martin and Hayman, 1967.
Sminthopsis crassicaudata (Gould, 1844)	12+XX/XY	A	24	Martin and Hayman, 1967.
Sminthopsis froggatti (Ramsay, 1887)	12+XX/XY			Hayman and Martin, 1969.
Sminthopsis leucopus (Gray, 1842)	12+XX/XY			D. W. Cooper, pers. comm.
Sminthopsis macrura (Gould, 1845)	12+XX/XY			Sharman, 1961.
Sarcophilus harrisii (Boitard, 1841)	12+XX/XY	A	24	McIntosh and Sharman, 1953; Martin and Hayman, 1967.
Myrmecobiinae				
Myrmecobius fasciatus Waterhouse, 1836	12+XX/XY	A	22	Sharman, 1961.

TABLE III. Chromosome numbers, sex determination mechanisms and type of X chromosomes (A = acrocentric, M = metacentric) in Australian marsupials, family Peramelidea (Perameloidea). Count of autosome arms does not include short arms of acrocentrics unless equal to, or greater in length than, shortest long arm (excluding Y)

Species	2*n*	X shape	No. of autosome arms	References, etc.
Peramelinae				
Echymipera kalubu (Fischer, 1829)	12+XX/XY	M	20	Hayman, *et al.*, 1969 (1–5 supernumeraries in some tissues).
Echymipera rufescens (Peters and Doria, 1875)	12+XX/–	M	20	Sharman, 1973 (1–3 supernumeraries in some tissues).
Isoodon macrourus (Gould, 1842)	12+XX/XY	M	20	Hayman and Martin, 1965b.
Isoodon obesulus (Shaw 1797)	12+XX/XY	M	20	Hayman and Martin, 1965b.
Perameles gunnii Gray, 1838	12+XX/XY	M		McIntosh and Sharman, 1953.
Perameles bouganville Quoy and Gaimard, 1824	12+XX/XY	M		Sharman, 1961.
Perameles nasuta Geoffroy, 1804	12+XX/XY	M	20	Hayman and Martin, 1965b.
Peroryctes longicauda (Peters and Doria, 1875)	12+XX/XY			Hayman, *et al.*, 1969.
Thylacomyinae				
Macrotis lagotis (Reid, 1837)	16+XX/XY_1Y_2	compound M	20	Martin and Hayman, 1967 (autosome arms include an acrocentric pair translocated onto X chromosomes).

one member of each sex chromosome pair from most somatic tissues in both sexes (Section III E3).

The distribution of the perameline bandicoots is approximately the same as that of the Dasyurinae.

(b) Subfamily Thylacomyinae. There are two species of rabbit bandicoots (*Macrotis*) in the subfamily. The only species yet examined has the same number of autosome arms as have the perameline bandicoots but a higher diploid number: $2n = 18♀, 19♂$. A multiple sex chromosome mechanism of the XX♀, XY_1Y_2♂ type is found (Section III D3) and sex chromosome elimination from somatic tissues does not accur.

5. *Superfamily Phalangeroidea*

(a) Family Burramyidae. The unity and distinctness of the pigmy possums was recognized by Kirsch (1968a) from serological studies and by Gunson, *et al.* (1968) from cytological studies. The family contains but a single subfamily (Burramyinae) with about seven species, five of which have $2n = 14$ (Table IV). Sex determination is of the XX/XY type and the sex chromosomes bear a remarkable resemblance, in size, to those of some didelphid and all dasyurid marsupials. The pigmy glider (*Acrobates*) has two pairs of acrocentric autosomes, *Burramys* has three pairs and all the autosomes are metacentric or submetacentric in *Cercartetus* (2 spp.) and *Eudromicia*.

Some species of *Cercartetus* are amongst the most generalized of the phalangeroid marsupials and Bensley (1903) regarded their dentition as approximately like that of a (hypothetical) insectivorous phalangeroid prototype.

Species of the family occur in New Guinea, Australia and Tasmania.

(b) Family Tarsipedidae. This family includes but a single subfamily containing one species—the aberrant honey possum (*Tarsipes spencerae*) of southwestern Australia. On serological evidence Kirsch (1968b) considered *Tarsipes* so different from the remaining marsupials that he grouped so alone, in a separate superfamily. It is one of three marsupials thus far studied which has more than 22 chromosomes all of which are acrocentric (Table IV) as are the 20 chromosomes of the phalangeroids of the genus *Trichosurus*.

(c) Family Phalangeridae (i) Subfamily Trichosurinae. The brush possums are a homogeneous group, including three species of *Trichosurus*, all of which have been studied cytologically. The scaly-tailed possum (*Wyulda*) of

TABLE IV. Chromosome numbers, sex determination mechanisms and type of X chromosomes (A = acrocentric, M = metacentric) in some phalangeroid marsupials, subfamilies Burramyinae (Burramyidae), Tarsipedinae (Tarsipedidae), Trichosurinae, Phalangerinae (Phalangeridae), Petaurinae, Pseudocheirinae (Petauridae), Phascolarctinae (Phascolarctidae) and Vombatinae (Vombatidae). Count of autosome arms does not include short arms of acrocentrics unless equal to, or greater in length than, shortest long arm (excluding Y)

Species	$2n$	X shape	No. of autosome arms	References, etc.
Burramyinae				
Acrobates pygmaeus (Shaw, 1793)	12+XX/XY	A	20	Hayman and Martin, 1969.
Burramys parvus Broom, 1896	12+–/XY	A	18	Gunson, *et al.*, 1968.
Cercartetus nanus (Desmarest, 1818)	12+XX/XY	A	24	Martin and Hayman, 1967.
Cercartetus concinnus (Gould, 1845)	12+XX/XY	A	24	Martin and Hayman, 1967.
Eudromicia lepida (Thomas, 1888)	12+XX/XY		24	Sharman, 1961 (as *Cercaërtus lepidus*).
Tarsipedinae				
Tarsipes spencerae Gray, 1842	22+XX/–	A	22	Hayman and Martin (1969).
Trichosurinae				
Trichosurus vulpecula (Kerr, 1792)	18+XX/XY	A	18	Hayman and Martin, 1969.
Trichosurus caninus Ogilby, 1836)	18+XX/XY		18	Sharman, 1961.
Trichosurus arnhemensis (Collett, 1897)	18+XX/–			Hayman and Martin, 1969.
Phalangerinae				
Phalanger orientalis (Pallas, 1766)	12+XX/–	A	20	This paper.
Phalanger maculatus (Geoffroy, 1803)	12+–/XY	A	20	This paper (Hayman and Martin, 1969 state that "*Phalanger* sp." has $2n = 12 + XY$).
Petaurinae				
Petaurus breviceps Waterhouse, 1839	20+XX/XY		34	Hayman and Martin, 1969.

TABLE IV—*continued*

Species	2n	X shape	No. of autosome arms	References, etc.
Pseudocheirinae				
Schoinobates volans (Kerr, 1792)	20+XX/XY	M	24	Hayman and Martin, 1965a.
Hemibelideus lemuroides (Collett, 1884)	18+XX/–			Hayman and Martin, 1969.
Pseudocheirus peregrinus (Boddaert, 1785)	18+XX/XY	A	24	Koller, 1936.
Pseudocheirus archeri (Collett, 1884)	14+XX–			Hayman and Martin, 1969.
Phascolarctinae				
Phascolarctos cinereus (Goldfuss, 1817)	14+XX/XY	M	20	Martin and Hayman, 1967.
Vombatinae				
Lasiorhinus latifrons (Owen, 1845)	12+XX/XY	A	20	Martin and Hayman, 1967.
Vombatus ursinus (Shaw, 1800)	12+XX/XY	M	20	Martin and Hayman, 1967 (as *Phascolomis mitchelli*).

Northwestern Australia has not had its chromosomes studied. The brush possums are of wide Australian distribution but are not found in New Guinea. They have 18 acrocentric autosomes (Table IV) and a small X and a minute Y chromosome.

(ii) Subfamily Phalangerinae. This subfamily includes the cuscuses, a group of about six New Guinea and Cape York (north Queensland, Australia) species. Two New Guinea species (Table IV) have four pairs of metacentric and two pairs of acrocentric autosomes a small X and a minute Y chromosome. Hayman and Martin (1969) provide a karyotype of an unidentified species of *Phalanger*.

(d) Family Petauridae (i) Subfamily Petaurinae. This subfamily includes the squirrel and sugar gliders (*Petaurus*) only one species of which has been studied cytologically (Table IV). *Petaurus breviceps* has 22 chromosomes, a number found in other phalangeroids and in other marsupials, but most of the chromosomes are metacentric so that about 34 chromosome arms are found, a number exceeded in only one other marsupial. Volant adaptations are found in *Petaurus*, *Schoinobates* (subfamily Pseudocheirinae) and in *Acrobates* (family Burramyidae) and it is generally conceded that these are convergent evolutions of the gliding habit. This is supported by what is known of the chromosome cytology of the three genera, there being considerable differences even between *Petaurus breviceps* and *Schoinobates volans* both of which have 22 chromosomes but remarkably different chromosome morphology (Table IV).

The subfamily is distributed in, generally coastal, eastern and northern Australia, some adjacent islands and New Guinea.

(ii) Subfamily Pseudocheirinae. The Pseudocheirinae includes about fifteen species of ring-tailed possums of Australian and New Guinea distribution and the greater gliders (*Schoinobates*). The subfamily has not been extensively studied cytologically which is unfortunate because the lower chromosome number karyotypes (Table IV) may have been derived from $2n = 22$ karyotypes by centric fusion (Section III B1).

(e) Families Phascolarctidae and Vombatidae. The Phascolarctidae, containing a single subfamily and but one species of living marsupial (the koala), has been grouped with the Phalangeridae. Anatomical evidence, however, indicates affinities between the koala and the wombats (*Vombatidae*) another phalangeroid family with a single living subfamily containing no more than three or four species. The two species of wombat thus far studied have 14 chromosomes while the koala has 16. All three species have 20 autosomal arms (Table IV).

(f) Family Macropodidae (i) Subfamily Macropodinae. The Macropodinae (kangaroos) occupy many diverse habitats but are not the largest subfamily of marsupials. There are about forty-two species (cf. sixty-five Didelphinae, forty-seven Dasyurinae) distributed from Tasmania in the South, throughout continental Australia, and into equatorial regions in New Guinea and adjacent islands. The kangaroos (including euros, wallaroos, wallabies, etc.) range in adult weights from about 2 to a maximum 80 kg and include rain forest, open forest, grassland, desert and rock haunting forms. The tree kangaroos (*Dendrolagus*) have secondarily adopted an arboreal habit.

About three quarters of the species have been studied cytologically and chromosome numbers (Table V) range from $2n = 10♀, 11♂$ to $2n = 24$ in one species, *Lagostrophus fasciatus* (the banded hare wallaby).

Aside from having a chromosome number unique in the subfamily, and nearly unique among marsupials, the banded hare wallaby differs from the remainder of the subfamily in other respects. Thomas (1887), after an examination of its dentition, skull and external characters, concluded that it must be considered, as a group, equal in value to all other genera of the Macropodinae combined. He therefore proposed subfamily division into two sections one of which included only the banded hare wallaby. The anatomy of the female urogenital system resembles that of the rat kangaroos (subfamily Potoroinae) (Tyndale-Biscoe, 1965) and this also indicates a substantial phylogenetic distinction from the remaining kangaroos.

The kangaroos with lower chromosome numbers—*Lagorchestes conspicillatus*, *Wallabia bicolor*, *Dendrolagus goodfellowi*, and probably other species of *Dendrolagus*—pose challenging cytotaxonomic questions. They have but 16 autosomal arms (as defined here) and this is not necessarily a reflection of their low diploid numbers since the remaining kangaroos, with chromosome numbers ranging from $2n = 14$ to $2n = 22$, have 24 or 26 autosomal arms. All, either singly or grouped with other species, have hitherto been the subject of taxonomic speculation and have, for one or more reasons, been grouped aside from the majority of the kangaroos. Wood Jones (1924) stated that two divergent stocks occurred in the kangaroos and that the smaller of these (the brachyodont section) included *Dendrolagus* (also *Dorcopsis* and *Setonix*). *Wallabia bicolor* (the swamp wallaby) is usually classified apart from most of the remaining larger kangaroos (*Macropus*) because of its unique chromosome constitution, reproductive pattern and behaviour. The spectacled hare wallaby (*Lagorchestes conspicillatus*) has a sex determination mechanism unique amongst marsupials but may have had, as an intermediate stage in the evolution of its sex mechanism, an $XX♀/XY_1Y_2♂$ system like that of the swamp wallaby (Section III D3). There are three species of *Lagorchestes*, the only other studied cytologically having 22 chromosomes rather like those of the remaining 22 chromosome kangaroos (Sharman, 1961) so doubts must be held, in view of the generally demonstrated chromosomal stability within genera (Table V), about its inclusion in the same genus as is the spectacled hare wallaby.

The 16 autosome arm kangaroos may or may not form a natural group, distinct from the remainder of the subfamily, but this possibility should be explored at physiological and behavioural levels. Kirsch (1968c) distinguished four groups of the subfamily Macropodinae on serological evidence. He placed *Lagostrophus* with the 22 chromosome genera *Thylogale* and *Petrogale* (including probably also *Setonix*); *Lagorchestes conspicillatus* (*L. hirsutus* not studied serologically) with *Macropus* (the 16 chromosome group); and *Dendrolagus* as a separate group. Kirsch's remaining group contained only *Onychogalea* (species *unguifer* alone studied).

Most members of the 16 chromosome group of kangaroos are cytologically distinguishable but no differences are evident between the two most closely related species *Macropus giganteus* (Fig. 5) and *M. fuliginosus*. Kirsch and Poole (1972) provide a valuable account of the taxonomy of these two forms which were long confused as a single species.

(ii) Subfamily Potoroinae. This subfamily includes the rat kangaroos

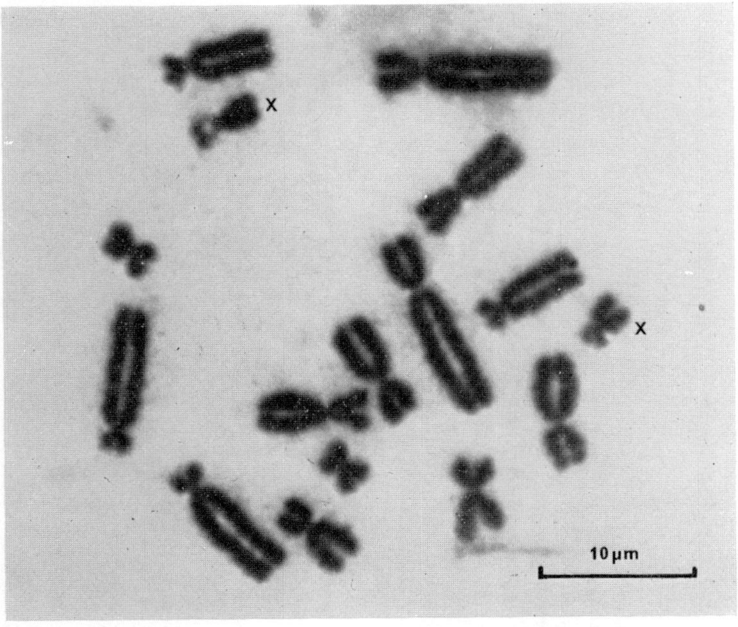

Fig. 5. Metaphase chromosomes from leucocyte culture of the Eastern grey kangaroo (*Macropus giganteus*). X — X chromosomes, the smaller of the two, without a distinct secondary constriction, is the late replicating (inactive) X chromosome (Graves, 1967).

TABLE V. Chromosome numbers, sex determination mechanisms and type of X chromosome (A = acrocentric, M = metacentric) in subfamilies Macropodinae and Potoroinae (Phalangeroidea, Macropodidae). Count of autosome arms does not include short arms of acrocentrics unless equal to, or greater in length than, shortest long arm (excluding Y)

Species	2*n*	X shape	No. of autosome arms	References, etc.
Macropodinae				
Lagostrophus fasciatus (Peron & Lesueur, 1807).	22+XX/XY	M	24	Sharman, 1961.
Lagorchestes hirsutus Gould, 1844	20+–/XY	M	ca. 24	Sharman, 1961.
Lagorchestes conspicillatus Gould, 1842	$12+X_1X_1X_2X_2/X_1X_2Y$	Compound M+A	16	Martin and Hayman, 1966 (autosome arms include two pairs incorporated in sex chromosomes).
Onychogalea unguifer (Gould, 1841)	18+–/XY		ca. 24	Sharman, 1961.
Petrogale rothschildi Thomas, 1904	20+XX/XY	A	26	Hayman and Martin, 1969.
Petrogale pearsoni Thomas, 1922	20+XX/XY		26	Sharman, 1961 (as *P. penicillata pearsoni*).
Petrogale xanthopus Gray, 1855	20+XX/XY			Hayman and Martin, 1969.
Petrogale hacketti Thomas, 1905	18+XX/XY	A	26	Hayman and Martin, 1969 (as *P. penicillata hacketti*).
Petrogale brachyotis signata Thomas, 1926	16+XX/–			Hayman and Martin, 1969 (as *P. penicillata signata*).
Petrogale inornata Gould, 1842	16+–/XY			Hayman and Martin, 1969 (as *P. penicillata inornata*).
Thylogale billardierii (Desmarest, 1822)	20+XX/XY	M	26	McIntosh and Sharman, 1953.
Thylogale stigmatica Gould, 1860	20+XX/XY	M	26	Sharman, 1961.
Thylogale thetis (Lesson, 1827)	20+XX/XY		26	Sharman, 1961.
Setonix brachyurus (Quoy and Gaimard, 1830)	20+XX/XY		26–28	Hayman and Martin, 1969.
Dorcopsis sp.	20+XX/XY			Hayman and Martin, 1969.
Megaleia rufa (Desmarest, 1822)	18+XX/XY	M	26	Martin and Hayman, 1965.

Macropus agilis (Gould, 1842)	14+XX/XY			Sharman, 1961 (as *Protemnodon agilis*).
Macropus dorsalis (Gray, 1837)	14+XX/XY			Sharman, 1961 (as *Protemnodon dorsalis*).
Macropus eugenii (Desmarest, 1817)	14+XX/XY	A	24	Martin and Hayman, 1965; Graves, 1967 (as *Protemnodon eugenii*) Sharman, *et al.*, 1970.
Macropus irma (Jourdan, 1837)	14+XX/XY	M		Sharman, 1961 (as *Protemnodon irma*).
Macropus parma Waterhouse, 1846	14+XX/XY		26	Hayman and Martin, 1969.
Macropus parryi Bennett, 1835	14+XX/XY	M	26	This paper.
Macropus rufogriseus (Desmarest, 1817)	14+XX/XY	M	26	Sharman, 1961 (as *Protemnodon rufogrisea*).
Macropus fuliginosus (Desmarest, 1817)	14+XX/XY			Sharman, 1961 (as *M. major fuliginosus*).
Macropus giganteus Shaw, 1790	14+XX/XY	M	24	This paper.
Wallabia bicolor (Desmarest, 1804)	8+XX/XY_1Y_2	compound M	16	Sharman, 1961 (as *Protemnodon bicolor*); Hayman and Martin, 1965c (as *Protemnodon bicolor*) (autosome arms include a pair incorporated in sex chromosomes).
Dendrolagus goodfellowi Thomas, 1908	12+–/XY	A	16	Hayman and Martin, 1969.
Dendrolagus lumholtzi Collett, 1884	12+–/XY			Hayman and Martin, 1969.
Dendrolagus matschiei Förster & Rothschild, 1907	12+XX/–			Hayman and Martin, 1969.
Dendrolagus ursinus Schlegel & Muller, 1839–44	10+–/XY			Hayman and Martin, 1969.
Potoroinae				
Hypsiprymnodon moschatus Ramsay, 1876	20+–/XY	A	22	Sharman, 1961.
Potorous tridactylus (Kerr, 1792)	10+XX/XY_1Y_2	compound M	22	Hayman and Martin, 1965c; Sharman and Barber, 1952 (autosome arms include a pair incorporated in sex chromosomes).
Aepyprymnus rufescens (Gray, 1837)	30+XX/XY	M	38	Hayman and Martin, 1969.
Bettongia gaimard (Desmarest, 1822)	20+XX/XY		ca. 36	McIntosh and Sharman, 1953 (as *B. cuniculus*).
Bettongia lesueuri (Quoy & Gaimard, 1824)	20+XX/XY			Sharman, 1961.
Bettongia penicillata Gray, 1837	20+XX/XY		ca. 34	Sharman, 1961.

containing only eight species, six of which have had their chromosomes studied. Woodburne (1967) included the genera *Dorcopsis* and *Dorcopsulus*, here included with the Macropodinae, in the Potoroinae. Hayman and Martin (1969) recorded, for an un-named species of *Dorcopsis*, $2n = 22$—a chromosome number found in both macropodid subfamilies. A karyotype or other illustration of the chromosomes was not provided so it is impossible to speculate about the cytotaxonomic affinities of *Dorcopsis*.

The range of chromosome numbers in the rat kangaroos is greater than the range in the true kangaroos. It is possible that those which have had their chromosomes studied fall into two groups—those with 22 autosome arms and those with 34–38 autosome arms (Table V). The group (as considered here) is of Australian, Tasmanian and some offshore island distribution but does not occur in New Guinea.

B. *Karyotype Evolution*

1. *Reciprocal translocation and centromere loss.*

Rearrangements following breaks in different chromosomes are translocations and special cases of reciprocal translocations involving two acrocentric chromosomes are called "centric fusions" (White, 1957). If one chromosome breaks in the long arm, very close to the centromere, while the other breaks in the minute "short arm", reunion of the long acentric and the other chromosome will give rise to one metacentric in place of two former acrocentric chromosomes. If the chromosome material in the small chromosome, formed by union of the centromere plus short arm with part of short arm, is genetically inert its rapid loss from the genome may be expected. Other types of rearrangements leading to centric fusions have been postulated (White, 1957).

The generally constant arm number in marsupials with different chromosome numbers but belonging to the same subfamily (Tables I, II, III) suggests that centric fusions may have been involved in karytoype evolution. Hayman and Martin (1969) state that "*Hemibelideus lemuroides* is probably derived by centric fusion from a chromosome complement like that of *Schoinobates volans* and we presently incline to the view that *Pseudocheirus archeri* also arose by chromosome reduction from $2n = 22$". This may be equivocal because there is a paucity of knowledge about chromosome number and morphology in other species of Phalangerinae but there seems little doubt that evolutionary chromosome reduction has occurred in the rock wallabies (*Petrogale*, Table V). Three species of *Petrogale* have $2n = 22$ as have three species of *Thylogale* and serological evidence (Kirsch, 1968b) indicates that the similar karyotypes found in the two genera (Sharman,

1961) are indicative of relationship. There is little doubt therefore that the primitive number in the group is $2n = 22$. *P. hacketti* has $2n = 20$, a karyotype undoubtedly derived from a $2n = 22$ ancestor by centric fusion (Fig. 6). *P. brachyotis* and *P. inornata* have different $2n = 18$ karyotypes "suggesting two separate instances of chromosome reduction from $2n = 22$" (Hayman and Martin, 1969).

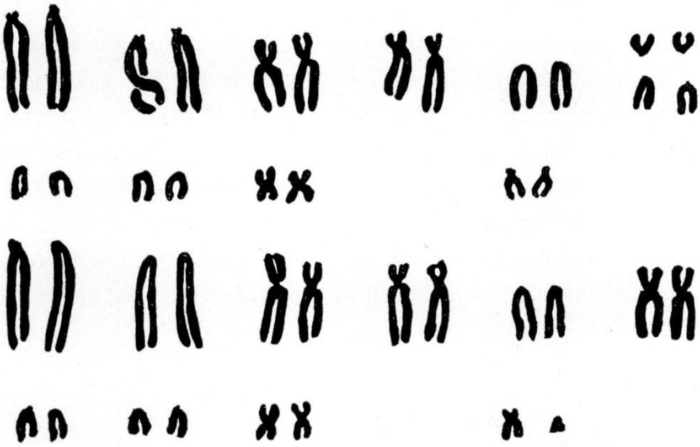

Fig. 6. Karyotypes of the rock wallabies *Petrogale rothschildi*, $2n = 22$ (♀), above and *P. hacketti* $2n = 20$ (♂) below. Two pairs of acrocentric chromosomes (at right) in *P. rothschildi* are represented by a pair of metacentric chromosomes of equivalent size indicating the occurrence of centric fusion in the evolution *P. hacketti*. Redrawn from Hayman and Martin (1969).

A male heterozygous for a translocation involving the centric fusion of two acrocentric chromosomes was found amongst seven individuals of the hare-wallaby *Lagorchestes conspicillatus* (Martin and Hayman, 1966). Amongst first metaphase cells studied 30% had linear trivalents which would have been expected to provide unbalanced gametes but no unbalanced spermatocytes were observed at second metaphase suggesting that unbalanced cells did not progress beyond interphase. The population from which the aberrant individual came has not been further studied so it is unknown whether the centric fusion was confined to a single individual or whether the population is polymorphic for chromosome number differences as are certain eutherians (Meylan, 1970; Ford and Hamerton, 1970). What is important, to the present theme, is that the animal bearing the centric fusion was viable and was producing apparently normal germ cells while being heterozygous for a fused and unfused chromosome pair.

Whereas centric fusions result from two proximal breaks, tandem fusions apparently result from a break in the long arm of an acrocentric chromosome near the centromere and a near-terminal break in one arm of a metacentric or submetacentric chromosome. Tandem fusions seem to have been responsible for at least one of the four separate formations of multiple sex chromosome systems in marsupials (Section III D3).

There are thus eight species of marsupials, and possibly at least a further two species, which have undergone evolutionary lowering of chromosome number following reciprocal translocations in ancestral forms. On the other hand no occurrence of "karyotypic fissioning" (Todd, 1970) during marsupial evolution has been clearly demonstrated. *Megaleia rufa*, with $2n = 20$, may be derived from a $2n = 16$ ancestor by the occurrence of two separate instances of chromosome fission (Fig. 7) but if this occurred there must have been further chromosome changes which shifted the centromeres of the new telecentric chromosomes from hitherto terminal positions. Alternatively if the region adjacent to the centromeres of acrocentric chromosomes is genetically inert its loss during reciprocal translocations which give rise to one metacentric, in place of two acrocentric chromosomes, might have no effect on viability.

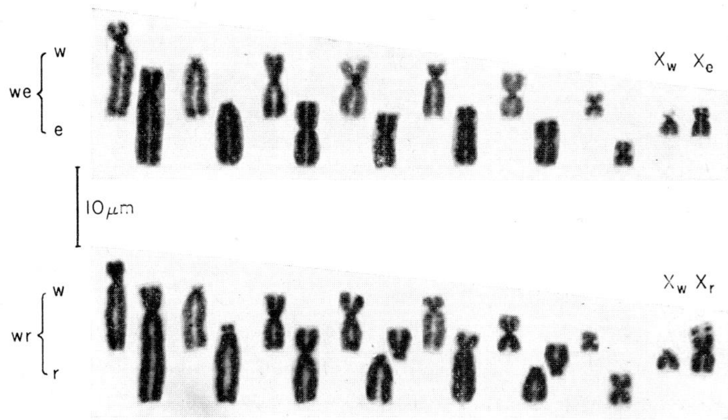

Fig. 7. Haploid karyotypes of wallaroo (*Macropus robustus robustus*) (w), euro (*M. r. erubescens*) (e) and red kangaroo (*Megaleia rufa*) (r). Female hybrids (we, wr) between these subspecies or species have an X chromosome of different morphology derived from each parent. Two submetacentric chromosomes in wallaroo and euro are represented by four acrocentric chromosomes in the red kangaroo.

2. *Evolutionary increase in chromosome number*

The didelphine marsupials have a fossil history extending back to the late Cretaceous (Clemens, 1966) and are the most generalized of the living marsupials. Clemens (1968, 1971) suggested that both South American and Australian radiations of marsupials are derived from *Alphadon*, or a closely related late Cretaceous didelphine. Living didelphines have $2n = 22$, 18 or 14 so the occurrence of three phalangeroid marsupials with chromosome numbers of $2n = 24$ (2) and $2n = 32$ may mean that evolutionary increase in chromosome number has occurred in the marsupials on at least three occasions. Unless a monocentric chromosome can fragment to form two chromosomes, each with a functional centromere, the problem associated with evolutionary chromosome number increase is to find a source of the extra centromeres. Supernumerary chromosomes are rare in mammals but have been reported in three species of marsupials. Hayman and Martin (1965a) found two, three, four, five and six supernumeraries in seven individuals of the greater gliding possum *Schoinobates volans*. The same authors found one to five supernumeraries in the testis and corneal epithelium of six out of eight individuals of the New Guinea bandicoot *Echymipera kalabu* (Hayman, *et al*., 1969). Supernumeraries also occur in another New Guinea bandicoot (Section III E3).

The particular significance of the, presumably genetically inert, supernumeraries is that they carry centromeres and may thus act as donors to acentric chromosome fragments thus providing a mechanism for evolutionary chromosome number increase. Such a mechanism has not, however, been demonstrated.

3. *Paracentric and pericentric inversions and interchanges*

Martin and Hayman (1967) measured chromosome arm lengths and DNA values in five species of dasyurine marsupials and from these constructed a "standard dasyurid" karyotype from which only 8 autosome arms (of a total of 60) differed significantly in length. In *Sarcophilus harrisii* chromosome 5 had a shorter long arm and a longer short arm than had the standard dasyurid and the long arm of chromosome 1 was shorter. These observations were compatible with the hypothesis of an interchange between chromosome 1 and 5 and a pericentric inversion in chromosome 5 of *Sarcophilus*. More pronounced differences between the chromosomes were found when the "standard dasyurid" karyotype was compared with those of various marsupials belonging to other superfamilies.

The $2n = 14$ karyotypes of the perameline bandicoots *Perameles nasuta*, *Isoodon macrourus* and *I. obesulus* are superficially very similar and presumably all three species had a common ancestor with 14 chromosomes. However only two whole chromosomes were "shared" (in the sense that they were not shown to be significantly different) by the two species of *Isoodon* so presumably considerable rearrangement of chromosome material, not resulting in change in chromosome number, has occurred during the evolution of the group.

A comparison of the karyotypes of the American marsupials *Didelphis marsupialis* and *D. virginiana* is particularly instructive (Fig. 8). These two opossums, both with $2n = 22$, are very similar, have been grouped in the one species (*D. marsupialis*) and are undoubtedly closely related, yet they have different chromosome morphology. The haploid lengths of the karyotypes of the two species do not differ significantly and corresponding auto-

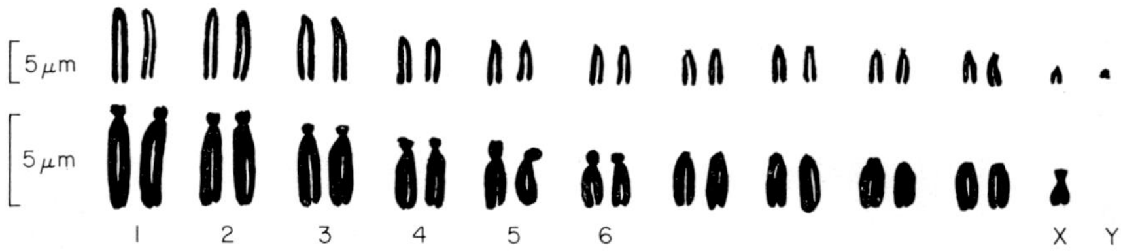

Fig. 8. Idiograms of the chromosomes of American opossums *Didelphis marsupialis*, redrawn from Perondini and Perondini (1965) and *D. virginiana*, redrawn from Shaver (1962) (below). These are closely related forms which have been grouped in a single species but the centromeres have shifted to definite subterminal positions in six pairs of chromosomes. (1–6) during the evolution of *D. virginiana*. The X chromosome (X), shown singly for female *D. virginiana*, is larger than that of *D. marsupialis* and submetacentric.

somes have about the same relative lengths (Perondini and Perondini, 1965). In addition to *D. marsupialis* there are four other species of South American $2n = 22$ marsupials which have seemingly terminal centromeres on all autosomes. From these *D. virginiana* alone differs in having definite subterminal centromeres on the six largest pairs of autosomes. It may be concluded that in the evolution of *D. virginiana*, perhaps during a post-Pliocene period of North American isolation (see p. 488), "centric shifts" occurred in chromosomes 1 to 6. Perondini and Perondini (1965) suggest that the morphological differences of the chromosomes of *D. virginiana* could be explained by the occurrences of chromosomic mutations, such as translocations and inversions.

C. *The Primitive Karyotype*

Chromosome numbers in one hundred species of marsupials have a bimodal distribution (Fig. 9). There may be some species, conservative in chromosomal evolution, which retain the chromosome number and morphology of the stem form from which the extant South American and Australian radiations arose. Alternatively none of the living marsupials have retained the primitive chromosome number and, if this could be shown to be so, further speculation would be pointless.

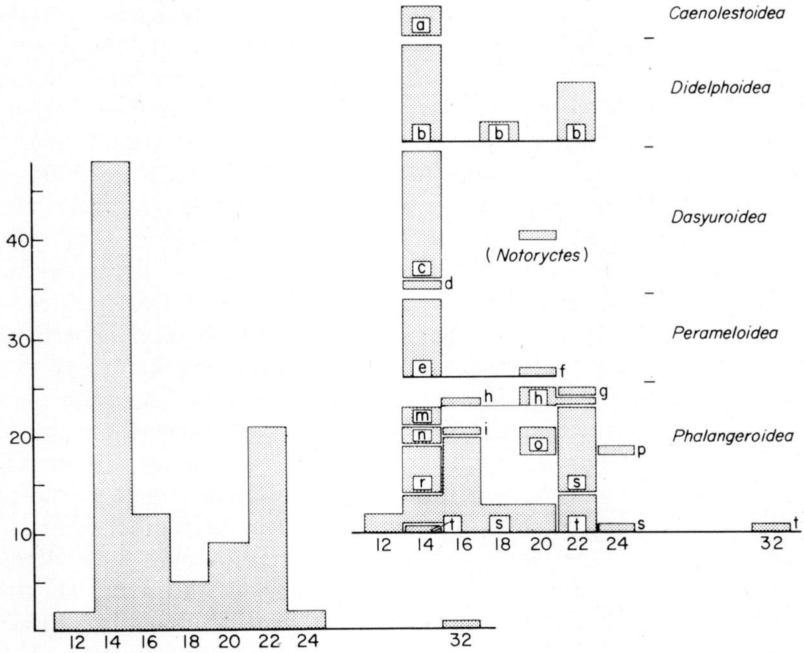

Fig. 9. Distribution of diploid chromosome numbers in marsupials and the distribution by superfamilies and subfamilies. The four marsupials with multiple sex chromosome mechanisms have been allotted the diploid chromosome number of the form from which they were derived before fusion of autosomes to sex chromosomes. a—Caenolestinae, b—Didelphinae, c—Dasyurinae, d—Myrmecobiinae, e—Peramelinae, f—Thylacomyinae, g—Petaurinae, h—Pseudocheirinae, i—Phascolarctinae, m—Phalangerinae, n—Vombatinae, o—Trichosurinae, p—Tarsipedinae, r—Burramyinae, s—Macropodinae, t—Potoroinae.

Most marsupials have 14 or 22 chromosomes and either of these numbers could be that of the stem form which gave rise to the extant radiations. Both numbers are found amongst the didelphoids which have the longest fossil history and which include, amongst extinct forms, the late Cretaceous *Alphadon* from which existing radiations probably arose.

The subfamily Caenolestinae includes living and fossil South American forms and is the sole extant subfamily of a more extensive past radiation. Two of the three subfamilies of the family Caenolestidae are extinct and there is a further extinct family of the superfamily Caenolestoidea (Simpson, 1945). Three of the six living caenolestoids have $2n = 14$ and, although chromosome measurements are not available, it is probable that they share in Martin and Hayman's (1967) sense, one or more pairs of chromosomes with the $2n = 14$ didelphoids (Fig. 4).

The Australian dasyuroids, of which two subfamilies have been studied cytologically (Table II), have a remarkably uniform $2n = 14$ karyotype (Fig. 4). However if the marsupial mole, *Notoryctes typhlops* (family Notoryctidae), which has $2n = 18 + XY$ (J. H. Calaby, L. Corbett, P. G. Johnston and G. B. Sharman, unpublished) is admitted to be a dasyuroid then this superfamily is not entirely composed of $2n = 14$ species. Ride (1964) regarded the marsupial mole as constituting a family *incertae sedis* while Kirsch (1968b) placed it as order Polyprotodonta *incertae sedis* (Polyprotodonta of Kirsch = Didelphoidea + Dasyuroidea + Perameloidea). Simpson (1945) included Notoryctidae as a family of the Dasyuroidea but later (Simpson, 1970) suggested that it might eventually merit superfamily distinction. The dasyuroids also have chromosomes remarkably like those of the $2n = 14$ didelphoids and the caenolestoids. In all dasyuroids critically studied, and in the caenolestoids, secondary constrictions, presumably the nucleolar organizers, are found on the short arms of the fifth or sixth largest pair of autosomes. The location of the nucleolar organizer has not been described for any $2n = 14$ (or other) South American didelphoid.

The bandicoots (superfamily Perameloidea) were thought to be characterized by a uniform $2n = 14$ (Sharman, 1961) but Martin and Hayman (1967) found the rabbit bandicoot—*Macrotis lagotis*, subfamily Thylacomyinae—to have $2n = 18$♀, 19 ♂ and a karyotype derived from $2n = 20$ by incorporation of a pair of acrocentric autosomes into the sex determination mechanism (Fig. 10). The superfamily is thus comparable to the Didel-

Fig. 10. The chromosomes of male *Macrotis lagotis* redrawn from Martin and Hayman (1967). There are 19 chromosomes and the X of an ancestral form is joined to an acrocentric autosome which pairs with its homologue (Y_2) at meiosis.

phoidea, in which a range of chromosome numbers is found, and to the Dasyuroidea which also has a range of chromosome numbers if *Notorcytes* be included. As in the caenolestoids and dasyuroids the nucleolar organizers are found on the fifth or sixth pair of autosomes in the peramelines.

The superfamily Phalangeroidea includes forms with $2n = 14$, $2n = 20$, $2n = 22$ and other numbers (Fig. 9). No $2n = 22$ karyotype is exactly like the $2n = 22$ karyotypes of any didelphoid and there are at least three different $2n = 14$ karyotypes; that of the Burramyinae being most like that of $2n = 14$ didelphoids, caenolestoids and dasyuroids. Hayman and Martin (1969) state that, in the pigmy possum *Cercartetus nanus*, the satellited chromosomes (i.e. the chromosomes with secondary constrictions) are one of the two small autosomal pairs. However in another pigmy possum, *Eudromicia lepida* (= *Cercaërtus lepidus*), Sharman (1961) reported secondary constrictions in the long arms of one of the larger autosomal pairs.

It has been suggested that $2n = 14$, with the form of chromosome complement found in $2n = 14$ didelphoids, caenolestoids, dasyuroids, perameloids and some phalangeroids is ancestral for the marsupials (Hayman and Martin, 1969; Hayman, *et al.*, 1971). Reig and Bianchi (1969) and Reig, *et al.* (1972) believed $2n = 14$ to be the primitive number for the family Didelphidae. This implies that evolutionary increase in chromosome number must have occurred independantly in the didelphoids, perameloids, phalangeroids and in the dasyuroids if *Notoryctes* be included in the latter group. However no mechanism for chromosome number increase in marsupials has yet been demonstrated although it must, of course, have occurred even if the other common number, $2n = 22$, be regarded as primitive because two marsupials have $2n = 24$, and one has $2n = 32$. On the other hand it has clearly been demonstrated that reduction in chromosome number has occurred during the evolution of at least 8% of extant marsupials (p. 506).

Reig and Bianchi (1969) suggested the homologies between the chromosomes of the $2n = 14$, $2n = 18$ and $2n = 22$ didelphines shown diagrammatically in Figure 11. They favoured "centromeric fission" as the mechanism by which the $2n = 18$ and $2n = 22$ karyotypes arose from $2n = 14$ karyotypes.

Hayman, *et al.* (1971) point to the similarities between the $2n = 14$ karyotype of the superfamily Dasyuroidea and those of some members of Didelphoidea, Phalangeroidea and Perameloidea. However Martin and Hayman's (1967) careful measurements of the chromosomes in $2n = 14$ species from three of the superfamilies, although demonstrating conservatism in dasyuroid chromosome evolution, failed to demonstrate homology between superfamilies. Thus no chromosome of the $2n = 14$ perameloids was "shared" (in the sense that it was not shown to be different) with the dasyuroid standard karyotype. With one exception all chromosome arms of the combined karyotype of two of the $2n = 14$ phalangeroids (*Cercartetus*) were significantly different from the standard dasyurid. When total chromosome lengths were compared four chromosomes of *Cercartetus* did not differ from the standard dasyurid and the authors considered that each of the four changed arm ratios could be the result of one or more pericentric inversions. *Perameles nasuta* (Perameloidea) shared at least two chromosomes with *Lasiorhinus* (Phalangeroidea). The satellited number 5 or 6 chromosomes of the $2n = 14$ species were regarded as evidence of comparability of karyotype (Hayman and Martin, 1969; Hayman, *et al.*, 1971).

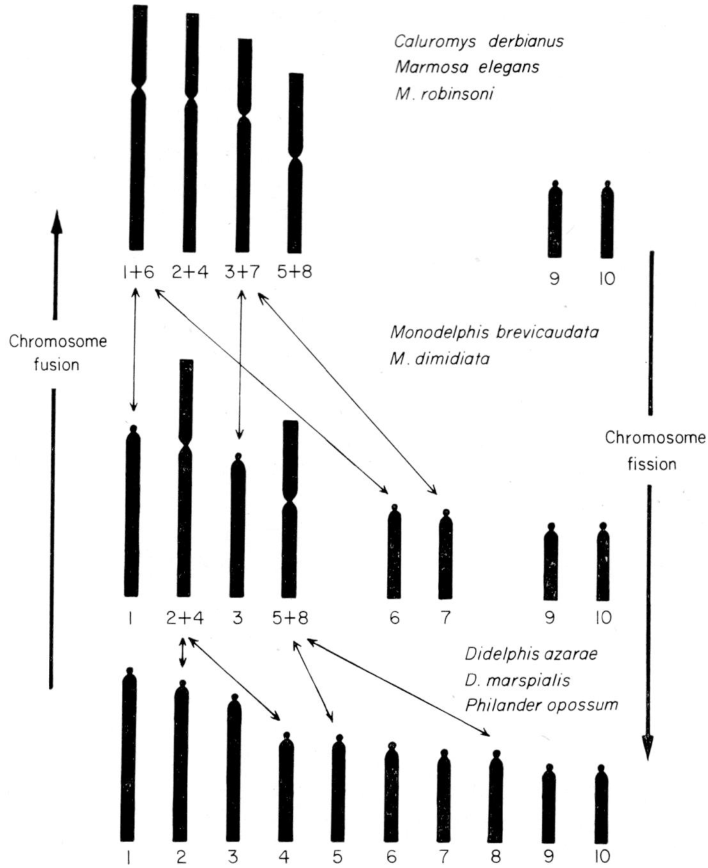

Fig. 11. Acrocentric and metacentric autosomes in some species of the superfamily Didelphoidea based on karyotypes from authors given in Table I. Sex chromosomes not shown. For further explanation see text. Reproduced from Sharman (1973).

These were the same relative lengths in two species of *Isoodon* (Perameloidea) but were of significantly different lengths in other comparisons, both within and between superfamilies.

It must be concluded that if $2n = 14$ is the primitive marsupial chromosome number then considerable rearrangements, resulting from inversions and translocations, have occurred within $2n = 14$ species during their evolution. Alternatively marsupials are descended from $2n = 22$ ancestors and the lower chromosome numbers in extant forms are the products of evolutionary lowering of the chromosome number by reciprocal translocation and centromere loss (centric and tandem fusions). In this context

Martin and Hayman (1966) presented evidence for a basic chromosome number of $2n = 22$ in macropodid marsupials with several of the lower chromosome numbers derivable from $2n = 22$. They also state (Martin and Hayman, 1967) that the most acceptable explanation of the relationships between *Macrotis* (= *Thylacomys*) *lagotis* and the $2n = 14$ perameloids is "that *Thylacomys* is the more primitive and, by Robertsonian fusions, it gave rise to a fourteen chromosome species". If the $2n = 18$ and $2n = 14$ didelphoids be also regarded as derived by Robertsonian fusions from $2n = 22$ ancestors (although Reig and Bianchi (1969) favoured the reverse hypothesis) then chromosome number reduction to $2n = 14$ may be presumed to have occurred in three superfamilies of marsupials. Given that *Alphadon*, regarded by Clemens (1966, 1968, 1970) as being ancestral to existing marsupial radiations, is found as a Lance formation (late Cretaceous) fossil there would appear to have been ample time, during seventy million years of evolution, to allow for the possibility that the primitive number in each of the superfamilies was $2n = 22$. If this is so the numerous $2n = 14$ karyotypes found in all superfamilies (as considered here) must be regarded as separate instances of chromosome reduction in immediately ancestral forms. Further evidence of this possibility may be found when chromosome arms are compared using specific staining techniques.

D. *Evolution of Sex Chromosomes*

1. *Sex determination*

In man male sex is normally determined by the presence of a Y chromosome and in the absence of a Y female sex develops. At least one intact X chromosome is necessary for life itself (Ford, 1970). This rule probably prevails throughout both eutherian and marsupial mammals while the situation in monotremes is uncertain due to the possibility that sex determination is of the XO male: XX female type (Bick and Jackson, 1967c). The occurrence of XO: XX sex determination in monotremes is however unlikely (p. 487).

Marsupials with sex chromosome aneuploidy have been reported (Sharman, *et al.*, 1970). An XXY male tammar wallaby (*Macropus eugenii*) was of female body phenotype and had a pouch (marsupium) containing teats with underlying mammary tissue but undescended, non-functional testes were present, the accessory reproductive structures were of male type and the penis was well developed. A 14 + XO tammar intersex was of female body phenotype and had a non-functional reproductive system of female type. These two animals essentially resembled, respectively, Klinefelter's and Turner's syndromes which are the outcome of XXY and XO sex chromosome aneuploidy in man. It was concluded that, in the tammar at least, the Y chromosome is strongly male determining as in man and other eutherian mammals.

2. *X and Y sex chromosomes*

In all members of the superfamilies Didelphoidea, Caenolestoidea and Dasyuroidea which have been studied cytologically the X chromosome is smaller (usually much smaller) than the smallest autosome. The Y chromosome in these superfamilies is usually even smaller than the X and in some species is only just above the limit of resolution by the light microscope (Ford, 1970). If the basic mechanism of male sex determination by presence of Y chromosome is common to all marsupials then, as Ford points out, the genetic information responsible for switching sex development in the male direction may be carried in a chromosome segment of negligible size.

Since no smaller X chromosomes than those of the didelphoids, caenolestoids and dasyuroids occur in marsupials it may be presumed that these superfamilies carry "primitive" type marsupial X chromosomes. In the dasyuroids such X chromosomes constitute 2%, or less, of the haploid genome (Martin and Hayman, 1967) that is they are rather smaller than the basic X chromosome of the mammals as postulated by Ohno (1969). Despite its small size the marsupial "primitive" type X chromosome occurs in both acrocentric (A) and metacentric (M) forms. Didelphoids have either acrocentric or (sub) metacentric X chromosomes (Table I). The X chromosomes of such dasyurids and caenolestoids as have been critically studied however appear to be uniformly acrocentric (Table II). X and Y sex chromosomes which do not appear to differ greatly in size from those of the didelphoids, caenolestoids and dasyuroids are found in several groups of the superfamily Phalangeroidea including the subfamilies Burramyinae, Trichosurinae and Phalangerinae.

In the remainder of the Phalangeroidea and in the Perameloidea the sex chromosomes are larger; the X being usually equivalent in size to, or larger than, the smallest autosome and the Y of more substantial size than in the remaining three superfamilies. In both these groups instances of X chromosome autosome fusion resulting in the production of multiple sex chromosome systems occur (Section III D3). Hayman and Martin (1969) reported preliminary observations on the X and Y chromosomes of the subfamilies Macropodinae and Potoroinae in which the nucleolar organizer is found on the X chromosome (Figs. 5, 7, 12, 13) in contrast to its autosomal location in the remaining marsupials. In *Lagostrophus fasciatus* the nucleolar organizer has not been identified on the X chromosome with certainty and this was attributed to the poorer quality of the preparations studied. However in view of the unique (amongst Macropodidae) chromosome number, uncertain taxonomic position and relatively small size of the X chromosome of this species it would certainly be worthwhile to ascertain that the nucleolar organizer is not of autosomal location as in non-macropodid marsupials. The remaining macropodids which have been studied show pronounced variation in the position of the nucleolar organizer on the X chromosome and, in the groups with larger X chromosomes (Fig. 12), the

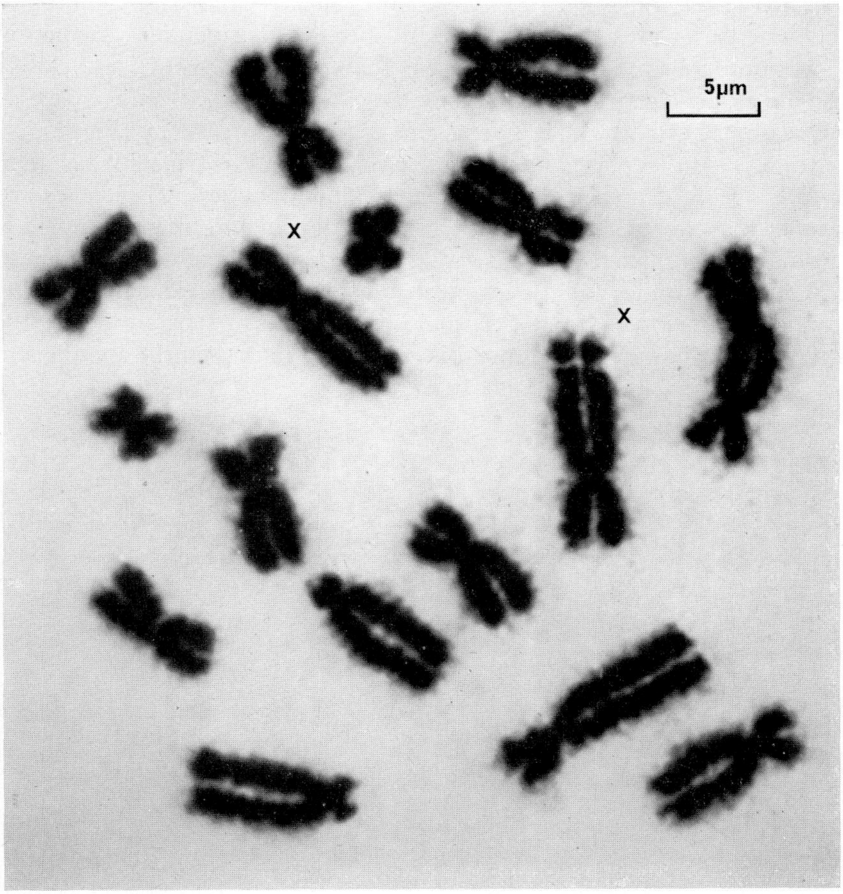

Fig. 12. Diploid chromosome complement of the whiptail wallaby, *Macropus parryi*. A secondary constriction (the presumed nucleolar organizer) is clearly visible in one of the two X chromosomes.

X may be two to four times larger than the average smaller X chromosomes. The origin of the extra material which has contributed to the evolution of the larger X chromosomes in marsupials with XX♀: XY♂ sex determination systems remains unknown. In general species with large X chromosomes also have comparatively large Y chromosomes although the correlation is by no means absolute.

The sex chromosomes do not appear to be of cytotaxonomic significance. Wallaroos and euros are subspecies of the same species (*Macropus robustus*) and will cross in captivity and produce fertile F1 hybrids (Richardson, *et al.*, 1971) but the X chromosome of the euro is one and a half times larger than that of the wallaroo (Fig. 7). Within the genus *Macropus* sex chromosome size varies greatly from species to species. Considerable variation in X chromosome size and morphology is found in other groups and closely related species such as *Didelphis marsupialis* and *D. virginiana* have X chromosomes of both different size and different morphology (Fig. 8).

Koller (1936) reported that the sex chromosomes of marsupials were involved more frequently in numerical abnormalities of chromosome number than were the autosomes. In spermatogonial mitoses the single X chromosome of male marsupials shows a different cycle of condensation to that of the autosomes and the sex bivalent is precociously condensed during prophase of meiosis. X and Y chromosomes begin to separate before the autosomes at first division of meiosis but thereafter lag on the spindle and

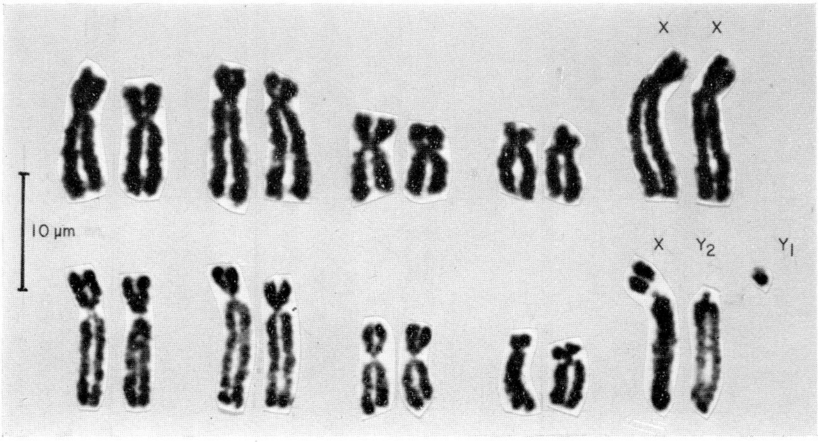

Fig. 13. Female (above) and male (below) somatic chromosomes of the swamp wallaby, *Wallabia bicolor*. The compound X chromosome (X) was formed by the fusion of a small acrocentric X chromosome, with its nucleolar organizer near the centromere, and an acrocentric autosome. The nucleolar organiser appears as a non-staining gap near the centromere of the compound X chromosome of the male and in favourable preparations may be seen in one, or both, compound X chromosomes of the female. During meiosis in the male the neo Y chromosome (Y_2) pairs with its homologue (the long arm of the compound X) and the ancestral Y chromosome (Y_1) pairs with the short arm.

reach the poles after anaphase separation of autosomes is complete. Lagging of X and Y chromosomes was also observed during second anaphase (McIntosh and Sharman, 1953). Failure of metaphase association between X and Y chromosomes occurs in *Dasyurus* (Koller 1936) and *Macropus rufogriseus* (= *Wallabia rufogrisea*) (McIntosh and Sharman, 1953). These observations, at the time they were made, suggested that the X chromosome of male marsupials was at some disadvantage during call division. In retrospect they also suggest that the single X chromosome of male marsupials is genetically active in somatic cells but becomes inactivated during spermotogenesis and is inherited in inactive form by the zygote (Section III E2).

Koller (1936) believed that the post-diplotene association of marsupial X and Y chromosomes was achieved without the occurrence of chiasmata between them but McIntosh and Sharman (1953) believed that the X and Y were associated by a single chiasma. There is general agreement, amongst the few authors who have studied marsupial sex chromosomes during meiosis, that the X and Y pairing segments do not include the centromere.

3. *Multiple sex chromosome mechanisms*

Agar (1923) reported that during meiosis the XY bivalent of *Macropus ualabatus* (= *Wallabia bicolor*) was "possibly sometimes independant but more often attached to one of the autosomes". His findings were reinterpreted by Darlington (1937) and Matthey (1949) who considered that a multiple sex chromosome mechanism existed. Sharman, *et al.*, (1950) and Sharman and Barber (1952) re-examined the chromosomes of *Potorous tridactylus* in which XX♀ : XY♂ sex determination had been hitherto reported and found that the male of this species had thirteen chromosomes and the female twelve. The sex determination mechanism is of the XX♀ : XY_1Y_2♂ type due to reciprocal translocation between the X chromosomes of an ancestral XX♀ : XY♂ system and a pair of acrocentric autosomes. Subsequently *Wallabia bicolor* was re-examined and found to have a similar sex-determination mechanism (Fig. 13) and one autosome pair less than *Potorous* (Sharman, 1961). Another XX♀ : XY_1Y_2♂ system (Fig. 10), formed in a similar manner to the multiple sex chromosome systems of *Potorous* and *W. bicolor*, was subsequently found in the rabbit bandicoot, (*Macrotis* = *Thylacomys*) *lagotis* (Martin and Hayman, 1967). In *Lagorchestes conspicillatus* both X and Y chromosomes are translocated onto autosomes (Fig. 14) resulting in the formation of an $X_1X_1X_2X_2$♀ : X_1X_2Y♂ sex determining mechanism (Martin and Hayman, 1966). Ohno (1969) objected to the terminology used by the authors who described the XX♀ : XY_1Y_2♂ systems of marsupial and eutherian mammals and preferred to call the Y_2 chromosome "an intact autosome which is seen only in the male".

However, a chromosome which appears only in the heterogametic (male) sex is, by definition, a Y chromosome.

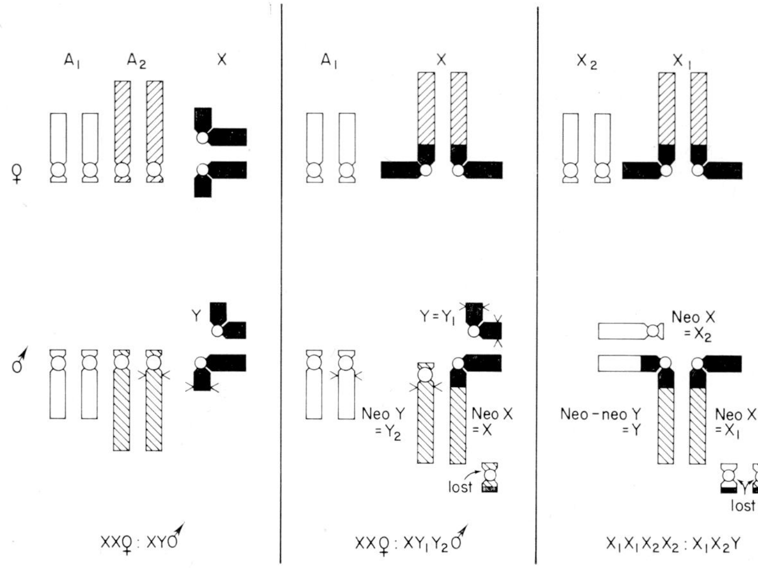

Fig. 14. (Left) The chromosomes of a marsupial with an XX♀/XY♂ sex-determining mechanism and two pairs of autosomes (A_1, A_2). Further autosomal pairs not shown. (Centre) A reciprocal translocation between one autosome and the X chromosome converts the system to an XX♀/XY_1Y_2♂ sex determining mechanism and reduces the number of autosome pairs by one. (Right) Further reciprocal translocations between Y_1 and Y_2 chromosomes of the XY_1Y_2 system and between Y_1 and another autosome convert the system to an $X_1X_1X_2X_2$♀/X_1X_2Y♂ system reducing the number of autosomes by a further pair. From Sharman (1970).

E. *Sex-Linked Inheritance and Dose Compensation for X Chromosomes*

1. *Sex-linked inheritance*

Ohno (1969) listed a series of homologous X linked genes found amongst a variety of eutherian mammals and implied an extensive homology of the X chromosome amongst mammalian species. The inheritance of glucose-6-phosphate dehydrogenase (G6PD) variation was subsequently shown to be X linked in red kangaroos (*Megaleia rufa*), euros (*Macropus robustus erubescens*) and wallaroos (*M. r. robustus*) (Richardson, *et al.*, 1971) and also in rednecked wallabies (*M. rufogriseus*) (P. G. Johnston, personal communication). Similarly the genes controlling phosphoglycerate kinase (PGK) variation are X linked in eastern (*M. giganteus*) and western (*M.

fuliginosus) grey kangaroos and in whiptail wallabies (*M. parryi*) (Cooper, *et al.*, 1971). It would thus appear that the marsupial X chromosome may be basically homologous with the X of eutherian mammals on which the G6PD and PGK loci are also carried. During the early study of X linked genes in marsupials it became apparent however that their pattern of inheritance differed from the basic pattern of X linked gene inheritance in eutherians. This suggested a different form of dose compensation to that prevailing in eutherian mammals where it is a random matter, from cell to cell, as to whether the X chromosome of maternal or paternal origin is genetically inactive (Lyon, 1961, 1966, 1972).

2. *Dose compensation*

Red kangaroos, euros and wallaroos have X chromosomes of different sizes (Fig. 15) so in female hybrids between these different species or sub-

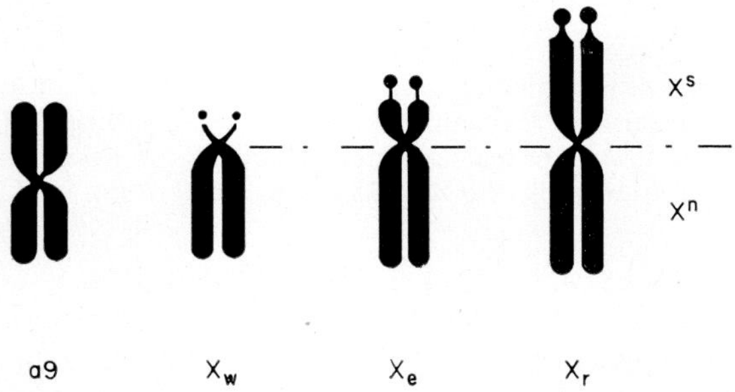

Fig. 15. The X chromosomes of the wallaroo *Macropus robustus robustus* (X_w), euro, *M. r. erubescens* (X_e) and red kangaroo, *Megaleia rufa* (X_r) compared to an autosome (a9) common to all three forms. X^n non-satellited arm of X chromosome; X^s satellited arm of X chromosome.

species it is possible to positively identify each of the two X chromosomes as to parental origin (Fig. 7). In thymidine H^3 labelled dividing leucocytes from hybrids the X of paternal origin universally, or almost universally, underwent DNA replication later than the X of maternal origin (Sharman, 1971). It was subsequently shown that the X chromosome of paternal origin was late replicating irrespective of whether it was of euro, wallaroo or red kangaroo origin (Fig. 16). Euros and wallaroos are subspecies of the same species (*Macropus robustus*) so their hybrids are fertile and female hybrids have produced F2 offspring when backcrossed to euros or wallaroos. The F2 generation may inherit from its mother either of her morphologically distinguishable X chromosomes and, in keeping with results in F1 hybrids, the X of maternal origin is the active X as determined by labelling studies

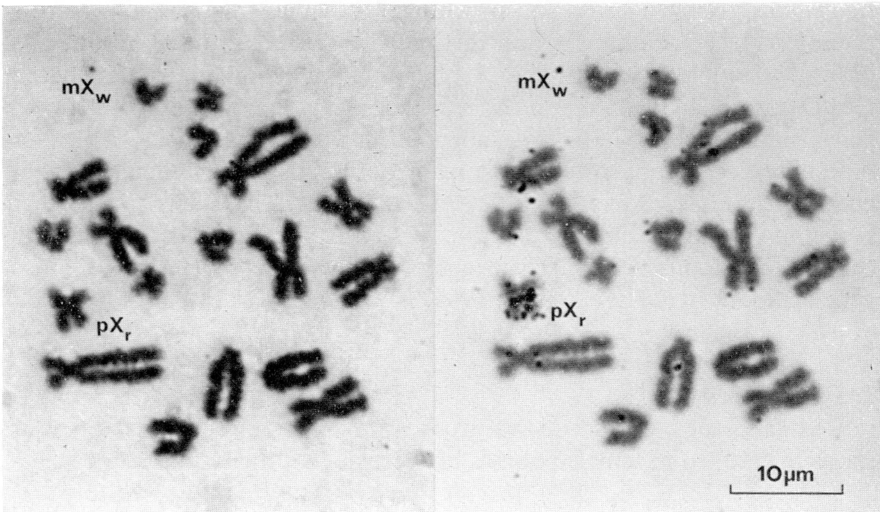

Fig. 16. The chromosomes of a female hybrid derived from crossing a female wallaroo ($2n = 16$) with a male red kangaroo ($2n = 20$) before addition of stripping film to labelled preparation (left) and three days after addition of stripping film (right). The X chromosome of male parental origin (pX_r) is heavily labelled indicating late DNA replication while the X of female parental origin (mX_w) is unlabelled.

(P. G. Johnston and G. B. Sharman, unpublished data, see Fig. 17). This implies that both X chromosomes, of maternal and paternal origin, are active in the female germ line and that the zygote inherits an active X from its mother and an inactive X from its father. Reasons for believing that the X of paternal origin becomes inactivated during meiosis in the male parent were outlined above (p. 517). The mechanism by which the X of paternal origin becomes re-activated in the female germ line is unknown.

Electrophoresis of red blood cell haemolysates on cellulose acetate revealed that red kangaroo and euro populations were characterized by a single slow moving band (G6PD-S) while wallaroos had a single fast moving band (G6PD-F). Female hybrids derived from crosses between euros and wallaroos, and between wallaroos and red kangaroos also showed a single band on electrophoresis and in every case this was that characteristic of the mother in spite of the offspring having inherited an X from each parent. By analogy with results obtained from intraspecies hybrids of eutherian mammals (Trujillo, *et al.*, 1965; Hamerton, *et al.*, 1969) the female marsupial hybrids should have had electrophoretic phenotypes with both slow and fast moving bands. Electrophoretic studies thus confirmed inactivity of the X chromosome of paternal origin in so far as the G6PD locus was concerned. Since G6PD is found in all cells it was possible to show inactivation of the

G6PD locus of paternal origin in a variety of tissues other than red blood cells. In keeping with the chromosome data backcross hybrids inherit one or other of the electrophoretic phenotypes determined by the X inherited from their hybrid female parent (Fig. 17). This data also supports the hypothesis of activity of both X chromosomes in the female germ line but is, as yet, incomplete.

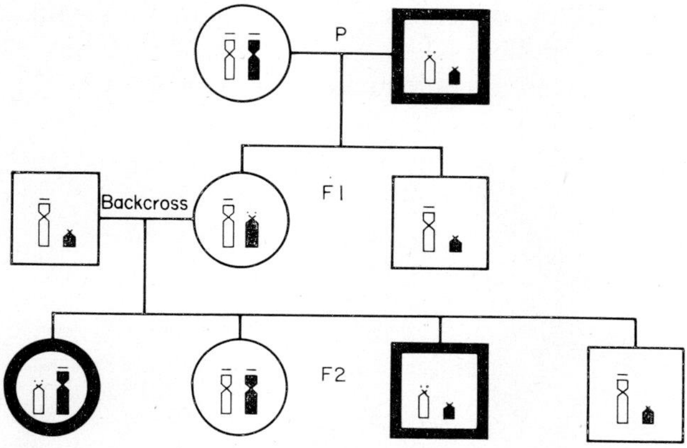

Fig. 17. Inheritance of glucose-6-phosphate dehydrogenase (G6PD) variation and active and inactive X chromosomes in euros and wallaroos. Circles—females; squares—males; thick outline—G6PD-F; thin outline—G6PD-S; largest chromosome—euro X; second largest—wallaroo X; smallest—Y; chromosomes drawn in outline—active X chromosomes; solid chromosomes—inactive X and Y. Note inheritance in F_2 generation of maternal inactive X as an active chromosome with associated electrophoretic phenotype (female and male with G6PD-F).

The inheritance of PGK variation in eastern and western grey kangaroos and whiptail wallabies also indicated that inactivation of the X chromosome of paternal origin is the rule in marsupials (Cooper, *et al.*, 1971). All results obtained from electrophoresis of blood were in accord with the hypothesis that inheritance of the gene controlling the enzyme is sex linked and the paternal X inactivated. A variety of tissues, besides blood, were studied in one 35 day old pouch young the product of a mating between a female of normal PGK phenotype and a male with a variant form of the enzyme. All tissues except the heart and embryonic upper reproductive tract (including ovaries) had only the maternal electrophoretic phenotype. The reproductive organs showed normal and variant bands on electrophoresis indicating the presence of two active X chromosomes while the heart had a strong normal band and slight variant activity. The occurrence of two active X chromosomes in the ovaries, or at least in the germ cells, is in accord with the hypothesis of universal inactivation of the X of paternal origin in somatic tissues but the occurence of slight activity in the heart suggests the possibility of reactivation of the X of paternal origin in some tissues. Investigations are continuing.

3. *Sex chromosome elimination*

In all bandicoots of the subfamily Peramelinae which have been adequately investigated the chromosome number in ovarian tissue is 12 + XX and, in the male germ line, 12 + XY (Hayman and Martin, 1965b; Hayman, *et al.*, 1969). Chromosome numbers in bone marrow (Fig. 18) liver,

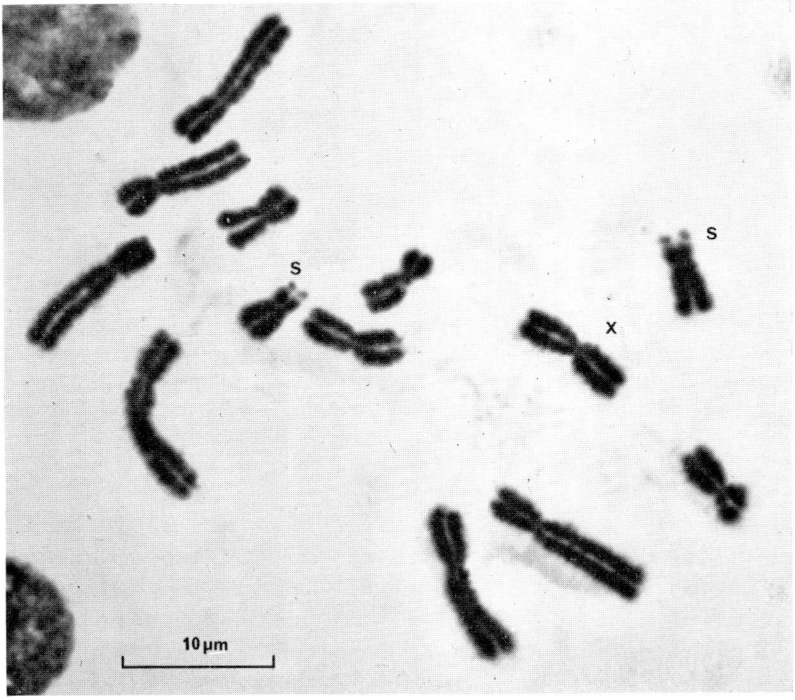

Fig. 18. Chromosomes of a female bandicoot, *Perameles nasuta* from bone marrow. X X chromosome; S satellited chromosomes.

spleen, leucocyte cultures, etc. are generally one less than the gonadal chromosome numbers, an X being eliminated from female and the Y from male somatic tissues. In *Perameles nasuta* both sex chromosomes are retained in corneal epithelium (Fig. 19) and Jackson and Ellem (1968) found both sex chromosomes to be retained in cultures derived from skin.

Walton (1971) studied sex chromosome elimination during development

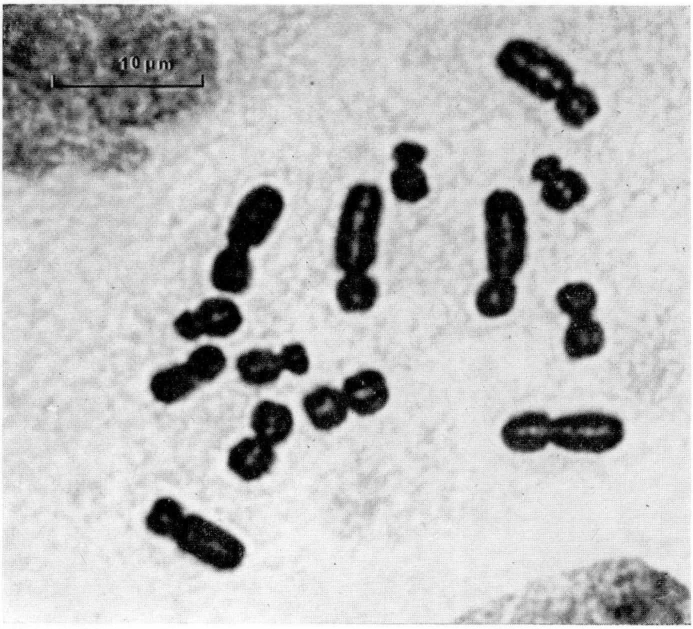

Fig. 19. Chromosomes from cornea of the same female *Perameles nasuta* as illustrated in Fig. 18. There are fourteen chromosomes including two (metacentric) X chromosomes.

in *Perameles nasuta* and *Isoodon macrourus*. Gut and skin cells of newborn animals retained both sex chromosomes and elimination was not complete in the spleen of 10 day old *P. nasuta* or the liver of 24 day old *I. macrourus*. All cells of the cornea of 55 day old *I. macrourus* retained both sex chromosomes. Sex chromosome elimination thus occurs over a considerable period and if, as Hayman and Martin (1965b) suggest, it is to be explained in terms of a dose compensation mechanism then elimination presumably occurs some time after inactivation. In eutherian mammals the sex chromatin body appears at the late blastula or early primitive streak stage and X inactivation is presumed to occur at, or near, this stage (Lyon, 1966). No sex chromatin difference between male and female nuclei has yet been found in Australian marsupials although Graham and Barr (1959), Ohno *et al.* (1959) and Perondini and Perondini (1966) reported a larger sex chromatin body in the female of some American marsupials.

The supernumerary chromosomes of the bandicoots *Echymipera kalubu* and *E. rufescens* are eliminated from somatic cells as is one of the sex chromosomes. Hayman *et al.* found that *E. kalubu* retains both sex chromosomes in corneal tissue (as does *Perameles nasuta*) and that the supernumeraries are present in adult corneal epithelium as in the testis or ovary. Since both supernumeraries and sex chromosomes exhibit negative heteropycnosis during meiosis it is argued that their cycle of compaction may be associated with the recognition of the elimination system. It is possible therefore that the supernumerary chromosomes are derived from sex chromosomal material and while such an origin would be in keeping with certain of their properties there is no evidence for such an origin in *E. kalubu* (Hayman, *et al.*, 1969). The supernumeraries of *E. kalubu* are mediocentric but those of *E. rufescens* are acrocentric (Fig. 20) and bear a remarkable resemblance,

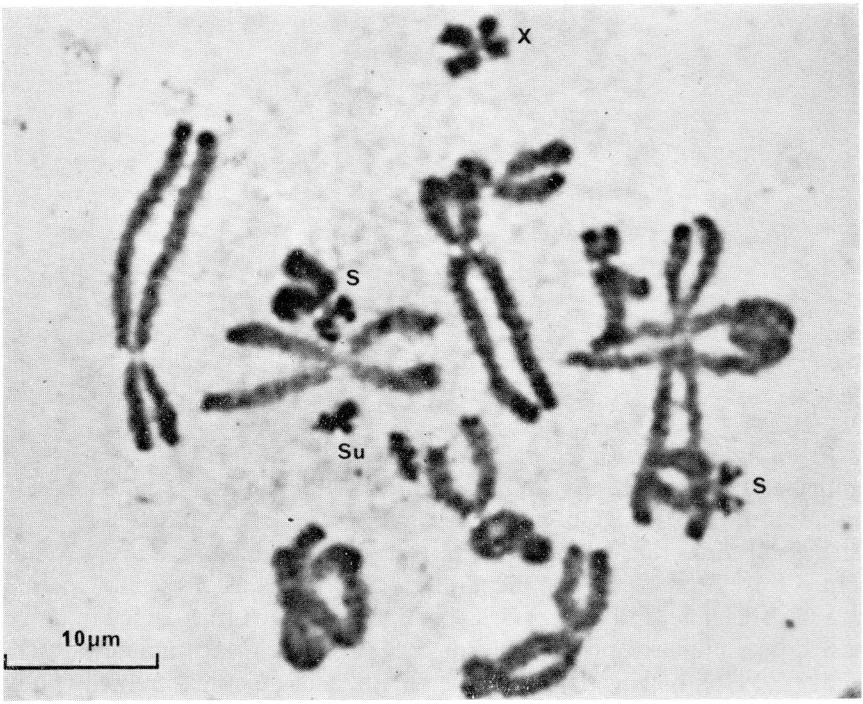

Fig. 20. Chromosomes of a female bandicoot, *Echymipera rufescens* from bone marrow. There are fifteen chromosomes but only one X chromosome (X). S—satellited chromosomes; Su—supernumerary chromosomes.

both in size and shape, to the acrocentric X chromosomes of dasyuroids. In *E. rufescens* elimination of the supernumerary elements does not exactly parallel sex chromosome elimination since bone marrow cells may lack one of the sex chromosomes yet have one or more supernumeraries still present (Fig. 20).

IV. CHROMOSOMAL EVOLUTION IN NON-EUTHERIAN MAMMALS

The oldest fossils referable to the order Monotremata are of Pleistocene age so palaeontological evidence of their affinities is not available. The monotremes have some obvious reptilian characteristics but their karyotypes are not firmly divisible into macro and microchromosomes as are the karyotypes of most modern reptiles. Bick and Jackson (1967c) regard the chromosomes of the echidna as "an extreme mammalian type strongly resembling that found in crocodiles and turtles". No definite evidence of differentiated sex chromosomes has yet been found in either of these reptilian groups. The Squamata, containing the lizards (Lacertilia) and their Cretaceous offshoot, the snakes (Ophidia), are modern reptiles which evolved coincident with the main period of mammalian evolution and radiation. Those lizards in which differentiated sex chromosomes are found show male heterogamety (Pennock, *et al.*, 1969) while the snakes apparently have female heterogamety (Beçak, *et al.*, 1964). The occurrence of an uneven chromosome number in the male echidna and platypus strongly suggests male heterogamety but the probability that this is an $X_1X_1X_2X_2$♀/X_1X_2Y♂ system, or some even more complicated system, is not excluded. We have, through the courtesy of Dr. Y. A. E. Bick, examined preparations of echidna meiosis which indicate a considerable degree of multivalent formation.

If the similar DNA values of monotremes to those of eutherians and marsupials (Bick and Jackson, 1967a) are to be interpreted as evidence of the unity of the mammalian group then it may possibly be assumed that the common mammal-like reptile ancestors of the three groups also had high DNA values compared to those of other contemporary reptiles. Since eutherians and monotremes, as well as modern reptiles and birds, have generally high chromosome numbers it is also reasonable to assume that the ancestral amnioted vertebrates were characterized by a high number of generally small chromosomes and that, amongst the living descendants of the ancestral amniote stock, only the marsupials and occasional eutherians such as the Indian muntjac (Wurster and Benirschke, 1970) have evolved very much lower chromosome numbers. Viewed in this light the monotremes might be regarded as surviving mammal-like reptiles which have evolved, separated from the therian mammal line, since earliest Jurassic times. Such an early separation of prototherian and therian mammal lines was postulated by Kermack (1965). McIntyre (1969) proposed classification of the monotremes with therapsid reptiles.

Gregory (1947) proposed that monotremes were "derived, probably within the Australasian region by relatively rapid divergence from the ancestors of some of the Australian marsupials". This conclusion is untenable on cytological grounds. The monotremes show no chromosomal affinities with any other marsupial, American or Australasian and although some monotreme features may be thought to be neotenic or degenerate marsupial features the karyotype cannot be thus interpreted.

The occurrence of sex-linked inheritance has not yet been demonstrated in monotremes and the method of dose compensation for X chromosomes (p. 519), if such occurs, is unknown. The occurrence of homologous sex-linked genes born on the X chromosomes of both marsupial and eutherian mammals however suggests basic homology of the X chromosomes in both groups of therian mammals. Similarly the Y chromosome of both marsupials and eutherians is male determining. The methods of dose compensation (p. 519) however differ between such eutherians and marsupials as have been investigated. The active X chromosome in female marsupial somatic tissues is invariably of female parental origin while in eutherians it is a random matter, from cell to cell, as to whether the X of maternal or of paternal origin is carried in the active state. Few species of either eutherians or marsupials have yet had their dose compensation mechanisms investigated but the results to date indicate a primary dichotomy of marsupials and eutherians in this respect. Cooper (1971) suggested a mechanism to account for the derivation of the random X inactivation system of eutherians from an ancestral paternal X inactivation system like that of marsupials.

Reorganization of DNA into fewer and larger chromosomes presumably occurred early in marsupial evolution and the "primitive" marsupial chromosome number is assumed to be $2n = 14$ or $2n = 22$. At first sight $2n = 14$ appears to have the greater claim to be regarded as basic to the marsupial radiation because all superfamilies (as considered here) have at least some $2n = 14$ species and these are, generally speaking, the most generalized forms. However, if this be the case, it remains to be explained how chromosome number increase to $2n = 20$ or $2n = 22$ occurred in three, or perhaps four, superfamilies (as considered here). No conclusive demonstration of chromosome number increase during marsupial evolution has been provided although there is conclusive evidence of reduction during the evolution of at least eight species (p. 506).

The basic sex chromosomes of marsupials appear to be an X smaller in size than the smallest autosome and a very small Y chromosome. Evolution of sex chromosome mechanisms has been accompanied by shifts in centromere position, by increase in sex chromosome size and by the incorporation of the nucleolar organizer into the X chromosomes of macropodid marsupials. Generally speaking X chromosome size and shape has no taxonomic significance although the occurrence of the nucleolar organizer on the X chromosomes of macropodids appears to be specific for the family with one possible exception.

Incorporation of autosomal material into the sex chromosome mechanism has occurred on a number of occasions during marsupial evolution. The multiple sex chromosome mechanisms thus formed result in lowering of the diploid chromosome number but this has occurred in three distinct superfamilies and is not indicative of close relationship of the species concerned.

To date there is some information about the chromosomes of about one hundred of a total of about two hundred and thirty species of marsupials. In some instances cytological study has given a remarkably clear insight into the processes of chromosome evolution. Study of the remaining one hundred and thirty species, in the light of information already gained and taking into account demonstrated parallelisms, would undoubtedly contribute substantially to knowledge of chromosome evolution and cytotaxonomy in the group.

Acknowledgements

Hitherto unpublished work reported in this chapter is the results of research supported by the Australian Research Grants Committee, Macquarie University Research Fund and the R/V Alpha Helix New Guinea Program, 1969 (Scripps Institution of Oceanography). I am indebted to my colleagues Dr. E. S. Robinson and P. G. Johnston for helpful discussion and to Mrs. Caroline Murtagh who drew the figures and assisted with photography. I thank Dr. D. W. Cooper for the gift of fixed material from a female echidna, Dr. Angel Spotorno O. for providing a karyotype of *Dromiciops australis* and J. H. Calaby for critically reading the manuscript.

References

Agar, W. E. (1923). *Q. Jl. microsc. Sci.* **67**, 183–202.

Beçak, W., Beçack, M. L., Nazareth, H. R. S. and Ohno, S. (1964). *Chromosoma, Berlin* **15**, 606–617.

Bensley, B. A. (1903). *Trans. Linn. Soc. Lond. (Zool).* **9**, 82–217.

Bick, Y. A. E. and Jackson, W. D. (1967a). *Nature Lond.* **215**, 192–193.

Bick, Y. A. E. and Jackson, W. D. (1967b). *Nature Lond.* **214**, 600–601.

Bick, Y. A. E. and Jackson, W. D. (1967c). *Am. Nat.* **101**, 79–86.

Biggers, J. D., Fritz, H. I., Hare, W. C. D. and McFeely, R. A. (1965). *Science N.Y.* **148**, 1602–1603.

Cestari, A. N. and Perondini, A. L. P. (1964). *In* "Mammalian Cytogenetics and Related Problems in Radiobiology" (C. Pavan, C. Chagas, O. Frost-Pessoa and L. R. Caldas, eds.) pp. 157–162. Pergamon Press, Oxford.

Clemens, W. A. (1966). *Univ. Calif. Publs. geol. Sci.* **62**, i–vi, 1–122.
Clemens, W. A. (1968). *Evolution Lancaster P.A.* **22**, 1–18.
Clemens, W. A. (1970). *Annu. Rev. ecol. Systematics* **1**, 357–390.
Clemens, W. A. (1971). *In* "Early Mammals" (D. M. Kermack and K. M. Kermack, eds.) pp. 165–180. *Zool. J. Linn. Soc.* **50**, suppl. 1.
Cooper, D. W. (1971). *Nature Lond.* **230**, 292–294.
Cooper, D. W., VandeBerg, J. L., Sharman, G. B. and Poole, W. E. (1971). *Nature New Biology*, **230**, 155–157.
Darlington, C. D. (1937). "Recent Advances in Cytology". University Press, Cambridge.
Dreyfus, A. and Campos, J. E. (1939). *Biol. Fac. Filos Cienc. S. Paulo* **17**, 3–28.
Drummond, F. H. (1938). *Cytologia* **8**, 343–352.
Ford, C. E. (1970). *J. biosoc. Sci., Suppl.* **2**, 7–30.
Ford, C. E. and Hamerton, J. L. (1970). *Symp. zool. Soc. Lond.* No. **26**, 223–236.
George, W. and Weir, B. J. (1972). *Nature New Biology* **236**, 205–206.
Graham, M. A. and Barr, M. L. (1959). *Archs Anat. microsc. Morph. exp.* **48**, 111–121.
Gregory, W. K. (1947). *Bull. Am. Mus. nat. Hist.* **88**, 5–52.
Graves, J. A. Marshall (1967). *Expl. Cell Res.* **46**, 37–57.
Gunson, M. M., Sharman, G. B. and Thomson, J. A. (1968). *Aust. J. Sci.* **31**, 40–41.
Hamerton, J. L., Gianelli, F., Collins, F., Hallett, J., Fryer, A., McGuire, V. M. and Short, R. V. (1970). *Nature Lond.* **222**, 1277–1278.
Hayman, D. L., Kirsch, J. A. W., Martin, P. G. and Waller, P. (1971). *Nature Lond.* **231**, 194–195.
Hayman, D. L. and Martin, P. G. (1965a). *Aust. J. biol. Sci.* **18**, 1081–1082.
Hayman, D. L. and Martin, P. G. (1965b). *Genetics N.Y.* **52**, 1201–1206.
Hayman, D. L. and Martin, P. G. (1965c). *Cytogenetics* **4**, 209–218.
Hayman, D. L. and Martin, P. G. (1969). *In* "Comparative Mammalian Cytogenetics" (K. Benirschke, ed.) pp. 191–217. Springer-Verlag, New York.
Hayman, D. L., Martin, P. G. and Waller, P. (1969). *Chromosoma, Berlin* **27**, 371–380.
Jackson, L. G. and Ellem, K. A. O. (1968). *Cytogenetics* **7**, 183–189.
Kermack, K. A. (1965). *Sci. J.* Sept., 66–72.
Kirsch, J. A. W. (1968a). *Aust. J. Sci.* **31**, 45–46.
Kirsch, J. A. W. (1968b). *Nature Lond.* **217**, 418–420.
Kirsch, J. A. W. (1968c). Ph.D. Thesis, University of Western Australia.
Kirsch, J. A. W. and Poole, W. E. (1972). *Aust. J. Zool.* **20**, 315–339.
Koller, P. C. (1936). *J. Genet.* **32**, 452–472.
Legator, M., Jacobsen, C., Perry, M. and Dolimpio, D. (1966). *Life Sci.* **5**, 397–402.
Lyon, M. F. (1961). *Nature Lond.* **190**, 372–373.
Lyon, M. F. (1966). *In* "Advances in Teratology" **1** (D. H. M. Woolam, ed.) pp. 25–54. Logos Press, London.
Lyon, M. F. (1972). *Biol. Rev.* **47**, 1–35.
McIntosh, A. J. and Sharman, G. B. (1953). *J. Morph.* **93**, 509–532.

MacIntyre, G. T. (1967). *Evolution* **21**, 834–841.
Martin, P. G. and Hayman, D. L. (1965). *Evolution* **19**, 157–161.
Martin, P. G. and Hayman, D. L. (1966). *Chromosoma, Berlin* **19**, 159–175.
Martin, P. G. and Hayman, D. L. (1967). *Chromosoma, Berlin* **20**, 290–310.
Matthey, R. (1949). "Les Chromosomes des Vertebres". F. Rouge, Lausanne.
Matthey, R. (1958). *Arch. Julius Klaus-Stift. Vererb-Forsch.* **33**, 253–297.
Matthey, R. and van Brink, J. M. (1957) *Evolution* **11**, 163–165.
Meylan, A. (1970). *Symp. zool. Soc. Lond.* No. **26**, 211–222.
Ohno, S. (1969). *Annu. Rev. Genet.* **3**, 495–524.
Ohno, S., Kaplan, W. D. and Kinosita, R. (1959). *Expl. Cell Res.* **19**, 417–419.
Pennock, L. A., Tinkle, D. W. and Shaw, M. W. (1969). *Cytogenetics* **8**, 9–19.
Perondini, A. L. P. and Perondini, D. R. (1965). *Revta. bras. Biol.* **25**, 377–386.
Perondini, A. L. P. and Perondini, D. R. (1966). *Cytogenetics* **5**, 28–33.
Reig, O. A. (1955). *Investnes zool. chil.* **2**, 121–130.
Reig, O. A. (1968). *Experientia* **24**, 185–187.
Reig, O. A. and Bianchi, N. O. (1969). *Experientia* **25**, 1210–1211.
Reig, O. A., Fernandez, R. and Spotorno, A. (1972). *Sonderdr. Z. Säugetier.* **37**, 37–42.
Reig, O. A. and Lobig, I. (1970). *Acta cient. Venez.* **21**, Suppl. 1, 35.
Reig, O. A. and Sonnenshein, C. (1970). *Experientia* **26**, 199–201.
Rendel, J. M. and Kellerman, G. M. (1955). *Nature Lond.* **176**, 829.
Richardson, B. J., Czuppon, A. and Sharman, G. B. (1971). *Nature New Biology* **230**, 154–155.
Ride, W. D. L. (1964). *J. Proc. Roy. Soc. West Aust.* **47**, 97–131.
Saez, F. A. (1931). *Am. Nat.* **65**, 287–288.
Saez, F. A. (1938). *Revta. Soc. argent. Biol.* **14**, 156–161.
Sharman, G. B. (1961). *Aust. J. Zool.* **9**, 38–60.
Sharman, G. B. (1970). *Science N.Y.* **167**, 1221–1228.
Sharman, G. B. (1971). *Nature Lond.* **230**, 231–232.
Sharman, G. B. (1973). *Aust. Mammalogy* **2**, in press.
Sharman, G. B. and Barber, H. N. (1952). *Heredity* 6, 345–355.
Sharman, G. B., McIntosh, A. J. and Barber, H. N. (1950). *Nature Lond.* **166**, 996.
Sharman, G. B., Robinson, E. S., Walton, S. M. and Berger, P. J. (1970). *J. Reprod. Fert.* **21**, 57–68.
Shaver, E. L. (1962). *Can. J. Genet. Cytol.* **4**, 62–68.
Simpson, G. G. (1929). *Mem. Peabody Mus. Yale.* **3**, i–xv, 1–171.
Simpson, G. G. (1945). *Bull. Am. Mus. nat. Hist.* **85**, v–xvi, 1–350.
Simpson, G. G. (1970). *Bull. Mus. comp. Zool. Harvard* **139**, 1–80.
Tate, G. H. H. (1947). *Bull. Am. Mus. nat. Hist.* **88**, 101–155.
Thomas, O. (1887). *Proc. zool. Soc. Lond.* 1886, 544–547.
Todd, N. B. (1970). *J. theor. Biol.* **26**, 455–480.
Trujillo, J. M., Walden, B., O'Neill, P. and Anstall, H. B. *Science N.Y.* **148**, 1603–1605.
Tyndale-Biscoe, C. H. (1965). *Aust. J. Zool.* **13**, 255–267.
van Brink, J. M. (1959). *Chromosoma, Berlin* **10**, 1–72.

Walton, S. M. (1971). *Cytogenetics* **10,** 115–120.
White, M. J. D. (1957). *In* "Survey of Biological Progress" pp. 109–147. Academic Press, New York and London.
Woodburne, M. O. (1967). *Commonwealth of Australia, Bureau of Mineral Resources Bull.* **87**, vii–ix. 1–187.
Wood Jones, F. (1924). "The Mammals of South Australia" Part II. Government Printer, Adelaide.
Wurster, D. H. and Benirschke, K. (1970). *Science N.Y.* **168,** 1364–1366.
Yunis, E., Ramirez, E., Cayon, J. and Hernandez, J. (1972). *Aust. J. Zool.* 265–269.

15

The Chromosome Formulae of Eutherian Mammals

ROBERT MATTHEY
Institute of Animal Biology and Zoology, University of Lausanne, Switzerland.

I. INTRODUCTION

Since 1930 I have been compiling a file on the chromosome formulae of mammals. For quite a number of years, say from 1930 to 1952, this file was further extended by contributions provided almost entirely by S. Makino, myself and our pupils. At that time, the material after fixation in Champy's Flemming-Heitz's fluid was sectioned, then stained by iron hematoxylin, and occasionally by the Feulgen reaction.

In 1949 I published "Les Chromosomes des Vertébrés", which was the sum of our knowledge then acquired. Since 1952, Makino and Nishimura, resorting to the "squash" technique, whose introduction by La Cour had proved to be most promising, succeeded in improving it considerably by hypotonic treatment. Further development of this method, by applying Colcemid to tissue cultures, made the analysis of the karyotype of mammals, which had been so difficult until then, a relatively easy task. Moreover, discoveries such as the exact chromosome number in Man (Tjio and Levan, 1956) and also its changes causally correlated with pathological conditions, mongolism or intersexuality, conferred a practical interest to the study of chromosomes, so that it became a routine procedure in medicinal analyses.

Meanwhile, the accumulation of exact data concerning the karyotype of mammals—the number of investigated species or subspecies having increased in the course of seventeen years from 104 to 1165—led to the development of modern mammalian Cytogenetics and thus new items of information, among which the diploid number 2N, and the "nombre fondamental" (*N.F.* of Matthey, 1945) play an essential role were made available to taxonomists and theoreticians of evolution.

Thus, it seemed to me that any zoologist or cytologist would be happy to find in a single list all the chromosome formulae of the eutherians known up to 1970 and that, to this end, my file built up over forty years should become available to them. I have avoided assembling a full Bibliography on this subject since a whole book would be required, but I have given *one* reference for each species; this is, as a rule, the earliest to supply exact determination of the 2N number of the species under study. It is not always easy to establish this priority when investigations are carried out simultaneously by several authors or teams of workers as was the case for Primates.

Some clue to the planning of this list is needed:

(1) In keeping with modern trends in Taxonomy, I have separated the *Cricetidae* from the *Muridae*, raising the sub-family to the rank of family. Other sub-families, e.g. the *Dendromurinae* and *Gerbillinae*, whose classic position was maintained, would similarly have deserved being accorded a higher status.

(2) When there are more than two writers for any paper, only the name of the first author is mentioned, followed by "*et al.*"

(3) Among the generally adopted abbreviations, two require some explanation: (a) *Mlekop.* This is a Russian publication whose transcription from cyrillic into latin is *Mlekopitayushchie*. In this publication important contributions by cytologists and mammalogists from the USSR to the "2nd All-Union Mammalogy Conference" are brought together. It was prepared under the direction of N. N. Vorontsov and printed at Novo-Sibirsk in 1969. (b) *MCNL.* The valuable *Mammalian Chromosomes Newsletter* edited by Dr. T. H. Hsu should not have been used, in principle, as a bibliographic source. But, owing to the great importance that this Journal has attained, it cannot be ignored; indeed, it often contains preliminary reports antedating by a year or two the publication of the relative definitive papers. For this reason, its bearing on questions of priority, although immaterial in my opinion, must be acknowledged.

II. TRANSFORMATIONS OF THE KARYOTYPE

When considering a given taxon, a genus for example, the observer is faced by three entirely different conditions:

(1) At the optical level all the species of this genus have identical or almost identical karyotypes; this uniformity may involve a whole family: this is found in *Felidae* where eighteen species have 38 chromosomes and five species 36.

(2) Some morphologically closely related species may display altogether diverging 2*N* numbers; *Microtus socialis* has 62 chromosomes, while

Microtus oregoni has 18. Three possibilities are then open: (a) the *N.F.*'s corresponding to these different 2N numbers are the same for all species, or at least much closer among themselves than are the 2N numbers; for example, in *Gerbillinae* the *N.F.* varies from 68 to 80, and the 2N values range from 23 to 72; even more schematic is the case of the majority of the African *Leggada* in which the *N.F.* is consistently 36, whereas their diploid numbers, varying from 18 to 34, show all intermediate combinations; this type of chromosome variation is the commonest; (b) the 2N number is constant and the *N.F.* differs from species to species; a classical example is found in the genus *Peromyscus* (2N = 48) the different species or subspecies of which have *N.F.*'s ranging from 56 to 96; in *Mus* (*Leggada*) *oubanguii* the *N.F.* may be 30, 31, 32, 33 and 34 while the 2N is 28 and (c) even less often, no relationship seems to exist between the diploid number and the *N.F.*; the most striking example is provided by two cervids, *Muntiacus reevesi* and *Muntiacus muntjack*, the first having 46 chromosomes, the second 6 (♀) or 7 (♂), or a pair of metacentrics, one of acrocentrics and a third pair of acrocentric-like chromosomes, but whose short arms correspond to the two X-chromosomes (♀) translocated on two acrocentrics, thereby forming long arms. In the case of the male, this pair is heteromorphic, and a short Y is also present. Prior to this discovery, the lowest 2N numbers lay between 16 and 20 (*Ellobius lutescens*, *Microtus oregoni*, and several Chiroptera of the *Phyllostomatidae* family); these appeared in taxa displaying diploid numbers between 40 and 56.

(3) One final, quite unique example may be mentioned, namely, that of the genus *Reithrodontomys*, within which two species possess a varying number of supernumerary minute chromosomes resulting in 2N numbers of 42, 43, 44, 45 and 46.

Cytological interpretations of the foregoing different kinds of transformation are as follows:

(a) (variable 2N, constant or slightly variable *N.F.*) Here, we have cases of Robertsonian processes i.e. fusions/fissions. Fusion (Figs 1 and 2) consists in the proximal union of two nonhomologous chromosomes resulting in a metacentric element. Its mechanism is based either upon reciprocal translocation (Fig. 2) involving the elimination of two fragments, one centric and the other acentric, or, possibly more frequently, on the fusion of both centromeres of the two acrocentrics into a single one (Fig. 1).

Fission, which is harder to envision cytologically, since it implies either the formation, starting from a metacentric, of two telocentric chromosomes each being furnished with a hemi-centromere, or the occurrence of centromere "donors" in the form of supernumerary chromosomes, seems rarer than fusion.

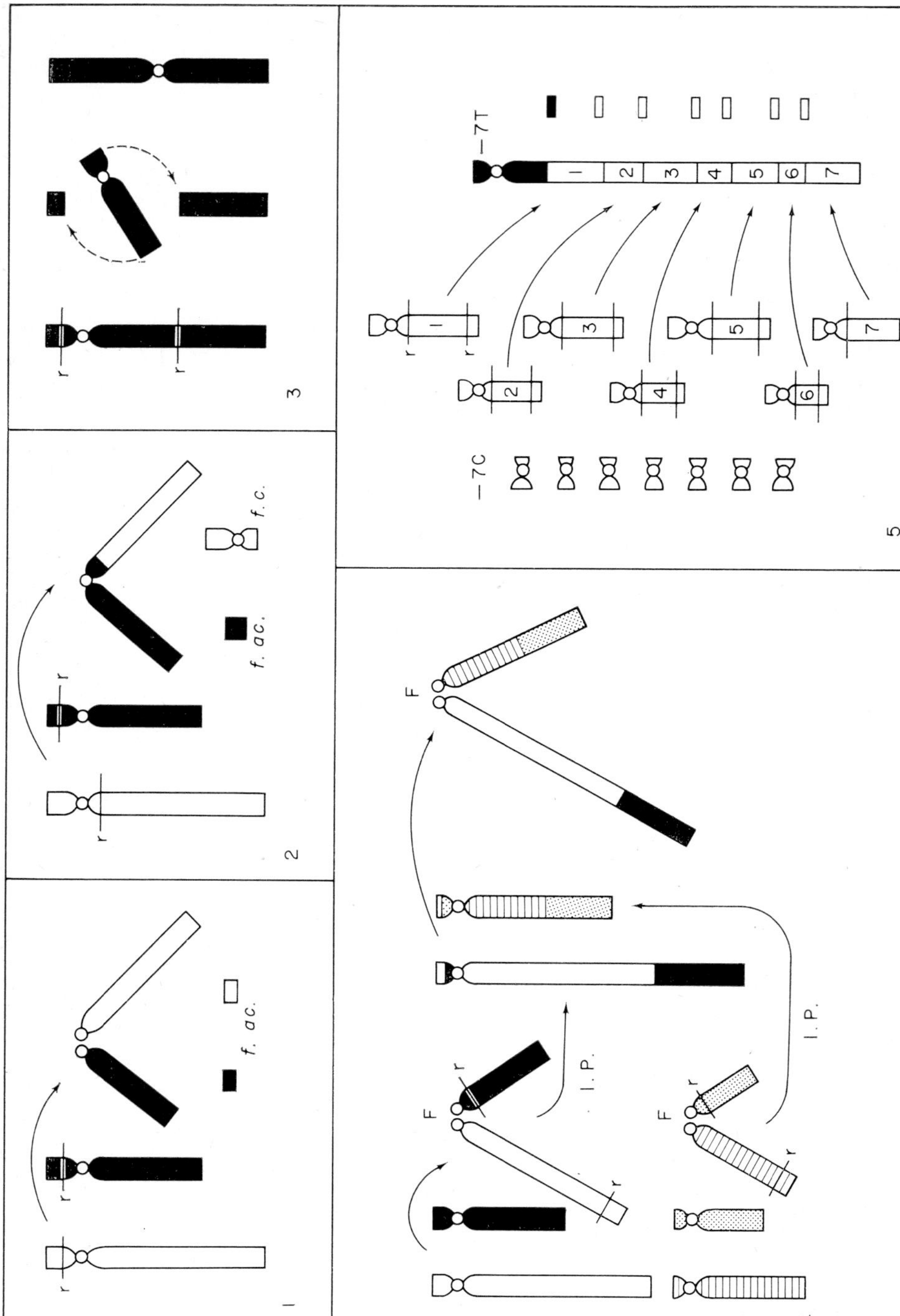

Figs. 1–5. Chromosomal mutations. Fig. 1. "Robertsonian fusion". Fig. 2. Fusion through reciprocal translocation. Fig. 3. Pericentric inversion. Fig. 4. Alternation of fusions and pericentric inversions as theoretical explanation of the lowest diploid numbers. Fig. 5. A possible explanation for the case of *Muntiacus muntjack*.

(b) (uniform $2n$, variable *N.F.*) Here, pericentric inversion takes place (Fig. 3). One acrocentric chromosome undergoes two breaks, one in the short arm, the other in the long arm, the latter being divided into two subequal segments. The fragment between these two breaks, which contains the centromere, is then rotated by 180° and the centromere is shifted to a submedian position: the acrocentric chromosome has thus become metacentric and the *N.F.* is increased by one unit.

(c) The establishment of a very low diploid number (16 to 20) may be explained by many theoretical possibilities. In the case of *Ellobius lutescens* ($2n = 17$), belonging to a genus which contains three other species with 54, 52 and 36 chromosomes, I suggested an evolutionary process involving four nonhomologous acrocentric chromosomes: these may give rise to two metacentrics by centric fusions. Each of these metacentrics undergoes a pericentric inversion, and once they become acrocentric, fuse again (Fig. 4). In the case of *Muntiacus muntjack* ($2n = 6$) where one chromosome corresponds on the average to seven chromosomes of the congeneric species *Muntiacus reevesi*, the only possible mechanism I can visualize is repeated translocations harnessing the larger part of from five to seven chromosomes to a single centromere (Fig. 5). The final translocation concerns an element whose telomere is preserved, while the other chromosomes are inserted in the chain after losing both centromeres and telomeres. This hypothesis agrees with the fact that, according to Wurster and Benirschke (1970), in *M. muntjack* the nuclear DNA content amounts to 70% of that characteristic of *M. reevesi*. This 30% difference may represent the amount of non-genetically active DNA on either side of the centromere and in the telomeres making up the lost fragments.

All these interpretations are complete conjectures and will remain so until the nature of the centromere, the determinism of chromosomal mutations and the mechanism of mitosis are better understood. It is obvious that the *sense*—in the dual meaning of this term, i.e. *significance* and *direction*—of the changes undergone by the karyotype is still a riddle. Within a single population these changes can ensure the reproductive isolation of individuals in which mutations have arisen affecting the sex chromosomes, for example the Leggadas *PR* and *TR*. On the other hand, the idea that in a polymorphic Robertsonian series disjunction of a trivalent formed by a metacentric and two acrocentrics is bound to occur in a normal way and centric fusions fail to promote reproductive isolation among members of a series having different diploid numbers, can no longer be considered as generally valid after Tettenborn and Gropp's (1970) observations on hybrids between *Mus musculus* ($2N = 40$, all acrocentrics) and *M. poschiavinus* ($2N = 26$, twelve acrocentrics and fourteen metacentrics). In these F_1 hybrids, formation of aneuploid gametes, resulting in much reduced fertility, is brought about by meiotic nondisjunctions.

Another problem concerns the "timing" of chromosome mutations. Let us consider two related species with the same *N.F.* though with different diploid numbers, e.g. the above mentioned *M. musculus* and *M. poschiavinus*, or else *M. setulosus* (2N = 36, all acrocentrics), and *M. minutoides* (2N = 18, all metacentrics). Classic concepts regarding the delicate adjustment of the genic balance make it mandatory that consecutive fusions should have occurred sufficiently spaced out over time to allow balance to be re-established.

Each of these mutations being improbable, it follows that the time needed to pass from 36 to 18 or from 40 to 26 could far exceed the time allotted to the diversification of the *Mus* genus. Anyway, as has often been stressed by White, chromosome mutations involved in the evolution of a groups are never casual, but, as far as this group is concerned, they depend upon the same type. This argues in favour of some characteristics being shared by all the chromosomes of a single karyotype and rules out the notion that each mutation is a phenomenon independent of past or future mutations.

I am increasinly inclined to admit, contrary to the classical point of view, that moments exist when, within a given group, fusions and other karyotype changes may simultaneously affect a high number of chromosomes.

However this may be, chromosome mutations seem to be associated with speciation processes which may be referred to as sympatric or, according to White's term (1969), stasipatric.

III. LIST AND TABLES

A. *General list*

The general list contains all the even and odd diploid numbers so far known. Over the years I have published the data relative to 104 (1953), 240 (1958), 267 (1966) and 769 (1969) eutherians. In the present list 1165 chromosome formulae, corresponding to 1143 species and subspecies, are set out. The difference between these figures is accounted for by polymorphism and also by those cases which have a 2N number which differs between the sexes. Thus, an odd chromosome number characterizes the male sex, when one of the following exceptional types are encountered:

$$\text{♂}: X_1X_2/Y \quad — \quad \text{♀}: X_1X_2/X_1X_2$$
$$\text{♂}: X/Y_1Y_2 \quad — \quad \text{♀}: X/X$$
$$\text{♂}: X/O \quad — \quad \text{♀}: X/X$$ (existence of the last is not proved).

B. *General Table* (Fig. 6.)

In this table, drawn from the general list, only even numbers are given.

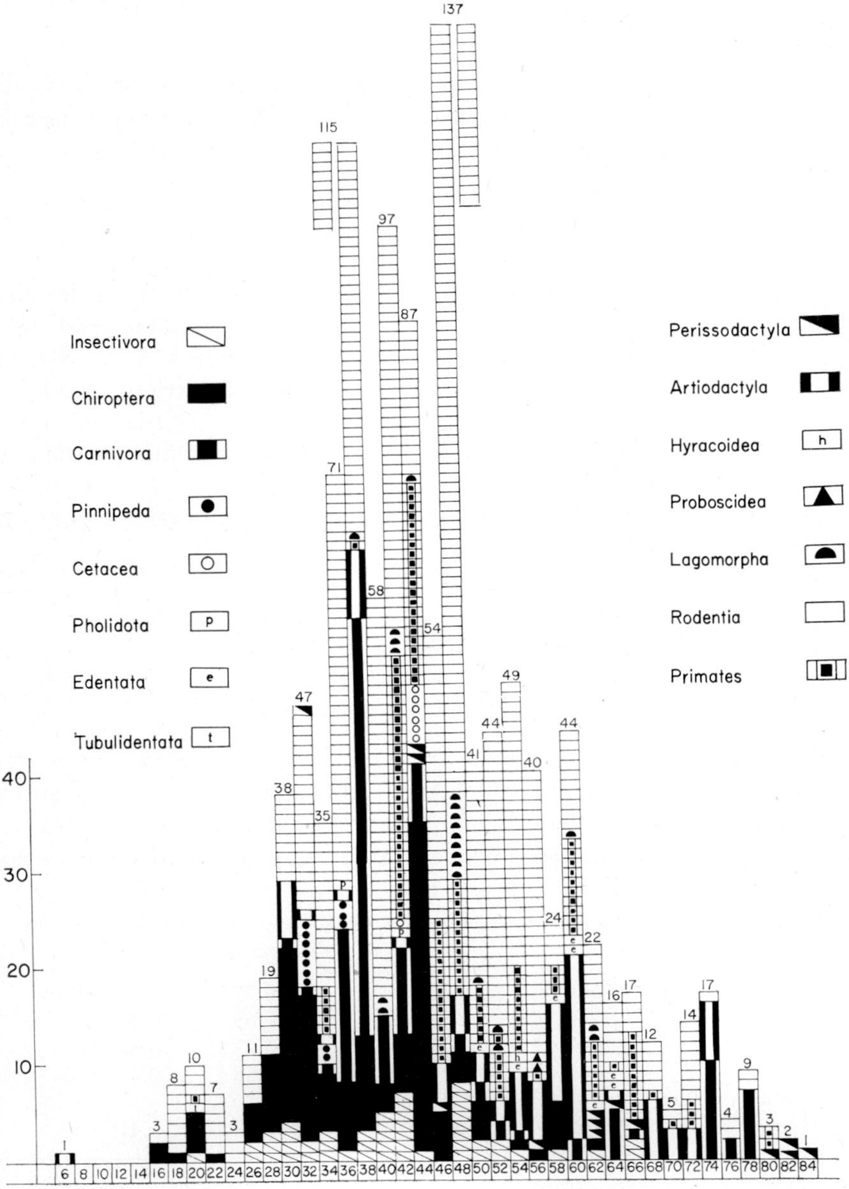

Fig. 6. The diploid numbers by 1063 species and subspecies of eutherian mammals.

Accordingly, mammals in which both sexes possess an odd 2N number (e.g. *Ellobius lutescens*, 2N = 17; *Choleopus hoffmanni*, 2N = 49; Okapia *johnstoni*, 2N = 45) and those in which the only sex examined has an odd chromosome number (e.g. *Alcephalus boselaphus* ♀, 2N = 39; *Ctenomys tuconax* ♂, 2N = 61; *Taterillus gracilis* ♂, 2N = 23) are excluded.

An odd number is also found among heterozygotes or double heterozygotes of the polymorphic Robertsonian series, for instance 34, 33, 32 and 31 (*Mus musculoides*). Therefore, in the general Table only the extreme even numbers, e.g. 34 and 32 in the present case, have been reported.

C. *Special Tables* (Figs 7–22)

These are devoted to groups in which a large number of species have been analysed, which endows them with particular statistical interest. Here also only even values (including the extreme even limits of the Robertsonian series) are reported. These Tables supply the data relative to Orders (Insectivora, Chiroptera, Carnivora, Primates, Perissodactyla and Artiodactyla,) some Families (*Sciuridae*, *Cricetidae*) and some sub-families (*Murinae*, *Gerbillinae*, *Microtinae*).

In five Tables I have gathered the data concerning eutherian groups, the survey of which is far less advanced.

IV. DISCUSSION

A. *Statistical considerations*

From 1953 onwards, increasingly richer samples of species were submitted for statistical analysis (method of confidence limits). The conclusion was drawn that over half of the *Eutheria* must consist of species characterized by diploid numbers ranging from 40 to 56, as shown by the following Table:

Year	n	x	limits at 5%	limits at 1%
1953	104	72	0.59–0.77	0.56–0.80
1958	240	150	0.55–0.68	0.53–0.70
1960	267	163	0.54–0.67	0.52–0.68
1966 (published in 1969)	769	401	0.48–0.55	0.47–0.55
1970	1165	606	0.46–0.58	0.44–0.60

n = number of known formulae
x = number of the sc-called modal formulae, lying between 40 and 56

Hence, there is only a slight probability, from 5 to 1%, that at least half of the eutherian species are not endowed with 2N numbers ranging from 40 to 56.

The agreement between these consecutive results is most gratifying, since the richer the *n* sample, the more significant the results obtained. Now, let us examine a question which has a direct bearing on the problem of evolution by fusion or fission, already mentioned in Section I, and to which we will come back again in the next Section. Considering the modal values 40–56 as primitive, what are the relative proportions of the categories with less than 40 or more than 56 chromosomes ? For the samples of 1966 (1969) and 1970 we have the following values, expressed as percentages:

	Below 40	*40 to 56*	*Above 56*
1966 (1969)	30%	52.4%	17.6%
1970	31.5%	52.1%	16.4%

According to our hypothesis, these results lend support to the contention that fusions were twice as frequent as fissions. When we turn to more homogeneous groups than the eutherian complex, more particularly analysing taxa with a large number of already investigated species, examination of Fig. 7 shows that the ratio 2/1 in favour of a predominance of fusion mechanisms is only an average value, since some of our taxa exhibit a greater number of formulae above 56 than below 40.

In this respect, the *Rodentia* are very interesting: among *Sciuridae* the most extreme case of chromosomal evolution is found, tending towards "frozen" formulae (see Section IV B where this concept will be dealt with), because 63.8% of species and subspecies studied have less than 40 chromosomes and 1.1% more than 56. The other families and sub-families have from 62.6 to 72.9% of species included in the range of modal values (40–56). The *Heteromyidae* display more than 56 chromosomes in 19.2% of cases, numbers lower than 40 appearing only in 15.3% of them. Among *Dipodidae* the ratio 18.1/9% is even more strongly in favour of karyotypes with more than 56 chromosomes, and the same is true for *Gerbillinae* (16.2/11.6%).

However, it should be noted that the study of these groups in which fission seems to prevail upon fusion concerns a relatively small body of species (26, 22, 43) and the possibility cannot be ruled out that if richer samples were available, these deviations might decrease or even be reversed In fact, when the number of analysed species is larger, predominance of fusions over fissions becomes apparent, although two notable exceptions are found, namely the artiodactyles and the primates. Let us go back to the general case, concerning the 94 *Sciuridae*, 87 *Cricetidae*, 78 *Microtinae* and 144 *Murinae*, where the ratios $<40/>56$ are: 28.7/3.4%; 25.6/7.6% and 31.9/5.5%.

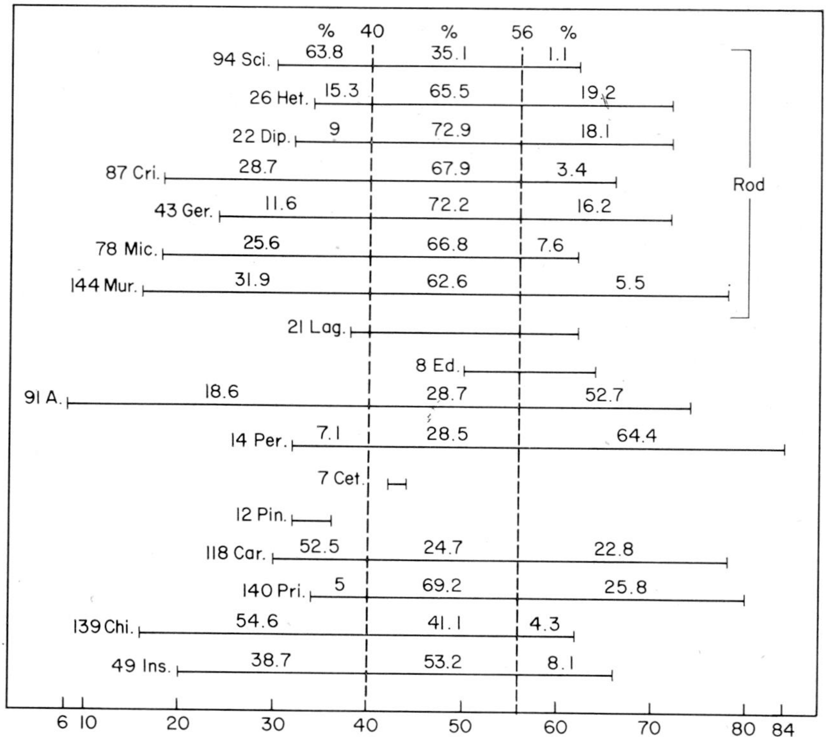

Fig. 7. Distribution of modal (40–56), infra-modal (6–38) and supra-modal (58–84) diploid numbers in the Orders and some Families of eutherian mammals.

As to the 21 *Lagomorpha*, 8 *Edentata*, 7 *Cetacea* and 12 *Pinnipeda*, the available assembly is too reduced for definite conclusions to be drawn from inspection of the figures which are slightly in favour of fission in the two first Orders and of fusion in the last one, while the *Cetacea* fall in the modal class range.

In *Perissodactyla* the ratios observed are not very significant. This Order consists of only 17 species, 14 of which are included in the sample. It is clear, however, that the kinship of *Equidae* with *Rhinocerontidae* and *Tapiridae* is one of the most remote.

The 91 *Artiodactyla* attain only the 28.7% of the formulae lying between 40 and 56; 18.6% of the species having less than 40 chromosomes and 52.7% more than 56. This augmentation obviously points to an evolution requiring repeated fissions as its major mechanism.

Like the *Sciuridae*, the 118 *Carnivora* show the peculiarity of poorly represented modal classes, namely, 24.7% while 52.5% of species possess less than 40 chromosomes and 22.8% more than 56. Among the 49 *Insectivora* a little more than half of them fall into the modal classes. 4.3% display more than 56 chromosomes and 38.7% less than 40.

We are now left with the *Primates*. This Order, of which 140 species and sub-species have been analysed, exhibits a very peculiar pattern of 2N distribution, in that 69.2% of forms belong to the modal classes. Only in 5% of them is the chromosome number below 40, while in 25.8% it is above 56. Here, since a considerable body of investigations has been carried out on all the families, it is evident that fissions have largely prevailed upon fusions.

To sum up, the best represented Orders, Families and Sub-families in our sample, lend support to the assumption of a direction of evolution towards the narrowing down of the 2N number. An exception is represented by *Artiodactyla* and *Primates* which, instead of entering the Robertsonian blind alley leading to "frozen" chromosome formulae, work a change by resorting to the fissioning process.

Figures 8–22 call for few comments. Some peaks, e.g. 44 for *Chiroptera* (Fig. 9), 38 for *Carnivora* (Fig. 13), 42 for *Primates* (Fig. 11) and 48 for *Cricetidae* (Fig. 13), may be explained by extensive investigations, within the same Family, upon genera with a large number of species and subspecies, the taxonomic value of the latter being often questionable. This is particularly true for the genera *Peromyscus*, *Myotis* and *Macaca*.

Glancing at Fig. 7 once more, it can be seen that the deviation between minimum and maximum 2N numbers—not considering the aberrant and unique case of *Muntiacus muntjack*—ranges from 16 to 84. The conclusion I had already drawn in 1953 seems to be confirmed by fresh acquisitions: the primitive eutherians, when monophyletic arise from a stock displaying 48 chromosomes, when polyphyletic they are derived from a pool in which the chromosome formulae ranged from 40 to 56. Evolution towards karyotypes with a chromosome number below 40 or above 56 is all the more unlikely, the more marked the deviation is from the modal values. Finally, it seems doubtful that the diploid number is a character subject to natural selection by environmental factors.

Fusions and fissions have played their part, and the relative importance of each of them will be taken again into consideration. I have not spoken about the possible part played by deletions or duplications in the chromosome evolution; owing to the fact that the DNA-content is almost the same in all species of mammals, this part can be neglected.

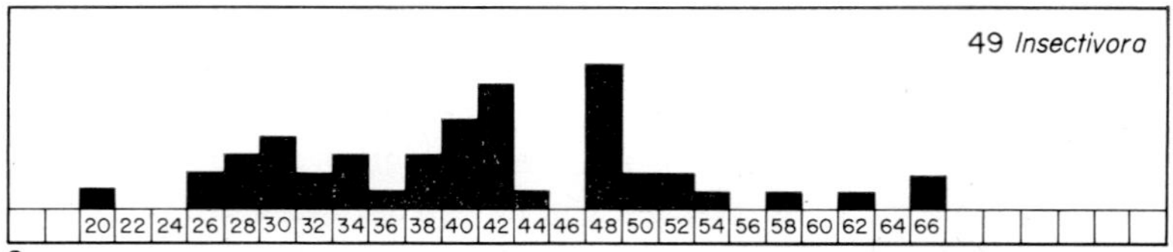

8

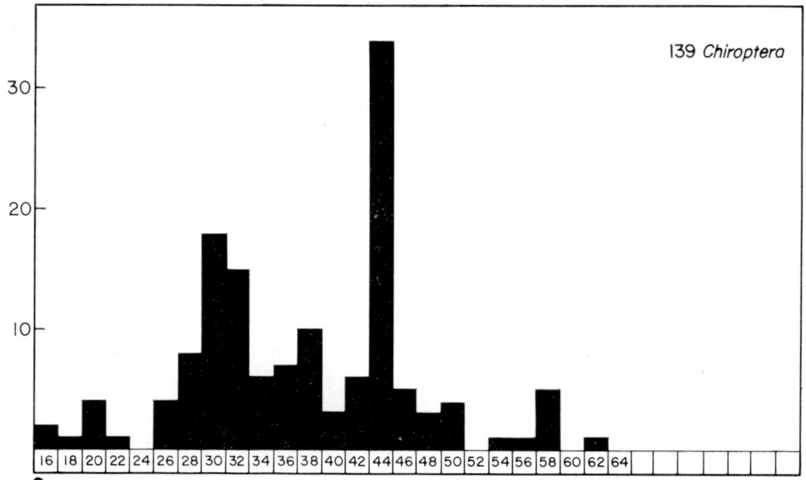

9

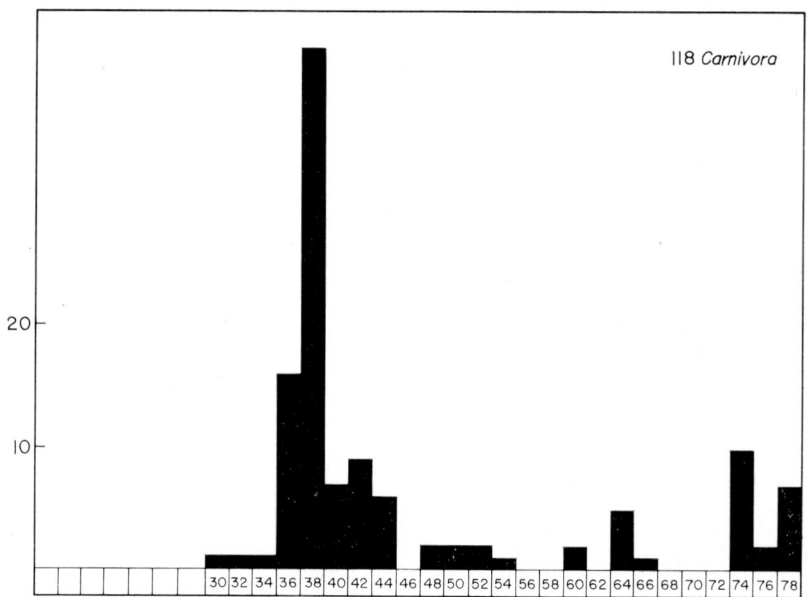

10

Figs. 8–10.

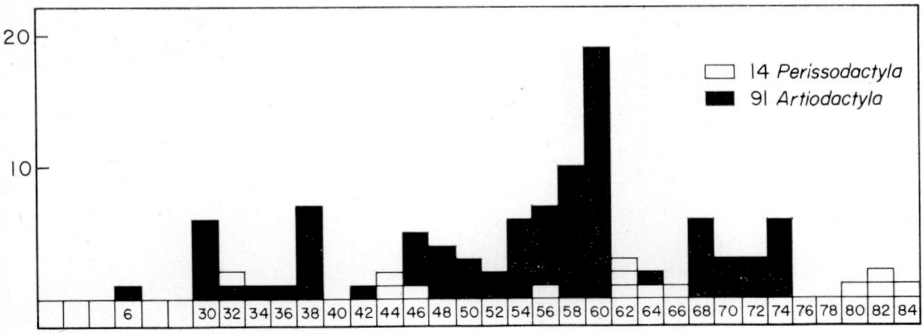

11

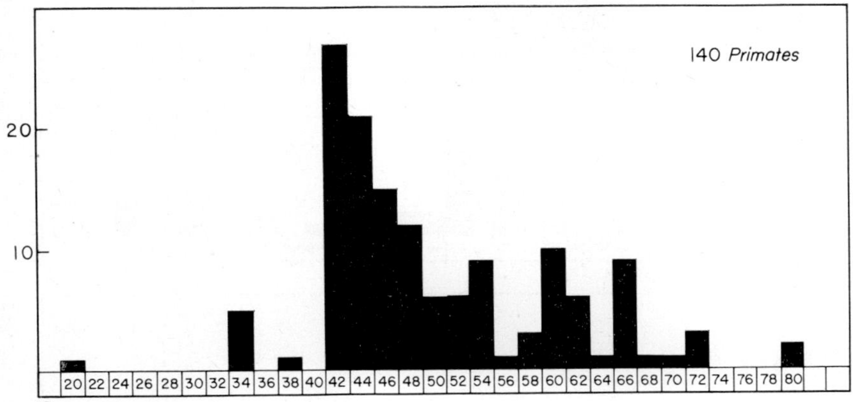

12

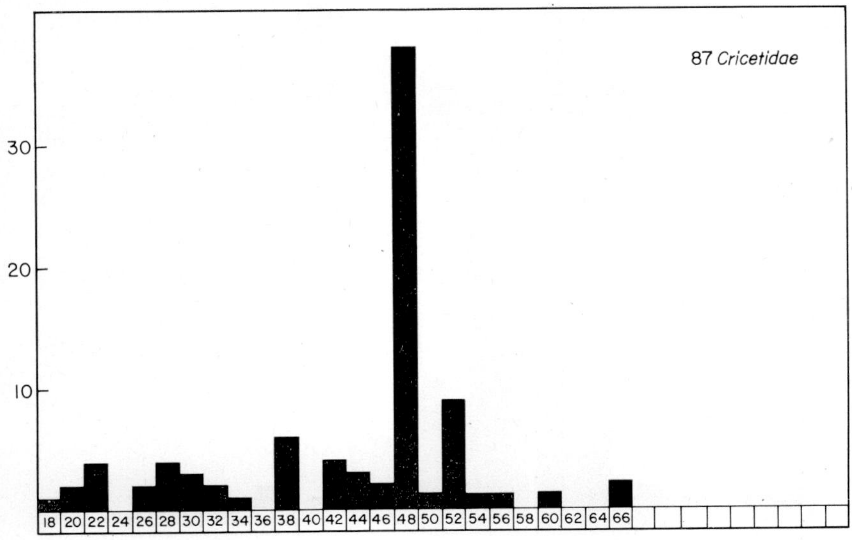

13

Figs. 11–13.

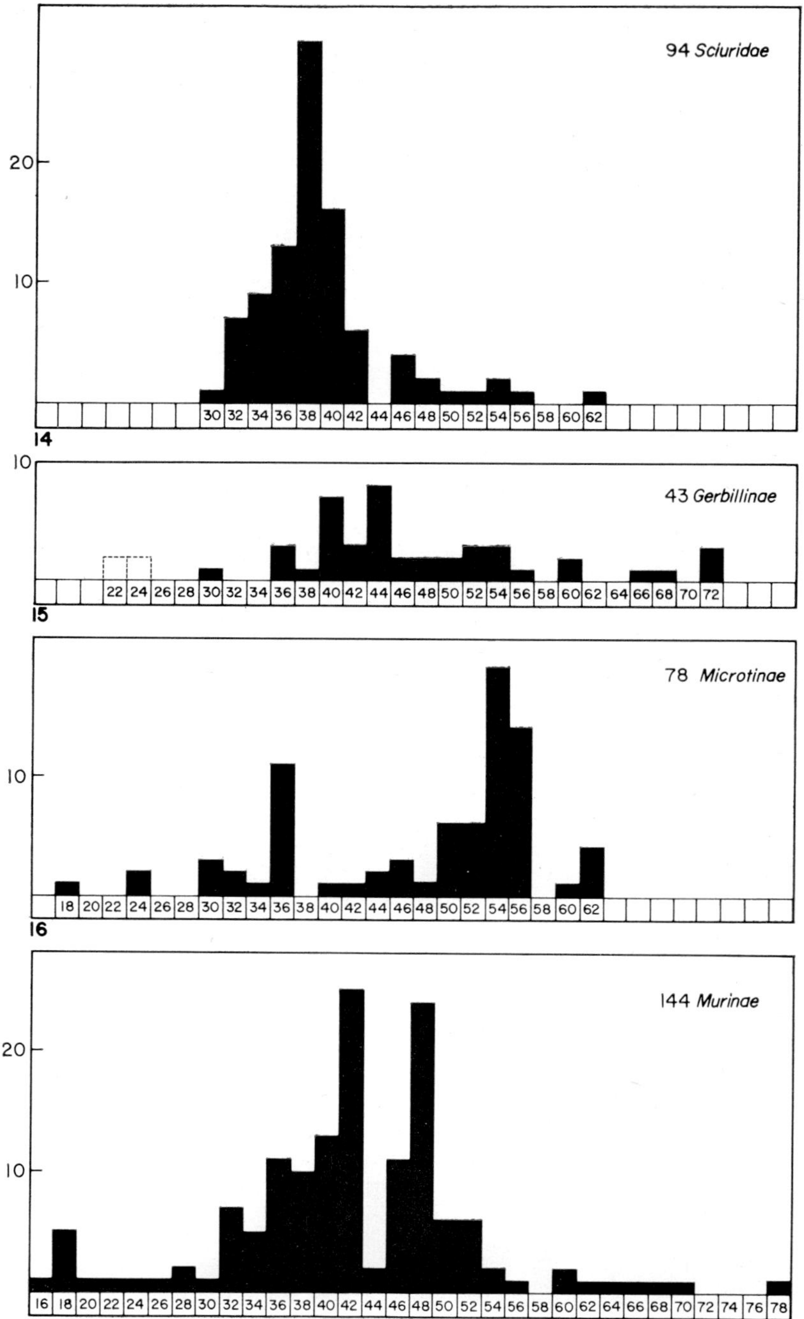
94 Sciuridae
20
10
14
10
43 Gerbillinae
15
78 Microtinae
10
16
144 Murinae
20
10
17

Figs. 14–17.

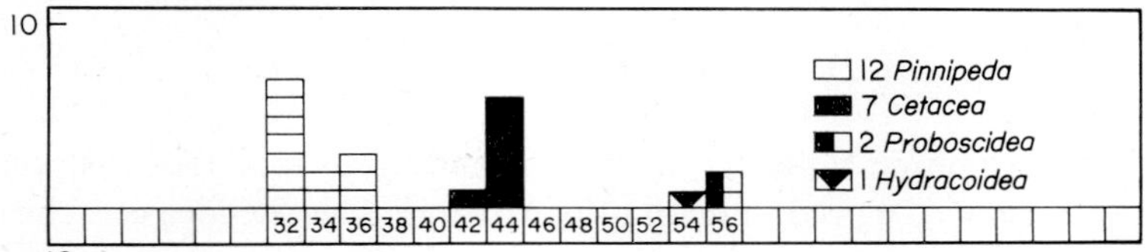

18

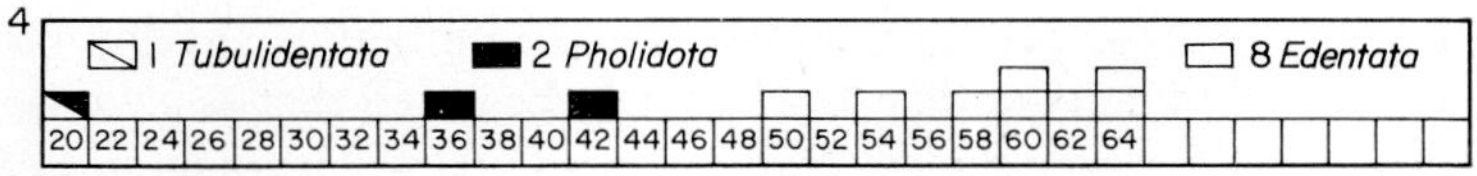

19

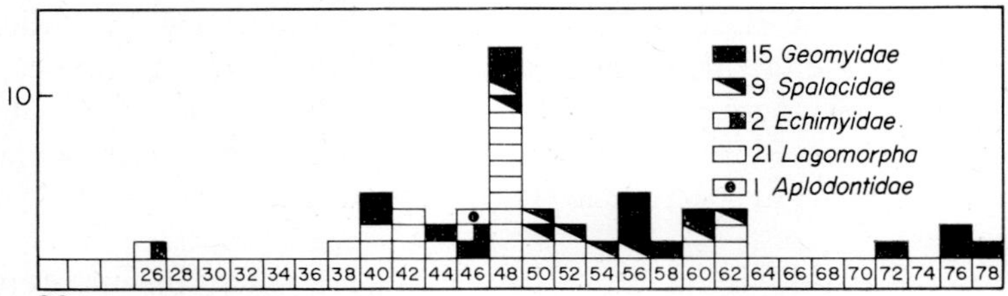

20

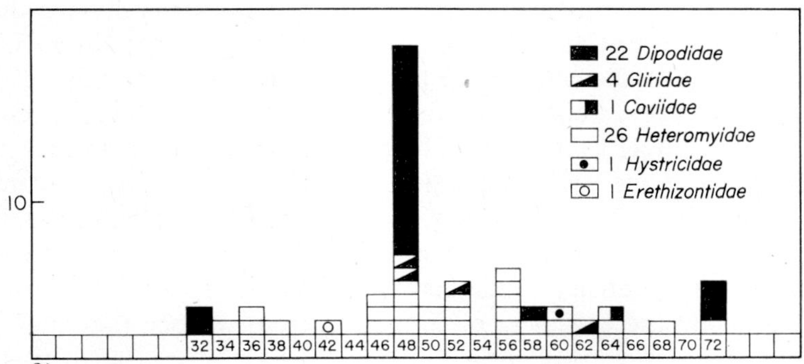

21

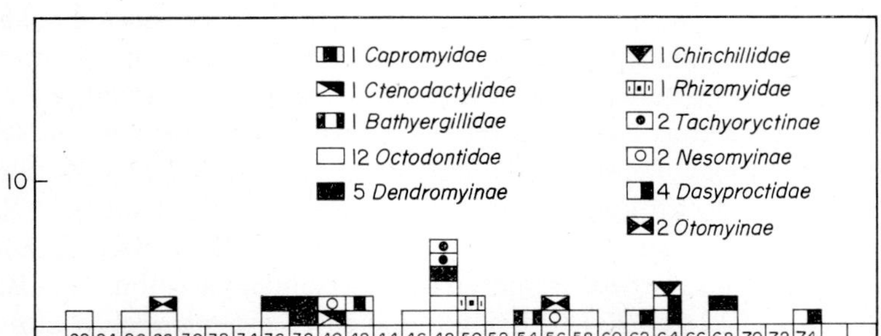

22

Figs. 18–20.

B. *Direction of chromosomal evolution*

We must examine some recent works tending to maintain that centric fissions have played a predominant role, which implies the idea of a chromosomal evolution arising from formulae with a small number of chromosomes.

Hayman and Martin (1969) found, when investigating the cytogenetics of marsupials, that the distribution of diploid numbers is sharply bimodal with one peak of 14 and another of 22. It may be noted that as a whole these karyotypes show a rather typical Robertsonian series. Hayman and Martin advance two hypotheses, either an ancestral number of 14, or an ancestral number of 22. The reason why they endorse the first view (evolution by fission) lies in the fact that among the seven families where number 14 is found (33 species) there is a high karyotype uniformity, a common scheme, which is not observed in the six families (31 species) whose representatives have 2N's of 20 (eight species), 22 (twenty species), 24 (two species) and 32 (one species).

In contrast to this viewpoint, we have strong arguments in support of the second hypothesis. We know (Thenius and Hofer, 1960) that marsupials and eutherians, which issued from the pantotherian stock, divided during the Lower or Middle Cretaceous and were sharply demarcated in the Upper Cretaceous. While allowing for the present limited documentation of Australian paleontology, we must not forget that the oldest marsupial remains date from the Upper Cretaceous in North America and do not differ in the least from the living *Didelphidae* which, in the genus *Didelphis* contains species displaying 22 chromosomes. According to Viret (1955), the didelphoid ancestors, widespread in North America and Asia, possibly since the end of the Jurassic, would have reached the southern hemisphere travelling along the cordillera of East Asia, from whence they may have migrated to Australia, populating it, therefore, at a relatively late period.

In my opinion, the uniformity of the karyotypes with 14 chromosomes is in favour of the second hypothesis rather than the first, since it is known that in many quite distantly related lineages of the same group evolution by centric fusion often resulted in identical or very similar karyotypes which should be regarded as final ("frozen" formulae) rather than initial stages. That is why among *Lacertilia* the karyotype with 12 metacentric macrochromosomes and 24 microchromosomes is characteristic of species belonging to families as far apart as *Agamidae*, *Chamaleontidae*, *Iguanidae*, *Gerrhosauridae* and *Amphisbaenidae*, its Robertsonian variations, e.g. 10 *V*/4 *I*; 8 *V*/8 *I*; 6 *V*/12 *I*; 4 *V*/16 *I* being found in *Helodermatidae*, *Varanidae*, *Xantusiidae* and *Anguidae*. Beyond doubt the same karyotype has established itself within taxa which have started to diverge since the Jurassic.

There is a further objection, from a cytological standpoint, to Hayman and Martin's hypothesis, which is that four marsupial species, a *Peramelidae* and three *Macropodidae*, respectively endowed with 18, 12, 16 and 10 chromosomes in the female, show multiple sex chromosome of the following type: ♂: X/Y_1Y_2, ♀: X/X. Now, among eutherians, this type or another aberrant type appear most often in species with a very low diploid number. It seems that chromosomal mutations affecting the sex chromosomes have little likelihood of establishing themselves, owing to the genetic balance which is particularly delicate at this level, while, only after all possible autosomal fusions have taken place, heterochromosomes themselves may be involved in transformations resulting in sex determining mechanisms of an aberrant type. Figure 9 shows that out of 139 *Chiroptera* species, eleven of which possess 30 to 50 chromosomes with a peak of 44 (34 species), multiple sex chromosomes are found in seven species with 30 chromosomes and in seven out of eight species with less than 24 chromosomes. The situation is even clearer in the *Microtinae* (Fig. 16): in this subfamily 32 species out of 78 have 54 or 56 chromosomes and all of them belong to the type referred to by taxonomists as primitive. *Ellobius lutescens* (2N = 17) and *Microtus oregoni* (2N = 17 and 18) with unusual and also poorly understood sex determining mechanisms raise the question of how could it be alleged that they show the primitive conditions of the sub-family.

Biochemical evidence also lends support to the second hypothesis. Atkin *et al.* (1965) have brought to light that in *Microtus oregoni* (2N = 17, 18) the nuclear DNA content is some 10% lower than in Eutherians which display a higher number of chromosomes. They interpret this deficiency as the result of the elimination of centric and acentric fragments during fusion, whereby the diploid number has shifted from higher to very low levels.

The same reasoning applies to marsupials. Ohno (quoted by Bick and Jackson, 1967) remarks that in *Didelphis* (2N = 22) nuclei have a DNA content which is 94% of that of the eutherians. Bick and Jackson (1967) evaluate this amount as 81% in *Potorous* (2N = 12 in the ♀, 13 in the ♂). Here again the 10% deficiency is likely to be due to fragment losses during fusion.

Hence, in marsupials the primitive formula is 22. The interesting fact is that either side of this value of 22 the distribution of diploid numbers compared to eutherians is very variable: while 27 species (33.7%) show 20 or 22 chromosomes, 50 species (62.5%) have less than 20 and only 3 species (3.75%) have more than 22 chromosomes. This means that the karyotypes of the majority of existent marsupials may be ascribed to centric fusions leading to these final formules (i.e. "frozen") shown by *Lacertilia*. In eutherians, while 31.5% of the karyotypes have 2N's below the modal values (40–56), 16.4% have higher; this must be due to fissioning processes.

Accordingly, I would suggest that the marsupials in their chromosomal evolution have attained a more advanced stage than the eutherians. This fits in with Grassé's (1955) suggestion that "the marsupials, richer in archaisms, are more than liable to have appeared earlier than the eutherians, though the latter are not necessarily descended from them."

Theoretical studies were made on chromosomal evolution in the eutherians by Todd (1967, 1970) and Staton (1967).

Evolution is envisaged by Todd as a series of cycles each being characterized by alternating phases: first, one of speciation followed by an intense adaptive radiation and a second phase in which forms undergo gradual transformation without proliferation from a variety of types starting from the primitive stock. Todd speculates that the first of these phases coincides with generalized fissioning processes, while the second is correlated to chromosomal mutations such as fusions and pericentric inversions, the latter resulting in the shifting of a terminal centromere to an intermediate position, at least in some chromosomes. Three of these cycles are supposed to have taken place in eutherian evolution, the first phases being referred to as "potentiation" phases. Before going further let us turn again to the terminology. Todd calls *ST* all the chromosomes with two arms, regardless of whether they are acrocentric, submetacentric or metacentric, whereas for the cytologist a subterminal chromosome (*ST*)is acrocentric. In the present paper the term *ST* will be replaced by *DB* (dibrachial).

In eutherian evolution the starting point would be the same as envisioned by Hayman and Martin in marsupials, namely 2N = 14. An example lending support to this hypothesis would be the case of an insectivore, *Myurus elephantulus*, in which 14 chromosomes had been counted by Brenner (1946). But, as I showed in 1954, Brenner interpreted haploid images as diploid, the actual diploid number being 30. From this 14 chromosome stage, called pantotherian, we can, as an example, follow the representation proposed by Todd of the evolution undergone by three lineages, namely the *Carnivora*, the *Artiodactyla* and the *Perissodactyla*. The earliest "potentiation" phase leads from:

$$2N = 10\,DB + 2\,AC + 2\,X\,(\text{or XY}) = 14,$$
$$\text{to } 2N = 22\,AC + 2\,X\,(\text{or XY}) = 24$$

coincidental with *Miacidae* and *Condylarthra* differentiation. Out of the 22 *AC*, 10 or 14 having become *DB* during the progressive transformation phase, a second fission cycle leads from:

$$2N = 14\,(10)\,DB + 8\,(12)\,AC + 2\,X\,(\text{or XY}) = 24$$
$$\text{to } 2N = 36\,AC + 2\,X\,(\text{or XY}) = 38$$
$$\text{or } 2N = 32\,AC + 2\,X\,(\text{or XY}) = 34.$$

The first of these formulae characterizes *Carnivora* and *Artiodactyla*, the second one *Perissodactyla*. Starting the second phase of the second cycle, the *Carnivores* will have given origin to the *Canoidea* and *Feloidea*; the *Artiodactyla* will have split into the *Ruminantia* and *Suina* and the *Perissodactyla* into the *Hippomorpha* and *Ceratomorpha*.

The third "potentiation" cycle will divide the *Canoidea* into two branches; the formula:

$$2\text{N} = 38\,DB + 2\,\text{X (or XY)} = 40 \text{ is changed into:}$$
$$2\text{N} = 76\,AC + 2\,\text{X (or XY)} = 78\ (\textit{Ursidae, Canidae})$$

or remains equal to 36 or 38 (*Felidae, Herpestidae*).

The *Ruminantia*, starting with the formula:

$$2\text{N} = 34\,DB + 4\,AC + 2\,\text{X (or XY)} = 40$$

diverge into the *Camelidae* (2N = 74) and *Suidae*, the latter retaining the initial formula.

As for *Perissodactyla*, which arise from the karyotype:

$$2\text{N} = 32\,AC + 2\,\text{X (or XY)} = 34$$

They undergo fissions involving 12 *AC*, whereby:

$$2\text{N} = 12\,DB + 20\,AC + 2\,\text{X (or XY)} = 34, \text{ which turns into:}$$
$$2\text{N} = 44\,AC + 2\,\text{X (or XY)} = 46.$$

Equus zebra (2N = 32) was not drawn into the fission phase, whereas appropriate modulations of the formula 46 make it possible to infer the 2N numbers (44, 46, 48, 54, 64, 66) displayed by the other *Equidae*.

This concept was applied by Staton (1967) to the New World Primates, and Todd (1970) developed it focusing his attention in particular on the analysis of the Order *Carnivora*. In this last work a considerable portion is devoted to the examination of paleontologic and zoogeographic data. My lack of first hand knowledge in this field does not allow any appraisal on my part of the value of this documentation. and I must keep to the level of cytogenetics and comparative cytology.

Todd's hypotheses can be tested by fitting them into the general picture: the problem of the chromosome number in primitive eutherians admits *a priori* three solutions which we intend discussing here:

(1) Chromosome evolution was brought about essentially by fission, with secondary rearrangements triggering pericentric inversions, fusions, translocations, starting from a very low primitive number: 2N = 14, according to Todd. As far as Hayman's and Martin's work on marsupials is concerned, I have presented the arguments that seem to contradict their hypothesis.

These arguments to me are for the most part valid and others concerning the eutherians are being added to them.

(a) The scarcity of low numbers, regarded as primitive (eutherians).

(b) The frequency of aberrant mechanisms in the process of sex determination. This is particularly high when the 2N is extremely low (marsupials and Eutherians).

(c) The morphologically most primitive and paleontologically most ancient forms (*Didelphis*) have a higher number of chromosomes (marsupials).

(d) It is generally agreed that the monotremes derived, starting in the Jurassic, from a Docodont stem, which itself issued from a triasic trunk from which has also sprung the pantotherian branch destined to diverge, by the Cretaceous, into marsupials and placentals. If this is true, all mammals have a common origin. Now, the *Ornithorinchus* has 54 chromosomes (Bick and Jackson, 1967), the *Echidna* has 64 (van Brink, 1959; Bick and Jackson, 1967) with *N.F.*'s of about 100.

However, among present day reptiles, no groups exhibit a chromosome number below 36. Many Squamata (*Lacertilia* and *Serpentes*) have between 36 and 50 chromosomes, the chelonians between 50 and 66, the crocodilians between 30 and 42. Though none of these orders have a direct kinship with mammals, it would be surprising if from reptiles and monotremes to marsupials and eutherians a massive reduction of the 2N number had occurred, which, if true, would support fusions rather than fissions.

(e) The contributions afforded by Biochemistry cannot be ignored here (Atkin *et al.*, 1965; Beçak *et al.*, 1964; Ohno *et al.*, 1968). Ohno and his team of workers have measured the nuclear DNA content in various vertebrate classes. I would also draw attention here to the sauropsides and mammals.

From this standpoint, reptiles divide into two sharply defined categories: the *Squamata* have a DNA content which is about 50% only of that characteristic of placental mammals, while birds have still lower DNA values. On the other hand, chelonians and crocodiles are very close to the eutherians. It is of interest to note that the marsupials, if the DNA content in eutherians is equated to 100, are scarcely less rich in DNA than the latter; i.e. 94% (*Didelphis*) and 81% (*Potourus*). The same is true for monotremes: 93.3% (*Echidna*) and 98.6% (*Platypus*).

In other words, the mammals of the three sub-classes share the same DNA content in their chromosomes, whereby their common origin seems to be testified. In passing, we might remark that we can discard straightway the hypothesis which might be suggested by the comparison between primitive diploid numbers, 22 in marsupials and 48 in eutherians, namely, that of polyploidy in the former with respect to the latter. Such a process in fact also played a not unimportant role in teleost evolution, and some cases are also known among amphibians and reptiles.

This bulk of arguments, all of which run counter to the hypothesis of the existence of very low diploid numbers in primitive eutherians, is irrefutable in my opinion. Indeed, the method used by Todd, starting from the alleged pantotherian formula 2N = 14 to infer the subsequent chromosomal evolution, has no cytological basis, rather it directly contradicts it. As far as I am concerned, this concept is simply a clever product of the imagination. It is clear that by fitting, at the time of origin of the "potentiation" cycles, the number of the *DB* chromosomes destined to fission to the actually observed karyotypes one can always ascertain the composition of these karyotypes.

(2) The primitive 2N numbers were identical to the highest *N.F.*'s found by us in the eutherians, that is between 80 and 100. This hypothesis removes a major obstacle, in that a centromere increase becomes useless. A considerable difficulty, however is raised. If chromosomal evolution occurred by the reduction of the 2N number, why did it attain the final result as shown in Fig. 6, namely an overwhelming number of forms displaying between 40 and 56 chromosomes, and on either side of this peak a rapid increase or decrease in the diploid numbers ? On the contrary, in view of the fact that chromosome mutation is rare and hardly ever maintained, one would expect a curve in which the majority of forms have a high number of chromosomes, with this curve flattening down exponentially to the lowest values.

(3) Starting with the primitive number 48, or near it, chromosomal evolution has occurred through a variety of mutations, the most frequent of which are fusions. Less frequently are pericentric inversions and fissions involved.

This concept is the most plausible, since it emerges from the statistical data reported above, which reveal a high probability of the existence of 40 to 56 chromosomes in half the eutherian species. The ancestral forms were likely to possess 48 $\pm$ 8 chromosomes, the numerous Robertsonian series known to us suggesting fusion processes which account for diploid numbers below 40.

From 56 to the highest values we must admit that fissions have been at work. I know only of a single case which can show the intervention of a fissioning process, namely that of *Microtus longicaudus*. Figure 16 shows the distribution of the 2N numbers in the *Microtinae* sub-family. Out of 78 species analysed, 32 have 54 or 56 chromosomes, these figures being exclusively found in the genera regarded by taxonomists as the most primitive on the basis of their morphology. Hence, a striking parallelism exists between the chromosome constitution and systematic position of these voles within the sub-family. Species displaying 54 and 56 chromosomes are generally composed of acrocentrics, their *N.F.*'s being the same as the 2N numbers. In *M. longicaudus* (2N = 56) 20 to 30 meta- or submetacentric chromosomes are found (this animal was the object of study prior to the use of Colcemid, a more precise count was not therefore feasible). These dibrachial elements are the size of the acrocentrics from other species, therefore origin by fusion may be discarded. It is believed that the fission of some of these chromosomes, which are likely to arise by pericentric inversions, might result in a 2N number well above 56.

These are the theoretical considerations I wished to add, as a commentary, to the list of species. I apologise for any errors or omissions, which due to the complexity of a work of this nature, are inevitable.

References

Atkin, N. B., Mattinson, G., Beçak, W. and Ohno, S. (1965). *Chromosoma, Berlin* **17**, 1–10.

Beçak, W., Beçak, M. L., Nasareth, H. R. and Ohno, S. (1964). *Chromosoma, Berlin* **15**, 606–617.

Bick, Y. A. E. and Jackson, W. D. (1967). *Amer. Nat.* **101**, 79–86.

Brenner, S. (1946). *Sth Afr. J. Med. Sci.* **11**, 71–78.

Brink, J. van (1959). *Chromosoma, Berlin* **10**, 1–72.

Grassé, P. (1955). *In* "Traité de Zoologie", (Grassé, P. ed.) **142**, Masson Paris 17.

Hayman, D. L. and Martin, P. G. (1969). *In* "Comparative Mammalian Cytogenetics, (ed. K. Benirschke), 191–217. Springer, New York.

Matthey, R. (1945). *Experientia* **1**, 50–56, 78–86.

Matthey, R. (1949). *Les chromosomes des Vertébrés*, Ec. Rouge, Lausanne.

Matthey, R. (1953). *Rev. suisse Zool.*, **60**, 225–283.

Matthey, R. (1954). *Caryologia* **6**, 1–44.

Matthey, R. (1958). *Arch. J. Klaus Stift.* **33**, 253–297.

Matthey, R. (1969). *In* "Traité de Zoologie", (Grassé, P. ed.) Masson, Paris 16[6, 855–909.

Ohno, S., Wolf, U. and Atkin, N. B. (1968). *Hereditas* **59**, 169–187.

Staton, D. (1967). *Mamm. Chrom. Newslet.* 8/**3**, 203–223.

Tettenborn, U. and Gropp, A. (1970). *Cytogenet.* **9**, 272–283.

Thenius, E. and Hofer, H. (1960). *In* "Stammesgeschischte der Säugetiere" Springer Verlag, Berlin.

Tjio, J. H. and Levan, A. (1956). *Hereditas* **42**, 1–6.

Todd, N. B. (1967). *Mamm. Chrom. Newslet.* **8**/**4**, 268–272.
Todd, N. B. (1970). *Mamm. Chrom. Newslet.* **11**/**4**, 136–140.
Viret, J. (1955). *In* "Traité de Zoologie", (Grassé, P. ed.), Masson, Paris 17, 172.
White, M. J. D. (1969). *Ann. Rev. Genet.* **3**, 75–98.
Wurster, D. H. and Benirschke, K. (1970). *Science N.Y.* **168**, 1364.

	INSECTIVORA	
	Erinaceidae	
Erinaceus amurensis Schr.	44	Kang, Y. S. and Kim, Y. J. (1963) *Zoologica*, **2**, 1–8.
Erinaceus concolor Martin	48	Kral, B. (1967) *MCNL.*, **8**, 201.
Erinaceus europaeus L.	48	Painter, T. (1925) *Am. Nat.*, **59**, 385–392.
	48	Bovey, R. (1949) *Revue Suisse Zool.*, **56**, 371–460.
Erinaceus europaeus roumanicus, B.H.	48	Gropp, A. and Geisler, M. (1966) *MCNL.*, **22**, 207.
Hemiechinus auritus aegyptius Fisch.	48	Gropp, A. *et al.*, (1969) *Chromosoma*, **27**, 288–307.
Hemiechinus megalotis Pallas	48	Gropp, A. *et al.*, (1969) *Chromosoma*, **27**, 288–307.
Paraechinus h. hypomelas Brandt	48	Boulatova, N. S. and Vorontsov, N. N. (1969) *Mlekop.*, Novo-Sibirsk.
	Talpidae	
Condylura cristata L.	34	Meylan, A. (1968) *Revue Suisse Zool.*, **75**, 691–696.
Galemys pyrenaicus K.	42	Peyre, A. (1957) *Bull. Soc. Zool. Fr.* **82**, 434–437.
Mogera insularis Swinh.	32	Tateishi, S. (1938) *Annot. Zool. Jap.*, **17**, 515–521.
Talpa caeca Savi	36	Meylan, A. (1966) *Revue Suisse Zool.*, **73**, 548–558.
Talpa europaea L.	34	Bovey, R. (1949) *Revue Suisse Zool.*, **56**, 371–460.
	Solenodontidae	
Solenodon paradoxus Brandt	34	Borgaonkar, D. S. (1966) *MCNL.*, **19**, 16.
	Centetidae = Tenrecidae	
Centetes ecaudatus Schr.	38	Borgaonkar, D. S. and Gould, E. (1965) *MCNL.*, **17**, 74.
Echinops telfairi Martin	♀:40; ♂:41	Borgaonkar, D. S. (1967) *J. Hered.*, **58**, 211–213.
Hemicentetes nigriceps Günther	38	Borgaonkar, D. S. (1967) *MCNL.*, **8/1**, 27–28.
Hemicentetes semispinosus Cuv.	38	Borgaonkar, D. S. (1966) *MCNL.*, **19**, 16.
Microgale cowani Th.	54	Borgaonkar, D. S. and Gould, E. (1969) *MCNL.*, **10/1**, 31.
Microgale dobsoni Th.	30	Borgaonkar, D. S. and Gould, E. (1968) *Experientia*, **24**, 506–509.
Microgale talazaci Fors. Maj.	30	Borgaonkar, D. S. and Gould, E. (1968) *Experientia*, **24**, 506–509.
Microgale thomasii Th.	30	Borgaonkar, D. S. (1966) *MCNL.*, **22**, 182.

Setifer setosus Schr.	40	Borgaonkar, D. S. (1967) *MCNL.*, **8/1**, 27–28.
	Soricidae	
Blarina brevicauda Say	48, 49, 50	Meylan, A. (1968) *Revue Suisse Zool.*, **75**, 691–696.
Crocidura horsfieldi Th.	26	Yosida, T. H. *et al.* (1968) *A. Rep. Nat. Inst. Genet. Jap.*, **18**, 24–25.
Crocidura leucodon Herm.	28	Meylan, A. (1966) *Revue Suisse Zool.*, **73**, 548–558.
Crocidura murina L. (= *Suncus murinus* L.)	40	Tateishi, S. (1938) *Annotnes. Zool. Jap.* **17**, 515–521.
Crocidura occidentalis kiwu Osg.	50	Meylan, A. (1967) *Can. J. Zool.*, **45**, 1119–1127.
Crocidura russula Herm.	42	Bovey, R. (1949) *Revue Suisse Zool.*, **56**, 371–460.
Crocidura russula dsinezumi Temm.	40	Yosida, T. H. *et al.* (1968) *A. Rep. Nat. Inst. Genet. Jap.*, **18**, 24–25.
Crocidura suaveolens Pallas	40	Meylan, A. (1966) *Revue Suisse Zool.*, **73**, 548–558.
Microsorex hoyi Baird	62	Meylan, A. (1968) *Revue Suisse Zool.*, **75**, 691–696.
Neomys fodiens Pallas	52	Bovey, R. (1949) *Revue Suisse Zool.*, **56**, 371–460.
Neomys anomalus milleri Mottaz	52	Meylan, A. (1966) *Revue Suisse Zool.*, **73**, 548–558.
Sorex alpinus Schinz	58	Meylan, A. (1966) *Revue Suisse Zool.*, **73**, 548–558.
Sorex araneus L.	♂:23	Bovey, R. (1949) *Revue Suisse Zool.*, **56**, 371–460.
	22–25	Sharman, G. B. (1956) *Nature (Lond)*, **177**, 941–942.
	♀:22–25; ♂:22–27	Ford, C. E. *et al.* (1957) *Nature (Lond)*, **180**, 392.
Type A	♀:22; ♂:23	Meylan, A. (1965) *Revue Suisse Zool.*, **72**, 636–646.
Type B	♀:20–32; ♂:21–31	
Sorex arcticus Kerr	♀:28; ♂:29	Meylan, A. (1968) *Revue Suisse Zool.*, **75**, 691–696.
Sorex caecutiens Laxm.	42	Skarren, U. and Halkka, O. (1966) *Hereditas*, **54**, 376–378.
Sorex cinereus Kerr	66	Meylan, A. (1968) *Revue Suisse Zool.*, **75**, 691–696.
Sorex fumeus Miller	66	Meylan, A. (1968) *Revue Suisse Zool.*, **75**, 691–696.
Sorex minutus L.	42	Meylan, A. (1965) *Experientia*, **21**, 268.
Sorex shinto saevus Th.	42	Takagi, N. and Fujimaki, Y. (1966) *MCNL.*, **20**, 68.
Sorex unguiculatus Dobson	42	Tagaki, N. and Fujimaki, Y. (1966) *MCNL.*, **20**, 68.
Suncus etruscus Savi	42	Meylan, A. (1968) *Bull. Soc. Vaud. Sci. Nat.*, **70**, 85–90.

	Macroscelidae	
Elephantulus rupestris jamesoni Chubb	30	Ford, C. E. and Hamerton, J. L. (1956) *Nature* (*Lond*), **177**, 140–141.
Macroscelides rozeti Duv.	28	Matthey, R. (1954) *Revue. Suisse Zool.*, **33**, 669–677.
Nasilio b. brachyrhynchus Smith	26	Stimson, C. and Goodman, M. (1966) *MCNL.* **22**, 188.
	CHIROPTERA	
	Pteropidae	
Cynopterus sphinx gangeticus And.	34	Pathak, S. (1965) *MCNL.*, **17**, 80.
Eidolum helvum Kerr	34	Matthey, R. (1962) *MCNL.*, **8**, 22.
Nycteris sp.	42	Bovey, R. (1949) *Revue Suisse Zool.*, **56**, 371–460.
Pteropus dasymallus inopinatus Kuroda	38	Makino, S. (1948) *Biol. Bull. Woods Hole.*, **94**, 275–282.
Pteropus giganteus Brun.	38	Manna, G. K. and Talukdar, M. (1965) *MCNL.*, **17**, 77.
Roussettus leschnaulti Desm.	36	Pathak, S. (1966) *Chrom. Inform. Serv.*, **7**, 5.
	Emballonuridae	
Balantiopteryx plicata Burt.	32	Patton, J. L. and Baker, R. J. (1966) *MCNL.*, **20**, 66–67.
Taphozous l. longimanus Harw.	42	Ray Chaudhuri, S. P. and Pathak, S. (1966) *MCNL.*, **22**, 206.
Taphozous melanopogon Temm.	42	Ray Chaudhuri, S. P. and Pathak, S. (1966) *MCNL.*, **22**, 206.
	Hipposideridae	
Hipposideros b. bicolor Temm.	32	Ray Chaudhuri, S. P. and Pathak, S. (1966), *MCNL.*, **22**, 206.
	Megadermatidae	
Megaderma l. lyra Geoff.	54	Ray Chaudhuri, S. P. and Pathak, S. (1966) *MCNL.*, **22**, 206.
	Molossidae	
Promops nasutus ancilla Th.	40	Wainberg, R. L. (1966) *Arch. Biol.*, 77, 411–423.
Tadarida brasiliensis St-Hilaire	48	Patton, J. L. and Baker, R. J. (1966) *MCNL.*, **20**, 66–67.
Tadarida femorosacca Merr.	48	Patton, J. L. and Baker, R. J. (1966) *MCNL.*, **20**, 66–67.
Tadarida molossa Pallas	48	Patton, J. L. and Baker, R. J. (1966) *MCNL.*, **20**, 66–67.

	Natalidae	
Natalus stramineus Miller	36	Patton, J. L. and Baker, R. J. (1966) *MCNL.*, **20**, 66–67.
	Noctilionidae	
Noctilio leporinus L.	34	Beçak, M. L. *et al.* (1969) *Experientia*, **25**, 81–83.
	Phyllostomatidae	
Caroliinae		
Carollia castanea Allen	♀:20; ♂:21	Baker, R. J. and Bleier, W. J. (1971) *Experientia*, **27**, 220–222.
Carollia perspicillata azteca Saus.	♀:20; ♂:21	Hsu, T. C. *et al.* (1968) *Cytogenetics*, **7**, 27–38.
Carollia subrufa Hahn	♀:20; ♂:21	Hsu, T. C. *et al.* (1968) *Cytogenetics*, **7**, 27–38.
Rhinophylla fischerae Carter	34	Baker, R. J. and Bleier, W. J. (1971) *Experientia*, **27**, 220–222.
Rhinophylla pumilio Peters	36	Baker, R. J. and Bleier, W. J. (1971) *Experientia*, **27**, 220–222.
Chilonycterinae		
Mormops megalophylla Peters	38	Baker, R. J. and Hsu, T. C. (1970) *Cytogenetics*, **9**, 131–138.
Pteronotus davyi Gray	38	Hsu, T. C. *et al.* (1968) *Cytogenetics*, **7**, 27–38.
Pteronotus parnellii Allen	38	Hsu, T. C. *et al.* (1968) *Cytogenetics*, **7**, 27–38.
Pteronotus psilotus Dobson	38	Hsu, T. C. *et al.* (1968) *Cytogenetics*, **7**, 27–38.
Pteronotus rubigonosa Allen	38	Patton, J. L. and Baker, R. J. (1966) *MCNL.*, **20**, 66–67.
Desmodinae		
Desmodus rotundus murinus Tr.	28	Hsu, T. C. and Benirschke, K. (1967–68) *Atlas mamm. chrom.*, New York.
Glossophaginae		
Anoura geoffroyi Gray	30	Hsu, T. C. *et al.* (1968) *Cytogenetics*, **7**, 27–38.
Choeroniscus godmani Th.	♂:19	Hsu T. C. *et al.* (1968), *Cytogenetics*, **7**, 27–38.
Choeronycteris mexicana Tsch.	♀:16	Hsu, T. C. *et al.* (1968), *Cytogenetics*, **7**, 27–38.
Chrotopterus auritus Peters	28	Yonenaga ,Y. *et al.* (1969), *Caryologia*, **22**, 63–80.
Glossophaga alticola Davis	32	Makino, S. (1948), *Biol. Bull. Woods Hole*, **94**, 275–282.

Glossophaga commissarisi Gardner	32	Hsu, T. C. *et al.* (1968), *Cytogenetics*, 7, 27–38.
Glossophaga soricina Pallas	32	Hsu, T. C. *et al.* (1968), *Cytogenetics*, 7, 27–38.
Leptonycteris sanborni Hoffm.	32	Patton, J. L. and Baker R. J. (1966), *MCNL.*, **20**, 66–67.
Phyllostomatinae		
Lonchorina aurita Th.	32	Baker, R. J. and Hsu T. C. (1970), *Cytogenetics*, **9**, 131–138.
Macrotus waterhousi Gray	46	Hsu, T. C. *et al.* (1968), *Cytogenetics*, 7, 27–38.
Macrotus californicus Baird	40	Nelson-Rees, W. A. and Kniazeff, A. J. (1968), *MCNL*, **9**/**1**, 40.
Macrotus mexicanus Sauss.	46	Nelson-Rees, W. A. and Kniazeff, A. J. (1968), *MCNL.*, **9**/**1**, 40.
Mimon crenulatum St-Hilaire	32	Baker, R. J. and Hsu ,T. C. (1970), *Cytogenetics*, **9**, 131–138.
Micronycteris megalotis Gray	40	Hsu, T. C. *et al.* (1968), *Cytogenetics*, 7, 27–38.
Micronycteris nicefori Sanborn	28	Baker, R. J. and Hsu, T. C. (1970), *Cytogenetics*, **9**, 131–138.
Phylloderma stenops Peters	32	Baker, R. J. and Hsu, T. C. (1970), *Cytogenetics*, **9**, 131–138.
Phyllostomus discolor Wagner	32	Hsu, T. C. *et al.* (1968), *Cytogenetics*, 7, 27–38.
Phyllostomus hastatus Allen	32	Kibliski, P. (1969), *Experientia*, **25**, 1203–1204.
Tonatia bidens Spix	16	Baker, R. J. and Hsu, T. C. (1970), *Cytogenetics*, **9**, 131–138.
Tonatia minuta Goodwin	30	Baker, R. J. and Hsu, T. C. (1970), *Cytogenetics*, **9**, 131–138.
Trachops cirrhosus Spix	30	Hsu, T. C. *et al.* (1968), *Cytogenetics*, 7, 27–38.
Vampyrum spectrum L.	32	Baker, R. J. and Hsu, T. C. (1970), *Cytogenetics*, **9**, 131–138.
Sternoderminae		
Ametrida centurio Gray	♀:30; ♂:31	Baker, R. J. and Hsu, T. C. (1970), *Cytogenetics*, **9**, 131–138.
Artibeus cinereus Gerv.	♀:30; ♂:31	Baker, R. J. and Hsu, T. C. (1970), *Cytogenetics*, **9**, 131–138.

Artibeus jamaicensis Leach	♀:30; ♂:31	Hsu, T. C. *et al.* (1968), *Cytogenetics*, 7, 27–38.
Artibeus litturatus Licht.	♀:30; ♂:31	Hsu, T. C. *et al.* (1968), *Cytogenetics*, 7, 27–38.
Artibeus phaeotis Miller	♀:30; ♂:31	Hsu, T. C. *et al.* (1968), *Cytogenetics*, 7, 131–138.
Artibeus toltecus Sauss.	♀:30; ♂:31	Hsu, T. C. *et al.* (1968), *Cytogenetics*, 7, 131–138. 131–138.
Artibeus turpis nanus And.	♀:30; ♂:31	Patton J. L., and Baker, J. R. (1966), *MCNL.*, **20**, 66–67.
Centurio senex Gray	28	Hsu, T. C. *et al.* (1968), *Cytogenetics*, 7, 27–38.
Chiroderma trinitatum Goodw.	26	Baker, R. J. and Hsu, T. C. (1970), *Cytogenetics*, **9**, 131–138.
Chiroderma villosum Peters	26	Hsu, T. C. *et al.* (1968), *Cytogenetics*, 7, 27–38.
Enchistenes hartii Th.	30	Hsu, T. C. *et al.* (1968), *Cytogenetics*, 7, 27–38.
Mesophylla macconnelli Th.	♀:22; ♂:21	Baker, R. J. and Hsu, T. C. (1970), *Cytogenetics*, **9**,
Sturnira lilium Geoff.	30	Patton, J. L. and Baker, J. R. (1966), *MCNL.*, **20**, 66–67.
Sturnira ludovici Anth.	30	Hsu, T. C. *et al.* (1968), *Cytogenetics*, 7, 27–38.
Sturnira tildae De la Torre	30	Baker, R. J. and Hsu, T. C. (1970), *Cytogenetics*, **9**, 131–138.
Uroderma bilobatum Peters	44	Hsu, T. C. *et al.* (1968), *Cytogenetics*, 7, 27–38.
Vampyrodes caraccioloi Th.	30	Baker, R. J. and Hsu, T. C. (1970), *Cytogenetics*, **9**, 131–138.
Vampyrops helleri Peters	30	Hsu, T. C. *et al.* (1968), *Cytogenetics*, 7, 27–38.
	Rhinolophidae	
Rhinolophus blasii Peters	58	Dulic, B. (1967), *Bull. Scient. Cons. Acads. RPF. Yougosl.*, **A**/**12**, 63–65.
Rhinolophus c. cornutus Temm.	62	Sasaki, M. (1970), *MCNL.*, **11**/**1**, 22.
Rhinolophus euryale Blasius	58	Bovey, R. (1949), *Revue Suisse Zool.*, **56**, 371–460.
Rhinolophus ferrum-equinum Schr.	58	Bovey, R. (1949), *Revue Suisse Zool.*, **56**, 371–460.
Rhinolophus ferrum-equinum nippon Temm.	58	Makino, S. (1948), *Biol. Bull. Woods Hole*, **94**, 273–282.
Rhinolophus hipposideros Bechst.	56	Capanna, E. *et al.* (1967), *Atti Accad. Naz. Lincei Rc.*, **63**, 125–128.
Rhinolophus mehelyi Matschie	58	Dulic, B. and Soldatovic, B. (1969), *Caryologia*, **22**, 1–6.
	Rhinopomatidae	
Rhinopoma hardwicki Gray	36	Ray-Chaudhuri, S. P. and Pathak, S. (1966), *MCNL.*, **22**, 206.

	Vespertilionidae	
Antrozous pallidus Le Conte	46	Patton, J. L. and Baker, R. J. (1966), *MCNL.*, **20**, 66–67.
Barbastella barbastellus Schr.	32	Bovey, R. (1949), *Revue Suisse Zool.*, **56**, 371–460.
Eptesicus andinus Allen	50	Patton, J. L. and Baker, R. J. (1966), *MCNL.*, **20**, 66–67.
Eptesicus furinalis d'Orb.	50	Patton, J. L. and Baker, R. J. (1966), *MCNL.*, **20**, 66–67.
Eptesicus fuscus Beauv.	50	Patton, J. L. and Baker, R. J. (1966), *MCNL.*, **20**, 66–67.
Eptesicus serotinus turcomanus Eversm.	50	Vorontsov, N. N. *et al.* (1969), *Mlekop.*, Novo-Sibirsk.
Lasiurus borealis Miller	28	Patton, J. L. and Baker, R. J. (1966), *MCNL.*, **20**, 66–67.
Lasiurus cinereus Beauv.	28	Patton, J. L. and Baker, R. J. (1966), *MCNL.*, **20**, 66–67.
Lasiurus ega Gerv.	28	Patton, J. L. and Baker, R. J. (1966), *MCNL.*, **20**, 66–67.
Lasiurus intermedius Allen	26	Patton, J. L. and Baker, R. J. (1966), *MCNL.*, **20**, 66–67.
Lasionycteris noctivagans Le Conte	20	Patton, J. L. and Baker, R. J. (1966), *MCNL.*, **20**, 66–67.
Miniopterus schreibersii Natt.	46	Bovey, R. (1949), *Revue Suisse Zool.*, **56**, 371–460.
Myotis austroriparius Rhoads	44	Baker, R. J. and Patton, J. L. (1967), *J. Mamm.*, **48**, 270–286.
Myotis californicus Audub. and Bachm.	44	Baker, R. J. and Patton, J. L. (1967), *J. Mamm.*, **48**, 270–286.
Myotis capaccinii Bonap.	44	Capanna, E. *et al.* (1968), *Caryologia*, **21**, 225–240.
Myotis dasycneme Boie	44	Strelkov, P. P. and Volobujev, V. T. (1969), *Mlekop.*, Novo-Sibirsk.
Myotis daubentonii Leisler	44	Strelkov, P. P. and Volobujev, V. T. (1969), *Mlekop.*, Novo-Sibirsk.
Myotis elegans Hall	44	Patton, J. L. and Baker, R. J. (1966), *MCNL.*, **20**, 66–67.
Myotis emarginatus Geoff.	44	Bovey, R. (1949), *Revue Suisse Zool.*, **56**, 371–460.
Myotis e. turcomanus Bobr.	44	Strelkov, P. P. and Volobujev, V. T. (1969), *Mlekop.*, Novo-Sibirsk.
Myotis evotis Allen	44	Patton, J. L. and Baker, R. J. (1966), *MCNL.*, **20**, 66–67.
Myotis fortidens Mill. and Allen	44	Osborne, J. (1965), MSc. Thesis Univ. Ariz.
Myotis grisescens Howell	44	Patton, J. L. and Baker, R. J. (1966), *MCNL.*, **20**, 66–67.
Myotis keenii Merr.	44	Patton, J. L. and Baker, R. J. (1966), *MCNL.*, **20**, 66–67.
Myotis leibii Audub. and Bachm.	44	Patton, J. L. and Baker, R. J. (1966), *MCNL.*, **20**, 66–67.
Myotis lucifugus Le Conte	44	Patton, J. L. and Baker, R. J. (1966), *MCNL.*, **20**, 66–67.

Myotis macrodacylus Temm.	44	Sasaki, M. (1970), *MCNL.*, **11**/**1**, 22.
Myotis myotis Borkh.	44	Bovey, R. (1949), *Revue Suisse Zool.*, **56**, 371–460.
Myotis mystacinus Leisler	44	Bovey, R. (1949), *Revue Suisse Zool.*, **56**, 371–460.
Myotis nattereri Kuhl	44	Strelkov, P. P. and Volobujev, T. (1969), *Mlekop.*, Novo-Sibirsk.
Myotis nigricans Schinz	44	Patton, J. L. and Baker, R. J. (1966), *MCNLl.*, **20**, 66–67.
Myotis oxygnathus Montic.	44	Radjabli, S. I. and coll. (1969), *Mlekop.*, Novo-Sibirsk.
Myotis sodalis Miller	44	Patton, J. L. and Baker, R. J. (1966), *MCNL.*, **20**, 66–67.
Myotis subflavus leibii Audub. and Bachm.	44	Baker, R. J. and Patton, J. L. (1967), *J. Mamm.*, **48**, 270–286.
Myotis subulatus leibii Miller and Allen	44	Baker, R. J. and Patton, J. L. (1967), *J. Mamm.*, **48**, 270–286.
Myotis thysanodes Miller	44	Patton, J. L. and Baker, R. J. (1966), *MCNL.*, **20**, 66–67.
Myotis velifer Allen	44	Patton, J. L. and Baker, R. J. (1966), *MCNL.*, **20**, 66–67.
Myotis volans Miller	44	Patton, J. L. and Baker, R. J. (1966), *MCNL.*, **20**, 66–67.
Myotis yumanensis Allen	44	Patton, J. L. and Baker, R. J. (1966), *MCNL.*, **20**, 66–67.
Nyctalus noctula Schr.	42	Dulic, B. *et al.* (1967), *Experientia*, **23**, 945.
Nycticeus humeralis Raf.	46	Patton, J. L. and Baker, R. J. (1966), *MCNL.*, **20**, 66–67.
Pipistrellus affinis Dobson	36	Ray-Chaudhuri, S. P. and Pathak, S. (1966), *MCNL.*, **22**, 206.
Pipistrellus hesperus Allen	28	Patton, J. L. and Baker, R. J. (1966), *MCNL.*, **20**, 66–67.
Pipistrellus kuhli Kuhl	44	Capanna, E. and Civitelli, M. V. (1966), *Caryologia*, **19**, 231–240.
Pipistrellus mimus, Wr.	34	Manna, G. K. and Talukdar, M. (1965), *MCNL.*, **17**, 77.
	38	Ray-Chaudhuri, S. P. and Pathak, S. (1966), *MCNL.*, **22**, 206.
Pipistrellus mordax Peters	34	Ray-Chaudhuri, S. P. and Pathak, S. (1966), *MCNL.*, **22**, 206.
Pipistrellus nathusii Kays. and Blasius	44	Bovey, R. (1949), *Revue Suisse Zool.*, **56**, 371–460.
Pipistrellus pipistrellus Schr.	42	Bovey, R. (1949), *Revue Suisse Zool.*, **56**, 371–460.
Pipistrellus subflavus Cuv.	30	Patton, J. L. and Baker, R. J. (1966), *MCNL.*, **20**, 66–67.
Pipistrellus tralatitius abramus Horf.	26	Takayama, S. (1959), *Jap. J. Genet.*, **34**, 107–110.

Pipistrellus savii Bonap.	44	Capanna, E. and Civitelli, M. V. (1967), *Caryologia*, **20**, 265–272.
Pizonyx vivesi Men.	44	Patton, J. L. and Baker, R. J. (1966), *MCNL.*, **20**, 66–67.
Plecotus auritus L.	32	Bovey, R. (1949), *Revue Suisse Zool.*, **56**, 371–460.
Plecotus phyllotus Allen	30	Baker, R. J. and Patton, J. L. (1966), *MCNL.*, **20**, 66–67.
Plecotus townsendii Cooper	32	Baker, R. J. and Patton, J. L. (1966), *MCNL.*, **20**, 66–67.
Rhogoessa gracilis Miller	44	Patton, J. L. and Baker, R. J. (1966), *MCNL.*, **20**, 66–67.
Rhogoessa parvula Allen	44	Patton, J. L. and Baker, R. J. (1966), *MCNL.*, **20**, 66–67.
Rhogoessa tumida Allen	42	Patton, J. L. and Baker, R. J. (1966), *MCNL.*, **20**, 66–67.
Scotophilus kuhli Leach	36	Ray-Chaudhuri, S. P. and Pathak, S. (1966), *MCNL.*, **22**, 206.
Scotophilus temmincki wroughtoni T.	36	Chaudhuri, S. P. and Pathak, S. (1966), *MCNL.*, **22**, 206.
Vespertilio murinus L.	38	Vorontsov, N. N. *et al.* (1969), *Mlekop.*, Novo-Sibirsk.
Vespertilio superans Th.	38	Vorontsov, N. N. *et al.* (1969), *Mlekop.*, Novo-Sibirsk.
	CARNIVORA	
	Viverridae	
Atilax paludinosus Cuv.	♀:36; ♂:35	Todd, N. B. (1967), *MCNL.*, **8**/**1**, 21–22.
Arctitis binturong Raffles	42	Hsu, T. C. and Arrighi, F. E. (1966), *MCNL.*, **21**, 155–160.
Bdeogale sp.	36	Wurster, D. H. and Benirschke, K. (1967), *MCNL.*, **8**/**3**, 195–196.
Civettictius civetta Schr.	38	Todd, N. B. (1967), *Carn. Genet. NL.*, **3**, 49–51.
Cryptoprocta fossa Benn.	42	Wurster, D. H. and Benirschke, K. (1967), *MCNL.*, **8**/**3**, 195–196.
Cynictis sp.	36	Todd, N. B. (1967), *Carn. Genet. NL.*, **2**, 31.
Fossa fossa Schr.	42	Wurster, D. H. and Benirschke, K. (1967), *MCNL.*, **8**/**3**, 195–196.
Galidia elegans Geoff.	44	Wurster, D. H. and Benirschke, K. (1968), *Chromosoma*, **24**, 336–382.
Genetta genetta L.	54	Matthey, R. (1965), *MCNL.*, **17**, 74.
Genetta genetta neumanni Matschie	52	Wurster, D. H. and Benirschke, K. (1968), *Chromosoma*, **24**, 336–382.

Hemigalus derbyanus Gray	42	Wurster, D. H. and Benirschke, K. (1968), *Chromosoma*, **24**, 336–382.
Herpestes auropunctatus Hodgson	♀:36; ♂:35	Fredga, K. (1964), *Hereditas*, **52**, 411–420.
Herpestes edwardsi edwardsi Geoff.	♀:36; ♂:35	Ray-Chaudhuri, S. P. *et al.* (1968), *MCNL.*, **9**/**2**, 82–84.
Herpestes ichneumon L.	48	Wurster, D. H. and Benirschke, K. (1968), *Chromosoma*, **24**, 336–382.
Herpestes javanicus Geoff.	♀:36; ♂:35	Todd, N. B. (1966), *MCNL.*, **21**, 154–155.
Ichneumia albicauda Cuv.	36	Wurster, D. H. and Benirschke, K. (1967), *MCNL.*, **8**/**3**, 195–196.
Mungos mungo Gmelin	36	Wurster, D. H. and Benirschke, K. (1967), *MCNL.*, **8**/**3**, 195–196.
Nandinia binotata Gray	38	Todd, N. B. (1967), *Carn. Genet. NL.*, **3**, 49–51.
Paguma larvata H.S.	44	Wurster, D. H. and Benirschke, K. (1967), *MCNL.*, **8**/**3**, 195–196.
Paradoxurus hermaphroditus Pallas	42	Ray-Chaudhuri, S. P. *et al.* (1966), *Experientia*, **22**, 740–741.
Prionodon linsang Hardw.	34	Wurster, D. H. and Benirschke, K. (1968), *Chromosoma*, **24**, 336–382.
Suricatta suricatta Desm.	36	Todd, N. B. (1966), *MCNL.*, **21**, 154–155.
Viverricula indica Desm.	36	Wurster, D. H. and Benirschke, K. (1968), *Chromosoma*, **24**, 336–382.
	Canidae	
Alopex lagopus L.	50	Lande, O. (1960), *Nature* (*Lond*), **188**, 170.
	48	Wurster, D. H. and Benirschke, K. (1968), *Chromosoma*, **24**, 336–382.
Atelocynus microtus Schlater	76	Wurster, D. H. (1969), *In* "Comparative Mammalian Cytogenetics", Springer-Verlag, New York.
Canis aureus L.	78	Ranjini, P. V. (1966), *MCNL.*, **19**, 5.
Canis dingo Blbch.	78	Valenti, C. and Levy, L. (1965), *MCNL.*, **18**, 147.
Canis familiaris L.	78	Minouchi, O. (1928), *Jap. J. Zool.*, **1**, 255–268.
Canis latrans L.	78	Benirschke, K. and Low, R. J. (1965), *MCNL.*, **15**, 102.
Canis lupus L.	78	Hungerford, D. A. and Snyder, R. L. (1966), *MCNL.*, **20**, 72.

Canis niger Bartram	78	Hsu, T. C. and Arrighi, F. E. (1966), *MCNL.*, **21**, 155–159.
Chrysocyon brachyurus Illig.	76	Newnham, R. E. and Davidson, W. M. (1966), *Cytogenetics*, **5**, 152–153.
Dusicyon vetulus Lund	74	Wurster, D. H. and Benirschke, K. (1968), *Chromosoma*, **24**, 336–382.
Lycaon pictus Temm.	78	Wurster, D. H. and Benirschke, K. (1968), *Chromosoma*, **24**, 336–382.
Megalotis (Fennecus) zerda Zimm.	64	Matthey, R. (1954), *Mammalia*, **18**, 225–230.
Nyctereutes viverrinus Temm.	42	Minouchi, O. (1929), *Cytologia*, **1**, 88–108.
Speothos venaticus Lund	74	Wurster, D. H. (1969), *In* "Comparative Mammalian Cytogenetics", Springer Verlag, New York.
Urocyon cinereoargenteus Schr.	66	Wurster, D. H. and Benirschke, K. (1968), *Chromosoma*, **24**, 336–382.
Vulpes bengalensis Shaw	60	Ranjini, P. V. (1966), *MCNL.*, **22**, 216.
Vulpes ruppelli Schinz	40	Matthey, R. (1954), *Mammalia*, **18**, 225–230.
Vulpes, vulpes L.	38	Makino, S. (1947), *J. Fac. Sci. Hokkaido Univ.*, **IV/9**, 345–357.
	Felidae	
Acinonyx jubatus Erxl.	38	Hsu, T. C. *et al.* (1962), *Am. Nat.*, **97**, 225–234.
Felis aurata Temm.	38	Malouf, N. and Schneider, T. G. (1965), *MCNL.*, **15**, 107.
Felis bengalensis Kerr	38	Makino, S. and Tateishi, S. (1952), *J. Morph.*, **90**, 93–102.
Felis caracal Schr.	38	Hsu, T. C. and Arrighi, F. E. (1966), *MCNL.*, **21**, 155–160.
Felis chaus Guld.	38	Manna, G. K. and Talukdar, M. (1965), *MCNL.*, **17**, 77.
Felis colocolo Smith	36	Wurster, D. H. and Benirschke, K. (1969), *MCNL.*, 10/1, 20.
Felis concolor L.	38	Hsu, T. C. (1960), *MCNL.*, **2**, 4.
Felis domestica Schr.	38	Minouchi, O. and Ohta, T. (1934), *Cytologia*, **5**, 355–362.
Felis geoffroyi d'Orb.	36	Wurster, D. H. and Benirschke, K. (1968), *Chromosoma*, **24**, 336–382.

Felis (Panthera) leo L.	38	Makino, S. and Tateishi, S. (1952), *J. Morph.*, **90**, 93–102.
Felis lynx (Canadian lynx) = *Lynx canadensis* Kerr	38	Benirschke, K. *et al.* (1966), *MCNL.*, **21**, 148–149.
Felis (Neofelis) nebulosa Griff.	38	Wurster, D. H. (1969), *In* "Comparative Mammalian Cytogenetics", Springer Verlag, New York.
Felis nigripes Burch.	38	Hsu, T. C. and Arrighi, F. (1966), *MCNL.*, **21**, 155–160.
Felis (Panthera) onca L.	38	Hsu, T. C. (1960), *MCNL.*, **3**, 4.
Felis pardalis L.	36	Hsu, T. C. (1962), *MCNL.*, **8**, 4.
Felis (Panthera) pardus L.	38	Hsu, T. C. (1960), *MCNL.*, **3**, 4.
Felis serval Schr.	38	Wurster, D. H. and Benirschke, K. (1968), *Chromosoma*, **24**, 336–382.
Felis temmincki Vig.	38	Wurster, D. H. and Benirschke, K. (1968), *MCNL.*, **9**/**4**, 236–237.
Felis tigrina Schr.	36	Leyhausen, P. (1967), *MCNL.*, **8**/**4**, 287.
Felis (Panthera) tigris L.	38	Hsu, T. C. and Rearden, H. (1965), *Chromosoma*, **16**, 365–371.
Felis viverrina Benn.	38	Wurster, D. H. and Benirschke, K. (1968), *MCNL.*, **9**/**1**, 19.
Felis wiedii Schinz	36	Hsu, T. C. (1962), *MCNL.*, **8**, 4.
Felis yagouaroundi Lacép.	38	Hsu, T. C. and Luquette, G. F. (1963), *Am. Nat.*, **97**, 225–234.
	Ursidae	
Euarctos americanus Pallas	74	Low, R. J. *et al.* (1964), *MCNL.*, **13**, 3.
Helarctos malayanus Raffles	74	Wurster, D. H. (1969), *In* "Comparative Mammalian Cytogenetics", Springer Verlag, New York.
Melursus ursinus Shaw	74	Hard, W. L. (1968), *MCNL.*, **9**, 242.
Selenarctos thibetanus Cuv.	74	Newnham, R. E. and Davidson, W. M. (1966), *Cytogenetics*, **5**, 152–163.
Thalarctos maritimus Phipps.	74	Low, R. J. *et al.* (1964), *MCNL.*, **13**, 3.
Tremarctos ornatus Cuv.	52	Newnham, R. E. and Davidson, W. M. (1966), *Cytogenetics*, **6**, 156–163.
Ursus arctos syriacus Hemp. and Ehrenb.	74	Hungerford, D. A. and Snyder, R. L. (1966), *MCNL.*, **21**, 150.

Ursus horribilis Ord.	74	Wurster, D. H. and Benirschke, K. (1968), *Chromosoma*, **24**, 336–382.
Ursus middendorfi Merr.	74	Low, R. J. *et al.* (1964), **13**, 3.
	Procyonidae	
Ailuropoda melanoleuca David	42	Newnham, R. E. and Davidson, M. W. (1964), *MCNL.*, **14**, 52–53.
Ailurus fulgens Cuv.	36	Todd, N. B. (1968), *Carn. Genet. NL.*, **5**, 105–108.
Bassaricyon gabbii Allen	38	Wurster, D. H. and Benirschke, K. (1966), *MCNL.*, **8**, 195.
Bassariscus astutus Leht.	38	Hsu, T. C. and Arrighi, F. E. (1966), *MCNL.*, **21**, 155–160.
Nasua narica L.	38	Todd, N. B. (1966), *MCNL.*, **21**, 153.
Nasua nasua solitaria Schinz	38	Panzetta, P. and Alaimo, I. (1967), *MCNL.*, **8**/**2**, 97.
Potos flavus Schr.	38	Fredga, K. (1966), *MCNL.*, **21**, 145–151.
Procyon lotor L.	38	Benirschke, K. *et al.* (1966), *MCNL.*, **21**, 149.
	Mustelidae	
Amblonyx cinerea Illig.	38	Hsu, T. C. and Arrighi, F. E. (1966), *MCNL.*, **21**, 155–160.
Aonyx cinerea Illig.	38	Hsu, T. C. and Arrighi, F. E. (1966), *MCNL.*, **21**, 155–159.
Eyra (Tayra) barbara L.	38	Hsu, T. C. and Arrighi, F. E. (1966), *MCNL.*, **21**, 150–160.
Grison vittatus Bell	38	Fredga, K. (1966), *MCNL.*, **21**, 145–151.
Gulo gulo L.	42	Wurster, D. H. and Benirschke, K. (1967), *MCNL.*, **8**/**3**, 195.
Helictis subaurantiaca Swinhoe	38	Makino, S. (1952), *Cytologia*, **16**, 288–301.
Lutra canadensis Schr.	38	Benirschke, K. *et al.* (1966), *MCNL.*, **21**, 148–149.
Lutra perspicillata Geoff.	38	Manna, G. K. and Talukdar, M. (1965), *MCNL.*, **17**, 78.
Martes americana Miller	38	Wurster, D. H. and Benirschke, K. (1968), *Chromosoma*, **24**, 336–382.
Martes flavigula Boddaert	40	Fredga, K. (1966), *MCNL.*, **21**, 146–151.
Martes foina Erxl.	38	Ehrlich, I. (1949), *Revue Suisse Zool.*, **56**, 621–626.

Martes martes L.	38	Wurster, D. H. and Benirschke, K. (1967), *MCNL.*, **8**/**3**, 195.
Martes pennanti Erxl.	38	Benirschke, K. and Young, E. (1966), *MCNL.*, **21**, 150.
Martes zibellina L.	38	Orlov, V. N. and Malygen, V. M. (1969), *Mlekop.*, Novo-Sibirsk.
Meles meles L.	44	Omodeo, P. and Renzoni, A. (1966), *Caryologia*, **19**, 219–226.
Melogale moschata Gray	38	Wurster, D. H. and Benirschke, K. (1967), *MCNL.*, **8**/**3**, 195.
Mephitis mephitis Schr.	50	Fredga, K. (1966), *MCNL.*, **21**, 145–151.
Mustela erminea L.	44	Fredga, K. (1967), *Hereditas*, **57**, 295.
Mustela erminea cicognanii Bonap.	44	Meylan, A. (1967), *Can. J. Genet.*, **9**, 569–574.
Mustela itatsi itatsi Temm.	38	Makino, S. (1947), *J. Fac. Sci. Hokkaido Univ.*, **VI**, 345–357.
Mustela nigripes Audub. and Bachm.	38	Hsu, T. C. and Arrighi, F. E. (1966), *MCNL.*, **21**, 155–158.
Mustela sibirica coreana Doman.	38	Kang, Y. S. and Kim, Y. J. (1964), *Zoologica*, **3**, 21–32.
Mustela nivalis nivalis L.	42	Omodeo, P. and Renzoni, A. (1966), *Caryologia*, **19**, 219–226.
Mustela putorius putorius L.	40	Fredga, K. (1966), *MCNL.*, **21**, 146–151.
Mustela putorius furo L.	40	Hsu, T. C. and Arrighi, F. E. (1966), *MCNL.*, **21**, 155–158.
Mustela vison Peale	30	Lande, O. (1957), *Hereditas*, **43**, 578–582.
Mustela rixosa Bangs	44	Meylan, A., *in litt.*
Spilogale gracilis latifrons Merr.	60	Hsu, T. C. (1966), *MCNL.*, **21**, 177.
Spilogale putorius L.	64	Hsu, T. C. and Arrighi, F. E. (1966), *MCNL.*, **21**, 155–160.
Spilogale putorius interrupta Raf.	64	Hsu, T. C. and Arrighi, F. E. (1966), *MCNL.*, **21**, 155–160.
Spilogale putorius ambarvalis Bangs	64	Hsu, T. C. and Arrighi, F. E. (1966), *MCNL.*, **21**, 155–160.
Spilogale putorius phenax Merr.	64	Hsu, T. C. and Arrighi, F. E. (1966), *MCNL.*, **21**, 155–160

Taxidea taxus Schr.	32	Wurster, D. H. and Benirschke, K. (1969), *MCNL.*, **10**/7, 20.
	Hyaenidae	
Crocuta crocuta L.	40	Wurster, D. H. and Gray, C. (1967), *MCNL.*, **8**/**3**, 197.
Hyaena hyaena L.	40	Hsu, T. C. and Arrighi, F. E. (1966), *MCNL.*, **21**, 155–160.
Proteles cristata Sparr	40	Wurster, D. H. *et al.* (1968), *MCNL.*, **9**/**1**, 4–5.
	PINNIPEDA	
	Phocidae	
Erignathus barbatus Erxl.	34	Fay, F. H. *et al.* (1967), *Can. J. Zool.*, **45**, 773–778.
Leptonychotes weddelli Lesson	34	Fay, F. H. *et al.* (1967), *Can. J. Zool.*, **45**, 773–778.
Pagophilus groelandicus Erxl.	32	Anbinder, E. M. (1969), *Mlekop.*, Novo-Sibirsk.
Phoca hispida krascheninikovi Nan. and Smirn.	32	Corfman, P. A. and Richart, R. M. (1964), *Nature* (*Lond*), **204**, 502–503.
Phoca vitulina L.	32	Fay, F. H. *et al.* (1967), *Can. J. Zool.*, **45**, 773–778.
Pusa caspica Gmel.	32	Anbinder, E. M. (1969), *Mlekop.*, Novo-Sibirsk.
Pusa hispida Schr.	32	Fay, F. H. *et al.* (1967), *Can. J. Zool.*, **45**, 773–778.
Pusa sibirica Gmel.	32	Anbinder, E. M. (1969), *Mlekop.*, Novo-Sibirsk.
	Otariidae	
Callorhinus ursinus L.	36	Fay, F. H. *et al.* (1967) *Can. J. Zool.*, **45**, 773–778.
Eumetopias jubata Schr.	36	Fay, F. H. *et al.* (1967), *Can. J. Zool.*, **45**, 773–778.
Zalophus californicus Lesson	36	Hungerford, D. A. and Snyder, R. L. (1964), *Am. Nat.*, **98**, 125–127.
	Odobenidae	
Odobenus rosmarus L.	32	Fay, F. H. *et al.* (1967), *Can. J. Zool.*, **45**, 773–778.
	CETACEA	
	Delphinidae	
Globicephala scammonii Cope	44	Walen, K. H. and Madin, S. H. (1965), *Am. Nat.*, **99**, 349–354.

Lagenorhynchus obliquidens Gill	44	Duffield, D. A. *et al.* (1967), *Nature* (*Lond*), **213**, 189–190.
Orcinus orca L.	44	Horrall, J. F. *et al.* (1968), *MCNL.*, **9**/**4**, 244.
Orcinus rectispinna Cope	44	Carr, D. H. *et al.* (1966), *MCNL.*, **22**, 208.
Phocoenoides dallii True	44	Makino, S. (1948), *Chromosoma*, **3**, 220–231.
Tursiops truncatus Montagne	44	Walen, K. H. and Madin, S. H. (1965), *Am. Nat.*, **99**, 349–354.
	Physeteridae	
Physeter catodon L.	42	Atwood, R. P. and Razavi, L. (1965), *Nature* (*Lond*), **207**, 328–329.
	EDENTATA	
	Dasypodidae	
Cabassous centralis Miller	62	Benirschke, K. *et al.* (1969), *In* "Comparative Mammalian Cytogenetics", Springer Verlag, New York.
Chaetophractus villosus Desm.	60	Benirschke, K. *et al.* (1969), *In* "Comparative Mammalian Cytogenetics", Springer Verlag, New York.
Dasypus hybridus Desm.	64	Saez, F. A. *et al.* (1964), Proc. Symposium Mammalian Cytogenetics, Brazil, 1962, pp. 163–170. Pergamon Press.
Dasypus novemcinctus L.	64	Beath, M. M. *et al.* (1962), *Chromosoma*, **13**, 27–38.
Euphractus sexcinctus L.	58	Benirschke, K. *et al.* (1969), *In* "Comparative Mammalian Cytogenetics", Springer Verlag, New York.
Priodontes giganteus Geoff.	50	Benirschke, K. *et al.* (1969), *In* "Comparative Mammalian Cytogenetics", Springer Verlag, New York.
	Bradypodidae	
Choloepus hoffmanni Peters	♀:49; ♂:49 XO XY	Corin-Frederic, J. (1969), *Chromosoma*, **27**, 268–287.
	Myrmecophagidae	
Myrmecophaga tridactylus L.	60	Hsu, T. C. (1965), *MCNL.*, **15**, 108.
Tamandua tetradactyla, L.	54	Hsu, T. C. (1965), *MCNL.*, **15**, 108.

	TUBULIDENTATA	
	Orycteropidae	
Orycteropus afer Pallas	20	Wurster, D. H. and Benirschke, K. (1970), *MCNL.*, **11**/**1**, 29.
	PHOLIDOTA	
	Manidae	
Manis pentadactyla L.	42 (?)	Makino, S. and Tateishi, S. (1951), *J. Fac. Sci. Hokkaido Univ.*, **6**, 319–322.
	36	Ray-Chaudhuri, S. P. *et al.* (1969), *Experientia*, **25**, 1167–1168.
	HYRACOIDEA	
	Procaviidae	
Procavia capensis Pallas	54	Hungerford, D. A. and Snyder, R. L. (1969), *Experientia*, **25**, 870.
	PROBOSCIDEA	
	Elephantidae	
Elephas indicus L.	56	Sasaki, M. and Veomett, R. (1963), *MCNL.*, **11**, 124.
Loxodonta africana Blumenb.	56	Hungerford, D. A. *et al.* (1966), *Cytogenetics*, **5**, 243–246.
	PERISSODACTYLA	
	Rhinocerotidae	
Ceratotherium simum Burch.	82	Wurster, D. H. and Benirschke, K. (1968), *Experientia*, **24**, 511.
Rhinoceros unicornis L.	82	Wurster, D. H. and Benirschke, K. (1968), *Experientia*, **24**, 511.
Diceros bicornis Gray	84	Hungerford, D. A. *et al.* (1967), *Am. Nat.*, **101**, 357–358.

		Tapiridae	
Tapirus terrestris columbianus L.		80	Nelson-Rees, W. A. *et al.* (1967), *MCNL.*, **8**/**3**, 229–230.
		Equidae	
Equus asinus L.		62	Trujillo, J. M. *et al.* (1962), *Chromosoma*, **13**, 243–248.
Equus asinus africanus Fitz.		62	Benirschke, K. *et al.* (1962), *J. Reprod. Fertil.*, **4**, 319–326.
Equus asinus somalicus Noack		62	Benirschke, K. *et al.* (1962), *J. Reprod. Fertil.*, **4**, 319–326.
Equus burchelli böhmi Matschie		44	Benirschke, K. and Brownhill, L. E. (1963), *MCNL.*, **10**, 82.
Equus burchelli antiquorum Smith		44	Benirschke, K. and Brownhill, L. E. (1963), *MCNL.*, **10**, 82.
Equus caballus L.		64	Trujillo, J. M. *et al.* (1962), *Chromosoma*, **13**, 243–248.
Equus grevyi Oustal.		46	Mutton, D. E. *et al.* (1964), *MCNL.*, **13**, 7–9.
Equus onager Pallas		56	Benirschke, K. and Malouf, N. (1967), "Equus", Vol. 1–2, pp. 253–284, Tierpark Berlin.
Equus przewalskii Poliak.		66	Benirschke, K. (1965), *Science*, **148**, 382–383.
Equus zebra hartmannae Matschie		32	Benirschke, K. (1967), *MCNL.*, **8**/**3**, 231–232.
		ARTIODACTYLA	
		Suidae	
Pecari tajacu L.		30	Krallinger, H. F. (1936), *Z. Zellforsch.*, **24**, 1–10.
Pecari tajacu angulatus Cope		30	Spalding, J. F. and Berry, R. O. (1956), *Cytologia*, **21**, 81–84.
Phacochoerus aethiopicus Pallas		34	Wallace, C. and Fairall, N. (1967), *S. Afr. J. Med. Sci.*, **32**, 51.
Sus cristatus Wagner		38	Ray-Chaudhuri, S. P. *et al.* (1968), *MCNL.*, **9**/**1**, 82–84.
Sus scrofa L.	wild	38, 36	Gropp, A. *et al.* (1969), *Experientia*, **25**, 778.
Sus scrofa domesticus L.	domestic	38	Muramoto, J. *et al.* (1965), *Proc. Jap. Acad.*, **41**, 236–239.
Sus vittatus leucomystax, Temm. and Schl.		38	Muramoto, J. *et al.* (1965), *Proc. Jap. Acad.*, **41**, 236–239.
		Camelidae	
Camelus bactrianus L.		74	Hungerford, D. A. and Snyder, R. L. (1966), *MCNL.*, **20**, 71.

Camelus dromedarius L.	74	Taylor, K. M. *et al.* (1968), *Cytogenetics*, **7**, 8–15.
Lama glana L. = *L. peruana* Tiedm.	74	Benirschke, K. *et al.* (1965), *MCNL.*, **15**, 100.
Lama huanachus Mol.	72	Capanna, E. and Civitelli, M. V. (1965), *MCNL.*, **17**, 75.
—	74	Taylor, K. M. *et al.* (1968), *Cytogenetics*, **7**, 8–15.
Lama pacos L.	72	Capanna, E. and Civitelli, M. V. (1965), *MCNL.*, **17**, 75.
—	74	Taylor, K. M. *et al.* (1968), *Cytogenetics*, **7**, 8–15.
Lama vicugna Mol.	74	Taylor, K. M. *et al.* (1968), *Cytogenetics*, **7**, 8–15.
	Cervidae	
Alces alces L.	68	Aula, P. and Kääriäinen, L. (1963), *Hereditas*, **51**, 274–278.
Alces alces americana Clinton	70	Wurster, D. H. and Benirschke, K. (1967), *MCNL.*, **8**/**3**, 226–229.
Axis porcinus Zimm.	68	Wurster, D. H. and Benirschke, K. (1967), *MCNL.*, **8**/**3**, 226–229.
Capreolus capreolus L.	70	Gustavsson, I. (1965), *MCNL.*, **18**, 149.
Cervus elaphus L.	68	Gustavsson, I. (1965), *MCNL.*, **18**, 149.
Cervus nippon hortulorum Swinh.	64, 65, 66, 67, 68	Gustavsson, I. and Sundt, C. O. (1969), *Chromosoma*, **28**, 245–254.
Cervus unicolor Kerr	58	Chandra, H. S. *et al.* (1967), *Chromosoma*, **21**, 211–220.
Cervus yesoensis Heude	68	Makino, S. and Muramoto, J. (1966), *MCNL.*, **22**, 197.
Dama dama L.	68	Gustavsson, I. (1965), *MCNL.*, **18**, 149.
Muntiacus muntjak Zimm.	♀:6; ♂:7	Wurster, D. H. and Benirschke, K. (1970), *Science*, **168**, 1364.
Muntiacus reevesi Ogilby	46	Wurster, D. H. and Benirschke, K. (1967), *MCNL.*, **8**/**3**, 226–229.
Odocoileus virginianus borealis Miller	70	Benirschke, K. *et al.* (1963), *MCNL.*, **10**, 82–83.
Rangifer phylarchus Holl.	72	Makino, S. (1944), *Zool. Mag.* (*Tokyo*), **56**.
Rucervus duvauceli Cuvier	56	Chandra, H. S. *et al.* (1967), *Chromosoma*, **21**, 211–220.
	Giraffidae	
Giraffa cameleopardis giraffa Bodd.	30	Wallace, C. and Fairall, N. (1965), *J. Sci. Res. Nat. Park, S. Afr.*, **8**, 97–103.
Okapia johnstoni	45	Hösli, P. and Lang, E. M. (1970), *MCNL.*, **11**/**3**, 109–110.

Bovidae

Antilopinae		
Aepyceros melampus Licht.	60	Wurster, D. H. and Benirschke, K. (1967), *MCNL.*, **8**/**3**, 226–229.
Antidorcas marsupialis Zimm.	56	Wurster, D. H. and Benirschke, K. (1967), *MCNL.*, **8**/**3**, 226–229.
Antilope cervicapra Pallas	30	Wurster, D. H. and Benirschke, K. (1967), *MCNL.*, **8**/**3**, 226–229.
Gazella dama Pallas	38	Wurster, D. H. and Benirschke, K. (1968), *Chromosoma*, **25**, 152–171.
Gazella thomsoni Günther	58	Wurster, D. H. and Benirschke, K. (1967), *MCNL.*, **8**/**3**, 226–229.
Litocranius walleri Brooke	60	Wurster, D. H. and Benirschke, K. (1968), *Chromosoma*, **25**, 152–171.
Rhaphicerus campestris Thunb.	30	Wurster, D. H. and Benirschke, K. (1968), *Chromosoma*, **25**, 152–171.
Rhynchotragus kirki Günther	46	Chandra, H. S. *et al.* (1967), *Chromosoma*, **21**, 211–220.
Bovinae		
Anoa depressicornis Smith	48	Nelson-Rees, W. A. *et al.* (1968), *MCNL.*, **9**/**2**, 87–89.
Bison bison L.	60	Nichols, W., after Nelson-Rees, W. A. (1968), *MCNL.*, **9**/**2**, 87–89.
Bison bonasus L.	60	Melander, Y. (1959), *Hereditas*, **45**, 649–664.
Bos banteng Raffles	60	Wurster, D. H. and Benirschke, K. (1967), *MCNL.*, **8**/**3**, 226–229.
Bos gaurus Smith	58	Heck, L. *et al.* (1968), *Z. Säugetierk.*, **33**, 172–179.
Bos indicus L.	60	Makino, S. (1944), *Cytologia*, **13**, 247–264.
Bos taurus L.	60	Makino, S. (1944), *Cytologia*, **13**, 247–264.
Boselaphus tragocamelus Pallas	46	Chandra, H. S. *et al.* (1967), *Chromosoma*, **21**, 211–220.
Bubalus buffelus L. = *Bubalus bubalis* Pallas	48	Makino, S. (1944), *Cytologia*, **13**, 247–264.
—	50	Chandra, H. S. (1968), *MCNL.*, **9**/**4**, 231.
Poephagus grunniens L.	60	Wurster, D. H. and Benirschke, K. (1967), *MCNL.*, **8**/**3**, 226–229.

Syncerus caffer Sp. (*ex* Belgian Congo)	54	Wurster, D. H. and Benirschke, K. (1967), *MCNL.*, **8**/**3**, 226–229.
— (Kenya)	52	Ulbrich, F. and Fischer, H. (1966), *Z. Tierzücht. Zücht. Biol.*, **83**, 219–223.
Tetracerus quadricornis Blainv.	38	Wurster, D. H. and Benirschke, K. (1967), *MCNL.*, **8**/**3**, 226–229.
Tragelaphus angasi Gray	♂:55	Wurster, D. H. and Benirschke, K. (1968), *Chromosoma*, **25**, 152–171.
Tragelaphus spekei Sclater	30	Wurster, D. H. *et al.* (1968), *Chromosoma*, **23**, 317–323.
Tragelaphus s. strepsiceros Pallas	♀:32; ♂:31	Wallace, C. and Fairall, N. (1969), *MCNL.*, **10**/**1**, 11–12.
Caprinae		
Ammotragus lervia Pallas	58	Wurster, D. H. and Benirschke, K. (1967), *MCNL.*, **8**/**3**, 226–229.
Capra falconeri Wagner	60	Wurster, D. H. and Benirschke, K. (1968), *Chromosoma*, **25**, 152–171.
Capra hircus L.	60	Sokolov, I. (1930), *Bull. Bureau genet. Acad. Sc. URSS.*, **8**, 39–54.
Capra ibex L.	60	Hauschteck-Jungen, E. and Meili, R. (1967), *Chromosoma*, **21**, 198–210.
Hemitragus jemlahicus Smith	48	Chandra, H. S. *et al.* (1967), *Chromosoma*, **21**, 211–220.
Oreamnos americanus Blainv.	42	Wurster, D. H. and Benirschke, K. (1968), *MCNL.*, **9**/**2**, 80–82.
Ovibos m. moschatus Zimm.	48	Heck, H. *et al.* (1968), *Z. Säugetierk.* **33**, 172–179.
Ovis ammon cycloceros Hutton	58	Schmitt, J. and Ulbrich, F. (1968), *Z. Säugetierk.*, **33**, 180–186.
Ovis ammon laristanica Nason.	54	Schmitt, J. and Ulbrich, F. (1968), *Z. Säugetierk.*, **33**, 180–186.
Ovis ammon nigrimontana Severt.	56	Schmitt, J. and Ulbrich, F. (1968), *Z. Säugetierk.*, **33**, 180–186.
Ovis aries L.	54	Berry, J. H. (1941), *J. Hered.*, **32**, 261–267.
Ovis canadensis Shaw	54	Wurster, D. H. and Benirschke, K. (1968), *Chromosoma*, **25**, 152–171.

Ovis musimon Schr.	54	Schmitt, J. and Ulbrich, F. (1968), *Z. Säugetierk.*, **33**, 180–186.
Ovis orientalis Gmel.	54	Manna, G. K. and Talukdar, M. (1965), *MCNL.*, **17**, 78.
Rupicapra rupicapra L.	58	Gropp, A. and Giers, D. (1969), *MCNL.*, **10**/**1**, 19.
Saiga tatarica L.	60	Wurster, D. H. and Benirschke, K. (1968), *Chromosoma*, **25**, 152–171.
Cephalophinae		
Cephalophus dorsalis Gray	60	Hard, W. L. (1969), *MCNL.*, **10**/**4**, 216–217.
Cephalophus grimmi L.	60	Koulischer, L. *et al.* (1967), *Acta Zool. Path.* (*Antwerp*)., **43**, 135–141.
Cephalophus jentinki Thos.	60	Hard, W. L. (1969), *MCNL.*, **10**/**4**, 216–217.
Cephalophus maxwelli H.Sm.	60	Hard, W. L. (1969), *MCNL.*, **10**/**4**, 216–217.
Cephalophus niger Gray	60	Hard, W. L. (1969), *MCNL.*, **10**/**4**, 216–217.
Cephalophus rufilatus Gray	60	Hard, W. L. (1969), *MCNL.*, **10**/**4**, 216–217.
Hippotraginae		
Addax nasomaculatus Blainv.	58	Wurster, D. H. and Benirschke, K. (1968), *Chromosoma*, **25**, 152–171.
Alcelaphus boselaphus Pallas	♀939	Wurster, D. H. and Benirschke, K. (1967), *MCNL.*, **8**/**3**, 226–229.
Connochaetes gnu Zimm.	58	Wurster, D. H. and Benirschke, K. (1968), *Chromosoma*, **25**, 152–171.
Connochaetes taurinus Burck.	58	Wurster, D. H. and Benirschke, K. (1968), *Chromosoma*, **25**, 152–171.
Damaliscus dorcas Pallas	38	Wurster, D. H. and Benirschke, K. (1967), *MCNL.*, **8**/**3**, 226–228.
Hippotragus niger Harris	60	Wurster, D. H. and Benirschke, K. (1968), *Chromosoma*, **25**, 152–171.
Kobus ellipsiprymnus Ogilb.	50	Wurster, D. H. and Benirschke, K. (1968), *Chromosoma*, **25**, 152–171.
Kobus kob Erxl.	50	Taylor, K. M. *et al.* (1967), *MCNL.*, **8**/**3**, 233–234.
Kobus megaceros Heugl.	52	Wurster, D. H. and Benirschke, K. (1968), *Chromosoma*, **25**, 152–171.

Oryx leucoryx Pallas	58	Newnham, R. and Davidson, W. M. (1967), *MCNL.*, **8**/**1**, 15.
—	56	Wurster, D. H. and Benirschke, K. (1968), *Chromosoma*, **25**, 152–171.
Antilocaprinae		
Antilocapra americana Ord.	56	Wurster, D. H. and Benirschke, K. (1967), *MCNL.*, **8**/**3**, 226–228.
	PRIMATES	
	Lemuridae	
Cheirogaleus major Geoff.	66	Chu, E. H. Y. and Swomley, B. A. (1961), *Science*, **133**, 1925–1926.
Hapalemur griseus Link	54	Chiarelli, B. (1961) *MCNL.*, **6**, 3–4.
Hapalemur griseus olivaceus Geoff.	58	Chu, E. H. Y. and Swomley, B. A. (1961), *Science*, **133**, 1925–1926.
Lemur catta L.	56	Chiarelli, B. (1961), *MCNL.*, **6**, 3–4.
Lemur fulvus Geoff.	58	Chiarelli, B. (1961), *MCNL.*, **6**, 3–4.
Lemur f. fulvus Geoff.	48	Chu, E. H. Y. and Swomley, B. A. (1966), *MCNL.*, **20**, 55–64.
Lemur fulvus albifrons Geoff.	60	Chu, E. H. Y. and Swomley, B. A. (1961), *Science*, **133**, 1925–1926.
Lemur fulvus collaris Geoff.	48	Chu, E. H. Y. and Bender, M. A. (1962), *Ann. N.Y. Acad. Sci.*, **102**, 253.
Lemur fulvus rufus Aud.	60	Chu, E. H. Y. and Swomley, B. A. (1961), *Science*, **133**, 1925–1926.
Lemur macaco L.	44	Chu, E. H. Y. and Bender, M. A. (1961), *Science*, **133**, 1399.
Lemur mongoz L.	60	Chu, E. H. Y. and Swomley, B. A. (1961), *Science*, **133**, 1925–1926.
Lemur sp. nova	52	Chu, E. H. Y. and Swomley, B. A. (1961), *Science*, **133**, 1925–1926.
Lemur variegatus Kerr	46	Chu, E. H. Y. and Swomley, B. A. (1961), *Science*, **133**, 1925–1926.

Microcebus murinus Miller	66	Chu, E. H. Y. and Swomley, B. A. (1961), *Science*, **133**, 1925–1926.
	Indriidae	
Propithecus v. verreauxi Grand.	48	Chu, E. H. Y. and Bender, M. A. (1962), *Ann. N.Y. Acad. Sci.*, **102**, 253.
Propithecus v. coquereli M.E.	48	Chu, E. H. Y. and Bender, M. A. (1962), *Ann. N.Y. Acad. Sci.*, **102**, 253.
	Lorisidae	
Arctocebus calabarensis Smith	52	Egozcue, J. and Vilarasau, M. (1966), *MCNL.*, **20**, 53–54.
Galago crassicaudatus Geoff.	62	Chu, E. H. Y. and Bender, M. A. (1961), *Science*, **133**, 1399.
Galago senegalensis Geoff.	38	Matthey, R. (1955), *Revue Suisse Zool.*, **62**, 190–197.
Loris tardigrada L.	62	Manna, G. K. and Talukdar, M. (1966), *MCNL.*, **17**, 78.
Nycticebus coucang Bodd.	50	Bender, M. A. and Mettler, L. E. (1958), *Science*, **128**, 186–190.
Nycticebus coucang bengalensis Fisch.	50	Chu, E. H. Y. and Bender, M. A. (1962), *Ann. N.Y. Acad. Sci.*, **102**, 253.
Nycticebus pygmaeus Bonh.	50	Chu, E. H. Y. and Bender, M. A. (1962), *Ann. N.Y. Acad. Sci.*, **102**, 253.
Perodicticus potto Geoff.	62	Chu, E. H. Y. and Bender, M. A. (1961), *Science*, **133**, 1399.
	Tarsiidae	
Tarsius bancanus Horsf.	80	Klinger, H. P. (1963), *Cytogenetics*, **2**, 140–151.
Tarsius syrichta L.	80	Chiarelli, B. and Egozcue, J. (1968), *MCNL.*, **9**/**2**, 85.
	Tupaiidae	
Tupaia chinensis Anders	62	Arrighi, F. E. (1969), *MCNL.*, **10**/**1**, 27.
Tupaia glis Diard.	62	Klinger, H. P. (1963), *Cytogenetics*, **2**, 140–151.
—	60	Hsu, T. C. and Johnson, M. L. (1963), *Am. Nat.*, **97**, 127–129.
Tupaia longipes Th	60	Arrighi, F. E. (1969), *MCNL.*, **10**/**1**, 27.

Tupaia minor Günth.	66	Arrighi, F. E. (1969), *MCNL.*, **10/1**, 27.
Tupaia montana Th.	68	Arrighi, F. E. (1969), *MCNL.*, **10/1**, 27.
Tupaia palawanensis Th.	52	Arrighi, F. E. (1969), *MCNL.*, **10/1**, 27.
Urogale everetti Th.	44	Chu, E. H. Y. and Bender, M. A. (1962), *Ann. N.Y. Acad. Sci.*, **102**, 253–266.
	Cebidae	
Alouatta caraya Humb.	52	Egozcue, J. and Vilarasau, M. (1965), *MCNL.*, **17**, 84.
Alouatta seniculus L.	44	Bender, M. A. and Mettler, L. E. *In* Chu, E. H. Y. and Bender, M. A. (1966), *MCNL.*, **20**, 56–65.
Alouatta villosa Gray	53	Egozcue, J. *In* "Comparative Mammalian Cytogenetics," Springer Verlag, New York.
Aotus trivirgatus Humb.	50	Chiarelli, B. and Barberis, L. (1966), *MCNL.*, **22**, 216.
Ateles arachnoides Geoff.	34	Egozcue, J. (1969), *In* "Comparative Mammalian Cytogenetics," Springer Verlag, New York.
Ateles belsebuth Geoff.	34	Egozcue, J. (1969), *In* "Comparative Mammalian Cytogenetics," Springer Verlag, New York.
Ateles fuscipes Gray	34	Wurster, D. H. and Benirschke, K. (1969), *MCNL.*, **10/1**, 3–4.
Ateles geoffroyi cucullatus Gray	34	Bender, M. A. and Mettler, L. E. (1958), *Science*, **128**, 186–190.
Ateles paniscus chamek Humb.	34	Bender, M. A. and Mettler, L. E. (1958), *Science*, **128**, 186–190.
Cacajo rubicundus Geoff.	46	Egozcue, J. (1969), *In* "Comparative Mammalian Cytogenetics," Springer Verlag, New York.
Callicebus cupreus Spix	46	Bender, M. A. and Mettler, L. E. (1958), *Science*, **128**, 186–190.
Callicebus moloch Hoffm.	46	Egozcue, J. (1969), *In* "Comparative Mammalian Cytogenetics," Springer Verlag, New York.
Callicebus torquatus Hoffm.	20	Egozcue, J. (1969), *In* "Comparative Mammalian Cytogenetics," Springer Verlag, New York.
Callimico goeldii Th.	48	Bender, M. A. and Mettler, L. E. (1960), *Cytologia*, **25**, 400–404.

Cebus albifrons Humb.	54	Egozcue, J. and Vilarasau, M. (1966), *MCNL.*, **20**, 66.
Cebus apella Erxl.	54	Bender, M. A. and Mettler, L. E. (1958), *Science*, **128**, 186–190.
Cebus capucinus Erxl.	54	Bender, M. A. and Mettler, L. E. (1958), *Science*, **128**, 186–190.
—	52	Egozcue, J. and Vilarasau, M. (1966), *MCNL.*, **20**, 71.
—	54	Chiarelli, B. and Barberis, L. (1966), *MCNL.*, **22**, 216.
Cebus fatuellus L. = *C. apella* Erxl.	54	Chiarelli, B. (1961), *MCNL.*, **6**, 3–4.
Lagothrix ubericola Elliot	62	Bender, M. A. and Eide, P. E. (1963), *MCNL.*, **9**, 40.
Pithecia pithecia L.	46	Egozcue, J. (1969), *In* "Comparative Mammalian Cytogenetics," Springer Verlag, New York.
Saimiri sciureus L.	44	Bender, M. A. and Mettler, L. E. (1958), *Science*, **128**, 186–190.
Saimiri madeirae juruanus Lönnb.	44	Egozcue, J. *et al.* (1967), *MCNL.*, **8**/**1**, 14.
Saimiri boliviensis nigriceps Th.	44	Egozcue, J. *et al.* (1967), *MNCL.*, **8**/**1**, 14.
	Callithricidae-Hapalidae	
Callithrix argentata L.	44	Egozcue, J. (1969), *In* "Comparative Mammalian Cytogenetics," Springer Verlag, New York.
Callithrix chrysoleucos Wagner	46	Bender, M. A. and Mettler, L. E. (1960), *Cytologia*, **25**, 400–404.
Callithrix humeralifer Geoff.	44	Egozcue, J. (1969), *In* "Comparative Mammalian Cytogenetics," Springer Verlag, New York.
Callithrix jacchus L.	46	Chiarelli, B. (1961), *MCNL.*, **6**, 3.
Callithrix pygmaea Spix = *Cebuella pygmaea*	44	Benirschke, K. and Brownhill, L. E. (1963), *MCNL.*, **9**, 6
Leontocebus illigeri Push = *Saguinus fuscicollis illigeri*	46	Bender, M. A. and Mettler, L. E. (1960), *Cytologia*, **25**, 400–404.
Leontocebus nigricollis Elliot = *Saguinus nigricollis*	46	Benirschke, K. and Brownhill, L. E. (1962), *Cytogenetics*, **1**, 245–257.
Leontocebus oedipus L. = *Saguinus oedipus* = *Oedipomidas oedipus*	46	Chiarelli, B. (1961), *MCNL.*, **6**, 3.
Leontideus rosalia L. = *Leontocebus rosalia*	46	Benirschke, K. and Brownhill, L. E. (1963), *MCNL.*, **9**, 6.

Tamarinus leucopus Günther = *Saguinus leucopus*	46	Benirschke, K. and Brownhill, L. E. (1962), *Cytogenetics*, **2**, 331–341.
Tamarinus mystax Spix = *Saguinus mystax*	46	Wohnus, J. F. and Benirschke, K. (1966), *Cytogenetics*, **5**, 94–105.
Tamarin tamarin Link = *Saguinus tamarin*	46	Benirschke, K. and Brownhill, L. E. (1963), *MCNL.*, **9**, 6
	Cercopithecidae	
Colobinae		
Colobus polycomos Zimm.	44	Chiarelli, B. (1962), *Experientia*, **18**, 405–406.
Colobus badius Kerr = *Procolobus kirkii* Gray	44	Kuhn, H. J. (1967), *In* "Progress in Primatology," Fischer V., Stuttgart.
Nasalis variatus Wurmb.	48	Chiarelli, B. (1966), *Experientia*, **22**, 797.
Presbytis cristatus Raffles	44	Wurster, D. H. and Benirschke, K. (1969), *MCNL.*, **10/1**, 3–4.
Presbytis entellus Dufr.	50	Makino, S. (1952), *Cytologia*, **16**, 288–301.
Presbytis e. entellus Dufr.	44	Sharma, T. and Kakati, S. (1966), *MCNL.*, **20**, 70.
Presbytis obscurus Reid.	44	Chiarelli, B. (1962), *Experientia*, **18**, 405–406.
Presbytis phayrei Blyth	44	Wurster, D. H. and Benirschke, K. (1969), *MCNL.*, **10/1**, 3–4.
Presbytis senex Erxl.	44	Goldstein, P. R. and Birdwell, T. R. (1967), *MCNL.*, **8/3**, 197–301.
Pygathrix nemaeus L.	44	Wurster, D. H. and Benirschke, K. (1969), *MCNL.*, **10/1**, 3–4.
Papinae		
Cercocebus albigena Gray	42	Chu, E. H. Y. and Bender, M. A. (1961), *Science*, **133**, 1399.
Cercocebus aterrimus Oudem.	42	Chiarelli, B. (1962), *Experientia*, **18**, 405–406.
Cercocebus galeritus Ptrs	42	Chu, E. H. Y. and Bender, M. A. (1961), *Science*, **133**, 1399.
Cercocebus torquatus lunulatus Temm.	42	Chu, E. H. Y. and Giles, N. H. (1957), *Am. Nat.*, **91**, 273–282.
Cercocebus t. torquatus Kerr	42	Chiarelli, B. (1962), *Experientia*, **18**, 405–406.

Macaca assamensis McClell.	42	Chiarelli, B. (1962), *Experientia*, **18**, 405–406.
Macaca cyclopis Swinh.	42	Makino, S. (1952), *Cytologia*, **16**, 288–301.
Macaca fuscata Blyth	42	Chiarelli, B. (1962), *Experientia*, **18**, 405–406.
Macaca irus Cuv.	42	Chu, E. H. Y. and Bender, M. A. (1961), *Science*, **133**, 1399.
Macaca maura Cuv.	42	Chiarelli, B. (1962), *Experientia*, **18**, 405–406.
Macaca mulatta L.	42	Darlington, C. D. and Haque, A. (1955), *Nature* (*Lond*), **175**, 32.
Macaca nemestrina L.	42	Darlington, C. D. and Haque, A. (1955), *Nature*. (*Lond*), **175**, 32.
Macaca radiata Geoff.	42	Chiarelli, B. (1962), *Experientia*, **18**, 405–406.
Macaca silenus L.	42	Chiarelli, B. (1962), *Experientia*, **18**, 405–406.
Macaca sinica L.	42	Chiarelli, B. (1962), *Experientia*, **18**, 405–406.
Macaca speciosa Cuv.	42	Chiarelli, B. (1962), *Experientia*, **18**, 405–406.
Macaca sylvana L.	42	Chiarelli, B. (1962), *Experientia*, **18**, 405–406.
Papio comatus Geoff.	42	Chiarelli, B. (1962), *Experientia*, **18**, 405–406.
Papio cynocephalus L.	42	Chiarelli, B. (1962), *Experientia*, **18**, 405–406.
Papio doguera Puch.	42	Chu, E. H. Y. and Giles, N. H. (1957), *Am. Nat.*, **91**, 273–282.
Papio hamadryas L.	42	Chiarelli, B. (1962), *Experientia*, **18**, 405–406.
Papio leucophaeus Cuv.	42	Chiarelli, B. (1962), *Experientia*, **18**, 405–406.
Papio papio Desm.	42	Darlington, C. D. and Haque, A. (1955), *Nature* (*Lond*), **175**, 32.
Papio sphinx L.	42	Bender, M. A. and Mettler, L. E. (1958), *Science*, **128**, 186–190.
Papio u. ursinus Kerr	42	Ismaïl, H. and Tobias P. V., (1956), *In* "Chromosomes, Sex-cells and Evolution in a Mammal," P. V. Tobias, ed., Lund and Co., London.
Theropithecus gelada Rüppel = *Papio gelada*	42	Chiarelli, B. (1962), *Experientia*, **18**, 405–406.
Cercopithecinae		
Cercopithecus aethiops L. = *pygerythrus*	60	Chu, E. H. Y. and Giles, N. H. (1957), *Am. Nat.*, **91**, 273–282.

C. a. sabaeus L.	60	Chu, E. H. Y. and Giles N. H. (1957), *Am. Nat.*, **91**, 73–282.
C. a. tantalus Ogilb.	60	Chu, E. H. Y. and Giles, N. H. (1957), *Am. Nat.*, **91**, 273–282.
Cercopithecus cephus L.	66	Chiarelli, B. (1963), *Riv. Antrop.*, **50**, 87–124.
Cercopithecus diana L.	60	Chu, E. H. Y. and Giles, N. H. (1957), *Am. Nat.*, **91**, 273–282.
Cercopithecus diana roloway Schr.	58	Chiarelli, B. (1962), *MCNL.*, **22**, 210.
Cercopithecus hamlyni Pocock	64	de Boer, L. E. M. (1970), *J. Soc. Neerl. Genet.*, in press.
Cercopithecus l'Hoesti Scl.	72	Chu, E. H. Y. and Bender, M. A. (1961), *Science*, **133**, 1399.
Cercopithecus mitis Wolf	72	Bender, M. A. and Chu, E. H. Y. (1963), *In* "Evolution and Genetical Biology of Primates," Buetthez-Janusch, New York.
Cercopithecus mitis albogularis Skyes	72	Chiarelli, B. (1963), *Riv. Antrop.*, **50**, 87–124.
Cercopithecus mona mona Schr.	66	Bender, M. A. and Mettler, L. E. (1958), *Science*, **128**, 186–190.
Cercopithecus mona campbelli Waterh.	66	Chu, E. H. Y. and Giles, N. H. (1957), *Am. Nat.*, **91**, 273–282.
Cercopithecus mona denti Th.	66	Tappen, N. (1960), *Symp. Zool. Soc.* (*Lond*), **10**, 267–276.
Cercopithecus mona preussi Matsh.	66	Chiarelli, B. (1962), *Experientia*, **18**, 405–406.
Cercopithecus nictitans L. = *C. petaurista* = *C. ascanius*	70	Kuhn, H. J. (1967), *In* "Progress in Primatology," Fischer V., Stuttgart.
Cercopithecus nictitans buttikoferi Jent.	66	Chu, E. H. Y. and Giles, N. H. (1957), *Am. Nat.*, **91**, 273–282.
Cercopithecus nigroviridis Pocock	60	Chiarelli, B. (1962), *Experientia*, **18**, 405–406.
Cercopithecus (*Allenopithecus*) *nigroviridis* Pocock	48 (?)	Wurster, D. H. and Benirschke, K. (1969), *MCNL.*, **10/1**, 3–4.
Cercopithecus niger Desm.	42	Chiarelli, B. (1962), *Experientia.*, **18**, 405–406.
Cercopithecus (*Erythrocebus*) *patas* Schr.	54	Chu, E. H. Y. and Giles N. H. (1957), *Am. Nat.*, **91**, 273–282.
Cercopithecus talapoin Schr.	54	Chiarelli, B. (1962), *Experientia*, **18**, 405–406.

	Hylobatidae	
Hylobates lar L.	44	Chiarelli, B. (1962), *Experientia*, **18**, 405–406.
Hylobates agilis Cuv.	44	Chiarelli, B. (1962), *Experientia*, **18**, 405–406.
Hylobates concolor Harlan	52	Wurster, D. H. and Benirschke, K. (1969), *MCNL.*, **10/1**, 3–5.
Hylobates hoolock Harlan	44	Chiarelli, B. (1962), *Experientia*, **18**, 405–406.
Hylobates moloch Aude.	44	Chiarelli, B. (1962), *Experientia*, **18**, 405–406.
Symphalangus syndactylus Raffles	50	Klinger, H. P. (1963), *Cytogenetics*, **2**, 140–151.
	Pongidae	
Gorilla g. gorilla Savage and Wyeman	48	Hamerton, J. L. *et al.* (1963), *Cytogenetics*, **2**, 240–263.
Gorilla gorilla beringei Matschie	48	Hamerton, J. L. *et al.* (1963), *Cytogenetics*, **2**, 240–263.
Pan tr. troglodytes Blumenb.	48	Chiarelli, B. (1962), *Experientia*, **18**, 405–406.
Pan troglodytes paniscus	48	Chiarelli, B. (1962), *Experientia*, **18**, 405–406.
Pongo pygmaeus Hopp.	48	Chiarelli, B. (1961), *Nature*, (*Lond*), **4799**, 285.
	Hominidae	
Homo sapiens L.	46	Tjio, J. H. and Levan, A. (1956), *Hereditas*, **42**, 1–6.

LAGOMORPHA

	Leporidae	
Lepus alleni alleni Mearns	48	Hsu, T. C. and Benirschke, K. (1967), *In* "Atlas of Mammalian Chromosomes," Springer Verlag, New York.
Lepus americanus Erxl.	48	Hsu, T. C. (1965), *MCNL.*, **15**, 114–115.
Lepus californicus Gray	48	Worthington, D. H. and Sutton, D. A. (1966), *MCNL.*, **22**, 194.
Lepus capensis tolai Pallas	48	Vorontsov, N. N. and Ivanitskaia, E. J. (1969), *Mlekop.*, Novo-Sibirsk.
Lepus europaeus occidentalis Winton	48	Muldal, S. (1950), *J. Innes Hort. Inst.*, *Ann. Rep.*, **41**, 39–41.
Lepus formosus Th.	48	Tateishi, S. (1936), *Zool. Mag.* (*Tokyo*), **48**, 622–624.
Lepus gichiganus ainu Allen	48	Makino, S. (1950), *Cytologia*, **16**, 288–301.
Lepus townsendii campanius Holl.	48	Jalal, S. M. and coll. (1967), *MCNL.*, **8/3**, 194.

Oryctolagus cuniculus L.	44	Painter, T. S. (1925), *Science*, **61**, 423–424.
Sylvilagus auduboni Baird	42	Worthington, D. H. and Sutton, D. A. (1966), *MCNL.*, **22**, 194.
Sylvilagus bachmani Waterh.	48	Hsu, T. C. (1965), *MCNL.*, **15**, 114–115.
Sylvilagus floridanus Allen	42	Palmer, C. and Armstrong, R. (1967), *MCNL.*, **8**/**4**, 282–283.
Sylvilagus nuttallii Bachm.	42	Worthington, D. H. and Sutton, D. A. (1966), *MCNL.*, **22**, 194.
Sylvilagus transitionalis Bangs	52	Holden, H. E. (1968), *MCNL.*, **9**/**4**, 230.
	Ochotonidae	
Ochotona d. daurica Pallas	50	Vorontsov N. N. and Ivanitskaia, E. J. (1969), *Mlekop.*, Novo-Sibirsk, 1969.
Ochotona hyperborea kolymensis All.	40	Vorontsov, N. N. and Ivanitskaia, E. J. (1969), *Mlekop.*, Novo-Sibirsk.
Ochotona hyperborea yesoensis Kishida	40	Hayata, I. and Shimba, H. (1969), *J. Fac. Sci. Hokkaido Univ.*, **17**, 393–396.
Ochotona macrotis sacana Th.	62	Vorontsov, N. N. and Ivanitskaia, E. J. (1969), *Mlekop.*, Novo-Sibirsk.
Ochotona pricei sushkini Th.	38	Vorontsov, N. N. and Ivanitskaia, E. J. (1969), *Mlekop.*, Novo-Sibirsk.
Ochotona rufescens Gray	60	Nadler, C. F. *et al.* (1969), *Experientia*, **25**, 774–775.
Ochotona rutila Sev.	62	Vorontsov, N. N. and Ivanitskaia, E. J. (1969), *Mlekop.*, Novo-Sibirsk.
	RODENTIA	
	Spalacidae	
Spalax ehrenbergi Nehr. (Israel)	52, 54, 58, 60	Wahrman, J. *et al.* (1969), *Science*, **164**, 82–84.
Spalax leucodon Nord. (Caucasus)	48	Matthey, R. (1959), *Revue Suisse Zool.*, **66**, 175–211.
Spalax leucodon Nord. (Bulgaria)	54	Walknowska, J. (1963), *Folia Biol.*, **11**, 293–309.
Spalax leucodon Nord. (Banat, north from Danube)	48	Soldatovic, B. *et al.* (1966), *Caryologia*, **19**, 261–265.
Spalax leucodon Nord. (Rumania, Constanta and Jucu)	56, 50	Raiçu, P. *et al.* (1968), *Caryologia*, **21**, 127–135.

Spalax microphtalmus Guld.	62	Raiçu, P. *et al.* (1968), *Caryologia*, **21**, 127–135.
Spalax nehringi Sat.	50	Orlov, V. N. (1969), *Mlekop.*, Novo-Sibirsk.
	Sciuridae	
Ammospermophilus h. harrisii Audub. and Bachm.	32	Nadler, C. F. and Sutton, D. A. (1962), *Proc. Soc. Exp. Biol.*, **110**, 36–38.
Ammospermophilus l. leucurus Merriam	32	Nadler, C. F. and Sutton, D. A. (1962), *Proc. Soc. Exp. Biol.*, **110**, 3638.
Callosciurus flavimanus Geoff.	40	Nadler, C. F. and Hoffmann, R. S. (1970), *Experientia*, in press.
Castor canadensis Kuhl	40	Hsu, T. C. and Benirschke, K. (1968), *In* "Atlas of Mammalian Chromosomes", Vol. 2. Springer, New York.
Cynomys ludovicianus Ord.	52 (?)	Makino, S. (1953), *Science*, **118**, 3073.
Cynomys ludovicianus Ord.	50	Nadler, C. F. and Harris, K. E. (1967), *Experientia*, **23**, 41.
Dremomys rufigenis Blanf.	38	Nadler, C. F. and Hoffmann, R. S. (1970), *Experientia*, in press.
Eutamias a. amoenus Allen	38	Nadler, C. F. (1964), *Am. Midl. Nat.*, **72**, 298–312.
Eutamias asiaticus uthensis Gmel.	38	Nakamura, T. (1953), *Kromosomo*, **16**, 602–604.
Eutamias d. dorsalis Baird	38	Nadler, C. F. (1964), *Am. Midl. Nat.*, **72**, 298–312.
Eutamias minimus operarius Merriam	38	Nadler, C. F. and Block, M. H. (1962), *Chromosoma*, **13**, 1–15.
Eutamias minimus caryi Allen	38	Nadler, C. F. (1964), *Am. Midl. Nat.*, **72**, 298–312.
Eutamias minimus consobrinus Allen	38	Nadler, C. F. (1964), *Am. Midl. Nat.*, **72**, 298–312.
Eutamias minimus neglectus Allen	38	Nadler, C. F. (1964), *Am. Midl. Nat.*, **72**, 298–312.
Eutamias q. quadrivittatus Say	38	Nadler, C. F. (1964), *Am. Midl. Nat.*, **72**, 298–312.
Eutamias quadrivittatus hopiensis Merriam	38	Nadler, C. F. and Block, M. H. (1962), *Chromosoma*, **13**, 1–15.
Eutamias sibiricus Laxm.	38	Nadler, C. F. *et al.* (1969), *Experientia*, **25**, 868–869.
Eutamias speciosus frater Allen	38	Nadler, C. F. and Block, M. H. (1962), *Chromosoma*, **13**, 1–15.
Eutamias t. townsendii Bachm.	38	Nadler, C. F. (1964), *Am. Midl. Nat.*, **72**, 298–312.
Eutamias townsendii cooperi Baird	38	Nadler, C. F. (1964), *Am. Midl. Nat.*, **72**, 298–312.

Eutamias townsendii senex Allen	38	Nadler, C. F. (1964), *Am. Midl. Nat.*, **72**, 298–312.
Eutamias unbrinus montanus White	38	Nadler, C. F. and Block, M. H. (1962), *Chromosoma*, **13**, 1–15.
Funambulus pennanti Wrough.	54	Venkatasubba Rao, S. R. and Sharda (1964), *Cytogenetics*, **3**, 342–346.
Funambulus pennanti Wrough.	56 (?)	Manna, G. K. and Talukdar, M. (1965), *MCNL.*, **17**, 77–78.
Funambulus pennanti Wrough.	38	Ray-Chaudhuri, S. P. *et al.* (1968), *MCNL.*, **9**/**2**, 82–84.
Glaucomys sabrinus lascivus Bangs	48	Nadler, C. F. and Sutton, D. A. (1967), *Experientia*, **23**, 249–251.
Glaucomys volans L.	48	Nadler, C. F. and Sutton, D. A. (1967), *Experientia*, **23**, 249–251.
Marmota baibacina Brandt	38	Vorontsov, N. N. *et al.* (1969), *Zool.Ƶ.* (Russian), **48** 317–334.
Marmota b. bobak Muller	38	Liapounova, E. A. and Vorontsov, N. N. (1969), *Mlekop.*, Novo-Sibirsk.
Marmota c. caligata Esch.	42	Rausch, R. L. and Rausch, V. R. (1965), *Chromosoma*, **16**, 618–623.
Marmota broweri Hall and Gilmore	36	Rausch, R. L. and Rausch, V. R. (1965). *Chromosoma*, **16**, 618–623.
Marmota c. camtschatica Pallas	40	Liapounova, E. A. and Vorontsov, N. N. (1969), *Mlekop.*, Novo-Sibirsk.
Marmota caudata Jaquem.	38	Vorontsov, N. N. *et al.* (1969), *Zool.Ƶ.* (Russian), **48**, 317–334.
Marmota flaviventris Audub. and Bachm.	42	Makino, S. (1953), *Science*, **118**, 3073.
Marmota marmota L.	38	Liapounova, E. A. and Vorontsov, N. N. (1969), *Mlekop.*, Novo-Sibirsk.
Marmota menzbieri Kashk.	38	Vorontsov, N. N. *et al.* (1969), *Zool.Ƶ.* (Russian), **48**, 317–334.
Marmota monax L.	38	Couser, W. *et al.* (1963), *Cytologia*, **28**, 108–111.
Marmota olympus Merriam	40	Rausch, R. L. and Rausch, V. R. (1965), *Chromosoma*, **16**, 618–623.

Marmota s. sibirica Radde	38	Liapounova, E. A. and Vorontsov, N. N. (1969), *Mlekop.*, Novo-Sibirsk.
Menetes berdmorei Blyth	62	Nadler, C. F. and Hoffmann, R. S. (1970), *Experientia*, in press.
Sciurus aberti Woodh.	40	Hsu, T. C. (1965), *MCNL.*, **15**, 114–115.
Sciurus anomalus Güldenst.	40	Nadler, C. F. and Hoffmann, R. S. (1970), *Experientia*, in press.
Sciurus c. carolinensis Gmel.	40	Nadler, C. F. and Sutton, D. A. (1967), *Experientia*, **23**, 249–251.
Sciurus carolinensis pennsylvanicus Ord.	40	Nadler, C. F. and Sutton, D. A. (1967), *Experientia*, **23**, 249–251.
Sciurus carolinensis Gmel. (*ssp* ? "eastern grey squirrel")	40	Singh, R. P. and Sergovich, F. R. (1966), *MCNL.*, **22**, 208.
Sciurus granatensis Humb.	42	Nadler, C. F. and Hoffmann, R. S. (1970), *Experientia*, in press.
Sciurus g. griseus Ord.	40	Nadler, C. F. and Hoffmann, R. S. (1970), *Experientia*, in press.
Sciurus niger rufiventer Geoff.	40	Nadler, C. F. and Sutton, D. A. (1967), *Experientia*, **23**, 249–251.
Sciurus (*Tenes*) *persicus anomalus* Gmel.	40	Liapounova, E. A. and Zholnerovskaia, E. I. (1969), *Mlekop.*, Novo-Sibirsk.
Sciurus vulgaris L.	40	Muldal, S. (1950), *J. Innes Hort. Inst. Ann. Rep.*, **40**, 18–19.
Spermophilus = *Citellus armatus* Kennic.	34	Nadler, C. F. (1966), *J. Mamm.*, **47**, 579–596.
Spermophilus (*Citellus*) *beecheyi douglasii* Rich.	34	Nadler, C. F. and Sutton, D. A. (1962), *Proc. Soc. Exp. Biol.*, **110**, 36–38.
Spermophilus (*Citellus*) *b. beldingi* Merriam	30	Nadler, C. F. (1962), *Proc. Soc. Exp. Biol.*, **110**, 785–788.
Spermophilus (*Citellus*) *citellus* L.	40	Vorontsov, N. N. and Liapounova, E. A. (1969), *Mlekop.*, Novo-Sibirsk.
Spermophilus (*Citellus*) *citellus lascarevi* Mart.	40	Vorontsov, N. N. and Liapounova, E. A. (1969), *Mlekop.*, Novo-Sibirsk.
Spermophilus (*Citellus*) *citellus balcanicus* Markov	40	Vorontsov, N. N. and Liapounova, E. A. (1969), *Mlekop.*, Novo-Sibirsk.

Spermophilus (*Citellus*) *columbianus* Ord.	32	Nadler, C. F. (1966), *J. Mamm.*, **47**, 579–596.
Spermophilus (*Citellus*) *d. dauricus* Brandt	36	Vorontsov, N. N. and Liapounova, E. A. (1969), *Mlekop.*, Novo-Sibirsk.
Spermophilus (*Citellus*) *erythrogenys* Brandt	36	Vorontsov, N. N. and Liapounova, E. A. (1969), *Mlekop.*, Novo-Sibirsk.
Spermophilus (*Citellus*) *franklini* Sabine	42	Nadler, C. F. (1966), *J. Mamm.*, **47**, 579–596.
Spermophilus (*Citellus*) *fulvus* Licht.	36	Vorontsov, N. N. and Liapounova, E. A. (1969), *Mlekop.*, Novo-Sibirsk.
Spermophilus (*Citellus*) *l. lateralis* Say	42	Nadler, C. F. and Sutton, D. A. (1962), *Proc. Soc. Exp. Biol.*, **110**, 36–38.
Spermophilus (*Citellus*) *m. major* Pallas	36	Vorontsov, N. N. and Liapounova, E. A. (1969), *Mlekop.*, Novo-Sibirsk.
Spermophilus (*Citellus*) *mexicanus parvidens* Mearns	34	Nadler, C. F. and Sutton, D. A. (1962), *Proc. Soc. Exp. Biol.*, **110**, 36–38.
Spermophilus (*Citellus*) *musicus saturatus* Ogn.	36	Vorontsov, N. N. and Liapounova, E. A. (1969), *Mlekop.*, Novo-Sibirsk.
Spermophilus (*Citellus*) *parryi kennicotti* Rich.	32	Vorontsov, N. N. and Liapounova, E. A. (1969), *Dokl. Akad. Nauk S SSR*, **187**, 207–210.
Spermophilus (*Citellus*) *pygmaeus* Pallas	36	Vorontsov, N. N. and Liapounova, E. A. (1969), *Mlekop.*, Novo-Sibirsk.
Spermophilus (*Citellus*) *relictus* Kaschk.	36	Vorontsov, N. N. and Liapounova, E. A. (1969), *Mlekop.*, Novo-Sibirsk.
Spermophilus (*Citellus*) *r. richardsonii* Sabine	36	Nadler, C. F. (1968), *J. Mamm.*, **49**, 312–314.
Spermophilus (*Citellus*) *richardsonii aureus* Davis	34	Nadler, C. F. (1968), *J. Mamm.*, **49**, 312–314.
Spermophilus (*Citellus*) *richardsonii elegans* Kennic.	34	Nadler, C. F. and Sutton, D. A. (1962), *Proc. Soc. Exp. Biol.*, **110**, 36–38.
Spermophilus (*Citellus*) *richardsonii nevadensis* Howell	34	Nadler, C. F. (1968), *J. Mamm.*, **49**, 312–314.
Spermophilus (*Citellus*) *spilosoma marginatus* Bailey	32	Nadler, C. F. and Hughes, C. E. (1966), *J. Mamm.*, **47**, 46–53.
Spermophilus (*Citellus*) *spilosoma canescens* Merriam	32	Nadler, C. F. and Hughes, C. E. (1966), *J. Mamm.*, **47**, 46–53.
Spermophilus (*Citellus*) *suslicus averini* Migulin	36	Vorontsov, N. N. and Liapounova, E. A. (1969), *Mlekop.*, Novo-Sibirsk.

Spermophilus (*Citellus*) *x. xanthoprymnus* Bennet	42	Vorontsov, N. N. and Liapounova, E. A. (1969), *Mlekop.*, Novo-Sibirsk.
Spermophilus (*Citellus*) *tereticaudus neglectus* Merriam	36	Nadler, C. F. (1962), *Proc. Soc. Exp. Biol.*, **110**, 785–788.
Spermophilus (*Citellus*) *t. townsendii* Bachm.	36	Nadler, C. F. (1964), *Am. Midl. Nat.*, **72**, 298–312.
Spermophilus (*Citellus*) *townsendii idahoensis* Merriam	38	Nadler, C. F. (1968), *Cytogenetics*, **7**, 144–157.
Spermophilus (*Citellus*) *townsendii mollis* Kenn.	38	Nadler, C. F. (1968), *Cytogenetics*, **7**, 144–157.
Spermophilus (*Citellus*) *townsendii nancyae, nova ssp.*	38	Nadler, C. F. (1968), *Cytogenetics*, **7**, 144–157.
Spermophilus (*Citellus*) *townsendii vigilis* Merriam	46	Nadler, C. F. (1968), *Cytogenetics*, **7**, 144–157.
Spermophilus (*Citellus*) *t. tridecimlineatus* Mitch.	34	Nadler, C. F. (1962), *Proc. Soc. Exp. Biol.*, **110**, 785–788.
Spermophilus (*Citellus*) *tridecimlineatus pallidus* Allen	34	Nadler, C. F. and Hughes, C. E. (1966), *J. Mamm.*, **47**, 46–53.
Spermophilus (*Citellus*) *undulatus* Pallas	34	Nadler, C. F. (1966), *J. Mamm.*, **47**, 579–596.
Spermophilus (*Citellus*) *undulatus* Pallas	32	Vorontsov, N. N. and Liapounova, E. A. (1969), *Dokl. Akad. Nauk SSSR*, **187**, 207–210.
Spermophilus (*Citellus*) *washingtoni* Howell	36	Nadler, C. F. (1966), *J. Mamm.*, **47**, 579–596.
Spermophilopsis leptoadctylus Licht.	38	Nadler, C. F. *et al.* (1969), *Experientia*, **25**, 774–775.
Tamasciurus hudsonicus streatori Allen	46	Hsu, T. C. (1966), *MCNL.*, **19**, 22–23.
Tamasciurus hudsonicus fremonti Aud. and Bachm.	46	Nadler, C. F. and Sutton, D. A. (1962), *Proc. Soc. Exp. Biol.*, **110**, 36–38.
Tamasciurus douglasii Bachm.	46	Hsu, T. C. (1966), *MCNL.*, **19**, 22–23.
Tamias sibiricus Laxm.	38	Sasaki, M. *et al.* (1969), *Chrom. Inform. Serv.*, **9**, 6–8.
Tamias striatus griseus Mearns	38	Nadler, C. F. (1964), *Am. Midl. Nat.*, **72**, 298–312.
Tamias striatus lysteri Rich.	38	Nadler, C. F. and Block, M. H. (1962), *Chromosoma*, **13**, 1–5.
	Geomyidae	
Geomys bursarius breviceps Swenk	72	Matthey, R. (1960), *Mammalia*, **24**, 165–170.
Thomomys bottae alienus Goldm.	76	Patton, J. L. and Dingman, R. E. (1968), *J. Mamm.*, **49**, 1–13.
Thomomys bottae modicus Goldm.	76	Patton, J. L. and Dingman, R. E. (1968), *J. Mamm.*, **49**, 1–13.
Thomomys talpoides attenuatus Hall	48	Thaeler, C. S. (1968), *Chromosoma*, **25**, 172–183.
Thomomys talpoides bridgeri Merriam	40	Thaeler, C. S. (1968), *Chromosoma*, **25**, 172–183.

Thomomys talpoides clusius Coues	46	Thaeler, C. S. (1968), *Chromosoma*, **25**, 172–183.
Thomomys talpoides fossor Allen	56, 60	Thaeler, C. S. (1968), *Chromosoma*, **25**, 172–183.
Thomomys talpoides kaibabensis Goldm.	40	Thaeler, C. S. (1968), *Chromosoma*, **25**, 172–183.
Thomomys talpoides meritus Hall	44	Thaeler, C. S. (1968), *Chromosoma*, **25**, 172–183.
Thomomys talpoides ocius Merriam	56	Thaeler, C. S. (1968), *Chromosoma*, **25**, 172–183.
Thomomys talpoides pygmaeus Merriam	58	Thaeler, C. S. (1968), *Chromosoma*, **25**, 172–183.
Thomomys talpoides rostralis Hall	48, 56	Thaeler, C. S. (1968), *Chromosoma*, **25**, 172–183.
Thomomys unbrinus Richard.	78	Patton, J. L. and Dingman, R. E. (1968), *J. Mamm.*, **49**, 1–13.
	Echimyidae	
Proechimys cherriei Th.	26	Reig, O. A. *et al.* (1970), *Experientia*, **26**, 201–202.
Proechimys guyannensis quairae Th.	46	Reig, O. A. *et al.* (1970), *Experientia*, **26**, 201–202.
	Aplodontidae	
Aplodontia rufa Raf.	46	Hsu, T. C. (1965), *MCNL.*, **15**, 114–115.
	Heteromyidae	
Dipodomys agilis simulans Merriam	60	Jackson, L. and Hunzaker, D. (1968), *MCNL.*, **9**/**1**, 33.
Dipodomys deserti Step.	64	Jackson, L. and Hunzaker, D. (1968), *MCNL.*, **9**/**1**, 33.
Dipodomys merriami Mearns	68	Matthey, R. (1956), *Chromosoma*, **7**, 670–692.
Dipodomys merriami Mearns	52	Jackson, L. and Hunzaker, D. (1968), *MCNL.*, **9**/**1**, 33.
Dipodomys spectabilis perblandus Goldm.	72	Jackson, L. and Hunzaker, D. (1968), *MCNL.*, **9**/**1**, 33.
Liomys irroratus Gray	58	Makino, S. (1953), *Science*, **118**, 3073.
Perognathus amplus pergracilis Goldm.	56	Patton, J. L. (1967), *S. West. Nat.*, **12**, 429–438.
Perognathus amplus rotundus Goldm.	56	Patton, J. L. (1967), *S. West. Nat.*, **12**, 429–438.
Perognathus amplus taylori Goldm.	56	Patton, J. L. (1967), *S. West. Nat.*, **12**, 429–438.
Perognathus baylei Merriam	48	Patton, J. L. (1967), *J. Mamm.*, **48**, 27–37.
Perognathus f. flavus Baird	50	Patton, J. L. (1967), *S. West. Nat.*, **12**, 429–438.
Perognathus formosus melanocaudus Cock.	36	Patton, J. L. (1967), *S. West. Nat.*, **12**, 429–438.
Perognathus formosus mohavensis Huey	36	Towner, J.W . (1965), *Hlth Phys.*, **11**, 1569–1571.
Perognathus goldmani Osg.	52	Patton, J. L. (1967). *J. Mamm.*, **48**, 27–37.
Perognathus hispidus Baird	34	Patton, J. L. (1967), *J. Mamm.*, **48**, 27–37.
Perognathus intermedius Merriam	46	Patton, J. L. (1967), *J. Mamm.*, **48**, 27–37.

Perognathus longimembris bombycinus Osg.	56	Patton, J. L. (1967), *S. West. Nat.*, **12**, 429–438.
Perognathus longimembris pimensis Huey	56	Patton, J. L. (1967), *S. West. Nat.*, **12**, 429–438.
Perognathus m. merriami Allen	50	Patton, J. L. (1967), *S. West. Nat.*, **12**, 429–438.
Perognathus p. penicillatus Woodh.	48	Patton, J. L. (1969), *Caryologia*, **22**, 351–358.
Perognathus penicillatus angustirostris Osg.	46	Patton, J. L. (1969), *Caryologia*, **22**, 351–358.
Perognathus penicillatus atrodorsalis Dalqu.	48	Patton, J. L. (1969), *Caryologia*, **22**, 351–358.
Perognathus penicillatus eremicus Mearns	46	Patton, J. L. (1969), *Caryologia*, **22**, 351–358.
Perognathus penicillatus pricei Allen	48	Patton, J. L. (1969), *Caryologia*, **22**, 351–358.
Perognathus p. pernix Allen	38	Patton, J. L. (1967), *J. Mamm.*, **48**, 27–37.
Perognathus pernix rostratus Osg.	52	Patton, J. L. and Soule, O. H. (1967), *MCNL.*, **8**/**4**, 263–264.
	Gliridae	
Dyromys nitedula intermedius Nehr.	48	Renaud, P. (1938), *Revue Suisse Zool.*, **45**, 349–383.
Eliomys quercinus L.	52	Renaud, P. (1938), *Revue Suisse Zool.*, **45**, 349–383.
Glis glis L.	62	Renaud, P. (1938), *Revue Suisse Zool.*, **45**, 349–383.
Muscardinus avellanarius L.	48	Renaud, P. (1938), *Revue Suisse Zool.*, **45**, 349–383.
	Dipodidae	
Alactagulus acontion Pallas	48	Vorontsov, N. N. *et al.* (1969), *Mlekop.*, Novo-Sibirsk.
Allactaga bobrinski Kolesn.	48	Vorontsov, N. N. *et al.* (1969), *Mlekop.*, Novo-Sibirsk.
Allactaga elater Licht.	48	Vorontsov, N. N. *et al.* (1969), *Mlekop.*, Novo-Sibirsk.
Allactaga jaculus Pallas	48	Vorontsov, N. N. *et al.* (1969), *Mlekop.*, Novo-Sibirsk.
Allactaga salvator Eversm.	48	Vorontsov, N. N. *et al.* (1969), *Mlekop.*, Novo-Sibirsk.
Allactaga severtzovi Vinogr.	48	Vorontsov, N. N. *et al.* (1969), *Mlekop.*, Novo-Sibirsk.
Allactaga williamsi Th.	48	Matthey, R. (1956), *Chromosoma*, **7**, 670–692.
Dipus sagitta Pallas	48	Vorontsov, N. N. *et al.* (1969), *Mlekop.*, Novo-Sibirsk.
Eremodius lichtensteini Vinogr.	48	Vorontsov, N. N. *et al.* (1969), *Mlekop.*, Novo-Sibirsk.
Jaculus jaculus L.	48	Matthey, R., unpublished.
Jaculus turkmenicus Vinogr. and Bond.	48	Voronstov, N. N. *et al.* (1969), *Mlekop.*, Novo-Sibirsk.
Napaeozapus insignis Miller	70, 72 (?)	Meylan, A. (1968), *Revue Suisse Zool.*, **75**, 691–696.
Paradipus ctenodactylus Vinogr.	48	Vorontsov, N. N. *et al.* (1969), *Mlekop.*, Novo-Sibirsk.
Pygerethmus platyurus Licht.	48	Vorontsov, N. N. *et al.* (1969), *Mlekop.*, Novo-Sibirsk.
Pygerethmus vinogradovi Vorontsov	48	Vorontsov, N. N. *et al.* (1969), *Mlekop.*, Novo-Sibirsk.

Pygerethmus zhitkovi Kuznetz	48	Vorontsov, N. N. *et al.* (1969), *Mlekop.*, Novo-Sibirsk.
Salpingonotus crassicauda Vinogr.	48	Vorontsov, N. N. *et al.* (1969), *Mlekop.*, Novo-Sibirsk.
Scirtopoda telum Licht.	58	Vorontsov, N. N. *et al.* (1969), *Mlekop.*, Novo-Sibirsk.
Sicista betulina Pallas	32	Walknowska, J. (1960), *Folia Biol.*, **8**, 65–70.
Sicista subtilis Pallas	32	Johansen W. (1959), unpublished.
Zapus hudsonius Zimm.	72	Meylan, A. (1968), *Revue Suisse Zool.*, **75**, 691–696.
Zapus princeps Allen	72	Meylan, A. (1968), *Revue Suisse Zool.*, **75**, 691–696.
	Hystricidae	
Hystrix cristata L.	60	Renzoni, A. (1967), *MCNL.*, **8**/**1**, 11–12.
	Erethizontidae	
Erethizon dorsatum L.	42	Benirschke, K. and Kessler, D. (1967), *MCNL.*, **8**/**1**, 17.
	Caviidae	
Cavia cobaya Pallas = *C. porcellus* L.	64	Makino, S. (1947), *J. Fac. Sci. Hokkaido Univ.*, **VI**/**9**, 345–357.
	Octodontidae	
Ctenomys australis Rusc.	46	Reig, O. A. and Kibliski, P. (1969), *Chromosoma*, **28**, 211–244.
Ctenomys azarae Th.	48	Reig, O. A. and Kibliski, P. (1969), *Chromosoma*, **28**, 211–244.
Ctenomys latro Th.	42	Reig, O. A. and Kibliski, P. (1969), *Chromosoma*, **28**, 211–244.
Ctenomys magellanicus Bennett	36	Cacheiro, N. *et al.* (1964), *MCNL.*, **14**, 67–68.
Ctenomys minutus Nehr.	50	Cacheiro, N. *et al.* (1964), *MCNL.*, **14**, 67–68.
Ctenomys occultus Th.	22	Reig, O. A. and Kibliski, P. (1968), *Experientia*, **24**, 274–276.
Ctenomys opimus luteolus Th.	26	Reig, O. A. and Kibliski, P. (1969), *Chromosoma*, **28**, 211–244.
Ctenomys porteousi Th.	48	Reig, O. A. and Kibliski, P. (1969), *Chromosoma*, **28**, 211–244.
Ctenomys t. talarum Th.	48	Kibliski, P. and Reig, O. A. (1966), *Nature* (*Lond*), **212**, 437–438.

Ctenomys torquatus Licht.	68	Cacheiro, N. *et al*. (1964), *MCNL*., **14**, 67–68.
Ctenomys tuconax Th.	♂:61	Reig, O. A. and Kibliski, P. (1968), *Experientia*, **24**, 274–276.
Ctenomys tucumanus Th.	28	Reig, O. A. and Kibliski, P. (1968), *Experientia*, **24**, 274–276.
Octodon degus M.	58	Donoso, R. F. (1966), *Biologica*, **38**, 30–37.
	Dasyproctidae	
Cuniculus paca L. = *Agouti paca* L.	74	Fredga, K. (1966), *MCNL*., **20**, 45–46.
Dasyprocta aguti L.	64	Fredga, K. (1966), *MCNL*., **20**, 45–46.
Dasyprocta variegata Tsch.	64	Hungerford, D. A. and Snyder, R. L. (1964), *Am. Nat*., **98**, 125–127.
Myoprocta acouchi Erxl.	62	Fredga, K. (1966), *MCNL*., **20**, 45–46.
	Chinchillidae	
Chinchilla laniger Molina	64	Makino, S. (1953), *Experientia*, **9**, 213.
	Capromyidae	
Myocastor coypus Molina	42	Makino. S. (1947), *J. Fac. Sci. Hokkaido Univ*., **VI**/**9**, 345–357.
	Bathyergidae	
Georychus capensis Pallas	54	Matthey, R. (1956), *Chromosoma*, 7, 670–692.
	Ctenodactylidae	
Ctenodactylus joleaudi H. de B.	40	Matthey, R. (1956), *Chromosoma*, **7**, 670–692.
	Rhizomyidae	
Rhizomys sumatrensis Raffles	50	Hsu, T. C. and Johnson, M. L. (1963), *Am. Nat*., **97**, 127–129.
	Cricetidae	
Akodon azarae Fischer	38	Bianchi, N. O. and Contreras, J. R. (1967), *Cytogenetics*, **6**, 306–313.
Akodon urichi Allen and Chapm.	18	Reig, O. A. and Kibliski, P. (1969), *Acta Cient. Venez*., 73.
Allocricetulus curtatus Allen	20	Matthey, R. (1960), *Caryologia*, **13**, 199–223.

Allocricetulus eversmanni Brandt	26	Matthey, R. (1958), *Arch. Julius Klaus-Stift. Vererb Forsch*, **32**, 385–404.
Allocricetulus eversmanni beljaevi Arg.	26	Vorontsov, N. N. and Krjukova, E. P. (1969), *Mlekop.*, Novo-Sibirsk.
Baiomys taylori Th.	48	Hsu, T. C. (1965), *MCNL.*, **15**, 114–115.
Calomys callosus Rengger	66	Matthey, R. (1968), *MCNL.*, **9/2**, 72–78.
Calomyscus bailwardi Th.	32	Matthey, R. (1961), *Revue Suisse Zool.*, **68**, 41–61.
Cricetulus b. barabensis Pallas	20	Matthey, R. (1960), *Caryologia*, **13**, 199–223.
Cricetulus griseus M. E.	22	Matthey, R. (1952), *Chromosoma*, **5**, 113–138.
Cricetulus migratorius Pallas	22	Matthey, R. (1953), *Revue Suisse Zool.*, **60**, 225–283.
Cricetus cricetus L.	22	Matthey, R. (1952), *Chromosoma*, **5**, 113–138.
Mesocricetus auratus Waterh.	44	Husted, L. *et al.* (1945), *J. Hered.*, **36**, 93–96.
Mesocricetus brandti Nehr.	42	Matthey, R. (1953), *Revue Suisse Zool.*, **60**, 225–283.
Mesocricetus newtoni Nehr.	38	Raiçu, P. and Bratosin, S. (1966), *Z. Säugetierk.*, **31**, 241–255.
Mesocricetus raddei Nehr.	44	Ivanov, V. G. (1969), *Mlekop.*, Novo-Sibirsk.
Mesocricetus raddei nigriculus Nehr.	42	Vorontsov, N. N. and Krjukova, E. P. (1969), *Mlekop.*, Novo-Sibirsk.
Mystromys albicaudatus Smith	32	Matthey, R. (1954), *Caryologia*, **6**, 1–44.
Neotoma albigula Hartley	52	Baker, R. J. and Mascarello, J. T. (1969), *Cytogenetics*, **8**, 187–198.
Neotoma cinerea Ord.	54	Meylan, A. (1968), *Revue Suisse Zool.*, **75**, 691–696.
Neotoma floridana Ord.	52	Cross, J. C. (1931), *J. Morph.*, **52**, 373–401.
Neotoma fuscipes Baird	56	Baker, R. J. and Mascarello, J. T. (1969), *Cytogenetics*, **8**, 187–198.
Neotoma lepida Th.	52	Baker, R. J. and Mascarello, J. T. (1969), *Cytogenetics*, **8**, 187–198.
Neotoma mexicana Baird	52	Baker, R. J. and Mascarello, J. T. (1969), *Cytogenetics*, **8**, 187–198.
Neotoma micropus Baird	52	Baker, R. J. and Mascarello, J. T. (1969), *Cytogenetics*, **8**, 187–198.
Neotoma phenax Merriam	38	Baker, R. J. and Mascarello, J. T. (1969), *Cytogenetics*, **8**, 187–198.

Neotoma stephensi Goldm.	52	Baker, R. J. and Mascarello, J. T. (1969), *Cytogenetics*, **8**, 187–198.
Onychomys leucogaster Merriam	48	Hsu, T. C. and Benirschke, K. (1967), *In* "Atlas of Mammalian Chromosomes", Vol. 1,
Onychomys torridus Coues	48	Hsu, T. C. and Benirschke, K. (1968), *In* "Atlas of Mammalian Chromosomes", Vol. 2. Springer, New York.
Oryzomys albigularis Tomes	66	Kibliski, P. (1969), *Experientia*, **25**, 1338–1339.
Oryzomys delicatus Allen and Chapm.	60	Kibliski, P. (1969), *Experientia*, **25**, 1338–1339.
Peromyscus boylei attwateri Allen	48	Cross, J. C. (1931), *J. Morph.*, **52**, 373–401.
Peromyscus boylei rowleyi Allen	48	Hsu, T. C. and Arrighi, F. E. (1966), *Cytogenetics*, **5**, 355–359.
Peromyscus c. californicus Gambel	48	Cross, J. C. (1931), *J. Morph.*, **52**, 373–401.
Peromyscus californicus insignis Rhoads	48	Cross, J. C. (1931), *J. Morph.*, **52**, 373–401.
Peromyscus collatus Burt.	48	Hsu, T. C. and Arrighi, F. E. (1966), *Cytogenetics*, **5**, 355–359.
Peromyscus crinitus Merriam	48	Hsu, T. C. and Arrighi, F. E. (1968), *MCNL.*, **9**/**2**, 95.
Peromyscus difficilis Allen	48	Hsu, T. C. and Arrighi, F. E. (1966), *Cytogenetics*, **5**, 355–359.
Peromyscus e. eremicus Baird	48	Hsu, T. C. and Arrighi, F. E. (1966), *Cytogenetics*, **5**, 355–359.
Peromyscus floridanus Chapm.	48	Hsu, T. C. and Arrighi, F. E. (1968), *MCNL.*, **9**/**2**, 95.
Peromyscus gossypinus palmarius Bangs	48	Cross, J. C. (1938), *Cytologia*, **8**, 408–419.
Peromyscus leucopus noveboracensis Fischer	48	Cross, J. C. (1938), *Cytologia*, **8**, 408–419.
Peromyscus leucopus texanus Woodh.	48	Cross, J. C. (1938), *Cytologia*, **8**, 408–419.
Peromyscus m. maniculatus Wagner	48	Makino, S. (1953), *Science*, **118**, 3073.
Peromyscus maniculatus bairdii Hoy and Kennic.	48	Makino, S. (1953), *Science*, **118**, 3073.
Peromyscus maniculatus blandus Osg.	48	Cross, J. C. (1938), *Cytologia*, **8**, 408–419.
Peromyscus maniculatus gambelii Baird	48	Cross, J. C. (1931), *J. Morph.*, **52**, 373–401.
Peromyscus maniculatus gracilis Le Conte	48	Singh, R. P. and McMillan, D. B. (1966), *MCNL.*, **19**, 6–7.
Peromyscus maniculatus hollesteri Osg.	48	Arakaki, D. T. and Sparkes, R. S. (1965), *MCNL.*, **17**, 79.

Peromyscus maniculatus keeni Rhoads	48	Hsu, T. C. (1965), *MCNL.*, **15**, 114–115.
Peromyscus maniculatus osgoodi Mearns	48	Cross, J. C. (1938), *Cytologia*, **8**, 408–419.
Peromyscus maniculatus rubidus Osg.	48	Arakaki, D. T. and Sparkes, R. S. (1965), *MCNL.*, **18**, 150–151.
Peromyscus maniculatus rufinus Merriam	48	Hsu, T. C. and Arrighi, F. E. (1966), *Cytogenetics*, **5**, 355–359.
Peromyscus maniculatus sonoriensis Le Conte	48	Cross, J. C. (1938), *Cytologia*, **8**, 408–419.
Peromyscus maniculatus artemisiae Rhoads	48	Cross, J. C. (1938), *Cytologia*, **8**, 408–419.
Peromyscus megalops Merriam	48	Hsu, T. C. and Arrighi, F. E. (1968), *MCNL.*, **9**/**2**, 95.
Peromyscus melanophris Coues	48	Hsu, T. C. and Arrighi, F. E. (1968), *MCNL.*, **9**/**2**, 95.
Peromyscus nasutus Allen	48	Moree, R. (1950), *Proc. Pa. Acad. Sci.*, **24**, 36–39.
Peromyscus nudipes Allen	48	Hsu, T. C. and Arrighi, F. E. (1968), *MCNL.*, **9**/**2**, 95.
Peromyscus nuttali Harl.	52	Patton, J. L. and Hsu, T. C. (1967), *J. Mamm.*, **48**, 637–639.
Peromyscus ochroventer Baker	48	Hsu, T. C. and Arrighi, F. E. (1968), *MCNL.*, **9**/**2**, 95.
Peromyscus pectoralis Osg.	48	Hsu, T. C. and Arrighi, F. E. (1968), *MCNL.*, **9**/**2**, 95.
Peromyscus polionotus leucocephalus Howell	48	Makino, S. (1953), *Science*, **118**, 3073.
Peromyscus p. polionotus Wagner	48	Cross, J. C. (1938), *Cytologia*, **8**, 408–419.
Peromyscus sitkensis Merriam	48	Hsu, T. C. and Arrighi, F. E. (1966), *Cytogenetics*, **5**, 355–359.
Peromyscus thomasi Merriam	48	Hsu, T. C. and Arrighi, F. E. (1968), *MCNL.*, **9**/**2**, 95.
Peromyscus t. truei Shuf.	48	Cross, J. C. (1938), *Cytologia*, **8**, 408–409.
Phodopus przhewalskii Vorontsov	34	Vorontsov, N. N. and Krjukova, E. P. (1969), *Mlekop.*, Novo-Sibirsk.
Phodopus roborovskii Sat.	28	Matthey, R. (1960), *Caryologia*, **13**, 199–223.
Phodopus sungorum Pallas	28	Matthey, R. (1958), *Arch. Julius Klaus-Stift. Vererb Forsch*, **32**, 385–404.
Phyllotis caprinus Pearson	38	Kibliski, (1967), *MCNL.*, **8**/**4**, 281–282.
Reithrodontomys fulvescens Allen	50	Hsu, T. C. (1965), *MCNL.*, **15**, 114–115.
Reithrodontomys megalotis distichlis von Bloeker	42, 43, 44, 45, 46 supernum.	Blanks, G. A. (1967), M.A. Thesis San José State Coll.
Reithrodontomys megalotis longicaudus Baird	42, 43, 44, 45, 46	Blanks, G. A. and Shellhammer, H. S. (1968), *J. Mamm.*, **49**, 726–731.

Reithrodontomys megalotis dychei Allen		44	Matthey, R. (1961), *Mammalia*, **25**, 145–161.
Reithrodontomys r. raviventris Dixon		38	Shellhammer, H. S. (1969), *Chromosoma*, **27**, 102–108.
Reithrodontomys raviventris halicoetes Dixon		38	Shellhammer, H. S. (1969), *Chromosoma*, **27**, 102–108.
Sigmodon fulviventer Allen		28, 29, 30	Lee, M. R. and Zimmerman, E. G. (1969), *J. Mamm.*, **50**, 333–339.
Sigmodon hispidus Say and Ord.	East populations	52	Zimmerman, E. G. and Lee, M. R. (1968), *Chromosoma*, **24**, 243–250.
	West populations	22	
Sigmodon minimus Mearns		29, 30	Hsu, T. C. (1969), *Experientia*, **25**, 205–206.
Sigmodon ochrognathus Bailey		52	Lee, M. R. and Zimmerman, E. G. (1969), *J. Mamm.*, **50**, 333–339.
Tscherkia albipes Ognev		28	Vorontsov, N. N. and Radajbli, S. I. (1969), *Mlekop.*, Novo-Sibirsk.
Tscherkia triton de Winton		30	Makino, S. (1951), *Kromosomo*, **8**, 311–312.
		Muridae	
Murinae			
Acomys cahirinus Desm.		38	Wahrman, J. and Zahavi, A. (1953), *Bull. Res. Coun. Israel*, **3**, 265.
Acomys cahirinus nesiotes Bate		38	Zahavi, A. and Wahrman, J. (1956), *Bull. Res. Coun. Israel*, **5**, 316.
Acomys ignitus Dollm.		50	Matthey, R. (1956), *Arch. Julius Klaus-Stift Vererb Forsch*, **31**, 294–306.
Acomys minous Bate		38, 40	Matthey, R. (1963), *Chromosoma*, **14**, 468–497.
Acomys percevali Dollm.		36	Matthey, R. (1968), *Mammalia*, **32**, 621–627.
Acomys russatus Wagner		66	Wahrman, J. (1953), *Bull. Res. Coun. Israel*, **3**, 265.
Acomys selousi de Winton		60	Matthey, R. (1965), *Revue Suisse Zool.*, **72**, 119–144.
Acomys seurati H. de B.		40	Matthey, R. and Baccar, H. (1967), *Revue Suisse Zool.*, **74**, 546–547.
Acomys subspinosus Waterh.		64	Matthey, R. (1965), *Mammalia*, **29**, 228–249.
Acomys wilsoni Th.		60	Matthey, R. (1968), *Mammalia*, **32**, 621–627.
Aethomys bocagei Th.		50	Matthey, R. (1963), *Mammalia*, **27**, 157–176.
Aethomys chrysophilus de Winton		44	Matthey, R. (1958), *Acta Tropica*, **15**, 97–117.
Aethomys granti Wrough.		32	Matthey, R. (1964), *Mammalia*, **28**, 403–418.

Aethomys kaiseri medicatus Wrough.	50	Matthey, R. (1963), *Mammalia*, **27**, 157–176.
Aethomys namaquensis Smith	24	Matthey, R. (1964), *Mammalia*, **28**, 403–418.
Apodemus a. agrarius Pallas	48	Matthey, R. (1936), *Z. Zellforsch.*, **25**, 501–515.
Apodemus agrarius insulaemus Tokuda	48	Makino, S. (1951), *J. Morph.*, **88**, 93–126.
Apodemus argenteus Temm.	46	Shimba, H. *et al.* (1969), *J. Fac. Sci. Hokkaido Univ.*, **17**, 257–262.
Apodemus flavicollis Melchior	48	Matthey, R. (1936), *Z. Zellforsch.*, **25**, 501–515.
Apodemus geisha Th. (pre-reduction of X and Y)	46	Makino, S. (1951), *J. Morph.*, **88**, 93–126.
Apodemus hebridensis de Winton	48	Koller, P. C. (1941), *J. Genet.*, **41**, 375–389.
Apodemus microps Kratochvil	48	Matthey, R. (1962), *Genetica*, **32**, 268–271.
Apodemus semotus Th.	48	Makino, S. (1951), *J. Morph.*, **88**, 93–126.
Apodemus s. speciosus Temm.	46 or 48	Shimba, H. *et al.* (1969), *J. Fac. Sci. Hokkaido Univ.* **17**, 257–262.
Apodemus speciosus ainu Th.	48	Shimba, H. and Kobayashi, T. (1969), *Jap. J. Genet.*, **44**, 117–122.
Apodemus sylvaticus L.	48	Matthey, R. (1936), *Z. Zellforsch*, **25**, 501–515.
Arvicanthis abyssinicus Rüppel	62	Matthey, R. (1959), *Revue Suisse Zool.*, **66**, 175–209.
Arvicanthis niloticus Desm.	56	Matthey, R. (1965), *Mammalia*, **29**, 228–249.
Bandicota bengalensis Gray and Hardw.	42	Matthey, R. (1956), *Arch. Julius Klaus-Stift. Vererb Forsch*, **31**, 294–306.
Bandicota i. indica Bechst.	42	Ray-Chaudhuri, S. P. *et al.* (1968), *MCNL.*, **9**/**2**, 82–84.
Bandicota indica setifera Horsf.	42	Matthey, R. (1956). *Arch. Julius Klaus-Stift. Vererb Forsch.*, **31**, 294–306.
Cricetomys gambianus Waterh.	78	Matthey, R. (1954), *Caryologia*, **6**, 1–44.
Dasymys incomtus rufulus Miller	38	Matthey, R. (1958), *Acta Trop.*, **15**, 97–117.
Golunda ellioti Gray	52	Matthey, R. (1956), *Arch. Julius Klaus-Stift. Vererb Forsch*, **31**, 294–306.
Hybomys univittatus Peters	48	Matthey, R. (1959), *Revue Suisse Zool.*, **66**, 175–209.
Hydromys chrysogaster Geoff.	48	Dartnall, J. A. (1969), *Proc. R. Soc. Tasm.*, **104**, 79–80.
Hylomyscus aeta Th.	52	Matthey, R. (1963), *Mammalia*, **27**, 157–176.
Hylomyscus stella Th.	46	Matthey, R. (1963), *Mammalia*, **27**, 157–176.
Leggadina delicatula Gould	48	Kennedy, J. A. (1969), *Aust. J. Zool.*, **17**, 465–471.

Leggadina h. hermannsburgensis Waite	48	Matthey, R. (1968), *Experientia*, **24**, 1160.
Lemniscomys barbarus L.	54	Matthey, R. (1954), *Caryologia*, **6**, 1–44.
Lemniscomys striatus L.	48	Matthey, R. (1959), *Revue Suisse Zool.*, **66**, 175–209.
Lophuromys aquilus laticeps Th. and Wr.	70	Matthey, R. (1967), *Mammalia*, **31**, 281–287.
Malacomys edwarsi Rochebr.	48	Matthey, R. (1958), *Acta Trop.*, **15**, 97–117.
Mastomys coucha Smith (= *natalensis*)	36	Matthey, R. (1954), *Caryologia*, **6**, 1–44.
Mastomys sp.	32	Matthey, R. (1955), *Revue Suisse Zool.*, **62**, 163–206.
Mastomys erythroleucus Temm.	38	Matthey, R. (1965), *Mammalia*, **30**, 105–109.
Melomys littoralis Lörnb.	52	Yosida, T. H. *et al.* (1968), *Ann. Rep. Nat. Inst. Genet.*, **19**, 9–13.
Millardia meltada pallidior Ryley	50	Matthey, R. (1956), *Arch. Julius Klaus-Stift Vererb. Forsch*, **31**. 294–306.
Millardia m. meltada Gray	52	Matthey, R., unpublished.
Micromys minutus Pallas	68	Makino, S. (1944), *Cytologia*, **13**, 237–245.
Mus booduga Gray	40	Matthey, R. and Petter, F. (1968), *Revue Suisse Zool.*, 75, 463–498.
Mus dunni Wrough.	40	Matthey, R. and Petter, F. (1968), *Revue Suisse Zool.*, 75, 463–498.
Mus musculus L.	40	Cox, E. K. (1926), *J. Morph.*, **43**, 45.
Mus musculus bactrianus Blyth.	40	Matthey, R. (1953), *Revue Suisse Zool.*, **60**, 225–283.
Mus musculus caroli Bonh. (= *formosanus* Kuroda ?)	40	Makino, S. (1941), *J. Fac. Sci. Hokkaido Univ.*, **VI**/7, 305–380.
Mus musculus molossinus Temm.	40	Oguma, K. (1935), *J.* **L***ac. Sci. Hokkaido Univ.* **V**/**4**, 35–37.
Mus musculus spretus Lataste	40	Matthey, R. (1955), *Revue Suisse Zool.*, **62**, 163–206.
Mus musculus wagneri Eversm.	40	Masui. K. (1923), *J. Coll. Agric. Imp. Univ. Tokyo*, **8**, 207.
Mus poschiavinus Fatio	26	Gropp, A. *et al.* (1970), *Cytogenetics*, **9**, 9–23.
Mus tahi Kloss	40	Yosida, T. H. *et al.* (1968), *Ann. Rep. Nat. Inst. Genet.* **19**, 9–13.

African pigmy-mice (*"Leggada"*)

In this group, the knowledge of the Karyotype does not fit with the views of the Taxonomists. There are too many original karyotypes which cannot be referred to a described *sp.* or *ssp.* and that I quote as *f.c.* (from the French, "formule chromosomique"). For the sake of clarity, I give here an enumeration of "Leggada", as may be established after two cytological criteria: (1) the fundamental number (N.F.) of the ♀. (2) The type of sex-chromosomes, either "primitive" (PR), X and Y acrocentric, or "translocated" (TR), X metacentric, Y submetacentric. The polymorphism rests more frequently on Robertsonian processes (*musculoides*) sometimes on pericentric inversions (*oubanguii*) or multiple translocations (*gundae*).

	N.F.	Sex-chr.	2N	
Mus haussa Th. and Hinton	38	PR	32, 33, 34	Matthey, R. (1967), *Genetica*, **38**, 311–226.
Mus bufo Th.	36	PR	36	Matthey, R. (1967), *Experientia*, **23**, 133.
Mus f.c. 1	36	PR	36	Matthey, R. (1964), *Arch. Biol.*, **75**, 169–206.
Mus indutus Th. (= *deserti*)	36	PR	36	Matthey, R. (1964), *Arch. Biol.*, **75**, 169–206.
Mus mattheyi Petter	36	PR	36	Matthey, R. (1966), *Genetica*, **37**, 171–180.
Mus setulosus Peters	36	PR	36	Matthey, R. (1964), *Arch. Biol.*, **75**, 169–206.
Mus f.c. 3	36	♀:X_1X_2/X_1X_2 ♂:X/Y	36 35	Matthey, R. (1965), *Chromosoma*, **16**, 351–364.
Mus f.c. 10	36	♀:X/X ♂:X/Y_1Y_2	18 19	Matthey, R. (1970), *Experientia*, **26**, 102.
Mus minutoides Smith	36	TR (Y acroc.)	18, 19	Matthey, R. (1963), *Cytogenetics*, **2**, 290–322.
Mus f.c. 6	36	TR (Y subm.)	18	Matthey, R. (1966), *Genetica*, **37**, 171–180.
Mus musculoides Smith (incl. f.c. 7 and f.c. 9)	36	TR	18, 19, 20, 21, 22, 23, 24, 25, 26, 27, 28, 29, 30, 31, 32, 33, 34	Matthey, R. (1963), *Revue Suisse Zool.*, **70**, 173–189. Matthey, R. (1970), *Revue Suisse Zool.*, **77** in press.
Mus f.c. 4	36	TR	32	Matthey, R. (1967), *Arch. Julius Klaus-Stift. Vererb Forsch.*, **42**, 21–30.

Mus bellus Th.	32	TR	30	Matthey, R. (1963), *Cytogenetics*, **2**, 290–322.
Mus gundae Petter	30	TR	16, 17, 18, 19	Jotterand, M. (1970), *Experientia*, in press.
Mus oubanguii Petter	30, 31, 32, 33, 34	TR	28	Matthey, R. and Jotterand, M. (1970), *Revue Suisse Zool.*, in press.
Mus triton Th. (Congo)	34	TR	32	Matthey, R. (1964), *Arch. Biol.*, **75**, 169–206.
Mus triton Th. (Tanzania)	♀:31; ♂:34	TR	♀:20, 21 ♂:20, 21, 22	Matthey, R. (1963), *Cytogenetics*, **2**, 290–322.
Myomys daltoni Th.			36	Matthey, R. (1964), *Mammalia*, **28**, 403–418.
Mylomys dubowskii Pous.			42	Matthey, R. (1970), *Mammalia*, in press.
Nesokia indica Gray and Hardw.			42	Matthey, R. (1953), *Revue Suisse Zool.*, **60**, 225–283.
Nesokia nemorivaga taiwanus Tokuda			46	Makino, S. (1944), *Cytologia*, **13**, 237–245.
Notomys alexis Th.			48	Kennedy, J. A. (1969), *Aust. J. Zool.*, **17**, 465–471.
Notomys cervinus Gould			48	Kennedy, J. A. (1969), *Aust. J. Zool.*, **17**, 465–471.
Oenomys hypoxanthus Puch.			32	Matthey, R. (1963), *Mammalia*, **27**, 157–176.
Pelomys campanae Huet.			48	Matthey, R. (1963), *Mammalia*, **27**, 157–176.
Praomys jacksoni de Winton			28	Matthey, R. (1959), *Revue Suisse Zool.*, **66**, 175–209.
Praomys morio Tr.			34	Matthey, R. (1970), *Mammalia*, in press.
Praomys tullbergi Th.			34	Matthey, R. (1958), *Acta Trop.*, **15**, 97–117.
Praomys sp. 1 (Pointe-Noire)			42	Matthey, R. (1963), *Mammalia*, **27**, 157–176.
Praomys sp. 2 (Boukoko)			42	Matthey, R. (1970), *Mammalia*, in press.
Praomys taitae Heller			48	Matthey, R. (1965), *Mammalia*, **29**, 228–249.
Praomys verreauxi Th.			42	Matthey, R. (1965), *Mammalia*, **29**, 228–249.
Pseudomys higginsi Tr.			48	Dartnall, J. A. (1969), *Proc. R. Soc. Tasm.*, **104**, 79–80.
Pseudomys minnie Trough.			48	Kennedy, J. A. (1969), *Aust. J. Zool.*, **17**, 465–471.
Rattus annandalei Bonh.			42	Yong Hoi Sen (1968), *Cytologia*, **33**, 174–180.
Rattus argentiventer Robins.			42	Yong Hoi Sen (1968), *Cytologia*, **33**, 174–180.
Rattus assimilis Gould			38	Martin, J. H. D. (1969), *Qd. J. Agr. Anim. Sci.*, **26**, 125–141.
Rattus (Stenomys) bowersii Anders.			40	Yong Hoi Sen, (1968), *Cytologia*, **33**, 174–180.
Rattus (Lenothrix) canus Miller			46	Yong Hoi Sen, (1968), *Cytologia*, **33**, 174–180.
Rattus conatus Th.			32	Yosida, T. H. *et al.* (1968), *Ann. Rep. Nat. Inst. Genet.*, **19**, 9–13.

Rattus concolor otteni Kopst. (= *R. exulans concolor* Blyth)	42	Matthey, R. (1956), *Arch. Julius Klaus-Stift. Vererb Forsch*, **31**, 294–306.
Rattus coxingi Swinh. (= *R. fulvescens coxinga* Gray)	46	Makino, S. (1949), *Cytologia*, **15**, 153–160.
Rattus (*Maxomys*) *cremoriventer* Miller	46	Yong Hoi Sen (1968), *Cytologia*, **33**, 174–180.
Rattus culturatus Th.	46	Makino, S. (1949), *Cytologia*, **15**, 153–160.
Rattus (*Leopoldamys*) *edwardsii* Th.	42	Yong Hoi Sen (1968), *Cytologia*, **33**, 174–180.
Rattus exulans Peale	42	Yosida, T. H. *et al.* (1968), *Ann. Rep. Nat. Inst. Genet.*, **19**, 9–13.
Rattus fuscipes Waterh.	38	Yosida, T. H. *et al.* (1968), *Ann. Rep. Nat. Inst. Genet.*, **19**, 9–13.
Rattus greyii Th.	38	Kennedy, J. A. (1969), *Aust. J. Zool.*, **17**, 465–471.
Rattus hung ?	42	Yosida, T. H. *et al.* (1968), *Ann. Rep. Nat. Inst. Genet.*, **19**, 9–13.
Rattus (*Lenothrix*) *inas* Bonh.	42	Yong Hoi Sen (1968), *Cytologia*, **33**, 174–180.
Rattus losea Swinh.	42	Makino, S. (1949), *Cytologia*, **15**, 153–160.
Rattus lutreolus Grey	42	Kennedy, J. A. (1969), *Aust. J. Zool.*, **17**, 465–471.
Rattus (*Stenomys*) *muelleri* Jent.	42	Yong Hoi Sen (1968), *Cytologia*, **33**, 174–180.
Rattus niveiventer Hodg.	46	Matthey, R. (1956), *Arch. J. Klaus Stift.*, **32**, 385–404.
Rattus norvegicus Berk.	42	Painter, T. S. (1926), *Science*, **64**, 336.
Rattus (*Lenothrix*) *rajah* Th.	36	Yong Hoi Sen (1968), *Cytologia*, **33**, 174–180.
Rattus r. rattus L. and *Rattus r. alexandrinus* Geoff. polymorphic.	38–42	Oguma, K. (1935), *J. Fac. Sci. Hokkaido Univ.*, **4**, 35–57. Pincus, G. (1927), *J. Morph.*, **44**, 515–528. Yosida, T. H. *et al.* (1965), *Chromosoma*, **16**, 70–78. Bianchi, N. O. *et al.* (1969), *Experientia*, **25**, 1111–1112. Capanna, E. and Civitelli, M. V. (1969), *MCNL.*, **10**/**4**, 220–221.
Rattus rattus brevicaudatus Horst	42	Matthey, R. (1956), *Arch. Julius Klaus-Stift.*, *Vererb Forsch*, **31**, 294–306.
Rattus rattus diardi Jent.	42	Matthey, R. (1956), *Arch. Julius Klaus-Stift.*, *Vererb Forsch*, **31**, 294–306.
Rattus rattus flavipectus M.E.	42	Yosida, T. H. *et al.* (1968), *Ann. Rep. Nat. Inst. Genet.*, **19**, 9–13.

Rattus rattus jalorensis Bonh.	42	Yong Hoi Sen (1968). *Cytologia*, **33**, 174–180.
Rattus rattus rufescens Gray	42	Tateishi, S. (1935), *Proc. Jap. Assoc. Adv. Sci.*, **10**, 1007–1012.
Rattus (Leopoldamys) sabanus Th.	42	Yong Hoi Sen (1968), *Cytologia*, **33**, 174–180.
Rattus sp. (Port-Moresby)	34	Yosida, T. H. *et al.* (1968), *Ann. Rep. Nat. Inst. Genet.*, **19**, 9–13.
Rattus (Lenothrix) surifer Miller	52	Yong Hoi Sen (1968), *Cytologia*, **33**, 174–180.
Rattus villosissimus Waite	50	Kennedy, J. A. (1969), *Aust. J. Zool.*, **17**, 465–471.
Rattus (Lenothrix) whiteheadi Th.	36	Yong Hoi Sen (1968), *Cytologia*, **33**, 174–180.
Rhabdomys pumilio Sparr.	48	Matthey, R. (1954), *Caryologia*, **6**, 1–44.
Saccostomus campestris Peters	44	Ford, C. E. and Hamerton, J. L. (1956), *Nature (Lond)*, **177**, 140–141.
Saccostomus campestris Peters	46	Matthey, R. (1958), *Acta Trop.*, **15**, 97–117.
Thallomys moggi lebomboensis Roberts	48	Matthey, R. (1959), *Revue Suisse Zool.*, **66**, 175–209.
Thallomys paedulcus Sundev.	48	Matthey, R. (1959), *Revue. Suisse Zool.*, **66**, 175–209.
Thamnomys rutilans Peters	50	Matthey, R. (1963), *Mammalia*, **27**, 157–176.
Uranomys ruddi Dollm.	52	Matthey, R. (1970), *Mammalia*, in press.
Uromys caudimaculatus Krefft	54	Yosida, T. H. *et al.* (1968), *Ann. Rep. Nat. Inst. Genet.*, **19**, 9–13.
Dendromyinae		
Dendromus insignis kivu Th.	48	Matthey, R. (1967), *Mammalia*, **31**, 281–287.
Dendromus melanotis Th.	36	Matthey, R. (1970), *Mammalia*, in press.
Dendromus mystacalis messorius Th.	38	Matthey, R. (1970), *Mammalia*, in press.
Malacothrix typica Smith	38	Allenbach, M. (1964), *Bull. Soc. Vaud. Sci. Nat.*, **68**, 493–494.
Steatomys pratensis Peters	68	Matthey, R. (1954), *Caryologia*, **6**, 1–44.
Gerbillinae		
Desmodillus auricularis Smith	52	Matthey, R. (1954), *Caryologia*, **6**, 1–44.
Gerbillus allenbyi Th.	40	Wahrman, J. and Zahavi, A. (1955), *Nature (Lond)*, **175**, 600.
Gerbillus calurus Th.	38	Wahrman, J. and Zahavi, A. (1955), *Nature (Lond)*, **175**, 600.

Gerbillus campestris Lev.		56	Matthey, R. (1953), *Revue Suisse Zool.*, **60**, 225–283.
Gerbillus (*Dipodillus*) *dasyurus* Wagner		60	Wahrman, J. and Zahavi, A. (1955), *Nature* (*Lond*), **175**, 600.
Gerbillus garamantis Lat.		54	Matthey, R. (1954), *Caryologia*, **6**, 1–44.
Gerbillus gerbillus Oliv.		♀:42; ♂:43	Matthey, R. (1954), *Experientia*, **10**, 464–465.
Gerbillus (*Dipodillus*) *nanus* Blanf.		52	Wahrman, J. and Zahavi, A. (1955), *Nature* (*Lond*), **175**, 600.
Gerbillus paeba Smith		36	Matthey, R. (1959), *Revue Suisse Zool.*, **66**, 175–209.
Gerbillus pyramidum Geoff.	(Algeria)	40	Matthey, R. (1952), *Arch. Julius Klaus-Stift. Vererb-Forsch.*, **27**, 163–166.
Gerbillus pyramidum Geoff.	(Algeria)	40	Wahrman, J. and Zahavi, A. (1955), *Nature* (*Lond*), **175**, 600.
	(Coast of Israel)	52	
	(Negev)	66	
Meriones crassus Sund.		60	Matthey, R. (1953), *Revue. Suisse Zool.*, **60**, 225–283.
Meriones hurrianae Jerdon		40	Nadler, C. F. and Lay, D. M. (1967), *Z. Säugetierk.*, **32**, 285–291.
Meriones l. lybicus Licht.		44	Matthey, R. (1953), *Revue Suisse Zool.*, **60**, 225–283.
Meriones lybicus erythroura Gray		44	Matthey, R. (1953), *Revue Suisse Zool.*, **60**, 225–283.
Meriones meridianus Pallas		50	Nadler, C. F. *et al.* (1969), *Experientia*, **25**, 774–775.
Meriones persicus Heptner		42	Matthey, R. (1957), *Säugetierk. Mitt.*, **5**, 145–150.
Meriones sacramenti Th.		46	Zahavi, A. and Wahrman, J. (1957), *Mammalia*, **21**, 341–380.
Meriones shawi Duv.		44	Matthey, R. (1953), *Revue, Suisse Zool.*, **60**, 225–283.
Meriones tamariscinus Pallas		40	Vorontsov, N. N. and Korobitsina, K. V. (1969), *Mlekop.*, Novo-Sibirsk.
Meriones tristrami Th.		72	Matthey, R. (1957), *Säugetierk. Mitt.*, **5**, 145–150.
Meriones u. unguiculatus M.E.		44	Awa, A. *et al.* (1959), *Jap. J. Zool.*, **12**, 257–265.
Meriones vinogradovi Heptner		44	Matthey, R. (1954), *Caryologia*, **6**, 1–44.
Pachyuromys duprasi Lat.		54	Matthey, R. (1954), *Revue Suisse Zool.*, **62**, 163–206.
Psammomys obesus Cretz.		48	Matthey, R. (1954), *Revue Suisse Zool.*, **62**, 163–206.
Rhombomys opimus Licht.		40	Matthey, R. (1953), *Revue Suisse Zool.*, **60**, 225–283.
Tatera afra Gray		44	Matthey, R. (1954), *Caryologia*, **6**, 1–44.

Lemmus lemmus L.	50	Matthey, R. (1957), *Revue Suisse Zool.*, **64**, 39–71.
Microtus abbreviatus Miller	50	Rausch, R. L. and Rausch, W. R. (1968), *Z. Säugetierk.*, **33**, 65–99.
Microtus afghanus Th.	54	Liapounova, E. A. and Mirokhanov, J. M. (1969), *Mlekop.*, Novo-Sibirsk.
Microtus agrestis L.	50	Matthey, R. (1950), *Cellule*, **53**, 163–184.
Microtus arvalis Pallas	46	Renaud, P. (1938), *Revue Suisse Zool.*, **45**, 349–383.
Microtus californicus eximius Kell.	54	Matthey, R. (1954), *Caryologia*, **6**, 1–44.
Microtus canicaudus Miller	24	Hsu, T. C. (1965), *in litteris*.
Microtus chrotorrhinus Miller	60	Meylan, A. (1967), *Can. J. Genet. Cyt.*, **9**, 569–574.
Microtus fortis Büchner	52	Meyer, M. *et al.* (1967), *Folia Biol.*, **15**, 251–258.
Microtus fortis pelliceus Th.	52	Liapounova, E. A. and Mirokhanov, J. M. (1969), *Mlekop.*, Novo-Sibirsk.
Microtus gregalis buturlini Ognev	36	Liapounova, E. A. and Mirokhanov, J. M. (1969), *Mlekop.*, Novo-Sibirsk.
Microtus gregalis dolguschini Afan.	36	Liapounova, E. A. and Mirokhanov, J. M. (1969), *Mlekop.*, Novo-Sibirsk.
Microtus gregalis eversmanni Polj.	36	Liapounova, E. A. and Mirokhanov, J. M. (1969), *Mlekop.*, Novo-Sibirsk.
Microtus g. gregalis Pallas	36	Liapounova, E. A. and Mirokhanov, J. M. (1969), *Mlekop.*, Novo-Sibirsk.
Microtus gregalis major Ognev	36	Fedik, S. (1970), *Acta Ther.* **15**, 142–151.
Microtus gregalis muriei Nelson	54	Rausch, R. L. (1964), *Z. Säugetierk.*, **29**, 343–358.
Microtus guentheri Danf. and Alston	54	Matthey, R. (1952), *Chromosoma*, **5**, 113–138.
Microtus hyperboreus Vinogr.	♀:50; ♂:52	Liapounova, E. A. and Krivosheyev, V. G. (1969), *Mlekop.*, Novo-Sibirsk.
Microtus irani Th.	54	Matthey, R. (1954), *Caryologia*, **6**, 1–44.
Microtus kikuchii Kuroda	30	Makino, S. (1950), *Annotnes Zool. Jap.*, **23**/**2**, 63–68.
Microtus longicaudus Merriam	56	Matthey, R. (1955), *Revue Suisse Zool.*, **62**, 163–206.
Microtus mexicanus Sauss.	44	Matthey, R. (1955), *Revue Suisse Zool.*, **64**, 39–70.
Microtus middendorffi Pol. and Lat.	50	Matthey, R. and Zimmermann, K. (1961), *Revue. Suisse Zool.*, **68**, 63–72.

Microtus mongolicus Radde	50	Meyer, M. and coll. (1967), *Folia Biol.*, **15**, 251–258.
Microtus montanus Peale	24	Matthey, R. (1954), *Caryologia*, **6**, 1–44.
Microtus montebelli Edw.	30	Utakoji, T. (1967), *MCNL.*, **8/4**, 283.
Microtus nivalis Martin	56	Matthey, R. (1953), *Revue Suisse Zool.*, **60**, 225–283.
Microtus ochrogaster haydenii Baird	54	Matthey, R. (1955), *Revue Suisse Zool.*, **62**, 163–206.
Microtus o. ochrogaster Wagner	54	Matthey, R. (1955), *Revue Suisse Zool.*, **62**, 163–206.
Microtus orcadensis Millais	46	Matthey, R. (1953), *Revue Suisse Zool.*, **60**, 225–283.
Microtus (*Chilotus*) *oregoni* Bachm.	♀: Soma, 17; ♂: Germen, 17	Matthey, R. (1957), *Revue Suisse Zool.*, **64**, 39–71.
Microtus (*Chilotus*) *oregoni* Bachm.	♂: Soma, 18	Ohno, S. *Cytogenetics*, **2**, 232–239.
Microtus pennsylvanicus Ord.	46	Matthey, R. (1952), *Chromosoma*, **5**, 113–138.
Microtus ratticeps K. and Bl. (= *M. oeconomus* Pallas)	30	Makino, S. (1950), *Annotnes. Zool. Jap.*, **23/2**, 63–68.
Microtus (*Aulacomys*) *richardsoni* de Kay	56	Matthey, R. (1958), *Arch. Julius Klaus-Stift. Vererb-Forsch.*, **32**, 385–404.
Microtus socialis irani Th.	62	Matthey, R. (1953), *Revue Suisse Zool.*, **60**, 225–283.
Microtus t. transcaspicus Sat.	54	Liapounova, E. A. and Mirokhanov, J. M. (1969), *Mlekop.*, Novo-Sibirsk.
Microtus ungurensis Kast.	42	Meyer, M. *et al.*, (1967), *Folia Biol.*, **15**, 251–258.
Myopus schisticolor Lillj.	32	Matthey, R. (1958), *Arch. Julius Klaus-Stift. Vererb-Forsch.*, **32**, 385–404.
Neofiber alleni True	52	Matthey, R. (1961), *Mammalia*, **25**, 145–161.
Ondatra zibethica L.	54	Makino, S. (1953), *Science*, **118**, 3073.
Phaiomys carruthersi Th.	54	Liapounova, E. A. and Fisher, T. B. (1969), *Mlekop.*, Novo-Sibirsk.
Phaiomys juldaschi Sever.	54	Liapounova, E. A. and Fisher, T. B. (1969), *Mlekop.*, Novo-Sibirsk.
Phenacomys ungava Merriam	56	Matthey, R. (1957), *Revue Suisse Zool.*, **64**, 39–71.
Pitymys duodecim-costatus S.L.	62	Matthey, R. (1955), *Revue Suisse Zool.*, **62**, 163–206.
Pitymys mariae Forsith Major	62	Meylan, A. (1970), *Revue Suisse Zool.*, in press.
Pitimys multiplex Fatio	48	Matthey, R. (1955), *Revue Suisse Zool.*, **62**, 163–206.
Pitymys pinetorum nemoralis Bailey	62	Matthey, R. (1955), *Revue Suisse Zool.*, **62**, 163–206.
Pitymys savii S.L.	54	Meylan, A. (1970), *Revue Suisse Zool.*, in press.

Pitymys subterraneus S.L.	54	Matthey, R. (1954), *Caryologia*, **6**, 1–44.
Pitymys tatricus Kratochvil	32	Matthey, R. (1964), *Z. Säugetierk.*, **29**, 235–242.
Prometheomys schaposchnikowi Sat.	56	Matthey, R. (1958), *Revue Suisse Zool.*, **65**, 377–383.

ADDENDUM 1

To this check-list of 1165 chromosomic formulae, 58 references must be added, some previously omitted, most of them to recently published papers. In the general list, some species, principally nearctic or siberian, are included without subspecific identification. Recent papers from Vorontsov's school in USSR and from Nadler and his co-workers in USA supply more accurate taxonomic determinations. These results are useful for a better understanding of the steps of the speciation.

These 58 references are not included in the general list and in the tables used as basis for the statistical analysis.

INSECTIVORA

Soricidae

Cryptotis parva Say	52	Hsu, T. C. and Benirschke, K. (1970), *In* "Atlas of Mammalian Chromosomes," Vol. 4, 155. Springer New York.
Sorex minutissimus Zimm.	38	Hallka, O. (1970), *Suomal. Tiedeakat Toim.*, **A/IV**, 1–5.

Talpidae

Parascalops breweri Bachm.	34	Gropp, A. (1969), *In* "Comparative Mammalian Cytogenetics," Springer Verlag, New York.

PRIMATES

Hylobatidae

Symphalangus brachytanites Schulz = *Hylobates klossii* Miller	50	Hsu, T. C. and Benirschke, K. (1970), *In* "Atlas of Mammalian Chromosomes," Vol. 4, 200. Springer New York.

RODENTIA

	Caviidae	
Hydrochaeris hydrochaeris urugauaensis L.	66	Saez, F. A. *et al.* (1971), *Experientia.* in press.
	Heteromyidae	
Perognathus arenarius Merriam	42	Patton, J. L. (1970), *MCNL.*, **11/1**, 3–5.
Perognathus californicus Merriam	44	Patton, J. L. (1970), *MCNL.*, **11/1**, 3–5.
Perognathus fallax Merriam	44	Patton, J. L. (1970), *MCNL.*, **11/1**, 3–5.
Perognathus nelsoni Merriam	46, 48	Patton, J. L. (1970), *MCNL.*, **11/1**, 3–5.
Perognathus spinatus Marriam	44	Patton, J. L. (1970), *MCNL.*, **11/1**, 3–5.
	Sciuridae	
Callosciurus flavimanus Geoff.	40	Nadler, C. F. and Hoffman, R. S. (1970), *Experientia*, **26**, 1383–1386.
Cynomys gunnisoni zuniensis Hollist.	40	Nadler, C. F. and Hoffman, R. S. (1969), *in litt.*
Cynomys leucurus Merriam	50	Nadler, C. F. and Hoffman, R. S. (1969), *in litt.*
Dremomys rufigenis Blanf.	38	Nadler, C. F. and Hoffmann, R. S. (1970), *Experientia*, **26**, 1383–1386.
Eutamias sibiricus asiaticus Gmel.	38	Nadler, C. F. *et al.* (1969), *Experientia*, **25**, 868–869.
Eutamias sibiricus jacutensis Ogn.	38	Liapounova, E. A. and Jolneroskia, E. I. (1969), *Mlekop.*, Novo-Sibirsk.
Eutamias sibiricus orientalis Bonh.	38	Liapounova, E. A. and Jolneroskia, E. I. (1969), *Mlekop.*, Novo-Sibirsk.
Marmota baibacina centralis Thom.	38	Vorontsov, N. N. *et al.* (1969), *Zool. J.*, **48**.
Marmota caligata nivaria Howell	42	Hoffmann, R. S. and Nadler, C. F. (1968), *Experientia*, **24**, 740–742.
Marmota vancouverensis Swarth	40	Rausch, R. L. and Rausch, V. R. (1965), *Chromosoma*, **16**, 618–623.
Menetes berdmorei Blyth	62	Nadler, C. F. and Hoffman, R. S. (1970), *Experientia*, **26**, 1383–1386.
Sciurus anomalus Güldenst.	40	Nadler, C. F. and Hoffmann, R. S. (1970), *Experientia*, **26**, 1383–1386.
Sciurus granatensis Humb.	42	Nadler, C. F. and Hoffmann. R. S. (1970), *Experientia*, **26**, 1383–1386.

Spermophilopsis leptodactylus schumakovi Satunin	38	Liapounova, E. A. and Jolneroskia, E. I. (1969), *Mlekop.*, Novo-Sibirsk.
Spermophilus (Citellus) beldingi crebrus Hall	30	Nadler, C. F. (1965), *Cytogenetics*, **4**, 37–44.
Spermophilus (Citellus) beldingi oregonus Merriam	30	Nadler, C. F. (1965), *Cytogenetics*, **4**, 37–44.
Spermophilus (Citellus) elegans Kenn.	34	Nadler, C. F. (1966), *J. Mamm.*, **47**, 579–596.
Spermophilus (citellus) erthrogenys brevicauda Brandt	36	Vorontsov, N. N. and Liapounova, E. A. (1970), *Bull. Soc. Moscov. Nat.*, **3**, 112–126.
Spermophilus (Citellus) erythrogenys carruthersi Thom.	36	Vorontsov, N. N. and Liapounova, E. A. (1970), *Bull. Soc. Moscov. Nat.*, **3**, 112–126.
Spermophilus (Citellus) fulvus oxianus Thom.	36	Vorontsov, N. N. and Liapounova, E. A. (1970), *Bull. Soc. Moscov. Nat.*, **3**, 112–126.
Spermophilus (Citellus) fulvus nigrimontanus Antipin	36	Vorontsov, N. N. and Liapounova, E. A. (1970), *Bull. Soc. Moscov. Nat.*, **3**, 112–126.
Spermophilus (Citellus) mollis Kenn.	36	Nadler, C. F. (1966), *J. Mamm.*, **47**, 579–596.
Spermophilus (Citellus) musicus boehmi Krass.	36	Tembotov, A. K. *et al.* (1969), *Bull. Soc. Moscov. Nat.*, **74**.
Spermophilus (Citellus) nevadensis Howell	34	Nadler, C. F. (1966), *J. Mamm.*, **47**, 579–596.
Spermophilus parryi buxtoni Allen	32	Liapounova, E. A. (1969), *Mlekop.*, Novo-Sibirsk.
Spermophilus (Citellus) parryi kodiacensis Allen	34	Nadler, C. F. (1966), *J. Mamm.*, **47**, 579–596.
Spermophilus (Citellus) parryi osgoodi Merriam	34	Nadler, C. F. (1966), *J. Mamm.*, **47**, 579–596.
Spermophilus (Citellus) parryi plesius Osgood	34	Nadler, C. F. (1966), *J. Mamm.*, **47**, 579–596.
Spermophilus (Citellus) parryi steinegeri Allen	32	Liapounova, E. A. (1969), *Mlekop.*, Novo-Sibirsk.
Spermophilus (Citellus) parvidens Mearns	34	Nadler, C. F. and Sutton, D. A. (1962), *Proc. Soc. Exp. Biol. Med.*, **110**, 36–38.
Spermophilus (Citellus) pygmaeus brauneri Martino	36	Denisov, V. *et al.*, (1969), *Folia Biol.*, **17**, 170–175.
Spermophilus (Citellus) pygmaeus planicola Satunin	36	Tembotov, A. K. *et al.* (1969), *Bull. Soc. Moscov. Nat.*, **74**.
Spermophilus (Citellus) relictus rallii Kuzn.	36	Vorontsov, N. N. and Liapounova, E. A. (1970), *Bull. Soc. Moscov. Nat.*, **3**, 112–126.
Spermophilus (Citellus) suslicus suslicus Güld.	34	Orlov, V. N. *et al.* (1969), *Mlekop.*, Novo-Sibirsk.
Spermophilus (Citellus) suslicus guttatus Pallas	34	Denisov, V. *et al.* (1969), *Folia Biol.*, **17**, 170–175.
Spermophilus (Citellus) undulatus eversmanni Brandt	32	Vorontsov, N. N. and Liapounova, E. A. (1969), *Dokl. Akad. Nauk SSSR*, **187**.

Spermophilus (*Citellus*) *undulatus stramineus* Obol.		32	Vorontsov, N. N. and Liapounova, E. A. (1969), *Dokl. Akad. Nauk. SSSR*, **187**.
Spermophilus (*Citellus*) *u. undulatus* Pallas		32	Vorontsov, N. N. and Liapounova, E. A. (1970), *Bull. Soc. Moscov. Nat.*, **3**, 112–126.
Spermophilus (*Citellus*) *variegatus grammurus* Say		38	Nadler, C. F. (1966) *Syst. Zool.*, **15**, 199– .
Spermophilus (*Citellus*) *vigilis* Merriam		46	Nadler, C. F. (1966), *J. Mamm.*, **47**, 579–596.
		Muridae	
Apodemus speciosus Temm.	polymorphic	48, 49, 50, 51	Vorontsov, N. N. (1970), *in litt.*
Mus platythrix Bennett		22	Pathak. S. (1970), *MCNL.*, **11**/**3**, 105.
Rattus rattus arboreus Horsf.		42	Ray-Chaudhuri, S. P. and Pathak, S. (1970), *MCNL.*, **11**/**4**, 135–136.
Rattus rattus gangutrianus Minton		42	Ray-Chaudhuri, S. P. and Pathak, S. (1970), *MCNL.*, **11**/**4**, 135–136.
Rattus ssp.		44	Ray-Chaudhuri, S. P. and Pathak. S. (1970), *MCNL.*, **11**/**4**, 135–136.
Rattus rattus thai Kloss	polymorphic	42, 44, 46, 48	Gropp, A. *et al.* (1970), *MCNL.*, **11**/**3**, 111.
		ARTIODACTYLA	
		Bovidae	
Taurotragus oryx Poryx Lydekker		♀:32	Taylor, K. M. and Taylor, B. K. (1970), *MCNL.*, **11**/**4**, 123–125.
Taurotragus oryx pattersonianus Lydekker		♂:31	Taylor, K. M. and Taylor, B. K. (1970), *MCNL.*, **11**/**4**, 123–125.

ADDENDUM 2

It seems appropriate to give references to 51 new papers, the most of them published in the first months of 1971. With the both addenda, the total number of references is brought up to 1274.

	INSECTIVORA	
	Talpidae	
Neurotrichus g. gibbsii Baird	38	Brown, R. M. and Waterbury, A. M. (1971). *MCNL.*, 12/**2**, 45.
	CHIROPTERA	
	Emballonuridae	
Cormura brevirostris Gray	22	Baker, R. J. and Jordan, R. G. (1970). *Caryologia*, **23**, 595–604.
Rhynchonycteris naso Wied-Neuwied	22	Baker, R. J. and Jordan, R. G. (1970). *Caryologia*, **23**, 595–604.
Saccopteryx bilineata Temm.	26	Baker, R. J. and Jordan, R. G. (1970). *Caryologia*, **23**, 595–604.
Saccopteryx leptura Schr.	28	Baker, R. J. and Jordan, R. G. (1970). *Caryologia*, **23**, 595–604.
	Hipposideridae	
Hipposideros ater Templ.	32	Baker, R. J. (1970). *In* "Biology of Bats", Academic Press, New York and London.
Hipposideros fulvus Gray	32	Baker, R. J. (1970). *In* "Biology of Bats", Academic Press, New York and London.
	Molossidae	
Eumops perotis Schinz	48	Baker, R. J. (1970). *In* "Biology of Bats", Academic Press, New York and London.
Molossops greenhalli Goodwin	34	Baker, R. J. (1970). *In* "Biology of Bats", Academic Press, New York and London.
Tadarida macrotus Gray	48	Patton, J. L. and Baker, R. J. (1966). *MCNL.*, **20**, 66–67.
Tadarida molossa Pallas (= *Molossus molossus* ?)	48	Baker, R. J. and Lopez, G. (1970). *Caryologia*, **23**, 465–472.
	Natalidae	
Natalus tumidirostris Miller	36	Baker, R. J. and Jordan, R. G. (1970). *Caryologia*, **23**, 595–604.

	Noctilionidae	
Noctilio labialis Kerr	34	Baker, R. J. and Jordan, R. G. (1970). *Caryologia*, **23**, 595–604.
Noctilio leporinus L.	34	Baker, R. J. and Jordan, R. G. (1970). *Caryologia*, **23**, 595–604.
	Phyllostomatidae	
Brachyphylla cavernarum Gray	32	Baker, R. J. and Lopez, G. (1970). *Caryologia*, **23**, 465–472.
Choeroniscus intermedius Allen	20	Baker, R. J. (1970). *In* "About Bats", Slaughter and Walton, Dallas.
Erophylla bombifrons Miller	32	Baker, R. J. and Lopez, G. (1970). *Caryologia*, **23**, 465–472.
Monophyllus redmani Leach	32	Baker, R. J. and Lopez, G. (1970). *Caryologia*, **23**, 465–472.
Pteronotus fuliginosa Gray	38	Baker, R. J. and Lopez, G. (1970). *Caryologia*, **23**, 465–472.
Stenoderma rufum St Hilaire	♀:30 ♂:31	Baker, R. J. and Lopez, G. (1970). *Caryologia*, **23**, 465–472.
Uroderma magnirostrum Davis	36	Baker, R. J. and Lopez, G. (1970). *J. Mamm.*, **51**, 786–789.
	Vespertilionidae	
Eptesicus brasiliensis Peters	50	Baker, R. J. and Jordan, R. G. (1970). *Caryologia*, **23**, 595–604.
Euderma maculatum Allen	30	Williams, D. F., *et al.* (1970). *J. Mamm.*, **51**, 602–606.
Myotis auriculus Baker and Stains	44	Baker, R. J. and Jordan, R. G. (1970). *Caryologia*, **23**, 595–604.
Myotis riparius Handley	44	Baker, R. J. and Jordan, R. G. (1970). *Carologia*, **23**, 595–604.
Myotis simus Th.	44	Baker, R. J. and Jordan, R. G. (1970). *Caryologia*, **23**, 595–604.
Tylonycteris pachypus Temm.	46	Yong, H. S. (1971). *Exper.*, in print.
Tylonycteris robustula Th.	32	Yong, H. S. (1971). *Exper.*, in print.

	CARNIVORA	
	Felidae	
Felis libyca Forster	38	Jotterand, M. (1971). *R. suisse Zool.*, in print.
Felis libyca ornata Gray	38	Jotterand, M. (1971). *R. suisse Zool.*, in print.
Felis marbarita Loche	38	Jotterand, M. (1971). *R. suisse Zool.*, in print.
Felis silvestris Schr.	38	Jotterand, M. (1971). *R. suisse Zool.*, in print.
	Mustelidae	
Enhydra lutris L.	38	Rausch, V. R. and Feltz, E. T. (1971). *MCNL.*, **12/2**, 47.
	Viverridae	
Viverra zibetha L.	38	Pathak, S. (1971). *MCNL.*, **12/2**, 61.
	CETACEA	
	Physeteridae	
Kogia breviceps Blainv.	42	Benirschke, K. *in litt* (1971).
	RODENTIA	
	Cricetidae	
Anotomys leander Th.	92	Gardner, A. L. (1971). *Exper.*, in print.
	Dasyproctidae	
Agouti taczanowskii stolzm.	42	Gardner, A. L. (1971). *Exper.*, in print.
	Heteromyidae	
Dipodomys a. agilis Gambel	62	Csuti, B. A. (1971). *Cytogenet.*, **52**, 202–206.
Dipodomys agilis perplexus Merr.	62	Csuti, B. A. (1971). *Cytogenet.*, **52**, 202–206.
Dipodomys d. deserti Stephens	64	Jackson, L. and Hunsaker, D. (1971). *MCNL.*, **12/2**, 50–51.
Dipodomys heermanni arenae Boulw.	64	Csuti, B. A. (1971). *Cytogenet.*, **52**, 202–206.
Dipodomys heermanni goldmani Merr.	64	Csuti, B. A. (1971). *Cytogenet.*, **52**, 202–206.
Dipodomys heermanni morroensis Merr.	64	Csuti, B. A. (1971). *Cytogenet.*, **52**, 202–206.
Dipodomys m. merriami Mearns	52	Jackson, L. and Hunsaker, D. (1971). *MCNL.*, **12/2**, 50–51.

Dipodomys m. microps Merr.	60	Csuti, B. A. (1971). *Cytogenet.*, **52**, 202–206.
Dipodomys nitratoides Merr.	54	Jackson, L. and Hunsaker, D. (1971). *MCNL.*, **12**/**2**, 50–51.
Dipodomys panamintius mohavensis Grinn.	64	Csuti, B. A. (1971). *Cytogenet.*, **52**, 202–206.
Dipodomys spectabilis perblandus Goldm.	72	Jackson, L. and Hunsaker, D. (1971). *MCNL.*, **12**/**2**, 50–51.
Dipodomys s. stephensi Merr.	70	Jackson, L. and Hunsaker, D. (1971). *MCNL.*, **12**/**2**, 50–51.
	Muridae	
Vandeleuria oleracea Bennett	♀:29	Pathak, S. (1971). *MCNL.*, **12**/**2**, 60.
	Sciuridae	
Marmota vancouverensis Swarth	42	Rausch, R. L. and Rausch, V. R. (1971). *Mammalia*, **35**, 85–101.

Section V

SEX CHROMOSOMES AND THE SEX DETERMINATION

16

The Evolution of the Sex Chromosomes

N. N. VORONTSOV

Department of Evolutionary Zoology, Institute of Biology and Pedology, Far East Science Centre, Academy of Sciences of the USSR, Vladivostok, USSR

I. INTRODUCTION

A. *General Remarks*

1. *Hermaphroditism and dioecism. (From polysexuality to bisexuality)*
Most plants, at least the higher ones, and a large number of animal species are hermaphrodites, i.e. the same individual is capable of elaborating either simultaneously or after some time both female and male gametes. As a rule, hermaphroditism (monoecism) does not result in self-pollination and self-fertilization caused by the asynchronous maturation of male and female gametes or by other mechanisms preventing inbreeding. Nevertheless, it is evident that in lines following a progressive evolutionary course animals, at any rate, have displayed a tendency towards the adaption of dioecism.

Dioecism first arose in unicellular organisms like *Chlamidomonas*; in some *Infusoria* conjugation may occur between cells with different "sex valency". In the genus *Clamidomonas* there may be as many as 8 lines differing in sex valency. This may be adduced as an argument that polysexuality has probably preceded bisexuality, and that bisexuality is the most optimal solution of the problem of dioecism. Evolution has given preference to bisexuality as the minimum number ensuring combinatory variability.

2. *From isogamy to oogamy. Independent arisal of oogamy and bisexuality in different types of animals*

The sex-producing, large and generally immobile gamete is referred to as female; the opposite sex-producing, smaller and mobile gamete is designated male. The passage from the conjugation of morphologically identical gametes (isogamy) to the fusion of morphologically and functionally differing gametes (aniso- and oogamy) occurs independently in various phyla of plants and animals. For this reason, the problem of the homology of gametes and sex is much more sophisticated than may appear at first sight.

From the very moment of transition from isogamy to oogamy both sexes of the dioecious forms should have been endowed with gene groups determining, at least, the development of the sex gland and the differentiation of germ cells into egg cells charged with nutritive material or into active spermatozoa almost devoid of cytoplasm. However, it is clear that general requirements set to the egg cell may lead to the convergent arisal of non-homologous genes determining the development of the female sex.

In this connection, there is no certainty as to whether Coleoptera males and Trichoptera males are homologous and as to whether there is a homology between *Drosophila* and human males.

3. *The determination of differentiation of sex*

A distinction should be made between the sex-determining processes and the differentiation of sexual characters in ontogeny. Sex reversal may occur in ontogeny under the effect of a wide variety of factors. Only the cytological mechanisms of sex-determination will be discussed here.

4. *Types of sex determination*

Three main modes of sex differentiation are distinguished:

(a) The progamic mode in which the sex of the progeny is determined as early as in the gametogenesis of the parent prior to meiosis or in the course of it. This mode of sex-determination is characteristic of many Rotatoria, Archiannelides, Cecidomyidae and Aphidodea (Vorontsov 1966).

(b) The syngamic mode by which the sex of the progeny is determined at the moment of gamete fusion. The syngamic mode of sex determination is most frequent in animals and has been studied more comprehensively than the other two modes of sex determination.

(c) The epigamic mode in which sex is determined in ontogeny under the effect of external factors as in the case of *Bonnelia*.

Both the progamic and syngamic types of sex determination are genetically conditioned. The syngamic mode of sex determination is characteristic of vertebrates, excluding some parthengenetic fishes, amphibia and reptiles in which sex is determined progamically.

It should be recalled that the sex-possessing genetically homologous (in each of the pairs) chromosome is known as the homogametic sex; that in which at least one pair of genetically non-homologous chromosomes are present is designated as heterogametic.

The presence of homo- and heterogametic sexes, being a requisite for animals with the syngamic mode of sex determination, is not necessarily related to the differentiation of sex-determining heterochromosomes.

B. "*What are sex chromosomes?*"

1. *Means of identifying sex chromosomes*

There are several means of identifying heterochromosomes.

(a) The morphological identification of chromosomes is the one commonly used. The detection of a non-homologous pair of chromosomes in the heterogamic sex, provided that there are homologous pairs in the homogametic sex, is considered adequate for the description of sex chromosomes.

However, it is known that genetically non-homologous sex chromosomes may be morphometrically indistinguishable. This is exemplified by isomorphic heterochromosomes inherent to the voles of the genus *Ellobius* (Cricetidae, Rodentia, Mammalia), which have been described by Vorontsov *et al.* (1969). In the case of isomorphic chromosomes, the cytological determination of sex chromosomes is possible only on the basis of meoitic studies.

(b) In contrast to autosomes, the homologous chromosomes of which conjugate side-by-side in pairs, the heterochromosomes in the heterogametic sex, being non-homologous, either conjugate terminally, or the homologous regions of the X and Y chromosomes conjugate laterally, whereas the non-homologous regions do not conjugate at all.

There is a large variety of mammals with terminal conjugation of the X and Y chromosomes, while in *Cricetulus migratorius* the homologous short arm of the Y and the distal segment of the long arm of the X chromosomes are able to undergo conjugation (Utakoji, 1966).

(c) It is generally believed that sex chromosomes are characterized by abundant chromatin. Sex chromatin has been found in the interphase nuclei of homogametic sex. This has been interpreted as evidence of the genetical inertness of one of the X chromosomes and an indication of an equilibrated genetical balance between the sexes owing to the inactivation of one of the X chromosomes in the somatic cells of homogametic females

(Galton 1966). However the phenomenon of "dosage compensation" has not been demonstrated in birds.

(d) DNA replication in one of the X chromosomes in the homogametic sex and in the Y of the heterogametic sex proceeds after the termination of DNA replication in autosomes. There have only been a few radioautographic studies carried out solely on mammals confirming this important observation.

It is, thus, obvious that the main method for the demonstration of the absence and presence of heterochromosomes remains the establishment of heteromorphism or isomorphism of mitotic chromosomes. Does the wide occurrence of chromosomal isomorphism among lower vertebrates testify to the absence of heterochromosomes?

2. *The trigger hypothesis of the origin of sex chromosomes*

A sceptic cannot fail to observe that studies on chromosome conjugation in meiosis have been restricted to a limited number of species and that conjugation has not been studied in groups with heterogametic females. Finally, a point of no less importance emerges as to whether there could have been a particular stage in the evolution of sex chromosomes when 90% of the length of the isomorphic X and Y chromosomes was occupied by homologous material and the sex-triggering genes were concentrated on negligibly small (relative to total length of the chromosome) non-conjugating portions of the X and Y chromosomes undiscernible cytomorphologically.

Sex-linked genes, which are irrelevant to the determination of sex and to the development of sexual characters in ontogeny, have been distinguished from sex genes. Likewise, two types should be recognized within sex genes: sex-determining genes controlling the development of the characters of a given sex in ontogeny on the one hand and sex-triggering genes, on the other hand, which determine along which of the possible developmental pathways the sexual characters will proceed.

It is quite obvious that the number of sex determining genes is much higher than the number of sex-triggering genes. Using the terms we have proposed, it may be said that in the *Drosophila* the trigger function is performed by the balance between the X chromosomes and autosomes, which has been established by Bridges. However, Diptera and Mammalia are the last stages of the evolution of sex chromosomes, which has occurred independently within Insecta and Amniota.

It is conceivable that at the early evolutionary stages of sex-triggering genes were concentrated at one of the pairs of autosomes, whereas the other sex-determining genes were distributed among the other autosomes. At this stage homozygosity or heterozygosity for a particular trigger gene was of importance. In some groups homozygotes have given rise to the subsequent development of females, which led, eventually, to the XX–XU system and its derivatives; in other groups homozygotes have triggered those genes into action, which determine the development of the male sex leading to the ultimate ZZ–ZW system and its derivates.

At first, the presence of trigger genes was no hindrance to the conjugation of chromosomes bearing these genes. Conjugation disturbances started to manifest only when both sex triggering and determining genes were concentrated on one and the same chromosome under the effect of translocations and the pressure of natural selection. Translocations involving the transfer of chromatin material from autosomes to heterochromosomes may have entrained genes having no bearing to the differentiation and determination of sex. These genes proved to be sex-linked.

The accumulation of single trigger genes together with a large number of sex determining genes may have led to disturbances in the conjugation between heterochromosomes along their length. It is only now that it is possible to identify heterochromosomes by means of cytological techniques in meiosis. Such are the conceptual stages of the evolvement of sex chromosomes.

3. *The relation between heterochromosomes and autosomes in the determination of sex*

It is evident that only the heterochromosomes are responsible for the determination and, moreover for the development of sexual characters. In man the Y chromosome always plays the male-determining role both in heterochromosomally normal individuals (XY) and in abnormal individuals (XXY, XXXY, XXXXY). However, individuals with abnormal chromosomal ratios are sterile. The absence of the Y chromosome in humans with the XX constitution causes it to pursue a normal female development and to develop into a phenotypical female. In contradistinction individuals with heterochromosomes of the XO, XXX, XXXX type lacking the Y chromosome deviate in sexual development and are sterile. Disturbances of the ratio of sex chromosomes to autosomes lead to sterility.

The general framework of the balance theory of sex determination (Bridges, 1932) based on studies of the *Drosophila* with abnormal numbers of heterochromosomes and autosomes holds for all animals with the syngamic mode of sex determination, although the ratio of chromosomes number to the ploidy of the autosomes, which lead to the development of sexual characters, and the role of separate sex chromosomes may differ largely between species.

II. SEX CHROMOSOMES IN LOWER VERTEBRATES (TAXONOMICAL REVIEW)

A. *Acrania* (*Amphioxiformes*) and *Agnata*

No sex chromosomes have been disclosed in *Amphioxus*.

All the ten karyologically studied species of lampreys of the genera *Lampetra*, *Entosphenus*, *Mordacia* and *Petromyzon* have large numbers of small chromosomes ($2n = 60$, 94–96, 142, 156, 168; $1 = 0.7\mu$–2.7μ). Heteromorphic chromosomes were not found in Cyclostomata (Potter and Rothwell, 1970).

B. *Anamnia*

1. *Pisces*

(a) Bisexual species: At present, the mitotic chromosomes of not less than 100 fish species have been described on the basis of data yielded by modern techniques. However, no proven cases of heteromorphic chromosomes have been described in either sex. It should be emphasized that in recent years the chromosome complements of such taxonomically remote species have been described as *Hydrolagus colliei* (Chondrichthyes, Elasmobranchia, Holocephala, Chimaeriformes), *Lepidosiren paradoxa* (Osteichthyes, Dipnoi, Ceratodiformes) *Scaphirhynchus platorhynchus* (Osteichthyes, Teleostomi, Chondrostei), *Lepidosteus productus* and *Amia calva* (Osteichthyes, Teleostomi, Holostei). Teleostei include the rather well studied Salmonoidei (*Salmo*, *Oncorhynchus*, *Coregonus*, *Osmerus*), Cyprinoidei (Cyprinidae: *Carassius*; Cobitidae: *Misgurnus*, *Barbatula*, *Lefua*, *Cobitis*), Cyprinodontoidei (Poeciliidae: *Xiphophorus*), Percoidei (Centrarchidae: *Lepomis*; Cichlidae: *Symphysodon*), Pleuronectoidei (Bothidae: *Xystreurys*; Pleuronectidae: *Pleuronichthys*) and Gasterosteiformes (*Gasterosteus*, *Pungitius*) (Ohno and Atkin, 1966; Ohno *et al.*, 1969; Ojima *et al.*, 1963; Rees, 1964; Sasaki *et al.*, 1968; Muramoto *et al.*, 1969).

Multiple sex chromosomes have been recently reported in an undetermined species of the Mexican Cyprinodotidae (Uyeno and Miller, 1971). In this fish females have $2n = 48 = 44A + X_1X_1X_2X_2$, whereas males possess $2n = 47 = 44A + X_1X_2Y$; Uyeno and Miller consider that multiple sex chromosomes originate from the karyotype of the initial species: ♂ $2n = 48 = 46A + X_1Y$; ♀ $2n = 48 = 46A + X_1X_1$.

In recent years T. R. Chen has described some cases of chromosome heteromorphism in different sexes and among diverse groups of fish. It was observed that among deep-sea fishes, *Bathylagus wesethi* (Bathylagidae) possess the male heterogamety whereas *Gambusia affinis* bear the female heterogamety (Chen and Ebeling, 1968); the X-chromosome is slightly hypertrophied in *Bathylagus* (Chen and Ebeling, 1966; Chen, 1969). Some species of the killifish genus *Fundulus* (Cyprinodontinae) bear the isomorphic sex chromosomes (*F. heteroclitus*, *F. majalis*) while XY–XX system is found in *F. diaphanus* and *F. parvipinnus* (Chen and Ruddle, 1970).

It should be noted that relatively clear-cut heteromorphism of XY male chromosomes was found in *F. parvipinnus* but the presence of heteromorphism of chromosomes in *F. diaphanus* males requires further investigations. Meiosis of these species has not been studied.

(b) Unisexual species: Gynogenetic reproduction has been established in a number of fish species. Hubbs and Hubbs (1932) found gynogenesis in *Poecilla* (= *Molliensia*) *formosa* (Poeciliidae), which live in the waters of Mexico and Texas. It has been shown that *P. formosa* females are derived from crosses between bisexual species of *P. sphenops* and *P. latipinna*. Subsequently, gynogenesis has been discovered in two species of the genus *Poeciliopsis* (Schultz, 1967) belonging to the same family. Gynogenesis has been established in several races of *Carassius auratus* within the family Cyprinidae (Golovinskaja and Romashov, 1947; Lieder, 1959; Cherfas, 1966). In all the cases described in the literature the gynogenetic females are triploid, triploidy being a device for cytogenetical sex differentiation.

2. *Amphibia*

Amphibia with their small numbers of large chromosomes are very convenient tools for karyological studies. Conclusions concerning the presence or absence of chromosome heteromorphism based on Amphibia are more reliable than those made on the basis of Reptilia with their small microchromosomes. There are no data on the sex chromosomes in Gymnophiona.

(a) Caudata (= Urodela): The XY system has been described in ♂♂ of the Plethodontidae family within the North American genera *Oedipina* (Matthey, 1969a) and *Thorius* and also in the tongueless salamandra *Chiropterotriton abscodens* (Kezer, personal communication; Manchino and Nardi, 1971).

As a consequence of investigations on lampbrush chromosomes of female oocytes in *Triturus marmoratus* (Salamandridae), Manchino and Nardi (1971) inferred that there exist heterogametic females of the ZW–ZZ type. However, thorough karyometric analysis of representatives of Salamandridae—*Salamandra salamandra* and *Euproctus asper*—did not reveal heteromorphic chromosomes (Jaylet, 1966; Kessous *et al.*, 1968).

Hauschka and Brunst (1964) have described heterogametic females (ZW) in *Ambystoma mexicanum* (Ambystomidae). It is astonishing that heteromorphic sex chromosomes are encountered in the genus *Ambystoma* considering the finding of gynogenesis and triploidy in females of the unisexual complex *Ambystoma jeffersonianum* (Thomas and Uzzell, 1964).

(b) Salienta (= Anura): Comparatively detailed examination of the karyotype of all the 17 families of Anura has been carried out. Among almost 200 karyologically studied Anura (Rabello, 1970) sex chromosomes, or chromosomal heteromorphism, in one of the sexes have been demonstrated only in single cases.

Heteromorphism of sex chromosomes has been shown in the African frog *Xenopus laevis* (Pipidae). *Xenopus* females are heterogamous (ZW), the Z being an acrocentric and the W being a large metacentric (Weiler and Ohno, 1962). Morescalchi (1964) has distinguished sex chromosomes in *Discoglossus pictus* (Discoglossidae).

The idea that the Japanese tree frog *Hyla arborea* has sex chromosomes of the XY type (Yoshida, 1957) was abandoned (Seto, 1964); there is not a single proven case of chromosomal heteromorphism among the 40 karyologically studied species in the genus *Hyla* (Stephenson and Stephenson, 1970).

Surveying the literature on the karyology of Anura, Rabello stated (1970 p. 57): "When both sexes were available, male and female presented identical karyotypes."

Brum-Zorrilla and Saez (1971) have analysed a large number of South American Anura species belonging to the families Bufonidae, Ceratophrydidae, Leptodactylidae and Hylidae. They have found no evidence for cytologically differentiated sex chromosomes. The results of Brum-Zorrilla and Saez cast doubt on the data of Yoshida (1957) concerning the XX–XY system in *Hyla arborea* and those of Weiler and Ohno (1962) on the ZW–ZZ in *Xenopus laevis*; the results also make the description of the ZW–ZZ system in *Discoglossus pictus* doubtful (Morescalchi, 1964) and speak against the ZW system in ♀ *Bufo melanostictus* as indicated by Manna and Bhunya. Karyometric studies of ♀ and ♂ *Bufo arenarium* have not established heteromorphic chromosomes (Brum-Zorrilla and Saez, 1971).

The body of evidence points directly to the fact that there are no cytogenetically differentiated heterochromosomes in all or, at least, in the majority of Anura.

It may be argued that the probable widespread polyploidy in the evolution of Anura (and, perhaps, in all Amphibia) is related to the absence of sex chromosomes in Amphibia. It should be noted that the first proven instances of bisexual polyploid species among vertebrates belong to Anura *Odontophrynus americanus* ($4n = 44$), *Pelomedusa burmeisteri* ($4n = 52$) and *Ceratophrys dorsata* ($8n = 104$) and, furthermore, that cytophotometric studies of DNA indicate that polyploidy was a common occurrence in the evolution of Amphibia (Beçak *et al.*, 1967, 1970).

What is the cause of the almost complete absence of heteromorphic sex chromosomes in Amphibia? Can the absence be accounted for by the fact that the first attempts of cytogenetical sex differentiation are made in this class? Perhaps the cause may reside in the physiological determination of sex, which is characteristic of Amphibia and occasionally counteracts the genetic determination of sex.

Recent data indicate that polyploidy and the duplication of gene material have played a significant role in the evolution of Amphibia. Variations in DNA content support the idea that gene duplications have occurred in Bufonidae, Ceratophrydidae (true polyploidy has been observed in this family), Hylidae and Leptodactylidae (Beçak *et al.*, 1970). Polyploidy, which has included separate groups of Amphibia in the course of evolution, removed the advantages which cytogenetic sex determination could have conferred.

C. *Sauropsida*

The definite relationship of all Sauropsida is reflected by the clear-cut distribution of their chromosome complements into relatively large macrochromosomes and small ($\approx 1\mu$) microchromosomes. However, the sex chromosomes in Reptilia and Aves are different. By contrast to birds with their distinctive ZZ–ZW system or its derivatives, Reptilia display a variegated pattern ranging from the absence of heteromorphic chromosomes to heterogamety in males and females.

1. *Reptilia*

A brief review covering the subject of sex chromosomes in Reptilia has been recently presented in Russian (Kuprijanova, 1969; Sokolovsky, 1973).

In the thirties and forties the Japanese school of cytologists reached the conclusion that sex chromosomes are heteromorphic. The establishing of the fact that in *Lacerta vivipara* (Lacertidae) the males possess $2n = 36$ and the females have $2n = 35$ led to the suggestion of the ZZ–ZO system (Oguma, 1934). It was considered that heterogamety of the same type is inherent to *Calotes versicolor* (Agamidae) (Makino and Asana, 1948). However, the cytogeneticists working in Europe have produced arguments against the interpretation of the Japanese colleagues.

For instance, Margot (1946) showed that both ♂ and ♀ *L. vivipara* have $2n = 36$, which refuted the ZZ–ZO system proposed by Oguma. In their reviews of karyotypic data Matthey (1949), Matthey and van Brink (1956, 1957) and van Brink (1959) deny the existence of cytologically identifiable sex chromosomes in Reptilia.

van Brink (1959) concludes her study on sex chromosomes in Vertebrata by: "Fishes: no cytologically recognizable sex chromosomes; Amphibians: no cytologically recognizable sex chromosomes; Reptiles: no cytologically recognizable sex chromosomes; Birds: female heterogamety (XO, XY?); Other mammals: male heterogamety of the XY type, rarely, multiple sex chromosomes" (p. 66).

In spite of all the previous data, heteromorphic ZW was found in ♀ *Vipera berus* (Kobel, 1962) and in ♀ *Bothrops jaracaca* (Beçak *et al.*, 1962). This has called for the reconsideration of the problem of the existence of sex chromosomes in Reptilia. At present, heteromorphism of sex chromosomes has been primarily established in various sexes of some groups of Reptilia, mainly in snakes.

(a) Rhinchocephalia: No information has been derived from the early studies of Keenen (1932) concerning heterochromosomes in *Sphenodon*.

(b) Chelonia: Of the 32 karyologically studied tortoise species 6 were investigated after 1951 (Sokolovsky, 1972). Recent data do not make it possible to discern heteromorphism in the six species studied in the families Pelomedusidae, Emydidae, Testudinidae and Trionchidae (Huang and Clark, 1967, 1969).

(c) Squamata: (i) Sauria: Karyotypic data have been presented for 19 of the 22 lizard families (Sokolovksy, 1972).

Geckonidae. Sex chromatin was not found in the interphase nuclei of *Gekko gecko* and neither were heteromorphic chromosome pairs observed in both sexes (Cohen *et al.*, 1967). There is evidence indicative of parthenogenesis in Geckonidae (Hall, 1970).

Pygopodidae. Heteromorphism of the X_1X_2Y–$X_1X_1X_2X_2$ type has been shown in *Lialis burtonis* (Gorman and Gress, 1970b). Trivalents have been shown in males in meiosis I, when cells with $n = 14$ or $n = 17$ appear, $2n$ o $= 33$, $2n$ o $= 34$.

Iguanidae. The karyotype of *Iguanidae* has been quite thoroughly investigated (see the review by Kral, 1969). It is of interest that the transition from isomorphic chromosomes to the XY–XX system and its derivative system X_1X_2Y–$X_1X_1X_2X_2$ is observed not only within the family Iguanidae, but also in the range of separate genera (*Anolis*, *Sceloporus*). It is suggested that the large Y chromosome in *Polychrus marmoratus* is the result of the fusion of the X_1 and X_2 (Gorman *et al.*, 1967). That the X_1X_2Y–$X_1X_1X_2X_2$ mechanism operates in *Uma inornata* has been confirmed by mitotic studies and by the existence of spermatocytes II of two types: having $n = 10$ and $n = 11$ (Král, 1969). 13 bivalents and 1 trivalent in diakinesis have been demonstrated in some *Anolis* males. The trivalent forms an unpaired metacentric (Y), a large acrocentric (X_1) and a microchromosome (X_2) (Gorman and Atkins, 1966). In *Uma inornata* (Král, 1969), *Sceloporus poissetti*, *S. jarrovi* (Cole and Lowe, 1967) and *Anolis gingivinus* (Gorman and Atkins, 1966) the Y chromosome is the largest microchromosome. The intergradation from isomorphic to multiple sex chromosomes indicates that the chromosomal mechanism determining sex has evolved independently in different Iguanidae groups. Against this background the independent origin and the wide occurrence of the X_1X_2Y–$X_1X_1X_2X_2$ should attract attention; it seems that in this group multiple sex chromosomes had a certain selective value.

Agamidae. Makino and Asana (1948) thought it possible that the ZZ–ZO system existed in Agamid *Sitana ponticeriana* and *Calotes versicolor*. According to Japanese workers, the males of latter species have $2n = 34$ (12M + 22m) and the females have $2n = 33$ (12M + 21m).

New karyotype data on 4 Agamidae species do not reveal any heteromorphic chromosomes in this family which has been scantily studied from the karyological view-point (Arronet, 1965; Dutt, 1969; Hall, 1970). Hall (1970) has raised the possibility of parthengenesis in Agamidae.

Lacertidae. The data produced by Oguma (1934) on the heterogamety of *Lacerta vivipara* have been disputed by many workers (van Brink, 1959 and others). However, recently Chevalier (1969) maintained that she had established heterogamy of the $Z_1Z_1Z_2Z_2$–$Z_1Z_2W_1$ type, but other workers have not succeeded in identifying heterochromosomes in *L. vivipara* (Orlova and Orlov, 1969). A $Z_1Z_1Z_2Z_2$–Z_1Z_2W mechanism was reported in *L. strigata* (Ivanov and Fedorova, 1970) and in *Eremias* (Ivanov, personal communication). Conversely, most studies on other *Lacerta* (Darevsky and Kulikova, 1962, 1964; Dallai and Urbani, 1967; Kuprianova, 1968, 1969; Orlova and Orlov, 1969; Gorman, 1969) and on other genera of this family including *Ophisops*, *Eremias* and *Acanthodactylus* (Arronet, 1968; Kuprianova and Arronet, 1969; Gorman, 1969) have not revealed heteromorphic chromosomes in Lacertidae of both sexes.

Within the superspecific group *Lacerta saxicola*, parthenogenesis was first discovered in Amniota (Darevsky, 1958). It is of interest that the karyotypes of the parthenogenetic *L. unisexualis*, *L. dahli* and *L. s. nairensis* females are indistinguishable from the isomorphic karyotypes of both sexes of the bisexual forms *L. saxicola valentini*, *L. s. nairensis* and *L. s. darevskii*. In the nature triploidy arises in zones of contact between parthenogenetic and bisexual forms (Darevsky and Kulikova, 1964).

Teiidae. The occurrence of parthenogenesis has also been observed in the American *Cnemidophorus*, which occupies the niche of *Lacerta* in the Nearctic. Unisexual species have proved to be diploid (*C. neomexicanus*) as well as triploid (*C. uniparens*) (Lowe *et al.*, 1966, 1970). In this connection, heteromorphism of the XY–XX type in *C. tigris* (Cole *et al.*, 1969) and in other species belonging to this genus is an unexpected finding. The heteromorphism is so weakly expressed in these cases that polykaryograms are necessary to differentiate them.

The problem of the actual existence of sex chromosomes in Lacertidae and Teiidae remains obscure.

Cytogenetically differentiated heterochromosomes were not demonstrated in other lizard families such as Scincidae (Dallai and Talluri, 1969; Dutt, 1969), Varanidae (Dutt, 1968; Gorman and Gress, 1970a), Anguidae (Bury *et al.*, 1969) and Amphi baenidae (Huang *et al.*, 1967).

(ii) Serpentes: Three karyologically studied snake families are definitely known to possess sex chromosomes. The initial studies were by Kobel (1962) who described the heterochromosomes in *Vipera berus*.

Colubridae. This family exhibits all the morphological intergrades of a chromosome which ranks fourth in size, encompassing isomorphism and heteromorphism (Beçak, and Beçak, 1969). It is considered that the acrocentric W is an element which has arisen from the initial (primary) metacentric due to pericentric inversion. In *Thamnodynastes strigatus* and *Liophis miliaris* the W chromosome has become smaller in size. In the genera *Thomodon* and *Xenodon* the W chromosome is a metacentric, which is, probably, of secondary origin and the result of a secondary pericentric inversion. In *Clelia occipitolutea* the W is bigger than the Z chromosome, which is the probable consequence of a duplication (Beçak and Beçak, 1969; Itoh *et al.*, 1970). In the three species of the genus *Elaphe–E. subocularis*, *E. guttata* and *E. obsoleta*—the Z-chromosome is the third in size of four submeta- or subtelocentric chromosome pairs, the W is a subtelocentric equal in size to the fourth pair of chromosomes (Baker *et al.*, 1971).

Hydrophiidae. The system ZZ–ZW_1W_2 was described in sea snake *Enhydrina schistosa* (Singh, 1972).

Elapidae. Multiple sex chromosomes of Z_1Z_1 Z_2Z_2 type in males and Z_1Z_2W in females were found in Indian krait *Bungarus caeruleus* (Singh *et al*., 1970).

Viperidae. Heterogamety of the ZW–ZZ type has been proven for representatives of this family. Furthermore, it has been shown that both in *Vipera* and *Crotalus*, just as in the above mentioned families, the heterochromosomes are equal in size to the fourth chromosome pair (Kobel, 1962, 1963, 1967; Baker *et al*., 1971).

Constant female heterogamety and size of the Z chromosome are characteristic of Serpentes.

(d) Crocodilia: All the 21 species of the order have been examined to determine their karyotypes, but no sex chromosomes were found (Cohen and Clark, 1967; Cohen and Gans, 1970).

Thus, the impression is gained that the heterochromosomes in Reptilia are highly heterogeneous. Leaving the problem of heterochromosomes in Rhinchocephalia open; it may be stated that probably in Chelonia and certainly in Crocodilia cytogenetically distinguishable sex chromosomes are absent.

The sex-determining mechanisms are evolving in Squamata, but heteromorphic chromosomes may be absent in Sauria, and XX–XY devices and its $X_1X_1X_2X_2$–X_1X_2Y derivatives may be elaborating (female heterogamety of the ZW–ZZ type cannot be accepted as proven in Sauria). Serpentes is characterized by a clear-cut ZW–ZZ mechanism and its Z_1Z_2W–$Z_1Z_1Z_2Z_2$ derivative.

Heteromorphic sex chromosomes seem to be absent in these groups of Sauria, in which parthenogenesis is encountered which is due in some lizard species to triploidy. However, parthenogenesis cannot account for the absence of sex chromosomes in tortoises and crocodiles. Morphological divergence, which is widest in Reptilia in comparison to all the Amniota, has undoubtedly been mirrored by the divergence of sex chromosomes in various groups of Reptilia.

2. *Aves*

Homoiothermal animals belong to those in which sex chromosomes are not the exception, but rather the rule.

As early as 1922 Zawadowsky showed that the castration of the heterogametic sex results in the adoption of the phenotype of the heterogametic sex. For example, a castrated hen acquires the colour of the cock's plumage, whereas the castrated cock retains the colour of his sex. The castration of male antelopes (*Portax pictus* and *Antelope cervicapra*) is associated with their acquirement of the female coat colour. The experiments of Zawadowsky have led to the conclusion that homogamety plays a decisive role in the development of secondary sexual characters and that the characters of the heterogametic sex are realized in ontogeny only under the control of sex hormones which inhibit the effect of the large X and Y chromosomes in the heterogametic sex.

With respect to birds it has been suggested that the evolution of sex chromosomes occurred at the expense of the gradual reduction of the size of one of the chromosomes of the homologous pairs of autosomes. Here, the sex-determining factors are limited to one of the sex chromosomes of unequal size. It is believed that birds lack the effect of "dosage compensation" characteristic of mammals. Sex determination of females depends on the locus of the W chromosomes and in males it is related to the combined effect of both Z chromosomes (Galton, 1966; Galton and Bredbury, 1966).

The karyotypes of more than 200 studied bird species are surprisingly uniform, when compared with the karyotypes of mammals. Their chromosome complements usually exhibit 7–8, rarer 10–12 macrochromosome pairs and a large number of microchromosome pairs (22–37). Only *Burhinus oedicnemus* ($2n = 40$), which has been studied in our laboratory (Bulatova *et al.*, 1971) has but 8 pairs of microchromosomes.

The karyotype of *Struthio camelus* (Ratitae) is very similar to that of the typical Carinatae (Itoh *et al.*, 1969), but its sex chromosomes have not been described in as much as animals of undetermined sex have been investigated.

In Carinatae female heterogamety of the ZW type may be considered as established in females and males homogamety of ZZ type. As in snakes, the Z chromosome occupies a relatively stable karyotype position: most frequently it occurs fourth and more rarely third, fifth or sixth. The sixth chromosome is usually situated between the macro- and microchromosomes and is ranged most commonly eighth, more rarely sixth, seventh and ninth in the karyotypes (Itoh *et al.*, 1969; Bulatova *et al.*, 1971). The structure of sex chromosomes in *Calandrella cinerae* (Alaudidae, Passeriformes) is unusual in that the Z chromosome is the largest in karyotype and the W chromosome is the second in the complement (Bulatova *et al.*, 1971). The presence of the ZW–ZZ system has been established in Falconiformes, Galliformes, Charadriiformes (including Oedicnemidae), Lariformes, Columbiformes, Cuculiformes, Psittaciformes, Strigiformes and Passeriformes (Ohno *et al.*, 1964b; Takagi and Makino, 1966; Renzoni and Vegni-Talluri, 1966; Ray-Chaudhury *et al.*, 1966; Ray-Chaudhury, 1967; Jovanovič and Atkins, 1969; Itoh *et al.*, 1969; Bulatova *et al.*, 1971). The homogeneity of the types of sex chromosomes, in spite of the few cytogenetical studies performed on birds, is in accordance with the morphological uniformity of the Aves class and contrasts with the variety of chromosome types in Reptilia and Mammalia.

III. THE SEX CHROMOSOMES OF MAMMALS

In comparison with all the other vertebrates our knowledge of mammalian sex chromosomes is one most extensive. A comprehensive review of the subject would require a volume. The literature on the abnormal structure of the sex chromosomes in mammals has been aptly summarized by Fredga (1970). In all mammals males are the heterogametic sex.

A. *Monotremata*

The presence of the sex chromosomes in Monotremata was denied (Matthey, 1949; van Brink, 1959), until Bick and Jackson (1967a, b) established the XX system in *Tachyglossus aculeatus* ($2n = 64$) and *Ornithorynchus anatinus* females ($2n = 54$) and the XO system in males. The XO system has been confirmed for somatic cells and spermatogonia of *Tachyglossus* ($2n = 63$), but only for the somatic cells of the *Ornithorhynchus* male. Since there are no thorough meiotic studies, Fredga (1970) was compelled to refer this species to the group with "unknown mechanisms of sex determination".

B. *Eutheria*

It has been proved that the XX–XY mechanism is inherent to the great majority of cytologically studied Eutheria. The sizes of both the X- and Y-chromosomes may vary in this "classical" mechanism.

A system with multiple sex chromosomes may evolve on the basis of such a mechanism in consequence of the fragmentation of either the X or the Y as well as the fusion of one of the heterochromosomes with an autosome.

1. *The XX–XY system. The hypothesis of Ohno, Beçak and Beçak. Matthey's hypothesis*

Ohno *et al.*, (1964a) called attention to the fact that in a large number of mammalian species the length of the X-chromosome makes up only 5% of the length of the haploid complement of autosomes. A larger X-chromosome may attain about 10%, 15% or 20% of the length and then is called a duplicate, triplicate or quadruplicate type of the X-chromosome, respectively. According to the hypothesis advanced by Ohno, Beçak and Beçak, in all the cases the functioning part of the X-chromosome always makes up 5% of the length of the haploid complement and the rest of the chromatin is represented by heterochromatin. The hypothesis implies that the large X-chromosomes arise by means of duplication of the X-chromosome of the initial type. The hypothesis of Ohno–Beçak appeared singularly attractive against a background where descriptive works predominated and it was approved almost unreservedly (Lewis and John, 1968) and included in reviews.

Matthey (1961, 1965) suggested that large mammalian chromosomes are the result of the translocation of a single or several autosomes to the sex chromosomes. If Matthey's assumption is correct, then the terminal association of the large sex chromosomes would not be observed, but rather a side–side conjugation with the formation of chiasmata between the homologous regions of the X- and Y-chromosomes.

Modern data indicate that the evolution of the sex chromosomes in mammals is along different pathways involving the duplication of heterochromosomic material and its subsequent inactivation, on the one hand, and the inclusion or loss of the autosome material by the heterochromosomes, on the other.

The X- and Y-chromosomes may both be telocentric or one of them may be meta- or submetacentric, or they may both be metacentric. There is no relation between the shape of the autosomes and heterochromosomes: karyotypes are known where all the chromosomes are acrocentrics with the exception of the sex chromosomes (*Canis familiaris*; *Bos taurus*); whereas in the vole *Microtus montanus*, among entirely metacentric autosomes, telocentric heterochromosomes are distinguished (Schmidt, 1967; after Fredga, 1970).

In most species the X-chromosome is smaller in size than the first pair of autosomes. The content of the sex chromosomes in the haploid complement of autosomes varies and the transition from the "initial" type to the "duplicate" type described by Ohno is actually not so distinctive (Table I).

An increase in the size of the X-chromosome accompanied by a small size of the Y has been observed in *Balaenoptera physalus* (Cetacea), *Leggadina hermannsburgensis* (Muridae, Rodentia), *Chinchilla laniger* (Chinchilidae, Rodentia); it has been shown that in the latter species the large X and the small Y are associated end-to-end (Fredga, 1970).

Three different meiotic patterns are distinguished among species having a large X- and a large Y-chromosome:

Species	X-ratio	Y-ratio	Reference
Capra ibex	2.6	—	Hauchtek-Jung *et al.*, 1967
Rattus rattus (2*n* = 38)	4.5	1.1	Badr and Badr, 1970
R. rattus (2*n* = 42)	4.6	2.4	Badr and Badr, 1970
Sorex minutus	4.8	0.8	Kozlovsky, personal communication
Talpa europaea	5.0	0.7	Kozlovsky, personal communication
Sorex raddei	5.4	2.4	Kozlovsky, personal communication
S. isodon	5.5	2.1	Kozlovsky, personal cummunication
Rhinophylla pumilio	5.4–6.3	—	Baker and Bleier, 1971
Pipistrellus savii	5.6	1.6	Capanna and Civitelli, 1967
Myotis capaccini	5.6	1.7	Capanna *et al.*, 1968
Rattus (*Mastomys*) *natalensis*	5.8	4.2	Huang, 1968
Arvicola terrestris	5.8	2.2	Fredga, 1968
Barbastella barbastellus	5.9	1.8	Capanna *et al.*, 1968
Myotis myotis	6.0	1.0	Capanna *et al.*, 1968
M. oxyganthus	6.5	1.5	Radjabli *et al.*, 1970
Arvicola scherman exitus	6.8	—	Schmidt and Leppert, 1968
Rhinophylla fischerae	6.9	—	Baker and Bleier, 1971
Marmota caudata	7.4	—	Vorontsov *et al.*, 1969
Tatera indica cuvierii	8.0 $(X)^{sm}$ 7.3 $(X)^{m}$	2.8	Rao *et al.*, 1968
Neotoma cinerea	8.1	6.7	Wurster *et al.*, 1971
Dolichotis patagonicum	8.4	—	Wurster *et al.*, 1971
Hydrochoerus hydrochoerus	8.5	4.1	Wurster *et al.*, 1971
Leggada minutoides musculoides	8.6	4.4	Matthey, 1966a
Chinchilla laniger	9.0	—	Galton *et al.*, 1965
Sorex volnuchini	9.1	1.2	Kozlovsky, personal communication
Suncus murinus	11.2	—	Rao *et al.*, 1970
Tragelaphus spekei	13.1	7.3	Wurster *et al.*, 1968
Antelope cervicapra	15.0	—	Wurster *et al.*, 1968
Sorex araneus			
USSR—Tula	15.1	1.0(Y_1) 8.4(Y_2)	Kozlovsky, personal communication
USSR—Kemerovo	15.5	1.9$(Y)_1)$ 9.5(Y_2)	Kozlovsky, personal communication
USSR—Moscow	16.8	2.2(Y_1) 8.5(Y_2)	Kozlovsky, personal communication
Carrolia castanea	16.4	—	Baker and Bleier, 1971
C. subrufa	17.3	—	Baker and Bleier, 1971
C. perspicillata	16.5–18.2	—	Baker and Bleier, 1971

(a) An end-to-end association is observed in *Mesocricetus* and in *Phodopus sungorus* (Cricetidae, Rodentia), *Rangifer tarandus* (Cervidae, Artiodactyla) and in the vole with giant sex chromosomes *Microtus agrestis* (Cricetidae, Rodentia) (Matthey, 1949, 1952; Fredga and Santesson, 1964; Fredga, 1970; Fraccaro *et al.*, 1969; Radzhabli and Kryukova, 1971).

(b) The giant sex chromosomes were found in *Microtus agrestis* (Cricetidae, Rodentia) (Matthey, 1950). The concept of their end-to-end association which appears in the literature (Fredga, 1970) seems to be erroneous. Zenzes and Wolf (1971) have established that no conjugation of the sex chromosomes independent of one another has been observed in *M. agrestis*, either in zygotene or in diplotene. Thus, in *M. agrestis* the change among the heterochromatin, supposedly homologous regions of the sex chromosomes does not occur in meiosis.

(c) The presence of chiasma between the X- and Y-chromosomes in the male meiosis. It is of interest that unlike *Mesocricetus*, *Cricetulus griseus* and *Cricetus cricetus*, which belong to the same Cricetini tribe (Fredga and Santesson, 1964; Utakoji, 1966), undergo the conjugation among the homologous regions of the X- and Y-chromosomes. This indicates that the evolution of the sex chromosomes runs in different directions among relatively cognate genera. The similar type of association was described in Muridae, *Leggada minutoides* (Matthey, 1965), *Rattus* (*Mastomys*) *natalensis* (Hamerton, 1958; Huang and Strong, 1962) and in Erethizontidae *Erethizon dorsatum* (Benirschke, 1968).

2. *Isomorphism of X- and Y-chromosomes*

The problem of the true existence of isomorphism of the sex chromosomes requires special proof. Morphological comparisons make it difficult to distinguish the X- and Y-chromosomes in the cases where the karyotype consists of gradually decreasing rows of chromosomes. Isomorphism of the heterochromosomes in mammals occurs rarely if ever.

Isomorphism of the sex chromosomes has been found in *Ellobius talpinus* (Vorontsov *et al.*, 1967), *E. alaicus* (Vorontsov *et al.*, 1969), *E.* "*talpinus*" with $2n = 32$ (Liapunova *et al.*, 1972) (all Microtinae, Cricetidae, Rodentia) and in *Uranomys ruddi* (Muridae, Rodentia) (Matthey, 1970). The heteromorphic sex chromosomes have not been found in *Spalax* (*Microspalax*) *ehrenbergi* (Wahrman *et al.*, 1969) whereas in the other Spalacinae the sex chromosomes readily differentiate (Soldatovič *et al.*, 1966; Liapunova *et al.*, 1972). The isomorphism of heterochromosomes in mammals can be expected to be of a secondary nature and to arise in different groups of mammals independently.

3. *Gonosomic mosaicism in Marsupialia. Sex chromosome mosaicism in Marsupials. XO–XX, XO–XY system*

Of 5 species of Marsupials of Peramelidae family (in which the somatic and reproductive tissues were examined) 4 species bore mosaicism of sex chromosomes (Hayman and Martin, 1965, 1969). It has been supposed that there is early elimination at the initial stages of the embryogenesis of one of the X-chromosomes in ♀ and of the Y-chromosomes in ♂. This suggestion was confirmed by the analysis of chromosomes from the different tissues (Table II). At present we have no data on the reasons for elimination of the second sex chromosome in a number of somatic tissues.

TABLE II. The diploid number and sex chromosome complements in different tissues of adult of Marsupials *Peramelidae* (compiled from Hayman and Martin, 1969 and Hayman *et al.*, 1969).

Species	Sex	Tissue	2*n*	Sex chromosome
		bone marrow	13	XO
		leucocytes	13	XO
		spleen	13	XO
Isoodon macrourus		corneal epithelium	13	XO
and *I. obesulus*	♂ & ♀			
		intestinal epithelium	13[a]	XO
	♂	testis	14	XY
	♀	ovarian tissue	14	XX
		bone marrow	13	XO
		leucocytes	13	XO
Perameles nasuta	♂ & ♀	spleen	13	XO
and *P. gunni*		corneal epithelium	14	XX and XY
		intestinal epithelium	14[b]	XX and XY
	♂	testis	14	XY
		ovarian tissue	14	XX
Peroryctes longicauda	♀	corneal epithelium	14	XX
		spleen	13	XO
Echymipera kalabu	♂ & ♀	bone marrow	13	XO
		spleen	13	XO
		corneal epithelium	14	XX and XY
	♂	testis	14	XY

[a] Only *I. obesulus* examined.
[b] Only *P. nasuta* examined.

TABLE III. Species with the sex chromosomes of XX–XY_1Y_2 type (after Fredga, 1970, with supplements)

Order / Family / Species	2n ♀	2n ♂	Reference
Marsupialia			
Peramelidae			
Macrotis (Thylacomys) lagotis	18	19	Hayman and Martin, 1965
Macropodidae			
Wallabia (= *Protemnodon*) *bicolor apicalis*	10	11	Sharman, 1961
Potorous tridactylus	12	13	Sharman *et al.*, 1950
Insectivora			
Centetidae			
Echinops telfairi	40	41, 40	Borgaonkor, 1967a, b
Soricidae			
Sorex araneus (many regions of Palearctic)	20–32	21–33	Bovey, 1948 Sharman, 1956 Kozlovsky, 1971
S. arcticus (Canada)	28	29	Meylan, 1968
S. arcticus (Siberia, Kemerovo)	—	33	Kozlovsky, 1971
S. arcticus (Siberia, Irkutsk)	36	37	Kozlovsky, 1971
S. caucasicus	—	25	Kozlovsky, 1971
S. gemellus	22	23	Meylan, 1964
Chiroptera			
Phyllostomatidae			
Choeroniscus godmani	—	19	Hsu *et al.*, 1968
Carrolia perspicillata	20	21	Hsu *et al.*, 1968
C. brevicauda	20	21	Patton and Gardner, 1971
C. subrufa	20	21	Hsu *et al.*, 1968
C. castanea (Colombia, Costa Rica)	20	21	Baker and Bleier, 1971
Artibeus jamaicensis	30	31	Hsu *et al.*, 1968
A. lituratus	30	31	Hsu *et al.*, 1968
A. cinereus	30	31	Baker and Hsu, 1970
A. talectus	30	31	Hsu *et al.*, 1968
Enchisthenes hartii	30	31	Baker and Hsu, 1970
Ametrida centurio	30	31	Baker and Hsu, 1970
Stenoderma refum	30	31	Baker and Lopez, 1970
Rodentia			
Cricetidae			
Gerbillus gerbillus	42	43	Matthey, 1954
Taterillus gracilis	36	37	Matthey and Jotterand, 1972
T. pygargus	36	37	Matthey and Jotterand, 1972
Arthiodactyla			
Cervidae			
Muntjacus muntjak	6	7	Wurster and Benirschke, 1970

4. *Multiple sex chromosomes*

Until recently the cases of discovery of XX–XY_1Y_2 or $X_1X_2X_2X_2$–X_1X_2Y systems were treated as fragmentation of heterochromosomes. It is generally believed now that the multiple sex chromosomes arise at the expense of translocation of X- or Y-chromosome to autosome. An autosomic segment with the translocated sex chromosome (A^Y or A^X) has an affinity with the genetically homologous nontranslocated autosome (A), with a proper sex chromosome. So trivalents A^XYA or A^YXA are produced under a conjugation in meiosis of males. Thus there is a possibility of the existence of two versions of multiple sex chromosomes.

(a) XX–XY_1Y_2 (A^XA^X–A^XYA) system: The presence of two Y-chromosomes was first found in the shrew *Sorex araneus* (Bovey, 1948) and in the gerbil *Gerbillus gerbillus* (Warhman and Zahavi, 1955). Table III gives the review of known material.

This system, as generally accepted, arises at the expense of X-chromosome translocation to autosome. In this case a large Y-chromosome (Y_2) is a homologue of the autosomic part of X-chromosome whereas a small Y-chromosome (Y_1) bears normal genetical material for a Y-chromosome. This is true not only of species which have one large Y-chromosome (phyllostomid bat *Carrolia perspicillata*) but also of species with both small Y-chromosomes such as a bat *Artibeus lituratus* (Beçak *et al.*, 1969).

It is of interest that XY_1Y_2 system arises in various subfamilies of Phyllostomatidae (Carollinae, Stenoderminae), although many species with a normal XY–XX constitution are among the latter one (Baker, 1970; Baker and Hsu, 1970). There are many species with XY–XX system in genus *Sorex* together with the species group of XY_1Y_2-XX system. The XY_1Y_2–XX system seems to have an advantage over phyllostomid bats since it occurs independently in different genera of subfamily Stenoderminae. A convincing proof of the translocated origin of the multiple sex chromosome has recently been obtained. It was found that geographical chromosomic polymorphism was inherent in the bat *Carrolia castanea*: bats from Costa Rica and Colombia have ♀ $2n = 20$, ♂ $2n = 21$, while in

TABLE IV. Comparison of the relative sizes (% of the haploid genome) of the X-chromosome and acrocentric autosome of *Carollia castanea* with $2n = 20$, 21 and $2n = 22$ (Patton and Gardner, 1971)

XX_1Y_2 system Costa Rica, ♂ $2n = 21$		XY system Peru, ♂ $2n = 22$	
orig. part X	6.07 ± 0.37	X	6.17 ± 0.52
transloc. A	8.56 ± 0.54	A	8.82 ± 0.61
neo–X(A^X)	14.63 ± 0.62	X + A	14.99 ± 0.45

Peru ♂♂ and ♀♀ of this species have $2n = 22$ (Baker and Bleier, 1971; Patton and Gardner, 1971). The large X-chromosome in ♂ with $2n = 21$ has the secondary constriction in the long arm whereas ♀ $2n = 22$ has a small X-chromosome without a secondary constriction with the length equal to a centromeric (remaining) part of X-chromosome of the previous form. The measurements showed that the hypothesis of A + X translocation giving rise to XY_1Y_2 is true.

The giant X- and Y_2-elements arise in a paradoxical karyotype of the Indian muntjac *Muntjacus muntjak* at the expense of translocation with autosomes. As a rule the long arm of X-chromosome is morphologically identical to a large acrocentric Y_2-chromosome (Wurster and Benirschke, 1970).

(b) $X_1X_1X_2X_2$–X_1X_2Y (XXAA–XAA^Y) system: This system is related to the translocation of the Y-chromosome to one of the autosomes and as reported by Fredga (1970), it can be designated as XXAA–XAA^Y or simply XX–XA^Y. There is a normal (initial) X-chromosome in all these

TABLE V. Mammals with the sex chromosomes of $X_1X_1X_2X_2$–X_1X_2Y type (after Fredga, 1970, with changes).

Order / Family / Species	2n ♀	2n ♂	References
Marsupialia			
Macropodidae			
Lagorchestes conspicillatus	16	15	Martin and Hayman, 1966
Rodentia			
Muridae			
Leggada minutoides	36	35	Matthey, 1965
Carnivora			
Viverridae			
Herpestes ichneumon	44	43	Fredga, 1970
H. sanguineus	—	41	Fredga, 1970
H. javanicus (?)	36	35	Todd and Pressman, 1966
H. auropunctatus	36	35	Fredga, 1965
H. edwardsi	36	35	Ray-Chaudhuri *et al.*, 1968
H. fuscus	36	35	Fredga, 1970
H. brachyurus	36	35	Fredga, 1970
Atalax paludinosus	36	35	Hsu, 1966
Artiodactyla			
Antilopidae			
Tragelaphus angasi	—	55	Wurster and Benirschke, 1968
T. strepsiceros	32	31	Wallace and Fairall, 1968

cases; X_2X_2-chromosomes of females are the typical autosomes. The Y-element was translocated to one of the X_2-autosomes in ♂♂ and it results in a large Y-chromosome associated with the X_2-chromosome (originally autosome). Table III presents a review of data on mammals with $X_1X_1X_2X_2$–X_1X_2Y type of chromosomes. It is noteworthy that unusually large sex chromosomes (Table I) have been observed in the antelope genus *Tragelaphus* when the sex structure is XX–XY and the multiple sex chromosomes are present (Table V); it has been pointed out that the XO system exists in this genus (Wurster and Benirschke, 1968).

It is interesting to note the occurrence of $X_1X_1X_2X_2$–X_1X_2Y system in Viverridae (in different species of *Herpestes* and *Atilax*) (Fredga, 1970).

(c) $X_1X_1X_2X_2$–$X_1X_2Y_1Y_2$ system: Liapunova and Krivoshejev (1969) advanced the hypothesis of the existence of the $X_1X_2X_1X_2$–$X_1X_1X_2X_2$ heteromorphism on the basis of finding clean-cut sex polymorphism in a limited number of voles *Microtus middendortii* and *M. hyperboreus*. At the same time, Raiçu *et al.* (1969) who studied *M. arvalis* chromosomes put forward the same suggestion. Matthey (1969c), who examined the karyotype of the same species, denied not only the conclusion of Raiçu *et al.* (1969) but also the possibility of the existence of the $X_1X_2X_2X_2$–$X_1X_2Y_1Y_2$ system. Liapunova and Krivoshejev (1969) do not insist on the existence of such a system of sex chromosomes since they have not investigated meiosis. However, an additional study of the paradoxical chromosomal heteromorphism in *M. middendorfii* and *M. hyperboreus* must be carried out.

5. *XX–XO system*

If a small Y-chromosome was translocated to a large autosome the karyotype would become XO.

Recently two cases showing the existence of the XX–XO type were described but no results of meiosis in either of them have been shown for

TABLE VI. Mammals with the sex chromosomes of XX–XO type

Order / Family / Species	2n ♀	2n ♂	References
Chiroptera			
Phyllostomatidae			
Mesophylla macconelli	22	21	Baker and Hsu, 1970
Primates			
Callithricidae			
Callimico goeldii	48	47	Hsu and Hampton, 1970

the present (Table VI).* It should be noted that the karyotype *Callimico goeldii* (Hsu and Hampton, 1970) was described for the animals bred in capture.

The numerous cases, between 1920 and 1940, describing the XX–XO system in many mammalian species still require further verification.

6. *Deviations from the XX–XY system. Unpaired sex chromosomes in females (or in both sexes)*

Matthey (1953) found that *Ellobius lutescens* shows the lowest diploid number among placental mammals and an odd number of chromosomes in males and females ($2n = 17$). The paradoxical type of sex determination and karyotypic peculiarities stimulated further investigations (Matthey, 1954, 1956).

White (1957) suggested the following structure of the karyotype in *E. lutescens*: 16 autosomes + a fused XX element in females or + a fused XY element in males. The finding of the additional 17 chromosome species among the placental burrowing North American voles *Microtus* (*Chilotus*) *oregoni* has prompted the suggestion that common, but as yet unclear, mechanisms underlie sex determination in *M.* (*Ch.*) *oregoni* and in *E. lutescens* (Matthey, 1958).

If Wright's hypothesis were true, then the 17 chromosome species would produce gametes of two types with 8 and 9 chromosomes. Then the following combinations would be possible:

$$[8 \text{ autosomes (a)} + (XY)] \times (8a + 0) = ♂ (2n = 17) \quad (1)$$

$$[8a + (XX)] + (8a + 0) = ♀ (2n = 17) \quad (2)$$

$$(8a + 0) + (8a + 0) = \text{lethal} (2n = 16) \quad (3)$$

$$[8a + (XX)] + [8a + (XY)] = \text{lethal} (2n = 18) \quad (4)$$

The lethal combinations 3 and 4 should lead to the death of 50% of zygotes.

As a result of the recent studies *Choloepus* (Xenarthra) was included in this rather heterogenous group of species.

(a) Gonosomic mosaicism and XY/OY–XO system, the case of Chilotus: It has been shown (Brown and Coneway, 1964) that in *M.* (*Ch.*) *oregoni*

* Reig and Kiblisky (1968) examined somatic chromosomes 1 ♂ *Ctenomys tuconas* (Ctenomyidae, Hystricomorpha, Rodentia), $2n = 61$. The unpaired chromosome is a large metacentric. This case should be thoroughly examined.

the number of ovulated egg cells (3.90 ± 0.26) is close to that of implanted zygotes (3.82 ± 0.25). Thus, the hypothesis of Wright does not hold for *M.* (*Ch.*) *oregoni*. Detailed studies have confirmed that only males in *M.* (*Ch.*) *oregoni* with spermatocytes I (16 + OY) have the diploid number $2n = 17$, as estimated earlier by Matthey (1958) and have shown that the male somatic cells have the usual XY constitution ($2n = 18$).

The non-disjunction of X-chromosomes in the course of germ cell development leads to the formation of the germ cells of two types: 16a + XXY ($2n = 19$) and 16a + OY ($2n = 17$). Of the two types only 16a + OY produces definite spermatogonia. Spermatozoa 8a + O fusing with the egg cell 8a + X determines the sex of the female 16a + XO ($2n = 17$) and the fusion of spermatozoa 8a + Y with egg cell 8a + X gives rise to a male with somatic formula 16a + XY ($2n = 18$). Females have the somatic complement 16a + XO ($2n = 18$), but in the process of the germ cell differentiation the X-chromosome gives rise to two lines: (1) 16 A + XX ($2n = 18$) which differentiates into definite oogonia and (2) 16a + OO ($2n = 16$) which is eliminated (Ohno *et al.*, 1963).

(b) Unpaired number of chromosomes in males and females: (i) Case of *Ellobius lutescens*, sex chromosomes as regulators of fetal mortality: Another hypothesis was put forward to account for sex determination in *E. lutescens*. Autoradiographic studies on DNA replication in the ninth unpaired chromosome of *E. lutescens* of both sexes have shown that it has identical sequences (Castro-Sierra and Wolf, 1967). In size the unpaired chromosome in *E lutescens* makes up about 5% of the haploid chromosome complement, which corresponds with the percentage of the X-chromosome in the karyotype of most placental mammals (Ohno *et al.*, 1964a). For this reason, Castro-Sierra and Wolf (1967) have raised objections against the hypothetical concept of the direct fusion of the X + Y elements in males and X + X elements in females as formulated by White and Matthey.

According to the hypothesis of Castro-Sierra and Wolf (1968) the Y segment bearing the sex-determining factor in *E. lutescens* males may be translocated to the terminal of the autosome; this autosome (A^Y) retains its capacity not only to conjugate with the normal autosome, but also to associate terminally with the X-chromosome. Accordingly, the male has the XAA^Y constitution and the female has the XAA. Accepting the hypothesis, the association of the X-chromosome with the terminal region of the Y of the autosome pair should form a trivalent, which has been observed in meiosis in males.

On the basis of the meiotic analysis Castro-Sierra and Wolf (1968) suggested that one (unpaired) autosome of the *E. lutescens* male bears a part of the Y element.

Granting that the interpretation is correct, a male should produce the following 4 types of gametes:

$$7a + A, 7a + A^Y, 7a + A + X, 7a + A^Y + X$$

and female should produce 2 types of gametes:

$$7a + A, 7a + A + X.$$

Random conjugation of gametes gives 6 types of zygotes:

$$2\,(14a + AA^Y + X) = ♂\,(2n = 17) \quad (5)$$

$$2\,(14a + AA + X) = ♀\,(2n = 17) \quad (6)$$

$$1\,(14a + AA + XX) = \text{lethal, "superfemale"}\,(2n = 18) \quad (7)$$

$$1\,(14a + AA^Y + XX) = \text{lethal, atypical } ♂\,(2n = 18) \quad (8)$$

$$1\,(14a + AA) = \text{lethal, intersex (?)}\,(2n = 16) \quad (9)$$

$$1\,(14a + AA^Y) = \text{lethal, atypical } ♂\,(2n = 16) \quad (10)$$

Only combinations 5 and 6 are viable and combinations 7–10 are lethal. It follows that the hypothesis, inasmuch as it implies a 50% zygote mortality is essentially correct.

The studies we have carried out (Vorontsov *et al.*, 1973) have shown that, in contrast to *Chilotus*, the somatic complements of both sexes in *E. lutescens* have $2n = 17$. Studies on the fetal mortality in *E. lutescens* have indicated that the number of ovulated egg cells in 17 females is 12.9 ± 1.7 and that 24 females had 4.8 ± 0.4 embryos (Vorontsov *et al.*, 1972). These data indicate the existence of a mechanism for cytogenetical regulation of offspring number. It appears that much larger numbers of young should die among the underground burrowing species *E. lutescens* than among the terrestrial species *Microtinae*. Decrease in the number of young in *E. lutescens* is obtained through the cytogenetical regulation of offspring number. It is of interest that all the other *Ellobius* species have the usual even chromosome number (Vorontsov *et al.*, 1969). Thus, if zygote mortality does occur in all the other *Ellobius* species, it should be determined by other than cytogenetical factors.

(ii) Case of *Choloepus hoffmanni:* Corin-Frederic (1969) has found a quite different situation in the sloth *Choloepus hoffmanni*. The two sexes have $2n = 49$ (the tissue culture was studied in 6 ♀♀ and 4 ♂♂). One lately replicated X_1-chromosome was found in ♀. The same chromosome (X) was inherent to ♂. There are two autosomes in ♀ which were not involved in the sex determination of ♀, but the Y element was translocated to the distal end of one of these autosomes in ♂♂. So the karyotype of ♂ *Choloepus* is similar to the animals with the XY_1Y_2-chromosomes. It has been found that corresponding to XAX^Y hypothesis the trivalent occurs in meiosis in ♂ and thus the principal mechanism of the sex determination is not distinguishable from the males with the XY structure.

7. *Polymorphism of the sex chromosomes*

In comparison with the usual morphological characters variability is, however, inherent to the chromosome in spite of its high stability. The cases of autosomic polymorphism are rather numerous. But the number of known cases of heterochromosome polymorphism is not large.

Matthey (1966a) has established that *Leggada triton* (Muridae) females are heterozygous by the X-chromosomes, one of which was a normal submetacentric and the second, called X_{dc}, was acrocentric and coincided in size with the length of the long arm in the normal X. He has also (1967) described similar phenomenon in *L. minutoides musculoides*.

Polymorphism of the X-chromosomes has been revealed in *Phodopus sungorus* and *Ph. campbelli* (Cricetidae) in our laboratory; there are 4 types of the X-chromosomes ranging from subtelocentric to metacentric and X_1, X_2, X_3 and X_4-chromosomes have a constant length of a long arm and only the length of the short arm changes discretely (Vorontsov *et. al.*, 1967; Radzhabli and Kryukova, 1971).

It was found that the South American hamsters *Akodon azarae* (Cricetidae, Rodentia) bear both the heteromorphism of X-chromosomes in females and somatic mosaicism in females (XX–XO) (Bianchi and Contreras, 1967). Heteromorphism of the X-chromosomes has been also observed in the other species of the genus *Akodon*. Korobytsina (1969) established, in our laboratory, that there are clear-cut interpopulational differences in the morphology of X-chromosome in the gerbils *Meriones meridianus*, which are located along the Volga shores. It was pointed out that the interpopulational differences in the Y form occurred in *M. unguiculatus* (Vorontsov and Korobitsyna, 1970). The occurrence of the polymorphism and heteromorphism of the X-chromosomes in females prompted the suggestion that some combinations of the sex chromosomes, which have deletions under the homozygous state in ♀ and the hemyzygous in ♂, can be lethal (Vorontsov *et al.*, 1967). Polymorphism of the X-chromosomes may be related to the regulation of the sex ratio and the number in populations.

IV. CONCLUSION

Thus the gradual formation of mechanisms of the cytogenetical sex determination which lead to female (Serpentes, Aves) or male (Eutheria) heterogamety has been observed in Vertebrata line. We have only found the probing attempts of the formation of the mechanisms among the lower Vertebrata from Pisces to Reptilia.

Just as the transition from isogamy to anisogamy and to oogamy took place independently of each other in the various phyla of plants so the formation of mechanisms of the cytogenetical sex determination with differentiated heterochromosomes follows the same pattern in various kingdoms and phyla and results in an independent occurrence of the XX–XY system in *Melandrium* as well as in many Insecta and Mammalia, whereas the ZW–ZZ system evolved independently in Trichoptera, Lepidoptera, Serpentes and in Aves. Against the background of these facts it is unclear whether the male species of different groups are homologous to each other or not; they appear to be nonhomologous.

REFERENCES

Arronet, V. N. (1965). Description of the Karyotypes of *Agama caucasica* and *Phrynocephalus helioscopus* (Agamidae, Reptilia) *Cytology* (*Leningrad*), 7, 237–239 (Russian with English summary).

Arronet, V. N. (1968). The karyotype of lizard *Ophysops elegans* Menetries. *Cytology* (*Leningrad*), **10**, 120–122 (Russian with English summary).

Badr, F. M. and Badr, R. S. (1970). The somatic chromosomes of a wild population of rats: Numerical polymorphism. *Chromosoma, Berlin* **30**, 465–475.

Baker, R. Y. (1970). The role of karyotypes in phylogenetic studies of bats. In "About Bats" (Slaughter and Walton, eds.) pp. 303–312. South. Methodist University Press, Dallas.

Baker, R. Y. and Bleier, W. Y. (1971). Karyotypes of bats of the subfamily Carolliinae (Mammalia; Phyllostomatidae) and their evolutionary implications. *Experientia*, **27**, 220–222.

Baker, R. Y. and Hsu, T. C. (1970). Further studies on the sex-chromosome systems of the American leaf nosed bats (Chiroptera, Phillostomatidae). *Cytogenetics*, **9**, 131–138.

Baker, R. Y. and Lopez, G. (1970). Karyotypic studies of the insular population of bats on Puerto Rico. *Caryologia*, **23**, 465–472.

Baker, R. Y., Bull, Y. Y. and Mengden, G. A. (1971). Chromosomes of *Elaphe subocularis* (Reptilia: Serpentes), with the description of an *in vivo* technique for preparation of Snake chromosomes. *Experientia*, **27**, 1228–1292.

Beçak, M. L., Batistic, R. F., Vizotto, L. D. and Beçak, W. (1969). Sex determining mechanism XY_1 Y_2 in *Artibeus lituratus litupatus* (Chiroptera–Phyllostomidae) *Experientia*, **25**, 81–83.

Beçak, W. (1967). Chromosomes numbers and sex chromosomes in Serpentes. *Mamm. Chrom.* **8**, 4–10.

Beçak, W. and Beçak, M. L. (1969). Cytotaxonomy and chromosomal evolution in Serpentes. *Cytogenetics* **8**, 247–262.

Beçak, W., Beçak, M. L. and Nazareth, H. R. S. (1962). Karyotypic studies of two species of South American snakes. *Cytogenetics*, **1**, 305–313.

Beçak, W., Beçak, M. L. Nazareth, H. R. S. and Ohno, S. (1964). Close karyological kinship between the reptilian suborder Serpentes and the class Aves. *Chromosoma*, **15**, 606–617.

Beçak, W., Beçak, M. L., Lavalle, D. and Schreiber, G. (1967). Further studies on polyploid amphibians (Geratophydidae). 2. DNA content and nuclear volume. *Chromosoma*, **23**, 14–23.

Beçak, W., Beçak, M. L., Schreiber, G., Lavalle, D. and Amorim, F. O. (1970). Interspecific variability of DNA content. *Experientia*, **26**, 204–206.

Benirschke, K. (1968). The chromosome complement and meiosis of the North American porcupine. *J. Hered.* **59**, 71–76.

Bianchi, N. O. and Contreras, Y. R. (1967). The chromosomes of the field mouse *Akodon azarae* (Cricetidae, Rodentia) with special reference to sex chromosome anomalies. *Cytogenetics*, **6**, 306–313.

Bianchi, N. O. and Marta, S. A. de Bianchi (1969). Origin of the pattern and chronology of chromosome replication in vertebrates. *Genetics*, **61**, 275–287.

Bick, Y. A. E. and Jackson, W. D. (1967a). A mammalian X-O sex chromosome system in the monotreme *Tachyglossus aculeatus* determined from leucocyte cultures and testicular preparations. *Amer. Nat*, **101**, 79–86.

Bick, Y. A. E. and Jackson, W. D. (1967b). Karyotype of the monotremes *Ornithorhynchus anatinus* (Platypus) and *Tachyglossus aculeatus* (Echiana), *Nature, Lond.*, **214**, 600–601.

Borgaonkar, D. S. (1967a). Additions to the list of chromosome number in the Insectivores et Primates. *J. Hered.*, **58**, 211–213.

Borgaonkar, D. S. (1967b). Cytotaxonomic studies on the Tenrecoidea Simpson. *Mamm. Chrom. Newsl.*, **8**, 27–28.

Bovey, R. (1948). Un type nouveau d'heterochromosomes chez un Mammifére Le trivalent sexual de *Sorex araneus*. *Arch Julius Klaus—Stift. Vererbungsforsch.* **23**, 506–509.

Brown, L. M. and Coneway, C. H. (1964). A balanced lethal sex-chromosome system. *J. Hered.*, **25**, 7–8.

Brum-Zorrilla, N. and Saez, F. A. (1971). Chromosomes of South American Bufonidae. *Experientia*, **27**, 470–471.

Bulatova, N.Sh., Panov, E. N. and Rajabli, S. I. (1971). Description of karyotypes of some species of birds of the USSR fauna. *Doklady Akad. Nauk SSR, Ser. Biol.*, **199**, **N4–6**, pp. 1420–1423.

Bury, R. B., Gorman, G. C. and Lunch, J. F. (1969). Karyotypic data for the five species of anguid lizards. *Experientia*, **25**, **N3**, 314–316.

Capanna, E. and Civitelli, M. V. (1967). I cromosomi di *Pipistrellus savii*. *Caryologia (Firenze)*, **20**, 256–272.

Capanna, E., Civitelli, M. V. and Spagnuolo, C. (1968). Contributo alla cariologia del genero *Myotis*. Considerazioni sull evoluzione del cariotipo dei Vespertilionidi (Mammalia-Chiroptera). *Caryologia (Firenze)*, **21**, 225–240.

Castro-Sierra, E. and Wolf, U. (1967). Replication patterns of the chromosome No. 9 of the rodent *Ellobius lutescens*. *Cytogenetics*, **6**, 268–275.

Castro-Sierra, E. and Wolf, U. (1968). Studies on the male meiosis of *Ellobius lutescens*. Thom. *Cytogenetics*, **7**, **4**, 241–248.

Chen, T. R. (1969). Karyological heterogamety of deep-sea fishes. *Postilla*, **130**, 1–29.

Chen, T. R. and Ebeling, A. W. (1966). Probable male heterogamety in the deep-sea fish *Bathylagus wssethi* (Teleostei, Bathylagidae). *Chromosoma, Berlin* **18**, 88–96.

Chen, T. R. and Ebeling, A. W. (1968). Karyological evidence of female heterogamety in the mosquitofish *Gambusia affinis* (Baird and Girard). *Copeia*, 1968, 70–75.

Chen, T. R. and Ruddle, F. H. (1970). A chromosome study of four species and a hybrid of the killifish genus *Fundulus* (Cyprinodontidae). *Chromosoma, Berlin* **29**, 255–267.

Cherfas, N. B. (1966). Natural triploidy in females of the uni-sexual form of the goldfish (*Carassius auratus gibelio* Bloch). *Genetics* (*USSR*), **2**, 5, 16–24 (in Russian with English summary).

Chevalier, M. (1969). Données nouvelles sur le caryotype du lézard vivipare (reptile, lacertililu). Existe-t-il une hitérogamétee femalle de type Z_1Z_2W? *C.R. Hebd. Séanc Acad. Sci. Paris.* **D268** N 16, 2098–2100.

Cohen, M. M. and Clark, H. F. (1967). The somatic chromosome of five Crocodilian species. *Cytogenetics*, **6**, 193–203.

Cohen, M. M. and Gans, C. (1970). The chromosomes of the order Crocodilia. *Cytogenetics*, **9**, 2, 81–105.

Cohen, M. M., Huang, Ch.C. and Clark, H. F. (1967). The somatic chromosomes of 3 Lizard species: *Gekko gecko*, *Iguana iguana* and *Crataphytus collaris*. *Experientia*, **23**, 769–770.

Cole, C. I. (1970). Karyotypes and evolution of the *spinosus* group of lizards in the genus *Sceloporus*. *Am. Mus. Novit.* **2431**, 1–47.

Cole, C. I. and Lowe, C. H. (1967). Sex chromosomes in lizards. *Science, N.Y.* **155**, 1028–1029.

Cole, C. I., Lowe, C. H. and Wright, I. W. (1969). Sex chromosomes in whiptail lizards (genus *Chemidophorus*). *Am. Mus. Novit.*, **2395**, 1–19a.

Corin-Frederic, I. (1969). Les formules gonosomiques dites aberrantes chez les Mammiféres Eutheriens. Exemple particulier du paresseux *Choloepus hoffmanni* Peters (Edente, Xenarthre, famille des Bradypodidae). *Chromosoma, Berlin* **27**, 268–287.

Dallai, R. and Tallari, M. V. (1969). A karyological study of three species of Scincidae (Reptilia). *Chromosoma, Berlin* **27**, 1, 86–94.

Dallai, R. and Urbani, C. B. (1967). Fine resolution of the karyogram of *Lacerta sicula campestris* (di Betta). *Caryologia*, **20**, 4. 347–353.

Darevsky, I. S. (1958). Natural parthenogenesis in some subspecies of lizards *Lacerta saxicola. Dokl. Akad. Nauk SSSR*, **122**, 4, 730–732 (in Russian).

Darevsky, I. S. and Kulikova, V. N. (1962). Taxonomical characters and some features of oogenesis of hybrids between bisexual and parthenogenetical forms of *Lacerta saxicola. Cytology (Leningrad)*, **4**, 160–170. (Russian with English summary).

Darevsky, I. S. and Kulikova, V. N. (1964). Natural triploidy in polymorphic groups of Caucasic lizards (*Lacerta saxicola* Eversmann) as evidence of hybridization between bisexual and parthenogenetical forms of this species. *Dokl. Akad. Nauk SSSR*, **158**, 1, 202–205 (in Russian).

Dutt, K. (1968). A karyotype study of *Varanus monitor* Linh. *Caryologia*, **21**, 1, 1–10.

Dutt, K. (1969). Study of chromosomes in two species of Indian lizards. *Microscope*, **17**, 3, 213–218.

Fraccaro, M., Gustavsson, I., Halten, M., Lindsten, I. and Tiepolo, L. (1968). Chronology of DNA replication in sex chromosomes of the reindeer (*Rangifer tarandus* L.). *Cytogenetics*, **7**, 196–211.

Fraccaro, M., Gustavsson, I., Hultén, M., Lindsten, I. and Tiepolo, L. (1969). Late-replicating Y-chromosome in spermatogonia of the Chinese hamster (*Cricetulus griseus*). *Cytogenetics*, **8**, 263–271.

Fredga, K. (1970). Unusual sex chromosome inheritance in mammals. *Phil. Trans. R. Soc. Lond. B.* **259**, 15–36.

Fredga, K. (1965). A new sex determining mechanism in mammal chromosomes of Indian mongoose (Herpestes auropunctatus). *Hereditas* **52**, 411–420.

Fredga, K. and Santesson, B. (1964). Male meiosis in the Syrian, Chinese and European hamsters. *Hereditas*, **52**, 36–48.

Galton, M. (1966). Evolution of sex chromosomes in mammals and birds. *Lancet*, **7478**, 1397–1398.

Galton, M. and Bredbury, P. R. (1966). DNA replication pattern of the sex chromosomes of the pigeon (*Columba livia domestica*) *Cytogenetics*, **5**, 295–306.

Galton, M., Benirschke, K. and Ohno, S. (1965). Sex chromosomes of the chinchilla: allocycly and duplication sequence in somatic cells and behavior in meiosis. *Chromosoma, Berlin* **16**, 668–690.

Golovinskaya, K. A. and Romashov, D. D. (1947). Studies of the gynogenesis in *Carassius auratus. Trudy Vses. Inst. Prudov. Ryb. Khoz.*, **4**, 73–85 (in Russian).

Gorman, G. C. (1969). New chromosome data for 12 species of Lacertid lizards. *J. Herpetology*, **3**, (1–2), 49–54.

Gorman, G. E. and Atkins, L. (1966). Chromosome heteromorphism in some male lizards of the genus *Anolis. Am. Nat.*, **100**, 579–583.

Gorman, G. C. and Atkins, L. (1968). Confirmation of a X-Y sex determining mechanism in lizards (*Anolis*). *Copeia*, **1**, 159–160.

Gorman, G. C. and Atkins, L. (1969). The Zoogeography of lesser antillean *Anolis* lizards—an analysis based upon chromosomes and lactic dehydrogeneses. *Bull. Mus. Comp. Zool. Han.* **138**, 3, 53–80a.

Gorman, G. C. and Gress, F. (1970a). A chromosome cytology of four boid snakes and a varanid lizard, with comments on the karyosystematics of primitive snakes. *Herpetologica*, **26**, 3, 308–317.

Gorman, G. C. and Gress, F. (1970b). Sex chromosomes of a pygopodid lizard, *Lialis burtenis*. *Experientia*, **26**, 2, 206–207.

Gorman, G. C., Atkins, L. and Holzinger, T. (1967). New karyotypic data on 15 genera of lizards in the family *Iguanidae*, with a discussion on taxonomic and cytological implications. *Cytogenetics*, **6**, 286–299.

Hall, W. P. (1970). Three probable cases of parthogenesis in Lizards (Agamidae, Chamaeleontidae, Gekkonidae). *Experientia*, **26**, 11, 1271–1273.

Hamerton, J. L. (1958). Mammalian sex chromosomes. *In*: "Symp. on nuclear sex". (eds.: D. R. Smith and W. M. Davidson) Heinemann: London.

Hammar, B. (1966). The karyotypes of nine birds. *Hereditas*, 55, 367–385.

Harris, G. B. and Edwards, R. G. (1970). A Discussion on determination of sex. *Phil. Trans. R. Soc. Lond. B*. **259**, 828, 1–206.

Hauschteck-Jung., E. and Meili, R. (1967). Vergleich der Chromosomensätze von Steinwild (*Capra ibex*) und Hausziege (*Capra hircus*). *Chromosoma, Berlin* **21**, 198–210.

Hayman, D. L. and Martin, P. G. (1965). An autoradiographic study of DNA synthesis in the sex chromosomes of two marsupials with an XX/XY_1Y_2 sex chromosomes mechanism. *Cytogenetics*, **4**, 209–218.

Hayman, D. L. and Martin, P. G. (1969). Cytogenetics of marsupials. *In* "Comparative Mammalian Cytogenetics" pp. 191–217. (K. Benirschke, ed.) Springer-Verlag, Berlin, Heidelberg and New York.

Hayman, D. L., Martin, P. G. and Waller, P. F. (1969). Parallel mosaicism of supernumerary Chromosomes and sex chromosomes in *Echymipera kalabu* (Marsupialia). *Chromosoma, Berlin* **27**, 371–380.

Hitotsumachi, S., Sasaki, M. and Ojima, Y. (1969). A comparative karyotype study in several species of Japanese loaches (Pisces, Cobitidae). *Jap. J. Genet.*, **44**, 3, 157–161.

Hsu, T. C., Baker, R. J. and Utakoji, T. (1968). The multiple sex chromosome system of American leaf-nosed bats (Chiroptera, Phyllostomidae). *Cytogenetics*, 7, 27–38.

Huang, C. C. (1968). Karyological and autoradiographic studies of the chromosomes of *Rattus* (*Mastomys*) *natalensis*. *Cytogenetics*, **7**, 97–107.

Huang, C. C. and Clark, H. F. (1967). Chromosome changes in cell lines of the box turtle (*Terrapene carolina*) grown at two different temperatures. *Can. J. Genet. Cytol.*, **9**, 449–461.

Huang, C. C. and Clark, H. F. (1969). Chromosome studies of the cultured cells of two species of side-necked turtle (*Podocnemis unifilis* and *P. expansa*). *Chromosoma, Berlin* **26**, 3, 245–253.

Huang, C. C., Clark, H. F. and Gans, C. (1967). Karyological studies of fifteen formes of *Ampisbaenus* (Amphisbaenia-Reptilia). *Chromosoma, Berlin* **221**, 1-15.

Huang, C. C. and Strong, L. C. (1962). Chromosomes of the African-mouse, *J. Hered.*, **53**, 95–99.

Hubbs, C. L. and Hubbs, L. C. (1932). Apparent partenogenesis in nature, in a form of fish of hybrid origin. *Science, N.Y.* **76**, 628–630.

Itoh, M., Ikeuchi, T., Shimba, H., Mori, M., Sasaki, M. and Makino, S. (1969). A comparative karyotype study in fourteen species of birds. *Jap. J. Genet.*, **44**, 163–170.

Itoh, M., Sasaki, M. and Makino, S. (1970). The chromosomes of some Japanese snakes, with special regard to sexual dimorphism. *Jap. J. Genetics*, **45**, 2, 121–128.

Ivanov, V. G. and Fedorova, T. A. (1970). Sex heteromorphism of chromosomes in *Lacerta strigata* Eichwald. *Cytology* (*Leningrad*), **12**, 12, 1582–1585. (Russian with English summary).

Jaylet, A. (1966). Le caryotype de L'amphibien Urodéle *Euproctus asper* (Dugés). *Chromosoma, Berlin* **18**, 79–87.

Jovanovič, V. and Atkins, L. A. (1969). Karyotypes of four passerine birds belonging to the families Turdidae, Mimidae and Corvidae. *Chromosoma*, **26**, 388–394.

Keenan, R. D. (1932). The chromosomes of *Sphenodon punctatum*. *J. Anat.*, **67** (cited by Matthey, 1942).

Kessous, A., Caussinus, H., Jaylet, A. and Beetschen, J. C. (1968). Essai d'analyse biometrique du caryotype de l'Amphibien Urodéle *Salamandra salamandra* L. *Chromosoma, Berlin* **23**, 324–332.

Kobel, H. R. (1962). Heterochromosome by *Vipera berus* L. (Viperidae, Serpentes). *Experientia*, **18**, 173–174.

Kobel, H. R. (1963). Vergleich der Chromosomensatze von *Vipera berus* and *Vipera aspis* L. (Viperidae, Serpentes). *Arch. Julius Klaus-Stift. VererbForsch*, **38**, 68–75.

Kobel, H. R. (1967). Morphometrische Karyotypanalyse einigen Schlangen-Arten, *Genetics*, **38**, 1, 1–31.

Korobitsina, K. V. (1969). Intraspecific variability of chromosomes of the some species gerbills (Meriones, Gerbillinae, Cricetidae, Rodentia). *In*: "The Mammals: (Evolution, Karyology, Taxonomy, Fauna)" (ed. N. N. Vorontsov,) 117–120, Novosibirsk.

Kozlovsky A., (1971). Chromosomal complements and taxonomy of some palaearctic Insectivores. Thesis. Moscow University (in Russian).

Král, B. (1969). A karyotype study of the Fringe-told Lizard *Uma inornata* Cope, 1895. *Zool Listy*, **18**, 2, 185–194.

Kupriyanova, L. A. (1968). Description of the karyotypes of three species of Lacertidae family. *Cytology* (*Leningrad*), **10**, 7, 892–895. (Russian with English summary).

Kupriyanova, L. A. (1969). Description of the karyotypes of the lizards subgenera *Archaeolacerta*. *Cytology* (*Leningrad*), **11**, 7, 803–814. (Russian with English summary).

Kupriyanova, L. A. and Arronet, V. N. (1969). The karyotype of the lizards *Eremias velox* Pall. (Reptilia, Lacertidae). *Cytology* (*Leningrad*), **11**, 8, 1057–1060 (Russian with English summary).

Lewis, K. B. and John, B. (1968). The chromosomal basis of sex determination. *Int. Rev. Cytol.*, **23**, 277–379.

Liapunova, E. A. and Krivoshejev, V. G. (1969). The heteromorphism of the chromosomes *Microtus hyperboreus* and *M. middendorffi* and taxonomic position of those species. *In*: The Mammals (Evolution, Karyology, Taxonomy, Fauna) (ed. N. N. Vorontsov), 146–149, Novosibirsk. (In Russian).

Liapunova, E. A., Vorontsov, N. N. and Martynova, L. Ya. (1972). Cytogenetical differentiation of Palearctic burrowing mammals. *In*: *Proc. Intern. Symp. on Species and Zoogeogr. of European Mammals. Praha.*

Leider, U. (1959). über die Entwicklung bei mänchenlosen Stämmen der Silberkarausche *Carassius auratus gibelio* (Bloch) (Vertebrata, Pisces). *Biol. Zbl.*, **78**, 2, 284–298.

Lowe, C. H., Wright, I. W. and Cole, C. I. (1966). Chromosomes and karyotypes of sceloporine iguanid lizards in the North American Southwest. *Mammal. Chrom. Newsl.*, **22**, 201–203.

Lowe, C. H., Wright, I. W., Cole, C. I. and Berzg, R. L. (1970). Natural hybridization between the teiid lizards *Chemidoporus sonorae* (parthenogenetic) and *Chemidoplorus tigris* (bisexual). *Syst. Zool.*, **19**, 2, 114–127.

Makino, S. (1951). "An Atlas of Chromosome Numbers in Animals." Iowa University Press.

Makino, S. and Asana, I. (1948). Sexual differences in the chromosomes of two species of Agamid lizards. *Chromosoma, Berlin* **3**, 208–219.

Manchino, G. and Nardi, I. (1971). Chromosomal heteromorphism and female heterogamety in the marbled new *Triturus marmoratus* (Latreille, 1800). *Experientia*, **27**, 7, 821–822.

Margot, A. (1964). Démonstration de l'absence d'hétérochromosomes morphologiquement differéncies chez deux espéces de Sauriens: *Anguis fragilis* L. et *Lacerta vivipara* Jacquin. *Revue. Suisse. Zool.*, **53**.

Martin, P. G. and Hayman, D. L. (1966). A complex sex-chromosomes system in the hare-wallaby *Lagorchestes conspieillatus* Gould. *Chromosoma, Berlin* **19**, 159–175.

Matthey, R. (1949). "Les chromosomes des Vertébrés". F. Rouge, Lausanne.

Matthey, R. (1950). Les chromosomes sexuals géants de *Microtus agrestis* L. *Cellule*, **53**, 162–184.

Matthey, R. (1953). La formule chromosomique et le probleme de la détermination sexuelle chez *Ellobius lutescens* (Thomas) (Rodentia-Muridae-Microtinae). *Arch. Julius Klaus-Stift. Vererbungsforsch.*, **28**, 65–73.

Matthey, R. (1952). Chromosomes de Muridae (Cricetinae et Microtinae). *Chromosoma, Berlin* **5**, 113–138.

Matthey, R. (1954). Nouvelles recherches sur chromosomes des Muridae. *Caryologia (Firenze)*, **6**, 1–44.

Matthey, R. (1956). Cytologie comparee des Muridae. L'origine des Ellobii. *Experientia*, **12**, 9, 337.

Matthey, R. (1958). Un nouvelu type de determination chromosomique du sexe chez mammiferes *Ellobius lutescens* Th. et *Microtus* (*Chilotus*) oregoni Bachm. (Muridés-Microtinés). *Experientia*, **14**, 7, 240.

Matthey, R. (1959). La formule chromosomique et le probleme de la determination sexuelle chez *E. lutescens* Thom. (Rodentia-Muridae-Microtinae). *Arch. Julius. Klaus-Stift. VererbForsch.* **28**, 65–73.

Matthey, R. (1961). Cytologie comparée des Cricetinae Palearctiques et d'Americanes. *Revue Suisse Zool.*, **68**, 41–61.

Matthey, R. (1964). Etudes sur les chromosomes d'*Ellobius lutescens* (Mammalia-Muridae-Microtinae). II. Informations complementaries sur les divisions meiotique. *Revue Suisse Zool.*, **71**, 401–410.

Matthey, R. (1965). Un nouveau de chromosomes sexuels multiples chez un souris africains du group *Mus* (*Leggada*) *minutoides* (Mammalia-Rodentia). Male X_1X_2Y : female X_1X_2/X_1X_2. *Chromosoma*, 16, 351–364.

Matthey, R. (1966a). Présence dans une population congolaise de *Mus* (*Leggada*) *triton* Th. de femelles hétérozygotes pour une délétion caractérisée la supression du bras court de l'un des chromosomes X métacentriques. *Z. Vererbungslehre.*, **97**, 361–369.

Matthey, R. (1966b) La probléme de la détermination du sexe chez *Acomys selousi* de Winton. Cytogénétique du genre *Acomys*. *Revue Suisse Zool.*, **73**, 4.

Matthey, R. (1967). Etude de deux femelles hétérozygotes pour une délétion partielle portant sur un bras chromosome X chez *Mus* (*Leggada*) *minutoides musculoides* Temm. *Cytogenetics*, **6**, 168–177.

Matthey, R. (1969a). Les chromosomes et l'evolution chromosomique des mammiféres. *Traite de Zoologie* **16**, 855–909.

Matthey, R. (1969b). Chromosomes de Gerbillinae, Genres *Tatera* et *Taterillus*. *Mammalia* (*Paris*), **33**, 3, 522–528.

Matthey, R. (1969c). Existe-t-il des chromosomes sexuels multiples chez le campagnol, *Microtus arvalis* Pallas ? *Genetica*, **40**, 4, 517–526.

Matthey, R. (1970). Caryotypes des Muridés et de Dendromuridés originaires de République Centrafricaine, *Mammalia* (Paris), **34**, 3, 459–466.

Matthey, R. and van Brink, J. M. (1956). La question des hétérochromosomes chez les Sauropsides. I. Reptiles. *Experientia* **12**, 53–55.

Matthey, R. and van Brink, J. M. (1957). Sex chromosomes in Amniota. *Evolution*, **11**, 163–165.

Meylan, A. (1964). Le polymorphisme chromosomique de *Sorex araneus* L. (Mammalia, Insectivora). *Revue Suisse Zool.* **71**, 903–983.

Morescalchi, A. (1964). Il corredo cromosomico di *Discoglossus pictus* Otth.: Cromosomi sessuali, spiralizzazione cromosomica e zone eterocromatiche. *Caryologia* (*Firenze*), **17**, 327–345.

Muramoto, J., Igarashi, K. Itoh, M. and Makino, S. (1969). A study of the chromosomes and enzymatic patterns of Sticklebacks of Japan. *Proc. Jap. Acad.*, **45**, 9, 803–807.

Oguma, K. (1934). Studies on the Sauropsid chromosomes. II. The cytological evidence proving female heterogamity in the lizard (*Lacerta vivipara*). *Arch. Biol.*, **45**, 27–46.

Ohno, S. and Atkin, N. B. (1966). Comparative DNA values and chromosome complements of eight species of fishes. *Chromosoma, Berlin* **18**, 455–456.

Ohno, S., Jainchill, J. and Stenius, C. (1963). The creeping vole (*Microtus oregonus*) as a gonosomic mosaic. I. The OY/XY constitution of the male. *Cytogenetics*, **2**, 232–239.

Ohno, S., Beçak, W. and Beçak, M. (1964a). X-autosome ratio and the behaviour pattern of individual X-chromosomes in placental mammals. *Chromosoma, Berlin* **15**, 14–30.

Ohno, S., Stenius, C., Christian, L. C., Beçak, W. and Beçak, M. L. (1964b). Chromosomal uniformity in the avian subclass Carinatae. *Chromosoma, Berlin* **15**, 280–288.

Ohno, S., Stenius, C. and Christian, L. (1966). The XO as the normal female of the creeping vole (*Microtus oregoni*). *In* "Chromosomes Today" (C. D. Darlington and K. R. Lewis, eds.). pp. 182–187. Oliver and Boyd, Edinburgh and London.

Ohno, S., Muramoto, J., Stenius, C., Christian, L. and Kittrell, W. A. (1969). Microchromosomes in Holocephalian, Chondrostean and Holostean Fishes. *Chromosoma, Berlin* **27**, 35–40.

Ojima, Y., Maeki, K., Takayama, S. and Nogusa, S. (1963). A cytotaxonomic study on the Salmonidae. *Nucleus*, **6**, 91–98.

Orlova, V. F. and Orlov, V. N. (1969). Chromosomal complements and some problems of taxonomy of lizards genus *Lacerta*. *Zool. Zh.* **48**, 7, 1056–1060 (Russian with English summary).

Patton, J. L. and Gardner, A. L. (1971). Parallel evolution of multiple sex-chromosome system in phyllostomatid bats, *Carollia* and *Choeroniscus*. *Experientia*, **27**, 105–106.

Piccini, E. and Stella, M. (1970). Some avian kariograms. *Caryologia (Firenze)*, **23**, 2, 189–202.

Pogosianz, H. E. (1970). Meiosis in the djungarian hamster. I. General pattern of male meiosis. *Chromosoma, Berlin* **31**, 392–403.

Potter, I. C. and Rothwell, B. (1970). The mitotic chromosomes of the Lamprey, *Petromyzon marinus* L. *Experientia*, **26**, 429–430.

Rabello, M. N. (1970). Chromosomal studies in Brasilien Anuras. *Caryologia (Firenze)*, **23**, 1, 45–59.

Radzhabli, S. I. and Kryukova, E. P. (1971). X-chromosome polymorphism in djungarian hamster *Phodopus sungorus* (Rodentia, Mammalia). *Cytology (Leningrad)*, **13**, 6, 710–790. (Russian with English summary).

Radzhabli, S. I., Vorontsov, N. N. and Volobuev, V. T. (1970). The karyological analysis of three *Myotis* species (Chiroptera-Vespertilionidae). *Cytology (Leningrad)*, **12**, 6. 767–773, V–VII (Russian with English summary).

Rao, S. R. V., Shah, V. C. and Seshadry, C. (1968). Studies on rodent chromosomes. II. Autoradiographic study of the sex chromosomes of the Indian gerbil, *Tatera indica cuverii* (Waterhouse) and its bearing on the Lyon hypothesis. *Chromosoma, Berlin* **23**, 309–316.

Rao, S. R. V., Sharma, V. K. and Shah, V. C. (1970). DNA synthesis in duplicate-type sex chromosomes of the Indian house shrew, *Suncus murinus* (Insectivora). *Cytogenetics*, **9**, 384–395.

Raiçu, P., Kirillova, M. and Hamar, M. (1969). A new chromosomal sex-determining mechanism in *Microtus arvalis* Pallas. *Genetica*, **40**, 97–102.

Ray-Chaudhuri, R. (1967). Mitotic and meiotic chromosomes of the koel, *Eudynamys scolopacea scolopacea*. *Nucleus*, **2**, 179–189.

Ray-Chaudhuri, S. P., Ranjini, P. V. and Sharma, T. (1968). Karyological studies of 16 species on Indian mammals. *Mamm. chrom. Newsl.*, **9**, 82–84.

Ray-Chaudhuri, S. P., Ray-Chaudhuri, R. and Sharma, T. (1966). The W chromosome in the females of two Indian species of Birds. *Chromosoma, Berlin* **20**, 151–154.

Rees, H. (1964). The question of polyploidy in the Salmonidae. *Chromosoma, Berlin* **15**, 275–279.

Reig, O. A. and Kiblisky, P. (1968). Chromosomes in fair species of rodents of the genus *Ctenomys* (Rodentia, Octodontidae) from Argentina. *Experientia*, **24**, 274–275.

Renzoni, A. and Vegni-Talluri, M. (1966). The karyograms of some Falconiformes and Strigiformes. *Chromosoma, Berlin* **20**, 133–150.

Rott, N. N. (1963). Sex determination in Amphibians. *Byull. Mosk. Obshch. Ispyt. Prir.*, **68**, 4, 118–134 (Russian with English summary).

Sasaki, M., Hitotsumachi, S., Makino, S. and Terao, T. (1968). A comparative study of the chromosomes in the chum salmon, the kokanee salmon and their hybrids. *Caryologia (Firenze)*, **21**, 4, 389–394.

Schmidt, W. (1967). Heterochromatin in mammals. *Arch. Julius Klaus-Stift. Vererb Forsch.*, **42**, 1–60.

Schmidt, W. and Leppert, M. (1968). Mammalian X-chromosomes: a new kind of composite type X in the vole *Arvicola scherman exitus* Miller. *Experientia*, **24**, 277–279.

Schultz, R. (1967). Gynogenesis and triploidie in the viviparous fish *Poecilopsis*, *Science, N.Y.* **197**, 1562–1567.

Seto, T. (1964). The karyotype of *Hyla arborea japonica* with some remarks on the heteromorphism of the sex chromosome. *J. Fac. Sci. Hokkaido Univ.*, **15**, 366–373.

Sharman, G. B., McIntosh, A. J. and Barber, N. V. (1950). Multiple sex chromosomes in the marsupials. *Nature (Lond.)*, **166**, 996.

Sharman, G. B. (1956). Chromosomes of the common shrew. *Nature (Lond.)*, **177**, 941/942.

Sharman, G. B. (1961). The mitotic chromosomes of marsupials and their bearing on taxonomy and phylogeny. *Austr. J. Zool.* 9: 38–60.

Singh, L. (1972). Multiple W-Chromosome in a sea snake, *Enhydrina schistosa* Daudin. *Experientia*, **28**, 95–96.

Singh, L., Sharma, T. and Ray-Chaudhuri, S. D. (1970). Multiple sex-chromosomes in the common Indian krait *Bungarus caeruleus* Schneider. *Chromosoma*, **31**, 4, 386–391.

Sokolovsky, V. (1973). The karyotaxonomy of reptiles with a list of the reptilian chromosome numbers. In press.

Soldatovič, B., Garžičič, B. and Živkovič, S. (1966). Chromosome analysis of *Spalax leucodon*. *Cariologia*, **19**, 3, 261–263.

Stephenson, E. M. and Stephenson, N. G. (1970). Karyotypes of two Australian hylids. *Chromosoma, Berlin* **30**, 38–50.

Takagi, N. and Makino, S. (1966). A revised study on the chromosomes of three species of birds. *Caryologia (Firenze)*, **19**, 4, 443–455.

Thomas, M. and Uzzell, T. M. (1964). Relations in the diploid and triploid species of the *Ambystoma jeffersonianum* complex (Amphibia, Caudata). *Copeia*, **2**, 257–300.

Todd, N. B. and Pessman, S. R. (1966). The karyotype of the marsh mongoose (*Atilax paludinosus*) and remarks on the phylogeny of the mongoose (Herpestidae). *Mamm. chrom. Newsl.* **8**, 21–22.

Utakoji, T. (1966). On the homology between the X and the Y chromosomes of the chinese hamster. *Chromosoma, Berlin* **18**, 449–454.

Uyeno, T. and Miller, R. R. (1971). Multiple sex chromosomes in a Mexican Cyprinodontid fish. *Nature (Lond)*, **231**, 452–453.

van Brink, J. M. (1959). L'expression morphologique de la digamétie chez les sauropsidés et les monotrèmes. *Chromosoma, Berlin* **10**, 1–72.

Vorontsov, N. N. (1966). Evolution of karyotype. *In* "Rukovodstvo po Cytologii" (Handbook of Cytology) Vol. 2, 359–389.

Vorontsov, N. N. and Korobitsina, K. V. (1970). Materials on a comparative karyology of Gerbillinae, *Cytology (Leningrad)*, **12**, 2: 152–157.

Vorontsov, N. N., Radjabli, S. I. and Liapunova, K. L. (1967). Karyological differentiation of allopatric forms of hamsters superspecies *Phodopus sungorus* and heteromorphism of sex-chromosomes in females. *Dokl. Acad. Sci. SSR*, **172**, 703–705.

Vorontsov, N. N., Liapunova, E. A., Zakarjan, G. G. and Ivanov, V. G. (1969). The karyology and taxonomy of the genus *Ellobius* (Microtinae, Rodentia). *In* "The Mammals (Evolution, Karyology, Taxonomy, Fauna) pp. 127–129. (N. N. Vorontsov, ed.) Novosibirsk.

Vorontsov, N. N., Liapunova, E. A. and Zakarjan, G. G. (1973). The chromosomes of genus *Ellobius* (in press).

Wahrman, J. and Zahavi, A. (1955). Cytological contributions to the phylogeny and classification of the rodent genus Gerbillus. *Nature (Lond.)*, **175**, 600–602.

Wallace, C. and Fairall, N. (1968). Chromosome analysis in the Kruger national park; a rare translocation chromosome in the kudu. *S. Afr. J. med. Sci.*, **33**, 113–118.

Weiler, C. and Ohno, S. (1962). Cytological confirmation on female heterogamety in the African water frog (*Xenopus laevis*). *Cytogenetics*, **1**, 173–184.

White, M. D. C. (1957). An interpretation of the unique sex-chromosome mechanism of the rodent *Ellobius lutescens* Thom., *Proc. Zool. Soc. Calcutta*, Mookerjee Mem. Vol., 113–114.

Wurster, D. H. and Benirschke, K. (1968). Chromosome studies in the superfamily Bovoidea. *Chromosoma, Berlin* **25**, 152–171.

Wurster, D. H. and Benirschke, K. (1970). Indian muntjac, *Muntiacus muntjak*: a deer with a low diploid chromosome number. *Science, N.Y.* **168**, 1364.

Wurster, D. H., Benirschke, K. and Noelke, H. (1968). Unusually large sex chromosomes in the sitatunga (*Tragelaphus spekei*) and the blackbuck (*Antilope cervicapra*). *Chromosoma, Berlin* **23**, 317–323.

Wurster, D. H., Snapper, J. R. and Benirschke, K. (1971). Unusually large sex chromosomes: new methods of measuring and description of karyotypes of six rodents (Myomorpha and Hystricomorpha) and lagomorph (Ochotonidae). *Cytogenetics*, **10**, 153–176.

Yosida, T. H. (1957). Sex chromosomes of the tree frog *Hyla arborea japonica*. *J. Fac. Sci. Hokkaido Univ.*, **13**, 352–358.

Zawadowsky, M. (1922) ."Das Geschlecht und die Entwickelung der Geschlechtsmerkmale." pp. 255, Vol. I–XX. Verlag "Gosisdat", Moskau. (In Russian with German summary.)

Zenzes, M. T. and Wolf, U. (1971). Paarungsverhalten der Deschlechtschromosomen in der männlichen Meiose von *Microtus agrestis*. *Chromosoma, Berlin* **33**, 41–47.

17

Evolution and the Mode of Action of the Genetic Regulatory Mechanisms which Determine Sex*

SUSUMU OHNO

Department of Biology, City of Hope Medical Centre, Duarte, California, U.S.A.

INTRODUCTION

The fact that a male and a female of the same species may have different chromosome constitutions was discovered in the last century. In 1891, Henking observed that at first meiotic prophase in the testis of a heteropteran insect, *Pyrrhocoris apterus* ($2n = 23/24$, XO/XX), each nucleus of primary spermatocytes contained one densely stained round body. Observing a pair of secondary spermatocytes, he noted that this densely stained body was always incorporated into one of the two daughter secondary spermatocytes. Thus, there were two kinds of haploid germ cells; one with 11 chromosomes and one with this body in addition to the 11 chromosomes. He was not quite certain as to the chromosomal nature of the dense body and, therefore, labelled it "X" for unknown. It is in Henking's honour that today the term X-chromosome denotes the sex chromosome which is present in the disomic state in the female and the monosomic state in the male. By 1902, the chromosomal basis of sex determination was firmly established by McClung.

Although enormous progress has been made since then in the fields of endocrinology, embryology and cytogenetics, the exact genetic mechanism which manifests bisexuality is still beyond our comprehension. While no one doubts the existence of sex determining genes, we have no idea as to the nature and function of the products specified by these genes.

In this day and age of molecular biology, the field of sex determination remains as one of the most underdeveloped areas in genetics. A drastic departure from conventional thinking appears to be in order. In this

* This work was supported in part by a grant (CA 05138) from the National Institutes of Health, U.S. Public Health Service.

Chapter, I shall attempt to equate the sex-determining genes with the Jacob and Monod type of regulatory genes (Jacob and Monod, 1963).

II. THE DOMINANT-RECESSIVE RELATIONSHIP BETWEEN MALE AND FEMALE DETERMINING GENES

Although synchronous as well as asynchronous hermaphroditism can be found in certain groups of teleost fish (Atz, 1964), and the all female parthenogenic or gynogenic species can be found not only among teleost fish (Hubbs, 1955), but also among reptiles (Pennock, 1965), the great majority of vertebrates are gonochorists; individual members of the species being born either as male or female. Among the gonochorists, a one-to-one sex ratio prevails.

With regard to this sex ratio which is a rule among gonochorists, it is well to remember an elementary law of genetics—that a mating between homozygotes and heterozygotes reproduces the two parental genotypes in the exact one-to-one ratio. Thus, the minimal requirement for becoming a gonochorist species is the acquisition of a pair of alleles at a single gene locus. In the case of the XX/XY scheme of male heterogamety, the Y-linked male determining allele M should be dominant over the X-linked female determining allele f. Then, the one-to-one sex ratio would perpetuate itself from generation to generation, since a mating would always be between f/f homozygotes (female) and f/M heterozygotes (male). Needless to say, when F is dominant over m, the female heterogamety of the ZW/ZZ scheme would develop.

It becomes clear that bisexuality as such does not require morphological differentiation of the sex chromosomes. Indeed, among vertebrates, the morphologically distinct X- and Y-chromosomes characterize only the class *Mammalia*, and the morphologically distinct Z- and W-chromosomes only the class *Aves*. In a great majority of reptiles, amphibians and fish, a pair of opposing sex-determining chromosomes apparently constitute a homomorphic pair, although exceptions are found among reptiles (Beçak *et al.*, 1964) and teleost fish (Chen and Ebeling, 1970). Furthermore, the analysis of sex-linked body colour genes of several species of aquarium fish revealed the extensive genetic homology between the X- and Y-chromosomes (Yamamoto, 1961) as well as the Z- and W-chromosomes (Winge and Ditlevsen, 1947; Gordon, 1951) of these lower vertebrates. Sex is determined not by the whole sex chromosome but by an allele or alleles of a particular gene locus or loci on the sex chromosome.

Regardless of the number of X-chromosomes present, the possession of a single Y-chromosome is sufficient for the manifestation of the male phenotype by human individuals (Jacobs and Strong, 1959). The complete dominance of the Y-linked heterogametic sex-determining genes (alleles) over the homogametic sex-determining genes (alleles) is clearly indicated. Sex determination appears to be based upon the dominant-recessive relationship that exists between the determining genes of opposite sexes.

III. DOMINANT MUTATION OF THE REGULATORY LOCUS

Our failure in making headway with the genetic mechanism of sex determination may be viewed as a reflection of our general ignorance on the genetic regulatory mechanisms of higher animals. So far as the genetics of mammals and other higher animals is concerned, spectacular advances in the last 15 years have been limited to the products specified by structural genes such as various enzymatic and nonenzymatic proteins. If the male determining genes on the mammalian Y-chromosome were structural genes, the problem would have been solved long ago by identifying the gene products found only in the male but not in the female. Unfortunately, this is not the case. In the case of the pituitary gonadotropic hormones, the same peptide hormone serves as LH (luteinizing hormone) in the female and as ICSH (interstitial cell stimulating hormone) in the male (Woods and Simpson, 1961). Similarly, another peptide hormone which functions as FSH (follicle stimulating hormone) in the female serves to stimulate the growth of seminiferous tubules of the male testis (Simpson *et al.*, 1951).

If the testosterone of the male and estrogen and progesterone of the female are synthesized by separate pathways, one might expect the presence of male specific enzymes for testosterone synthesis. The fact is that male and female steroid hormones are metabolites of the same synthetic pathway in that progesterone serves as a precursor of testosterone and estrone is a metabolite of androstendione. Clearly, the structural genes for the same set of steroid metabolizing enzymes are present in the male as well as in the female genome.

In the apparent absence of the male specific structural gene products which contribute to gonadal development and the manifestation of secondary sex characteristics, we have to conclude that sexual difference develops as a result of differential regulation of the same set of structural genes during ontogenesis. The sex-determining genes must be the regulatory genes in the sense of Jacob and Monod (1963).

The colon bacillus, *Escherichia coli*, normally utilizes glucose as a carbohydrate source. However, in the colon of suckling mammalian infants, *E.*

coli no doubt encounters more lactose than glucose. Perhaps for this reason, it has acquired the capacity to metabolize lactose. As the enzymes for lactose metabolism are not needed until *E. coli* encounters lactose, these enzymes became *inducible* in that *E. coli* synthesizes them only when they are needed. The elucidation of the genetic mechanism involved in the induction of lactose metabolizing enzymes of *E. coli* by Jacob and Monod (1963) gave us a profound insight into the nature of genetic regulatory mechanisms which must exist in all forms of life regardless of whether they are prokaryotes or eukaryotes, unicellular organisms or multicellular organisms.

In the inducible system, a group of structural genes is placed under the control of a single regulatory gene. In order to be regulated, each structural gene is provided with the region of a specific base sequence *operator* region which serves as a receptor site for a regulatory gene product. A regulatory gene product also has binding affinity to a particular inducer supplied from outside. By having this affinity, it serves as a sensor of a specific environmental stimulus.

The beauty of the genetic regulatory mechanism elucidated by Jacob and Monod is in the exclusion of the circular argument, for there is no need to regulate the regulator gene. A regulatory gene product is made all the time in a constitutive manner, yet structural gene products under its control are made only in the presence of an inducer. The regulatory gene may specify either a repressor or an activator. In the repressive control exemplified by the *lac*-operon system of *E. coli*, a repressor protein specified by the regulatory gene, *i*, normally binds with the *operator* region of the *lac*-operon which is a cluster of three structural genes for β-galactosidase, transacetylase and permease; each concerned with lactose metabolism. This occupation of the *operator* region by a repressor prevents the attachment of RNA polymerase to the adjacent *promotor* region; hence, no transcription of these structural genes to a polycistronic *messenger* RNA occurs in the absence of the inducer which is a metabolite of the substrate (lactose). When *E. coli* encounters lactose, a binding occurs between an inducer and a repressor, and the inducer-bound repressor loses its affinity to the *operator* region of the *lac*-operon. As an inducer removes the transcriptional block, the three enzymes for lactose metabolism are made in quantity as long as the inducer persists. The *lac*-repressor protein postulated by Jacob and Monod has actually been isolated in purified form (Gilbert and Müller-Hill, 1966) and the kinetics of its binding affinity to an inducer on one hand and to a *lac*-operator DNA base sequence on the other have been worked out (Riggs and Bourgeois, 1968; Riggs *et al.*, 1968).

In the activating control exemplified by the *l*-arabinose system of *E. coli* (Englesberg *et al.*, 1969), the structural genes concerned with arabinose metabolism are not clustered together to form an *operon*. Although they may exist as separate entities, since each is equipped with an *operator* region of similar base sequence, they all come under the control of an activating protein specified by a regulatory gene (C). Thus, these structural genes constitute a *regulon* rather than an *operon*. The *promotor* regions of these structural genes which constitute an *l*-arabinose *regulon* are not readily recognized by RNA polymerase unless they are first modified by a binding with an activator protein and only an inducer-bound activator protein can bind with these *promotor* regions. The net result is the same as the repressive control in that transcription and translation of structural genes under the control of the C gene occur only in the presence of an inducer. An activating protein specified by the C gene may be similar to but more specific than sigma factor (Burgess *et al*., 1969) or CAMP-protein (Emmer *et al*., 1970; Zubay *et al*., 1970).

Inasmuch as sexual difference manifests itself not through a single metabolic pathway but through many pathways, it is likely that sex-determining genes have regulatory functions affecting the activities of not one but a number of structural genes. Provided that sex-determining genes are regulatory genes, how can the dominant-recessive relationship be established between allelic forms of the regulatory locus ? At a repressive regulatory locus such as i of the lactose system of *E. coli*, the i^s type of mutations are indeed dominant over the wild-type i^+.

A mutational loss by a repressor of the binding affinity to an inducer is represented by i^s (*repressor noninducible*). The structural gene products which are normally synthesized in the presence of an inducer would never be made by mutants carrying i^s, for the transcriptional block would persist even in the presence of that inducer. In the diploid organism, i^s behaves as a dominant with heterozygotes (i^s/i^+) manifesting the *noninducible* phenotype. An inducer introduced to heterozygous cells would remove wild-type repressors from the *operator* regions, but these vacated *operator* regions would be reoccupied by mutant repressors which cannot be removed. As a result, no induction occurs in the heterozygous cells.

At an activating regulatory locus such as the C-gene of the *l*-arabinose system of *E. coli*, the C^c (*activator constitutive*) type, but not the C^s (*activator noninducible*) type, behaves as a dominant over the wild-type C^+. It should be recalled that a wild-type activator can recognize *promotor* regions only if it is first bound with an inducer. A mutant activator specified by C^c has acquired the capacity to bind with a *promotor* region without prior binding with an inducer. The structural gene products which are normally made only in the presence of an inducer would be made all the time by C^c regardless of the presence or absence of an inducer. In heterozygous C^c/C^+ cells not exposed to an inducer, all *promotor* regions would be occupied by mutant

C^c activators; hence, they would manifest the *constitutive* phenotype indistinguishable from that of C^c-homozygotes.

The first act of sex determination in ontogenic development is in determining the fate of indifferent gonads which can develop either as the ovary or testis. For the sake of argument, let us assume that early embryos of both chromosomal sexes are exposed to only one kind of inducer and that, unless additional structural genes are activated, the indifferent gonad automatically develops into the testis. The differentiation toward ovaries, on the other hand, would require the selective activation of a group of structural genes. Under this circumstance, the XX/XY scheme of sex-determining mechanism can be put into action simply by placing the i^+ repressive regulatory gene either on the X-chromosome or on an autosome and its mutant allele i^s on the Y-chromosome. In the XX-embryo which is i^+/i^+ homozygous, a group of structural genes would be activated as soon as an inducer appears. Consequently, ovaries would develop from the indifferent gonad. The XY embryo is a heterozygote of either the i^+/i^s type (if i^+ is X-linked) or the $i^+/i^+/i^s$ (if i^+ is autosomally inherited). Either way, the XY embryo should manifest the *noninducible* phenotype. The corresponding group of structural genes would not be activated even when the indifferent gonad was exposed to an inducer. The testis would automatically develop.

Conversely, the same end can be accomplished by two alleles of the activating regulatory locus. In this case, C^+ should be put either on the X or on an autosome, while its mutant allele C^c should be put on the Y-chromosome. Let us assume that activation of the same set of structural genes is required for the development of testes as well as ovaries. Early activation leads to testicular differentiation while delayed activation results in ovarian formation. Since the XX is an *inducible* C^+-homozygote, the group of structural genes in the indifferent gonad would not be activated until an inducer appeared; hence, ovarian development. In the XY which is a *constitutive* heterozygote of either the C^+/C^c or $C^+/C^+/C^c$ type, products specified by the corresponding group of structural genes would be made long before the appearance of an inducer; hence, testicular differentiation.

This completes the theoretical demonstration that so far as the gonads are concerned the minimal genetic requirement for the manifestation of the one-to-one sex ratio is a pair of alleles at a single regulatory gene locus. Of the two, the dominant allele which gives either the *noninducible* or *constitutive* phenotype should be placed on the Y-chromosome in the case of male heterogamety and on the W-chromosome in the case of female heterogamety.

IV. HORMONES AS INDUCERS

In order to study the inducible system, an inducer should first be identified and enzymes and nonenzymatic proteins which are induced should next be catalogued. Only then can we analyse the genetics of the inducible system by studying mutations affecting the *i* or *C* regulatory gene on one hand and those affecting the *promotor* or *operator* regions of structural genes on the other.

In unicellular organisms such as *E. coli*, the culture fluid in which they are suspended constitutes the external environment. Perhaps for this reason, their *regulon* or *operon* consists of a group of structural genes which specify a series of enzymes concerned with the intracellular transport and metabolism of a particular substrate, and either that substrate itself or a derivative of that substrate serves as an inducer. In multicellular organisms such as mammals, on the other hand, individual somatic cells of the body are, with the exception of skin and intestinal epithelium, not directly exposed to the external environment. The composition of the body fluid to which they are exposed is regulated by the body as a whole. Under this circumstance, it is not surprising if hormones rather than substrates or derivatives of substrates serve as inducers of mammalian inducible systems. Consequently, structural gene products of diverse functions rather than those concerned with a single metabolic pathway are expected to be induced by a single hormone. Of the various hormones, peptide hormones cannot serve as inducers in a strict sense, simply because they are incapable of entering the target cells. A specific receptor on the cell surface of the target organ recognizes a particular peptide hormone, and this binding appears to activate adenyl cyclase which results in the increased intracellular concentration of cyclic AMP. Thus, the action of a peptide hormone on target cells appears to be mediated, at least in a large part, through cyclic AMP (Robinson *et al.*, 1968).

Steroid hormones, on the other hand, may serve as *bona fide* inducers, for they are transported into the nuclei of target cells (Tomkins and Martin, 1970).

Inasmuch as the gonads themselves are the producers of sex steroids, it is not likely that sex steroids serve as inducers of gonadal development. However, determination of the fate of indifferent gonads is but an initial step of sexual differentiation. Before an embryo becomes either a male or a female, the developmental direction of gonoduct progenitors and urogenital sinus has to be determined. The regulatory genes which govern the development of gonoducts may also be called sex-determining genes, and it is in this area where we expect to make the first breakthrough on the genetic regulatory mechanism of sex determination, because sex steroids are likely to be involved in the inducible system as inducers.

V. WOLFFIAN DUCTS AS AN *INDUCIBLE* ORGAN AND MÜLLERIAN DUCTS AS A *CONSTITUTIVE* ORGAN (Fig. 1)

In ontogeny, sex develops in intimate association with the kidney from the very beginning. Not only does the mammalian gonad develop as a fold on the surface of mesonephros, but also the mesonephric duct becomes the Wolffian duct which is the anlage of the male gonoduct and its accessories such as epididymis and seminal vesicles. A pair of Müllerian ducts which serve as the anlage of Fallopian tubes and uterus is said to be of pronephric origin (Witschi, 1959). Normally, ovarian development by the XX-embryo is accompanied by the regression of Wolffian ducts and the differentiation of Müllerian ducts, while testicular development by the XY embryo is accompanied by the development of Wolffian ducts and the disappearance of Müllerian ducts. Retention of Müllerian ducts and regression of Wolffian ducts, however, are automatic processes which do not require the presence of ovaries. In a series of brilliant experiments, Jost (1947a, b, c, d, 1967) proved that castration of male (XY) or female (XX) rabbit embryos *in utero* before the onset of sexual differentiation results in the manifestation of the female phenotype. In the absence of gonads, Müllerian ducts automatically develop and Wolffian ducts atrophy. This is the very reason that XO humans (Turner's syndrome), whose gonads are streaks of connective tissue, manifest the female phenotype. Jost has further shown that the administration of testosterone to castrated embryos results in the hermaphroditic phenotype. Wolffian ducts develop in response to externally administered testosterone, but Müllerian ducts are not destroyed by testosterone. On this basis, Jost reasoned that the destruction of Müllerian ducts requires the presence of another male hormone (factor "X") secreted by embryonic testes. The doubt that Müllerian ducts of castrated embryos are maintained by the maternally derived estrogenic steroids was removed by *in vitro* experiments (Price, 1970).

As the development of Müllerian ducts does not require an inducer, it appears that gene products pertinent to development of this duct system are made all the time in a *constitutive* manner. Wolffian ducts, on the other hand, develop in the presence of an inducer (testosterone) and regress in its absence. It must be that pertinent structural genes in this duct system constitute a *regulon* and that testosterone serves as an inducer of this *regulon*. If there is a *regulon*, there must be a regulatory locus which controls it. If it

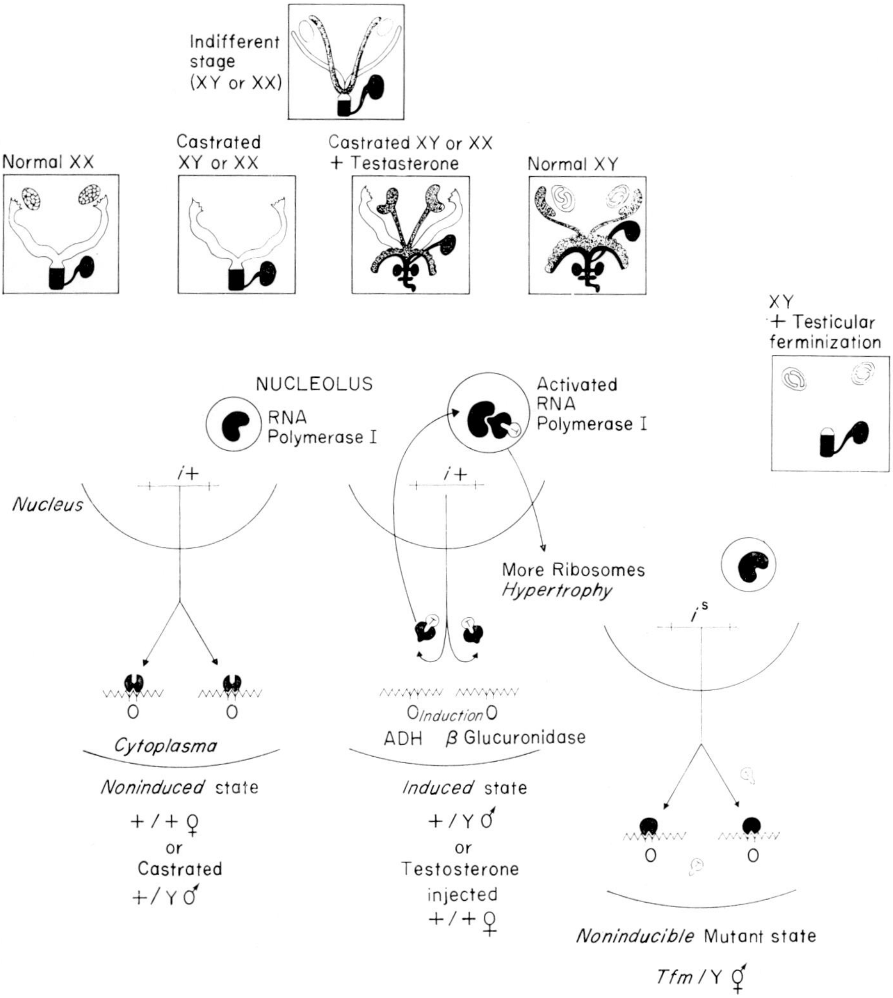

Fig. 1. A scheme of the minimum hypothesis on the genetic regulatory mechanism which determines the sexual phenotype of mammals. The status of Wolffian ducts (shaded), Müllerian ducts (outlined) and the urogenital sinus (solid black) is in each square. The pair of oblong bodies at the top signify gonads (modelled after Neumann *et al.*, 1969). As to the postulated genetic regulatory mechanism, i^+ represents the wild-type allele of the X-linked *Tfm* locus. Its product ( ) serves as a translational repressor of alcohol dehydrogenase (ADH) and β-glucuronidase in the cytoplasma until the arrival of testosterone ( ). When it becomes bound to testosterone or its metabolites (such as 5α-dihydrotestosterone), it moves into the nucleus and activates RNA polymerase I in the nucleolar region, thus causing increased production of ribosomes and hypertrophy. i^s represents a mutant *Tfm* allele; its product ( ) has lost the binding affinity to testosterone. The *o* (*operator*) region in each *messenger* RNA (wavy line) signifies a homologous base sequence shared by ADH and β-glucuronidase *messenger* RNAs which is recognized by the product of *i* gene.

exercises repressive control, the following two types of mutations are expected of this locus:

(a) The i^c (*repressor constitutive*) type represents a mutational loss by a repressor of the binding affinity to the *operator* regions of structural genes in a *regulon*. The structural gene products which were originally made only in the presence of testosterone would now be made all the time. In a diploid organism, i^c should behave as a recessive in that heterozygotes (i^+/i^c) should manifest the wild-type *inducible* phenotype, since, in the absence of testosterone, all the *operator* regions in a *regulon* of heterozygotes would be blocked by wild-type repressors. The phenotype of XY-males should not be affected by a *constitutive* mutation, because Wolffian ducts would develop as usual and Müllerian ducts would be destroyed by factor "X" secreted from the testis. The phenotype of XX-females homozygous for i^c should show a profound change. The hermaphroditic phenotype is expected since Wolffian ducts would persist and differentiate in the female equipped with ovaries and female genital tracts. Of various hermaphroditic conditions known in man and domestic animals (Overzier, 1963), none fits the expected phenotype of i^c. The absence of this class of mutant supports my view that the regulatory gene product has a dual role in being a translational repressor and a transcriptional activator as will become clear.

(b) The i^s (*repressor noninducible*) type represents a mutational loss by a repressor of the binding affinity to an inducer. Mutant repressors cannot be removed from the *operator* regions of structural genes in a *regulon* by an inducer. Thus, structural gene products which were originally inducible would not be made even in the presence of the inducer. In a diploid organism, a *noninducible* mutation should behave as a dominant over i^+ as already mentioned. This i^s mutation should not affect the XX-female, but should give the female phenotype to the genetic male. Despite the presence of testes, Wolffian ducts should regress, as testosterone cannot induce the production of pertinent structural gene products in this duct system. Müllerian ducts should regress in response to factor "X" secreted by testes. Thus, such a phenotypic female carrying this mutation should lack both the Wolffian duct and Müllerian duct derivatives. The inherited testicular feminization syndrome of man (Morris, 1953; Hauser, 1963) perfectly fits the predicted phenotype of an i^s mutant.

Since the postulated regulator of the testosterone-*regulon* serves as a translational repressor as well as a transcriptional activator, the question of whether the regulatory locus should be identified as *i* (repressive) or *c* (activating) becomes rather academic.

Figure 1 schematically illustrates the above view on the genetic regulatory mechanism which controls sexual phenotypes.

VI. X-LINKED TESTICULAR FEMINIZATION (*Tfm*) OF THE MOUSE AS A *NONINDUCIBLE* (i^s) MUTATION OF A REGULATORY LOCUS (Fig. 2)

The inherited testicular feminization syndrome which gives the phenotype expected of a *noninducible* regulatory mutation of a testosterone-*regulon* has been known to occur not only in man (Morris, 1953; Hauser, 1963), but also in cattle (Short, 1967) and in rats (Stanley and Gumbreck, 1964). In these species, the mode of inheritance of this defect has been compatible with either an X-linked mutation or an autosomal dominant mutation, both having sex limited expression. Only the demonstration of measurable linkage with other known X-linked genes would exclude autosomal inheritance.

The testicular feminization mutation (*Tfm*) has also been found in the mouse and the close linkage of this locus to two well-known X-linked marker genes has been established in this species. The *Tfm* locus is about one cross-over unit away from a coat structure gene locus, tabby (*Ta*), and about 4 cross-over units separate *Ta* from a coat colour gene locus, blotchy (*Blo*). The order of the three genes on the X is *Tfm*, *Ta*, *Blo* (Lyon and Hawkes, 1970). The discovery of *Tfm* in the mouse made the experimental analysis of the postulated testosterone-*regulon* possible.

If *Tfm* is indeed a *noninducible* mutation of the regulatory locus, structural gene products which are normally made only in the presence of testosterone should not be made even after the administration of a very large amount of an inducer to affected *Tfm*/Y ⚥. The so-called primary target organs of testosterone, such as seminal vesicles (Wolffian duct derivatives) and prostates (urogenital sinus origin), are entirely missing in *Tfm*/Y ⚥. Therefore, noninducibility cannot be studied on these organs. However, noninducibility can be studied on kidney of *Tfm*/Y ⚥. Unlike seminal vesicles and prostates, the kidney is present in both sexes. This permits a comparison to be made not only between *Tfm*/Y ⚥ and +/Y ♂ but also between *Tfm*/Y ⚥ and +/+ ♀ as well as *Tfm*/+ heterozygous ♀ with regard to the response of their kidney to testosterone. Probably reflecting its close ontogenic kinship with the Wolffian duct, the postnatal kidney (mesonephros) of the mouse is extremely responsive to testosterone. Normal male kidney weighs 20 to 40% more than normal female kidney, but the male kidney atrophies to the female level following castration. A number of enzymes in the mouse kidney of both sexes are known to be inducible by testosterone (Fishman, 1951; Shaw and Koen, 1964; Frieden, 1969; Ohno *et al.*, 1970b). This suggests that a testosterone-*regulon* consists of at least several structural gene loci. Of these, we have chosen alcohol dehydrogenase (ADH: E.C.

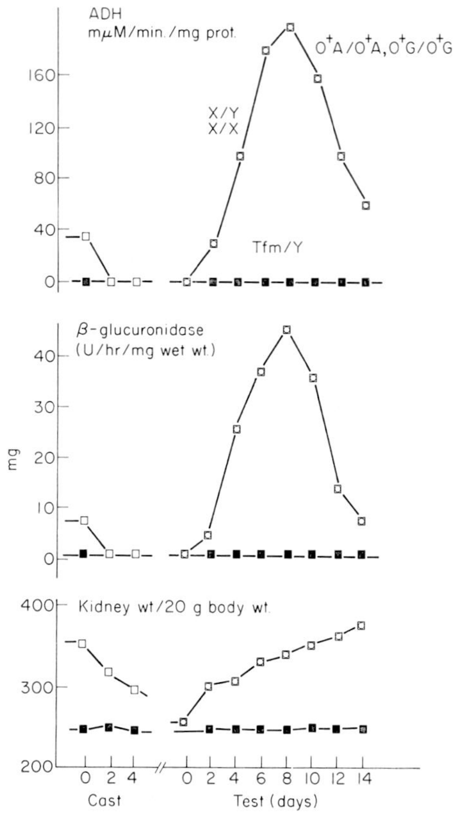

Fig. 2. Changes in kidney ADH (top) and β-glucuronidase (middle) levels as well as in kidney weight (bottom) of +/Y♂ and +/+♀ following castration and dihydrotestosterone administration are compared with those of *Tfm*/Y ☿ similarly treated. Each ADH value is expressed as mμM of reduced NAD oxidized in the presence of acetaldehyde per minute per mg protein (Ohno *et al.*, 1970b), while the β-glucuronidase value is expressed as units (micrograms) of phenolphthalein liberated per hour per mg wet tissue weight (Talalay *et al.*, 1946). Kidney weight is expressed as a combined wet weight of left and right kidney in mg per 20 g of body weight.
Cast. refers to the day of castration of +/Y♂. *Test.* refers to the day of administration of 10 mg of 5α-dihydrotestosterone. Each □ represents a mean value calculated from those obtained on five +/Y♂. Each ◘ represents a mean value calculated from three castrated +/Y♂ and two +/+♀. Each ■ represents a mean value from five *Tfm*/Y ☿.
The standard deviations are not given as those on ADH were given in a previous paper (Ohno *et al.*, 1970b). The β-glucuronidase values showed similar deviation from means.

1.1.1.1) and β-glucuronidase (E.C. 3.2.1.21) as marker enzymes partly because their assay methods are reliable but also because they are unrelated both in function and in intracellular location. ADH is found in the supernatant fraction while β-glucuronidase is found in lysosomes and ergastoplasm. Both enzyme loci are on autosomes, but the two are not linked. Although we do not know the exact location of the ADH locus (Ohno *et al.*, 1970a, b), the β-glucuronidase locus has been mapped on Linkage group XVII autosome (Paigen, 1963). The only thing that these two enzymes have in common is the fact that both are induced by testosterone in the kidney. Each of these two structural genes is probably equipped with an *operator* or *promotor* region of similar base sequence, and it is for this reason only that both belong to a testosterone-*regulon*.

Figure 2 compares the response of +/Y ♂ and +/+ ♀ kidney to 5α-dihydrotestosterone with that of *Tfm*/Y ⚥ kidney among the sibs. Observing Fig. 2, it should be noted that a sexual difference between +/Y ♂ and +/+ ♀ in the kidney ADH and β-glucuronidase levels disappears when the internal source of testosterone is removed from +/Y ♂ by castration. The following observations indicate that, in the absence of testosterone, a very small amount, if any, of each enzyme is present in the kidney: (1) if acetaldehyde used as a substrate for the ADH assay is uncontaminated, no oxidation of reduced NAD occurs and (2) in the case of β-glucuronidase, the small amount of phenolphthalein liberated from the substrate (phenolphthalein mono-β-glucuronide) does not increase lineally with the prolongation of incubation time.

From an undetectable level, both enzyme activities of +/Y and +/+ kidney begin a concordant rise following the administration of 10 mg of 5α-dihydrotestosterone, and after maintaining the fully induced state for 5 days (5th to 10th), both enzyme levels begin a gradual fall, again coordinately. This induction of kidney enzymes can be attributed to the marked increase in the net synthesis of ADH and β-glucuronidase rather than to slowing down of the degradation process. Concomitant administration of actinomycin D with an inducer prevents the induction of both enzymes (Frieden, 1969; Ohno *et al.*, 1970b). Furthermore, as kidney β-glucuronidase is induced, greater amounts of this enzyme are excreted into the urine (Fishman, 1951). The administration of 5α-dihydrotestosterone also causes hypertrophy of the +/Y and +/+ kidney which is accompanied by increased synthesis of all classes of RNA but not by cell division (Kochakian and Harrison, 1962). This hypertrophy, however, is the consequence but not a part of the specific induction as shall be shown later.

In sharp contrast to their normal +/Y and +/+ sibs, no induction of ADH and β-glucuronidase occurs in the kidney of *Tfm*/Y ⚥. A dose as large as 20 mg of testosterone or 5α-dihydrotestosterone was totally ineffective.

Tfm/Y kidney also fails to hypertrophy in response to administered dihydrotestosterone. Thus, *Tfm* which is an X-linked single gene mutation, affects the inducibility of two autosomal enzymes simultaneously satisfying the requirement to be classified as a *noninducible* regulatory mutation.

VII. TRANSLATIONAL REPRESSOR AND TRANSCRIPTIONAL ACTIVATOR (Fig. 3)

The current view is that, in order to be effective, testosterone has to be converted to 5α-dihydrotestosterone in target cells. This conversion is carried out by 5α-steroid reductase ($NADH_2$: Δ^4-5α-3-ketosteroid oxidoreductase) situated within target cell nuclei and 5α-dihydrotestosterone thus produced is found in the chromatin fraction of isolated nuclei (Bruchovsky and Wilson, 1968). This retention of 5α-dihydrotestosterone by the chromatin fraction appears to be mediated by a specific 3.5 S receptor protein (Fang *et al.*, 1969).

It would appear that the deficiency of either 5α-steroid reductase or 5α-dihydrotestosterone receptor protein could cause the testicular feminization syndrome. Indeed, the observation was made on suprapubic skin and pubic hair follicles of humans affected with the testicular feminization syndrome that 5α-steroid reductase was grossly deficient and this deficiency was suggested to be the cause of the syndrome (Heinrichs *et al.*, 1969; Northcutt *et al.*, 1969). However, preputial glands of rats affected with a homologous mutation were found to have a normal level of 5α-steroid reductase (Bullock *et al.*, 1970). We have also found the 5α-steroid reductase level of our *Tfm*/Y kidney cells to be normal (Bullock *et al.* 1971). This apparent discrepancy between man, on one hand, and rats and mice, on the other, has been reconciled as subsequent study has shown the 5α-steroid reductase of humans affected with the testicular feminization mutation to be normal (Jenkins and Ash, 1971).

The intracellular protein which binds the steroid hormone is customarily called "receptor" protein. The term "receptor", however, has an unfortunate connotation in that function as a permease is implied. Sixty minutes after an intravenous injection of 45 μc of tritiated-testosterone to effectively hepatectomized *Tfm*/Y ⚥, the kidney cytosol fraction contained 2000 cpm/mg protein of testosterone, 100 cpm of 5α-dihydrotestosterone and 400 cpm of androstandiols. As counts obtained on the cytosol fraction of +/+ ♀ and +/Y ♂ were no different, it was concluded that the *Tfm* mutation affected neither permease (if such exists) function nor intracellular metabolism of testosterone (Bullock *et al.*, 1971).

Influenced by a current view, we originally thought that 5α-dihydro-

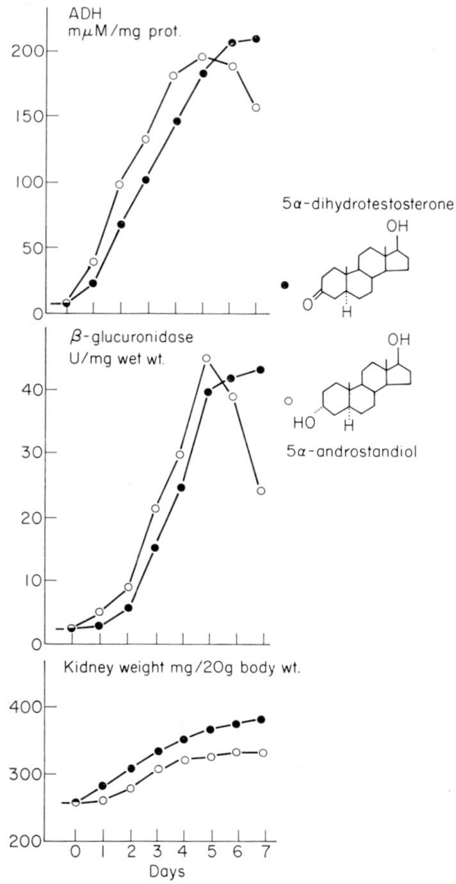

Fig. 3. The induction of kidney ADH and β-glucuronidase on castrated +/Y ♂ following a single intraperitoneal injection at day 0 of 10 mg 5α-androstan-3α-17β-diol (O) is compared with that following the administration of the same amount of 5α-dihydrotestosterone (●). The increase in kidney weight is also shown. Other details are the same as stated in Fig. 2.

testosterone was a *bona fide* inducer of the testosterone-*regulon*. But the suspicion that the role of 5α-dihydrotestosterone may have been overrated began to grow. The administration of 5α-dihydrotestosterone instead of its precursor (testosterone) to +/Y ♂ and +/+ ♀ did not result in faster induction of kidney ADH and β-glucuronidase. The only result was a more pronounced and prolonged hypertrophy of the kidney. This suspicion found further support when the most potent antiandrogen, cyproterone (1,2α-methyl-6-chloro-$\Delta^{4,6}$-pregnadiene-17α-o1,3,20-dione) acetate (Neumann *et al.*, 1970), totally failed as an anti-inducer of specific enzymes. Cyproterone acetate has been shown to specifically antagonize the action of 5α-dihydrotestosterone (Chandra *et al.*, 1967; Fang *et al.*, 1969).

When 10 mg of cyproterone acetate was administered 24 hours before the injection of 5 mg of 5α-dihydrotestosterone, the induction of kidney ADH and β-glucuronidase occurred according to schedule, but the prior administration of cyproterone acetate almost completely suppressed hypertrophy of the kidney (Ohno and Lyon, 1970). The above suggested that the induction of specific enzymes which appears to occur before hypertrophy need not be mediated by 5α-dihydrotestosterone. Accordingly, a search for another inducer began.

It has been shown that 5α-dihydrotestosterone released from nuclei of target cells is further converted by another enzyme to 5α-androstandiol (5α-androstan-3α-17β-diol) in their cytoplasm, and that androstandiol and its conjugated forms (glucuronides and sulphates) are excreted into urine. Although androstandiol has been regarded as a metabolite of no consequence, we recalled the experiments of Tomkins' group on the hydrocortisone induction of rat liver tyrosine aminotransferase and tryptophan pyrrolase. They observed that the steroid hormone induction of specific enzymes in mammals appears to involve the removal of a translational block rather than a transcriptional block (Tomkins *et al.*, 1969). If this is the case, a regulator specified by the wild-type allele of the *Tfm* locus must be recognizing the transcribed *operator* region of *messenger* RNAs in the cytoplasma and, therefore, 5α-androstandiol may indeed be able to substitute for testosterone and 5α-dihydrotestosterone as an inducer of specific enzymes (Ohno *et al.*, 1971).

Figure 3 compares the relative efficiency of the same dose of 5α-androstandiol and 5α-dihydrotestosterone as inducers. It should be noted that 5α-androstandiol causes a noticeably faster induction of kidney ADH and β-glucuronidase and that this fast induction is accompanied by only a slight degree of hypertrophy. This finding is compatible with the view that specific enzyme induction is the consequence of a cytoplasmic rather than a nuclear event. Here, we found a noticeable difference in *Modus operandi* between the testosterone *regulon* of the mouse and the *lac–operon* of *E. coli*.

Although cytoplasmic 3α-hydroxy steroid dehydrogenase (Tomkins, 1956) is capable of converting administered 5α-androstandiol back to 5α-dihydrotestosterone, if such conversion which is very inefficient is a prerequisite, androstandiol would not have behaved as a better inducer than dihydrotestosterone. Inducible kidney ADH can function as 3β-hydroxy steroid dehydrogenase but not as 3α-hydroxy steroid dehydrogenase (Waller *et al.*, 1965).

Subsequent hypertrophy must be caused by a rather unspecific stimulator of transcription. Increased production of ribosomes upon which translation of all *messenger* RNAs depends suffices as the cause of hypertrophy. It has been shown that testosterone stimulates synthesis of RNA specifically in the nucleolar region of target cell nuclei (Liao and Stumpf, 1968). A three-fold increase in activity of a class of RNA polymerase which functions at a low ionic strength has also been observed (Liao *et al.*, 1965). This is the RNA polymerase which was later identified as a specific RNA polymerase for transcription of *ribosomal* RNA genes in the nucleolar organizing region of the chromosome; nuclear RNA polymerase I (Roeder and Rutter, 1969). It is likely that testosterone and other steroid hormones cause hypertrophy of target cells by either directly or indirectly stimulating RNA polymerase I. The stimulation by estrogen of RNA polymerase I has actually been shown in rat uterus (Blatti *et al.*, 1970).

As already mentioned, 5α-dihydrotestosterone in target cell nuclei is bound to specific 3S protein and this complex exists in intimate association with the chromatin fraction. It may well be that a protein specified by the wild-type allele of the X-linked *Tfm* locus when bound with testosterone or 5α-dihydrotestosterone functions as a sigma-factor- (Burgess *et al.*, 1969) like subunit of RNA polymerase I. Our current study (to be reported elsewhere) strongly supports this possibility. Conversely, the testosterone-protein complex may function in a manner similar to cyclic AMP-binding protein of *E. coli* (Emmer *et al.*, 1970; Zubay *et al.*, 1970). The binding of the testosterone–protein complex to the *promotor* regions of *ribosomal* RNA genes may enhance their transcription by RNA polymerase I.

Neumann and his colleagues (1970) have shown that, in the presence of the wild-type allele of *Tfm*, the administration of a large dose of cyproterone acetate during embryonic development modifies +/Y ♂ to a phenocopy of *Tfm*/Y, although the complete regression of Wolffian ducts is unattainable. This suggests that in the presence of the wild-type allele of *Tfm*, cyproterone acetate can antagonize stimulation of RNA polymerase I by testosterone and that without this stimulation the manifestation of the male phenotype is impossible.

VIII. THE DUAL ROLE OF THE *TFM* REGULATOR

In affected *Tfm*/Y ⚥, neither specific enzyme induction nor hypertrophy of the kidney occurs in response to administered testosterone (Ohno and Lyon, 1970; Dofuku *et al.*, 1971). A mutational loss by the regulatory gene product (*Tfm* protein) of the binding affinity to testosterone and its metabolite must have occurred. Indeed, despite the unimpeded entrance of testosterone to *Tfm*/Y kidney cell cytoplasma (Bullock *et al.*, 1971), we found one class of cytoplasmic 5α-dihydrotestosterone-binding protein to be entirely missing (Gehring *et al.*, 1971). The *in vivo* uptake of testosterone by kidney cell nuclei of *Tfm*/Y ⚥ was grossly abnormal. No appreciable count was obtained 60 minutes after the intravenous injection of 45 μc of tritiated-testosterone, while kidney cell nuclei of +/+ ♀ and +/Y ♂ contained 500 cpm/10^7 nuclei of testosterone (Bullock *et al.*, 1971). Relative deficiency of nuclear testosterone-binding protein was also noted. At a tritiated-dihydrotestosterone concentration of 3×10^{-8} M, proteins in *Tfm*/Y kidney cell nuclei bound 24×10^{-15} moles/mg DNA of dihydrotestosterone after an incubation of two hours at 30°C, while a three times greater binding was observed in the kidney cell nuclei of +/+ ♀ and +/Y ♂ (Gehring *et al.*, 1971). A similar finding was reported on rats affected with an apparently homologous mutation (Bullock *et al.*, 1970). On the basis of the above findings, I have proposed (Ohno, 1971) that inside the target cells of testosterone a protein specified by the wild-type allele of the X-linked *Tfm* locus has a dual role. When it is not bound to testosterone, it stays in the cytoplasma and serves as a translational repressor of certain enzymes. A binding with testosterone or its metabolites changes its allosteric configuration, whereupon the bound form detaches itself from these *messenger* RNAs, releasing them from a translational block (specific enzyme induction) and moves into the nucleus where it activates RNA polymerase I inside the nucleolar organizing region (hypertrophy). In this way, a single gene product mediates the entire testosterone-induced response of target cells. Therefore, the mutational loss by this gene product of the binding affinity to testosterone simultaneously abolishes all responses from all of the target organs of testosterone (Fig. 1).

IX. SUMMARY AND FUTURE OUTLOOK

Studies on the X-linked testicular feminization mutation of the mouse led us to the conclusion that the wild-type allele of this X-linked regulatory locus alone controls the fate of the Wolffian duct and, therefore, the manifestation of the male phenotype (Ohno, 1971). Once developed, the testis merely serves as the supplier of the precursor (testosterone) of an inducer. When dissected, a mammalian regulatory system appears to be surprisingly simple. Apparent complexity is attributable to the fact that several regulatory systems often interlocked with each other operate within the same cell. For example, another regulatory system which uses aldosterone as an inducer also operates in the kidney (Edelman and Fimognari, 1968).

In a similar manner, the fate of indifferent gonads must also be controlled by either a single or a small number of regulatory loci. In goats, the mutation of an autosomal locus brings about testicular development in genetic females (Hamerton *et al.*, 1969). Cattanach *et al.* (1971) has recently recovered an autosomal dominant mutation (*Sxr*) in the mouse. The XX, *Sxr*/+ individuals develop as the phenotypically normal male complete with scrotal testes, although sterile. The dominance of *Sxr* suggests that this is either a *noninducible* (i^s) mutation of the repressive regulatory locus or a *constitutive* (c^s) mutation of the activating locus. If the fate of indifferent gonads is normally determined by the presence of i^+ on an autosome and the presence of its *noninducible* allele i^s on the Y-chromosome as mentioned at the beginning of this Chapter, mutation of the autosomal i^+ to i^s would surely give the testis to genetic females. If an inducer of testicular or ovarian development is identified and if structural gene products which are induced by that inducer become known, the genetic analysis of this regulatory system should become possible. We will finally understand the basis of sex determination.

REFERENCES

Atz, J. W. (1964). In "Intersexuality in Vertebrates Including Man." (C. N. Armstrong, and A. J. Marshall, eds.) pp. 145–232. Academic Press, London and New York.

Beçak, W., Beçak, M. L., Nazareth, H. R. S. and Ohno, S. (1964). *Chromosoma, Berlin* **15**, 606–617.

Blatti, S. P., Ingles, C. J., Lindell, T. J., Morris, P. W., Weaver, R. F., Weinberg, F. and Rutter, W. J. (1970). *Cold Spring Harb. Symp. Quant. Biol.*, **35**, 649–657.

Bruchovsky, N. and Wilson, J. D. (1968). *J. Biol. Chem.* **243**, 5953–5960.

Bullock, L., Schneider, G. and Bardin, C. W. (1970). *Life Sci.* **9**, 701–705.

Bullock, L. P., Bardin, C. W. and Ohno, S. (1971). *Biochem. Biophys. Res. Commn.* in press.

Burgess, R. R., Travers, A. A., Dunn, J. J. and Bautz, E. K. R. (1969). *Nature (Lond).* **221**, 43–46.

Cattanach, B. M., Pollard, C. E. and Hawkes, S. G. (1971). *Cytogenetics*, **10**, 318–337.

Chandra, P., Orii, H. and Wacker, A. (1967). *Hoppe Seyler's Z. Physiol. Chem.* **348**, 1085.

Chen, T. R. and Ebeling, A. W. (1970). *Trans. Am. Fish. Soc.* **99**, 131–138.

Dofuku, R., Tettenborn, U. and Ohno, S. (1971). *Nature New Biol.* **232**, 5–7.

Edelman, I. S. and Fimognari, G. M. (1968). *Recent Prog. Horm Res.* **24**, 1–34.

Emmer, M., deCrombrugghe, B., Pastan, I. and Perlman, R. (1970). *Proc. Natn. Acad. Sci. U.S.A.* **66**, 480–487.

Englesberg, E., Squires, C. and Meronk, F. (1969). *Proc. Natn. Acad. Sci. U.S.A.* **62**, 1100–1107.

Fang, S., Anderson, K. M. and Liao, S. (1969). *J. Biol. Chem.* **244**, 6584–6596.

Fishman, W. H. (1951). *Ann. N.Y. Acad. Sci.* **54**, 548–557.

Frieden, E. H. (1969). *In* "Metabolic Effects of Gonadal Hormones and Contraceptive Steroids". (H. A. Salhanick, D. M. Kipnis, and R. L. Vande Wiele, eds.) pp. 319–327. Plenum Press, New York.

Gehring, U., Tomkins, G. M. and Ohno, S. (1971). *Nature New Biol.* **232**, 106–107.

Gilbert, W. and Müller-Hill, B. (1966). *Proc. Natn. Acad. Sci. U.S.A.* **56**, 1891–1898.

Gordon, M. (1951). *Zoologica* **32**, 127–134.

Hamerton, J. L., Dickson, J. M., Pollard, C. E., Grieves, S. A. and Short, R. V. (1969). *J. Reprod. Fert.* **7**, 25–51.

Hauser, G. A. (1963). *In* "Intersexuality". (C. Overzier, ed). pp. 255–276. Academic Press, London and New York.

Heinrichs, W. L., Karsznia, R., Wyss, R. and Herrmann, W. L. (1969). *Clin. Res.* **17**, 143.

Henking, H. (1891). *Z. Wiss. Zool.* **51**, 685–736.

Hubbs, C. L. (1955). *Syst. Zool.* **4**, 1–20.

Jacob, F. and Monod, J. (1963). *In* "Cytodifferentiation and Macromolecular Synthesis". (M. Locke, ed). pp. 30–64. Academic Press, New York and London.

Jacobs, P. A. and Strong, J. A. (1959). *Nature* (*Lond*). **183**, 302–303.

Jenkins, J. S. and Ash, S. (1971). *J. Endocr.* **49**, 515–520.

Jost, A. (1947a). *C. R. Séanc. Soc. Biol.* **141**, 126–129.

Jost, A. (1947b). *C. R. Séanc. Soc. Biol.* **141**, 135–136.

Jost, A. (1947c). *C. R. Séanc. Soc. Biol.* **141**, 275–276.

Jost, A. (1947d). *Proc. Soc. Exp. Biol. Med.* **66**, 302–303.

Jost, A. (1967). *In* "Research on Steroids". Vol. 3, Rome.

Kochakian, C. D., and Harrison, D. G. (1962). *Endocrinology* **70**, 99–108.

Liao, S. and Stumpf, W. E. (1968). *Endocrinology* **83**, 629–632.

Liao, S., Leininger, K. R., Sagher, D. and Barton, R. W. (1965). *Endocrinology* **77**, 763–765.

Lyon, M. F. and Hawkes, S. G. (1970). *Nature* (*Lond*) **227**, 1217–1219.

McClung, C. E. (1902). *Kans. Univ. Sci. Bull.* **1**, 185–238.

Morris, J. M. (1953). *Am. J. Obstet. Gynec.* **65**, 1192–1211.

Neumann, F., Elger, W. and Steinbeck, H. (1969). *J. Reprod. Fert.* **7**, 9–24.

Neumann, F., von Berswordt-Wallrabe, R., Elger, W., Steinbeck, H., Hahn, J. D. and Kramer, M. (1970). *Recent Prog. Horm. Res.* **26**, 337–410.

Northcutt, R. C., Island, D. P. and Liddle, B. W. (1969). *J. clin. Endocr.* **29**, 422–425.

Ohno, S. (1971). *Nature* (*Lond.*), **234**, 134–137.

Ohno, S. and Lyon, M. F. (1970). *Clin. Genet.* **1**, 121–127.

Ohno, S., Stenius, C. and Christian, L. (1970a). *Clin. Genet.* **1**, 35–44.

Ohno, S., Stenius, C., Christian, L., Harris, C. and Ivey, C. (1970b). *Biochem. Genet.* **4**, 323–335.

Ohno, S., Dofuku, R. and Tettenborn, U. (1971). *Clin. Genet.* **2**, 1–13.

Overzier, C. (1963). *In* "Intersexuality" (C. Overzier, ed.) pp. 235–254. Academic Press, London and New York.

Paigen, K. (1963). *In* "Congenital Malformations". pp. 181–190. The International Medical Congress, Ltd., New York.
Pennock, L. A. (1965). *Science, N.Y.* **49**, 539–540.
Price, D. (1970). *Phil. Trans. R. Soc. (Lond). B.* **259**, 133–139.
Riggs, A. D. and Bourgeois, S. (1968). *J. Molec. Biol.* **34**, 361–364.
Riggs, A. D., Bourgeois, S., Newby, R. F. and Cohn, M. (1968). *J. Molec. Biol.* **34**, 365–368.
Robinson, G. A., Butcher, R. W. and Sutherland, E. W. (1968). *Ann. Rev. Biochem.* **37**, 149–174.
Roeder, R. G. and Rutter, W. J. (1969). *Nature (Lond)* **224**, 234–237.
Shaw, C. and Koen, A. (1964). *Histochem. Cytochem.* **13**, 431–433.
Short, R. V. (1967). *Ann. Rev. Physiol.* **29**, 373–400.
Simpson, M. E., Li, C. H. and Evans, H. M. (1951). *Endocrinology* **48**, 370–383.
Stanley, A. J. and Gumbreck, L. G. (1964). program, The Endocrinology Society, p. 40.
Talalay, P., Fishman, W. H. and Huggins, C. (1946). *J. Biol. Chem.* **166**, 757–772.
Tomkins, G. M. (1956). *J. Biol. Chem.* **218**, 437–447.
Tomkins, G. M. and Martin, D. (1970). *Ann. Rev. Genet.* **4**, 91–106.
Tomkins, G. M., Gelerhter, T. D., Granner, D., Martin, D., Samuels, H. H. and Thompson, E. B. (1969). *Science, N.Y.* **166**, 1474–1480.
Waller, G., Theorell, H. and Sjövall, J. (1965). *Arch. Biochem. Biophys.* **111**, 671–684.
Winge, O. and Ditlevsen, E. (1947). *Heredity* **1**, 65–83.
Witschi, E. (1959). *Ann. N.Y. Acad. Sci.* **75**, 412–435.
Woods, M. C. and Simpson, M. E. (1961). *Endocrinology* **69**, 91–125.
Yamamoto, T. (1961). *J. Exp. Zool.* **146**, 163–180.
Zubay, G., Schwartz, D. and Beckwith, J. (1970). *Proc. Natn. Acad. Sci. U.S.A.* **66**, 104–110.

18

Concluding Remarks

ERNESTO CAPANNA
Istituto di Anatomia Comparata, Rome, Italy

> "La chose la plus rare est de joindre la raison avec l'enthousiasme; la raison consiste à voir toujours les choses comme elles sont. Celui qui dans l'ivresse voit les objets doubles est alors privé de la raison. L'enthousiasme est précisément comme le vin; il peut exciter tant de tumulte dans les vaisseaux sanguins, et de si violents vibrations dans les nerfs, que la raison en est tout-à-fait détruite."
>
> VOLTAIRE, *Dictionnaire Philosophique*

I. ENTHUSIASMS

Can definite conclusions be drawn concerning the problem of chromosomal evolution in the vertebrates today? Or rather, from the tremendous body of available information that has been gathered in the present volume, is it possible nowadays to draw deductions of a general kind?

It might be said that the history of karyological studies in the vertebrates has been, and still is, an uninterrupted sequence of "great expectations" and "bitter disappointments".

Indeed, when Professor Matthey published "Les Chromosomes des Vertébrés" in 1949, we were in a period of great expectation as far as the karyology and cytotaxonomy of the vertebrates were concerned. In fact, the possibility was looming up at us of having to characterize univocally any animal species through a chromosomal formula which could reveal the physical basis of the genetic material, that is to say the biologically codified essence of the morphological characteristics of the species.

Moreover, Matthey in this book, a classic on vertebrate karyology, was able to illustrate, on the basis of his own observations and those of other authors, marvellous instances of Robertsonian changes in the Scindidae and Lacertidae (Matthey, 1931, 1939), in Phasianidae (Shokolov, 1936) and in

Bovidae (Makino, 1943), so that the concept of *Nombre Fondamental* acquired an ever increasing importance for the characterization of the karyotype of the species (Matthey, 1945).

Identification of highly exact mechanisms of chromosome transformations seemed, in fact, to bring to light those mechanisms of genetic transformations that lay at the basis of evolution in the neo-Darwinian sense.

Matthey's book is mentioned here because of its close correlation with the subject of the present volume. However, it should be remembered that at that time White's "Animal Cytology and Evolution", in which the same topic was dealt with, extended to the whole animal kingdom, had already been published (1945).

This similiarity of dates is not a coincidence. In that period the problem of chromosome evolution was ripening, and it even appeared that the technical methods devised for karyological analysis had attained a qualitative level beyond further improvement, although it was not entirely satisfactory at least as far as vertebrates were concerned.

Hence, when in the late 1950's new techniques were developed making use first of cytostatics, then of hypotonic shock (Ford and Hamerton, 1956) and followed by the techniques of somatic cell cultures (Hsu and Pomerat, 1953; Hsu and Kellog, 1960; Moorhead *et al.*, 1960). the hopes of those concerned with karyology were kindled, as the number of data and information on the vertebrate karyotype increased.

Matthey's findings reported here in Chapter 15 are thoroughly demonstrative. The eutherian species whose diploid number was known reached 104 in 1953; 240 in 1958; 267 in 1960; 769 in 1966 and no less than 1165 in 1970. I found it quite exciting to plot (in abscissae) the years between 1949, when "Les Chromosomes des Vertebrés" was published (in abscissae) against the number (in ordinates) of the eutherian species whose diploid number was known at that time (Fig. 1). An exponential curve was the result, interesting because of the fact that the point at which the curve starts climbing towards the vertical asymphote, namely the point after which the derivates are above 1, falls in the years 1958 and 1960, which is when Moorhead *et al.* (1960) devised an easy technique for leucocyte culture of peripheral blood for human cytogenetics. This technique was at once experimented with by all the mammalian cytogeneticists for eutherian cytotaxonomy. It must not be thought, however, that in the years following 1960 mammalian karyology was a mere "divertissement" for human geneticists: true, in some instances, this trend was recognized, but the reader of this book must have realized that an increase in the study of all vertebrate classes has actually been achieved. As a matter of fact, it would suffice to take reptiles into consideration in order to show that the species, for whom a diploid number was provided by Matthey in 1949, amounted to only 79, subdivided as follows: 49 saurians, 16 ophidians, 6 chelones and a single crocodile. In the review by Gorman in this work, we find over 500 diploid sets as follows: 330 saurians, 111 ophidians, 45 chelones and 21 crocodiles.

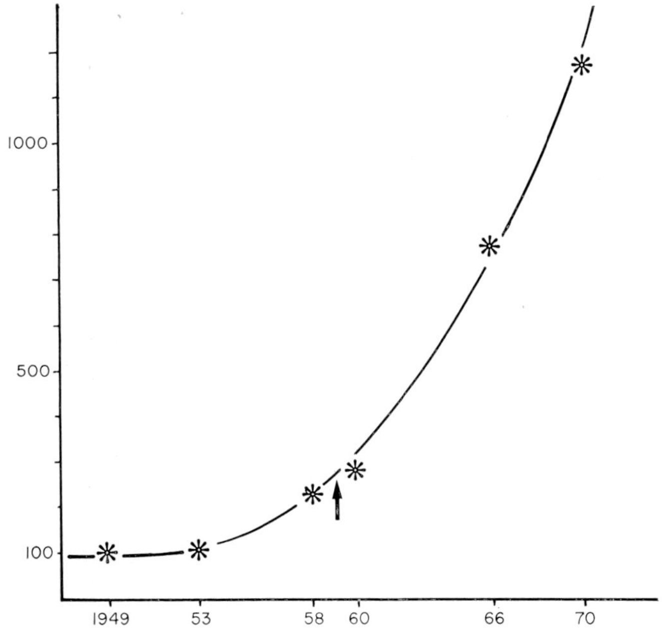

Fig. 1. Diagram showing the increase of the studies on Eutherian Mammal karyology; abscissae: years between 1949 and 1970; ordinatae: number of species whose diploid number was known in the corresponding year. Arrow indicates the point of the curve where the derivative is 1.

Furthermore, the difficulties in karyogram or metaphase plate interpretation, as well as tentatively suggested diploid numbers have been eliminated by modern techniques. Chromatid splitting by means of antiblastics reveals most accurately the centromere position thereby enabling characterization of each single metacentric chromosome through borrowed karyometric data, such as the centromeric index, arm ratio and so on. Likewise, other mechanisms of chromosome transformation are brought to light in the karyotype evolution in the vertebrates, e.g. pericentric inversions, deletions and so on.

However, contrary to our expectations, the first controversies within the chromosome establishment are now arising, causing conclusions of a general kind to become conjectural.

II. CHROMOSOME POLYMORPHISM

Beyond a doubt, chromosome polymorphism should be thought of as a much more widespread phenomenon than hitherto. As this problem is extensively discussed in this volume by White (Chapter 6), it need not be dealt with again by me here. I shall just call attention to one point. If, on the one hand, chromosome polymorphism reveals what may be the process of speciation *in fieri*, on the other hand it makes debatable the characterization univocally of one species through the morphology of its karyotype. This, however, would not be detrimental as regards the evolutionary interpretation of the changes in karyotype within a systematic homogeneous group, if it did not raise some misgiving as regards its purely taxonomic interpretation as well. In fact the suspicion often arises, that in reality chromosome polymorphism is but the expression of two sibling species which the traditional morphological characters fail to separate. Or, better, it may be assumed that such characters used as a first approximation had not been regarded as sufficient for a taxonomic separation.

Personal experience concerning the polymorphism of the black rat (*Rattus rattus*) has led me to the conclusion that here the coexistence of a real and a pseudo chromosome polymorphism is true. The form with 38 chromosomes would represent a species which is sharply demarcated from that with 42 chromosomes, which in its turn would be considerably polymorphic *per se* both from the chromosomal and morphological and taxonomic standpoints (see Capanna and Civitelli, 1971 for a recent summary on this subject.)

On the other hand, when considering *a posteriori* the morphological comparison between *R. rattus* with 38 and with 42 chromosomes, it is not difficult to find striking morphological differences between them by a traditional taxonomic approach, such as cranial indices and the pattern of dental cusps.

Which is more reliable: karyological evidence or morphological evidence? Which of them has the determinant importance for the separation of these species?

In 1945 White warned us that: "The morphology of the chromosome set is only one aspect of the general structure of the organism, and in the degree of divergence between two forms it should not be considered apart from differences in eye shape, wing venation, length of bristles, structure of genitalia or other recognized taxonomic characters" (White, 1945. p. 50).

Therefore, the two data, namely the karyological and the taxonomic, should be never discussed separately; but only by providing a valid morphological criterium for the identification of the species shall we be able to claim that we have solved the systematic problem, since an evolutionary problem is essentially a systematic one. This is indeed no easy task.

In this connection I should like to submit to your judgement a question that puzzles me at this moment. When Gropp *et al.* (1969), in the tobacco mouse (*Mus poschiavinus*), found an impressive Robertsonian reduction in the diploid number ($2n = 26$, NF 40) as compared to *Mus musculus*, the familiar house mouse ($2n = 40$, NF 40), it was universally surmised that this was a full confirmation of the debated divergence made by Fatio (1869) between *Mus musculus* and *Mus poschiavinus*. Today, however, Gropp *et al.* (1972) have found minor endemisms of *Mus musculus domesticus* in the Canton Grigioni (Switzerland) also characterized by Robertsonian reductions: $2n = 38$, NF 40 in a first case, and $2n = 34$, NF 40 in another. Following Professor Gropp's advice I devoted myself to the sampling of *Mus musculus* throughout Italy, from Sicily to the Alps. As well as absolutely normal populations ($2n = 40$), I was able to find two populations with a Robertsonially reduced number, the first one in the Central Appennine characterized by $2n = 22$, NF 40; the second in the western Dolomites (Eastern Alps) characterized by $2n = 32$, NF 40.

A Robertsonian fan ("eventail robertsonien") is thus delineated, which is similar to that demonstrated in another group of species from the same genus *Mus*, namely *Mus* (*Leggada*) (Matthey, 1963–1970; Matthey *et al.*, 1968; 1970).

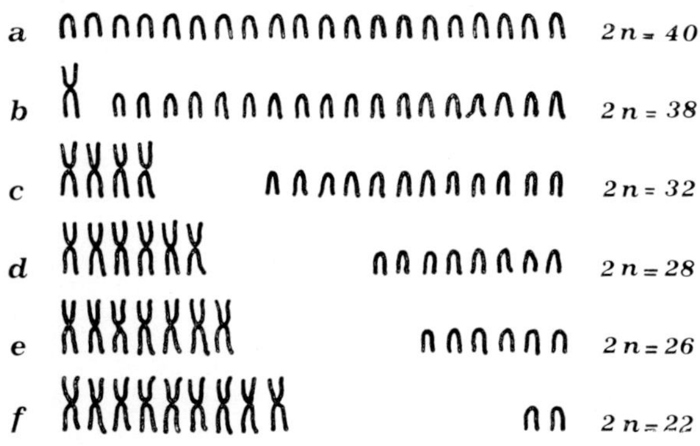

Fig. 2. The "Robertsonian fan" of *Mus musculus*. Haploid sets from different *Mus musculus* populations are represented: (a) the haploid set commonly shown by *Mus musculus* ($2n = 40$; NF $= 40$); (b) population from Val Bregaglia, Grisons, Switzerland (after Gropp *et al.*, 1972) $2n = 38$; (c) population from Val Mesolecina, Grisons, Switzerland (after Gropp *et al.*, 1972) $2n = 34$; (d) population from Val di Non, western Dolomites, Italy (personal observations) $2n = 32$; (e) *Mus poschiavinus* from Val Poschiavo, Grisons, Switzerland (after Gropp *et al.*, 1969) $2n = 26$ and (f) population from Greccio, Central Appennines, Italy (personal observations) $2n = 22$.

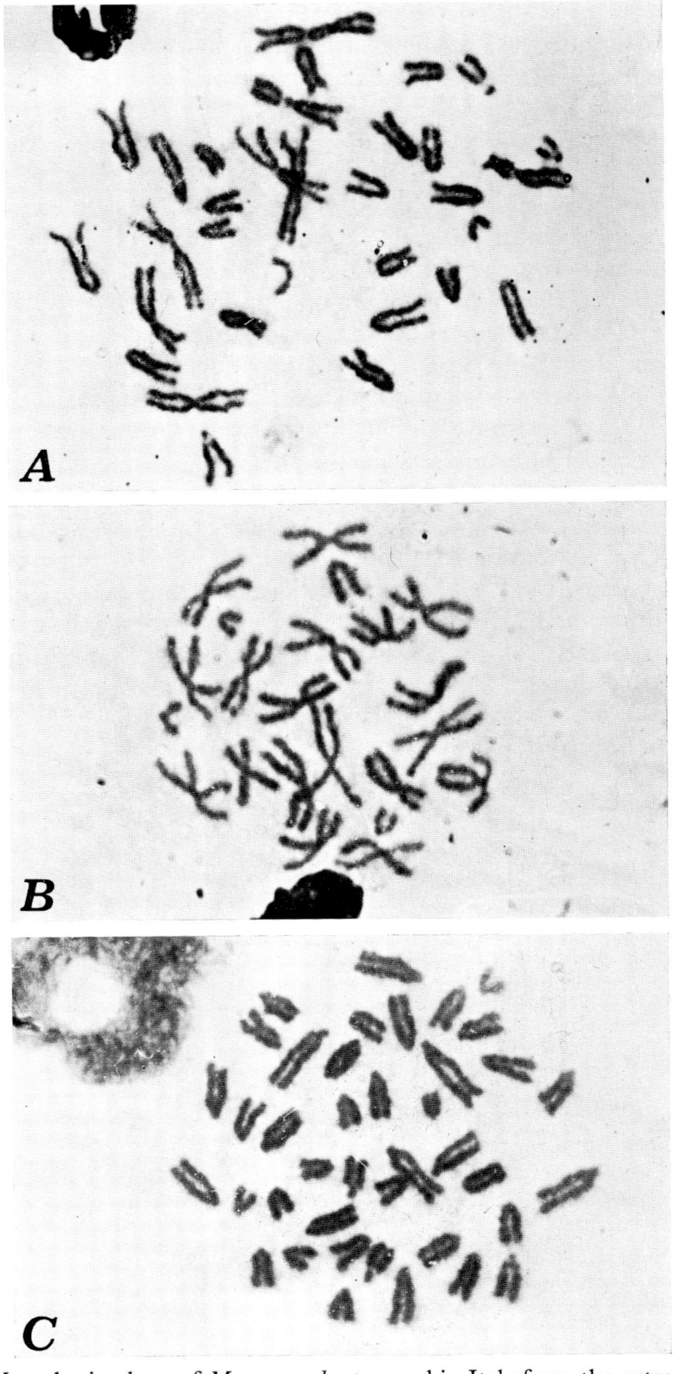

Fig. 3. Metaphasic plates of *Mus musculus* trapped in Italy from the extreme North to the deep South; results are ordered in this figure from North to South: (a) in the Alps, $2n = 32$; (b) in the Appennines, $2n = 22$ and (c) in Sicily, $2n = 40$.

Fig. 4. Reproduction of the original lithographic plate from Fatio's paper where *Mus poschiavinus* was described for the first time. Can gnawing cigars be assumed to be a better taxonomic character than seven Robertsonian fusions ? (see p. 685).

But, what makes us cautious as to the advisability of jumping to the conclusion that a taxonomic problem has been solved on the basis of the first karyological finding, is the evidence now that *Mus poschiavinus* contributes to this Robertsonian fan, being included among morphae that are with certainty related to *Mus musculus* (Figs. 2 and 3).

Related with certainty? Can we in the present state of knowledge speak of "certainty"? At all events there is no doubt that if Fatio had caught some mice in the Alpine Valleys crossing the Bernina Pass, south of the Poschiavo Valley, he would have found interesting morphae to compare with his *Mus poschiavinus*.

Beyond the limits of the above mentioned systematic interpretation, chromosomal polymorphism still remains the most fascinating problem that has been introduced in vertebrate karyology in the last decade of taxonomic investigations. Indeed, I would go as far as to say that the criticism expressed just now, as regards the interpretation of the most complex case, must be viewed as the most stimulating for further studies; all the more so if equipped with a careful evaluation of the systematic problems inherent in the material under examination.

III. THE MUNTJAC SCANDAL

For all passionate karyologists there always exists an unfailing axiom, some sort of absolute certainty, a lighthouse guiding the navigators across the stormy chromosome waters: *the karyotype morphology indicates the kinships between species*. It is quite logical to believe that if the affinity between species belonging to the same genus or between closely allied genera is the expression of a recent evolutionary derivation, the karyotype, due to the short time at its disposal for its own differentiation, will remain, as will the outer morphology, fairly similar, if not identical. On the basis of this truism a precise system has been built involving identical chromosomes, heterochromatic zones which characterize given elements of this or that karyotype. By means of the techniques of fluorescent banding or Giemsa-banding, the effort has been made in recent years to characterize homologous chromosomes between related species or the elements involved in Robertsonian transformations. All things considered, the Robertsonian transformation is among the most conservative of chromosome transformation processes, despite the spectacular reduction in the diploid number occasionally resulting from it. Needless to say, sometimes purely theoretical problems arise concerning the number and the nature of the centromere, but as a sceptical philosopher would put it: "Quid est centromerus?"

Fig. 5. The scandalous Muntjacs! On the right the Indian Muntjac (*Muntiacus muntjak*) and on the left the Reeve's Muntjac (*Muntiacus reevesii*) (see p. 690).

This assurance of ours, this apparatus consisting of chromosomal units either spontaneously evident or artificially visualized, is baffled by the scandalous Muntjac!

It was demonstrated unequivocally by Wurster and Benirschke (1970) that Reeve's muntjac (*Muntiacus reevesi*) possesses a diploid number $2n = 46$, NF 46, while the Indian muntjac (*Muntiacus muntjak*) has $2n = 7$ in the male and $2n = 6$ in the female (NF 12, sex complement X Y_1Y_2). The affinity between the two species is beyond dispute, as is demonstrated by the possibility of obtaining hybrids, between the two species (Gray, 1954), that are occasionally fertile. This report is fully known by mammalian karyologists and need not be commented upon here. However, an explanation seems unavoidable. Matthey, in Chapter 15, proposes as a hypothesis that the whole *Muntiacus reevesi* genome has undergone condensation as a result of subsequent translocations in the six chromosomes of *Muntiacus muntjak*; this hypothesis would be supported by the comparison between the DNA contents of interphase nuclei in the two species. Of course, if this challenging phenomenon truly exists, and it does, it calls for a rational explanation. At present, Matthey's assumption may serve as a working hypothesis. The scandal, however, persists!

The Muntjac is an extreme case, absurd in its incontestable reality; nonetheless it must not be regarded as a rarity in mammalian cytotaxonomy.

In his most brilliant work on Phyllostomidae bat chromosomes, Baker (1967) in fact proposes another puzzle. In the family Glossofaginae he distinguishes two cytotaxonomic groups; the first is comprised of *Glossophaga* and *Leptonycteris*, and possibly also *Anoura*, which are characterized by $2n = 30$–32 (NF 56–60), closely correlated in karyotype morphology to the other Phyllostomidae (Phyllostominae, Chilonycterinae, Sturnirinae and a part of the Stenoderminae). On the contrary, the second group represented by *Choeroniscus* and *Choeronycteris* ($2n = 16$–19, NF 26–32) is cytotaxonomically related to *Carollia* of the subfamily Carolliinae ($2n = 20$–21, NF 36). Further cytotaxonomical studies carried out by Baker and Bleier (1971) on Phyllostomidae bats of the subfamily Carolliinae revealed also that within the taxonomic field some species exist such as *Rhinophylla* ($2n = 36$, NF 62) which must be related to the main taxonomic group that contains most of the Phyllostomidae.

It is clear that Baker's observations have brought to light a mistake in traditional systematics. Obviously, *Carollia*, *Choeronycteris* and *Choeroniscus*, which have so far separated into two subfamilies, belong on the contrary to the same naturally occurring taxonomic group, in addition to the same cytological group. Phyllostomidae systematics should be revised in this sense.

Likewise, as it has also been stressed by the writer (Capanna and Civitelli, 1969), the systematics of the genus *Pipistrellus* (among Vespertilionidae) should be revised in the light of the sharp karyological divergences shown between the palearctic species on the one hand and the neotropic and paleotropic ones on the other, belonging to the same genus.

However, when we say "the systematics of this group should be revised" it does not mean that we have solved the problem of the evolution of that taxonomic entity. The cytologist, being aware of his absolute inability to solve this problem applies to the systematician for help. In fact, the systematicians do have adequate methodological tools to answer it.

It is evident also that for the study of the karyotype transformation within genera and families the same strictly taxonomic problems arise that have been discussed in relation to chromosomal polymorphism, namely that karyological findings provide further elements of judgment which must be of relevant weight in the restructuring of taxon systematics in natural terms.

Alas, how often, on the contrary, is the karyological item so laboriously obtained by devising new techniques or searching for rare zoological material, in reality devoid of any information?

Shifting our considerations from mammalian karyology to that of the Osteoichthyes I shall report what happened in a recent study of Mugilidae karyology (Cataudella and Capanna, unpublished). In three species from this family, *Liza ramada*, *Mugil cephalus* and *Chelon labrosus*, $2n = 48$, NF 48 were found. Obviously, we discussed this finding from a taxonomic standpoint taking the karyological evidence as supporting a systematic correlation of the Mugilidae with the Perciformes, thus strengthening the systematic outlook which considers this systematic unit as a family, against those who would treat Mugiliformes as an order *per se*. In fact the Perciformes almost always exhibit a constant karyotype $2n = 48$.

What is obvious, on the other hand, is that our conclusion stems from pre-existing—I would almost say preconceived—notions of a systematic order on morphological bases, but not from karyological evidence in itself. In fact, $2n = 48$, NF 48 is characteristic of many orders of Osteoichthyes, from the Pleuronectiformes (Ohno and Atkin, 1966) to the Mastacembeliformes (Kaur and Srivastava, 1965) to the Beloniformes (Post, 1964) and also of a number of Cypriniformes, Cyprinodontiformes and Clupeiformes (see Chapter 10).

IV. ANTICIPATIONS

Hence, if conclusions must be drawn at the end of a book, they should, in the present case, be anticipations. So far, the problem of chromosomal evolution is not fully understood in vertebrates: we can say that a satisfactory study has just begun for some of them. The foregoing considerations therefore should not give rise to criticisms or distrust as to the validity of taxonomic studies; but rather these studies should be further stimulated.

All things considered, what meaning is there in the fact that the karyotype is known in 1273 out of the 13000 Eutherian species?* Not to mention the fact that mammals are among the most thoroughly studied species from the karyological point of view. In fact what are the 500 reptilian known karyotypes when compared to the 5000 species hitherto undescribed? Our cytological information about fishes and birds is truly irrelevant, when we realize that the living fish species are about 20000 and that avian species are 28000 in number.

It is now entirely at one's discretion to attempt to build up "chromosomal cosmogonies" by trying to patch up the different evolutionary radiations into a unitary picture, even though restricted to the orders of a single Class.

In Chapter 15 written by Matthey, the ideas put forward by Todd (1967) and Staton (1967) are discussed at length and therefore need not be dwelt upon here. I should only like to suggest that, in my opinion, given the scarce and inhomogeneous data available, it is much too early to build up such a complex theory, based on a generalized chromosomal fissioning that would have occurred over and over in subsequent cycles. Some years hence, with the help of fresh findings, we may however recognize this theory to be valid. Nowadays we are objectively in a different position than lying waiting in doubt.

It might be objected that by so doing we run the risk of cutting the poor figure that formal science did towards Mendel's discovery! True, when Mendel published his investigations on the vegetal hybrids he was a forerunner, his interpretation lacked those logical prerequisites that only knowledge of meiotic processes could provide; all in all the monk from Brno stated only a hypothesis. Well, we hope that Todd's and Staton's hypothesis is as hazardous as that of Mendel and must stimulate us to persist in gathering more and more data.

The limits of reliability of the interpretations resting on the comparison between the karyotypes of different classes are still greater. When Bick and Jackson (1967) published their photos of *Echidna* chromosomes, I thought I could recognize in them the reptile chromosomes (either from a chelone or a crocodile), the likeness was so impressive; on the other hand, some saurian karyotypes are reminiscent because of the presence of micro-

* Here Hesse's (1929) finding is reported rather than Mayr's (1942), since the number of mammalian species was restricted by the latter worker to 3500 thereby degrading to polytypic species many of those used by Hesse. Hesse numerical data result, in fact, in polytypism phenomena which, as discussed earlier, lead to a huge complication of karyological data.

chromosomes, of the avian karyotype. This leads to the thought of chromosomal cosmogonies in which one goes back to the karyotype of *Euparkeria* as an ancestor common to both aves and Ornithischia (are not birds small feathered dinosaurs ?) as well as to the Lepidosauria.

Conversely, we may surmise the existence of a chromosomal differentiation appearing in the earliest Pennsylvanian, from whence the ancestral Cotilosauria karyotype and both the Anapsida and the mammalian karyological types would have descended. In this case it is more than evident that this is all the outcome of paleontological and morphological knowledge rather than of logically correct cytological inferences.

As regards the difference in the karyotype morphology between Monotremata and Therii (Placentals and Marsupials) evidence of mammalian diphyletism has been claimed; however, I endorse the opinion that the most obvious evidence of this diphyletism is the position of the epipterygoid (alisphenoid) in the constitution of the lateral wall of the brain case.

Diversification from the common Pantotherian ancestor of the two mammalian Theri radiations, a very ancient diversification dating back to the end of the Jurassic, would seem to be marked out by a sharp difference between the karyotypes characterized by few and large chromosomes in the Marsupials and small and numerous in the Eutherians. However, I might quote scores of instances of eutherian karyotypes with a marsupial-like karyotype (the notorious Indian muntjac or *Cheroniscus*). In contrast, *Aepyprymnus* with its diploid number $2n = 32$ exhibits a eutherian chromosome formula.

Available data are too meagre as yet for general deductions! Therefore, the following requirement stems from my concluding remarks: we must work for a unitary outlook of the evolutionary problem, namely of the systematic zoological problem in which cytological and taxonomic data, in the traditional sense, may equally be evaluated. The systematic problem together with karyological findings may provide a cytotaxonomic result; the karyogram in itself is just a string of chromosome cuttings.

It is fortunate that cytogenetic studies carried out on zoological material classified by the vulgar name alone or, still worse, erroneously classified are rapidly disappearing. At any rate they are justifiably ignored by the scientific literature. Regrettably enough, however, cases in which the name of the author or its abbreviation are omitted, may be frequently encountered. It should not be believed that (L.) or (PALLAS) or (CUVIER) are mere ornaments of the latin name, a sort of whim on the part of meticulous systematists. The everlasting rearrangement of the zoological nomenclature, on the basis of priority or of revision of non-natural groups, makes this specification essential for the exact knowledge of the animals of which the chromosomes were studied.

In addition, more must be done if we intend to set up a chromosomal *Systema Naturae*. It is growing increasingly important to record a full taxonomic card for each animal whose karyological study is undertaken. The highest possible number of elements for the diagnosis of the species should be collected (see van Gelder, 1969). This is the only way we can be sure of relating the karyotype to a sharply defined species even when its name may have changed because of some whimsical systematist concerned with priority and *nomina oblita et conservanda* or else because the karyological evidence has required the subdivision of a polytypic species into new *bonae species*.

The above is excellently demonstrated, as far as this methodological approach is concerned, by the work by Matthey and Petter (1970) who independently carried out some investigations on the genera Tatera and Tateryllus rodents both karyologically and taxonomically. Each worker ignored the findings of the other and they compared them only at the end of their respective studies. Truly encouraging from the point of view of the methodological approach is the perfect correspondence resulting from the taxonomic and chromosomal data.

To conclude, I should like to avail myself, of Ohno's final words (1969) although paraphrased in his "Conclusive Remarks" in Benirschke's "Comparative Mammalian Cytogenetics": "It would appear that in order to understand the role chromosomal changes played in evolution, we mammalian cytogeneticists should become less of a cytologist and more of a geneticist."

Since the present volume has the word Cytotaxonomy instead of Cytogenetics in its title, I hope Professor Ohno will allow me to change his sentence as follows: "It would appear that in order to understand the role chromosomal changes played in evolution, we vertebrate *cytotaxonomists* should become less of a cytologist and more of a *zoologist*."

REFERENCFS

Baker, R. J. (1967). *S. West. Nat.* **12**, 407–428.

Baker, R. J. and Bleier, W. J. (1971). *Experientia* **27**, 220–222.

Benirschke, K. (ed.) (1969). "Comparative Mammalian Cytogenetics." Springer-Verlag, Berlin, Heidelberg and New York.

Bick, Y. A. E. and Jackson, W. D. (1967). *Am. Nat.* **101**, 79–86.

Capanna, E. and Civitelli, M. V. (1969). *Caryologia*, **23**, 79–111.

Capanna, E. and Civitelli, M. V. (1971). *Boll. Zool.* **38**, 151–157.

Fatio, V. (1869). "Faune des Vetébrés de la Suisse". H. Georg, Geneva and Basle.

Ford, C. E. and Hamerton, J. L. (1956). *Stain Technol.* **31**, 247–251.

Gray, A. P. (1954). "Mammalian Hybrids. A Checklist with Bibliography"' Commonw. Agric. Bureaux, Farnham Royal, Bucks.
Gropp, A., Tettenborn, V. and Lehman, E. von (1969). *Experientia* **25**, 875–876.
Gropp, A., Winking, H., Müller, H. and Zech, L. (1972). *Chromosoma, Berlin* **39**, 265–288.
Hesse, R. (1929). *Naturwissenschaften (Berlin)* **17**, 570–575.
Hsu, T. C. and Kellog, D. S. (1960). *J. Nat. Cancer Res.* **25**, 221–235.
Hsu, T. C. and Pomerat, (1953). *J. Hered.* **44**, 23–29.
Kaur, D. and Srivastava, M. D. L. (1965). *Caryologia* **18**, 181–191.
Makino, S. (1943). *Cytologia* **13**, 39–54.
Matthey, R. (1931). *Revue Suisse Zool.* **38**, 117–183.
Matthey, R. (1939). *Cytologia* **10**, 32–39.
Matthey, R. (1945). *Experientia* **1**, 50–56 and 78–86.
Matthey, R. (1949). "Les Chromosomes des Vertébrés." Rouge, Lausanne.
Matthey, R. (1963). *Cytogenetics* **2**, 290–322.
Matthey, R. (1964). *Experienıia* **20**, 1–9.
Matthey, R. (1970). *Revue Suisse Zool.* 77, 625–629.
Matthey, R. and Jotterand, M. (1970). *Revue Suisse Zool.* 77, 630–635.
Matthey, R. and Petter, F. (1968). *Revue Suisse Zool.* **75**, 461–469.
Matthey, R. and Petter, F. (1970). *Mammalia* **34**, 585–597.
Mayr, E. (1942). "Systematic and Origin of Species" Columbia Univ. Press.
Moorhead, P. S., Nowell, P. C., Mellman, W. J., Battips, D. M. and Hungerford, D. A. (1960). *Exp. Cell Res.* **20**, 613–616.
Ohno, S. (1969). *In* "Comparative Mammalian Cytogenetics" (K. Benirschke ed.) pp. 451–453. Springer-Verlag, Berlin, Heidelberg and New York.
Ohno, S. and Atkin, N. B. (1966). *Chromosoma, Berlin* **23**, 79–111.
Post, A. (1964). *Z. Zool. Syst. Evol. Forsch.* **3**, 47–93.
Shokolov, N. N. (1936). *Cytologia* 7, (quoted by Matthey, 1949).
Staton, D. (1967). *Mamm. Chrom. Newsl.* **8**, 203–219.
Todd, N. B. (1967). *Mamm. Chrom. Newsl.* **8**, 268–279.
van Gelder, R. G. (1969). *In* "Comparative Mammalian Cytogenetics" (K. Benirschke, ed.) pp. 445–450. Springer-Verlag, Berlin, Heidelberg and New York.
White, M. J. D. (1945). "Animal Cytology and Evolution" Cambridge University Press.
Wurster, H. D. and Benirschke, K. (1970). *Science, N.Y.* **168**, 1364–1366.

AUTHOR INDEX

Numbers in italics are pages on which references are listed at the end of the paper

SUBJECT INDEX

B

H